Chemistry: Principles and Applications for Engineering Students

2nd Preliminary Edition

Lawrence S. Brown
Texas A&M University

Thomas A. Holme
University of Wisconsin - Milwaukee

THOMSON
BROOKS/COLE

Printed in the United States of America

Brooks/Cole
511 Forest Lodge Road
Pacific Grove, CA 93950
USA

For information about our products, contact us:
Thomson Learning Academic Resource Center
1-800-423-0563
http://www.brookscole.com

International Headquarters
Thomson Learning
International Division
290 Harbor Drive, 2nd Floor
Stamford, CT 06902-7477
USA

UK/Europe/Middle East/South Africa
Thomson Learning
Berkshire House
168-173 High Holborn
London WCIV 7AA

Asia
Thomson Learning
60 Albert Street, #15-01
Albert Complex
Singapore 189969

Canada
Nelson Thomson Learning
1120 Birchmount Road
Toronto, Ontario MIK 5G4
Canada
United Kingdom

0-534-65489-4

The Adaptable Courseware Program consists of products and additions to existing Brooks/Cole products that are produced from camera-ready copy. Peer review, class testing, and accuracy are primarily the responsibility of the author(s).

Overview

Chapter Topics and Template Themes

Detailed Contents

Chapter 13: Electrochemistry **515**

Appendices

Chapter 1: Introduction to Chemistry

In the future, nanoengineers may design and assemble tiny machines like the molecular scale pump shown in Figure 1.1. Their every decision will be guided by knowledge of the sizes and properties of the atoms of different elements. Such devices might be built up atom by atom: each atom would be specified based on relevant design criteria, and maneuvered into position using tiny tools. These nanomachines will be held together not by screws or rivets, but by the forces of attraction between the different atoms - by chemical bonds. Clearly these futuristic engineers will need an understanding of atoms and the forces that bind them together – an understanding of chemistry.

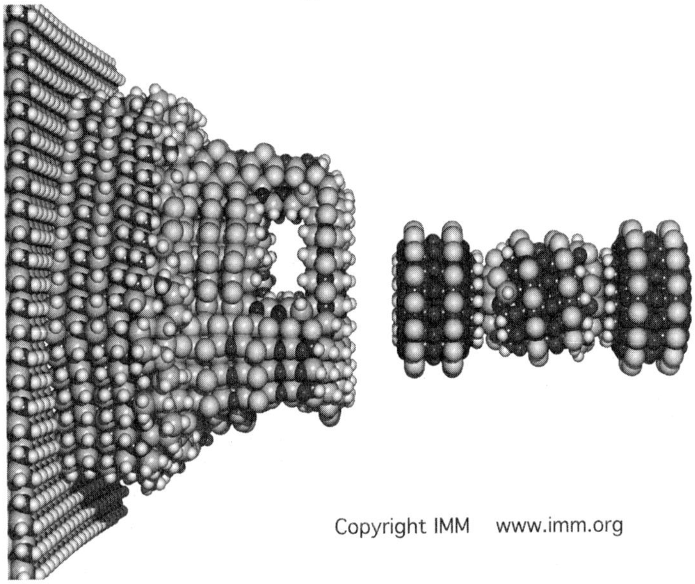

Figure 1.1 This drawing shows a molecular-scale pump, as conceived by the futurist K. Eric Drexler. The piece to the right is a rotor, and would be inserted into the housing at the left. Each sphere in the figure represents an individual atom, and the entire assembly shown here contains about 6,000 atoms. While the fabrication of such devices is not yet feasible, chemists and engineers have made impressive strides in the emerging field of nanoengineering. (Image courtesy of the Institute for Molecular Manufacturing.)

At least for now, though, this atomic level engineering remains in the future. But what about today's practicing engineers? How do their decisions depend on knowledge of chemistry? And from your own perspective as an engineering student, why are you required to take chemistry?

The Accreditation Board for Engineering and Technology, or ABET, is a professional organization that oversees engineering education. According to ABET's definition, "Engineering is the profession in which a knowledge of the mathematical and natural sciences gained by study, experience, and practice is applied with judgment to develop ways to utilize, economically, the materials and forces of nature for the benefit of mankind." So as one of the sciences, chemistry is clearly included in the realm of knowledge at the disposal of an engineer. Yet engineering students do not always recognize the role of chemistry in their chosen profession. One of the main goals of this textbook is to instill an appreciation for the role of chemistry in many areas of engineering and technology, and of the interplay between chemistry and engineering in a variety of modern technologies.

The study of chemistry involves a vast number of concepts and skills. The philosophy of this book is to not only present those basic ideas, but also to apply them to aspects of engineering where chemistry is important. Each chapter will begin with an example of chemistry related to engineering. Some of these examples, like the burning of fuels, will involve fairly clear applications of chemical principles and reactions. In other cases, the role of chemistry may be less immediately apparent. In Chapter 6 we will consider how evolving knowledge of chemical properties has fueled the design of different light sources, from the simple incandescent bulb through modern lasers and organic light emitting diodes (OLEDs). Other themes will involve the design and selection of materials for various uses or the importance of chemistry in environmental engineering problems. These chapter-opening sections all have titles that begin with "Insight into..." and the questions that are raised in them will guide our exploration of the relevant fundamentals of chemistry presented throughout that chapter. Our first case considers the production of aluminum and the history of aluminum as a structural material.

1.1 Insight into Aluminum

In 2002 over 100 billion aluminum cans were produced in the United States.

If you are thirsty you might ask yourself several questions about what to drink. In all likelihood you rarely ask, "Where did the can that holds this soda come from, and why is it made of aluminum?" The aluminum can has become such a common sight that it's easy to take for granted. What makes aluminum such an attractive material for this type of application, and how did it become such a familiar part of life?

You can probably identify a few properties of aluminum that make it suitable for use in a soda can. Compared with most other metals, aluminum is light but fairly strong. So a typical aluminum can is much lighter than a comparable tin or steel can. This means that the can does not add much weight when compared to the soda itself, so the cans are easier to handle and cheaper to ship. The fact that aluminum does not readily undergo chemical reactions that might cause it to degrade as the cans are transported and stored is also important. But while all of those features of the aluminum can are nice, they wouldn't be of much practical use if aluminum were not readily available and reasonably inexpensive.

The widespread availability of aluminum results from an impressive collaboration between the basic science of chemistry and the applied sciences of engineering. In the 19th century, aluminum was a rare and precious material. In Europe, Napoleon was emperor of a sizeable portion of the continent and he would impress guests by using extravagant aluminum tableware. In the United States, architects wanted a suitably impressive material for the capstone at the top of the Washington Monument, a tribute to the "father of our country"– and they chose aluminum. Why was aluminum so expensive then, and what changed to make it so affordable now?

Initial discussion of this question can be framed in terms of Figure 1.2, which looks at the interactions of human society with the Earth in a rather broad sense. Society, represented by the circle in the center, has needs for goods and materials. Currently, and for the foreseeable future, the raw materials needed to make these goods must somehow be extracted from the Earth. When the goods are used up, the leftovers become waste that must be disposed, completing the cycle in some sense by returning the exhausted materials to the ecosystem. Ultimately, the role of engineering in this cycle is to maximize the efficiency with which materials are extracted and minimize the amount of waste that is returned.

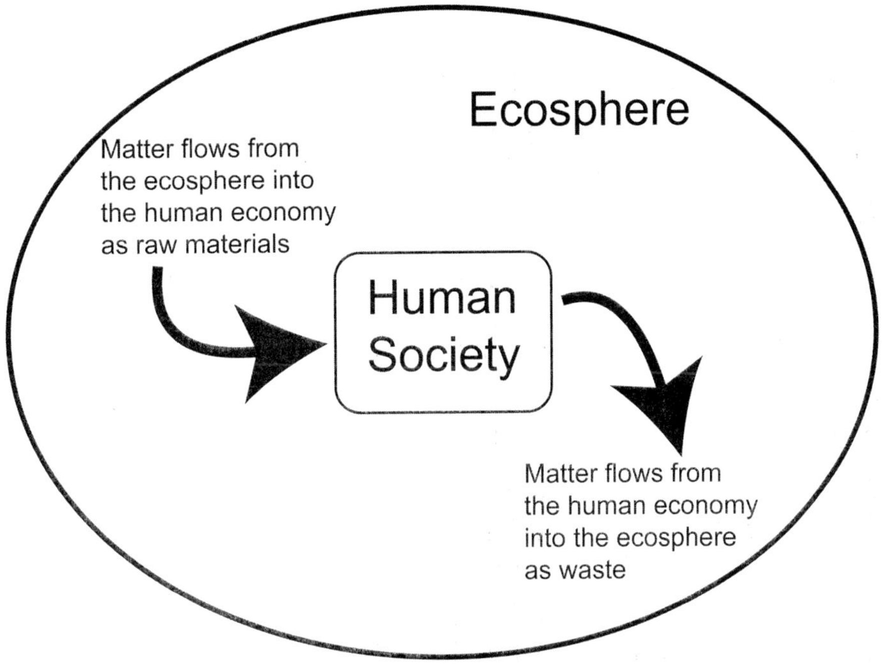

Figure 1.2 The interactions of human society with the Earth can largely be thought of in terms of the conversion of matter from raw materials into waste. Much of engineering consists of efforts to optimize the processes used in these conversions. And as the science of matter, chemistry is an important element of the knowledge exploited in engineering those processes.

Let's think about aluminum in this context. Pure aluminum is never found in nature. Instead, the metal occurs in ore composed of both useless rock and aluminum in combination with oxygen, called *alumina*. So before aluminum can be put to use in our soda can, it must first be extracted or "won" from its ore and purified. Because aluminum combines very readily with oxygen, this presents some serious challenges. Some of these challenges are chemical in nature and will be revisited in Chapter 13 of this text. Some of the early steps, however, can be solved by clever applications of physical properties, and we will consider a few of these as we investigate introductory material in this chapter. When confronted with a complex mixture of materials, such as an ore, how does a chemist think about separating the mixture?

To look into this type of question, we should adopt the approach that is commonly taken in science. The term **scientific method** has various possible definitions, and we'll look at this concept further in Section 1.4. But at this point we will consider it to be an approach to understanding that begins with the observation of nature, continues to hypothesis or model building in response to that observation, and ultimately includes further experiments that serve to either bolster or refute the hypothesis. In this definition, the hypothesis is an educated guess at ways to explain nature. We will see how this method relates to chemistry in general in this chapter and also to issues relating to materials like aluminum and their use in society.

1.2 The Study of Chemistry

Chemistry has been called the "central science" because of its importance to a number of other fields of scientific study. So, even if you have never taken a chemistry course before, chances are good that you have seen some chemistry before. This text and the course in which you are using it are designed to help you make connections between pieces of information you have already picked up, and to give you a more coherent and systematic picture of chemistry. The ultimate goal of introductory college chemistry courses is really to help you appreciate the chemical viewpoint and how it can help to understand the natural world. This type of perspective of the world is what enables chemists and engineers to devise strategies to refine metals from their ores, as well as to approach the many other applied problems we'll be exploring.

This coherent picture ultimately involves three levels of understanding or perspectives on the nature of chemistry: ***macroscopic, microscopic, and symbolic***. By the end of this course , you should be able to switch among these perspectives to look at problems involving chemistry in several ways. The things we can see about substances and their reactions provide the macroscopic perspective. We need to interpret these events considering the microscopic (or "particulate") perspective, where we focus on the smallest components of the system. Finally, we need to be able to communicate these concepts efficiently, so chemists have devised a symbolic perspective that allows us to do that. We can look at these three aspects of chemistry first, to provide a reference at the outset for framing our studies.

The Macroscopic Perspective

When we observe chemical reactions in the laboratory or in the world around us, we are observing matter at the macroscopic level. **Matter** is anything that has mass and can be observed. We are so often in contact with matter that we tend to accept our intuitive feel for its existence as an adequate definition. When we study chemistry, however, we need to be aware that some of what we observe in nature is not matter. For example, light is not considered matter, because it has no mass.

When we take a close look at matter - in this case aluminum - we can see that various questions arise. The behavior of the aluminum in a can is predictable. If the can is tossed into the air, little will happen except that the can will fall to the Earth under the force of gravity. Aluminum cans and other consumer goods like those shown in Figure 1.3 do not appear to decompose in the air or undergo other chemical reactions. If the aluminum from a soda can is ground into a fine powder and tossed into the air, however, it may ignite – chemically combining with the oxygen in the air. It is now believed that the Hindenburg airship burned primarily because it was covered with a paint containing aluminum powder and not because it was filled with hydrogen gas. (For a summary of the evidence, see http://www.aps.org/apsnews/0700/070004.html.)

One of the most interesting ways to observe matter is to allow it to change in some way. Two types of changes can be distinguished: physical changes and chemical changes. In a physical change the substances involved do not lose their chemical identity. **Physical properties** are variables that we measure while the substance being observed retains its identity. Mass and density are familiar physical properties. Mass is measured by making a comparison between the object given and some standard, using a balance. **Density** is a ratio of mass to volume. (This variable is sometimes called the mass density). To determine the density, both the mass and the volume must be measured. But these values can all be obtained without changing the material, so density is a physical property. Other familiar examples of physical properties include color, viscosity, hardness, and temperature. Some other physical properties, which will be defined later, include heat capacity, boiling point, melting point, malleability, and volatility.

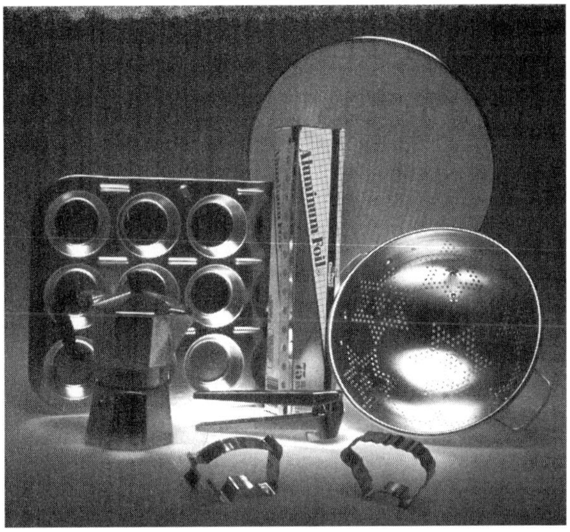

Figure 1.3 All of the common kitchen items shown here are made of aluminum. The metal's lightweight, non-corrosive nature, and low cost all make it a common choice for many consumer products.

Chemical properties are associated with the types of chemical changes a substance undergoes. For example, some materials burn readily while others do not. Burning is a chemical reaction called *combustion*. Corrosion is another commonly observed chemical change that can be hindered in a specific substance by treating it with a material, such as paint, that does not corrode. An important chemical property of paint is its resistance to corrosion. Chemical properties can be observed only by carrying out a measurement that examines how a substance changes its identity.

Aluminum is generally found second, behind gold, in rankings of metal malleability.

Both chemical and physical properties of aluminum are important to its utility. A structural material is useful only if it can be formed into desired shapes, which requires it to be malleable. **Malleability** is a measure of how much stress a material can take before it will break or crumble, and metals are valuable in part because of their malleability. It is a physical property because the substance remains intact – it is still the same metal, just in a different shape. An aluminum can is formed during its manufacturing process, but its shape can still be changed, as you have perhaps done many times when you crushed a can to put it into a recycling bin. Similarly the chemical properties of aluminum are important. Pure aluminum would be very likely to react with the acids found in many popular soft drinks. So aluminum cans are coated on the inside with a thin layer of polymer – a plastic - to keep the metal from reacting with the contents. This demonstrates how knowing chemical properties can allow product designs to account for and avoid potentially harmful processes.

When we make macroscopic observations of chemical reactions, there are three common forms, or *phases*, of matter that we encounter: solids, liquids and gases. At the macroscopic level, **solids** are hard and do not change their shape easily. If placed in a container, they retain their own shape rather than assuming that of the container. Even a powdered solid demonstrates this trait, since the individual particles still retain their shape, even though the collection of them fills the container.

Liquids can be distinguished from solids in the macroscopic perspective because unlike solids, liquids adapt to the shape of the container in which they are held. They may not fill the entire volume, but the portion they do occupy has its shape defined by the container. Finally, **gases** can be distinguished from both liquids and solids at the macroscopic level primarily because a gas expands to occupy the entire volume of its container. While many gases are colorless and thus invisible, the observation that a gas fills the available volume is a common experience; we are not concerned in walking through a large room that we will hit a pocket with no air.

The aluminum that we encounter on a daily basis is a solid, but during the refining process of the metal it must become molten, or liquid. Handling the molten metal, pouring it into containers, and separating impurities provide both chemical and engineering challenges for those who work to design aluminum production plants.

Chemical and physical properties are often difficult to distinguish at the macroscopic level. We can assert that boiling water is a physical change, but if you do nothing more than observe the water in a boiling pot disappear, how do you know if it has undergone a chemical or physical change? To answer this type of question we need to consider the particles that make up the water, or whatever we observe, and consider what is happening at the microscopic level.

Microscopic or Particulate Perspective

Perhaps the most fundamental tenet of chemistry is that all matter is composed of atoms and molecules. This is why chemists tend to think of everything as being "a chemical" of one sort or another. In many cases the matter we encounter is a complex mixture of chemicals and we refer to each individual component as a chemical substance. We will define these terms much more extensively as our study of chemistry develops, but we'll use basic definitions here. All matter comprises of a limited number of "building blocks" called **elements**. The elements are often associated with the periodic table of elements, shown inside the front cover of this textbook and no doubt hanging in the classroom where your chemistry lectures are held. **Atoms** are unimaginably small particles that cannot be made any smaller and still behave like a chemical system. When we study matter at levels smaller than an atom, we move into the areas of nuclear or elementary particle physics. In this sense, atoms are the smallest particles that can exist and retain the chemical identity of whatever element they happen to be. **Molecules** are groups of atoms held together in a manner so that they form a unit with a distinguishably different identity than the atoms alone. We will ultimately see how forces known as "chemical bonds" are responsible for holding the atoms in these molecules together.

The word atom comes from the Greek word "atomos" meaning indivisible.

The particulate perspective provides a more detailed look at the distinction between chemical and physical changes. Because atoms and molecules are far too small to observe, we will typically use simplified, schematic drawings to depict them in this book. Often, atoms and molecules will be drawn as circles to depict them and consider their changes.

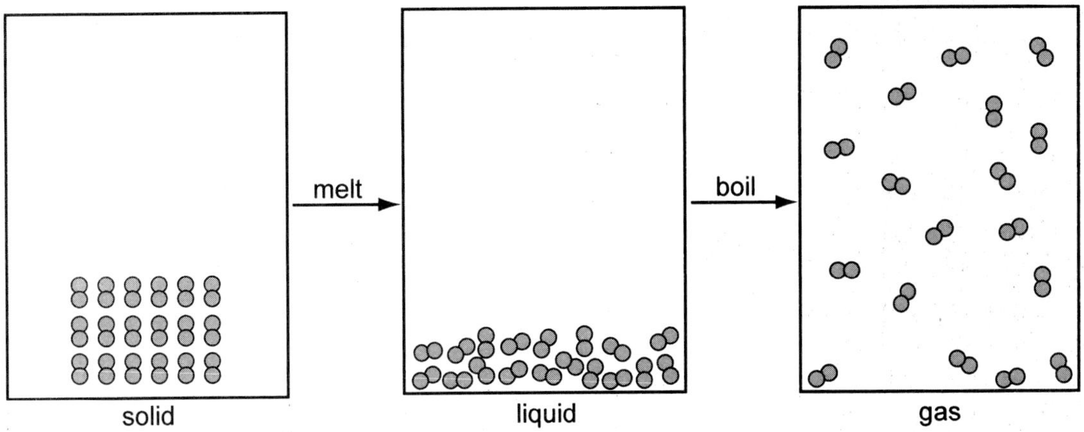

Figure 1.4 Particulate level views of the solid, liquid, and gas phases of matter. In the solid, the molecules maintain a regular ordered structure, so the sample maintains its size and shape. In the liquid, the molecules remain close to one another, but the ordered array breaks down. At the macroscopic level, this allows the liquid to flow and take on the shape of its container. In the gas phase, the molecules are very widely separated, and move independently of one another. This causes the gas to fill the available volume of the container.

If we consider solids, liquids and gases, how do they differ at the particulate level? Figure 1.4 provides a useful illustration of this perspective. Note that the solid is packed closely together and is depicted as maintaining its shape – here as a block or chunk. The liquid phase also has its constituent particles closely packed, but they are shown filling the bottom of the container rather than maintaining their shape. Finally, the gas is shown with much larger distances between the particles, and the particles themselves fill the entire volume of the container. These pictures have been inferred from experiments that have been conducted over the past 100 years. Many solids, for example, have well-ordered structures, called crystals, so the particulate perspective of solids usually includes this sense of order.

To correctly depict the relative density of a gas and a liquid, much more space would need to be shown between particles in a gas than can be shown in a drawing like Figure 1.4

How can we distinguish between a chemical and a physical change in this perspective? The difference is much easier to denote at this level, though it is often no more obvious to observe. If a process is a physical change, the atoms or molecules themselves do not change at all. To look at this idea, we turn to a "famous" molecule – water. Many people who have never studied chemistry can tell you that the chemical formula of water is "H two O." We depict this molecule using different sized circles; the slightly larger circle represents oxygen and the smaller circle represents hydrogen. In Figure 1.5 we see that when water boils the composition of the individual molecules is the same in the liquid phase and the gas phase. Water has not been altered, and this fact is characteristic of a physical change.

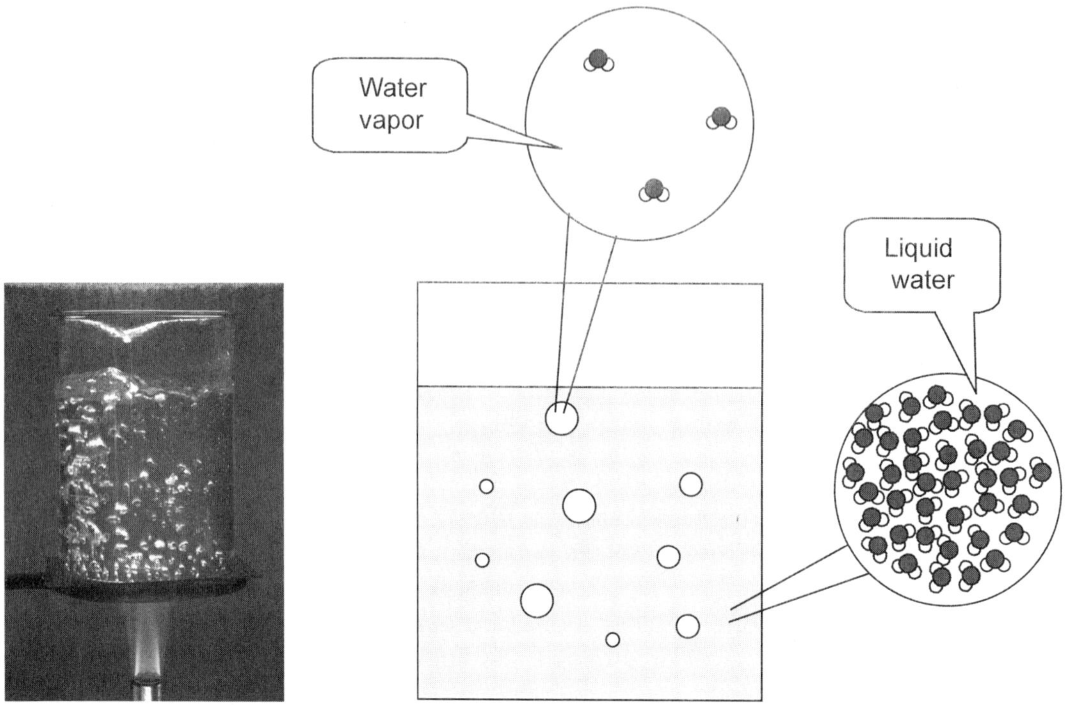

Figure 1.5 The boiling of water is a physical change, in which liquid water is converted into a gas. Both the liquid and gas phases are made up of water molecules, with each molecule containing two hydrogen atoms and one oxygen atom. The particulate scale insets in this figure emphasize that fact, and also show how the separation between the water molecules is much larger in the gas than in the liquid.

Contrast this with Figure 1.6, in which we show a particulate level depiction of a process called electrolysis, which occurs when water is exposed to an electric current. Notice that the molecules themselves change in this depiction. The water molecules are gone and the result is hydrogen and oxygen molecules; here, then, we have a chemical change.

If we observe these two reactions macroscopically, what would we see and how would we know the difference? In both cases we would see bubbles forming, only in one case the bubble will contain water vapor (gas) and in the other they contain hydrogen. Despite this similarity, we can make observations at the macroscopic level to distinguish between these two possibilities. Example Problem 1.1 poses an experiment that could be set up to make such an observation.

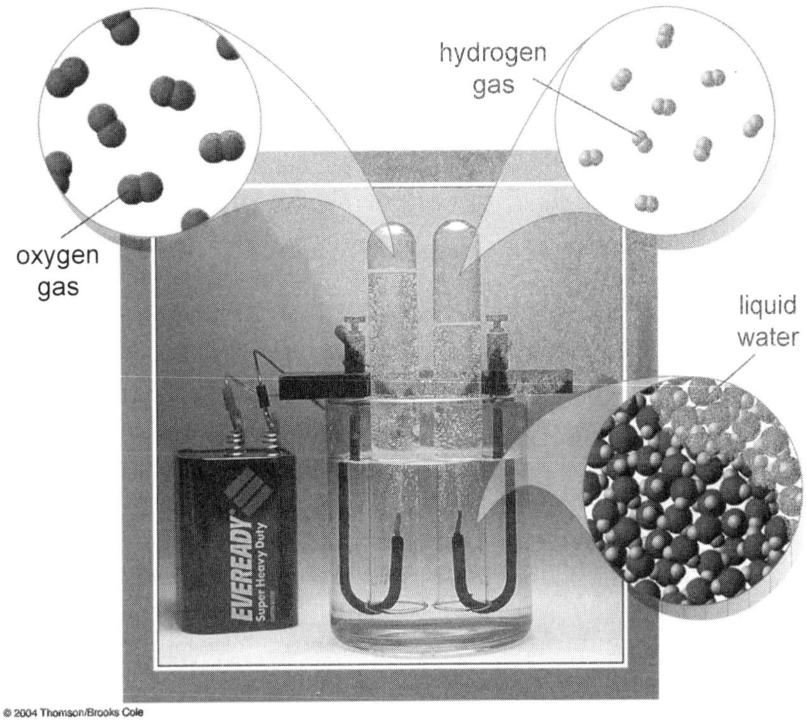

Figure 1.6 If a suitable electric current is passed through liquid water, a chemical change known as electrolysis occurs. In this process the water molecules are converted into molecules of hydrogen and oxygen gases, as shown in the particulate scale insets in the figure.

Example Problem 1.1

Consider the experimental apparatus shown in the diagram below. This equipment could be used to test a hypothesis about the chemical composition of the bubbles that rise from boiling water. What would the observation be if the bubbles were composed of (a) water, (b) hydrogen, or (c) oxygen?

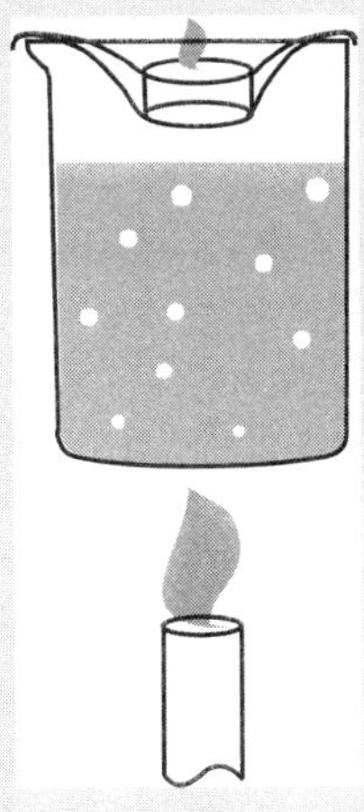

Strategy

This problem asks you to think about what you expect to observe in an experiment and alternatives for different hypotheses. At this stage, you may need to do a little research to answer this question – find out how hydrogen gas behaves chemically in the presence of a flame. We also have to remember some basic facts about fire that we've seen in science classes before. Fire, in order to be sustained, requires both a fuel and an oxidizer - usually the oxygen in the air.

Solution

(a). If the bubbles coming out of solution were water, we would expect that the flame would diminish in size or be extinguished. Water does not sustain the chemical reaction of combustion (as oxygen does), so if the bubbles are water the flame should not burn as brightly.

(b). Looking into hydrogen, you should have been able to find (on the web, for example) that it tends to burn explosively. If the bubbles coming out of the solution are hydrogen, we should expect to see the flame ignite the gas with some sort of an explosion. (Hopefully a small one.)

(c). If the bubbles are oxygen, the flame should burn brighter. The amount of fuel would remain the same, but the bubbles would increase the amount of oxygen present and make the reaction more intense.

Check Your Understanding

Work with students in your class or with your instructor to construct this apparatus and see whether or not your observations confirm any of these hypotheses. Draw a picture showing a particulate level explanation for what you observe.

Symbolic Representation

The third way that chemists perceive their subject is to use symbols to represent the atoms, molecules and reactions that make up the science. We will wait to introduce this perspective in detail in the next two chapters – but here we point out that you have certainly encountered the symbolism in your previous studies. The famous "H two O" molecule we have noted is never depicted as we have done here in the quotations. Rather, you have seen the symbolic representation of water, H_2O. We will look at chemical formulas in more detail in Chapter 2 and how we use them in describing reactions using chemical equations in Chapter 3. For now, we simply note that this symbolic level of understanding is very important, because it provides us a way to discuss some of the most abstract parts of chemistry. We need to think about atoms and molecules, and the symbolic representation provides a convenient way to keep track of these particles we'll never actually see. These symbols will be one of the key ways that we interact with ideas at the particulate level.

Geologically, bauxite is typically found in one of three forms, Gibbsite, Bohmite and Diaspore.

How can we use these representations to help us think about aluminum ore or aluminum metal? The macroscopic representation is perhaps the most familiar, especially to the engineer. From a practical perspective, the clear differences between unrefined ore and usable aluminum metal are immediately apparent. The principle ore from which aluminum is refined is called bauxite, and bauxite looks pretty much like an ordinary rock. There's no mistaking the fact that it is a different substance than aluminum metal. (A photo of a sample of bauxite, along with more information on aluminum and bauxite, can be found on the web at http://www.mii.org/Minerals/photoal.html.) At the molecular level, we might focus on the aluminum oxide (also called alumina) in the ore and compare it to aluminum metal, as shown in Figure 1.7. This type of drawing depicts the different atoms making up the alumina in the ore, while in the metal only one type of atom is seen. (Note that in reality, metals normally contain small amounts of impurities, sometimes intentionally introduced to provide specific, desirable properties. But in this case we have simplified the illustration by eliminating any impurities.) Finally, Figure 1.7 also shows the symbolic representation for aluminum oxide - its chemical formula. This formula is slightly more complicated than that of water, and we'll look at this type of symbolism more closely in Chapter 2.

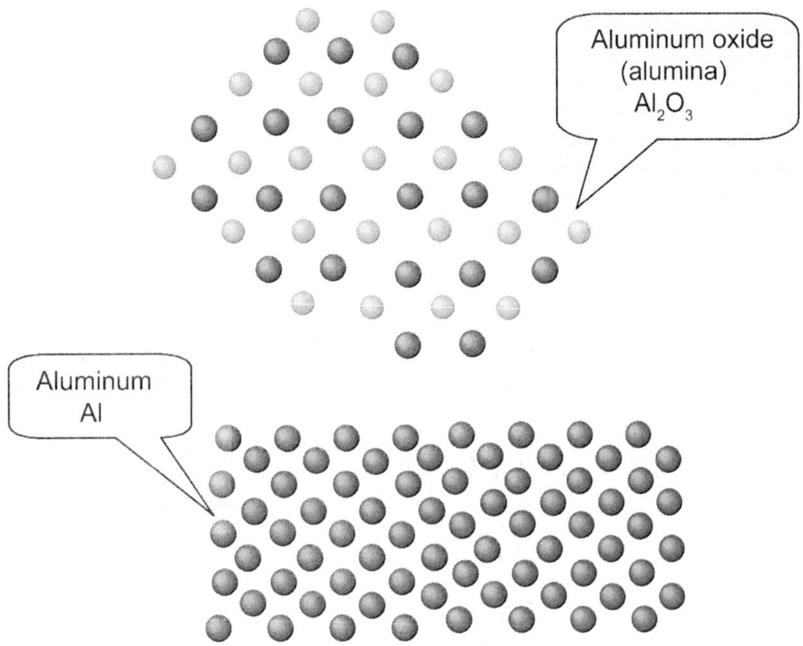

Figure 1.7 A particulate level representation of aluminum oxide (top) and pure aluminum (bottom). The darker circles represent aluminum atoms, while the lighter colored balls represent oxygen.

1.3 The Science of Chemistry: Observations and Models

Chemistry is an empirical science. In other words, scientists who study chemistry do so by making measurements of nature, observing chemical substances and chemical reactions. Once observations have been made, models are then created to help organize and explain the data. This structure of observations and models provides the backdrop of the science we'll explore throughout this book. In some sense, one difference between an engineer and a chemist is that chemists use their intellect and creativity to create models for understanding nature. The product of the intellect and curiosity in engineering is usually a design that somehow exploits or constrains nature. Ultimately, both fields must begin with the observation of nature.

Observations in Science

Observations in chemistry are made in a wide variety of ways for a wide variety of reasons. In some cases, the observations are made because of a need for materials with certain properties. For example, containers that hold liquids such as soft drinks need to be strong enough to hold the liquid, but lightweight so they don't increase the cost of transporting the product too much. Before aluminum cans were widely used, steel cans provided the containers that society demanded, but steel is relatively heavy, so there was incentive to find a different packaging material. Scientists and engineers worked together to make observations that confirmed the desirability of aluminum for this use.

A single aluminum can has a mass of roughly 14 grams.

Observations of nature involve some level of uncertainty in most cases. As an analogy, consider the attendance at a football game. We may be able to count with complete accuracy how many people use a ticket to attend a football game, which gives the paid attendance. But that number is not really how many people are there because it does not include members of the press, vendors, and coaching staffs, among others. This example points out one characteristic of making observations: we must be careful to define what we intend to observe. Carefully defining the measurement to be made does not, however, eliminate all sources of uncertainty. Virtually any scientific measurement must be made more than once in order to be valid.

Because we cannot observe nature with complete certainty, we need to establish the types of uncertainty we encounter in making observations. To do this, two terms are used: accuracy and precision. While these words may be synonymous in everyday speech, they each have their own meaning in scientific or engineering usage. **Accuracy** refers to how close the observed value is to the "true" value. **Precision** refers to the spread in values obtained from the measurement. A precise observation has several measurements close in value. Figure 1.8 illustrates the concepts of accuracy and precision in measurement, and shows how the two terms differ.

Error in measurement is unavoidable. Again characteristics of error fall into categories. **Random error** is fundamental to any measurement. Random error may cause the measurement to be either too high or too low, and is associated with the limitations of the equipment with which the measurement is made. **Systematic error** causes measurements to be consistently either too high or too low. This type of error is often associated with the existence of some unknown bias in the measurement apparatus. Impurities in metals provide one example of possible error sources. Suppose that an aluminum alloy contains very small amounts of another element, such as silicon. If we attempt to determine the amount of silicon in a sample, we might expect to have significant error in the measurement if the impurity is not uniformly distributed. This will be especially true if we carry out our measurement using a small sample. Systematic errors are also possible. Consider what would happen if we estimated the density of aluminum using soda cans and did not account for the thin layer of polymer that is applied to the can. Because the polymer coating probably has a different density than the aluminum, the measurement made would be systematically incorrect.

Figure 1.8 The dartboards illustrate the concepts of precision and accuracy, and show the difference between the two terms. The 'bulls-eye' on the target represents the true value of a quantity, and each dart represents a measurement of that quantity. The darts in the left hand panel are widely scattered and far away from the target; this is neither precise nor accurate. In the center panel, the darts are closely bunched but are not close to the target; this is good precision but poor accuracy. In the right hand panel we see good precision and good accuracy, as the darts are closely clustered and close to the target.

Interpreting Observations

Not all experiments provide direct information about the questions they ask. In many cases we must infer answers from the data that are obtained. The types of reasoning that are often needed are referred to as inductive and deductive reasoning.

Deductive reasoning is the talent required to solve most mysteries. Taking pieces of information and putting them together to reach a conclusion is important in science as well as detective work. **Inductive reasoning** is also important in science. In this type of reasoning, one imagines a particular answer to a puzzle or question and then constructs the implications of that answer. Inductive reasoning is essentially what was used in the Example Problem 1.1 to decide what might be observed in the proposed experiments.

To look at these types of reasoning, consider the developments that led to the affordable refining of aluminum. Even in the 19th century, aluminum ore was available. Metallurgists and other scientists were aware of the fact that the combination of aluminum with oxygen found in the ores was very stable. Observations allowed them to deduce that the chemical bonding forces in the alumina must be strong and would be difficult to overcome. Initial hypotheses on how to refine the metal involved heating and adding materials like carbon that would react with oxygen that might be liberated. This method was already in use for refining iron, so inductive reasoning suggested that it was a reasonable hypothesis, but it doesn't work for aluminum. As new discoveries were made about electrochemistry, people inferred that this method might be applied to the refining of aluminum and they devised experiments to prove their point. Eventually, Charles Hall devised a method that reduced the needed energy input for electrochemical refining of aluminum ore. Later, when this approach was coupled with the relatively low cost of hydroelectricity, the availability of aluminum was greatly increased. Ultimately, both deductive and inductive reasoning played vital roles in this technological advance.

Models in Science

The number of observations made in the history of science is tremendous. In order to organize this vast amount of information, many models have been devised in all fields of science. Models are important for a number of reasons. First, they allow us to summarize a large number of observations. Second, they allow us to predict behavior in circumstances that we haven't previously encountered. Third, they represent examples of creative thinking and problem solving.

In popular culture, to indicate something is just a theory indicates uncertainty. In science a theory has significant credibility gained by many observations that confirm its validity.

Models usually take time to develop. Utilizing the scientific method, a creative scientist will offer an explanation of why some pattern exists. This insight is considered tentative until additional observations confirm the hypothesis. The way that science develops, then, involves a cyclic process. Data are obtained, a hypothesis is advanced and more data are gathered to confirm or refute the hypothesis. A refuted explanation can be modified to fit the new data and tested again. Eventually, this process of proposing explanations and testing the implications of those proposals leads to a model that explains nature.

Models are typically dynamic in nature, changing as new information is obtained. Some models become sufficiently refined and explain so many aspects of nature that they are labeled **theories**. Theories often include mathematical explanations of phenomena. Again, theories can be overturned by new observations that invalidate their explanations of nature. Theories are, however, supported by a substantial number of experimental measurements and once designated tend to be in place for a long period of time. Although we hear phrases like "it's just a theory" in casual conversation, scientists only use the word to describe a very well established model.

The process just described is commonly called the scientific method as we defined in the "Insight" section of this chapter. The word "method" perhaps implies a more structured approach than actually exists in most scientific advancement. Many of the advances of science happen in coincidental ways, as products of serendipity. The stops and starts that are characteristic of scientific development are, however, guided by the process of hypothesis formation and observation of nature. Skepticism is a key component of this process. Explanations are accepted only after they have held up to the scrutiny of experimental observations.

The scientific method provides the philosophy by which science advances. In order to advance, however, every science must have a "beginning." For natural sciences, such as chemistry, biology and physics, the bedrock of advancement lies in the observation of some aspects of nature that seem self-evident. These tenets are collectively referred to as **laws of nature**. For example, the observation that objects that are less dense float on or rise above objects that are denser seems an irrefutable fact. We are not likely to observe any situation on this planet where an observation is counter to this expectation. That a less dense object will float upon a denser object is therefore a law of nature. Some tenets of science are designated a law, even though they are not self-evident as in the case of a natural law. But, typically, if we encounter a law we can expect it to summarize nature reliably enough so that we need not be concerned that it will be "repealed."

The models that played an important role in the discovery of affordable ways to refine aluminum were largely associated with the nature of chemical bonding. We'll revisit this topic in Chapters 6 and 7 and see how the models of chemical bonding were developed.

1.4 Numbers and Measurements in Chemistry

We can observe the world in a variety of ways. For example, we may see a professional basketball player and comment that he is tall, a reasonable statement. If we consider that same person on the basketball court, however, we may want to know *how* tall he is. In this case the answer we receive to the question might be "six-ten." To a basketball fan in the United States this answer makes sense because that person ascribes additional meaning to the answer, specifically that the numbers are in feet and inches: six feet 10 inches. A fan from another country might not find any meaning in the two numbers, six and ten, since they wouldn't make sense in terms of metric units. No one is six meters tall, after all.

Units can have consequences far more important than confusion over height. NASA attributed the loss of a Mars probe in 1999 to errors associated with the confusion of different units used in the design.

The example of the basketball player offers a good illustration of the difference between qualitative and quantitative information, and also points toward the need to be very specific in how we communicate information. Science and engineering are usually regarded as very quantitative professions, and in many ways this reputation is correct. But scientists also look at the world around us in general ways, akin to the assessment that the player is "tall." Such general or qualitative observations can be crucial in establishing a systematic understanding of nature. As that understanding deepens, though, we usually invoke quantitative or numerical measurements and models.

When discussing chemistry in qualitative or quantitative terms, though, it is always important that we communicate our observations and results as clearly as possible. In the case of quantitative observations, this usually means making sure that we carefully define the terms we use. As we move through this course, you will encounter many examples of "everyday words" that take on much more specific meanings when used in a scientific context. Similarly, when we talk about numerical measurements we will need to be very careful about the ways we use numbers and units.

Units

The possible misunderstanding between basketball fans from different countries about the height of a player provides an analogy to an unfortunate reality in studying science and engineering. Science has grown over the centuries and has seen contributions from a number of cultures and individuals. The legacy of centuries of development is the existence of a large number of units for virtually every basic measurement that can be made in science. Energy, for example, has long been important for human civilization and accordingly there are many units of energy. Distance and mass have also been measured for literally millennia.

The internationalization of science and engineering led to the establishment of a standard system that provides the needed flexibility to handle a wide array of observations. In this International System of Units (*Systeme International d' Unites*, or SI), carefully defined *units* are combined with a set of *prefixes* that designate powers of ten. This allows quantities of any size to be reported and understood, as illustrated in Figure 1.9.

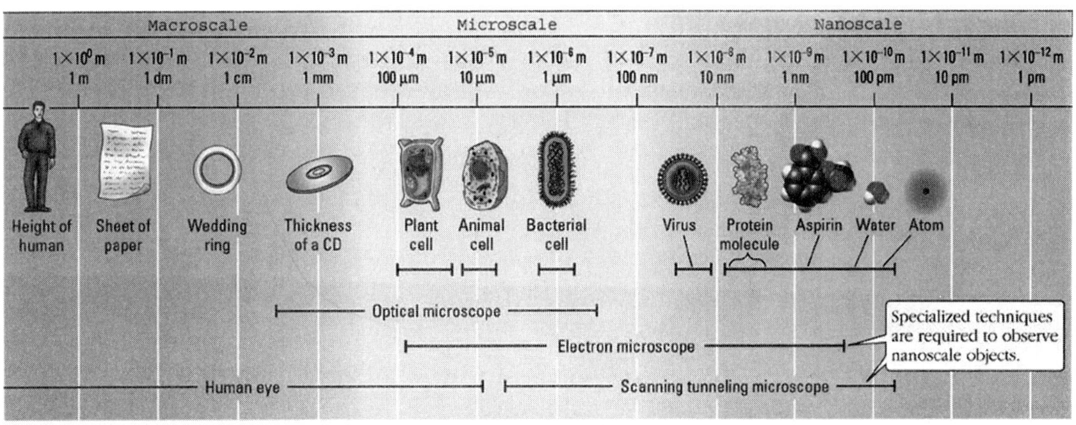

Figure 1.9 The typical dimensions of the objects shown extend over many orders of magnitude, and help point out the usefulness of the prefixes in the SI system.

Property	Unit, with abbreviation
mass	kilogram, kg
time	second, s
distance	meter, m
electric current	ampere, A
temperature	kelvin, K
number of particles	mole, mol
light intensity	candela, cd

Table 1.1 The seven base quantities of the SI system of units are listed, along with the names and abbreviations for their recognized units.

Due to the importance of powers of two in computer science, a kilobyte is actually defined as 1024 bytes.

When observations are reported in this system, the base unit designates the type of quantity measured. We can immediately recognize any quantity reported in meters (m), for example, as a distance. Table 1.1 shows the base units for a number of the quantities that chemists may want to measure. These base units, however, are not always conveniently matched to the size of the quantity to be measured. When we consider the sizes of single atoms or molecules, we will see that they are on the order of 0.0000000001 meters. Dealing with so many decimal places invites confusion, so it is usually preferable to choose a unit that is better matched to the scale of the quantity. The standard way to do this is to use prefixes that alter the "size" of any given base unit. A familiar example is the concept of the "kilo" unit. A kilometer is 1000 meters and a kilobyte, familiar in computer technology, is roughly 1000 bytes. The prefix *kilo-* implies that there are 1000 of whatever the base unit is. Prefixes such as this exist for numbers large and small, and these prefixes are provided in Table 1.2. The distance of 0.0000000001 meters could be reported as either 0.1 nm or 100 pm. Measuring time provides another interesting example of units. Laser based experiments are capable of measuring the progress of chemical reactions on a time scale of 10^{-15} s. Papers that report such experiments therefore use femtosecond, fs, to report time. Corrosion is also studied to determine its rate and reactions often occur at rates that take microseconds, μs, or milliseconds, ms, to complete.

Not every quantity of interest can be measured directly in terms of just the seven base units shown in Table 1.1, of course. Some units comprise combinations of these base units, and are therefore termed *derived units*. The SI unit for energy, for example, is the Joule (J), and 1 J is defined as 1 kg m^2 s^{-2}.

In principle any quantity of interest could be expressed in terms of appropriate combinations of the SI base units. But in practice many other units are so firmly entrenched that they remain in common use. One simple example of this is the measurement of times that are longer than seconds. Minutes, days and years are typically used in these circumstances, rather than kiloseconds, etc. In chemistry labs – and on soda containers – volumes are most often reported in terms of liters (L) or milliliters (mL) rather than the SI unit of m^3. Chemists also use a wide variety of units when describing concentration, which measures how much of a particular substance is present in a mixture. Metals often contain minor impurities, and while in some cases the units used are simply percentages, other units used include parts per million (ppm) or parts per billion (ppb). The **ppm** unit tells how many particles of a particular substance are present for every 1,000,000 particles in the sample. In **ppb** the sample size is one billion particles. Impurity concentrations on the order of even a few ppm may cause problems in some applications, or they may be intentionally added to impart some desirable property. Later in this chapter we will take up the issue of converting a measurement from one unit to another, as is frequently required in scientific and engineering calculations.

Factor	Name	Symbol
10^{24}	yotta	Y
10^{21}	zetta	Z
10^{18}	exa	E
10^{15}	peta	P
10^{12}	tera	T
10^{9}	giga	G
10^{6}	mega	M
10^{3}	kilo	k
10^{2}	hecto	h
10^{1}	deka	da

Factor	Name	Symbol
10^{-1}	deci	d
10^{-2}	centi	c
10^{-3}	milli	m
10^{-6}	micro	μ
10^{-9}	nano	n
10^{-12}	pico	p
10^{-15}	femto	f
10^{-18}	atto	a
10^{-21}	zepto	z
10^{-24}	yocto	y

Table 1.2 The prefixes listed here allow extremely large or extremely small numbers to be reported easily and interpreted correctly.

While many of these SI units have gradually found their way into everyday use, the units for temperature may be the least familiar. You are probably used to seeing temperatures in either degrees Fahrenheit or degrees Celsius, but even in "fully metric" countries the weather report does not generally use the Kelvin temperature scale. In general, temperature scales arise from the choice of two standard reference points that can be used to calibrate temperature with the use of a thermometer. The familiar Fahrenheit scale originally chose body temperature as one reference and set it to 100°F. (Accuracy of measurement was clearly not a priority when this temperature scale was proposed.) The second reference point was the coldest temperature that could be achieved by adding salt to water, a practice that lowers the melting point of ice. This established 0°F, and the temperature range between the two points was divided into 100 equal units. The scale is now defined by setting the freezing point of water as 32°F and the boiling point of water as 212°F.

Gabriel Daniel Fahrenheit invented the alcohol thermometer in 1709 and the mercury thermometer in 1714.

The Celsius scale was developed in a similar way, but the freezing point of pure water was chosen to be 0°C and the boiling point of water 100°C. Figure 1.10 shows the relationship between these scales. Conversions between the two scales are given by the following expressions.

$$°F = (1.8 \times °C) + 32 \qquad (1.1)$$

$$°C = (°F - 32) \, / \, 1.8 \qquad (1.2)$$

Scientific uses of temperature require yet another temperature scale. The choice of the kelvin as the standard reflects mathematical convenience more than familiarity. The Kelvin scale is similar to the Celsius scale but draws its utility from the fact that the lowest temperature theoretically possible is taken as zero kelvin. It violates the laws of nature to go below 0 K, as we will see in Chapter 12. The mathematical importance of this definition is that we are assured we will not divide by zero when we use a formula that has temperature in the denominator of an expression. Conversions between Celsius degrees and kelvin are common in science, and are also more straightforward.

$$K = °C + 273.15 \qquad (1.3)$$

$$°C = K - 273.15 \qquad (1.4)$$

Engineers in some disciplines use the Rankine temperature scale (°R), which is an absolute scale whose degrees are the same size as those of the Fahrenheit scale.

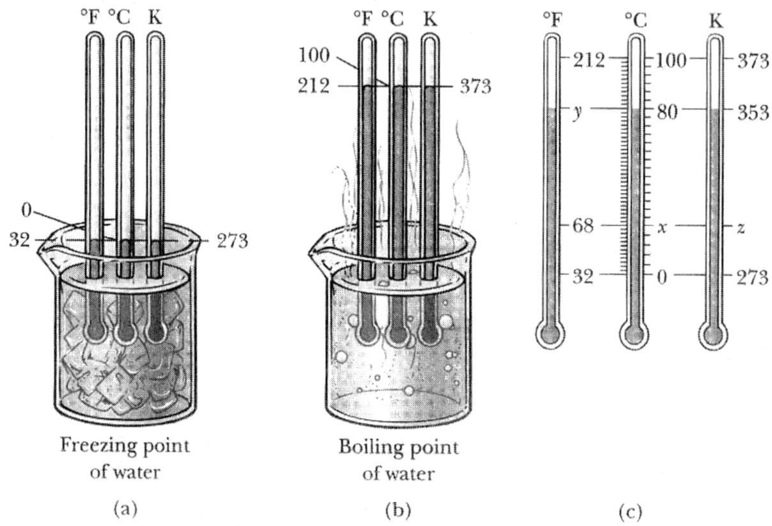

Figure 1.10 The Fahrenheit, Celsius, and Kelvin temperature scales are compared. The freezing point of water can be expressed as 32°F, 0°C, or 273 K. The boiling point of water is 212°F, 100°C, or 373 K. You should be able to find the temperatures represented by x, y, and z on the thermometers in the right hand panel.

Numbers and Significant Figures

We often encounter very small and very large numbers in dealing with chemistry problems. For example, pesticide production in the world exceeds millions of tons, while residues that may cause harm to animals or humans can have masses as small as nanograms. For either type of number, **scientific notation** is useful. Numbers written using scientific notation factor out all powers of ten and write them separately. Thus the number 54,000 is written as $5.4\text{x}10^4$. This notation is equivalent to 5.4 x 10,000, which is clearly 54,000. Small numbers can also be written in scientific notation using negative powers of ten, because 10^{-X} is identical to $1/10^X$. The number 0.000042 is $4.2\text{x}10^{-5}$ in scientific notation.

When numbers are derived from observations of nature, we need to report them with the correct number of significant figures. **Significant figures** are used to indicate the amount of information that is reliable when discussing a measurement. Numbers can be manipulated in a mathematical sense without accounting for how much information is reliable. When the integer number 5 is divided by the integer number 8, for example, the answer is 0.625. When numbers associated with an observation are used, however, we must be more careful about reporting digits. To understand why, consider whether or not to accept a wager about some measurement. For example, an almanac lists the population of Canada as 31,110,600. Suppose a study concluded that 24% of the people who live in Canada speak French. Based on this information alone, the wager is that 7,466,544 in Canada speak French. Should we accept the bet? What if the wager was that roughly 7.5 million Canadians speak French? Is this a better bet? This scenario shows the importance of significant figures. The figure of 7,466,544 is not attractive because we don't really know if 24% is *exactly* 24.000000%. It could well be 23.95% and the correct answer would then be 7,450,989. Both answers might qualify as "roughly 7.5 million," an answer with a more reasonable number of significant figures.

This type of reasoning has been formalized into rules for significant figures (digits) in numbers reported from scientific observations. When a measurement is reported numerically, we generally consider each digit given to be known accurately, with one important exception. There are special rules for zero such that it is sometimes significant and sometimes not. When it establishes the place for the decimal in the number, a zero is not significant. Thus the measurement 51300 m has two zeros that are not significant, as does the measurement 0.043 g. A zero is significant when it is the final digit *after* a decimal place and when it is *between* other significant digits. Thus the zeros in both 4.30 mL and 304.2 kg are significant. When numbers are written properly in scientific notation all of the digits provided are significant. Example Problem 1.2 provides some additional practice with determining the number of significant figures in measurements.

Scientific notation provides an advantage here, because any digits shown in the number are significant. There are no zeros needed for placement of the decimal.

Example Problem 1.2

An alloy contains 2.05% of some impurity. How many significant figures are reported in this value?

Strategy

Use the general rule for significant figures: digits that are reported are significant unless they are zeros whose sole purpose is to position the decimal place.

Solution

In this case, all digits reported are significant, so there are three significant figures in the number.

Check Your Understanding

How many significant figures are reported in each of the following measurements? (a). 0.000403 s (b). 200,000 g

The other aspect of significant figures in the use of numbers is in how they are used in calculations. Our earlier story with the number of French-speaking Canadians provides some insight into this issue as well. The wager that roughly 7.5 million such people live in Canada was the more attractive. The reason: 7.5 million has the same number of significant figures as the value 24%, so it provides a similar quality of information. Three key rules will be required to determine the number of significant figures for results of calculations.

Rule 1: For multiplication and division, the number of significant figures in a result must be the same as the number of significant figures in the factor with the fewest significant figures. When 0.24 kg is multiplied by 4,621 m the result on a calculator reads 1109.04, but if significant figures are correctly reported the result will be $1.1\text{x}10^3$ kg m. The value 0.24 kg has only two significant figures, and so the result should also have just two significant figures.

Rule 2: For addition and subtraction, the rules for significant figures center on the digits to the right of the decimal place. The result (a sum or difference) must have the same number of digits to the right of the decimal point as are present in the measurement that has the least number of digits to the right of the decimal point. The number 51.3

m added to 4.882 m yields 56.182 on a calculator, but should be reported as 56.2 m. In this case the last allowed digit was "rounded up" because the first non-significant digit, 8, was greater than 5. There are two possible conventions for rounding in such cases. In this text we will round down for numbers 4 and smaller and round up for numbers 5 and larger. Example Problem 1.3 provides some practice in working with significant figures.

Example Problem 1.3

Report the result for the indicated arithmetic operations using the correct number of significant figures. Consider all values to be measurements.

(a) 4.30 x 0.31 (b) 4.033 + 88.1 (c) 5.6 / 17.32

Strategy

Check the rules for significant figures as needed and determine which rules apply. Carry out the calculation and express the result in the correct number of significant figures.

Solution

(a) 4.30 x 0.31 = 1.3

(b) 4.033 + 88.1 = 92.1

(c) 5.6 / (17.32 x 10^4) = 3.2 x 10^{-5}

Check Your Understanding

Determine the value of the following expressions using the correct number of significant figures. (a) 7.10 m + 9.003 m (b) 0.004 g x 1.13 g

The rules above apply to any numbers that result from most of the measurements that we might make. But in the special case of countable objects, we must also consider one additional rule.

Rule 3: When we count discrete objects, the result has no ambiguity. Such measurements use exact numbers and they have essentially infinite significant figures. Thus if we need to use information such as four quarts in a gallon, there is no limitation on significant figures. Note that this rule also applies when we work with the various prefixes in the SI system. There are *exactly* 100 centimeters in a meter, so the factor of 100 would never limit the number of significant figures in a calculation.

1.5 Problem Solving in Chemistry and Engineering

Calculations play a major role in the practice of chemistry and in its application to real world issues and problems. And engineering designs routinely rely on a tremendous number of calculations. The types of exercises we will introduce in this text, though based in chemistry, will provide practice with techniques that can also be utilized in engineering applications. To a chemist, the questions associated with aluminum ore require looking into the nature of chemical bonding and how to overcome the stability associated with strong bonds between aluminum and oxygen. To an engineer, the problems associated with refining the ore might focus on how to deliver enough electricity when and where it is needed. What both disciplines need, though, is the careful use of quantitative or numerical reasoning.

Using Ratios

We encounter and use ratios regularly. We discuss the speed of our cars in miles per hour and buy our fruit at \$1.09 a pound. Most of us use such ratios intuitively. If we look at how we do this, we can deduce a formal set of rules to apply to chemical observations and situations. Let's say your grocery bill indicates you paid \$4.45 for a 5.0-pound bag of apples. How much did you pay per pound? While you may not think about it, you obtain the answer by creating an appropriate ratio and completing some simple arithmetic.

$$\text{price} = \frac{\$4.45}{5.0 \text{ pounds}} = \$0.89 \text{ per pound}$$

The notation "per" means for each of the specified units. Price per pound is the cost of ONE pound. Miles per hour tell the distance traveled in ONE hour. This is useful because one is a convenient number in multiplication. We obtain this type of information by creating and simplifying a ratio.

With the information we have about the bag of apples, we have another option to create a ratio. We could obtain the pounds of apples per dollar.

$$\frac{5.0 \text{ pounds}}{\$4.45} = 1.1 \text{ pounds of apples per dollar}$$

For this example such a manipulation may not be particularly useful – we would probably prefer to buy our apples in whole numbered quantities. Still, it provides useful insight. When we form the ratio and carry out the indicated arithmetic, the result tells us how much of the numerator is equivalent to ONE unit of the denominator. So, in general, given two equivalent quantities A = B, we can write either of two ratios.

A / B (tells how much A is in ONE unit of B)

or

B / A (tells how much B is in ONE unit of A).

Example Problem 1.4 shows how this type of manipulation fits into a problem solving strategy with non-chemistry examples before we look at it in a chemistry context.

Example Problem 1.4

Shrimp are usually labeled with a "count" that indicates the average number of shrimp per pound. The bigger the shrimp are, the smaller the count would be. Suppose your supermarket is offering 20-count shrimp for \$5.99 per pound. How much should you expect to pay for one dozen shrimp?

Strategy

We can determine the average cost per shrimp, then multiply by the number of shrimp needed to find the total price.

Solution

$$\frac{\$5.99}{20 \text{ shrimp}} \times 12 \text{ shrimp} = \$3.59$$

Check Your Understanding

A farmer purchases pesticide in 5-gallon drums that cost $23.00 each. If 65 gallons of the pesticide is applied to a field, what was the total cost? How might the situation be more complicated if the amount needed were not a multiple of five gallons?

For some problems in chemistry the units of measurements can be used to determine how to write the appropriate ratio. In the previous example, by considering the units in an algebraic sense, the "shrimp" (with the 20 in the denominator are "cancelled" by the same unit (with the 12) in the numerator.

$$\frac{\$5.99}{20 \, \cancel{\text{shrimp}}} \times 12 \, \cancel{\text{shrimp}} = \$3.59$$

This type of reasoning is referred to as **dimensional analysis** or the **factor-label** method for calculation. In some cases, using the dimensions can provide clues about what ratio to define. We will point out this type of reasoning in problems where it may be used.

Ratios in Chemistry Calculations

We use ratios in a variety of common calculations in chemistry. One that you will undoubtedly also encounter in engineering is the need to convert between units of different sizes. Example Problem 1.5 provides an introduction to this type of manipulation.

Example Problem 1.5

Visible light is commonly measured by considering its wavelength, in units of nanometers. In subsequent calculations this measurement often needs to be expressed in units of meters. If we are considering green light of wavelength 615 nm, what is its wavelength in meters?

Strategy

We need to establish a ratio that relates nanometers and meters. Then we can use that ratio to determine the needed answer.

Solution

$$1 \text{ m} = 1 \text{x} 10^9 \text{ nm}$$

We can write this as a ratio. Because we want to convert *from* nm and *to* m, we'll need m in the numerator and nm in the denominator.

$$\frac{1 \text{ m}}{10^9 \text{ nm}}$$

Then we just complete the calculation.

$$615 \text{ nm} \times \frac{1 \text{ m}}{10^9 \text{ nm}} = 6.15 \times 10^{-7} \text{ m}$$

Discussion

Writing the units on all of the quantities involved helps assure that we use the correct ratio. We should also check that the answer makes sense. Since nanometers are much smaller than meters, the fact that we got a much smaller number for our answer seems correct. Note that conversions between SI units are between exact numbers, when considering significant figures.

Check Your Understanding

Large amounts of electrical power, measured in watts (W), are used in the production of aluminum from ore. If a production plant uses 4.3 GW in a certain period of time, how many W are used?

Another common type of problem that uses ratios occurs when two quantities are related to each other either by some chemical or physical property. One such property is **mass density**. The mass density is defined as the mass of a substance per unit volume. This definition is itself a ratio, and allows us to convert between mass and volume as shown in Example Problem 1.6.

Example Problem 1.6

The mass density of water at 25°C is 0.975 g per mL. A child's swimming pool holds 346 L of water at this temperature. What mass of water is in the pool?

Strategy

We want to use the density as a ratio, but must first express the volume in appropriate units. We can convert the volume of the pool from liters to milliliters, and then use the density as given.

Solution

1 L = 1000 mL, so 1000 mL / 1 L provides a conversion from L to mL.

The given density is 0.975 g / 1 mL, and this provides the connection between volume and mass.

$$346 \text{ L} \times \frac{1{,}000 \text{ mL}}{1 \text{ L}} \times \frac{0.975 \text{ g}}{1 \text{ mL}} = 3.37 \times 10^5 \text{ g}$$

Discussion

Here we have written our solution as a single two-step process, in which we first converted the volume from liters to milliliters and then used the density to find the requested mass. Students sometimes feel more comfortable breaking these steps down into separate calculations; we could have explicitly found the volume as 346,000 mL and then multiplied that by the density. As long as the operations are carried out correctly, the result will be the same. If you do choose to break calculations down into smaller pieces, though, be careful not to round your answers until you reach the final result.

Check Your Understanding

A liquid pesticide has a density of 1.67 g per mL. What is the mass of liquid in a full 20.0-L container?

Calculations in chemistry use ratios very often. Example Problem 1.6 also shows how they are used in combinations. Learning how to carry out these sequential manipulations will be a key part of solving many of the problems we encounter in chemistry. One guide that often helps determine how to construct the appropriate ratio is to note the units being used. For example, the equality between 1000 mL and 1 L leads to two ways to express the ratio, 1000 mL/1L or 1L/1000mL. Which ratio did we need? Because we started with liters, we required the former ratio, with 1 L in the denominator. We will often include discussions about the ratios we form in the "strategy" part of the sample exercises we carry out.

Mass density also is important in determining bouyancy, because less dense objects float on more dense ones.

Mass density is more than a convenient way to define a ratio for carrying out calculations. It is also an important characteristic of materials. Among the reasons why aluminum is a popular structural material, its relatively low density is often important. When a design calls for a strong but lightweight material, aluminum is a good candidate. We'll look at this aspect further in the closing insight in Section 1.6 when we look at various materials used in bicycle frames.

Conceptual Chemistry Problems

While carrying out numerical calculations will always be an important component of chemistry, it is only a part of the field. Sometimes, in order to make sure that the concepts involved in chemistry are understood, we'll also work problems that focus on the particulate representation and other concepts. The strategies employed in this type of question are often different than those we have just described.

When we try to visualize a chemical concept using drawings, we will start from the conventions that were introduced for the particulate perspective of chemistry in Section 1.3. In this type of problem, we could be asked, for example, to draw a diagram that depicts what happens to the molecules when steam condenses into liquid water. Example Problem 1.7 illustrates this type of thinking.

Example Problem 1.7

Dry ice is solid carbon dioxide and it is called "dry" in large part because it goes directly from a solid to a gas without becoming a liquid under everyday conditions. Draw a picture that shows what the carbon dioxide molecules might look like before and after this process.

Strategy

To answer conceptual questions such as this one, we must depict the differences between the two phases in some way. Our drawing can be schematic, but it must imply the differences clearly. In this case, we know that a solid will have closely packed molecules, while a gas will exhibit significant space between the particles.

Solution

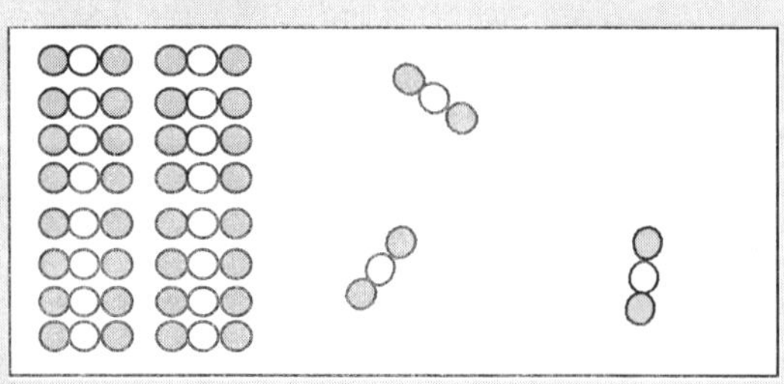

Our picture shows several things. First, we have included some information about the chemical composition of the molecules by using two different colors for the circles representing the atoms. Second, the solid is indicated on the left and is distinguishable because the molecules are shown closely packed together. Finally the gas is depicted with only a few molecules, and they are widely spaced.

Discussion

Obviously the solutions to conceptual questions like this are less exact than numerical answers. So your drawing might look somewhat different while still being "correct." Neither the gas nor the solid needs to look exactly like the picture we have drawn, but the drawing itself should impart the essential concepts we need to understand when thinking about the particulate perspective of this process.

Check Your Understanding

Draw a picture that shows a molecular scale view of steam condensing into liquid water.

We will continue to use conceptual problems throughout the text to help provide practice in thinking about how molecules and atoms behave.

Visualization in Chemistry

One area where conceptual understanding in chemistry is different than in most engineering pursuits is the nature of how we visualize the systems being studied. Chemistry, when it is applied in its most powerful way, provides multiple simultaneous ways to view problems, including the completely abstract perspective of atoms and molecules that will never be directly observed. Methods to visualize this level of chemistry provide an important tool in the way that chemistry is taught and learned. We can look at this aspect of learning chemistry by thinking about the refining process of aluminum.

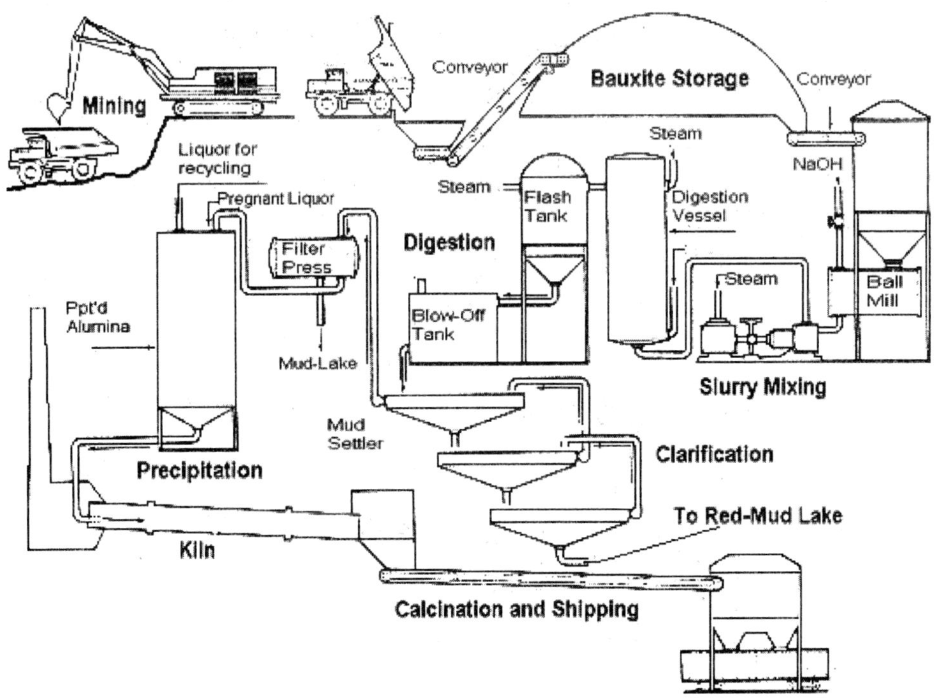

Figure 1.11 Several steps in the processing of bauxite are represented in this diagram. (Drawing courtesy of Prof. Robert J. Lancashire, University of the West Indies)

The most common aluminum ore is bauxite, which consists of alumina and rock. The first step in the production of aluminum metal is to separate the alumina from the remainder of the rock present. As shown in Figure 1.11, this actually entails several steps, the first being a digestion of the ore. Process engineers design "digesters" in which crushed ore and sodium hydroxide (known as NaOH or lime) are mixed at high temperatures to create a slurry. If we look at Figure 1.12, we can visualize this process at the microscopic level. The molecular level pictures show alumina as atoms of aluminum and oxygen, where the rock is shown as silicon and oxygen (or silica). This is essentially a simplified depiction based on the fact that much of the Earth's crust of contains a large percentage of these elements. (Bauxite ore also contains significant amounts of iron oxide and other minerals, but these behave much like the silica in the digestion process.) Using this level of visualization to consider the reactions that take place in the digester, we see that the alumina reacts with the NaOH while the silica does not. In this way we can look at a large-scale industrial process while thinking at the microscopic perspective.

Note that this type of conceptual drawing is schematic. It does not attempt to impart molecular detail, but rather it aims to provide a sense of the particles involved in the chemcical process. In this perspective, both Figures 1.12 and 1.13 are schematic.

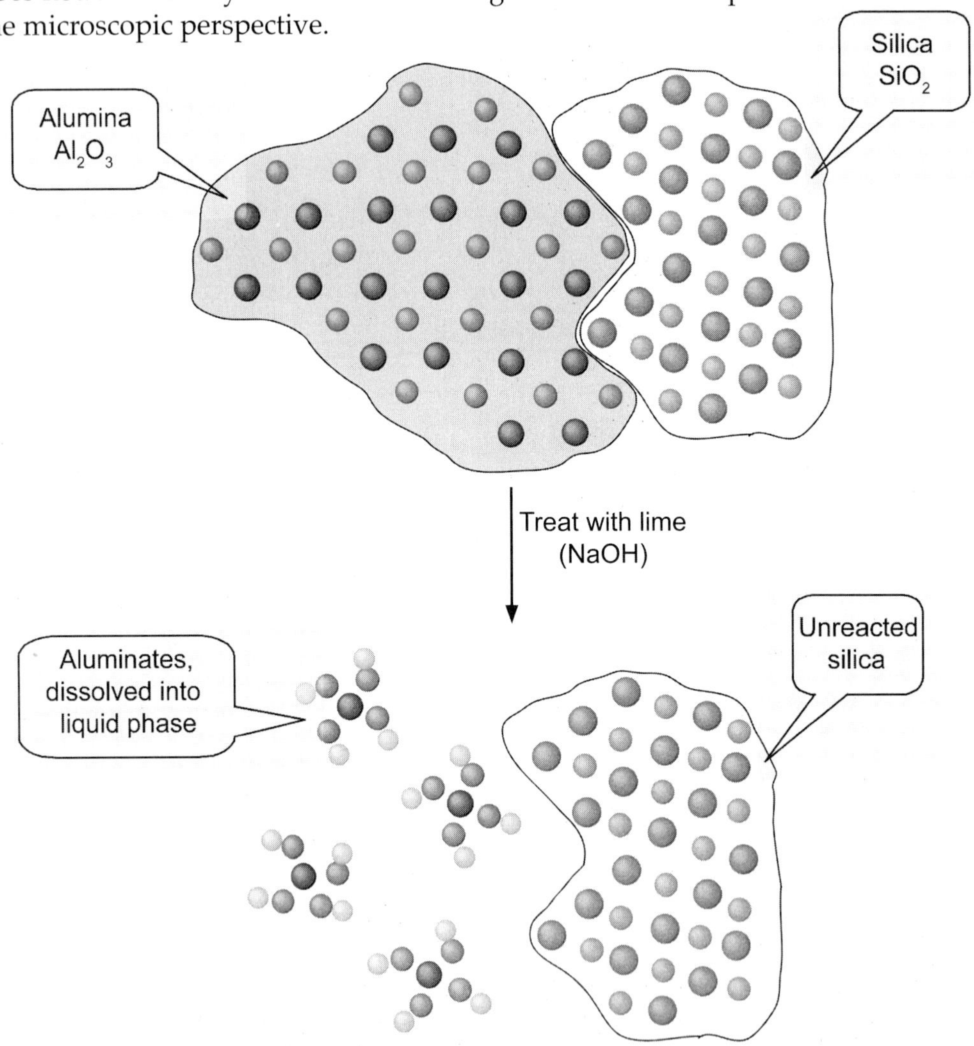

Figure 1.12 Bauxite is treated with lime in a process called *digestion*. The alumina in the ore reacts with the lime, and dissolves. This step achieves the separation of the aluminum from the rest of the ore, which contains a variety of other minerals.

The next step in aluminum refining is smelting. At this stage a chemical reaction is induced where the aluminates, now dissolved in a material called cryolite (Na_3AlF_6), lose oxygen atoms to a carbon rod, forming relatively pure aluminum and carbon dioxide. This portion of the process is shown in Figure 1.13. While this molecular level visualization again shows how the chemical reaction is rearranging atoms at the particulate level, it is worth remembering that this process is carried out on a truly massive scale: the U.S. aluminum industry produces about 2.7 million metric tons of aluminum per year.

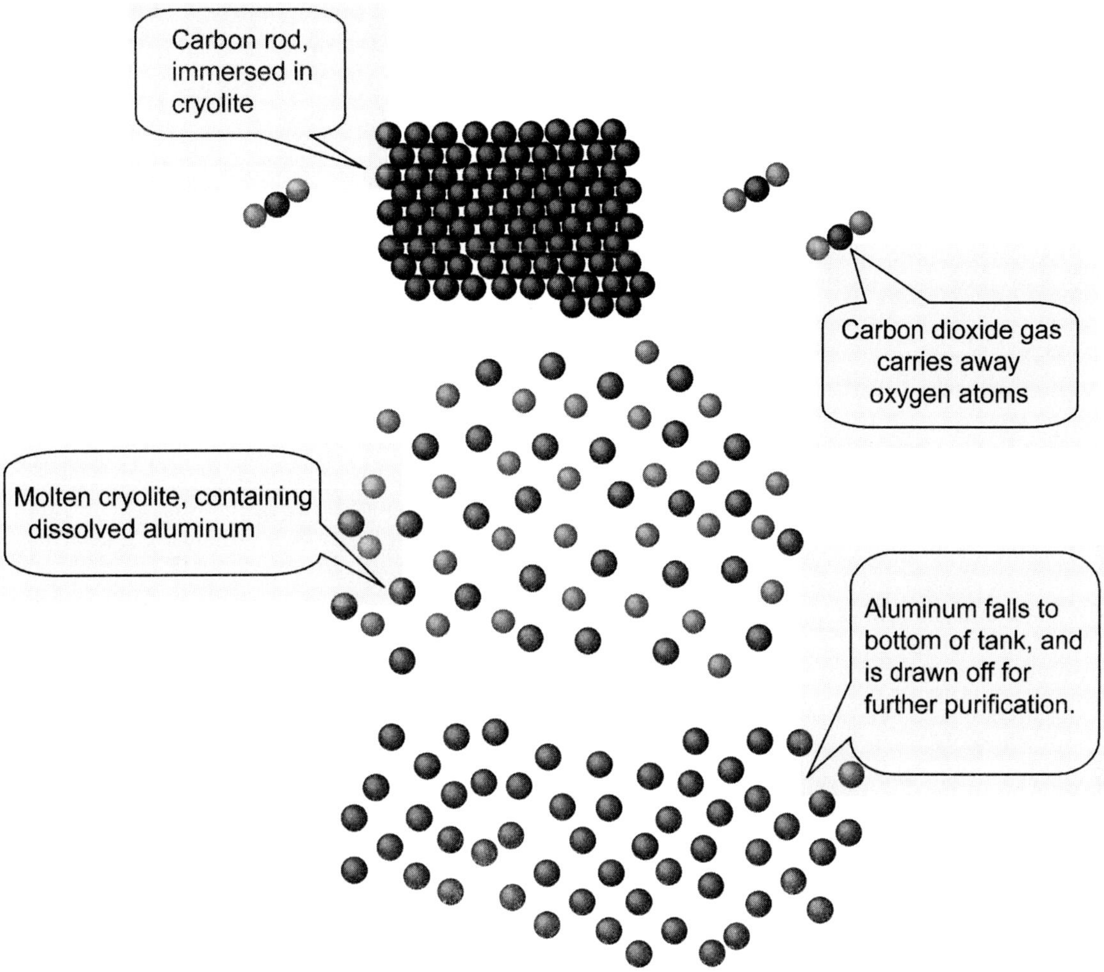

Figure 1.13 These particulate level illustrations provide a simplified view of the atomic scale process involved in the smelting of aluminum.

The visualization techniques introduced in this section will help us to develop a particulate level understanding of many concepts we encounter throughout the text. At this point let's take some time to look at one application where the properties of aluminum metal make it a popular material choice for engineering a consumer product – a bicycle.

1.6 Insight into Material Selection and Bicycle Frames

For the average rider who doesn't wish to spend a great deal of money on a bike, the frame is likely to be some alloy of steel. Using other materials for the frame can improve certain aspects of performance, but also tends to increase the cost of the bicycle. The choices a consumer has are confronted first by the engineer who designs the frame.

Among the properties that an engineer must consider in choosing a material for a frame are strength, density (which affects the weight of the frame), and stiffness.

While we are already familiar with density, talking about the strength and stiffness of materials requires some additional vocabulary. Stiffness is related to a property called the **elastic modulus** of the material. The elastic modulus measures the amount of stretch or compression a material experiences when it is stressed. Something with a high modulus will stretch very little even when experiencing large forces, and a bicycle frame composed from such a material would seem stiff. We will see this concept again when we discuss chemical bonding, because the ability of a material to stretch is related to how closely and strongly the atoms are connected to each other in the material.

Measurements such as elastic modulus or yield strength depend on temperature. For most engineering designs, however, this dependence is more critical in materials like plastics than in metals.

The strength of a material is formally measured by a property called the yield strength. The **yield strength** measures the amount of force required to produce a specified deformation of the material. A stronger material can withstand greater forces before it deforms, and so has a higher yield strength. Finally, because there are benefits to having a lighter frame, materials with low densities are also desirable.

The three most common materials for bicycle frames are aluminum, steel and titanium. Table 1.3 shows typical values for the elastic modulus, yield strength and density of these materials.

Material	Elastic Modulus (psi)	Yield Strength Range (psi)	Density (g/cm^3)
Aluminum	10.0×10^6	$5 \times 10^3 - 6 \times 10^4$	2.699
Steel	30.0×10^6	$4.5 \times 10^4 - 1.6 \times 10^5$	7.87
Titanium	16.0×10^6	$4 \times 10^4 - 1.2 \times 10^5$	4.507

Table 1.3 The elastic modulus, yield strength, and density are related to the stiffness, strength, and weight of a material. These are just a few of the many engineering properties that would be considered in the design of a bicycle frame.

Looking at this table, we can see that the physical properties of these materials afford tradeoffs in types of performance in the frame. Aluminum, while it is lightweight, is neither as strong nor as stiff as steel or titanium. Steel is the strongest, stiffest and heaviest of the materials. It's important to note the range of yield strengths in this table, because the details of how one can manipulate the strength of aluminum or steel depend on both the chemical composition and the manner in which the material is treated in fabricating the frame.

Of course, the engineer has more than the choice of material available in making a design for a bicycle frame. The size and connectivity of the tubing can also be adjusted, for example. Aluminum frames generally feature tubing with a much larger diameter than that found in steel or titanium frames. These larger diameter tubes give the frame itself a much stiffer feel – so much so that many cyclists find aluminum frames too stiff to give a comfortable ride. Many modern frames also use tubing that is not round, but rather is oval-shaped to specifically withstand certain types of stress that are commonly encountered in cycling. Some very expensive frames for competitive cycling are made from more exotic materials. New aluminum alloys containing trace amounts of scandium, for example, can achieve lower frame weights without requiring the large tubing diameters that cause the jarring ride aluminum is usually known for. And the extremely aerodynamic bikes preferred for high-speed time trial racing are most often made from carbon fiber composites, which can easily be molded into exotic shapes that reduce air resistance.

Chemistry and chemists have played an essential role in helping engineers develop and exploit these new materials. And the intersection between chemistry and engineering we've illustrated in this chapter for bicycles and soda cans is also apparent in looking at many other familiar products. Throughout this course, we'll look at places where the chemistry we learn has an impact on engineering designs. At this point, we know enough to say that if a lightweight bicycle frame is your desire, you may want to take advantage of aluminum and its low mass density. In Chapter 2 we take the next step and look at atoms and molecules in more detail and simultaneously introduce another important class of engineering materials – polymers.

Focus On Problem Solving

Engineering students often question wonder whether their college chemistry courses will prove useful later in their career. Through the opening and closing "Insight" sections of the chapters of this book, we'll make connections between the content of the chemistry course and places where that content is relevant in engineering applications. But this type of connection isn't the only place where things learned in chemistry courses help future engineers.

Because engineering is a problem solving discipline, the exposure to problem solving techniques in courses like this one help students to diversify and enhance their problem solving abilities. We will emphasize this connection after many of the chapters in this book with a special section like this one. In these "Focus on Problem Solving" sections, we will look at chemistry problems related to the chapter just presented. The difference will be that the "correct" answer will not be a number, but rather the identification of an appropriate strategy for solving the problem. We begin, here in Chapter 1, with the following question.

Question

Describe how you would determine which has a greater mass, a sphere of mercury with a radius of 2.00 cm or a cube of lead with an edge length of 2.00 cm. What formulas are needed and what other information would have to be looked up?

Strategy

Like many "real world" problems, this question does not have enough information given to provide an answer. To devise a strategy therefore, we have to think about what it takes to compare two materials – in this case mercury and lead. Moreover, we are asked to compare masses of the two materials, so that provides a clue. The dimensional information we are provided, along with formulas that are either familiar or can be looked up, provides information about volume. Mass and volume were related in this chapter by the density.

Solution

To answer this problem we would first need to look up the density of both mercury and lead. Then, we would need to use the data provided to determine the volume of each sample. For a cube, we can calculate the volume by taking the given measure for the side (s) and cubing it (s^3). For a sphere the volume is given by, $(4/3)\pi r^3$. Multiplying the density times volume provides the mass, so with the data we looked up for density and the calculated volumes we could answer the question.

Chapter 1
Problems & Exercises

Insight into Aluminum

1.1 Use the web to determine the mass of a steel beverage can from the 1970's and find the mass of a modern aluminum can. How much more massive would a 12-pack of soda weigh if aluminum were not used?

1.2 What properties of aluminum might cause you concern if you had to use the aluminum tableware that Napoleon employed to impress his guests?

1.3 Where does the scientific method "start?" What is the first step?

1.4 Use the web to determine the amount of aluminum used in the US in a single year. What is the primary use for this material?

1.5 Use the web to find current prices being offered for aluminum for recycling. Is there variation in the price based on where in the US the aluminum is returned?

1.6 Use the web to determine the difference in how much aluminum is recycled in states where there are deposits on aluminum cans versus states where recycling is voluntary? What is the most reliable way to estimate this value? What uncertainty is there in the estimate?

The Study of Chemistry

1.7 When we make observations in the laboratory, which perspective of chemistry are we normally utilizing?

1.8 Which of the following items are matter and which are not? (a) a flashlight (b) sunlight (c) an echo (d) air at sea level (e) air at the top of Mount Everest

1.9 What macroscopic characteristics differentiate solids, liquids and gases? (List as many as possible.)

1.10 How can a liquid be distinguished from a fine powder? What type of experiment or observation might be undertaken?

1.11 Some farmers use ammonia, NH_3, as a fertilizer. This ammonia is stored in liquid form. Use the particulate perspective to show the transition from liquid ammonia to gaseous ammonia.

1.12 Do the terms element and atom mean the same thing? If not how do they differ?

1.13 Which of the items listed describe physical changes and which describe chemical changes? (a) a banana ripens (b) water evaporates (c) an object falls out a window (d) food is digested

1.14 Why do physical properties play a role in chemistry if they do not involve any chemical changes?

1.15 Physical properties may change because of a chemical change. For example, the color of an egg "white" changes from clear to white when it is cooked because of a chemical change. Think of another common situation when a chemical change also leads to a physical change.

1.16 Identify the physical and chemical properties noted in the following statement: When a piece of iron is exposed to air it corrodes, forming a layer of red-brown powder that flakes off and is denser than the original material.

1.17 Use a molecular level description to explain why gases are less dense than liquids or solids.

1.18 All molecules attract each other to some extent, and the attraction decreases as the distance between particles increases. Based on this idea, which state of matter would you expect has the strongest interactions between particles, solids, liquids or gases?

Observations and Models

1.19 We used the example of attendance at a football games to emphasize the nature of observations. Describe another example where deciding how to "count" the subjects of interest could affect the observation.

1.20 Complete the following statement: Data that have a large random error may be (accurate, precise, or neither).

1.21 Complete the following statement: Data that have a large systematic error may be (accurate, precise, or neither).

1.22 Two golfers are practicing shots around a putting green. Each golfer takes 20 shots. Golfer 1 has 7 shots within 1 meter of the hole, and the other 13 shots are scattered around the green. Golfer 2 has 17 shots go into a small sand trap near the green and three just on the green near the trap. Which golfer is more precise? Which is more accurate?

1.23 Use your own words to explain the difference between deductive and inductive reasoning.

1.24 Suppose you are waiting at a corner for a bus. Three different routes pass this particular corner. You see a bus pass by from the two routes you are not interested in taking. When you say to yourself, "My bus must be next," what type of reasoning (deductive or inductive) are you using? Explain your answer.

1.25 When a scientist looks at an experiment and then predicts the results of other related experiments, what type of reasoning is being used? Explain your answer.

1.26 What is the difference between a hypothesis and a question?

1.27 What must happen in order for a model to come to be considered a theory?

1.28 What is a law of nature? Are all scientific laws examples of a law of nature?

Numbers and Measurements

1.29 Describe an example of a miscommunication that can arise because units are not included as part of the information.

1.30 What is the difference between a qualitative and a quantitative measurement?

1.31 Identify which of the following units are base units in the SI system: grams, meters, joules, liters, amperes.

1.32 What is a "derived" unit?

1.33 Rank the following prefixes in order of increasing size of the number they represent, centi-, giga-, nano-, kilo-

1.34 The largest computers now include disk storage space measured in terabytes. How many bytes are in a terabyte? (Recall that in computer terminology, the prefix is only "close" to the value it designates in the metric system.)

1.35 How many μg are equal to one gram?

1.36 Convert the value 0.120 ppb into ppm.

1.37 How was the Fahrenheit temperature scale calibrated? Describe how this calibration process reflects the measurement errors that were evident when the temperature scale was devised.

1.38 Oxygen boils at –186°C. What is this temperature in kelvin?

1.39 Express each of the following temperatures in kelvin. (a) -10°C (b) 0°C (c) 280°C (d) 1.4×10^3 °C

1.40 Express each of the following temperatures in °C. (a) 177 K (b) 298 K (c) 4 K (d) 1500 K

1.41 Express each of the following numbers in scientific notation. (a) 62.13 (b) 0.000414 (c) 0.0000051 (d) 871000000 (e) 9100

1.42 How many significant figures are present in each of the following experimental measurements? (a) 0.003 m (b) 6030 kg (c) 400.3 s (d) 0.000701 L (e) 31.624 km

1.43 How many significant figures are present in each of the following experimental measurements? (a) 4.15×10^3 km (b) 4.15×10^6 km (c) 3.10×10^{-5} s (d) 1.1×10^{-9} g

1.44 Assuming the numbers given are measurements, carry out the indicated arithmetic operations and give the answer with the correct number of significant figures.

(a) 7.132 / 6 (b) $3.65 \times 10^2 + 8.1 \times 10^3$

(c) 18.13 - 1.3 (d) $9 \times 10^{-6} \times 1.33 \times 10^{-3}$

1.45 Assuming the numbers given are measurements, carry out the indicated arithmetic operations and give the answer with the correct number of significant figures.

(a) 34.1 + 0.34 (b) $3.4 \times 10^{-4} \times 1.4 \times 10^3$

(c) 5.2 x 6.13 (d) $5.102 \times 10^3 / 4.3 \times 10^5$

1.46 In an attempt to determine the velocity of a person on a bicycle, an observer uses a stopwatch and times the length of time it takes to cover 25 "squares" on a sidewalk. A measurement of one of the squares shows it to be 1.13 m long. The bicycle takes 4.82 seconds to travel this far. What velocity, in m/sec, should the observer report?

1.47 A student measures the mass of an object to be 4.131 g and its volume to be 7.1 mL. What density should be reported in g/mL?

1.48 Measurements indicate that 23.6% of the residents of a city with a population of 531,314 are college graduates. Considering significant figures, how many college graduates are estimated to reside in this city?

Problem Solving in Chemistry and Engineering

1.49 A package of 8 apples has a mass of 1.00 kg. On average what is the mass of one apple?

1.50 If a 1.00-kg bag containing 8 apples costs $1.48, how much does one apple cost? What mass of apples costs $1.00?

1.51 A person is measured to be 173 cm tall. What is this height in meters? In feet and inches?

1.52 The distance between two atoms in a molecule is 148 pm. What is this distance in meters?

1.53 3.47×10^{-6} g to μg (b) 2.73×10^{-4} L to mL (c) 725 ns to s (d) 1.3 m to km

1.54 Carry out the following unit conversions. (a) 82 μg to mg (b) 6.73×10^{-4} mm to nm(c) 1.37×10^{6} nA to mA (d) 4.8×10^{18} mW to GW

1.55 What is the mass of 4.3×10^{6} lbs expressed in kg? 1 pound = 454 g.

1.56 If a vehicle is traveling 92 m/s what is its velocity in miles per hour? 0.60 miles = 1.00 km.

1.57 A load of asphalt contains 115 kg and occupies a volume of 220 L. What is the density of this asphalt in g/L?

1.58 A tube of toothpaste occupies a volume of 75 mL and contains toothpaste with a mass of 230 g. What is the density of the toothpaste?

1.59 A sample of crude oil has a density of 0.87 g/mL. What volume does a 3.6-kg sample of this oil occupy?

1.60 Mercury has a density of 13.6 g/mL. What is the mass of 4.72 L of mercury?

1.61 Draw a molecular scale picture that depicts the difference between a dense liquid such as mercury (density of 13.6 g/mL) and a less dense liquid such as water (1.00 g/mL).

1.62 Draw a molecular scale picture to show how a crystal differs from a liquid.

1.63 Draw a molecular scale picture that distinguishes between alumina and silica. Is your picture structurally accurate or schematic?

1.64 In the description of the refining of aluminum the molecular scale depiction of precipitation was not included. Use the web to determine what chemicals are involved in the precipitation process and draw a molecular scale drawing of this reaction.

Insight into Materials Selection and Bicycle Frames

1.65 Suppose a new material has been devised for a bicycle frame with an elastic modulus of 22.0×10^{6} psi. Is this bike frame likely to be more or less stiff than an aluminum frame?

1.66 Rank aluminum, steel and titanium in order of increasing stiffness.

1.67 Compare the strengths of aluminum, steel and titanium. If high strength is needed for a particular design, would aluminum be a good choice?

1.68 Aluminum is not as strong as steel. What other factor should be considered when comparing the desirability of aluminum versus steel if strength is an important consideration for a design?

1.69 Use the web to research the differences in the design of steel framed bicycles versus aluminum-framed bicycles. Write a report that details the similarities and differences you discover.

1.70 Use the web to research the elastic modulus and yield strength of carbon fiber composites. How do these materials compare to aluminum, steel and titanium?

1.71 Use the web to research the relative cost of aluminum, steel and titanium frames for bicycles. Speculate about how much of the relative cost is due to the costs of the materials themselves.

Focus on Problem Solving Exercises

1.72 A student was given two metal cubes that looked similar. One was 1.05 cm on an edge and had a mass of 14.32 grams; the other was 2.66 cm on a side and had a mass of 215.3 grams. How can the student determine if these to cubes of metal are the same material using only the data given?

1.73 The diameter of a wire is sometimes referred to by its American wire gauge number. A 16-gauge wire has a diameter of 0.0582 in. Suppose you have a 1-pound spool of this wire, (a) what would you need to look up in order to determine the length of the wire on the spool and (b) once you had that information, how would you determine the length of the wire on the spool?

1.74 Battery acid has a density of 1.285 g/mL and contains 38.0% sulfuric acid by mass. Describe how you would determine the mass of pure sulfuric acid in a car battery, noting what item(s) you would have to measure or look up.

1.75 Unfermented grape juice used to make wine is called a "must." The sugar content of the must determines whether the wine will be dry or sweet. The sugar content is found by measuring the density of the must. If the density is lower than 1.070 g/ml sugar syrup is added until the density reaches 1.075 g/ml. Suppose you have a sample taken from a must whose mass is 47.28 g and whose volume is 44.60 mL. Describe how you would determine whether or not sugar syrup needs to be added and if so how would you estimate how much sugar to add?

1.76 A solution of ethanol in water has a volume of 54.2 mL and a mass of 49.6 g. What information would you need to look up and how would you determine the percentage of ethanol in this solution?

1.77 Legend has it that Archimedes, a famous scientist of Ancient Greece was once commanded by the king to determine if a crown he was presented was pure gold or a gold silver alloy. He was not, however, allowed to damage the crown (by slicing off a piece for example). If you were assigned this same task, what would you need to know about both gold and silver and how would you make a measurement that would tell you if the crown was pure gold?

1.78 Mercury is traded by the "flask," a unit that has a mass of 34.5 kg. How would you determine the volume of one of these flasks? What would you have to look up to solve the problem?

1.79 The chemical commonly referred to as dioxin has the formula, $C_{12}H_4O_2Cl_4$. This chemical is lethal to many laboratory animals; an oral dose of just 0.6 μg per kg of body mass will kill half of a test population of guinea pigs. Because animals are composed of primarily water, assume a 1 kg guinea pig is entirely water and use this assumption in a description of how you would determine the percentage of molecules that are dioxin in a guinea pig after such a dose is administered.

1.80 Imagine you place a cork measuring 1.30 cm x 4.50 cm x 3.00 cm in a pan of water. On top of this cork you place a small cube of lead measuring 1.15 cm on a side. Describe how you would determine if the combination of the cork and lead cube will still float in the water. Note any information you would need to look up to answer the question.

1.81 A calibrated flask was filled to the 25.00 mL mark with ethyl alcohol and it was found to have a mass of 19.7325 g. In a second experiment 25.0920 g of metal beads are put into the container and the flask was again filled to the 25.00 mL mark. The total mass of the metal plus the alcohol was 43.0725 g. Describe how to determine the density of the metal sample.

Chapter 2: Atoms and Molecules

Atoms and molecules are the building blocks of chemistry. You've probably been hearing this since middle school, so the existence of atoms is not something that you are likely to question or challenge. Chances are, however, that you rarely think about atoms or molecules when you come across most of the items in your day-to-day life. When chemists want to understand some aspect of the world around us, though, they tend to focus their attention at the level of atoms and molecules. So an important part of studying chemistry is learning how to interpret nature by thinking about what atoms and molecules are doing. It can certainly take some time to get comfortable with this type of thinking. Individual atoms and molecules are not usually observed, of course, so we often must infer what they are doing from indirect evidence. Chemistry has matured tremendously over the past century by building an increasingly thorough understanding of atoms and molecules. We will introduce some of the basic concepts of atoms and molecules here, and then refine these ideas as we progress through the text.

A variety of techniques known as scanning probe miscroscopy are now able to produce images of single atoms under some conditions.

2.1 Insight into Polymers

As human civilization and technology have progressed, eras have been closely associated with materials from which important objects have been made. The Stone Age gave way to the Bronze Age, which in turn was followed by the Iron Age. These labels were chosen much later, through the lens of history, and it may be dangerous to try to characterize our own time period. But it isn't hard to imagine that future archaeologists or historians might label the late 20th Century as the polymer age. As you go about your life, the plastics and synthetic fibers you encounter are all examples of what chemists call *polymers*. The properties and applications of these polymers are so diverse that you may not even recognize them as having anything in common with one another. Hard and durable plastics are routinely used as structural materials for things like computer cases or casual furniture. Softer, flexible plastics give us sandwich bags and Saran® wrap, while other polymers make up the nylon and rayon that are found in our carpets and clothing. Still more polymer materials, like the filling in many bulletproof vests, offer incredible combinations of light weight and high strength. The diversity of polymer properties is truly impressive. And yet if we take the chemist's approach, and turn our attention to the atomic and molecular level, we will see that all of these polymers share crucial common traits.

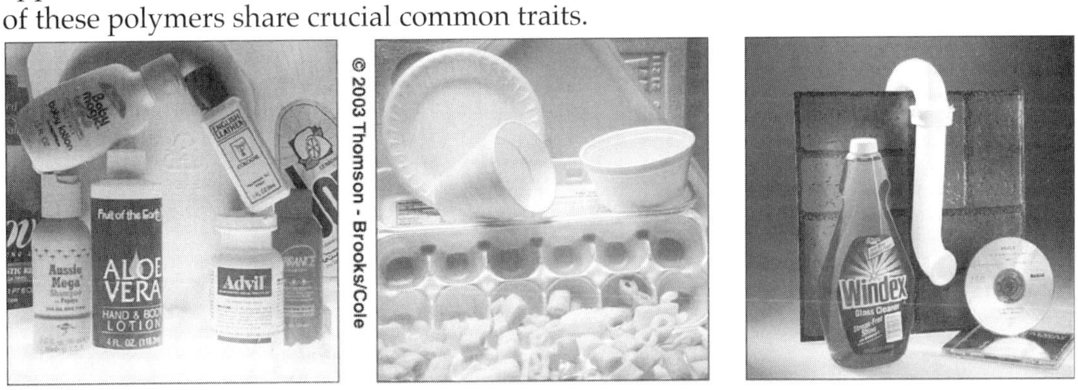

(a) (b) (c)

Figure 2.1 Polymers are the materials of choice for a host of everyday objects. All of the items shown in the left hand panel are made from high-density polyethylene. The objects in the center are made from polystyrene, and those in the right hand panel are made from poly(vinyl chloride).

Because these molecules are so large, polymers are also often referred to as macromolecules.

Traced back to its Greek origins, the word polymer literally means "many parts." And that definition offers us a clue as to what these seemingly disparate substances all have in common. All polymers are made up of very large molecules. And these large molecules, in turn, can be thought of as being made up of many smaller molecules, all linked end to end. A typical polymer molecule might contain hundreds or even thousands of these smaller constituent molecules, which are referred to as *monomers*. And if we look deeper into the composition of the polymer, we will find that these monomers are themselves made up of an assembly of atoms. But because polymer molecules are so large, it is often helpful to think of them as chains of monomers rather than collections of atoms.

Other uses of polyethylene are discussed in Section 2.7 at the end of this chapter.

The observable macroscopic properties of any particular polymer depend intimately on the identity of its constituent monomers, the number of monomers present, and the way that the monomers are connected to one another. To illustrate how dramatically the properties of a polymer can depend on its atomic composition, let's consider three common items and look at the polymers they are made from. The plastic bottles that your milk, juice, or shampoo come in are usually made of a plastic called polyethylene. Polyethylene molecules are composed entirely of just two elements, carbon and hydrogen. The carbon atoms are linked together in a long chain that is often referred to as the "backbone" of the polymer molecule, and there are two hydrogen atoms attached to each carbon. The molecular model at the top of Figure 2.2 shows a portion of a polyethylene molecule. If one of the two hydrogen atoms on every other carbon is replaced with a chlorine atom, as shown in the middle of Figure 2.2, we will have poly(vinyl chloride). This polymer is commonly referred to as PVC. Plastic pipe made from PVC has been widely used in plumbing for many years, so you have probably seen PVC pipe in either your house or a hardware store. As you might guess, the PVC used for pipes is much harder and stronger than the polyethylene found in soda bottles. Yet the chemical composition and structures of these two materials are very similar. Suppose we replace the second hydrogen on every other carbon atom with chlorine, too? Then we will have poly(vinylidene chloride), which is the "plastic wrap" used in almost every kitchen to cover leftovers.

These three common examples show how strongly the physical properties of a polymer are influenced by its chemical composition. To begin to explore the world of polymers in a systematic way, then, we will need to know some things about atoms, which are the simple building blocks of these giant molecules.

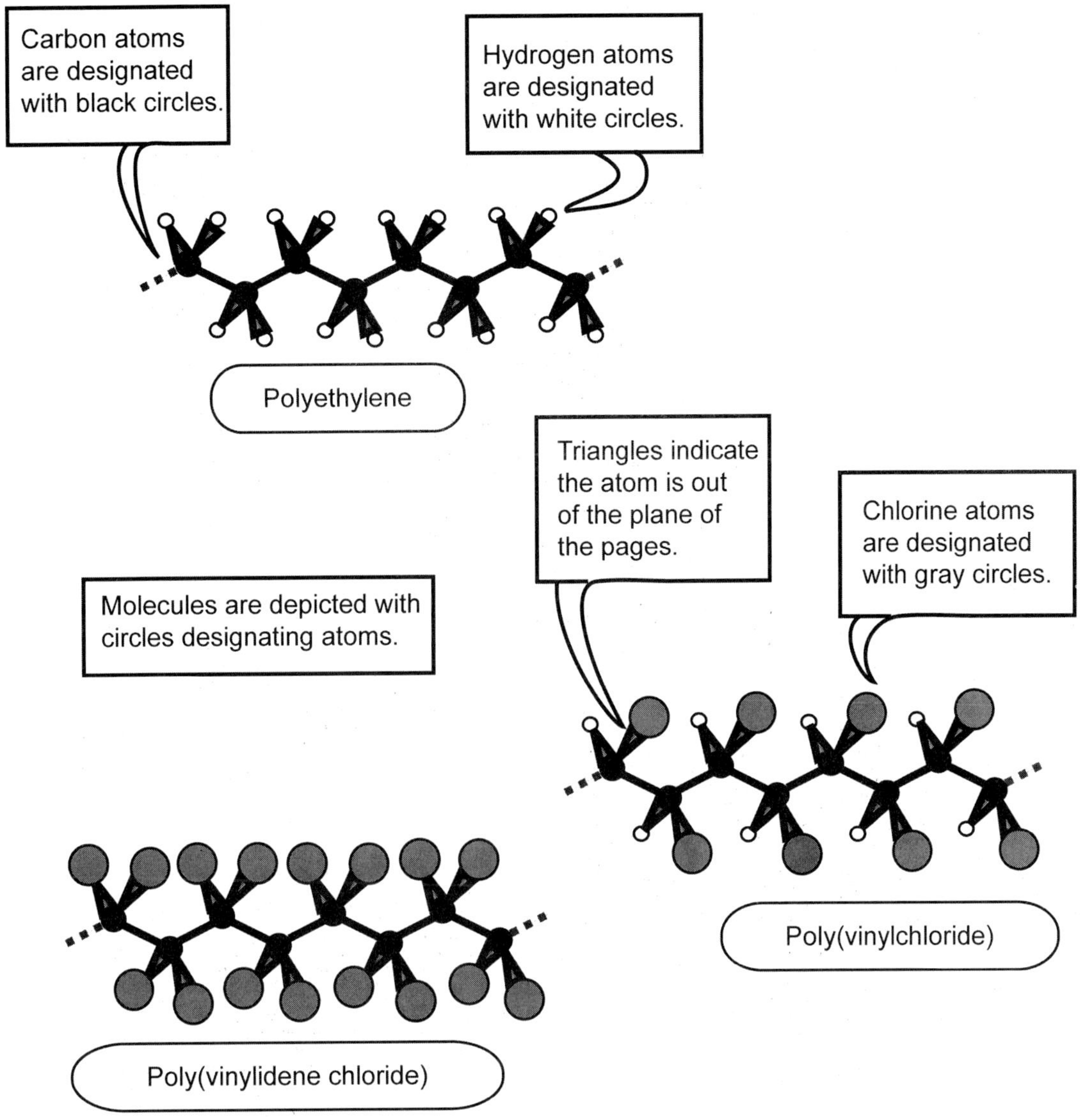

Figure 2.2 Models showing how atoms are arranged in molecules of polyethylene, poly(vinyl chloride), and poly(vinylidene chloride).

2.2 Atomic Structure and Mass

The comparison of polyethylene, poly(vinyl chloride) and poly(vinylidene chloride) demonstrates that the identity of the atoms in a molecule can have a tremendous impact on that molecule's properties. So let's begin our exploration by examining the structure of atoms, so that we can address the question of how a chlorine atom differs from a hydrogen atom. To do that we'll need to dig one layer deeper, into the realm of subatomic particles.

Fundamental Concepts of the Atom

We will examine the structure of atoms in much greater detail in Chapter 6.

Our current model of the structure of atoms has been accepted for nearly a century, but it took great creativity and many ingenious experiments to develop. The atom is composed of a small, compact core called the **nucleus** surrounded by a disperse cloud of **electrons**. The nucleus is composed of two types of particles: **protons** and **neutrons**. There is a vast amount of space between the electrons and the nucleus, so much that it is impossible to show to scale in an illustration. Consider Figure 2.3, which is almost certainly similar to pictures you've seen before in high school chemistry or physical science books. The figure shows the relative positions of the protons, neutrons, and electrons. But if the protons and neutrons were actually the size shown, then the electrons would be hundreds of meters away. Another misunderstanding promoted by this type of illustration is the picture of electrons following regular orbits around the nucleus. We will learn later that we need to view the electrons more in terms of clouds that surround the nucleus as opposed to particles that orbit around it in an orderly way.

Protons and neutrons are themselves made up of even smaller particles, known as quarks.

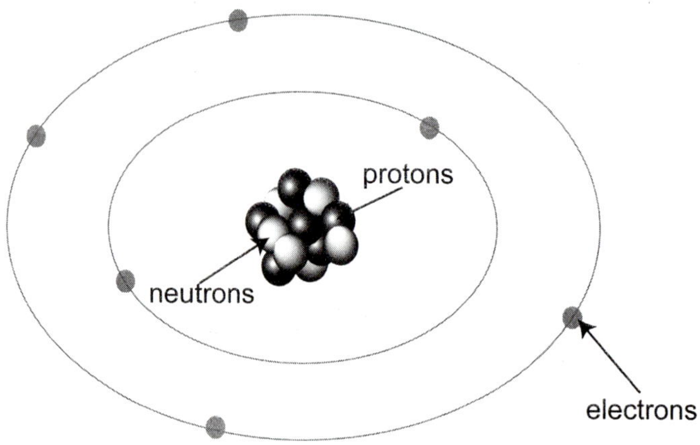

Figure 2.3 Atoms have often been depicted as resembling a solar system, with the nucleus at the center and the electrons orbiting around it. Here we see such a representation of ^{12}C, with six protons, six neutrons, and six electrons. While such pictures do help to emphasize the way that protons, neutrons, and electrons are distributed in the atom, they cannot accurately illustrate the currently accepted model of atomic structure.

We generally depict the charges in units of the electron charge, so that the charge of an electron is written as –1 and that of a proton is written as +1.

Now we turn our attention to the numbers of protons, neutrons, and electrons in the atom. Electric charge provides an important constraint on these numbers. Protons are positively charged, electrons are negatively charged, and neutrons are neutral. Atoms themselves are also electrically neutral, so the numbers of protons and electrons present must be such that their charges will cancel one another by adding up to zero. You may know from physics that the SI unit of charge is the Coulomb (C). Experiments have shown that the electrical charges on a proton and an electron are equal and opposite. Every electron carries a charge of -1.602×10^{-19} C, while every proton carries a charge of $+1.602 \times 10^{-19}$ C. So for an atom to remain neutral, the number of electrons and protons must be equal. Since neutrons have no charge, the number of neutrons present is not restricted by the requirement for electrical neutrality. And for most elements, the number of neutrons can actually vary somewhat from one atom to another, as we'll see.

Atomic Number and Mass Number

The number of protons in a particular atom is referred to as the **atomic number**, and identifies what element that atom is. Carbon atoms make up the backbone of nearly all polymers, so we will consider them as a first example. The atomic number of carbon is six, which tells us that a neutral carbon atom has six protons. And the requirement for electrical neutrality means a carbon atom must also have six electrons. The great majority of carbon atoms - roughly 99% - also contain six neutrons. But some carbon atoms contain seven or even eight neutrons. Atoms of the same element having different numbers of neutrons are called **isotopes.** The protons and electrons govern nearly all of the important chemical properties of atoms, so isotopes generally cannot be separated chemically. But the existence and even the relative abundance of isotopes can be proven by very careful examinations of the mass of atoms.

Protons and neutrons have quite similar masses, and are each nearly 2000 times more massive than the electron. So the mass of any atom is concentrated in its nucleus. Individual atoms are extremely small and light, so reporting their masses in conventional units like kilograms or grams is not convenient. So we use a unit that is appropriate to the atomic scale: the **atomic mass unit** or **amu.**

*The atomic mass unit is also referred to as a dalton, and is sometimes abbreviated as **u**.*

$$1 \text{ amu} = 1.6605 \times 10^{-24} \text{ g}$$

Both the neutron and the proton have masses very close to one amu. The mass of a neutron is 1.009 amu, and that of a proton is 1.007 amu. The mass of an electron, in contrast, is just 0.00055 amu. So for many practical purposes, we can determine the mass of an atom simply by counting the number of protons and neutrons. That number will be the mass in amu, to a fairly reasonable approximation. Because of this, the combined total of protons and neutrons is called the **mass number** of the atom. Since isotopes are atoms of the same element with different numbers of neutrons, they will have the same atomic number but different mass numbers.

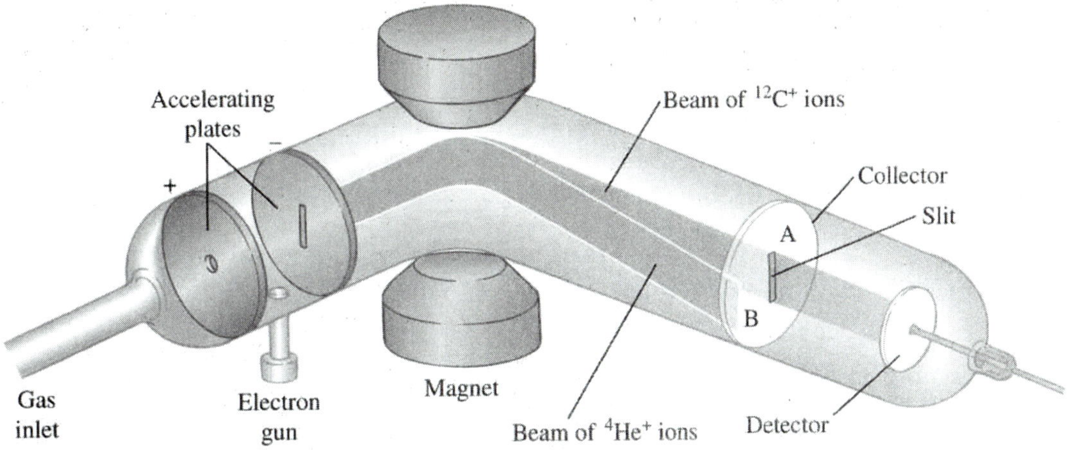

Figure 2.4 The schematic diagram shown here illustrates the key principles in the function of a mass spectrometer. A stream of gas to be analyzed enters at the left, and an electron gun causes some of the atoms to lose an electron, forming charged particles called *ions.* These ions are then accelerated to the right by an electric field, so that a beam of ions passes into a magnetic field. The magnetic field deflects the ions, and the extent of that deflection depends on the charge to mass ratio of each ion. For a given charge, lighter particles are deflected more severely than heavier ones. So if the sample contained both $^{4}He^+$ and $^{12}C^+$ ions, as shown here, the helium ions would be deflected much more than the carbon ions. This allows a slit to select ions of a particular charge to mass ratio, which then strike a detector. The current at this detector produces a signal that is proportional to the number of ions found with the desired charge to mass ratio, and this in turn is related to the amount of the parent gas molecule that entered the spectrometer.

Isotopes

How do we know that these isotopes exist? Modern experimental instruments called mass spectrometers provide direct evidence of the difference. The first important function of a mass spectrometer is to take a stream of microscopic particles - atoms or molecules - and "sort them out" according to mass. (Figure 2.4 shows a simplified explanation of how the instrument does this.) Once the particles have been separated by mass, the second key function of the mass spectrometer is to accurately measure the number of particles with a given mass. The data are usually presented as a "mass spectrum." Any time we refer to a spectrum we will be noting an observation that is made over a range of values in some variable. In this case that variable is mass, so the mass spectrum is really just a plot showing the number of particles detected as a function of mass. When a peak is seen at a particular mass, it means that the sample being analyzed has some component with that mass.

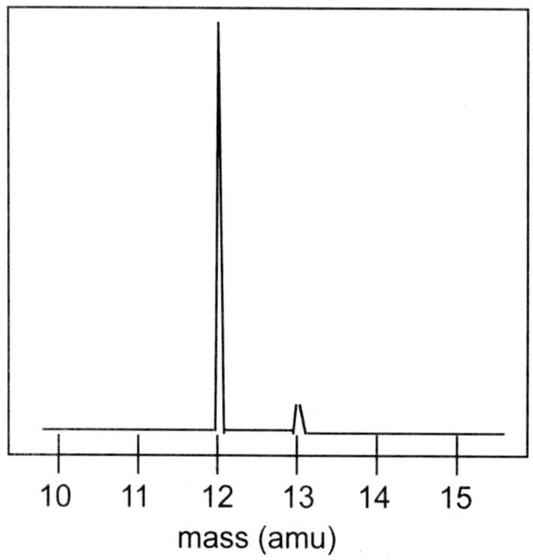

Figure 2.5 A sketch of the mass spectrum for elemental carbon is shown. The large peak is due to ^{12}C, while the smaller peak to the left is ^{13}C. The size of the ^{13}C peak here is actually somewhat exaggerated, as it would actually be just 1/99 the size of the ^{12}C peak.

A number of radioactive isotopes of carbon are aslo known. The most common of these is ^{14}C, the abundance of which is measured in the carbon dating of archaeological objects.

Figure 2.5 shows such a mass spectrum for a sample of carbon. Looking at the graph, we immediately see a large peak centered at mass 12. That represents the isotope called carbon-12, whose nucleus contains six protons and six neutrons. This isotope is actually used to define the amu: an atom of carbon-12 has a mass of *exactly* 12 amu. But if we look at the mass spectrum closely, we also see a much smaller peak centered near mass 13. This tells us that there is a small amount of a second isotope, carbon-13, with seven neutrons. Comparing the relative sizes of the two peaks, we could determine that the carbon-12 isotope accounts for roughly 99% of the carbon atoms. More accurate measurement gives a value of 98.93%, with just 1.07% of carbon-13. It is also possible to determine that the exact mass of the carbon-13 isotope is 13.0036 amu. Because any sample we measure will always contain vast numbers of atoms, these same percentages, or **isotopic abundances**, will be found in any naturally occurring sample of carbon.

Atomic Symbols

All the information about the structure of the atom that we have just discussed can be written in scientific shorthand, using atomic symbols. The general atomic symbol can be written as,

$$^{A}_{Z}\mathrm{E}$$

Here E represents the atomic symbol for the element in question, the superscript A is the mass number, and the subscript Z is the atomic number. Carbon-12, for example, would have the symbol $^{12}_{6}\mathrm{C}$.

Many atomic symbols are fairly obviously derived from the name of the element, such as the use of C for carbon in our example. For other elements, like iron, the symbol is based on the Latin name. The symbol for iron is Fe, derived from the Latin name of *ferrum*. An atom of iron with 26 protons and 30 neutrons is represented as $^{56}_{26}\mathrm{Fe}$. A listing of some common elements whose symbols are not based on the English name is provided in Table 2.1. A full list of elements and their symbols can be found in Appendix A at the back of this book.

Table 2.1 Names and symbols of some common elements whose symbols are not related to their English names.

Name	**Symbol (name origin)**
Gold	Au (Aurum)
Iron	Fe (Ferrum)
Lead	Pb (Plumbum)
Mercury	Hg (Hydrargyrum)
Silver	Ag (Argentum)
Sodium	Na (Natrium)

Atomic Masses

When you look at an entry on the periodic table you see some of the information we've just defined, such as the atomic symbol and atomic number. Periodic tables almost invariably incorporate additional information as well. In particular, the atomic mass is normally given. This number provides the average mass in amu of an atom of the element. If you look up carbon in the periodic table inside the front cover of this book, you will find the box shown in Figure 2.6. The atomic mass appears under the symbol: 12.011. But we have already said that the mass of an atom of carbon-12 is exactly 12 amu, and that of carbon-13 is 13.0036 amu. So the value of 12.011 does not seem to be the mass of any individual atom of carbon. Then how are atomic masses defined and determined?

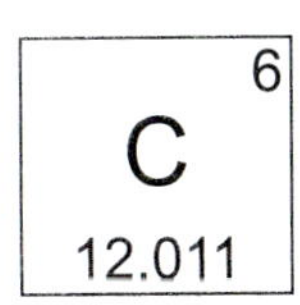

Figure 2.6 Entry for carbon from a periodic table. The atomic number (6) and the atomic mass (12.011) are shown, along with the symbol for the element (C). Some tables may display additional information, and the exact layout may vary from one table to another. But once you are familiar with the table itself it is usually easy to interpret whatever data are shown.

The atomic mass is defined as the *average* mass of an atom of a particular element. For carbon, we have two stable isotopes, with masses of 12.0000 and 13.0036 amu, respectively. So why is the average mass 12.011 and not something closer to 12.5? The answer is that when we take the average mass we must properly account for the relative abundance of each isotope. Suppose we could measure the mass of a 100-atom sample. Based on the isotopic abundances, we would expect to have 99 atoms of carbon-12 and only a single atom of carbon-13. And for any sample that we can actually weigh, the number of atoms will be far greater than 100. Even using the best available laboratory balances, the smallest quantity of matter that can generally be weighed is about a nanogram, or 10^{-9} g. And a nanogram of carbon would contain more than 10^{13} atoms. For such large numbers of atoms, it is safe to assume that the fraction of each isotope present will be determined by the natural abundances. For carbon, the fact that we only need to consider two stable isotopes makes the calculation fairly simple. We can multiply the mass by the fractional abundance to properly weight each isotope's contribution to the atomic mass.

carbon-12: $12.0000 \times 0.9893 = 11.8716$

carbon-13: $13.0036 \times 0.0107 = 0.1391$

weighted average mass $= 11.87 + 0.139 = 12.01$

The value of 12.011 found in the periodic table is obtained using additional significant figures on the isotopic abundance numbers.

Example Problem 2.1

The chlorine present in PVC has two stable isotopes. ^{35}Cl with a mass of 34.97 amu makes up 75.77% of the natural chlorine found. The other isotope is ^{37}Cl, whose mass is 36.95 amu. What is the atomic weight of chlorine?

Strategy

To determine the atomic weight we must calculate the average mass as weighted by the fractional abundance of each chlorine isotope. Since there are only two stable isotopes, their abundances must add up to 100%. So we can calculate the abundance of ^{37}Cl from the given abundance of ^{35}Cl.

Solution

First, we calculate the abundance of the chlorine-37 isotope.

abundance of ^{37}Cl = 100% - 75.77% = 24.23%

Now we can calculate the contribution of each isotope to the atomic mass.

^{35}Cl: $34.97 \times 0.7577 = 26.50$

^{37}Cl: $36.95 \times 0.2423 = 8.953$

weighted average mass $= 26.50 + 8.953 = 35.45$

So the atomic weight of chlorine is 35.45 amu.

Discussion

Some elements have several stable isotopes, but we can always do the same type of calculation by accounting for the mass and fractional abundance of each isotope.

Check Your Understanding
There are two naturally occurring isotopes of boron: ^{10}B has a mass of 10.0129 amu and ^{11}B has a mass of 11.0093 amu. Given that the atomic mass of boron is 10.811, find the percentage abundance of each isotope.

2.3 Ions

In developing our ideas about the composition of atoms in the previous section, we used the fact that atoms are electrically neutral to conclude that the numbers of protons and electrons must be equal. The operation of the mass spectrometer, however, depends critically on the instrument's ability to convert an atom into a charged particle, or **ion**. The separation of the particles by mass is generally based on the behavior of charged particles in a magnetic field, and the detector that counts the particles typically detects only ions, not neutral atoms. Ions can also play important roles in many chemical processes, including several that are important in the large-scale production of polymers.

Symbols for ions are written just like those of atoms, except that the charge appears as a superscript on the right of the symbol.

The fact that protons and electrons both carry an electrical charge plays an enormous role in the way atoms and molecules interact. By definition atoms are neutral species, requiring that the number of protons and electrons are the same. When the number of protons and the number of electrons does not match, the result is a species with a net charge, called an **ion**. Such species are no longer atoms, even if they contain only one particle, and they behave markedly different than atoms.

When an ion is derived from a single atom it is called a *monatomic ion*. When groups of atoms carry a charge they are called *polyatomic ions*. Ions, either monatomic or polyatomic, may carry either negative or positive charges. Negatively charged ions are called **anions**, and they contain more electrons than protons. Similarly a particle with more protons than electrons has a positive charge and is called a **cation**.

We write symbols for ions in a way analogous to those for atoms, adding the charge on the species as a superscript to the right of the atomic symbol. For monatomic ions, the number of protons still specifies the element whose symbol we choose. Table 2.2 provides some examples of monatomic ions. Notice that the monatomic anions have names ending in *–ide*, while the cations simply have "ion" added to the name of the element.

Cation Name	Symbol
Sodium Ion	Na^{+}
Lithium Ion	Li^{+}
Potassium Ion	K^{+}
Magnesium Ion	Mg^{2+}
Aluminum Ion	Al^{3+}

Anion Name	Symbol
Fluoride Ion	F^{-}
Chloride Ion	Cl^{-}
Bromide Ion	Br^{-}
Sulfide Ion	S^{2-}
Nitride Ion	N^{3-}

Table 2.2 Examples of monatomic Ions

The behavior and interaction of electrical charges are important topics in physics, but also provide a basis for thinking about many aspects of chemistry. For our current interests we will point out just two fundamental ideas about electric charge. First, *opposite charges attract each other and like charges repel one another.* And second, *electric charge is conserved*. These two ideas have important implications for the formation of ions in chemical processes. First of all, since charge is conserved we can say that if a neutral atom or molecule is to be converted into an ion, some oppositely charged particle - most likely an electron or another ion - must be produced at the same time. Moreover, since opposite charges attract one another, some energy input will always be needed to convert a neutral atom or molecule into a pair of oppositely charged particles.

Mathematical Description

The statement that "opposites attract and likes repel" can be quantified mathematically. **Coulomb's Law**, which you may recall from a physics class, describes the interaction of charged particles. The attraction of opposite charges and the repulsion of like charges are both described mathematically by one simple equation.

$$F = \frac{q_1 q_2}{4\pi\varepsilon_o r^2} \tag{2.1}$$

Here q_1 and q_2 are the charges, ε_0 is a constant called the permittivity of a vacuum, and r is the distance between the charges. F is the force the objects exert on one another as a result of their charges. Looking at this expression, when both charges have the same sign - either positive or negative - the resultant value for the force is a positive number. When the charges are opposite the value is negative. This is consistent with the usual sign conventions used throughout chemistry and physics for force and energy; a negative value of F from equation 2.1 indicates an attractive force and a positive value a repulsive one.

Now consider the effect of varying the distance, r, between two ions. If two positively charged particles are initially very far apart (effectively infinite distance), the r^2 term in the denominator of equation 2.1 will be very large and the particles will not interact with each other significantly. As the two like charges are brought closer together, the r^2 term in the denominator shrinks and so the (positive) force grows larger: the particles repel each other. If we somehow force the particles closer together, the repulsive force will continue to grow. The distance dependence of the Coulomb force is illustrated in Figure 2.7.

Ions and Their Properties

Many monatomic ions, both cations and anions, are known to exist. These ions can exist in the gas phase, and in fact many are important in atmospheric chemistry. But we encounter ions most frequently when dealing with the chemistry of substances dissolved in solution. For example, sodium atoms lose an electron relatively easily to form the sodium cation, Na^+. Because it still has 11 protons, this ion retains the symbol of sodium, yet it does not behave at all like an atom of sodium. Consider, for example, an order of french fries. You may have heard news stories about the high amount of sodium in an order of fries – and concerns have been raised about the possible health effects of too much sodium in our diets. This statement could be confusing, since here the word "sodium" does not refer to sodium metal. In fact, if we place sodium metal on freshly made french fries, the metal will burst into flames. The sodium we hear about in stories on diet and health is actually sodium ion, which is added to the fries when salt is sprinkled on. Too much salt might still be a health concern, but we certainly don't worry about the salt igniting when we add it. There is a big difference between ions and atoms, at least in this case.

In contrast to sodium, chlorine readily gains an extra electron forming an anion, Cl^-. Again, there is a noticeable difference between the ion and the atom of chlorine. The table salt we discussed above is sodium chloride, and contains chloride anions. Just like sodium, these chloride ions are present in your french fries or any other salted foods. Chlorine atoms, on the other hand, form a yellowish-green diatomic gas, Cl_2, which irritates the lungs and can be toxic. The behavior of the ion is clearly much different from that of the neutral atom or molecule.

Polymers do not generally contain ions. But ions are important in the chemical reactions used to produce many common plastics. As a result of their electric charge, ions are often much more reactive than neutral atoms or molecules. So small amounts of ions are often used to initiate or sustain the chemical reactions that join monomers together to form polymers.

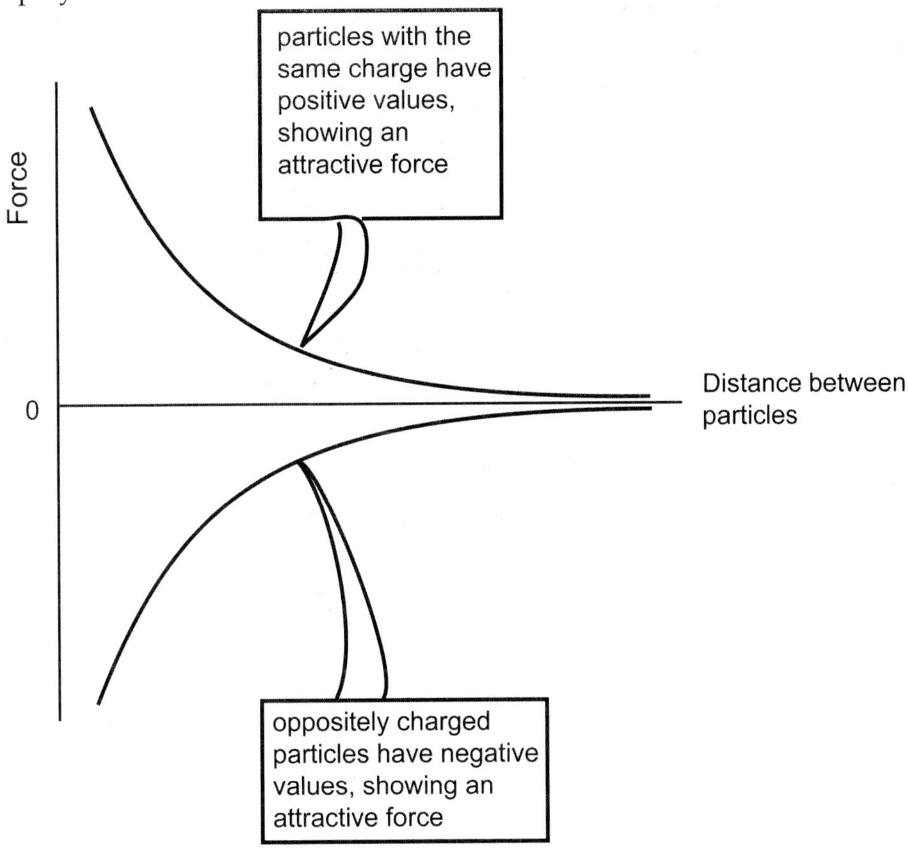

FIGURE 2.7 The figure shows how the Coulomb force (Equation 2.1) varies with the distance *d* between two particles with opposite or like charges. When the charges have the same signs, the particles will repel one another. So the value of the force is positive. If the charges have opposite signs, the particles will attract one another and the value of the force will be negative.

2.4 Compounds and Chemical Bonds

A basic picture of atoms is a good starting point for trying to understand the properties of various polymers. But to really begin to see how the observable properties of a polymer might be related to its atomic and molecular make-up we will need to consider the connections between the atoms. Which atoms are actually attached to one another? And what different types of connections - or **chemical bonds** - are involved? Once again we'll begin by trying to establish some vocabulary that will help us understand compounds and the chemical bonds that hold them together.

Chemical Formulas

Compounds arise from the combination of atoms in fixed ratios. In any such combination the resultant substance behaves differently than the atoms alone. For many compounds, atoms combine to form discrete particles called **molecules**. Molecules can be broken down into their constituent atoms, but the resulting collection of atoms no longer behaves like the original molecule. Other materials are composed of vast arrays or extended structures of atoms or ions, but do not really form discrete molecules. Alloys, metals and ionic solids (composed of paired ions) comprise this category of chemical compound. We've seen how we can use atomic symbols as a shorthand notation to designate atoms. That same idea can be extended to describe the composition of either molecules or extended compounds in a simple symbolic representation.

A **chemical formula** provides a description of a compound in terms of its constituent elements. We will actually encounter two distinct types of chemical formulas: **molecular formulas** and **empirical formulas**. The **molecular formula** for a compound can be seen as a kind of "parts list," describing the atomic composition of a molecule in an efficient way. The molecular formula of the ethylene monomer from which polyethylene is produced is C_2H_4; this tells us that there are two carbon atoms and four hydrogen atoms per molecule. The empirical formula tells us only the relative ratio between the numbers of atoms of the different elements present. Let's consider ethylene again. The ratio of carbon atoms to hydrogen is 1:2. So the empirical formula is CH_2. When dealing with an empirical formula, it is important to realize that no information is present about how large or small an individual molecule of the compound might be; only the relative numbers of atoms of each element are given. We often emphasize this fact by writing a subscript 'n' on the entire formula. For ethylene, this would give us $(CH_2)_n$, meaning that each molecule must contain some integer number of CH_2 units.

In Chapter 3 we will learn how to find the empirical formula for a compound from experimental data.

Empirical formulas, or a minor variation on them, are especially common when dealing with polymer molecules. Because polymer molecules are so large, the exact number of monomer units in a molecule is generally not terribly important. And in fact, the exact length of the polymer chains is often not the same for all of the molecules in a given sample. Instead, there is usually some range of chain lengths that will exist, depending on how the polymer was actually produced. But as long as the chains are all within some reasonable range of lengths, the macroscopic properties of the polymer will not be substantially affected. So polymer formulas are most often written like empirical formulas. The repeating unit contributed by each monomer molecule is written in parentheses or brackets, and a subscript 'n' is used to emphasize that a large number of these units will be found in any individual molecule. For polyethylene, we would write the formula as $-[CH_2CH_2]_n-$. Here the dashes are added to stress the fact that these units are to be attached end to end to build up the long chain of the polymer. For the most common forms of polyethylene, the number of monomer units (*i.e.*, the value of 'n') is in the tens of thousands. We could write similar formulas for the other polymers mentioned in the opening section.

In the closing Insight section for this chapter we will look more closely at how the properties of polyethylene are influenced by large changes in chain length.

Poly(vinyl chloride):	$-[CH_2CHCl]_n-$
Poly(vinylidene chloride):	$-[CH_2CCl_2]_n-$

There are four rules that allow us to write most of the formulas we will need in this textbook.

1. Indicate the types of atoms present in the substance by including their atomic symbols.
2. The number of each atom in the compound is indicated by a subscript to the right of the atomic symbol. For example the chemical formula of ethylene, C_2H_4, tells us that each molecule contains two carbon atoms and four hydrogen atoms.
3. Groups of atoms can be designated by using parentheses. Subscripts outside these parentheses mean that all atoms enclosed in the parentheses are multiplied by the value indicated in the subscript.
4. Water molecules associated with certain compounds called hydrates are indicated separately from the rest of the compound.

Example Problem 2.2 shows how to interpret chemical formulas by inverting some of these rules.

Example Problem 2.2

We cannot generally produce polymers by simply mixing up a large sample of the desired monomers. Instead, additional compounds called *initiators* or *catalysts* are almost always needed in order to get the polymerization started. One polymerization catalyst is $Al(C_2H_5)_2Cl$. How many of each type of atom are present in a molecule of this compound?

Strategy

The subscripts in a formula indicate how many atoms of each type are in the molecule. The parentheses designate a group of atoms, and the subscript associated with the parentheses multiplies each atom in the group.

Solution

In each molecule of $Al(C_2H_5)_2Cl$ there is one aluminum atom, one chlorine atom and two groups of C_2H_5. Each of the C_2H_5 groups contains two carbon atoms and five hydrogen atoms. We multiply those numbers by two since there are two C_2H_5 groups present, so we have four carbon atoms and ten hydrogen atoms.

Discussion

The number of atoms present might be easier to see if we wrote this formula as $AlC_4H_{10}Cl$. And right now you might feel that would be simpler. But when we write it as $Al(C_2H_5)_2Cl$, we are actually conveying some additional information about *how* the atoms are connected to one another. Specifically, we are showing that the carbon and hydrogen atoms are arranged as two C_2H_5 groups, and that each of these groups is attached to the aluminum atom. Later we will learn that such a C_2H_5 unit is called an *ethyl group*.

Check Your Understanding

A compound with the rather imposing name of 2,2′-azo- *bis* -isobutyrylnitrile is used to initiate the growth of some polymers, including poly(vinyl chloride). If the molecular formula is $C_8H_{12}N_4$, how many of each type of atom are present in a molecule of the compound? What is the empirical formula of this compound?

Chemical Bonding

Atoms combine to make compounds through the formation of chemical bonds. Several different types of chemical bonds are possible, and once we learn to recognize them these types of bonds will help us to understand at least some of the chemical properties of many substances.

All chemical bonds share two characteristics. First, all bonds involve some sort of exchange or sharing of electrons. We will return to this concept often in this text as we investigate chemical reactions and properties of molecules. Second, this exchange or sharing of electrons results in lower energy for the compound relative to the separated atoms. A chemical bond will not form, or will have only a fleeting existence, unless it lowers the overall energy of the collection of atoms involved.

There are three important classifications of chemical bonds. Some compounds are composed of collections of oppositely charged ions that form an extended array called a *lattice*. The bonding in these compounds is referred to as **ionic bonding**. In order to form the ions that make up the compound, one substance has lost an electron to become a cation while another has gained an electron to become an anion. We can view this as the donation of an electron from one species to another. Figure 2.8 shows this concept for an example of an ionic compound, NaCl.

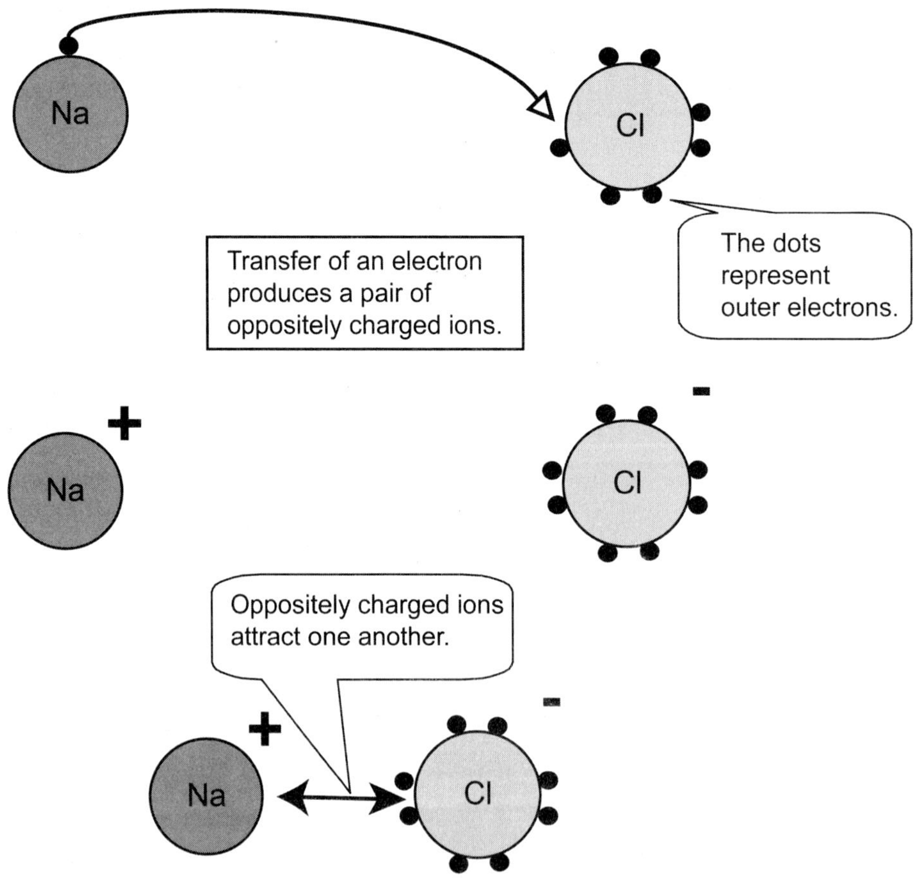

Figure 2.8 Conceptual illustration showing the transfer of one electron from a sodium atom to a chlorine atom, forming a pair of ions (Na^+ and Cl^-). Once the electron transfer takes place, the Coulomb force will draw the ions together.

Ionic compounds form extended systems of alternating positive and negative charges such as that shown in Figure 2.9. So while the formula NaCl does correctly describe the fact that the sodium and chlorine are present in a 1:1 ratio, we cannot really identify an individual "molecule" of NaCl. To emphasize this distinction, we sometimes refer to a **formula unit**, rather than a molecule, when talking about ionic compounds. The formula unit is the smallest whole number ratio of atoms in the ionic compound.

Lattices are often depicted as having shapes that are essentially cubic, but there are actually 17 different geometric shapes, not all of which are

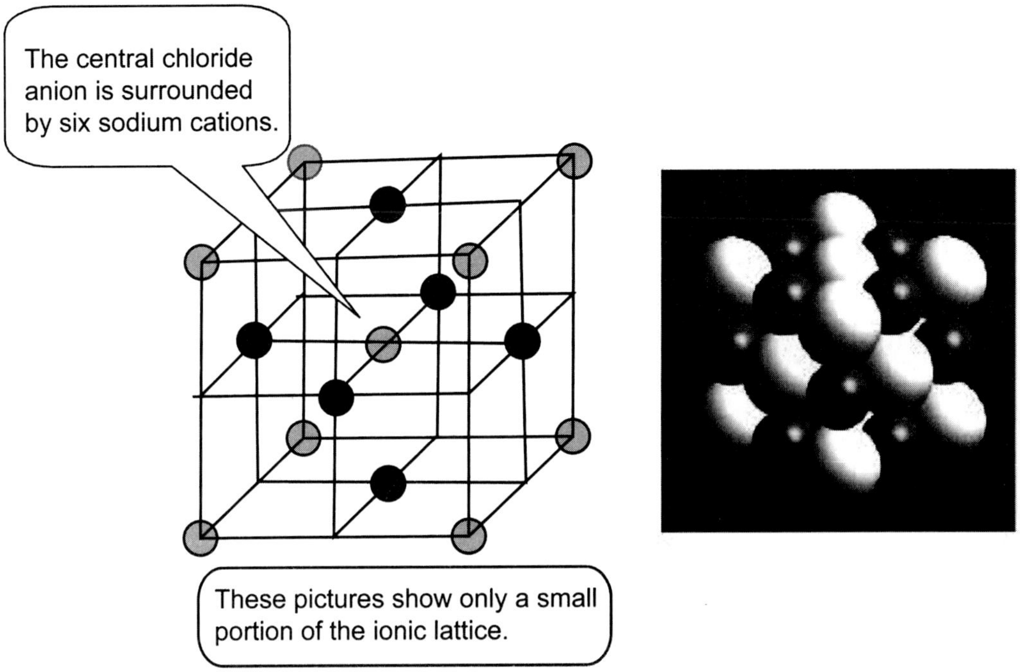

Figure 2.9 Two different representations of the NaCl crystal structure are shown. In the left hand panel, the lighter circles indicate the positions of chlorine ions, while the darker circles represent sodium ions. The lines are added to emphasize the cubic nature of the overall structure. In the right hand panel, the lighter spheres represent chlorine anions while the darker spheres denote sodium cations. In a macroscopic salt crystal, additional ions would simply extend this structure, repeating the same alternating pattern.

Metals represent another type of extended system, but here the chemical bonding is of a totally different nature. In metals the atoms are once again arranged in a lattice, but in this case there is not an alternation of positively and negatively charged species. Instead the nuclei and some fraction of their electrons comprise a positively charged "core" localized at these lattice points, and other electrons move more or less freely throughout the whole array. This is called **metallic bonding**, and leads to electrical conductivity because of the ease with which electrons can move through the bulk material. Figure 2.10 shows a schematic illustration of the concept of metallic bonding.

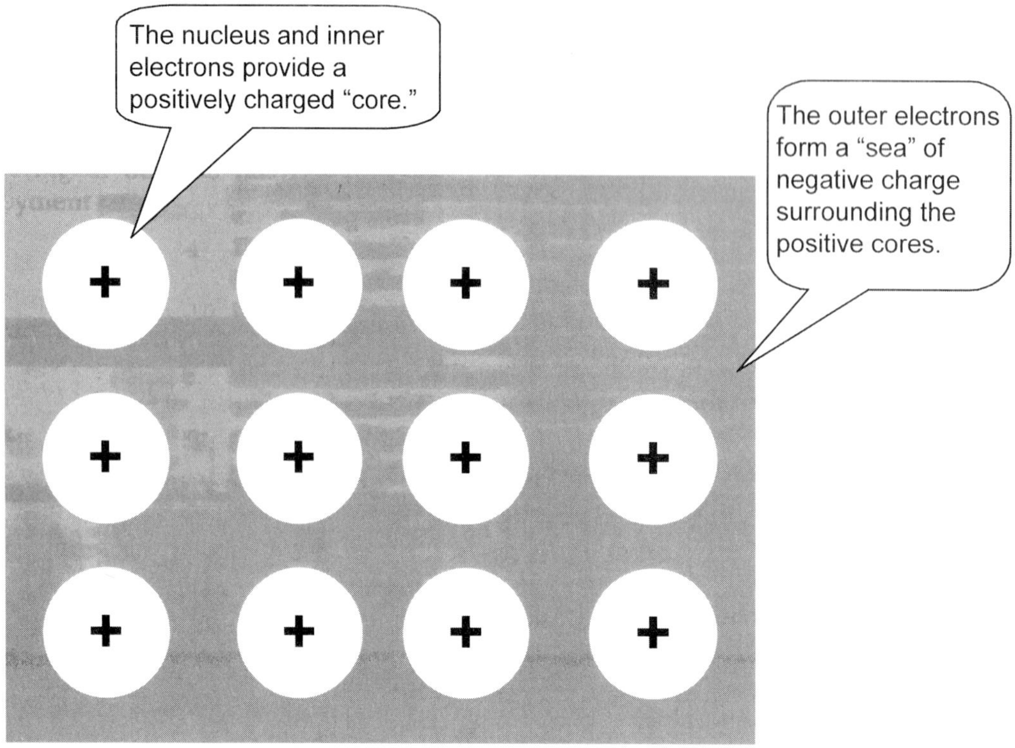

Figure 2.10 In this simple conceptual picture of metallic bonding, each metal atom contributes one or more electrons to a mobile "electron sea." The ability of the electrons to move freely through this "sea" allows the metal to conduct electricity. Here the gray shaded area depicts those mobile (or "delocalized") electrons, while the circles represent the positively charged "cores" of the individual atoms

Molecules in which electrons are shared between pairs of atoms rather than donated from one atom to another or mobile across an entire lattice are said to form **covalent bonds**. In covalent bonds electrons are usually shared in pairs. Two electrons (or four or six) are located between two nuclei and the sharing of them leads to an attraction between the nuclei. The long chains found in all polymers are formed by covalent bonds in which electrons are shared between adjacent carbon atoms. Smaller, more familiar molecules such as water, carbon dioxide or propane are simpler examples. All three types of chemical bonds will be discussed in much greater detail in Chapter 7.

Polymer molecules are built up by the successive addition of monomers to form the characteristic long chain backbone. The bonds that hold the monomers to one another, as well as the bonds between atoms within each monomer unit, are covalent bonds. But ionic bonding is important in many compounds that are used to initiate or sustain the reactions needed to grow the polymer.

2.5 The Periodic Table

One of the most recognizable symbols of chemistry is the periodic table. It is prominently displayed in practically every chemistry classroom, and many chemists own t-shirts, neckties, or coffee mugs on which the periodic table appears as a sort of professional badge of honor. Why do chemists hold this table in such high regard? Once you are familiar with it, the periodic table summarizes a wealth of information about the behavior of elements, organizing them simultaneously in ascending order of atomic number and in groups according to chemical behavior. An experienced chemist can get a rough idea of an element's properties simply by seeing where that element sits in the periodic table.

Today we can purchase or download artistic renderings of the periodic table, color-coded to display properties of particular interest for different applications. But like many developments in science, the emergence of what we now accept as the periodic table was accompanied by some degree of controversy. A number of scientists had devised various schemes for the arrangement of elements. These attempts to organize the understanding of the elements were not well received, however. One proposal by John Newlands in 1866 was literally laughed at during a scientific meeting where one critic questioned if a listing by alphabetical order had been tried, noting that "any arrangement would present occasional coincidences."

One of the difficulties that persisted in the development of the periodic table was the controversy of how to assign atomic masses. Without a unique scale for atomic mass scientists made different conclusions about how to arrange the elements "in order".

Despite the skepticism of the 19th century scientific community, efforts to organize the elements persisted. Numerous observations suggested a regularity, or periodicity, in the behavior of the elements known at that time. By 1869 Dmitri Mendeleev had published his first periodic table and enumerated the periodic law: the elements, when properly arranged, display a regular and periodic variation in their chemical properties. Perhaps the most significant and impressive feature of Mendeleev's work was his prediction of the existence of undiscovered elements. He left "holes" in his proposed table at positions where no known element seemed to fit. When the elements to fill in these holes were later identified, the scientific community came to accept Mendeleev's work. The discovery of the periodic law and construction of the periodic table represents one of the most significant creative insights in the history of chemistry. Prior to Mendeleev's time, chemists essentially had to learn the properties of each element individually. And as more and more elements were discovered, that task became increasingly daunting. By providing a simple, visual means to organize the elements in terms of their chemical and physical properties, the periodic table helped the study of chemistry to expand quickly.

Periods and Groups

The modern periodic table simultaneously arranges elements in two important ways. These are depicted by the horizontal rows of the table, called **periods**, and the vertical columns, called **groups**. The term "period" is used for the rows because many important properties of the elements vary in a systematic way as we move across a row. Figure 2.11 shows a plot of the density of elements, all in their solid state, as a function of atomic number. Looking at that graph, it is clear that the density varies according to a fairly regular pattern, going through a series of minima and maxima. Different symbols are used for the data points in this graph to show how the variation in density is correlated with position in the periodic table. Each symbol represents a period (row) in the table. And since the elements in the periodic table are arranged in order of increasing atomic number, moving across each segment of the graph corresponds to moving from left to right across the corresponding row of the periodic table. You can readily see that as we move across a row in this way, the density of the elements is initially small, increases until passing through a maximum, and then decreases again. Figure 2.12 shows the same data, with the density represented by the shading of each element's box. This representation clearly shows how the density of the elements varies in a regular way across each row of the table. The rows in the table are numbered sequentially from top to bottom as 1 through 7.

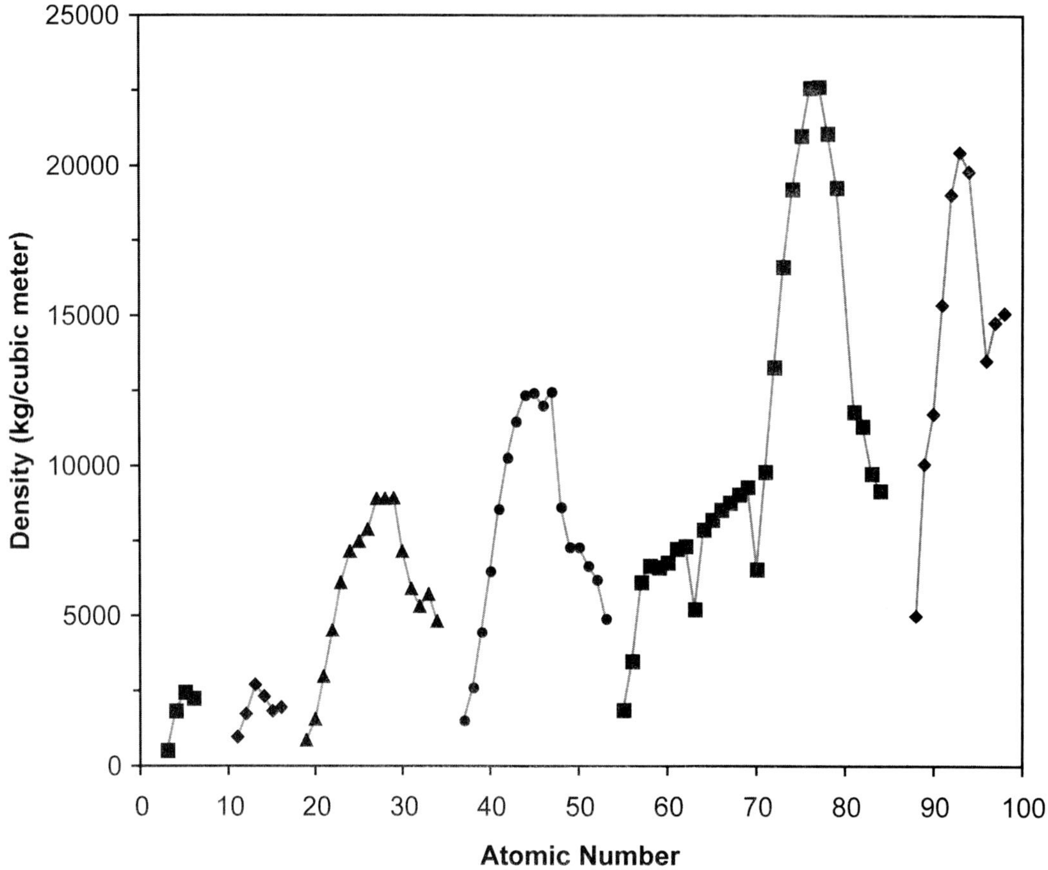

Figure 2.11 The density of elements in their solid states is plotted as a function of atomic number. Here density is in units of kg/m^3. The various data point symbols represent the different periods (rows) in the periodic table. Notice how the same general pattern repeats as we move across each row: the density is low at the left hand edge of the row (Group 1), increases through a maximum and then decreases as we continue to move from left to right across the table. (Only those elements that exist as solids under ordinary conditions are shown.)

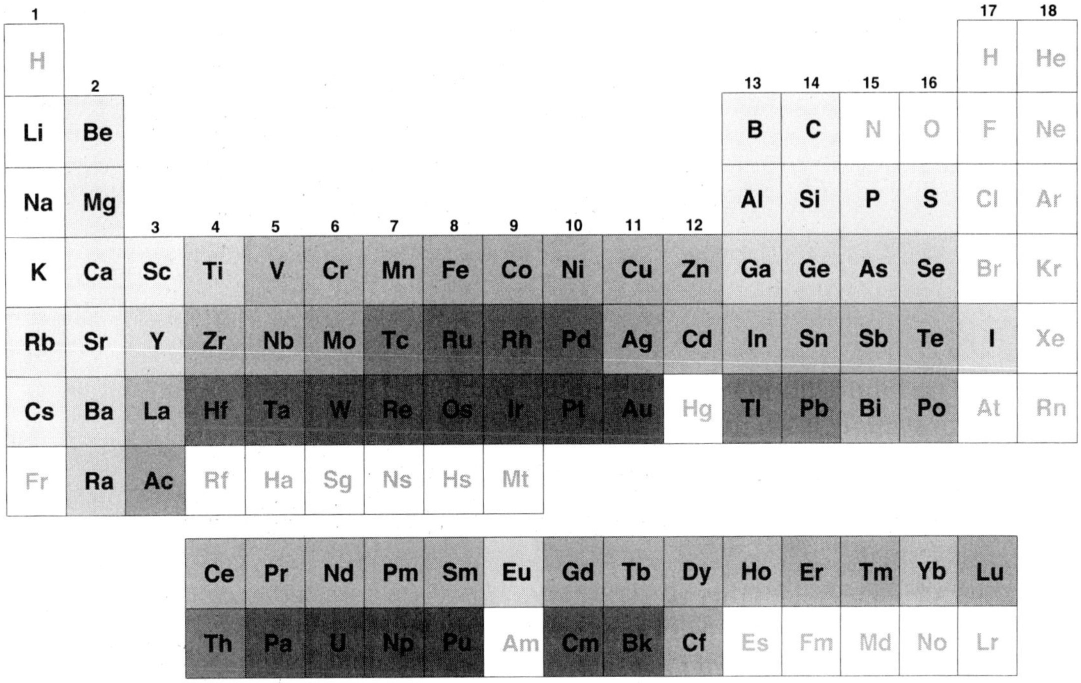

Figure 2.12 The data from Figure 2.11 are presented here in a different form. The shading of the boxes in the periodic table represents the density of each element, with darker shading indicating higher density. The general trends in density are readily seen as you look across a row or down a column.

While the properties of the elements can vary widely across a period, each column collects elements that have very similar chemical properties. Most elements can combine with hydrogen to form compounds. The graph in Figure 2.13 shows the number of hydrogen atoms with which an atom of each element will combine, and the regular variation in the plot clearly shows that this is a periodic property. Elements in a group (column) can be seen to combine with the same number of hydrogen atoms. Fluorine, chlorine, and bromine each combine with one atom of hydrogen for example, and all fall in the same group.

These types of chemical similarities were among the evidence that led to the development of the periodic table, so some of the groups predate the general acceptance of the table. These groups of elements were assigned names and those names have remained with them. Thus the elements of the far left-hand column (Li, Na, K, Rb, and Cs) are known collectively as the *alkali metals*. Similarly Be, Mg, Ca, Sr and Ba are called the *alkaline earths* and F, Cl, Br and I are referred to as the *halogens*. He, Ne, Ar, Kr and Xe were discovered much later than most of the other elements, and they too have been named the *rare gases* or the *noble gases*. Other groups are named but those names are less commonly used and won't be mentioned here.

The noble gas elements were once called inert gases, because they were thought to be completely unreactive. Several noble gas compounds are now known, so the term inert is no longer used.

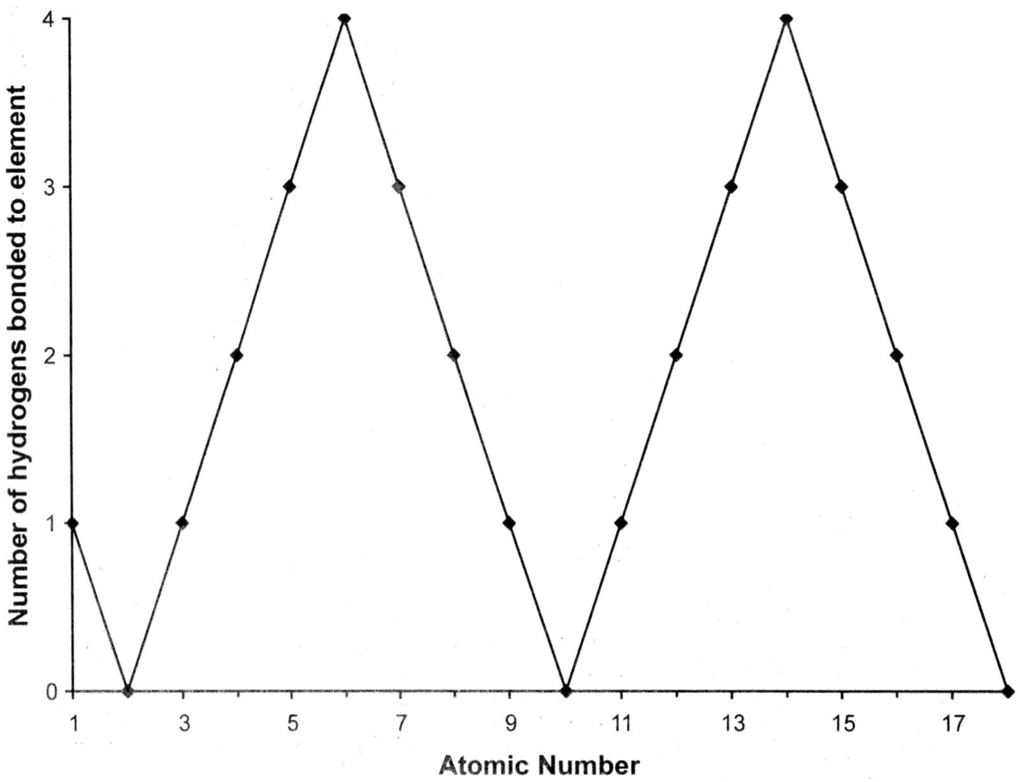

Figure 2.13 The graph shows the number of hydrogen atoms with which an individual atom of various elements will combine. The periodic nature of this chemical property is evident from the cyclic nature of the graph.

There are also names for different regions of the table. Elements in the two groups on the left side of the table and the six groups on the right side are collectively referred to as the **representative elements**, or main group elements. Elements that separate these two parts of the representative groups in the main body of the periodic table are called the **transition metals**. Iron is an example of a transition metal. The elements that appear below the periodic table are called the **lanthanides** (named after the element lanthanum, Z = 57) and the **actinides** (named after the element actinium, z = 89).

In addition to these names, several numbering systems have been used to designate the groups. Current convention dictates numbering from left to right starting with 1 and proceeding to 18. Thus, for example, the group containing C, Si, Ge, Sn and Pb is referred to as Group 14.

Metals, Nonmetals, and Metalloids

These labels have historical origins and are not without ambiguity.

Another way to classify an element is as a metal, nonmetal or metalloid. And once again the periodic table conveniently arranges elements so that one can easily place a given element into one of these categories.

Most of the elements are metals. Their general location in the periodic table is towards the left and bottom, as seen in the shading of the periodic table in Figure 2.14. Metals share a number of similarities in chemical and physical properties. Physically, metals are shiny and malleable, meaning they can be shaped easily by pounding or rolling.

They also conduct electricity and are ductile (can be pulled into wires), so they are used as wires. Chemical properties can also be used to distinguish metals. Metallic elements tend to form cations in most of their compounds, for example.

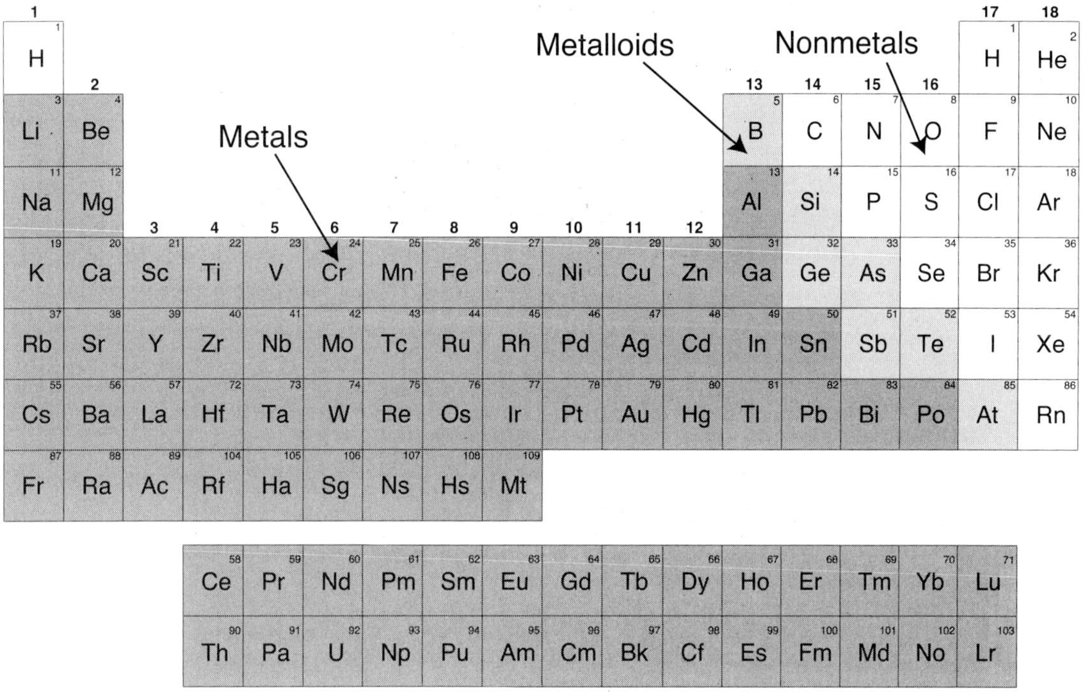

Figure 2.14 The shading in this periodic table identifies each element as a metal, nonmetal, or metalloid. Notice how the metals are grouped toward the left and bottom of the table, while nonmetals are towards the upper right.

Nonmetals occupy the upper right-hand portion of the periodic table. In terms of numbers, there are fewer nonmetals than metals. But when we consider the relative importance of elements, the nonmetals hold their own due to the role they play in the chemistry of living things. Most of the molecules that make up your body are composed predominantly or exclusively of the nonmetallic elements carbon, hydrogen, oxygen, nitrogen, sulfur, and phosphorous. In contrast to the metals, nonmetals are not shiny, malleable or ductile. They are not good conductors of electricity, either. These physical properties provide one means by which we can distinguish between metals and nonmetals.

The question of whether an element is a metal or nonmetal may seem like a simple one that we should be able to answer based on the physical properties cited above. Yet, some elements are not easily classified as either metal or nonmetal. The question of whether or not a substance conducts electricity, for example, does not always have a simple yes or no answer. Lacking a reliable means of drawing a clean boundary between the two categories, scientists have generally chosen to refer to intermediate cases as metalloids or semi-metals. In the periodic table, the metalloids are clustered along a diagonal path as shown in Figure 2.14. Using this categorization provides us with some useful flexibility, and also helps to emphasize that properties change slowly as one moves across or down the periodic table.

All of the polymer molecules we have mentioned are carbon-based, meaning that their skeletons consist entirely of carbon atoms. Since elements in the same group of the periodic table are supposed to have similar chemical behavior, you might be wondering whether similar polymers could be produced based on silicon, which appears right below carbon in Group 14. And while silicon-based polymers, known as silicones, do exist, they differ from carbon polymers in important ways. While covalent bonds can be formed between silicon atoms, they are not as strong as those between carbon atoms. So chains of silicon atoms become unstable beyond a length of around ten atoms, and the silicon analogues of polymers like polyethylene cannot be produced. Instead of pure silicon, the backbone chains in silicone polymers consist of alternating silicon and oxygen atoms. The Si-O bonding is strong enough to allow these chains to grow quite long. Additional atoms or groups of atoms are bound to the silicon atoms, and the identity of these influences the observed properties of the polymer. The range of polymers that can be produced is not nearly as diverse as for carbon, but silicone polymers are widely used in applications including greases, caulking materials, water repellants, and surfactants.

Proteins are natural polymers whose backbones include carbon, oxygen and nitrogen. The specific arrangment that allows this type of backbone is called a peptide bond.

2.6 Inorganic and Organic Chemistry

Just as engineering can be broken down into various specialties, so too can chemistry be viewed as a collection of subfields. Two of the most fundamental areas of chemistry are **organic chemistry** and **inorganic chemistry**. These names arise from the fact that organic chemistry would once have been defined as the chemistry of living things. A more modern definition is that organic chemistry is the study of compounds of the element carbon. As we've already seen, this includes not only naturally occurring biological molecules, but also nearly all synthetic polymers. And if organic chemistry deals only with carbon compounds, inorganic chemistry is then the study of all other elements and their compounds. At first glance this may well strike you as a bizarre distinction to make, since carbon is just one element out of a periodic table that contains more than 100 others. But the fact is that the chemistry of carbon is so rich, diverse, and important that organic chemists actually outnumber inorganic chemists. Because this text is intended to give a brief overview of important chemical principles, we will focus most of our attention on the behavior of molecules in general rather than delving too deeply into the details of either subfield. In this section we will provide a brief description of some important similarities and differences between organic and inorganic chemistry, and introduce a little bit of terminology and notation that we will need as we progress through the text.

The division of chemistry into sub-disciplines is an historic artefact. Much modern research in chemistry takes place at the interface of two or more sub-disciplines.

Inorganic Chemistry – Main Groups and Transition Metals

Many inorganic compounds exist as relatively small molecules. With their atoms joined together through covalent bonds. One such compound is aluminum trichloride, $AlCl_3$. Figure 2.15 shows some visual representations of $AlCl_3$ that we might use to illustrate how its atoms are actually arranged into molecules. Three chlorine atoms surround a central aluminum atom, with a pair of electrons shared between the aluminum and each chlorine. We will study chemical bonding in detail in Chapter 7, where we will learn that this structure leads the compound to be quite reactive. Because of this reactivity, $AlCl_3$ is used in the commercial preparation of poly(isobutylene), which is a synthetic rubber. The final polymer itself contains no aluminum or chlorine, but the reactive $AlCl_3$ serves as an *initiator*, helping to start the reactions that lead to polymer formation.

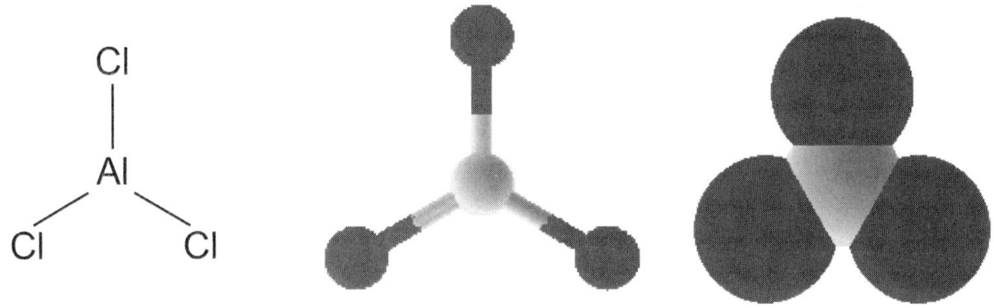

Figure 2.15 This figure presents three depictions of $AlCl_3$. In the drawing at the left, each atom is represented by its symbol and the lines between the symbols depict chemical bonds. In the center panel is a "ball and stick" model, where each atom is a ball and the bonds are shown as sticks connecting the balls. And in the right hand panel is a "space filling" model, where atoms are shown as balls that overlap strongly with one another. Each of these types of models is commonly used, and each has its strengths and weaknesses.

Aluminum and chlorine are both main group elements, found in Groups 13 and 17 of the periodic table, respectively. As mentioned in the previous section, elements from the same group tend to display similar chemical properties. Thus once we know that $AlCl_3$ exists, we might expect that other pairs of elements from the same groups might form similar compounds. And this prediction is correct: compounds such as BCl_3 and AlF_3 do exist, and exhibit structures and bonding analogous to what we saw in Figure 2.15 for $AlCl_3$.

Other compounds of the main group elements form extended ionic structures, such as we saw for NaCl in Figure 2.9. But despite the difference in the type of chemical bonds employed, we can still readily predict that similar compounds should exist for other pairs of elements from the same groups. Looking at the periodic table, we see that sodium is in Group 1 and chlorine is in Group 17. So we can expect other pairs of elements form these columns of the table to form ionic solids, too. And again, our prediction is accurate; compounds like LiCl, NaF, and KI all have structures that are analogous to that of NaCl. The reason for the existence of these similar compounds is simple. All of the metals in Group 1 form cations with a +1 charge, while all of the elements in Group 17 form anions with a –1 charge. Any of these cations can combine with any of the anions in a 1:1 ratio to form neutral compounds.

The chemistry of transition metals is somewhat more complicated than that of the main group elements, though, because most transition metals can form multiple cations with different charges. Iron commonly forms two different monatomic cations: Fe^{2+} and Fe^{3+}. As a result of this, iron can form a more diverse set of compounds than the Group 1 metals. It can combine with chlorine to form either $FeCl_2$ or $FeCl_3$, and these two compounds have significantly different physical properties. (Figure 2.16.) Largely because of the ability to form multiple cations, the chemistry of the transition metals does not vary as sharply from group to group. Regardless of their position in the periodic table, for example, most transition metals are able to form cations with a +2 charge. Thus predictions based simply on group number are not as reliable here as they are for the representative elements. When considering transition metals and their compounds we must rely more heavily on knowledge of the specific chemistry of each element.

The ionic charge for transition metals can have vitally important consequences. Some metals, such as chromium for example, are much more toxic in one ionic state than in others.

Iron(II) chloride, $FeCl_2$

greenish-yellow color
denisty 3.16 g cm^3
melts at 670°C

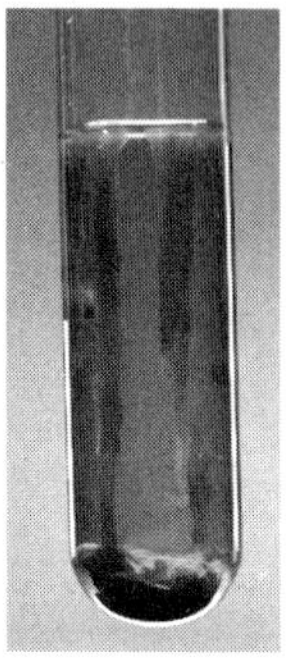

Iron(III) chloride, $FeCl_3$ (forming at bottom of test tube)
orange-brown color
denisty 2.96 g cm^3
melts at 306°C

Figure 2.16 Transition metals typically form more than one type of cation, giving them a very diverse chemistry. Iron, for example, forms cations with both +2 and +3 charges, and this allows it to form two different ionic compounds with chlorine. $FeCl_2$ and $FeCl_3$ have different appearances and properties.

Organic Chemistry

Organic compounds all feature carbon "skeletons," similar to what we have already seen in our introduction to polymers. Other elements frequently found in organic compounds include hydrogen, oxygen, and nitrogen. Despite this rather short list of elements, more than 18 million organic compounds are known to exist. This vast number of compounds arises due to some unusual aspects of the chemistry of carbon itself. Most importantly, carbon atoms readily attach to one another to form chains, and these chains can grow quite long. Many of the polymer molecules we have been discussing in this chapter contain thousands of carbon atoms. Furthermore, some of these long chains are straight while others are branched at one or more places. And finally, any pair of carbon atoms can bond to one another in three different ways, by sharing either one, two, or three pairs of electrons. When taken all together, these factors allow carbon to form a vast array of compounds.

The diversity of organic compounds presents some challenges. It is not uncommon for several different organic compounds to have the same molecular formula, for example, but to display different properties depending on exactly how the atoms are arranged into molecules. (Different compounds with the same molecular formula are called *isomers*.) So organic chemists must frequently depict molecules not only by their formulas, but in some way that also conveys important information about the arrangement of the atoms. This could be done using structural formulas of the sort we saw in Figure 2.15. But because organic chemistry often deals with very large compounds and complex structures, that option is somewhat unwieldy. So a shorthand notation known as a **line structure** has emerged, and is probably the most common method for describing organic compounds in a simple but unambiguous way. The line structure is really a modified version of the structural formula. As in any structural formula, lines are used to depict bonds between atoms. But in the line drawing, many of the elemental symbols are omitted. By definition an organic compound is based on carbon atoms. So to reduce clutter in a line drawing, the 'C' symbols for all carbon atoms are not written. Furthermore, because organic compounds almost always contain many hydrogen atoms, the 'H' symbol for any hydrogen atom that is attached directly to a carbon atom is also not written. Symbols are written for any elements other than carbon and hydrogen, as well as for any hydrogen atoms that are not directly attached to carbon. We can illustrate the relationship between a structural formula and a line drawing with an example.

Example Problem 2.3

Poly(methyl methacrylate) is widely known as Plexiglas®. The structural formula for the monomer, methyl methacrylate, is shown below. Write the corresponding line structure for methyl methacrylate.

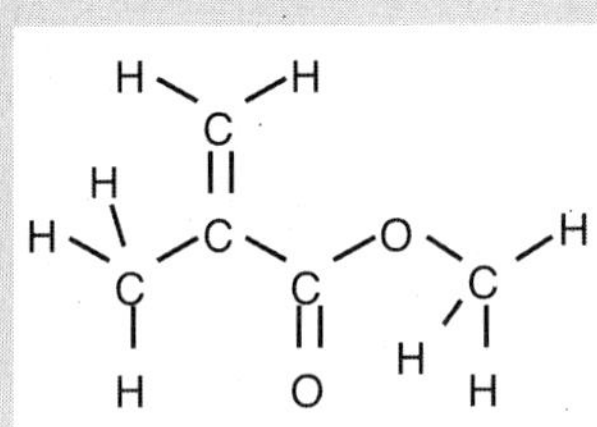

Strategy

We convert the structure into line drawing form by removing the symbols for all carbon atoms, and for hydrogen atoms attached directly to carbons. Bonds to or between carbon atoms remain, so carbon atom positions become either intersection points between lines or ends of lines. Bonds between carbon and hydrogen atoms are omitted.

Solution

First we will remove the symbols and bonds for all of the hydrogen atoms, since they are all bound directly to carbon.

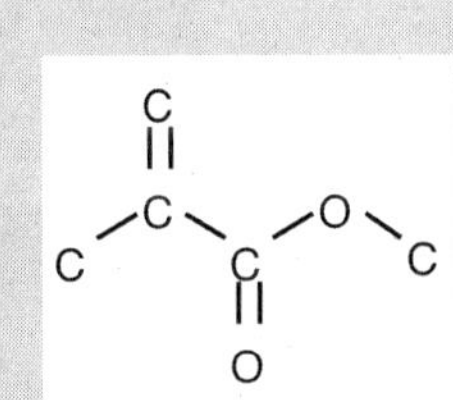

Next we remove the symbols for the carbon atoms, leaving the lines that depict the remaining bonds intact. This gives us the final line structure.

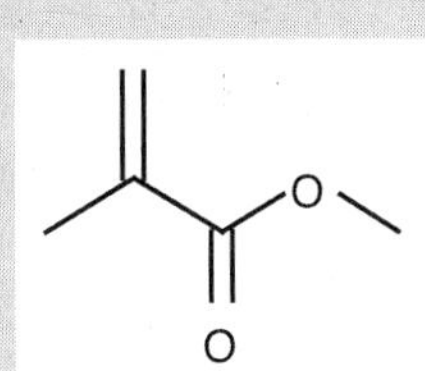

Discussion

The line structure is much more compact than the original structural formula. An experienced chemist quickly recognizes where the atoms whose symbols are not shown need to be.

Check Your Understanding

The structural formula for styrene, which is the monomer for the common plastic polystyrene, is shown below. Convert this to a line drawing.

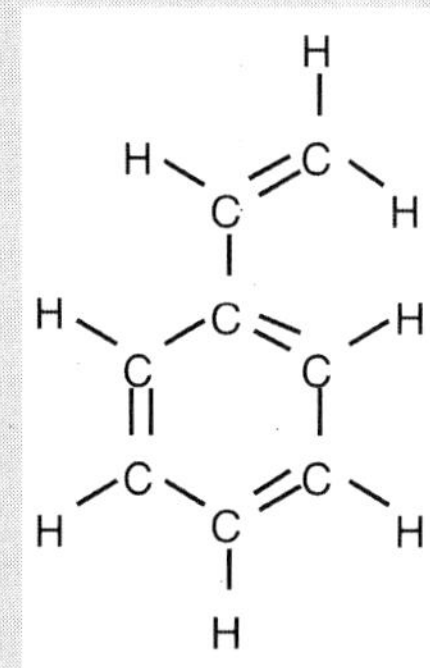

We will use these line structures throughout the rest of this textbook, and you may also encounter them in other places, such as the information sheets that accompany prescription drugs. In many instances it will be necessary to interpret the line drawing to determine the molecular formula, so we should develop a systematic way to do that. In addition to the rules we used above to go from the structural diagram to the line structure, we will need to introduce two important generalizations about chemical bonding.

1. A hydrogen atom in an organic molecule can form only one covalent bond to one other atom.
2. Every carbon atom in an organic molecule always forms exactly four covalent bonds.

In combination, these two facts allow us to fill in all of the carbon and hydrogen atoms that are not explicitly written in a line structure. First, we place a carbon anywhere that we see either an intersection between lines or the end of a line. Then we add hydrogen atoms as needed to bring each carbon's number of bonds up to four.

Example Problem 2.4

The line structure below represents 2,6-dimethylphenol, which is the monomer for a temperature resistant plastic called poly(phenylene oxide). What is the molecular formula for 2,6-dimethylphenol?

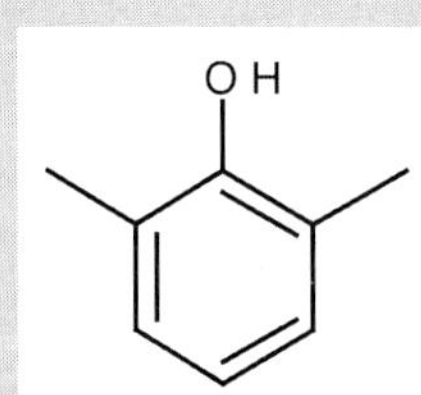

Strategy

First we can pencil in carbon atoms at the appropriate positions. Then we will add hydrogen atoms as needed. Once all of the atoms have been identified, it will be easy to count them to produce the needed formula.

Solution

We place a carbon atom at the end of a line or the intersection between lines.

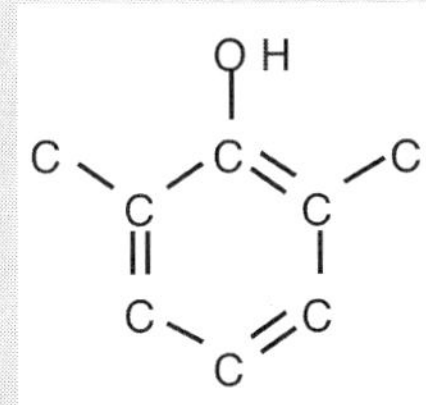

Next we count the number of bonds shown for each carbon. If that number is three or less, we add as many hydrogens as needed to bring it up to four.

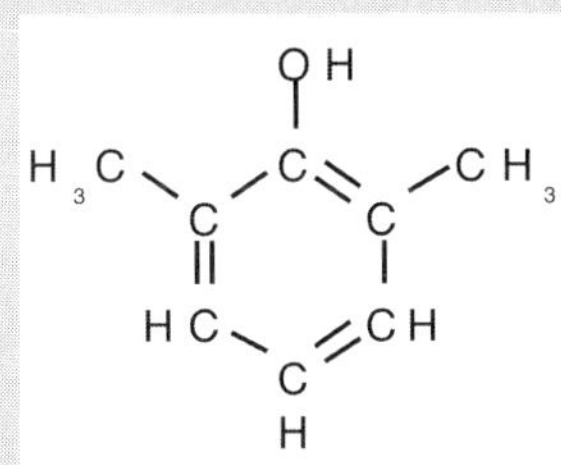

Now all of the atoms are shown explicitly. Counting, we arrive at the molecular formula as $C_6H_{10}O$.

Discussion

The double lines present in the ring in this structure represent double bonds, in which two pairs of electrons are shared between the two atoms. Notice that in locating the carbon atoms we treated these the same as we did the single lines: each intersection point represents a carbon atom, no matter how many lines meet.

Check Your Understanding

The line structure for vinylpyrrolidone, another monomer, is shown below. Find the corresponding molecular formula.

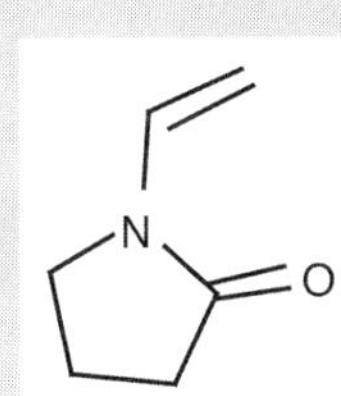

Given the vast number of organic compounds, the need for some systematic way to understand their chemistry should be apparent. One of the most important concepts for an organic chemist is the idea that certain arrangements of atoms tend to display similar chemical properties whenever they appear together. Such an arrangement of atoms is called a **functional group**. One of the simplest functional groups, and one that is central to many polymerization reactions, is a pair of carbon atoms joined by a double bond. If the double bond is converted to a single bond, then each carbon atom can form a new bond to another atom. Thus the characteristic reaction of a carbon-carbon double bond is addition, in which new atoms or groups of atoms are attached to a molecule. Line structures make it very easy to identify any C=C groups in a molecule, and thus to locate positions at which addition reactions might be feasible.

The role of addition reactions in producing polymers will be examined in Section 2.8.

The notion of functional groups often influences the way in which we choose to write chemical formulas. Compounds that contain an –OH functional group, for example, are collectively referred to as alcohols. The presence of the –OH group conveys certain properties to this class of molecules, including an ability to mix with water to a much greater extent than the corresponding hydrocarbons. If the chemical formula for an alcohol is written in such a way that the presence of the –OH group is emphasized, then it will be easier to recognize that this group is present. So the formulas for the simplest alcohol, methanol, is most often written as CH_3OH rather than as CH_5O. Similarly, ethanol is generally written as C_2H_5OH.

2.7 Chemical Nomenclature

Although only a limited number of elements are known to exist, the number of compounds that may be formed from those elements is virtually boundless. Given the vast number of molecules that can be made, we require a systematic means of assigning names to chemical compounds. This system should be sufficiently well defined that a person who knows the rules can draw the structure of any compound given its systematic name. This naming process for molecules is often referred to as **chemical nomenclature.** We will establish some of the basic premises of this system now, and supplement this initial set of rules as needed when we encounter new situations and types of compounds later.

Binary Systems

Compounds containing only two elements are referred to as *binary* compounds. Fe_2O_3, for example, is a binary compound. Many such compounds exist and they can be conveniently grouped according to their bonding tendencies. Thus, we encounter slightly different rules when we name binary molecules held together by covalent bonds than when we assign names to ionic compounds. This means it is important that you learn to recognize a compound as being ionic or covalent from its formula. This will become easier as you get more comfortable working with chemical formulas. A good way to start is to begin to recognize the elements as metals or nonmetals. When two nonmetals combine, they usually form a covalent compound. But when metals and nonmetals combine with one another, they frequently form ionic compounds. It is also handy to learn to recognize common polyatomic ions, like those listed below in Tables 2.4 and 2.5. The presence of these ions is a sign that a compound is probably ionic.

Naming Covalent Compounds

A given pair of elements can often form compounds in a number of different ways. Nitrogen and oxygen, for example, form NO, N_2O, NO_2, N_2O_4, and N_2O_5 all of which are stable enough to observe. So it is critical that our naming system is able to distinguish between these different molecules. To accomplish this, the nomenclature system uses a prefix to specify the number of each element present. The first ten of these prefixes, which arise from the Greek roots for the numbers, are listed in Table 2.3.

In a binary compound, the element that appears first in the formula also appears first in the name. The first element retains its full name, while the second element is described by replacing the ending from its name with the suffix -ide. Both elements will be preceded by a number designating prefix, the only exception being that when there is only one atom of the first element, it does not carry the prefix mono. An example of this procedure is provided in Example Problem 2.5.

Number	Prefix
one	mono-
two	di-
three	tri-
four	tetra-
five	penta-
six	hexa-
seven	hepta-
eight	octa-
nine	nona-
ten	deca-

Table 2.3 Greek prefixes for the first ten numbers

Example Problem 2.5

What are the systematic names of the following compounds? (a) N_2O_5, (b) PCl_3, (c) P_4O_6.

Strategy

The first element listed retains its full name and adds a prefix when more than one atom of it is in the compound. The second element will retain only the root of its name, followed by *-ide*, and it too takes a prefix to indicate the number of atoms.

Solution

(a) N_2O_5: dinitrogen pentoxide

(b) PCl_3: phosphorous trichloride (note: this is not called monophosphorous trichloride)

(c) P_4O_6: tetraphosphorous hexoxide

Check Your Understanding

What are the names of the following compounds? (a) CS_2 (b) SF_6 (c) Cl_2O_7

Naming Ionic Compounds

The iron chlorides shown earlier in Figure 2.16 serve as examples of binary ionic compound. Because ionic compounds must be neutral, the positive and negative charges of the ions must balance each other and only one formula unit is possible. In the name therefore, once one of the charges is specified, the entire formula is known. The rules of nomenclature specify that the positively charged species, the cation, be specified with enough information about its charge to indicate the complete formula. Unfortunately for novices in chemistry, some groups of the periodic table have only one possible cationic charge while others groups (such as Group 8 that contains iron) have more than one possibility.

The most common cations are summarized in Table 2.4. Note that the cations of all of the Group 1 metals have a charge of +1, and those of Group 2 have a charge of +2. Because the nomenclature system was designed by chemists to be used by chemists, this type of factual information is assumed to be known. The nomenclature system gives no indication of charge for groups that always have the same charge. When we look at the charges in transition metals, however, we note that they often form two cations and some even more than two. When we encounter such a situation, we indicate the charge of the cation in the name using a roman numeral in parenthesis following the elemental name. Thus Fe^{2+} has the name iron(II) while Fe^{3+} has the name iron(III). An older system was once used that distinguished these two ions by using a suffix. The smaller charge formed a name ending in *-ous*, (ferrous for Fe^{2+} in the case of iron whose root, *ferr* comes from the Latin word for iron) and the larger charge ending in *-ic*, (ferric for Fe^{3+}). We will not use this system in this book, but you may see these older names on some laboratory chemicals.

Sodium ion	Na^{+}	Potassium ion	K^{+}
Magnesium ion	Mg^{2+}	Calcium ion	Ca^{2+}
Iron(II) ion	Fe^{2+}	Copper(I) ion	Cu^{+}
Iron(III) ion	Fe^{3+}	Copper(II) ion	Cu^{2+}
Silver ion	Ag^{+}	Zinc ion	Zn^{2+}
Ammonium ion	NH_4^{+}	Hydronium ion	H_3O^{+}

Table 2.4 Common cations

We've already seen some examples of elements that form monatomic anions – the most important of which are the halogens. There are also several polyatomic anions that we will encounter often in this book. The ones listed in Table 2.5 are the most common anions that we will see throughout the text.

Halides	F^-, Cl^-, Br^-, I^-	Sulfate	SO_4^{2-}
Nitrate	NO_3^-	Hydroxide	OH^-
Phosphate	PO_4^{3-}	Cyanide	CN^-
Carbonate	CO_3^{2-}	Hydrogen carbonate	HCO_3^-

Table 2.5 Common anions

The naming convention for monatomic ions is familiar from the discussion of covalent molecules above: the name of a monatomic anion is the root of the element name with the suffix *–ide* added. Thus Cl^- is chloride, Br^- is bromide, and so on. We can now name $FeCl_2$ and $FeCl_3$ as iron(II) chloride and iron(III) chloride, respectively.

Many compounds contain polyatomic anions, including those shown in the table above. The names of these polyatomic ions are most often memorized rather than being obtained by a systematic nomenclature rule. There is, however, a system for polyatomic anions that contain oxygen and one other element, the oxoanions. The base name of the oxoanion is provided by the element that is not oxygen. If there are two possible groupings of the element with oxygen, the one with more oxygen atoms uses the suffix *-ate* and the one with fewer oxygens uses the suffix *-ite*. When there are four possible oxoanions we add a prefix *per-* to the *-ate* suffix for the most oxygens and a prefix *hypo-* to the *-ite* suffix for the least oxygens. Chlorine is the classic example of an element that forms four oxoanions, the names of which are provided in Table 2.6.

ClO^-	hypochlorite
ClO_2^-	chlorite
ClO_3^-	chlorate
ClO_4^-	perchlorate

Table 2.6 Oxoanions of chlorine

Once we know how to name both of the ions, naming an ionic compound is accomplished by simply combining the two names. The cation is listed first in the formula unit and in the name. Example Problem 2.6 provides some examples of how to determine the name of an ionic compound.

Example Problem 2.6

Determine the name of the following ionic compounds: (a) Fe_2O_3, (b) Na_2O, (c) $Ca(NO_3)_2$.

Strategy

We must determine the names of the constituent ions first. The anions will provide a hint about the cation charges if we need it.

Solution

(a) Fe_2O_3: As noted in Table 2.5, oxygen is always a -2 ion in these compounds, so there is a total of -6 represented in the formula unit. The two iron atoms must therefore have a total charge of +6, requiring +3 from each iron. So the name is iron(III) oxide

(b) Na_2O: Sodium from Group 1 is always +1 and oxygen is always -2. Therefore the name is sodium oxide. No roman numeral is needed for sodium, because it has only one common ion charge.

(c) $Ca(NO_3)_2$: The calcium ion is in Group 2 and always carries a charge of +2. NO_3^- is a common polyatomic anion called nitrate. The name is calcium nitrate.

Check Your Understanding

Name each of the following ionic compounds. (a) $CuSO_4$ (b) Ag_3PO_4 (c) V_2O_5

Occasionally throughout this text we will encounter new classes of chemical compounds that will need more rules to determine their names. We will introduce these nomenclature systems when necessary.

2.8 Insight into Polyethylene

We close this chapter by taking a closer look at polyethylene, which is probably the most common polymer in the world. US domestic production of all forms of polyethylene in 2001 totaled more than 33 billion pounds. And in its various forms, polyethylene can be found in items ranging from plastic grocery bags to children's toys, automobile gas tanks, and even bulletproof vests. Let's start at the beginning, with a look at how polyethylene is produced.

In Section 2.4 we pointed out that polyethylene is built up from a monomer called ethylene, and that the molecular formula for ethylene is C_2H_4. And we also noted that the formula for polyethylene itself is often written as $-[CH_2CH_2]_n-$. To begin to see how the monomers can be combined to form the polymer, we need to look at the structural formula for ethylene, which is shown below.

Free radicals are reactive wherever they form. Physiological processes like aging have been tied to the presence of free radicals, so their importance is wide ranging.

We know from our discussion of line structures that each carbon atom in a molecule always forms a total of four covalent bonds. For the carbon atoms in ethylene, two of those bonds are to hydrogen atoms while the other two make up a double bond between the two carbon atoms. In order to link the monomers together and form the polymer, then, we will need to convert this double bond into a single bond, leaving each carbon with one bond available for linking to the next monomer in the chain. The polymerization reaction begins with the addition of a small amount of an initiator molecule that easily breaks down, producing highly reactive fragments called **free radicals**. (The free radical is denoted as 'R•' below, where the '•' represents an unshared electron.) One of these free radicals attaches itself to a single ethylene molecule, opening its double bond and leaving one end "free."

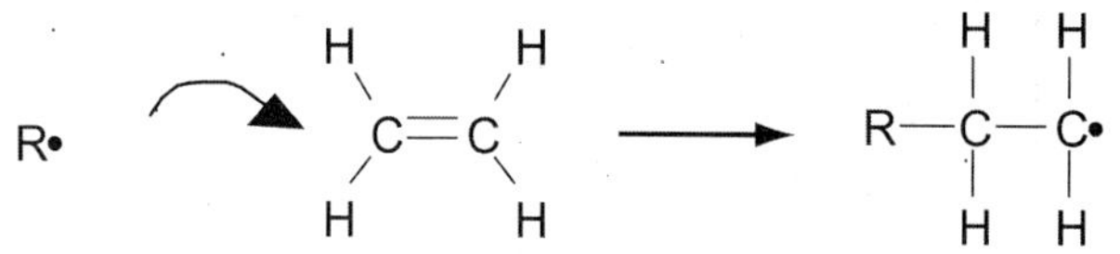

The free end of the ethylene now takes on the role of the free radical, attacking a second ethylene monomer and attaching it to the growing chain. As long as the number of available ethylene monomers remains large, the polymer can continue to grow in this fashion.

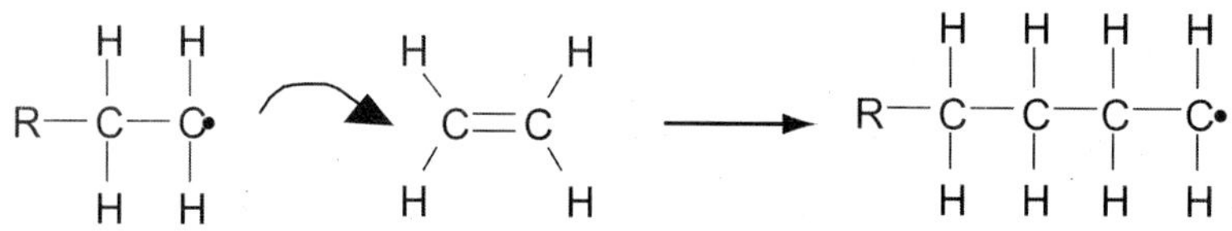

Linking together thousands of monomers in this way would generate a single polyethylene molecule, whose structural formula would look like that below, only extending over a much greater chain length.

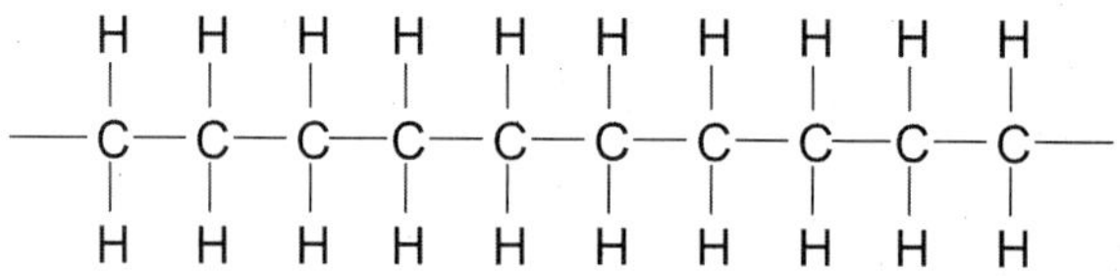

Eventually the free radical end of the growing chain will meet up with another free radical, either from an initiator molecule or from another growing chain. And when this happens, the chain will stop growing. When the monomer units grow end to end, as shown above, the result is known as linear polyethylene because all of the carbon atoms lie along a more or less straight backbone.

It is also possible to grow polyethylene under conditions that lead to branched chains, in which some of the hydrogen atoms along the backbone are themselves replaced by polyethylene-like chains. The contrasting forms of the linear and branched molecules are shown in Figure 2.17.

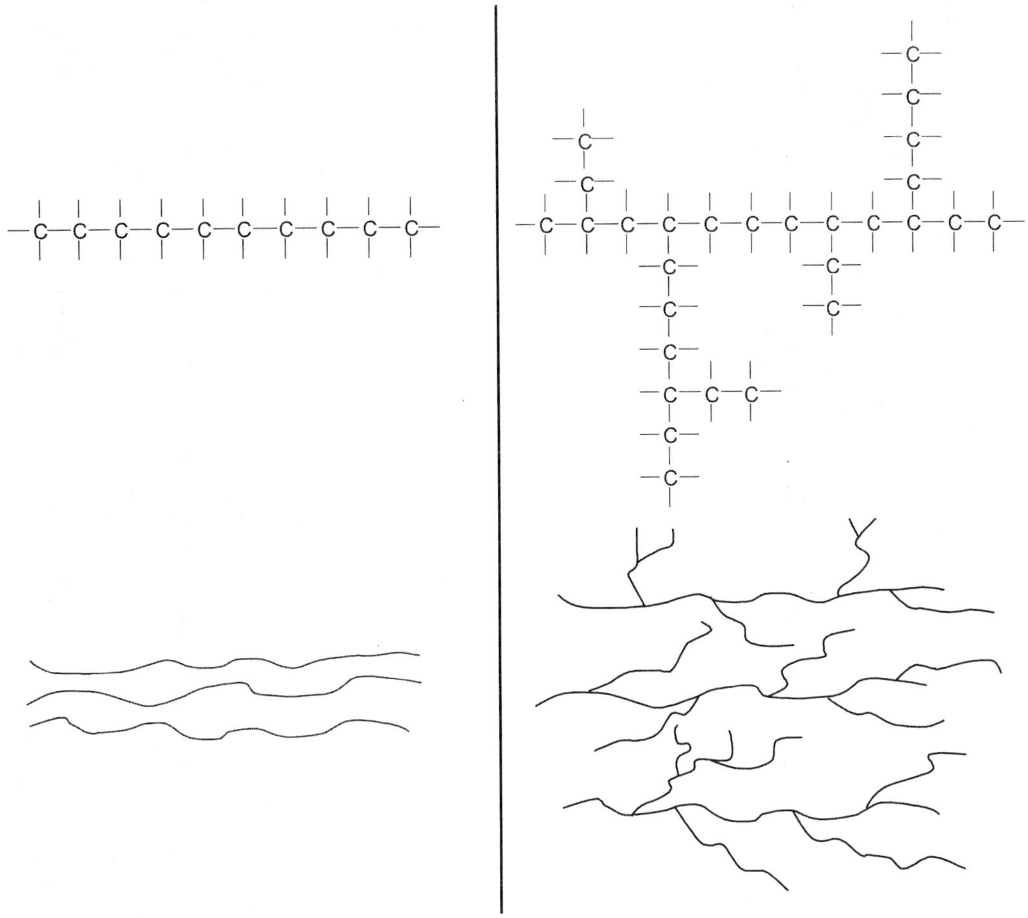

Figure 2.17 Differences between linear and branched polyethylene are illustrated. The left hand panel shows linear, or high-density polyethylene, while the right hand panel shows the branched, or low-density form. In each case the upper diagram illustrates the molecular structure for a part of a polymer chain. (Hydrogen atoms are omitted for clarity.) The lower diagrams show the way that polymer chains would pack together to form the solid plastic. The branched chains cannot approach one another as closely, so the resulting material has a much lower density.

Although linear polyethylene is conceptually simpler, branched polyethylene is actually cheaper and easier to produce. Perhaps surprisingly, the linear and branched forms of polyethylene are actually polymers with significantly different macroscopic properties. When arranged side by side, the long, linear polyethylene molecules can pack together very tightly, producing a relatively dense plastic. So linear polyethylene is also known as high-density polyethylene, or HDPE. It is a strong and hard material, used in bottles, kitchenware, and as a structural plastic in many children's toys. Because their chains cannot be stretched out straight, branched polyethylene molecules cannot be packed together as closely as those of linear polyethylene. This looser packing of the molecules produces a plastic with a much lower density, so branched polyethylene is usually referred to as LDPE, for low-density polyethylene. LDPE is commonly used in applications where relatively little strength is needed, such as plastic films, sandwich bags, and squeeze bottles.

Recent advances in polymerization technology have made it possible to grow extremely long linear polyethylene chains, stretching to hundreds of thousands of monomer units. Because the individual molecules are relatively large and heavy, this has been called "ultra-high molecular weight polyethylene," or UHMWPE. The very long chains are so strong that this material is replacing Kevlar® (another polymer) as the standard filling for bulletproof vests. UHMWPE can also be formed into large sheets, and these have been used as ice substitutes for skating rinks.

This quick glimpse at some of the many forms and uses for polyethylene shows how the observable properties of polymers are closely linked to the chemical structure of the individual molecules. It also illustrates how the interplay between chemistry and engineering can lead to polymers that are selectively produced to meet certain design specifications.

Focus on Problem Solving

Question

Naturally occurring boron has an average atomic mass of 10.811 amu. If the only isotopes of boron that are present are ^{10}B and ^{11}B, describe how you would determine the relative abundance of the two isotopes. Include in your description any information that you would need to look up.

Strategy

This problem is similar to those in the chapter for determining atomic masses. Those problems were essentially cases where we had two unknowns and two equations. There seems to be too little information provided here for that approach, though, so we need to consider any additional relationships that are implied in the problem. If we can write an extra formula that relates the amount of ^{10}B and ^{11}B that should provide the additional equation we need to solve this problem. We'll also need to know the mass of each isotope.

Solution

Because we need to know the mass of each isotope to solve this problem, that information would have to be looked up. We could get a rough approximation by assuming that ^{10}B has a mass of 10 amu and ^{11}B has a mass of 11 amu, but we probably need better data than that. The extra equation we need arises from the fact that we are told only two isotopes exist. That means that between the two of them there must be 100% of the abundance, so the (fraction of ^{10}B) + (fraction of ^{11}B) = 1.00. The fact that the resulting average atomic mass is 10.811 provides another equation, specifically,

(fraction of ^{10}B)x(mass of ^{10}B) + (fraction of ^{11}B)x(mass of ^{11}B) = 10.811 amu. We now have two equations in two unknowns (assuming we have looked up the needed masses) so the problem can be solved.

Chapter 2
Problems & Exercises

Insight into Polymers

2.1 Define the terms *polymer* and *monomer* in your own words.

2.2 How do polymers compare to their respective monomers?

2.3 Look around you and identify several objects that you think are probably made from polymers.

2.4 What one element forms the backbone of nearly all common polymers? What other elements are also found in common household polymers?

2.5 The fact that a polymer's physical properties depend on its atomic composition is very important in making these materials so useful. Why do you think this would be so?

2.6 Use the web to research the amount of PVC polymer produced annually in the United States. What are the three most common uses of this polymer?

2.7 Use the web to research the amount of polyethylene produced annually in the United States. What are the three most common uses of this polymer?

Atomic Structure and Mass

2.8 In a typical illustration of the atom such as Figure 2.3, what features lead to misunderstandings about the structure of atoms? Which ones give important insight?

2.9 Why is the number of protons called the atomic number?

2.10 If the amu is defined as 1/12 the mass of an atom of ^{12}C, which contains 6 protons and 6 neutrons, why don't both the proton and neutron have a mass of exactly 1.000 amu?

2.11 Define the term isotope.

2.12 Write the complete atomic symbol for each of the following isotopes. (a) carbon-13 (b) phosphorous-31 (c) sodium-23 (d) boron-10

2.13 How many protons, neutrons, and electrons are present in a neutral atom of the following isotopes? (a) iodine-126 (b) silicon-28 (c) chlorine-35 (d) niobium-93

2.14 What percent more massive is the average atom of potassium when compared to carbon-12?

2.15 Mercury is 16.716 times more massive than carbon-12. What is the atomic mass of mercury?

2.16 How can an element have an atomic mass that is not an integer?

2.17 Explain the concept of a "weighted" average in your own words.

2.18 The following table presents the abundances and masses of the isotopes of zinc. What is the atomic mass of zinc?

isotope	percent abundance	mass
^{64}Zn	48.6%	63.9291 amu
^{66}Zn	27.9%	65.9260 amu
^{67}Zn	4.10%	66.9271 amu
^{68}Zn	18.8%	67.9249 amu
^{70}Zn	0.60%	69.9253 amu

2.19 The atomic mass of copper is 63.55 amu. There are only two isotopes of copper, ^{63}Cu with a mass of 62.93 amu and ^{65}Cu with a mass of 64.93 amu. What is the percentage abundance of each of these two isotopes?

2.20 The element gallium, used in gallium arsenide semiconductors, has an atomic mass of 69.72 amu. There are only two isotopes of gallium, ^{69}Ga with a mass of 68.9257 amu and ^{71}Ga with a mass of 70.9249 amu. What are the isotopic abundances of gallium?

Ions

2.21 Define the term *ion* in your own words.

2.22 What is the difference between cations and anions?

2.23 Why is an atom NOT an ion?

2.24 Provide the symbol of the following ions given the number of protons and electrons in each.

(a) 8 protons, 10 electrons

(b) 20 protons, 18 electrons

(c) 53 protons, 54 electrons

(d) 26 protons, 24 electrons

2.25 How many protons and electrons are in each of the following ions?

(a) Na^+ (b) Al^{3+} (c) S^{2-} (d) Br^-

2.26 Identify each of the following species as an anion, a cation or a molecule.

(a). CO_3^{2-} (b). CO_2 (c). NH_4^+ (d). N_3^- (e). CH_3COO^-

2.27 Explain why the hydrogen ion, H^+, is sometimes referred to as a proton.

2.28 In what region of the periodic table are you likely to find elements that form more than one stable ion?

2.29 Use the web to identify which elements in groups 3 – 12 do not form more than one ion.

2.30 Use the web to find out how many ions are formed by the following elements (a) Fe, (b) V, (c) Pt, (d) Mo

2.31 Use the web to find a catalyst for a polymerization reaction that uses an ion. What are the apparent advantages of using this catalyst for creating the polymer?

2.32 Using Coulomb's Law explain how the difference between attractive and repulsive interactions between ions is expressed mathematically.

Compounds and Chemical Bonds

2.33 How many atoms of each element are represented in the formula, $Ba(OH)_2$?

2.34 Which of the following formulas shows the most hydrogen atoms?

C_2H_6, $(NH_4)_2CO_3$, H_2SO_4, $Fe(OH)_3$

2.35 In general, how are electrons involved in chemical bonding?

2.36 What is the difference between an ionic bond and a covalent bond?

2.37 When talking about the formula of an ionic compound, why do we typically refer to a formula unit rather than a molecule?

2.38 Which formula below is the correct one for an ionic compound? What is incorrect about each of the others? (a) $AlO_{3/2}$ (b) Ca_2Br_4 (c) $Mg(PO_4)_{3/2}$ (d) $BaCO_3$

2.39 Explain the differences between ionic and metallic bonding.

2.40 Conduction of electricity usually involves the movement of electrons. Based on the concept of metallic bonding, explain why metals are good conductors of electricity.

2.41 Describe how a covalently bonded molecule is different from compounds that are either ionic or metallic.

2.42 Explain the difference between a molecular formula and an empirical formula.

2.43 Why are empirical formulas generally preferred for describing polymer molecules?

2.44 The molecular formula for the ethylene monomer is C_2H_4. What is its empirical formula?

2.45 Poly(butadiene) is a synthetic elastomer, or rubber. The corresponding monomer is butadiene, which has the molecular formula C_4H_6. What is the empirical formula of butadiene?

The Periodic Table

2.46 What distinguished the work of Mendeleev that caused scientists to accept his concept of the periodic table when others before him were not believed?

2.47 How does the periodic table help to make the study of chemistry more systematic?

2.48 What is a period in the periodic table? From what does it derive its name?

2.49 How do binary compounds with hydrogen illustrate the concept of periodicity?

2.50 What part of the periodic table contains the representative groups or elements? Where are the transition metals?

2.51 Name the group to which each of the following elements belongs. (a) K (b) Mg (c) Ar (d) Br

2.52 What are some of the physical properties that distinguish metals from nonmetals?

2.53 What region of the periodic table contains the nonmetals?

2.54 Why are nonmetals important even though they account for only a very small fraction of the elements in periodic table?

2.55 What is a metalloid?

2.56 What elements are commonly referred to as metalloids?

2.57 Identify the following elements as metals, metalloids or nonmetals. (a) Si (b) Zn (c) B (d) N (e) K (f) S

2.58 List any metalloids that occur in the fourth period.

Inorganic and Organic Chemistry

2.59 The chemistry of the representative elements is generally simpler than that of the transition metals. Why is this so?

2.60 Calcium and fluorine combine to produce an ionic calcium fluoride, CaF_2. Use the periodic table to predict at least two other compounds that you would expect to have structures similar to that of CaF_2.

2.61 What is meant by the phrase *organic chemistry*?

2.62 Based on what you've learned in this chapter, would you classify the chemistry of polymers as organic or inorganic? Why?

2.63 What is a *functional group*? How does the concept of the functional group help to make the study of organic chemistry more systematic?

2.64 The molecule shown below is responsible for the smell of popcorn. Write the correct molecular formula for this compound.

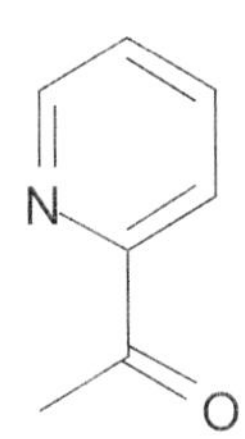

2.65 Not all polymers are formed by simply linking identical monomers together. Polyurethanes, for example, can be formed by reacting the two compounds shown below with one another. Write molecular and empirical formulas for each of these two substances.

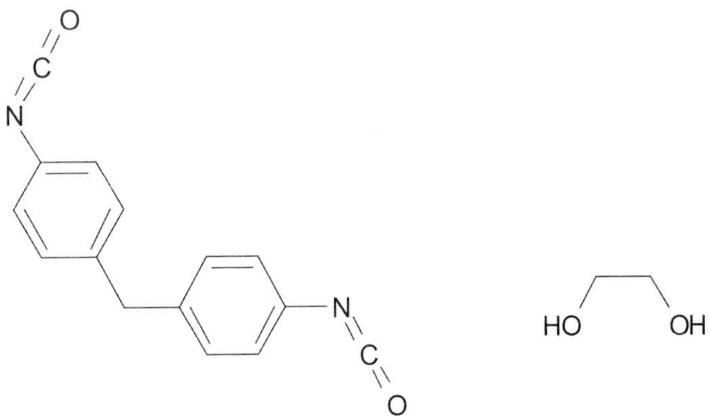

2.66 The compound shown below forms an amorphous solid (a glass) at room temperature, and has been used as a medium for the holographic storage of information. Write the correct molecular formula for this molecule.

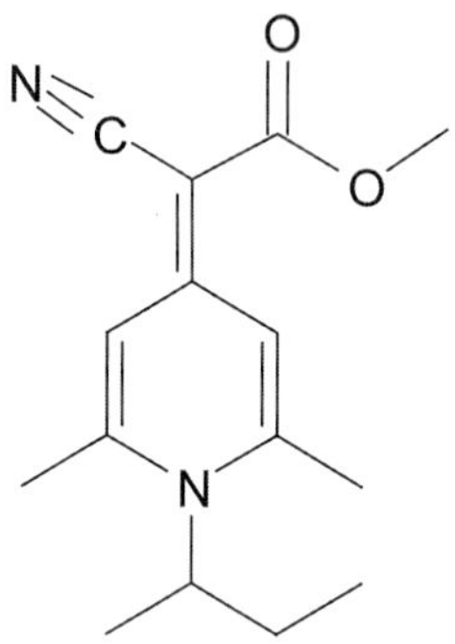

2.67 The figure below shows the structure gamma-aminobutanoic acid, or GABA. This molecule is a neurotransmitter. Some of the effects of alcohol consumption are due to the interaction between ethanol and GABA. Write the correct molecular formula for this compound.

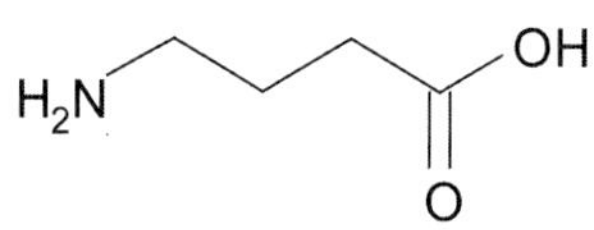

2.68 The figure below shows the structure of the adrenaline molecule. Write the correct molecular formula for this substance.

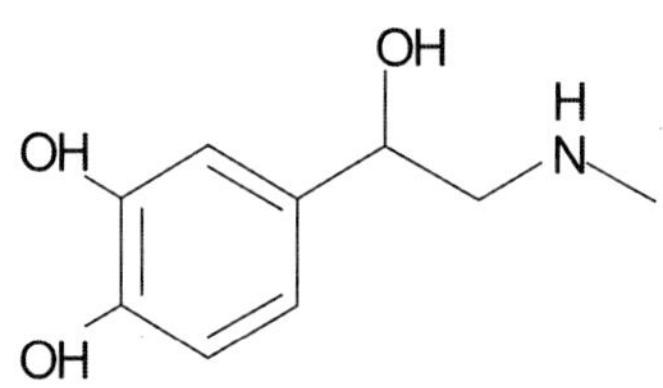

Chemical Nomenclature

2.69 Why are there different rules for naming covalent and ionic binary compounds?

2.70 What binary combinations of elements are most likely to give ionic substances?

2.71 Name the following covalent compounds. (a) N_2O_5 (b) S_2Cl_2 (c) Nbr_3 (d) P_4O_{10}

2.72 Write the molecular formula for each of the following covalent compounds. (a) silicon dioxide (b) tetraarsenic decaoxide (c) iodine heptafluoride (d) dinitrogen trioxide

2.73 Write the molecular formula for each of the following covalent compounds (a) sulfur hexafluoride (b) bromine pentafluoride (c) disulfur dichloride (d) tetrasulfur tetranitride

2.74 Name the following ionic compounds. (a) $NaHCO_3$ (b) $Fe(OH)_2$ (c) ZnS (d) Cr_2O_3

2.75 Name the following ionic compounds. (a) Cu_2O (b) Au_2S_3 (c) $AlCl_3$ (d) K_3N

2.76 Write the formula unit of the following ionic compounds. (a) copper(II)sulfide (b) barium chloride (c) iron(III) nitride (d) titanuim(IV) bromide

2.77 Name the following compounds. (a) PCl_5 (b) Na_2SO_4 (c) Ca_3N_2 (d) $Fe(NO_3)_3$ (e) SO_2 (f) Br_2O_5

Insight into Polyethylene

2.78 What is a free radical? How are free radicals important in the formation of polyethylene?

2.79 How do molecules of low-density polyethylene and high-density polyethylene differ? How do these molecular scale differences explain the differences in the macroscopic properties of these materials?

2.80 Why do you think an initiator molecule is needed to induce the polymerization of ethylene?

2.81 Use the web to determine the amount of Low-Density Polyethylene and High-Density Polyethylene produced annually in the US. What uses predominate the applications of these two materials?

Focus on Problem Solving Exercises

2.82 Describe how you can identify the isotope, X, in this puzzle. The nucleus contains one more neutron than protons and the mass number is nine times larger than the charge on the ion X^{3+}.

2.83 Many transition metals produce more than one ion. For example, iron has ions with charges of +2 and +3 that are both common. How could you use the compounds of a transition metal with oxygen to determine the charge of the metal ion? Use iron as your example.

2.84 Naturally occurring boron has an average atomic mass of 10.811 amu. If the only isotopes of boron that are present are ^{10}B and ^{11}B, describe how you would determine the relative abundance of the two isotopes. Include in your description any information that would need to be looked up.

2.85 Naturally occurring chlorine has an average atomic mass of 35.453 amu. If the only isotopes of chlorine that are present are ^{35}Cl and ^{37}Cl, describe how you would determine the relative abundance of the two isotopes. Include in your description any information that would need to be looked up.

2.86 Some candy manufacturer makes chocolate covered cherries. While all of the products look roughly the same, 3% of them are missing the cherry. The mass of the candy with a cherry is 18.5 g while those that are missing the cherry weigh only 6.4 g. (a) How would you compute the average mass of a box of 100 of these chocolate covered cherries from this manufacturer? (b) How is this question analogous to the determination of atomic masses?

2.87 Two common oxides of iron are FeO and Fe_2O_3. Based on this information tell how you would predict two common compounds of iron and chlorine.

CHAPTER 3: Molecules, Moles, and Chemical Equations

Today roughly twenty million chemical compounds are known. More are being discovered or created as you read this. While chemical nomenclature provides a systematic way to facilitate discussion of this tremendous diversity of substances, that is just a small step towards understanding chemistry. Much of the science of chemistry is aimed at understanding the reactions that produce or consume chemical substances. Although you probably don't realize it, chemical reactions are central to many of the technologies you use. Chemical reactions in batteries produce electricity for your portable appliances, while the burning of hydrocarbons powers your car. Other reactions, such as those designed to synthesize the light-emitting compounds used in LCD computer displays, play less obvious roles in technology. We'll begin our discussion by considering a class of reactions that can be either useful or destructive, but are always dramatic: explosions.

Figure 3.1 Explosions can be very large, as in the photo on the left, or quite small, as in the laboratory demonstration in the photo on the right. But all explosions are very rapid chemical reactions, releasing substantial amounts of energy in a short time.

3.1 Insight into Explosions

From the imposing sound to the frightening potential for destruction, nothing about an explosion is easily ignored. Whether it is a controlled detonation to bring down an old building, a violent industrial accident, or a deliberate bombing, an explosion always commands our attention. And if we look at an explosion scientifically, what we will find at its heart is a chemical reaction. There are actually many different chemical reactions that can produce explosions. Some are strikingly simple, at least in terms of the substances involved: a mixture of hydrogen and oxygen gases will explode with a distinctive popping sound if ignited with a spark or a flame. Others can be much more complex, involving several large molecules. What common characteristics can we identify that might contribute to a reaction's becoming explosive?

We will study the role of energy in chemical reactions in detail in Chapter 9.

Perhaps the most fundamental requirement for an explosive reaction is that it must release a large amount of energy. This release of energy is responsible for what might casually be described as the "force" of the explosion. The actual amount of energy released in a particular detonation will depend on both the identity and the quantity of the explosive used. In many cases, the energy release occurs as a fairly complex compound (the explosive) decomposes to form a collection of smaller, simpler molecules. The fact that explosives can decompose in a way that releases large amounts of energy means that these compounds are not very stable, and this raises obvious safety concerns for the storage or handling of what have come to be called "energetic materials." An explosive is most useful if it can be handled safely and then detonated readily on demand, but these objectives are not always mutually compatible. The same properties that lead to safe handling can also make detonation more difficult.

We will examine the factors that govern the rates of chemical reactions in Chapter 11.

Another common trait of explosive reactions is that they happen very fast; this is what makes an explosion much more dramatic and destructive than a slow burning fire that might release the same amount of energy. A careful examination of the progress of explosive chemical reactions reveals that they actually accelerate as they proceed. As a result, all of the available explosive is consumed in a very short period of time. And as that happens, the energy from the explosion is also released on that same short timescale. You may know from physics that power is defined as energy per unit time. So as the time required for the explosion to occur becomes shorter, the power of the explosion will increase.

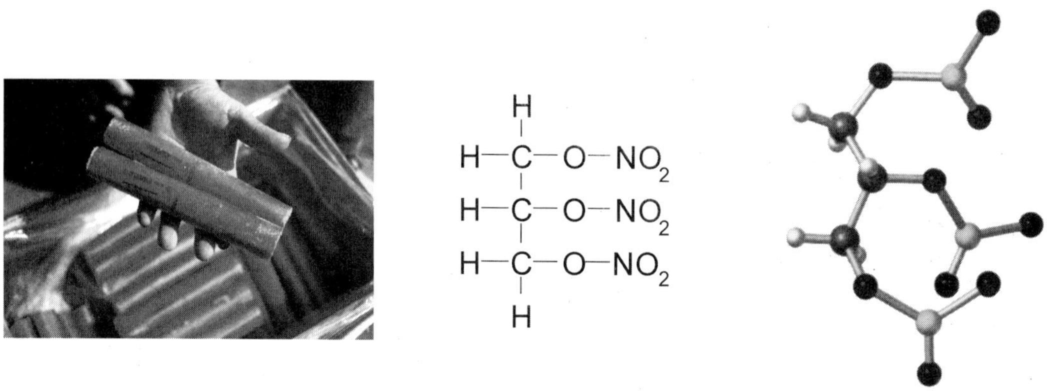

Figure 3.2 Alfred Nobel's invention of dynamite provided a powerful explosive in solid form. The actual explosive in dynamite is nitroglycerine, whose molecular structure is shown. Nitroglycerine is a liquid under ordinary conditions, so its use as a commercial explosive was extremely hazardous.

Modern explosives are generally solids. But that was not always true; liquid nitroglycerin was one of the first widely used explosives. Liquids are much harder to transport and handle than solids, though, and early use of nitroglycerin was plagued by many accidents due to unintentional detonation. The hazards associated with the use of liquid nitroglycerin prompted Alfred Nobel to develop dynamite, in which nitroglycerin is mixed with an inert binding agent to produce a solid material that is much easier and safer to handle. In the chemical reactions associated with explosions, the solid explosive is converted predominantly into gases. Far from being coincidental, the generation of gases is actually very important to the development of the explosion. You probably know from earlier chemistry or physical science classes that gases have much lower densities than solids. So when a chemical reaction transforms a solid explosive into a large number of gaseous molecules, those gases will tend to expand rapidly to reduce their density. This expansion typically results in the formation of a shock wave, which helps to deliver the blow of the explosion to its surroundings as shown in Figure 3.3.

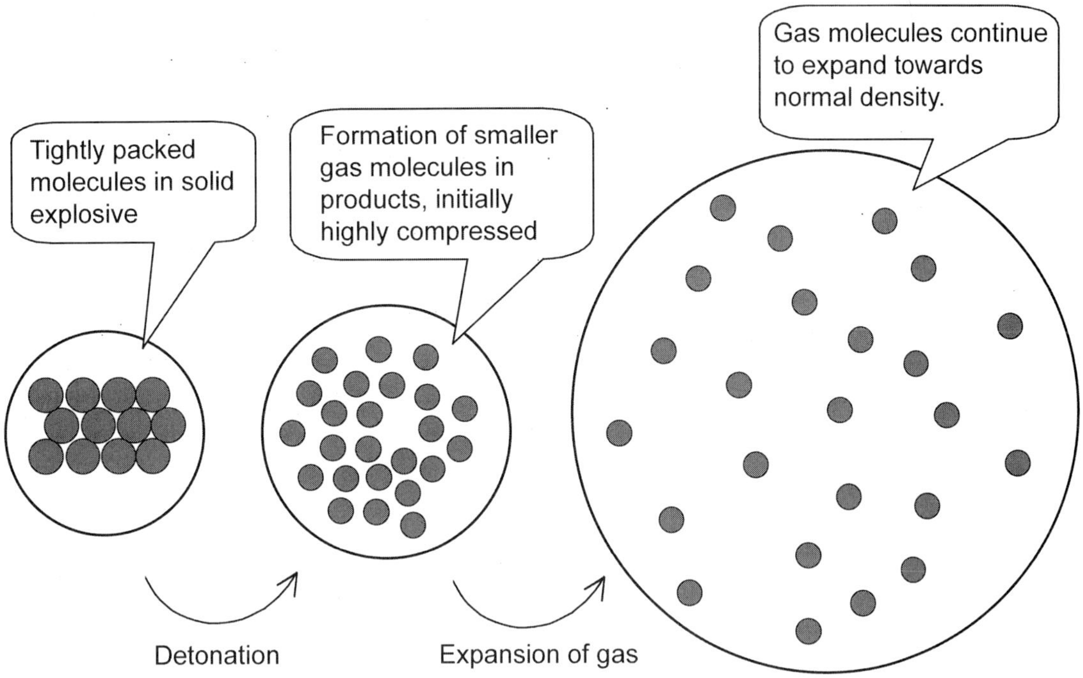

Figure 3.3 The destructive force of an explosion is due in part to the expansion of gases produced in the reaction. This molecular scale diagram illustrates the effect. In a solid explosive, relatively large molecules are packed very closely together. When the reaction occurs, the products are usually smaller gas molecules. These gases are initially formed in a volume similar to that of the solid. Since densities are typically hundreds of times smaller in gases than in solids, the products are very highly compressed. The gas must expand in order to lower its density, and this produces a shock wave.

3.2 Chemical Formulas and Equations

While the properties and uses of different explosives vary considerably, they all involve chemical reactions. Indeed, a realistic description of the chemistry of virtually any explosion would require us to consider many individual reactions. In order to discuss the science behind an explosion, than, we must be able to describe a chemical reaction in a concise manner. We've already seen how chemical formulas provide a concise way to represent chemical compounds, and we will now describe how chemical equations build on this to accomplish the same goal for chemical reactions.

Writing Chemical Equations

Chemical equations are designed to represent the transformation of one or more chemical species into new substances. In order to assure that the meaning is clear, we follow a set of conventions when writing chemical equations. Each chemical equation has two sides, and we usually envision the reaction as proceeding from left to right. The original materials are called the **reactants** and they appear on the left-hand side of the equation. The compounds that are formed as a result of the reaction are called

products and appear on the right-hand side of the equation. An arrow is used to represent the changes that occur during the reaction. Thus we can write a completely generic chemical equation.

$$\text{Reactants} \rightarrow \text{Products}$$

This would usually be read as "reactants go to products" or "reactants give products."

When a substance dissolves in water the result is called an aqueous solution, and the "aq" label is short for aqueous.

We use chemical formulas to identify the specific reactants and products. The physical states of the compounds are often designated, with (s) indicating a solid, (l) a liquid, (g) a gas, and (aq) a substance dissolved in water. We can write the reaction of hydrogen with oxygen, which is often done as a lecture demonstration, as a simple example.

$$2H_2(g) + O_2(g) \rightarrow 2H_2O(g)$$

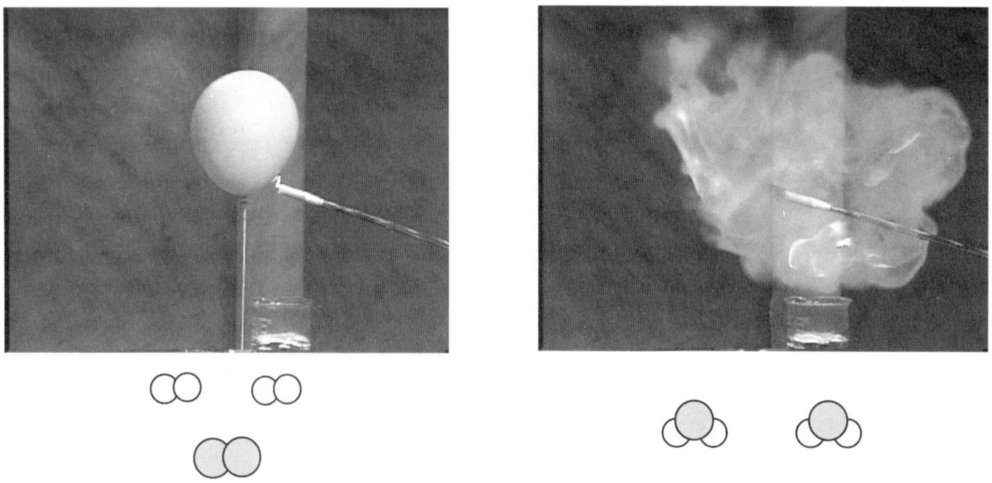

Figure 3.4 A balloon filled with a mixture of H_2 and O_2 explodes in a classroom demonstration. The gas mixture in the balloon is stable until ignited by the flame from a candle. Once ignited, the balloon explodes in a ball of flame as the reaction takes place. The accompanying microscopic scale illustration shows the molecular species present before and after reaction.

In this equation all of the substances involved are gases, as indicated by the "(g)" notation. One characteristic of this reaction is that it will only occur at high temperatures. As seen in the photos in Figure 3.4, the mixture of H_2 and O_2 is stable unless ignited. If our written equations are going to fully describe the chemical processes they represent, then we need ways to indicate such conditions in addition to identifying the chemicals involved. The way this is generally done is to place a symbol over the arrow. For now we will introduce two such symbols: one for heat and one for light. A reaction that requires heat (or high temperatures) is indicated with a Δ (pronounced delta) above the arrow, and one that is initiated by light energy is designated with an *hν*. We show two examples of these symbols below.

The symbol hν is used because it represents the energy of the light, as we will learn in Chapter 6.

$$2H_2(g) + O_2(g) \xrightarrow{\Delta} 2H_2O\,(g)$$
$$H_2(g) + Cl_2(g) \xrightarrow{h\nu} 2HCl\,(g)$$

The first equation is a more complete description of the reaction between hydrogen and oxygen. In the second equation, light energy is used to initiate the reaction of hydrogen with chlorine, producing hydrogen chloride. Such a light-induced process is called a *photochemical* reaction.

Balancing Chemical Equations

If you look carefully at the reactions we've been writing, you'll notice more than just the formulas and the states of the substances are present in the equation. Numerical information about the relative amounts of the substances involved is also evident. For example, we included the number '2' in front of both the $H_2(g)$ and $H_2O(g)$ in the equation for the reaction between hydrogen and oxygen. Why are these numbers there and what do they mean? If you've studied chemistry before in high school, you probably remember that these coefficients are often needed to *balance* a chemical equation.

The underlying premise of the chemical equation is that it is a written representation of a chemical reaction. So any reasonable representation must be consistent with the same constraints as nature. One of the most fundamental laws of nature is **the law of conservation of matter**: matter is neither created nor destroyed. If we specifically exclude nuclear reactions from our consideration, this law can be phrased more specifically. Atoms of one element are neither created nor destroyed in a chemical reaction. A chemical reaction simply rearranges the atoms present into new compounds. In its written representation of nature, therefore, the chemical equation must not "create or destroy" atoms. To uphold this condition, we must have the same number of atoms of each element on both sides of the chemical equation. (See Figure 3.5.) An equation that does not meet this condition cannot accurately represent the observed chemical reaction, and is said to be *unbalanced*.

If we also consider nuclear reactions, we must acknowledge that matter and energy can be interconverted.

In many cases, the most efficient way to balance the equations is "by inspection," which really means by trial and error. Even when this exploratory method is used, however, some systematic strategies can help make the balancing process easier. In all cases, the way we obtain balance is to introduce numbers in front of appropriate formulas in the equation. These numbers are called **stoichiometric coefficients**. The stoichiometric coefficient multiplies the number of atoms of each element present in the formula unit of the compound it precedes. Example Problem 3.1 shows one strategy for balancing a common class of reactions, the combustion of hydrocarbon fuels. Although these combustion reactions obviously release energy, in most instances they are carried out under conditions at which the reaction is not quite fast enough to produce a full-fledged explosion.

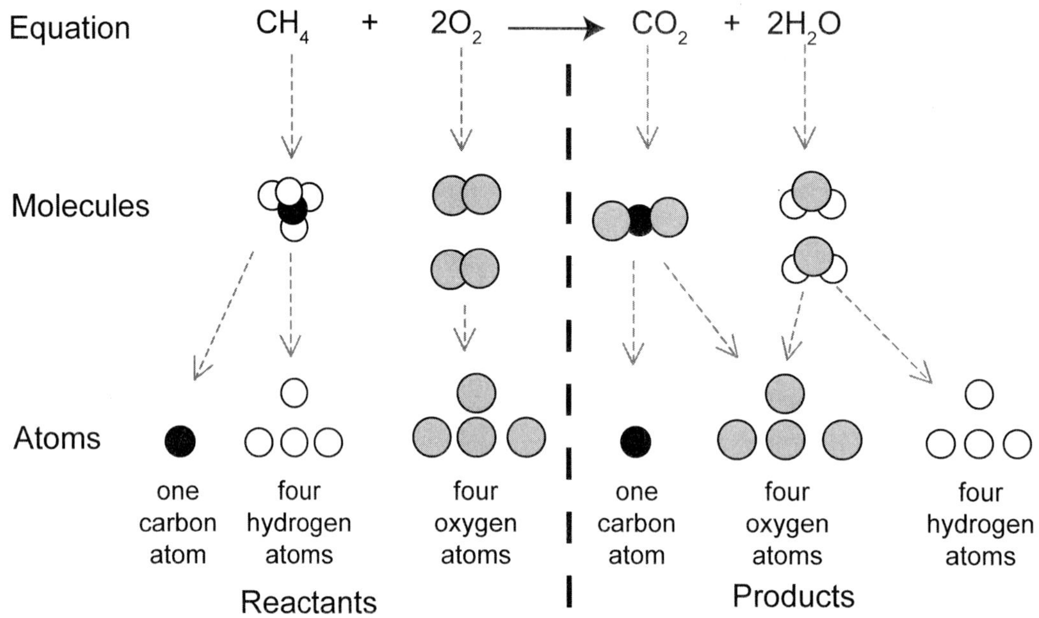

Figure 3.5 The chemical reaction for the burning of methane (CH_4) in oxygen illustrates the concept of atom balance for chemical equations. The equation is shown first in symbolic form, with each compound represented by its chemical formula. The individual molecules are pictured in the center row, and are then broken down into their constituent atoms. The number of atoms of each element on is the same on the left and right hand sides.

Example Problem 3.1

Propane, C_3H_8, is used as the fuel for gas barbecue grills, where it burns in a controlled fashion. But if a mixture of propane and air is ignited in a closed space, like a gas pipeline, an explosion can easily result. In either of these cases, the propane combines with oxygen, O_2, to form carbon dioxide and water. Write a balanced chemical equation describing this reaction.

Strategy

This problem requires two steps. First we must read the problem and determine what substances are reactants and which are products, which will allow us to write down an unbalanced "skeleton" equation. Then we must proceed to balance the equation, making sure the same number of atoms of each element appears on both sides.

Solution

Step 1: The problem notes that propane burns or explodes by combining with oxygen. The reactants are therefore C_3H_8 and O_2. We are also told that the products are CO_2 and H_2O. So we have enough information to write the unbalanced equation. (Here we will write "blanks" in front of each formula to emphasize that we still need to determine the needed coefficients.)

$$\_\_\_ C_3H_8 + \_\_\_ O_2 \rightarrow \_\_\_ CO_2 + \_\_\_ H_2O$$

Step 2: Balancing an equation like this one is aided by making some observations. In this case, both carbon and hydrogen appear in only one place on each side of the equation. Our first steps will be to balance these two elements, because there will be no other way to adjust them. Let's begin with carbon. (This choice is arbitrary.) To achieve the required balance we can use stoichiometric coefficients of one for C_3H_8 and three for CO_2, giving three carbon atoms on each side of the equation. At this point we have

$$1\ C_3H_8 + \_\_\_\ O_2 \rightarrow 3\ CO_2 + \_\_\_\ H_2O$$

(The coefficient of one in front of the propane would normally be omitted, but is explicitly written here for emphasis.) Next we will balance hydrogen. The propane already has its coefficient set from our work on carbon, so it dictates that there are eight H-atoms to be offset by water on the product side of the equation. We need to insert a coefficient of four to achieve this balance.

$$1\ C_3H_8 + \_\_\_\ O_2 \rightarrow 3\ CO_2 + 4\ H_2O$$

This leaves us with only oxygen to balance. We have saved this for last because it appeared in three compounds rather than two, and also because the fact that it appears as an element rather than a compound on the left side of the equation will make it easy to adjust the number of oxygen atoms without upsetting the balance of the other elements. Because there are ten O-atoms currently represented on the product side of the equation, we need ten on the reactant side. We can easily achieve this by using $10/2 = 5$ for the stoichiometric coefficient of the O_2 reactant. This gives us the balanced chemical equation.

$$1\ C_3H_8 + 5\ O_2 \rightarrow 3\ CO_2 + 4\ H_2O$$

This would normally be written without showing the coefficient of one in front of the propane.

$$C_3H_8 + 5\ O_2 \rightarrow 3\ CO_2 + 4\ H_2O$$

Check Your Understanding

The reaction we wrote above describes the "complete combustion" of propane. But under many conditions, hydrocarbons like propane may not burn cleanly or completely, and other products may be formed. One such possible reaction is the formation of formaldehyde (CH_2O) and water. Write a balanced chemical equation for a reaction in which propane and oxygen gas form CH_2O and H_2O as products.

MathConnections

We usually balance chemical equations by the sort of part systematic, part trial and error approach presented in Example 3.1. But it is actually fairly simple to transform that process into a more mathematical form, and doing so may help to emphasize the connection between balancing equations and the laws of conservation of matter. Let's illustrate the mathematical approach by reconsidering the equation we balanced in Example Problem 3.1.

$$\_\_\_\ C_3H_8 + \_\_\_\ O_2 \rightarrow \_\_\_\ CO_2 + \_\_\_\ H_2O$$

It will help to assign variable names to each of the coefficients in the balanced equation.

$$a\ C_3H_8 + b\ O_2 \rightarrow c\ CO_2 + d\ H_2O$$

When the correct values of the coefficients (a, b, c, d) are inserted, we know that the number of atoms of each element must be the same on both sides of the equation. We can express that idea as an equation for each element involved. Let's start with carbon. On the left hand side we will have 3 carbon atoms for each molecule of C_3H_8, for a total of $3a$ carbon atoms. And on the right side we will have c carbon atoms. So we can set these two terms equal to one another.

$$3a = c$$

Next we can write similar equations for the other elements. Considering hydrogen, we get

$$8a = 2d.$$

And for the oxygen we have

$$2b = 2c + d.$$

At this point we have three equations for the four unknown coefficients, so we don't appear to be in position to solve. The simplest way out of this predicament is just to set one coefficient arbitrarily equal to one. If we choose to set $a = 1$, we quickly get the following.

$$c = 3a = 3$$

$$2d = 8a = 8, \text{ and so } d = 4$$

$$2b = 2c + d = 6 + 4 = 10, \text{ and so } b = 5$$

Inserting these into the original equation gives us a balanced equation.

$$C_3H_8 + 5\,O_2 \rightarrow 3\,CO_2 + 4\,H_2O$$

Since all of the resulting coefficient values are integers, we have found the smallest whole number coefficients that balance the equation. And our result is identical to what we found in Example 3.1. If any fractional coefficients were obtained, we would simply multiply through by a constant to get whole numbers. This approach is quite general, and while it is not particularly popular with chemists it has been used to write computer algorithms for balancing chemical equations.

For most people the best way to learn how to balance chemical equations is just to practice until you gain confidence. The rules encountered for balancing equations are usually lists of steps that are prohibited rather than maps to be followed. For any reaction, there is more than one way to obtain the balanced equation, though some methods may be a dramatically more difficult than others. A look at some common difficulties encountered by students new to balancing chemical equations may prove useful in developing successful strategies.

One method that might appear to provide a route to the desired balanced chemical equation is to change the formulas of the molecules involved. Consider, for example, formaldehyde (CH_2O) that might be formed during the burning or explosion of a hydrocarbon like propane. As long as there is still oxygen available, any formaldehyde that is formed along the way can also burn. If CH_2O burns incompletely it will form carbon monoxide (CO) and water.

$$CH_2O + O_2 \rightarrow CO + H_2O$$

Counting the number of atoms on each side of the equation reveals that we need one more oxygen on the product side. Many students find it tempting to write

$$CH_2O + O_2 \rightarrow CO_2 + H_2O,$$

which is undeniably a balanced chemical equation. In making this change, however, the entire meaning of the equation has been altered. While this is a balanced equation, it does not depict the same reaction that we set out to describe. Carbon monoxide is no longer shown as a product, so this is *not* a valid way to balance the equation. Similarly, we can't write the oxygen as O rather than O_2, because the formaldehyde is reacting with oxygen molecules, not oxygen atoms. We may not change the chemicals involved in the equation, but only their stoichiometric coefficients. So let's take another look at our correct but unbalanced equation.

$$CH_2O + O_2 \rightarrow CO + H_2O$$

We still have two O-atoms on the right and three on the left. One possible way to balance this equation is to multiply the O_2 reactant by 1/2, so that there will be two oxygen atoms on each side.

$$CH_2O + 1/2\ O_2 \rightarrow CO + H_2O$$

It is generally preferable to avoid using fractional coefficients, since they could be interpreted as implying the existence of fractional molecules. But we can easily eliminate any fractions by multiplying the entire equation by a factor that removes them. In this case multiplying by two will do the trick, converting the 1/2 into one. This brings us to a final form of the balanced equation.

$$2\ CH_2O + O_2 \rightarrow 2\ CO + 2\ H_2O$$

Note that we must multiply *all* of the stoichiometric coefficients by two, *not* just the coefficient of 1/2 in front of the oxygen. Otherwise the balance we have worked to establish would be destroyed. Some of these ideas are depicted pictorially in Figure 3.6.

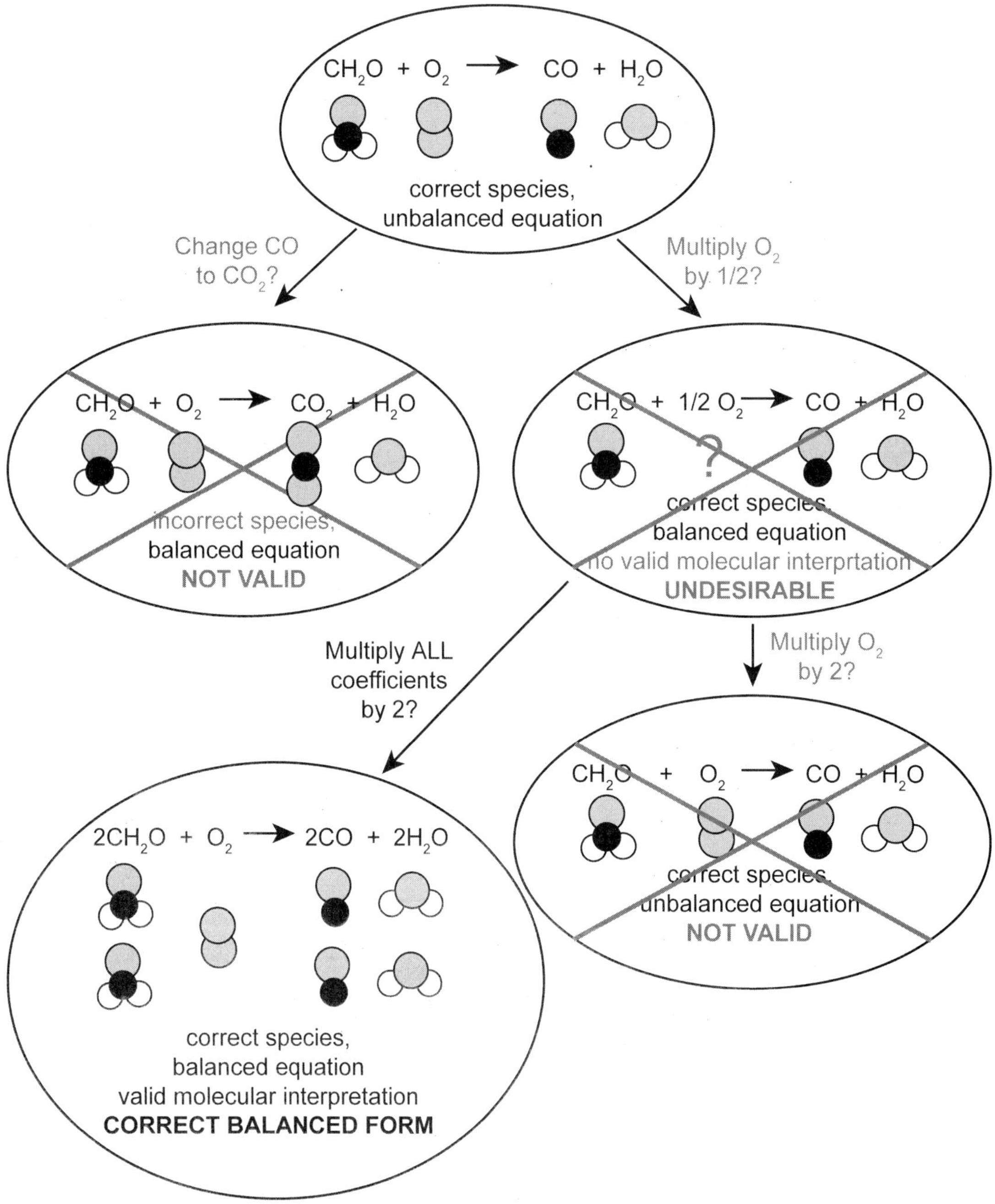

Figure 3.6 The molecular drawings here illustrate some common errors in the balancing of chemical equations.

3.3 Aqueous Solutions and Net Ionic Equations

In order for a chemical reaction to occur, the reactants involved must be able to come into contact with one another. In the gas or liquid phase, the molecules that make up the reactants move readily, allowing this contact to occur. In the solid phase, though, such motion is uncommon and reactions occur only very slowly if at all. One way to enable the needed contact between reactants that are solids under normal conditions is to dissolve them into solution. And the most common medium for producing such solutions is water. Reactions that occur in water are said to take place in **aqueous solution**. To describe this important group of chemical reactions, we will first need to define some key terms.

Solutions, Solvents, and Solutes

Although water accounts for the vast majority of all the molecules present in an aqueous solution, it is neither a reactant nor a product in many aqueous reactions. Water simply serves as the medium in which the reaction occurs. The entire liquid is called a **solution**, meaning that it is a homogeneous mixture of two or more substances. In any solution, the component present in greater amounts – usually much greater amounts – is called the **solvent**. In an aqueous solution, water is the solvent. Minor components of the solution are called **solutes**. The key feature of a solution is that the solutes *dissolve* in the solvent. When a substance dissolves, its particles are dispersed in the solvent, and typically many solvent molecules will surround any individual molecule or ion of solute. Although water is the most common solvent, it is by no means the only one. In fact, not all solutions are even liquids. The same ideas can be applied to homogeneous mixtures of gases or solids. Air is a good example of a gaseous solution, and alloys like brass can be described as solid solutions.

Figure 3.7 In this sequence of photos, one of the authors prepares aqueous solutions of $CuSO_4$. In the upper left panel, solid $CuSO_4$ – the solute – is transferred to a flask. In the upper right panel, water – the solvent – is added. The flask is shaken to speed the dissolution process (lower left). The final photo shows two $CuSO_4$ solutions of different concentrations. The solution on the left has the higher concentration, as seen from its darker color.

If we want to describe a pure substance, all we really need to specify is its identity. If you pick up a bottle in the chemistry lab that is labeled as (distilled) water, for example, you would know exactly what was in the container. But if the same container says only that it contains a solution of salt (NaCl) dissolved in water, you can't be as sure what the contents really are. Specifically, you would probably want to know something about the relative amounts of salt and water that were mixed to prepare the solution. So we need to specify at least one other key piece of information: the **concentration**. If there are many solute particles present, the solution is said to be *concentrated*. When there are few solute particles present the solution is called *dilute*. In Section3.5 we will discuss various units that can be used to report solution concentrations.

We will see in Chapter 12 that solubility is not an "either or" process, that there is continuous variability. These rules summarize the main quantitative tendencies we will learn then.

Some solutes can be dissolved in water to produce very concentrated solutions, while other substances do not dissolve to any measurable extent. So we can characterize various compounds in terms of their *solubility* in water. Those compounds that dissolve readily are said to *soluble*, while those that do not dissolve significantly are called *insoluble*. Chemists often use a series of *solubility rules* to predict whether or not a particular compound is likely to dissolve in water, and the most common of those rules are summarized in Table 2.1.

Usually Soluble	**Exceptions**
Group 1 cations (Li^+, Na^+, K^+, Rb^+. Cs^+), ammonium (NH_4^+)	No common exceptions.
Nitrates (NO_3^-), nitrites (NO_2^-)	Moderately soluble: $AgNO_2$
Chlorides, bromides, iodides (Cl^-, Br^-, I^-)	AgCl, Hg_2Cl_2, $PbCl_2$, AgBr, Hg_2Br_2, $PbBr_2$, AgI, Hg_2I_2, and PbI_2
Fluorides (F^-)	Insoluble: MgF_2, CaF_2, SrF_2, BaF_2, PbF_2
Sulfates (SO_4^{2-})	Insoluble: $BaSO_4$, $PbSO_4$, $HgSO_4$ Moderately soluble: $CaSO_4$, $SrSO_4$, Ag_2SO_4
Chlorates (ClO_3^-), perchlorates (ClO_4^-)	No common exceptions.
Acetates (CH_3COO^-)	Moderately soluble: $AgCH_3COO$
Usually Insoluble	**Exceptions**
Phosphates (PO_4^{3-})	Soluble: those of NH_4^+, Na^+, K^+
Carbonates (CO_3^{2-})	Soluble: those of NH_4^+, Na^+, K^+
Hydroxides (OH^-)	Soluble: LiOH, NaOH, KOH, $Ba(OH)_2$ Moderately soluble: $Ca(OH)_2$, $Sr(OH)_2$
Sulfides (S^{2-})	Soluble: those of NH_4^+, Na^+, K^+, Mg^{2+}, Ca^{2+}

Table 3.1 Solubility guidelines for ionic compounds in water at room temperature

Example Problem 3.2

Which of the following compounds would you predict to be soluble in water at room temperature? (a). $KClO_3$ (b). $CaCO_3$ (c). $BaSO_4$ (d). $KMnO_4$

Strategy

Solubility guidelines for common ions are given in Table 3.1. So we will identify the ions in each compound and consult the table as needed to determine the solubilities.

Solution

(a). $KClO_3$ is potassium chlorate. From the solubility guidelines in Table 3.1 we see that compounds containing K^+ and ClO_3^- tend to be soluble, and that no common exceptions are mentioned. So we predict that $KClO_3$ should be soluble.

(b). $CaCO_3$ is calcium carbonate. Again consulting the table, we see that carbonates are generally insoluble, and that Ca^{2+} is not listed among the exceptions. Thus $CaCO_3$ should be insoluble.

(c). $BaSO_4$ is barium sulfate. Although most sulfates are soluble, $BaSO_4$ is listed in the table as an exception to that rule. Therefore we expect $BaSO_4$ to be insoluble.

(d). $KMnO_4$ is potassium permanganate. Although the permanganate ion (MnO_4^-) is not listed in Table 3.1, the table does tell us that all compounds of K^+ are soluble. So we would predict that $KMnO_4$ should be soluble.

Discussion

Here we simply consulted the table to check the solubility of each compound. Chemists generally gain familiarity with many of these solubility rules, and learn to recognize soluble and insoluble salts without consulting such a table. You should check with your instructor to see whether you are expected to memorize any of these rules.

Check Your Understanding

Which of the following compounds would be soluble in water at room temperature? (a). NH_4Cl (b). KOH (c). $Ca(CH_3COO)_2$ (d). $Ba_3(PO_4)_2$

Although these rules imply that solubility is a simple yes or no question, the reality is more complicated than that. Saying that a compound is soluble does not mean that we could dissolve limitless amounts of it in a small beaker of water. If we keep adding more of the solute, eventually we will observe that the added material does not dissolve. When this occurs we have established a *saturated* solution. The concentration at which a given solution will become saturated depends on the identity of the solute, so it is useful to have a quantitative measure of solubility. The units used to report this number can vary, but one common choice is mass of solute per 100 g of solvent. For example, the solubility of table salt in water at room temperature is 35.7 g NaCl / 100 g H_2O.

One final observation to make about solutions is that many ionic compounds will dissociate into individual ions when they dissolve in water. Thus the salt solution above actually contains Na^+ and Cl^- ions, not NaCl molecules. The availability of freely moving charges causes these solutions to conduct electricity. Any substance that dissolves in water to produce an aqueous solution that conducts electricity is called an **electrolyte**. Substances whose solutions do not conduct electricity are called **non-electrolytes**. We can further divide electrolytes into two groups. **Strong electrolytes** dissociate completely, so that only the individual ions are present in the solution, with virtually no intact molecules. **Weak electrolytes**, in contrast, dissociate only partially; their solutions contain both intact molecules and individual ions in measurable quantities.

91

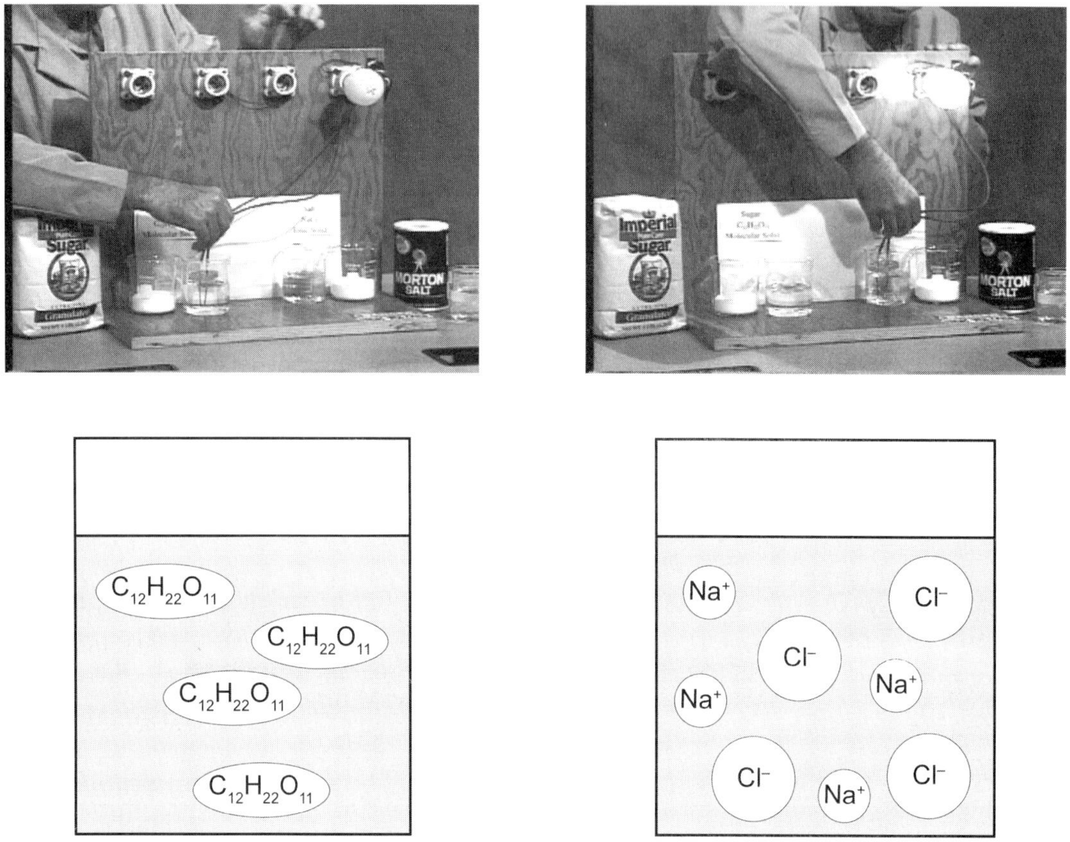

Figure 3.8 The photos show a classroom demonstration in which a pair of wires are dipped into different solutions. If the solution conducts electricity, a circuit is closed and the light bulb lights. In the photo on the left, a solution of ordinary sugar ($C_{12}H_{22}O_{11}$) dissolved in water does not conduct electricity. But in the right hand photo, a solution of table salt (NaCl) dissolved in water does conduct, and the bulb lights. The accompanying illustrations under each photo emphasize the solute species actually present in each solution.

Chemical Equations for Aqueous Reactions

We can describe the process by which a compound dissolves in water using a chemical equation. When a covalently bonded material like sugar dissolves in water, the molecules remain intact.

$$C_6H_{12}O_6(s) \rightarrow C_6H_{12}O_6(aq)$$

By contrast, when an ionic solid dissolves in water, it breaks into its constituent ions. This is called a **dissociation reaction**. The dissolving sodium chloride discussed above provides a common example of this process.

$$NaCl(s) \rightarrow Na^+(aq) + Cl^-(aq)$$

In both of these cases the water molecules are not shown explicitly in the chemical equation, although their presence is indicated by the "(aq)" on the product side. This omission reflects the general tendency of a solvent to disperse a solute but not to react chemically with it. When water appears in the chemical equation, the reaction actually involves water as either a reactant or product.

 92

In addition to these fairly simple reactions by which we describe compounds dissolving in water, many important reactions take place in water. The chemical equations we write to describe these reactions can be written in any of three forms, with the choice of which equation we would choose based mainly on the context in which the equation might be used. We'll illustrate these three types of equations for a reaction that is important in the production of commercial explosives: the synthesis of ammonium nitrate from ammonia and nitric acid. Ammonium nitrate is a precursor to many important explosives, and also has considerable explosive potential in its own right. Because ammonium nitrate is also widely used as a fertilizer, it is readily available for purchase. An unfortunate result of this availability is that it has often been used in bombings, including the 1995 attack on the Murrah Federal Building in Oklahoma City.

In the commercial preparation of ammonium nitrate, pure ammonia (NH_3) in the gas phase is combined with concentrated aqueous nitric acid (HNO_3). This produces a highly concentrated aqueous solution of ammonium nitrate, which is then dried and formed into small pellets (called prills). The **molecular equation** shows the complete formula of each compound involved. Both intact and dissociated compounds are shown, followed by the (aq) to designate an aqueous solution when appropriate.

$$HNO_3(aq) + NH_3(g) \rightarrow NH_4NO_3(aq)$$

Many times the compounds involved in aqueous chemistry are ionic in nature. In our example above, both HNO_3 and NH_4NO_3 are ionic. These ionic compounds dissociate when they dissolve in water, so the actual solution does not contain intact molecules of these compounds. This feature of ionic compounds is emphasized if we write the **total ionic equation** rather than the molecular equation. This form emphasizes what is actually present in the reacting mixture by writing dissociated compounds as separated ions in the solution. The total ionic equation for our example reaction from above is thus

$$H^+(aq) + NO_3^-(aq) + NH_3(g) \rightarrow NH_4^+(aq) + NO_3^-(aq)$$

The H^+ ion is a rather special one, as it is present in all aqueous solutions of acids. Strictly speaking, an H^+ ion would be nothing more than a proton. But as we will see in more detail later, such a bare proton in aqueous solution would actually be tightly surrounded by water molecules. Thus the H^+ ion is frequently written as H_3O^+ to remind us of this fact. If we write the ion in this way, our total ionic equation becomes

$$H_3O^+(aq) + NO_3^-(aq) + NH_3(g) \rightarrow NH_4^+(aq) + NO_3^-(aq) + H_2O(l)$$

Note that now we must include a molecule of water on the right hand side to balance the atoms that we added when we switched our representation of the first ion from H^+ to H_3O^+.

Careful inspection of this equation shows that one ion, $NO_3^-(aq)$, appears on both sides. Thus it is not really an active participant in the reaction. Ions that are uninvolved in the chemistry are referred to as **spectator ions**. In many instances we are not particularly concerned with these spectator ions and need not include them in the chemical equation. When the spectator ions are left out the result is called the **net ionic equation**. Using our same reaction as an example we would have either

The process of removing the spectator ions from the equation is similar to what you would do if you encountered an algebraic equation that had the same term on both sides.

$$H^+(aq) + NH_3(g) \rightarrow NH_4^+(aq)$$

or

$$H_3O^+(aq) + NH_3(g) \rightarrow NH_4^+(aq) + H_2O(l),$$

depending on whether we choose to write H^+ or H_3O^+ to represent the cation from the nitric acid. Because we have removed anions from both sides of the equation, we now have a net charge of +1 on each side. This is one way to recognize that you are looking at a net ionic equation: if there is a non-zero net charge on each side of the equation. If the equation is properly balanced, though, the net charge on each side must be the same. This rule reflects the fact that electrical charge must be conserved.

At this point you are probably wondering something like, "Which of these equations is the correct one?" The answer is that all of these forms provide valid descriptions of the reaction, so none is intrinsically "better" than the others. In this particular case, we would probably not choose the net ionic equation, since the identity of the anion (NO_3^-) is actually important if our aim is to produce solid ammonium nitrate for the explosives industry. But in many other cases, we might have little or no interest in the identity of a spectator ion, so the simplicity of the net ionic equation might make it attractive. The total ionic equation is more cumbersome, and is used less often.

Acid/Base Reactions

Two especially important categories of aqueous solutions are acids and bases, examples of which are quite common in our everyday lives as well as in the chemical industry. For our present purposes, we will define an **acid** as any substance that dissolves in water to produce H^+ (or H_3O^+) ions and a **base** as any substance that dissolves in water to produce OH^- ions. Table 3.2 lists some common acids and bases. Like other solutes, acids and bases can be either strong or weak electrolytes. Strong acids or bases dissociate completely in water, so that the resulting solution contains essentially no intact solute molecules. We can write a chemical equation for the dissociations, as illustrated below for HCl and NaOH.

$$\text{Strong acid: } HCl + H_2O(l) \rightarrow H_3O^+(aq) + Cl^-(aq)$$

$$\text{Strong base: } NaOH + H_2O(l) \rightarrow Na^+(aq) + OH^-(aq)$$

The only common strong acids and bases are those shown in Table 3.2.

Weak acids or bases undergo only partial ionization, so that their solutions contain intact molecules as well as dissociated ions. When we write equations for the dissolution of weak electrolytes like these, we use a "two-way" arrow that is intended to emphasize the fact that the reaction does not proceed completely from left to right. For the case of common weak acids, it is relatively easy to write the needed ionization equation. Many weak acids contaion the –COOH functional group, and it is the H-atom from that group that tends to be lost in solution. Acetic acid provides a good example.

We will revisit the concepts of acids and bases in Chapter 12 and provide additional detail at that time.

$$\text{Weak acid: } CH_3COOH(aq) + H_2O(l) \rightleftarrows H_3O^+(aq) + CH_3COO^-(aq)$$

The situation is slightly less obvious. Unlike the strong bases listed in Table 3.2, most weak bases do not contain –OH groups. So it may not be obvious that dissolving them will produce hydroxide ions in solution (*i.e.*, that they are bases at all). The most common weak base is ammonia (NH_3), which reacts in water according to the following equation.

$$\text{Weak base: } \quad NH_3(aq) + H_2O(l) \rightleftarrows NH_4^+(aq) + OH^-(aq)$$

Again we use the two-way arrow to show that intact NH_3 molecules as well as NH_4^+ and OH^- ions will be present in the solution. Many other weak bases are amines, which can be thought of as derivatives of ammonia in which one or more of the H-atoms have been replaced by methyl groups or longer hydrocarbon chains.

Strong Acids		Strong Bases	
HCl	Hydrochloric acid	$LiOH$	Lithium hydroxide
HNO_3	Nitric acid	$NaOH$	Sodium hydroxide
H_2SO_4	Sulfuric acid	KOH	Potassium hydroxide
$HClO_4$	Perchloric acid	$Ca(OH)_2$	Calcium hydroxide
HBr	Hydrobromic acid	$Ba(OH)_2$	Barium hydroxide
HI	Hydroiodic acid	$Sr(OH)_2$	Strontium hydroxide
Weak Acids		**Weak Bases**	
H_3PO_4	Phosphoric acid	NH_3	Ammonia
HF	Hydrofluoric acid	CH_3NH_3	Methylamine
CH_3COOH	Acetic acid		
HCN	Hydrocyanic acid		

Table 3.2 Strong and weak acids and bases. Note that *all* common strong acids and bases are shown, but only representative examples of weak acids and bases are listed.

Because many acidic and basic solutions occur in nature, observations about acids and bases date back hundreds of years. One of the most important observations is that a solution cannot be both acidic and basic at the same time. Mixing an acid and a base leads to a reaction known as **neutralization**, in which the resulting solution is neither acidic nor basic. To understand the origins of this neutralization, we might start with the definitions of acid and base. Acid solutions contain H_3O^+ ions, while bases contain OH^- ions. So for a solution to be both an acid and a base simultaneously, it would need to contain both of these species. Looking at those ions, it should be easy to see why this is not feasible. Hydronium ions and hydroxide ions combine readily to form water.

$$H_3O^+(aq) + OH^-(aq) \rightarrow 2H_2O(l)$$

This reaction will take place whenever an acid and a base are combined, and will always prevent a solution from being both acidic and basic simultaneously.

Example Problem 3.3

When aqueous solutions of acetic acid and potassium hydroxide are combined, a neutralization reaction will occur. Write molecular, total ionic, and net ionic equations for this process.

Strategy

This is the reaction of a weak acid with a strong base. The reaction between an acid and a base always produces water as a product, along with an ionic compound formed from the remaining ions. (This second product is often called a **salt**.) We can begin by using this idea to write the molecular equation. To generate the ionic equations we then account for the dissociation of strong electrolytes into their constituent ions.

Solution

A hydrogen atom from the acid and the hydroxide ion from the base will produce water. The remaining ions will be an acetate anion (CH_3COO^-, from the acid) and a potassium ion (K^+, from the base). Combining these gives us potassium acetate, KCH_3COO. This lets us write the molecular equation.

$$CH_3COOH + KOH \rightarrow H_2O + KCH_3COO$$

To produce the total ionic equation, we must now determine which of these species to write as dissociated ions. According to Table 3.3, acetic acid is a weak acid. This means that it is only partially dissociated in solution, and so we write it as an intact molecule. KOH, on the other hand, is a strong base, and so it will dissociate completely. So we write this as a pair of ions, K^+ and OH^-. According to Table 3.1, both potassium ion and acetate ion tend to produce soluble compounds. So KCH_3COO will be soluble, and hence will dissociate into its constituent ions. Putting all of this together lets us write the total ionic equation.

$$CH_3COOH(aq) + K^+(aq) + OH^-(aq) \rightarrow H_2O + K^+(aq) + CH_3COO^-(aq)$$

Looking at this equation, we see potassium ions on both sides. Thus potassium is a spectator ion, and can be deleted to give the net ionic equation.

$$CH_3COOH(aq) + OH^-(aq) \rightarrow H_2O + CH_3COO^-(aq)$$

Check Your Understanding

Write molecular, total ionic, and net ionic equations for the reaction between hydrochloric acid and calcium hydroxide.

Neutralization is only one example of aqueous reaction chemistry. Another common type of reaction is the **precipitation reaction**, in which a solid (called the precipitate) is formed and drops out of solution. Example Problem 3.4 provides an example of the three forms of chemical equations for precipitation reactions.

Example Problem 3.4

When aqueous sodium carbonate is added to barium chloride, the solution turns cloudy white with solid barium carbonate. Write the molecular, ionic and net ionic equations for this reaction.

Strategy

First we will use the nomenclature rules we presented in Chapter 2 to write formulas for all of the compounds involved. Then we construct a balanced chemical equation using complete formula units. Next we must identify any ionic compounds that are present, and write them as dissociated ions to give the total ionic equation. Finally, to obtain the net ionic equation we must eliminate any spectator ions from both sides of the equation.

Solution

Formulas for the substances listed are Na_2CO_3, $BaCl_2$, and $BaCO_3$. Although $BaCO_3$ is the only product listed in the problem statement, the sodium and chlorine atoms must also appear somewhere on the right hand side of our equation. So NaCl will also be produced as a product. We write reactants on the left and products on the right to set up the molecular equation:

$$Na_2CO_3(aq) + BaCl_2(aq) \rightarrow BaCO_3(s) + 2NaCl(aq)$$

Next we recognize that the three substances denoted as "(aq)" in that equation will actually exist as dissociated ions rather than intact molecules. This leads us to the total ionic equation:

$$2Na^+(aq) + CO_3^{2-}(aq) + Ba^{2+}(aq) + 2Cl^-(aq) \rightarrow BaCO_3(s) + 2Na^+(aq) + 2Cl^-(aq)$$

Finally, we can cancel the sodium and chloride ions, since they appear in equal numbers on both sides of the equation. This leads us to the net ionic equation:

$$CO_3^{2-}(aq) + Ba^{2+}(aq) \rightarrow BaCO_3(s)$$

Check Your Understanding

When sodium sulfate, Na_2SO_4, reacts with $Pb(NO_3)_2$, the products are solid lead sulfate and aqueous sodium nitrate. Write the molecular, ionic, and net ionic equations for this reaction.

Note that for a precipitation reaction, the net ionic equation does not have a net charge in either the reactants or products.

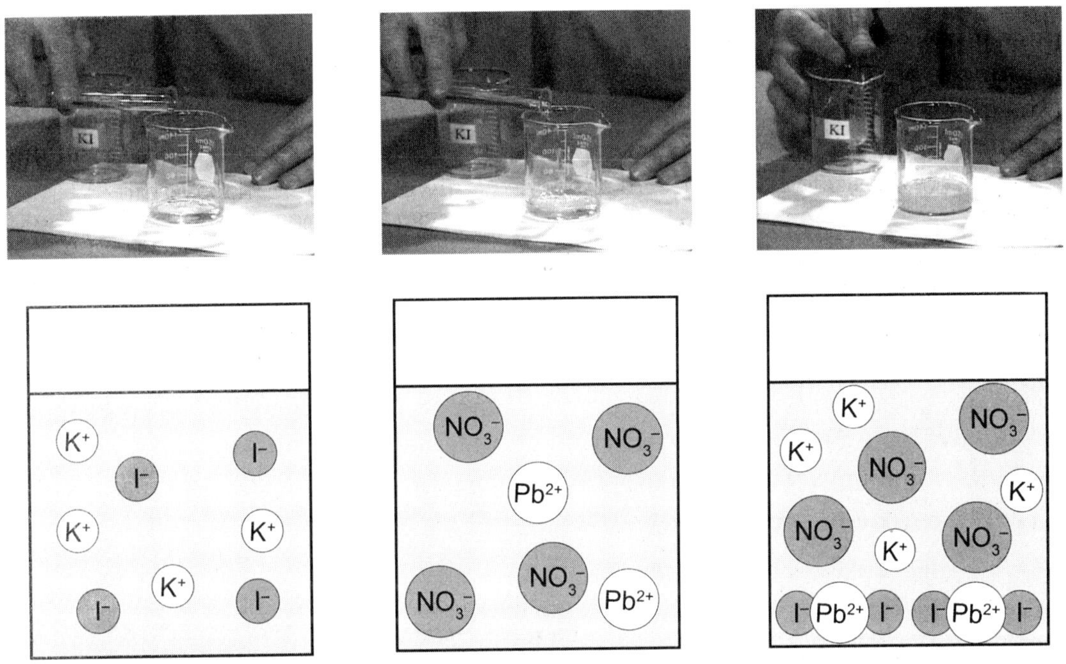

Figure 3.9 The photos show a demonstration in which clear and colorless solutions of KI and $Pb(NO_3)_2$ are mixed and react to form a precipitate of PbI_2. (The precipitate is bright yellow.) The accompanying illustration shows a molecular scale representation of the individual reactant solutions and the final product.

3.4 Interpreting Equations and the Mole

The chemical equations we have been learning to write are symbolic descriptions of chemical reactions. But to interpret these equations we must start to think about them from another point of view, in terms of the actual substances and processes they represent. As often happens in chemistry, we can do this at either the microscopic or the macroscopic level. The microscopic interpretation visualizes reactions between individual molecules, and that interpretation is the one we have used thus far. The macroscopic interpretation pictures reactions between bulk quantities. Neither perspective is inherently better than the other; we simply use the point of view that is best suited to whatever circumstances we are considering. To make the connection between the two worlds we need to look at how chemists count the molecules in a large sample.

Interpreting Chemical Equations

In describing how to write and balance a chemical equation we adopted a way of looking at reactions that seems natural. We spoke of a molecule reacting with another molecule to form some new compound we called the product. Real chemical reactions rarely involve just one or two molecules, though, so we will need to think of chemical equations in terms of larger quantities, too. Let's return to our first example, the explosive reaction between hydrogen and oxygen to produce water.

$$2H_2(g) + O_2(g) \rightarrow 2H_2O(g)$$

We have read this chemical equation as saying that "two H_2 molecules react with one O_2 molecule to form two H_2O molecules." But suppose we carried out the same reaction starting with 20 H_2 molecules rather than two. The balanced equation says that we need one oxygen molecule for every two H_2 molecules. So in this case we could say that 20 H_2 molecules react with 10 O_2 molecules to form 20 H_2O molecules.

Two key features of these statements must be emphasized. First, both statements are correct because they utilize the ratios established by the stoichiometric coefficients. There are twice as many hydrogen molecules as there are oxygen molecules in each case. Second, both statements refer to *numbers of particles*. When we interpret a chemical reaction we must remember that the stoichiometric coefficients *always* refer to numbers of molecules. In the example we just considered, the numbers were one and two or 10 and 20. We could just as easily have said 1,000,000 and 2,000,000 or 45,600,000 and 91,200,000. Provided we use numbers that satisfy the correct stoichiometric ratio, we will always have a reasonable description of an actual chemical reaction.

Avogadro's Number and the Mole

When we begin to consider chemical reactions with macroscopic amounts of materials, the numbers of molecules involved become staggeringly large. But people generally don't like to work with huge numbers, because it is very difficult to develop an intuitive feeling for them. So inevitably, when confronted with the fact that any real sample contains an almost preposterous number of molecules, scientists invented a way to make the numbers manageable. And the simple way to do this is just to count the molecules in terms of some large quantity rather than individually. The particular quantity that chemists have chosen to count by is called the **mole**. One mole is defined to be the number of atoms in exactly 12 grams of ^{12}C. This number is also referred to as Avogadro's number, and its value is 6.022×10^{23}. When chemists want to count molecules, they count by moles so that the resulting numbers will be easier to work with.

Because of its fundamental importance, Avogadro's number has been measured extremely carefully. The internationally accepted value is actually $6.02214199 \times 10^{23}$.

There are several reasons why this number is useful in chemistry. First, it is a large number that gives a useful unit for counting particles as small as atoms. In this sense, the mole is to a chemist what a dozen is to a baker. Second, one mole is the same number of particles no matter what substance we are talking about. So a mole of H_2 contains the same number of particles as a mole of TNT (trinitrotoluene, $C_7H_5N_3O_6$); both have 6.022×10^{23} molecules.

Figure 3.10 One mole samples of various elements are shown. Back row (*left to right*): bromine, aluminum, mercury, and copper. Front row (*left to right*): sulfur, zinc, and iron.

You may be thinking that the value of Avogadro's number seems like a rather peculiar choice for a counting unit. Why not just make it an even 10^{23}, for example, and simplify some calculations? The answer is that the chosen value of Avogadro's number produces a very convenient connection between the mass of an atom or molecule and the mass of a mole of those atoms or molecules. Our definition of the mole said that a mole of ^{12}C has a mass of exactly 12 g. And earlier we said that an atom of ^{12}C has a mass of exactly 12 amu. This is not merely a coincidence, but rather a very deliberate decision. And once we have defined the amu and the gram independently, Avogadro's number is fixed as the number of amu in one gram. It may help to envision a hypothetical experiment: suppose we had a balance that could accurately measure the mass of individual atoms and a way of loading atoms onto that ba*Section 3.4 - Interpreting Equations and the Mole*lance one at a time. We could then determine Avogadro's number by loading carbon-12 atoms onto the balance, one by one, until the mass read exactly 12 grams. Of course no such experiment is actually possible, so Avogadro's number has been measured by more indirect methods.

The mass of 6.022×10^{23} atoms of any element is the **molar mass** of that element, and its value in grams per mole is given in most periodic tables. The molar mass of silicon, for example, is greater than that of carbon because an atom of silicon is heavier than an atom of carbon. And the molar mass of each element takes into account the natural isotopic abundances. So the molar mass of carbon is 12.011 g/mol, reflecting the weighted average between the masses of ^{12}C and ^{13}C in the same way as we discussed in Chapter 2.

The mole is the key to the macroscopic interpretation of a chemical equation. Consider the same equation we have been discussing.

$$2H_2(g) + O_2(g) \rightarrow 2H_2O(g)$$

If we want to read this in terms of moles, we can say "two moles of H_2 and one mole of O_2 react to form two moles of H_2O." Since each mole contains the same number of molecules, the 2:1 **mole ratio** between the reactants is the same as the 2:1 ratio for the numbers of molecules. Chemical equations and their stoichiometric coefficients always provide ratios of numbers of particles, *not masses*.

Determining Molar Mass

Since balanced chemical equations are always expressed in terms of numbers of particles, it will be convenient to have some easy way to determine the number of particles in a given sample of some substance. But there is no simple laboratory instrument that can directly measure the number of moles in a sample. Instead we usually determine the number of moles indirectly from the mass of a sample.

Here we are using whole numbers for the molar masses of hydrogen and oxygen for simplicity. In general, you should use molar masses with at least as many significant figures as the actual data involved in the problem you are working.

The link between the mass of a sample and the number of moles present is the molar mass of the substance in question. To determine the molar mass of a compound we can exploit the idea of conservation of mass. Consider one mole of water as an example. We know that one mole of the compound must contain Avogadro's number of H_2O molecules. Furthermore, we know that each of those molecules must contain one O-atom and two H-atoms. Avogadro's number of oxygen atoms is one mole, and from the periodic table we know that one mole of O-atoms must have a mass of 16.0 g. Since each molecule contains two hydrogen atoms, the whole one mole sample will contain two moles of H-atoms. Again consulting the periodic table, we see that a mole of H-atoms has a mass of 1.0 g, so two moles must have a mass of 2.0 g. The masses of these atoms must be the same as the mass of the mole of H_2O, so we can simply add them to get 18.0 g as the mass of one mole of H_2O. In other words the molar mass of H_2O is 18.0 g/mol. We can do the same sort of thing for any compound; the sum of the molar masses of all of the atoms is the molar mass of the compound. In Example Problem3.5 we determine the molar mass of several explosive compounds.

Example Problem 3.5

Determine the molar mass of the following compounds, all of which have been used as explosives. (a) lead azide, PbN_6 (b) nitroglycerine, $C_3H_5N_3O_9$, (c) mercury fulminate, $Hg(ONC)_2$

Strategy

We must determine the mass contributed by each element and then add those to calculate the molar mass. When parentheses appear in the formula, each atom inside the parentheses must be multiplied by its own subscript *and* by the subscript appearing after the right hand parenthesis.

Solution

(a) PbN_6:	1 mole Pb:	1 x 207.2 g/mol	=	207.2 g/mol
	6 moles N:	6 x 14.007 g/mol	=	84.042 g/mol
		molar mass		291.2 g/mol
(b) $C_3H_5N_3O_9$:	3 moles C:	3 x 12.011 g/mol	=	36.033 g/mol
	5 moles H:	5 x 1.008 g/mol	=	5.040 g/mol
	3 moles N:	3 x 14.007 g/mol	=	42.021 g/mol
	9 moles O:	9 x 15.999 g/mol	=	143.991 g/mol
		molar mass		227.085 g/mol

(c) $Hg(ONC)_2$:	1 mole Hg:	1 x 200.59 g/mol	=	200.59 g/mol
	2 moles O:	2 x 15.999 g/mol	=	31.998 g/mol
	2 moles N:	2 x 14.007 g/mol	=	28.014 g/mol
	2 moles C:	2 x 12.011 g/mol	=	24.022 g/mol
		molar mass		284.62 g/mol

Discussion

The number of significant figures to which the molar masses of the elements are known varies from one element to another, as you can see if you examine the periodic table inside the front cover of this book. So the appropriate number of significant figures in the molar masses here varies according to the elements in each molecule.

Check Your Understanding

What are the molar masses of the following compounds, each of which are used in the preparation of various explosives? (a) H_2SO_4, (b) HNO_3, and (c) $(NH_2)_2HNO_3$

3.5 Using Molar Masses

Molar masses allow us to convert from mass, which can be measured easily, to number of moles, which is more useful in chemical calculations. The mass and the number of moles are really just two different ways of expressing the same information - the amount of a substance present. And the molar mass functions much like a unit conversion between them, allowing us to go from mass to moles or from moles to mass. The following examples illustrate typical uses of molar mass.

Example Problem 3.6

A sample of the explosive TNT ($C_7H_5N_3O_6$) has a mass of 650.5 g. How many moles of TNT are present in this sample? How many molecules is this?

Strategy

We are asked to convert from mass to moles. So we must determine the molar mass of the substance and then use it to carry out the conversion. Once we know the number of moles, we can easily find the number of molecules since we know that one mole contains Avogadro's number of molecules.

Solution

First we will calculate the molar mass of TNT.

$C_7H_5N_3O_6$:	7 moles C:	7 x 12.011 g/mol	=	84.077 g/mol
	5 moles H:	5 x 1.008 g/mol	=	5.040 g/mol
	3 moles N:	3 x 14.007 g/mol	=	42.021 g/mol
	6 moles O:	6 x 15.999 g/mol	=	95.994 g/mol
		molar mass		227.132 g/mol

Now we can use the molar mass to convert from mass to moles.

$$650.5 \text{ g TNT} \times \frac{1 \text{ mol TNT}}{227.132 \text{ g TNT}} = 2.864 \text{ mol TNT}$$

And finally, we can convert from moles to molecules using Avogadro's number.

$$2.864 \text{ mol TNT} \times \frac{6.022 \times 10^{23} \text{ molecules TNT}}{1 \text{ mol TNT}} = 1.725 \times 10^{24} \text{ molecules TNT}$$

Discussion

The mass of this sample corresponds to a little less than 1.5 pounds, so the sample would probably be roughly the size of a brick. Note the enormous number of molecules present in this fairly small amount of material.

Check Your Understanding

Liquid nitromethane was once widely used as an explosive, but is now much more common as a fuel for high performance engines in drag racing. How many moles are present in 1.00 gallon of pure nitromethane (CH_3NO_2), which has a density of 1.137 g/mL?

In some cases we might calculate the number of moles of a substance needed in a particular reaction. If we then want to actually carry out that reaction, we would need to know what mass of sample to prepare. We can easily invert the process above to convert from the number of moles to mass.

Example Problem 3.7

A demolition engineer is planning to use the explosive ethylenedinitramine ($C_2H_6N_4O_4$, also known as halite) to bring down an abandoned building. Calculations show that 315 moles of the compound will provide the necessary explosive force. How many pounds of $C_2H_6N_4O_4$ should be used?

Strategy

We are asked to convert from a number of moles to a mass. We know that the link between these two quantities is the molar mass of the compound, so we can start by calculating that. Once we know the molar mass, we can use it to find the required mass of explosive. Finally, we can easily convert that result from grams to pounds.

Solution

We will need to calculate the molar mass of the compound.

$C_2H_6N_4O_4$:	2 moles C:	2 x 12.011 g/mol	=	24.022 g/mol
	6 moles H:	6 x 1.008 g/mol	=	6.048 g/mol
	4 moles N:	4 x 14.007 g/mol	=	56.028 g/mol
	4 moles O:	4 x 15.999 g/mol	=	63.996 g/mol
		molar mass		150.094 g/mol

Now we can use the molar mass and the required number of moles to find the appropriate mass.

$$315 \text{ moles } C_2H_6N_4O_4 \times \frac{150.094 \text{ g } C_2H_6N_4O_4}{1 \text{ mole } C_2H_6N_4O_4} = 4.73 \times 10^4 \text{ g } C_2H_6N_4O_4$$

Finally, we can convert this from grams to pounds:

$$4.73 \times 10^4 \text{ g } C_2H_6N_4O_4 \times \frac{1 \text{ pound}}{454 \text{ g}} = 104 \text{ pounds } C_2H_6N_4O_4 \text{ needed}$$

Check Your Understanding

Approximately 80 billion moles of ammonium nitrate (NH_4NO_3) were produced in the U.S. in 2001. How many tons of ammonium nitrate is this?

Elemental Analysis: Determining Empirical and Molecular Formulas

When a new molecule is synthesized, an **elemental analysis** is routinely performed to help verify its identity. This test, which measures the mass percentage of each element in the compound, is also frequently done as part of the process of identifying any substance whose composition is unknown. The mass percentages describe the compound's composition, and so they must be related to its chemical formula. But the data obtained from elemental analysis describe the composition in terms of the mass of each element, while the formula describes the composition in terms of the number of atoms of each element. So these are two different representations of very similar information, and the molar masses of the elements provide a connection between them. The process of obtaining the empirical formula of a compound from its percent composition by mass is best illustrated by an example.

Example Problem 3.8

The explosive known as RDX contains 16.22% carbon, 2.72% hydrogen, 37.84% nitrogen and 43.22% oxygen by mass. Determine the empirical formula of the compound.

Strategy

The empirical formula is based on the mole ratios among the elements in the compound and the data given are in terms of mass. Molar mass provides the link between mass and moles, as usual. We can begin by choosing a convenient mass of the compound – usually 100 g - and use the given percentages to find the mass of each element in that sample. Then we will convert those masses into numbers of moles of each element. The ratios between those numbers of moles must be the same for *any* sample of the compound. Finally we will need to convert those ratios into whole numbers to write the empirical formula.

Solution

Consider a 100-g sample of RDX. From the given percentages, that sample will contain 16.22g C, 2.72g H, 37.84g N and 43.22g O. We will convert each of these masses into moles.

$$16.22\text{g C} \times \frac{1 \text{ mol C}}{12.011 \text{ g C}} = 1.350 \text{ mol C in 100 g RDX}$$

$$2.72\text{g H} \times \frac{1 \text{ mol H}}{1.008 \text{ g H}} = 2.70 \text{ mol H in 100 g RDX}$$

$$37.84\text{g N} \times \frac{1 \text{ mol N}}{14.007 \text{ g N}} = 2.702 \text{ mol N in 100 g RDX}$$

$$43.22\text{g O} \times \frac{1 \text{ mol O}}{15.998 \text{ g O}} = 2.701 \text{ mol O in 100 g RDX}$$

Those numbers tell us the ratio of C:H:N:O in the compound. We might be tempted to write a formula of $C_{1.35}H_{2.70}N_{2.70}O_{2.70}$. But we know that the correct empirical formula requires whole numbers of each element. In this case it may be easy to see the correct ratios, but we need to have a systematic way to approach such a problem. The usual strategy is to divide all of the numbers of moles we obtained above by the smallest number of moles. This assures that the smallest resulting number will always be one.

In this case we will divide all four number of moles by 1.350, since that is the smallest of the four.

$$\frac{1.350 \text{ mol C}}{1.350} = 1$$

$$\frac{2.70 \text{ mol H}}{1.350} = 2$$

$$\frac{2.702 \text{ mol N}}{1.350} = 2.001 \approx 2$$

$$\frac{2.701 \text{ mol O}}{1.350} = 2.001 \approx 2$$

In this case the result is a small whole number ratio: 1 mole C:2 moles of H:2 moles N:2 moles O. So the empirical formula is $CH_2N_2O_2$. The mass percentage data allow us to determine only the empirical formula, and not the molecular formula, because we do not have any information about how large or small a molecule of the compound might be.

Discussion

In working with this type of problem, it is a good idea to use accurate molar masses for the elements and to carry as many significant figures as possible throughout the calculation. This assures that no large rounding errors should occur, and helps make it easy to decide whether the final coefficients obtained are really whole numbers.

The steps we used above will produce whole numbers for all of the elements only if at least one of the subscripts in the empirical formula is a one. Otherwise the result for one or more elements will still not be an integer. In nearly all such cases, though, such non-integer results will be readily identified as involving small rational fractions (*i.e., 1.5 or 2.3*). We can obtain integer values by multiplying *all* of the coefficients by an appropriate integer. This is demonstrated in the "Check Your Understanding" exercise that follows.

Check Your Understanding

Nitroglycerine contains 15.87% C, 2.22% H, 18.50% N and 63.41% O by mass. Determine the empirical formula of the compound.

The procedure developed in Example Problem 3.8 allows us to find an empirical formula from mass percentage data. As we saw in Chapter 2, though, the empirical formula does not tell us the exact composition of a molecule of the compound. In the RDX example above, we now know the ratio of C atoms to H, N, and O atoms is 1:2, but we don't know anything about how many of these $CH_2N_2O_2$ units make up an actual molecule. To go from the empirical formula to the molecular formula we will need an additional piece of information - the molar mass of the compound. Fortunately modern instrumentation, like the mass spectrometer we discussed in Chapter 2, makes it fairly simple to measure the molar mass of a substance. Once we are armed with both the elemental analysis and the molar mass, we can find the molecular formula.

Example Problem 3.9

The molar mass of RDX is about 220 g/mol. Use this fact along with the result from Example Problem 3.8 to find the molecular formula of RDX.

Strategy
We already know that the empirical formula is $CH_2N_2O_2$. And now we also know the molar mass. We can find the mass of a single $CH_2N_2O_2$ unit, and then determine how many $CH_2N_2O_2$ units will give us the correct molar mass for the compound.

Solution
Molar masses of the elements involved are all close to whole numbers, and using those whole numbers will provide sufficient accuracy here. The mass of each empirical formula unit will be

$$12 + 2(1) + 2(14) + 2(16) = 74 \text{ g/mol}$$

We are told that the molar mass of the compound is "about 220 g/mol." If we divide that by the mass of the empirical formula, we should get the number of empirical formula units present in each molecule.

$$\frac{220 \text{ g/mol}}{74 \text{ g/mol}} = 2.97 \approx 3$$

So the molecular formula is $(CH_2N_2O_2)_3$, or $C_3H_6N_6O_6$.

Discussion
In this case we were given only an approximate value of the molar mass. But this has little impact on our ability to solve the problem. We require a whole number of empirical formula units in the molecular formula. So the possible molar masses for the compound are all multiples of 74 g/mol. The given molar mass of "about 220 g/mol" easily rules out the possibility that the formula might be either $(CH_2N_2O_2)_2$ (148 g/mol) or $(CH_2N_2O_2)_4$ (296 g/mol).

Note that if we had known an accurate molar mass at the outset of Example Problem 3.8 we could have saved ourselves some work by starting with a 1-mole sample instead of the 100-g sample we used.

Check Your Understanding
The explosive HMX has a molar mass of 296.2 g/mol and contains 16.22% C, 2.72% H, 37.84% N and 43.22% O by mass. Determine the molecular formula of this compound.

Molarity

Many different ways to express the concentration of a solute in a solvent have been developed. To define any unit of concentration we must know both the amount of solute and the quantity of solvent present in the solution of interest. The most commonly used concentration unit is **molarity** or **molar concentration**. **Molarity** is defined as the number of moles of solute per liter of solution, and is represented by the symbol **M**.

Molarity is another ratio between otherwise unrelated variables - similar to mass density. In calculations it can be used in the same ways that density is.

$$\text{Molarity (M)} = \frac{\text{moles of solute}}{\text{liter of solution}} \qquad (3.1)$$

The definition of molarity points to the ways in which it can be used. It provides a relationship among three things: molar concentration, moles of solute, and liters of solution. If we know any two of these quantities, we can determine the remaining unknown quantity. For any reaction stoichiometry calculation, we need to know the amounts of the substances involved in moles. We can readily measure the volume of a solution in the laboratory, and multiplying that volume (expressed in liters) by molarity readily gives moles.

The same relationship also allows us to find the number of moles of solute present if we know both the volume and molarity of any solution. If we use n to represent the number of moles, V for volume, and M for molarity, we can simply rearrange the definition of molarity.

$$n = \mathrm{M} \times \mathrm{V} \tag{3.2}$$

This relationship is central to most stoichiometric calculations for reactions in solution.

Example Problem 3.10

Aqueous solutions of sodium hypochlorite (NaClO) can be used in the synthesis of hydrazine (N_2H_4). Hydrazine has often been used as a rocket fuel, and a derivative of hydrazine is used as a fuel for engines in the orbital maneuvering system of the space shuttle. A solution is prepared by dissolving 45.0 g of NaClO in enough water to produce 750 mL of solution. What is the molarity of the solution?

Strategy

To obtain molarity we need two quantities—moles of solute and liters of solution. Neither of these quantities is given directly in the question, but both can be readily obtained from information given. We must use the molar mass of NaClO to convert from mass to moles, and convert the volume from milliliters to liters. Then we can use the definition of molarity to obtain the molarity of the solution.

Solution

First we compute moles of solute.

$$45.0 \text{ g NaClO} \times \frac{1.00 \text{ mol NaClO}}{74.442 \text{ g NaClO}} = 0.6045 \text{ mol NaClO}$$

Then convert the solution volume from mL to L.

$$750 \text{ mL} \times \frac{1.00 \text{ L}}{750 \text{ mL}} = 0.750 \text{ L}$$

Finally, calculate the molarity.

$$\text{Molarity} = \frac{\text{moles of solute}}{\text{liters of solution}} = \frac{0.6045 \text{ mol}}{0.750 \text{ L}} = 0.806 \text{ M NaClO}$$

Discussion

As usual, this problem could also be set up as single calculation including the same conversions.

If you choose this approach, you should always check to be sure that your units work out correctly. Here we have units of mol/L for our answer, which is appropriate for a molarity. Confirming this helps avoid careless errors.

Check Your Understanding

To produce hydrazine, the sodium hypochlorite must react with ammonia. How many moles of NH_3 would be present in a 4.0-L bottle of 14.8 M ammonia solution?

Dilution

One of the most common procedures encountered in any science laboratory is **dilution**, the process in which solvent is added to a solution to decrease the concentration of the solute. In dilution the amount of solute does not change. *The number of moles of solute is the same before and after dilution.* Since we already know that the number of moles of solute equals the product of molarity and volume, we can write the following equation, where the subscripts denote the initial and final values of the quantities involved.

This dilution formula can be used when there is only one solute present. If there is a reaction taking place, the problem must be approached as a stoichiometry problem.

$$M_i \times V_i = M_f \times V_f \tag{3.3}$$

The simplicity of equations such as this one is alluring. One of the most common errors made by students in chemistry is to use this relationship in situations where it is not valid. *The only place to use this relationship is in the process of dilution.* Example Problem 3.11 illustrates a common laboratory situation where this equation is valid and useful.

Example Problem 3.11

A chemist requires 0.15 *M* hydrochloric acid, HCl, for a series of reactions. The only solution available is a 6.0 *M* HCl. What volume of 6.0 *M* HCl must be diluted to obtain 5.0 L of 0.15 *M* HCl?

Strategy

We first need to recognize that this is the dilution of the concentrated solution to prepare the desired one. The underlying concept we must use is that the number of moles of HCl will be the same before and after dilution, and this means we can use Equation 3.3. We know the desired final molarity and final volume, as well as the initial molarity. So we can solve for the needed initial volume.

Solution

Initial concentration of HCl: $M_i = 6.0$ M

Final concentration of HCl: $M_f = 0.15$ M

Final volume of solution: $V_f = 5$ L

The unknown is the initial volume, V_i. Rearranging Equation 3.3 we have

$$V_i = \frac{M_f V_f}{M_i}$$

Inserting the known quantities on the right hand side

$$V_i = \frac{1.5 \text{ M} \times 5.0 \text{ L}}{6.0 \text{ M}} = 1.3 \text{ L}$$

To obtain the desired quantity of diluted HCl, the chemist should begin with 1.3 L of the concentrated solution and add enough water to bring the volume up to 5 L.

Check Your Understanding

If 2.70 mL of 12.0 M NaOH is diluted to a volume of 150.0 mL, what is the final concentration?

3.6 Insight into Fuel Cells

As you read through this chapter, you may have noticed some overlap between compounds we used as examples of explosives and compounds you may know are used as fuels. And this is more than coincidental. The first and foremost characteristic of an explosive reaction is that it must release large amounts of energy. But that is also the principle requirement for a reaction to be useful as a fuel. Indeed, the difference between a great explosive and a great fuel is largely one of control and timing. If an explosion is desired, than we want a reaction that accelerates to a runaway speed, delivering all of its energy in a very short time. But for a fuel, those circumstances spell disaster. What is needed in a fuel is a controlled and sustainable release of energy. In many cases the same reaction can play either of these roles, depending on the conditions under which it is carried out. The underlying chemistry is the same, and the difference is all in the engineering of a particular application.

In his 2003 State of the Union Address, President George W. Bush called for expanded research into hydrogen-fueled cars, calling attention to an area of research that has actually been active for many years. A hydrogen-fueled vehicle would almost certainly be powered by a *fuel cell* in which the energy is actually released through the same chemical reaction between hydrogen and oxygen that we used for our first example of an explosion.

$$2H_2(g) + O_2(g) \rightarrow 2H_2O(g)$$

The fact that this reaction is capable of producing an explosion demonstrates very clearly that it can indeed release considerable energy. And the equation illustrates one of the most attractive features of hydrogen as a fuel: the only product of this reaction is water vapor, so there would be no environmental concerns surrounding engine exhaust. The fact that we are not already driving hydrogen-powered cars is an indication that there are significant engineering obstacles to implementing this reaction in a controlled fashion. To operate safely, a fuel cell needs to carefully control the way that the hydrogen and oxygen react. Rather than allowing the reactants to mix in bulk quantities, the fuel cell keeps them physically separate. H_2 is converted into H_3O^+ ions in one part of the cell, while O_2 produces OH^- ions in another part of the cell. The resulting ions are allowed to come into contact, where they combine to form H_2O. The process is carried out in such a way that the energy liberated is in the form of electricity.

Just like for a gasoline engine, the necessary oxygen could easily be obtained from the air. But the vehicle would need to be able to carry enough hydrogen to cover a reasonable distance, and a convenient means of replenishing that hydrogen would need to be available. The storage and transportation of hydrogen is complicated by the fact that it is a gas under virtually any reasonable conditions. Gases have much lower densities than solids or liquids, so under normal ambient conditions an amount of hydrogen that was energetically equivalent to a typical modern gas tank would occupy a prohibitively large volume. Compressing the gas to high pressure could reduce that volume, but pressurized gases carry considerable safety risks. More novel solutions, including ways to store hydrogen by absorbing it into a porous metal "sponge" have been explored, but to date no viable approach has been identified.

Perhaps the most promising technology for hydrogen powered fuel cells actually involves using methanol, CH_3OH, as the initial fuel. The engine would then pre-process the methanol to produce the needed hydrogen (along with carbon dioxide), and the hydrogen would be passed on to the fuel cell. Because methanol is a liquid, it is more convenient for pumping and storing than gaseous hydrogen and this technology is in the advanced prototype stage of development. Figure 3.11 shows the components of such a two-stage fuel cell system along with the chemical reactions that are taking place in each stage.

The attractiveness of fuel cells as an energy source is by no means limited to the automobile industry. Active research programs also aim to develop fuel cells as replacements for batteries in electronic devices like laptop computers. Of course the amount of energy required to power a computer is so much less than that needed to propel a car. So it is very likely that you will see commercial fuel cells in your laptop before you see them in your car.

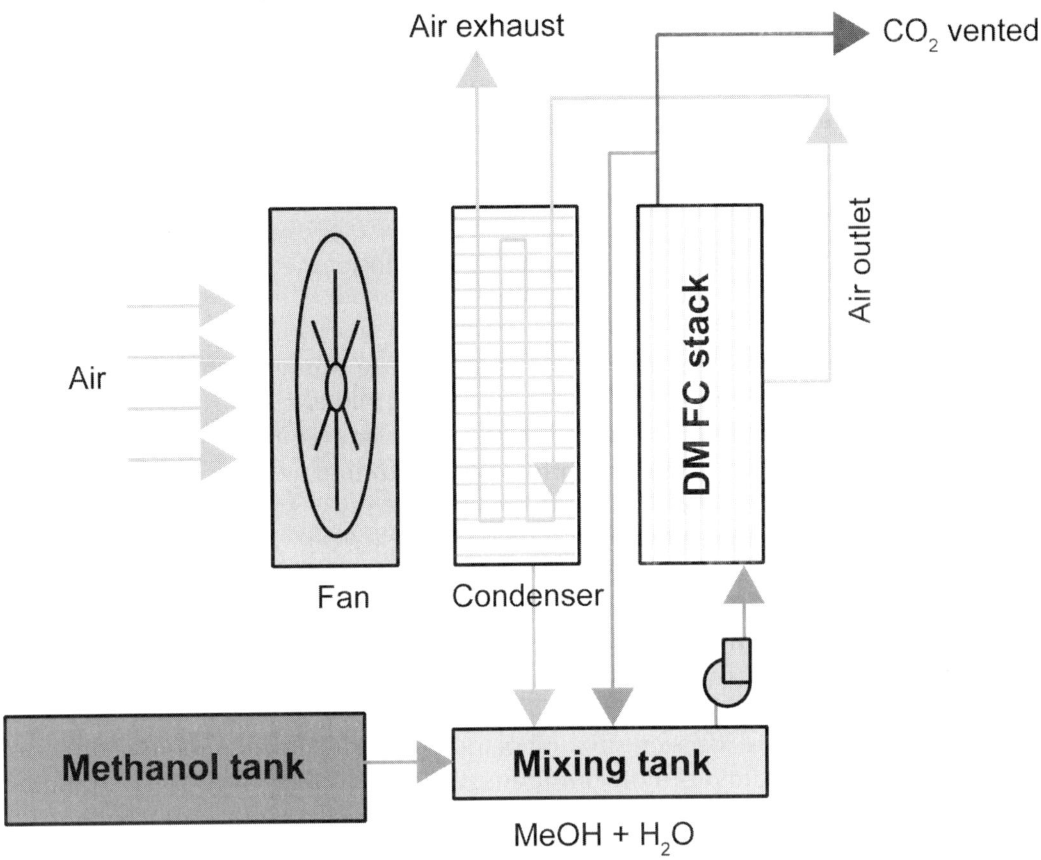

Figure 3.11 Schematic diagram of a direct methanol fuel cell system. (For Scientific and Technical Information Only. Copyright © 1998-2003 The Regents of the University of California. Figure from *Fuel Cells: Green Power,* available at http://education.lanl.gov/resources/fuelcells/)

Focus on Problem Solving

Question

When heated in air some element, X, reacts with the oxygen present to form an oxide with the chemical formula, X2O3. If a 0.5386-g sample of this unknown element yields an oxide with a mass of 0.7111 g, describe how you would determine the atomic mass of the element.

Strategy

Chemical formulas are expressed in terms of numbers of particles. In this problem we have to start with masses and somehow infer information about the numbers of particles. The problem we did that was similar in the chapter was the determination of an empirical formula from percentage by mass data. In this case, with the experiment described, we can determine the mass information for each element in the unknown compound – providing the same type of data we had from percentage mass calculations. Then, because we know the empirical formula, and we know the molar mass of the oxygen we can determine the molar mass of the unknown metal.

Solution

First we would determine the mass composition. The amount of the metal in the oxide sample must be assumed to be the 0.5386 g that was used to generate the oxide. Subtracting this value from the 0.7111 g of the oxide sample then provides the mass of the oxygen. The mass of oxygen can be converted to the moles of oxygen using the molar mass (which could be looked up if we don't know it). We could then use the mole ratio from the empirical formula given in the problem to assert that whatever value we obtain, the number of moles of metal would be 2/3 of this result. (This follows from the fact that there are two metal atoms for every three oxygen atoms.) We would then have a value for the number of moles of the metal and the mass of the metal in the sample – taking the ratio provides the molar mass, from which we could use a periodic table to look up the identity of the metal.

Chapter 3
Problems & Exercises

Insight into Explosives

3.1 List at least three traits shared by most explosives.

3.2 How does nitroglycerin differ from most modern explosives? Use the web to determine if any explosives based usage of nitroglycerin continues today.

3.3 In the chapter it was noted that a blast front forms when an explosive is used. Use the web to find more information about how a blast front propagates and draw a diagram that depicts this information.

3.4 How is an explosive similar to a fuel? How do explosives and fuels differ?

3.5 Many explosives contain nitrogen atoms. Why do you think this is?

Chemical Formulas and Equations

3.6 Define the terms reactants and products as they refer to chemical reactions.

3.7 What symbols are used to indicate solids, liquids, gases, and aqueous solutions in chemical equations?

3.8 How is the addition of heat symbolized in a chemical equation? The addition of light energy?

3.9 What law of nature underpins the concept of a balanced chemical equation?

3.10 Define the term *stoichiometric coefficient*.

3.11 The reactants and products for particular chemical reactions are shown below. Write a balanced chemical equation for each of these reactions.

(a). $CS_2 + Cl_2 \rightarrow CCl_4 + S_2Cl_2$
(b). $H_2S + O_2 \rightarrow H_2O + S_8$
(c). $SiO_2 + HF \rightarrow SiF_4 + H_2O$
(d). $C_2H_4O + O_2 \rightarrow CO_2 + H_2O$

3.12 Write balanced chemical equations for each of the following reactions.

(a). $P_4O_{10} + H_2O \rightarrow H_3PO_4$
(b). $N_2H_4 + H_2O_2 \rightarrow N_2 + H_2O$
(c). $HClO_3 \rightarrow ClO_2 + O_2 + H_2O$
(d). $C_{10}H_{14}N_2 + O_2 \rightarrow CO_2 + H_2O + N_2$

3.13 Ethanol, C_2H_5OH, is now found in gasoline blends in many parts of North America. Write a balanced chemical equation for the combustion of C_2H_5OH to form CO_2 and H_2O.

3.14 The reaction that starts combustion in a match involves the compound, P_4S_3. This chemical is made from phosphorous, P_4, and sulfur, S_8. Write a balance chemical equation for this reaction.

3.15 An explosive whose chemical formula is $C_3H_6N_6O_6$ produces, water, carbon dioxide and nitrogen gas when detonated. Write the chemical equation for the detonation reaction of this explosive.

3.16 Carbon dioxide is removed from the atmosphere by plants. This process combines CO_2 and H_2O to form a sugar, $C_6H_{12}O_6$ and O_2 gas. Write a balanced chemical equation for this reaction.

3.17 Sulfur oxides can form during the combustion of hydrocarbon fuels due to the presence of sulfur in those fuels. One important reaction in such processes is the reaction of SO_2 with O_2 to form SO_3. Write a balanced chemical equation to depict this reaction.

3.18 Diborane and related compounds were proposed as rocket fuels in the 1950's. A representative reaction for this class of molecules is $B_2H_6 + O_2$ to form $B_2O_3 + H_2O$. Write the balanced chemical equation for this process.

3.19 Butane, C_4H_{10}, is a common fuel in disposable lighters. Write a balanced chemical equation for the combustion of butane to form carbon dioxide and water.

3.20 The pictures below show a molecular scale view of a chemical reaction between the compounds AB_2 and B_2. (Shaded circles represent B atoms and open circles are A atoms). The box on the left represents the reactants at the instant of mixing, and the box on the right shows what is left once the reaction has gone to completion. Write a balanced chemical equation for this reaction. As usual, your equation should use the smallest possible whole number coefficients for all substances.

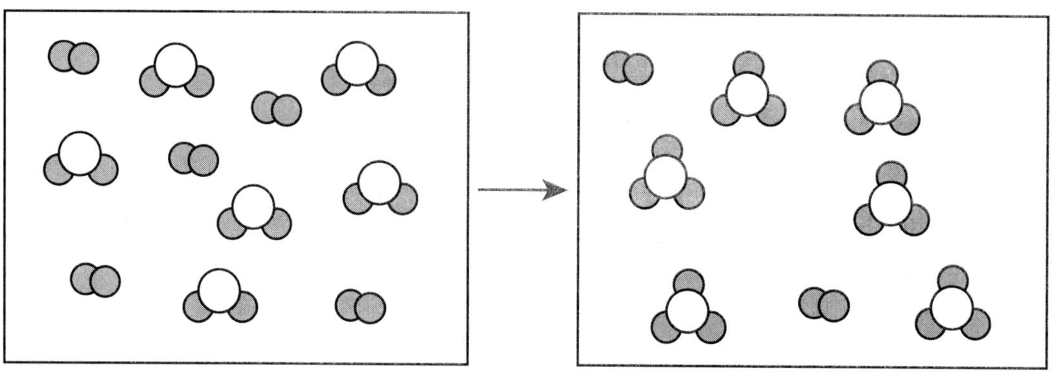

Aqueous Reactions and Net Ionic Equations

3.21 Define the terms solution, solute and solvent.

3.22 When a substance dissolves, does a chemical reaction always occur? What happens if it is not a reaction?

3.23 What is a concentrated solution? A dilute solution?

3.24 Define the term *saturated solution*.

3.25 Define the terms *electrolyte* and *nonelectrolyte*.

3.26 Under what circumstances does water appear explicitly in a chemical equation and when is it indicated only by an "(aq)" label on other species?

3.27 Write the chemical equation for the dissociation of KBr upon being dissolved in water.

3.28 The solubility of NaCl in water is 35.7 g NaCl / 100 g H_2O. If you have 2.0 L of water and 1.4 kg of salt, will all of the salt dissolve (assume the density of water to be 1.0 g / mL)?

3.29 The solubility of NaCl in water is 35.7 g NaCl / 100 g H_2O. Suppose you have 500 g of NaCl. What is the minimum volume of water you would need to dissolve it all (assume the density of water to be 1.0 g / mL)?

3.30 A student conducts an experiment to determine the solubility of K_2SO_4. With 215 mL of water in a beaker the student determines that 25 g of the salt dissolves. What is the solubility of K_2SO_4 according to this experiment (assume the density of water to be 1.0 g / mL)?

3.31 A student conducts a microscale experiment to determine the solubility of ammonium chloride, NH_4Cl. With 0.050 g of the salt, in a small test tube the student has to add 1.50 mL of water to have all of the solid dissolve. Based on this experiment, assuming a density of 1.0 g/mL for the water, calculate the solubility of ammonium chloride.

3.32 Define the term *spectator ion*.

3.33 What is the difference between the total ionic equation and the net ionic equation?

3.34 For each molecular equation given below, write the corresponding total ionic and net ionic equations.

(a). $AgNO_3(aq) + NaCl(aq) \rightarrow AgCl(s) + NaNO_3(aq)$

(b). $Na_2CO_3(aq) + BaCl_2(aq) \rightarrow 2NaCl(aq) + BaCO_3(s)$

(c). $HCl(aq) + NaOH(aq) \rightarrow H_2O(l) + NaCl$

3.35 Write the molecular, total ionic and net ionic equations for the reaction of aqueous solutions of Na_2S and $CuCl_2$ to form solid CuS and aqueous NaCl.

3.36 What is a precipitation reaction?

Interpreting Equations and the Mole

3.37 Explain the concept of the mole in your own words.

3.38 How many entities are present in

(a). 3.21 mol argon atoms

(b). 1.3×10^{-12} mol marbles

(c). 3×10^{6} mol water molecules

(d). 7.63×10^{-18} mol basketballs

3.39 If a typical grain of sand occupies a volume of 1.3×10^{-4} cm^3, what is the volume (in cm^3) of one mole of sand (ignoring space between grains)? What is the volume in L?

3.40 Estimate the size of particle for which one mole would fill your bedroom or dorm room.

3.41 Determine the molar mass of the following compounds. (a). NaCl (b). Al_2O_3 (c). $Mg(NO_3)_2$

3.42 Determine the molar mass of the following compounds. (a). H_2CO (b). $C_6H_{12}O_6$ (c). C_2H_5OH

3.43 Determine the molar mass of the following molecules involved in the formation of urban smog. (a). NO_2 (b). O_3 (c). $C_{16}H_{34}$ (hexadecane) (d). C_7H_5NS (benzothiazole)

3.44 Sodium azide, NaN_3, is used in air bags of automobiles. What is the molar mass of this compound?

3.45 Calculate the molar mass of the following compounds, (a). magnesium phosphate, (b). sodium sulfide, (c). dinitrogen tetroxide.

3.46 Calculate the molar mass of the following compounds, (a). potassium carbonate, (b). aluminum sulfate, (c). magnesium nitride.

Using Molar Masses

3.47 A chemist needs 2 moles of NaOH to make a solution. What mass of NaOH must be used?

3.48 What mass of O_3 gives 4.5 moles of the substance?

3.49 How many moles are present in the amount of material given? (a). 37.2 g SO_3 (b). 103.0 g H_2O (c). 3.1 mg N_2O (d). 8.37 kg Na_2SO_4

3.50 How many moles are present in each of the following samples? (a). 3.728 g H_2 (b). 2.38×10^4 g $C_6H_{12}O_6$ (c). 1.7×10^{-3} g AgCl (d) 93 g CO_2

3.51 How many moles are present in the given quantities of explosives?

(a). 358.1g trinitrotoluene (TNT), $C_7H_5N_2O_6$

(b). 82.6 g nitromethane, CH_3NO_2

(c). 1.68 kg RDX, $C_3H_6N_6O_6$

3.52 A test of an automobile engine's exhaust revealed that 3.7 g of NO_2 was emitted in 10 minutes of operation. How many moles of NO_2 would this engine release if it were used for a 45-minute commute, assuming this measured number is representative of the emission under all circumstances?

3.53 Modern instruments can measure a mass as small as 5 nanograms. If one observed 5.0 ng of CO_2, how many molecules were measured?

3.54 How many H atoms are present in 7.52 g of propane, C_3H_8?

3.55 How many N atoms are present in 214 g of TNT, $C_7H_5N_2O_6$?

3.56 A sample of $H_2C_2O_4{\cdot}2H_2O$ of mass 3.35 g is heated to drive off the waters of hydration (designated separately in the chemical formula). What mass of $H_2C_2O_4$ remains one the water has been removed by heating?

3.57 An average person inhales roughly 2.5 g of O_2 in a minute. How many molecules of oxygen are inhaled in (a) one minute (b) one hour (c) one day by an average person?

3.58 Determine the empirical formula of each of the following compounds, given the percentage composition by mass.

(a). 72.7% Sb, 28.3% S

(b). 23.1% Al, 15.4% C, 61.3% O

(c). 21.8% Mg, 27.9%P, 50.3% O

(d). 28.2% N, 8.0% H, 20.8% P, 43.0% O

3.59 The percent composition of morphine is 71.6% C, 6.7%H, 4.9% N and 16.8% O. Determine the empirical formula of morphine.

3.60 The percent composition of nicotine is 74.1% C, 8.8% H and 17.3% N. Determine the empirical formula of nicotine.

3.61 β-carotene is found in tomatoes and carrots. Its percentage composition is 89.5% C and 10.5% H. The molecular weight of the compound is 536 g/mol. Determine the empirical formula and the molecular formula of β-carotene.

3.62 Calculate the molarity of each of the following solutions.

(a). 1.45 mol HCl in 250. mL of solution

(b). 14.3 mol NaOH in 3.4 L of solution

(c). 0.341 mol KCl in 100 mL of solution

(d). 2.5×10^4 mol $NaNO_3$ in 350 L of solution

3.63 Given the masses of solute and solution volumes below, determine the molarities.

(a). 5.32 g HCOOH in 500 mL

(b). 1.5 kg $CuCl_2$ in 10 L

(c). 0.32 mg HNO_3 in 6.4 mL

(d). 3×10^5 g $NaHCO_3$ in 5000 L

3.64 How many moles of solute are present in each of these solutions?

(a). 48.0 mL of 3.4 M H_2SO_4

(b). 1.43 mL of 5.8 M KNO_3

(c). 3.21 L of 0.034 M NH_3

(d). 1.9×10^3 L of 1.4×10^{-5} M NaF

3.65 How many grams of solute are present in each of these solutions?

(a). 37.2 mL of 0.471 M HBr

(b). 113.0 L of 1.43 M Na_2CO_3

(c). 21.2 ml of 6.8 M CH_3COOH

(d). 1.3×10^{-4} L of 1.03 M H_2SO_3

3.66 Determine the final molarity for the following dilutions.

(a). 24.5 mL of 3.0 M solution is diluted to 100 mL

(b). 15.3 mL of 4.22 M solution is diluted to 1.00 L

(c). 1.45 mL of 0.034 M solution is diluted to 10 mL

(d). 2.35 L of 12.5 M solution is diluted to 100 L

3.67 Determine the initial volume needed to generate the desired solutions by dilution.

(a). 10 L of 0.45 M solution from 3.0 M solution

(b). 1.5 L of 1.00 M solution from 11.7 M solution

(c). 100 mL of 0.025 M solution from 1.15 M solution

(d). 50 mL of 3.3×10^{-4} M solution from 0.25 M solution

3.68 A scientist reads a research paper describing an experiment that takes place in a 0.035 M solution of sodium chloride. If 500 mL of this solution is required, describe how the solution could be prepared from a bottle of solid NaCl and water.

3.69 Magnesium is lighter than other structural metals, so it is increasingly important in the design of more efficient vehicles. Mg^{2+} ions are present in seawater, and the metal is often prepared by "harvesting" these ions and converting them to neutral Mg metal. The average magnesium content of the oceans is about 1270 g Mg^{2+} per ton of seawater, and the density of seawater is about 1.03 g/mL. What is the molarity of Mg^{2+} ions in seawater? The design for a concept car calls for 103 kilograms of magnesium per vehicle. How many gallons of seawater would be required to supply enough magnesium to build one of these cars?

Insight into Fuel Cells

3.70 List some advantages and disadvantages of hydrogen as a fuel.

3.71 Explain why methanol may be more attractive than hydrogen as a fuel.

3.72 Do a web search and see if you can locate any commercially available products using fuel cells as a source of energy.

3.73 Use the web to determine how fuel cells are used in space exploration. What are the advantages of fuel cells that make this application so common?

3.74 Search for information about fuel cells on the web. How many different possible fuels for these fuel cells can you find?

3.75 The Department of Energy estimates that 1 million tons of pollutants like nitrogen oxides, sulfur oxides, carbon monoxides and volatile organic chemicals (See Chapter 4) would be eliminated by the use of fuel cells in 10% of US automobiles. Why would the introduction of fuel cells have such a potentially large impact on pollutant gas release?

3.76 Some claim that converting 20% of US automobiles to fuel cells could reduce the amount of oil the US would need to import by 1.5 million barrels a day. What types of assumptions are likely built into this type of claim? Do these assumptions seem reasonable?

Additional Problems

3.77 Nitric acid (HNO_3) can be produced by the reaction of nitrogen dioxide (NO_2) and water. Nitric oxide (NO) is also formed as a product. Write a balanced chemical equation for this reaction.

3.78 One step in the enrichment of uranium for use in nuclear power plants involves the reaction of UO_2 with hydrofluoric acid (HF) solution. The products are solid UF_4 and water. Write a properly balanced chemical equation for this reaction.

3.79 Pyridine has the molecular formula C_5H_5N. When pyridine reacts with O_2, the products are CO_2, H_2O, and N_2. Write a properly balanced equation for this reaction.

3.80 Pyrrole has the molecular formula C_4H_5N. When pyrrole reacts with O_2, the products are CO_2, H_2O, and N_2. Write a properly balanced equation for this reaction.

3.81 Hydrogen cyanide (HCN) is extremely toxic, but it is used in the production of several important plastics. In the most common method for producing HCN, ammonia (NH_3) and methane (CH_4) react with oxygen (O_2) to give HCN and water. Write a balanced chemical equation for this reaction.

3.82 Many chemical reactions take place in the catalytic converter of a car. In one of these reactions, nitric oxide (NO) reacts with ammonia (NH_3) to give nitrogen (N_2) and water. Write a balanced equation for this reaction.

3.83 Adipic acid is used in the production of nylon, so it is manufactured in large quantities. The most common method for the preparation of adipic acid is the reaction of cyclohexane with oxygen. Balance the skeleton equation shown below.

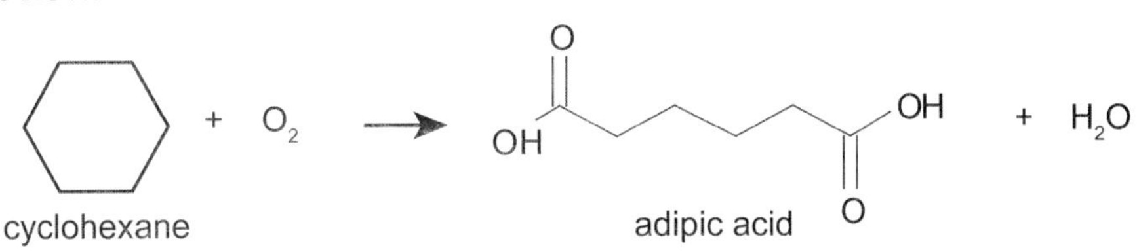

3.84 Calcium carbonate (limestone, $CaCO_3$) dissolves in hydrochloric acid, producing water and carbon dioxide. An *unbalanced* net ionic equation for this reaction is given below. Balance it.

$$CaCO_3\,(s) + H_3O^+\,(aq) \rightarrow H_2O\,(l) + CO_2\,(g) + Ca^{2+}\,(aq)$$

3.85 Answer each of the following questions. Note that none of these should require difficult calculations.

(a) How many molecules are present in 1 mole of octane (C_8H_{18})?

(b). How many moles of fluorine atoms are present in 4 moles of C_2F_6?

(c). What is the approximate mass (to the nearest gram) of 3×10^{23} carbon atoms?

3.86 Consider two samples of liquid: one mole of water (H_2O) and one mole of ethanol (C_2H_5OH). Answer each of the following questions. Note that none of these should require significant calculations.

(a). Which sample contains more atoms?

(b). Which sample contains more molecules?

(c). Which sample has a larger mass?

3.87 Liquid nitromethane was once widely used as an explosive, but is now much more common as a fuel for high performance engines in drag racing. How many moles are present in 1.00 gallon of pure nitromethane (CH_3NO_2), which has a density of 1.137 g/mL?

3.88 Kevlar is a lightweight, high strength polymer used in a variety of applications including bulletproof vests and bicycle tires. The material is produced by the reaction between terephthalic acid and phenylenediamine. Terephthalic acid has the structure shown below. Write the molecular formula for this compound, and calculate its molar mass.

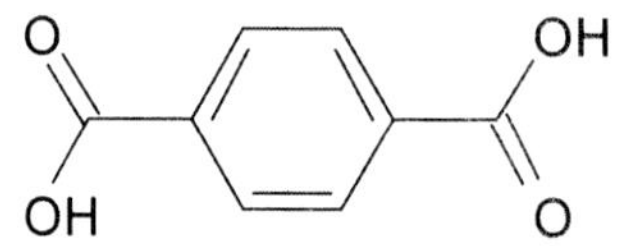

3.89 Cumene is a hydrocarbon, meaning that it contains only carbon and hydrogen. If this compound is 89.94% C by mass, and its molar mass is 120.2 g/mol, what is its molecular formula?

3.90 Methyl cyanoacrylate is the chemical name for the substance sold as Super Glue, and it has the chemical formula $C_5H_5NO_2$. Calculate the number of molecules of this substance in a 1.0-ounce tube of Super Glue, assuming that the glue is 80% methyl cyanoacrylate by mass.

3.91 A low grade form of iron is called taconite and the iron present in the ore is in the form Fe_3O_4. If a two ton sample of taconite pellets yields 1075 pounds of iron when it is refined, what is the mass percentage of Fe_3O_4 in taconite?

3.92 The characteristic odor of decaying flesh is due to the presence of various nitrogen-containing compounds. One such compound, called putrescine, was analyzed and found to contain 54.49% carbon, 13.72% hydrogen, and 31.78% nitrogen by mass. If the molar mass of putrescine is known to be between 85 and 105, what is its molecular formula?

3.93 The characteristic odors of garlic and onions are due to the presence of various sulfur compounds. One such compound was isolated from onions and purified. Analysis of a sample of this compound reveals that it contains 48.60% C, 8.15% H, and 43.25% S by mass. Another test shows that the molar mass of the compound is between 140 and 160 g/mol. Find the molecular formula for this substance.

3.94 Caproic acid is the chemical name for the substance that is responsible for the characteristic odor of gym socks and old running shoes. Analysis of a sample of caproic acid reveals that it contains 62.04% C, 10.41% H, and 27.55% O by mass. Another test shows that the molar mass of caproic acid is between 100 and 125 g/mol. Find the molecular formula for caproic acid.

3.95 Bay leaves are often used to add flavor to stews and other dishes. This flavoring effect is due in part to a substance called eugenol that is found in the bay leaf. Given that eugenol contains 73.15% C, 7.37% H, and 19.48% O by mass, find the empirical formula for this compound.

3.96 The flavor of spearmint comes mainly from a compound called carvone. Given that carvone contains 79.96% C, 9.39% H, and 10.65% O by mass, find the empirical formula for this compound.

3.97 The scent of citronella (used in candles as an insect repellant) comes mainly from a compound called geraniol. Given that geraniol contains 77.87% C, 11.76% H, and 10.37% O by mass, find the empirical formula for this compound.

3.98 A compound frequently used as a high temperature lubricant has the formula MS_2, where 'M' represents a metallic element, and contains 40.06% sulfur by mass. Identify the metal M.

3.99 Nitric acid is often sold and transported as a concentrated 16 M aqueous solution. How many gallons of such a solution would be needed to contain the entire 1.87×10^7 pounds of HNO_3 produced in the US in 1998?

3.100 If 200.0 mL of 0.053 M KBr and 550.0 mL of 0.078 M KBr are mixed, what is the molarity of KBr in the resulting solution? (Assume that the volumes are additive, so that the final volume is 750 mL.)

Focus on Problem Solving Exercises

3.101 The protein that carries oxygen in the blood is called hemoglobin. It is 0.335% Fe by mass. Given that a molecule of hemoglobin contains four iron atoms, describe how you can calculate the molar mass of hemoglobin. Do you need to look up anything to do it?

3.102 The chlorophyll molecule responsible for photosynthesis in plants has 2.72% Mg by mass. There is only one Mg atom per chlorophyll molecule. How can you determine the molar mass of chlorophyll based on this information?

3.103 In one experiment the burning of 0.614 g of sulfur produced 1.246 g of sulfur dioxide as its only product. In a second experiment run to verify the outcome of the first, a sample was burned and 0.813 g of sulfur dioxide was obtained, but the student running the experiment forgot to determine the initial mass of sulfur burned. Describe how the data already obtained could be used to calculate the initial mass of the sulfur burned in the second experiment.

3.104 $MgCl_2$ is often found as an impurity in table salt (NaCl). If a 0.5200 g sample of table salt is found to contain 61.10% Cl by mass, describe how you could determine the percentage of $MgCl_2$ in the sample.

3.105 As part of the clean air act companies are able to purchase the right to emit certain pollutants from other companies that emit less than they are allowed by law. Suppose the cost to purchase a "pollution credit" for 1 ton of sulfur dioxide is $300. Describe how you would determine how many SO_2 molecules may be emitted per dollar at this price.

3.106 The average person exhales 1.0 kg of carbon dioxide in a day. Describe how you would estimate the number of CO_2 molecules exhaled per breath for this "average" person.

3.107 The simplest approximate chemical formula for the human body could be written as $C_{728}H_{4850}O_{1970}N_{104}Ca_{24}P_{16}K_4S_4Na_3Cl_2Mg$. Based on this formula describe how you would rank **by mass** the 10 most abundant elements in the human body.

3.108 For the following oxides of iron, FeO, Fe_2O_3 and Fe_3O_4, describe how you would determine which has the greatest percentage by mass of oxygen. Would you need to look up any information to solve this problem?

Chapter 4: Stoichiometry

A tremendous number of chemical compounds exist in nature, and undergo myriad reactions. By building and exploiting systematic understanding of reactivity, chemists have also been able to produce an impressive array of manmade compounds. Many of the explosives and polymers we've encountered in the last two chapters are handy examples, as are most modern pharmaceuticals. Other practical applications of chemistry involve the development of new routes to synthesize existing natural compounds. As you might imagine, making any of these applications commercially feasible requires a detailed and *quantitative* understanding of the reactions involved. The economics of any chemical process obviously depend on the amounts of each reactant needed to produce a given amount of product. And for processes carried out on an industrial scale, even very small changes in efficiency can have enormous impact on profitability. The quantitative relationships between the amounts of reactants and products in a chemical reaction are referred to as **stoichiometry**. In this chapter we'll use a variety of chemical reactions associated with various fuels to illustrate important stoichiometric concepts.

The word stoichiometry derives from two Greek words: stoicheion (meaning "element") and metron (meaning "measure").

4.1 Insight Into Gasoline and Other Fuels

Gasoline is perhaps the most widely used fuel in our modern society: daily consumption in the U.S. averages more than 350,000 gallons. To explore gasoline's chemistry, we must begin by examining its composition. The fuel we know as gasoline is actually a rather complex mixture, typically containing more than 100 different chemical compounds. The exact composition of gasoline varies somewhat depending on factors including the grade of gas, geographic location, and time of year, but in any case the predominant compounds are hydrocarbons. Most of the hydrocarbon molecules in gasoline are **alkanes**, meaning their carbon atoms are all linked together by single bonds, and most contain between 6 and 11 carbon atoms.

When gasoline is burned in an engine, all of these various compounds undergo combustion simultaneously, reacting with oxygen from the air. As you might imagine, an accurate description of the combustion of such a complex mixture of compounds would be quite a challenge. So for our present purposes it will be helpful to use a simplifying model. The simplest possible model is one in which a single compound is used to represent the gasoline mixture; and the most common choice for that compound is octane, C_8H_{18}. If we start by assuming that gasoline can reasonably be represented by octane, then it is fairly easy to write and balance a chemical equation for its combustion. If we further assume complete combustion, then our equation should show octane and oxygen as reactants, and carbon dioxide and water as products.

Gasoline actually contains several different compounds with the molecular formula C_8H_{18}, and the equations we write would apply to any of those substances.

$$2\ C_8H_{18} + 25\ O_2 \rightarrow 16\ CO_2 + 18\ H_2O$$

Because we made two significant assumptions, this equation is a highly idealized model for the actual combustion of gasoline. The use of octane to represent all of the hydrocarbons in gasoline is mainly for simplicity. If we chose to, it would not be very difficult to write similar combustion equations for each hydrocarbon that is actually present. But the assumption of complete combustion is perhaps more drastic. You probably know that typical automobile exhaust contains a number of compounds besides carbon dioxide and water vapor. Most states require periodic emissions testing to measure the levels of carbon monoxide and hydrocarbons in a car's exhaust, and some states or local areas also require additional tests for other types of compounds.

Since these compounds don't appear among the products in our equation above, their presence implies that our simple model is not showing the full picture. What additional factors could we consider to get a more complete description of engine chemistry?

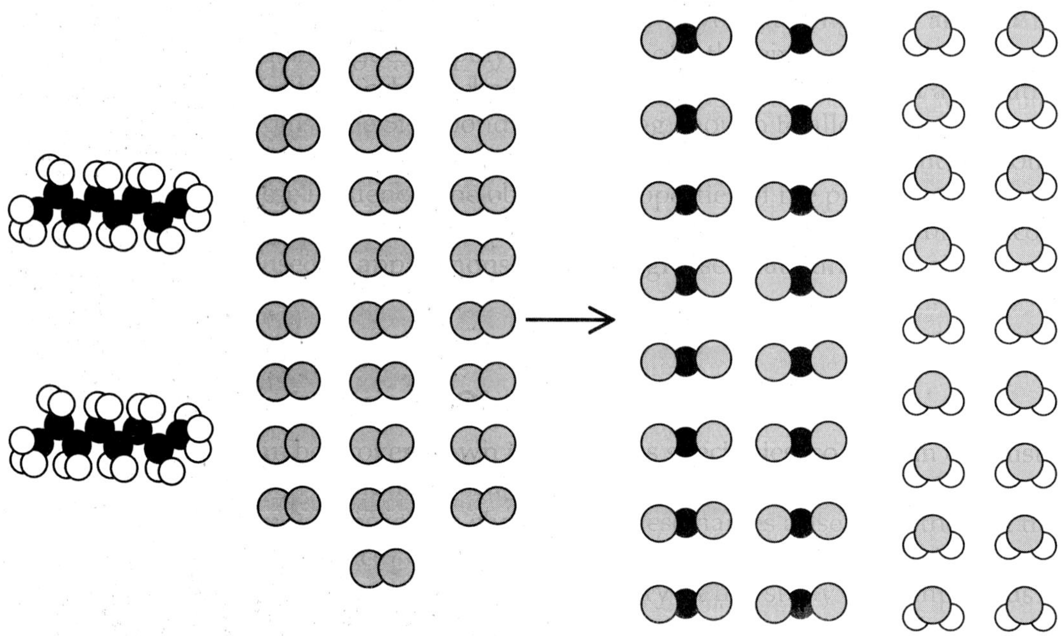

Figure 4.1 This molecular scale picture of the combustion of octane to give carbon dioxide and water shows the relative number of molecules of each compound involved.

Additional products of incomplete combustion are also possible. These include formaldehyde (CH_2O) and soot (particles of pure carbon).

We know that gasoline itself contains a wide array of hydrocarbons. So the presence of hydrocarbons in the exhaust most likely indicates that some of the gas has made it through the engine and into the exhaust stream without burning. This could mean that some hydrocarbon molecules have simply not reacted at all before leaving the engine. Or it could mean that some larger molecules have broken down into smaller fragments and some of those fragments have made it into the exhaust. What about the presence of carbon monoxide? We wrote our original equation based on the assumption of complete combustion, meaning that all of the carbon atoms are converted to CO_2. If insufficient oxygen is present in the engine cylinders, then we will see incomplete combustion, with CO produced instead of CO_2.

$$2\,C_8H_{18} + 17\,O_2 \rightarrow 16\,CO + 18\,H_2O$$

Notice that we have now written two chemical equations with the same reactants but different products. How do the molecules "know" which reaction to undergo? Under most conditions, both reactions will occur to some extent. But the relative importance of each reaction depends on the reaction conditions. The ratio of fuel to oxygen is of particular importance, and is closely monitored by the fuel injection system in modern car engines. Other factors, including engine temperature, are also important. A properly tuned engine assures that the extent of complete combustion is maximized, limiting hazardous CO emissions and also improving gas mileage.

The equations we have written provide us with a decent starting point in our consideration of the chemistry of gasoline or other hydrocarbon fuels. But many of the important questions surrounding the chemistry of fuels are quantitative. How much oxygen is needed to allow complete combustion to occur? How is the amount of CO_2

emitted related to the amount of fuel burned? To begin to answer questions like these, we will need to consider the stoichiometry of the underlying chemical reactions.

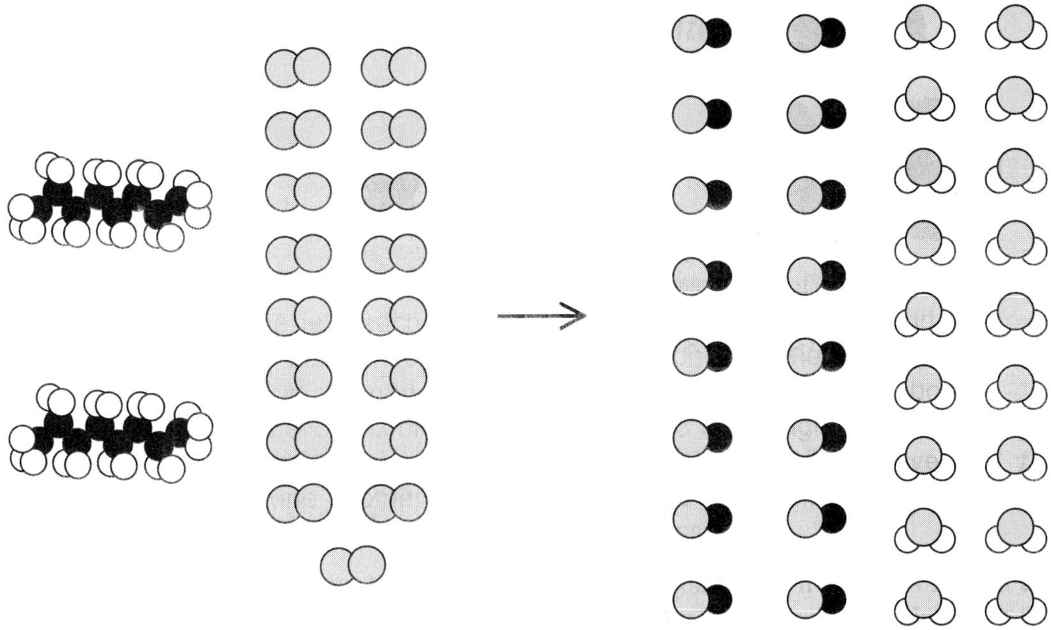

Figure 4.2 Here we show the same sort of molecular picture as in Figure 4.1, but this time for the incomplete combustion of octane to give carbon monoxide and water. Note that both equations balance, but they clearly show different stoichiometry.

4.2 Fundamentals of Stoichiometry

Stoichiometry is a term used to describe quantitative relationships in chemistry. Any chemistry question that asks "How much?" of a particular substance will be consumed or formed in a given chemical reaction is a stoichiometry problem. And at the heart of every such stoichiometry problem you'll always find a balanced chemical equation.

We have already seen that chemical equations are always written in terms of the numbers of particles involved. Whether we interpret them in terms of individual molecules or moles of molecules, the stoichiometric coefficients balancing a chemical equation refer to numbers of particles and not to masses. Unfortunately, we usually can't directly measure the number of particles in the laboratory; masses and volume of liquids are the quantities that are more likely to be measurable. Thus if we want to make quantitative calculations about a chemical reaction, we frequently need to convert between the measured value of a mass or volume and the desired value of a number of moles. Because such calculations are common and important, chemists have developed a standard approach to overcome this "variable mismatch." While you might think of this approach as an algorithm for solving a particular class of chemistry problems, it is instructive to understand its conceptual underpinnings. And the key concept is the use of the balanced chemical equation to establish ratios between moles of the various materials involved in the reaction.

Obtaining Ratios from the Balanced Chemical Equation

The simplest combustible hydrocarbon – and therefore one of the simplest of all fuels - is methane, CH_4. So the combustion of methane will provide a good place for us to begin our exploration of reaction stoichiometry. We start by writing a balanced chemical equation.

The physical state of water will depend on the reaction conditions.

$$CH_4(g) + 2\,O_2(g) \rightarrow CO_2(g) + 2\,H_2O(l)$$

Reading this equation aloud from left to right, we might say: "one molecule of methane gas plus two molecules of oxygen gas yields one molecule of carbon dioxide gas plus two molecules of liquid water." But what if we started with two molecules of methane and four molecules of oxygen? Simple proportional reasoning tells us that the products would be two molecules of carbon dioxide and four molecules of water. In almost any practical reaction, a much larger number of molecules will be used. So if we start with 1×10^{23} methane molecules and 2×10^{23} oxygen molecules, we would expect to obtain 1×10^{23} molecules of carbon dioxide and 2×10^{23} molecules of water. In each case we must have a 1:2 ratio of methane molecules to oxygen molecules, a 1:1 ratio of methane molecules to carbon dioxide molecules, a 1:2 ratio of methane molecules to water molecules, and a 2:2 (or 1:1) ratio of oxygen molecules to water molecules.

As we discussed in Chapter 3, the coefficients in a chemical equation can be interpreted as molar relationships as well as molecular relationships. So we can say that the equation shows that one mole of methane reacts with two moles of oxygen to produce one mole of carbon dioxide and two moles of water. From the chemical equation we can write the following set of *mole ratios*.

$$1 \text{ mol } CH_4 : 2 \text{ mol } O_2$$

$$1 \text{ mol } CH_4 : 1 \text{ mol } CO_2$$

$$1 \text{ mol } CH_4 : 2 \text{ mol } H_2O$$

$$2 \text{ mol } O_2 : 2 \text{ mol } H_2O$$

We could, of course, also write that last one as 1:1.

When doing stoichiometric calculations, mole ratios are very often written to look like fractions. They are used in a way analogous to unit conversion factors, relating the amount of one substance to that of another.

$$\frac{1 \text{ mol } CH_4}{2 \text{ mol } O_2} \qquad \frac{1 \text{ mol } CH_4}{1 \text{ mol } CO_2} \qquad \frac{1 \text{ mol } CH_4}{2 \text{ mol } H_2O} \qquad \frac{2 \text{ mol } O_2}{2 \text{ mol } H_2O} \quad \left(\text{or } \frac{1 \text{ mol } O_2}{1 \text{ mol } H_2O}\right)$$

Just like unit conversion factors, these mole ratios can be inverted as needed to carry out a particular calculation. Example Problem 4.1 shows how to use mole ratios in a stoichiometry problem.

Example Problem 4.1

In the combustion of methane, how many moles of O_2 are required if 6.75 mol of CH_4 is to be completely consumed in the reaction?

Strategy

We start with the balanced chemical equation, and use the stoichiometric coefficients to establish the mole ratio between methane and oxygen. We can then use that ratio to relate the amount of methane involved in the reaction to the amount of oxygen needed.

Solution

The balanced equation was given earlier in the chapter.

$$CH_4(g) + 2\,O_2(g) \rightarrow CO_2(g) + 2\,H_2O(l)$$

The coefficients from this equation give us the ratio between CH_4 and O_2, which can be expressed in either of the following forms.

$$\frac{1 \text{ mol } CH_4}{2 \text{ mol } O_2} \text{ or } \frac{2 \text{ mol } O_2}{1 \text{ mol } CH_4}$$

In order to go from the known amount of CH_4 to the desired amount of O_2, we should use the second form. This gives us the needed result.

$$6.75 \text{ mol } CH_4 \times \frac{2 \text{ mol } O_2}{1 \text{ mol } CH_4} = 13.5 \text{ mol } O_2$$

Discussion

The answer we obtained is a direct result of the 2:1 mole ratio, and so is reasonably intuitive. You might think of the choice of which form of the mole ratio to use as an application of dimensional analysis. Many stoichiometry problems seem more complicated than this, because the known quantity is usually not in moles. But the simple use of the mole ratio illustrated here is always the pivotal step in any reaction stoichiometry calculation.

Check Your Understanding

How many moles of H_2O are formed in the reaction considered above?

The balanced chemical equation provides all the mole ratios needed to relate the amounts of the compounds involved in a reaction. To carry out the reaction in the laboratory, however, we often use a balance to measure out the materials needed, giving us a measurement of mass in grams. The ratios from the chemical equation are in numbers of molecules or moles. So we need to be able to convert between grams and moles, and the molar mass provides the way to do this.

The use of ratios to convert from one unit to another is not any more complicated in stoichiometry than in simple physical measurements. The important difference, however, is that in many stoichiometry problems ratios are used three or more times. Each individual use of the ratio follows principles you already know, and if you keep that in mind you should be able to find your way through even the most complex stoichiometry problems. The main challenge is to determine which ratio to use and when to use it. One way to keep track of the steps needed is to create a flow chart using blocks to indicate when information is derived from a relationship, such as the balanced chemical equation or the molar mass.

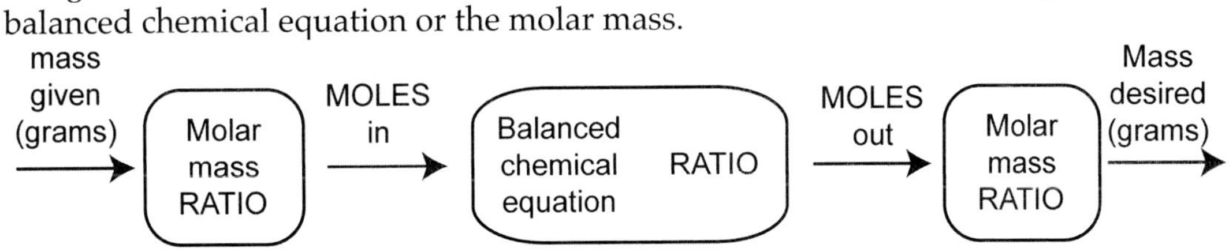

Figure 4.3 This flow diagram illustrates the various steps involved in solving atypical reaction stoichiometry problem.

In Figure 4.3, we can see that if we begin with the mass of one species involved in the reaction and we want to know the mass of some other species consumed or produced, we have to use three ratios: a ratio that uses molar mass of the substance given, a mole ratio from the balanced chemical equation, and a ratio using molar mass of the substance whose mass we wish to find. The simplest and most reliable way to identify which ratio to use for each step is generally to write all quantities with their complete units attached. In setting up the simple calculation in Example Problem 4.1, we wrote the initial quantity with units of "mol CH_4," and this helped us choose the correct form of the mole ratio to use in the calculation. Of course, keeping proper units attached to all the quantities involved in a calculation is *always* a good idea, whether it is a chemical stoichiometry problem or an engineering calculation. The following examples apply this idea to slightly more common stoichiometry calculations, where amounts are measured in mass rather than numbers of moles.

Example Problem 4.2

We saw in Chapter 3 that hydrogen has considerable potential as a fuel. How many grams of water can be produced if sufficient hydrogen reacts with 26.0 g of oxygen?

Strategy

First, it will help if we recognize that this is a reaction stoichiometry problem. Two signs of this are that it asks "How much ..." and that there is obviously a chemical reaction involved. Once we see that, we should immediately realize that we need a balanced chemical equation to give us the mole ratio between water produced and oxygen reacted. We will convert the mass of oxygen given to moles of oxygen, use the mole ratio between oxygen and water to provide moles of water, and then use the molar mass of water to find the mass of water produced.

Solution

Recall that both hydrogen and oxygen are diatomic gases. So the balanced equation is

$$2\,H_2(g) + O_2(g) \rightarrow 2\,H_2O(l)$$

The molar mass of oxygen is 32.0 g/mol, and the molar mass of water is 18.0 g/mol. Using the diagram in Figure 4.3, we write sequential ratios, starting with the mass of oxygen given and keeping in mind that the answer should be in grams of water:

$$26.0\text{ g } O_2 \times \frac{1\text{ mol}}{32.0\text{ g } O_2} \times \frac{2\text{ mol } H_2O}{1\text{ mol } O_2} \times \frac{18.0\text{ g } H_2O}{1\text{ mol } H_2O} = 29.2\text{ g } H_2O$$

Discussion

Note that we start with the substance whose quantity was given (26.0 g O_2). The first and last steps are just conversions between mass and moles, with which you should be comfortable. The use of the mole ratio from the balanced equation is the only new concept here.

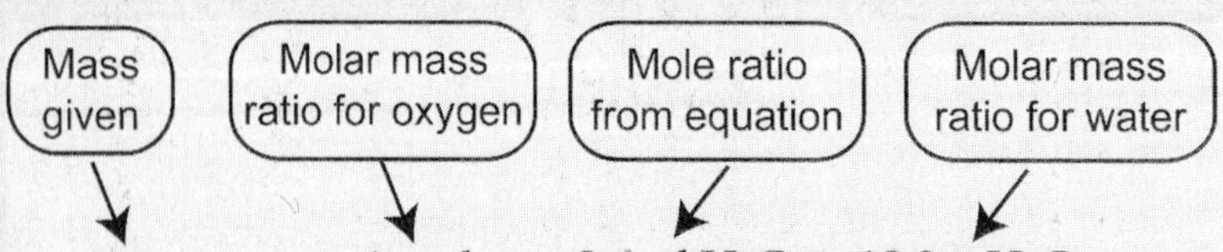

$$26.0\text{ g } O_2 \times \frac{1\text{ mol}}{32.0\text{ g } O_2} \times \frac{2\text{ mol } H_2O}{1\text{ mol } O_2} \times \frac{18.0\text{ g } H_2O}{1\text{ mol } H_2O} = 29.2\text{ g } H_2O$$

This calculation can also be carried out in a more stepwise fashion, in which you might first convert the 26.0 g O_2 to 0.8125 mol O_2, and so forth. The result will be the same so long as all of the calculations are carried out correctly.

Check Your Understanding

Calculate the mass of hydrogen needed to produce 29.2 grams of water.

Example Problem 4.3

The reaction that starts a match burning occurs between potassium chlorate, $KClO_3(s)$, and tetraphosphorus trisulfide, $P_4S_3(s)$. The equation for the formation of P_4S_3 from its constituent elements is

$$8\ P_4(s) + 3\ S_8(s) \rightarrow 8\ P_4S_3(l)$$

If we have 153 g of S_8 and an excess of phosphorus, what mass of P_4S_3 can be produced by this reaction?

Strategy

The heart of any stoichiometry problem is the balanced chemical equation that provides the mole ratio we need. We must convert between masses and the number of moles in order to use this ratio, first for the reactant, S_8, and at the end of the problem for the product, P_4S_3. Molar mass is used to provide the needed conversion factors. The phrase "excess of phosphorus" tells us that we have more than enough P_4 to completely consume 153 g of S_8.

Solution

The molar mass of S_8 is 256.6 g/mol, and that of P_4S_3 is 220.1 g/mol. Along with the balanced equation, those masses allow us to set up the needed relationships.

$$153\ \text{g}\ S_8 \times \frac{1\ \text{mol}}{256.6\ \text{g}\ S_8} \times \frac{8\ \text{mol}\ P_4S_3}{3\ \text{mol}\ S_8} \times \frac{220.1\ \text{g}\ P_4S_3}{1\ \text{mol}\ P_4S_3} = 350\ \text{g}\ P_4S_3$$

Check Your Understanding

The smell of a burning match is due to the formation of sulfur oxides. We can represent the relevant reaction as $S(s) + O_2(g) \rightarrow SO_2(g)$. If 4.8 g of sulfur are burned in this manner, what mass of sulfur dioxide is produced?

We can now relate the amount of carbon dioxide emitted by an automobile engine and the amount of gasoline burned. What mass of CO_2 will be produced for every gallon of gasoline used? While this may sound much different than a typical chemistry book stoichiometry problem, that's really all it is. Start by assuming that our simple model using the complete combustion of octane to represent the burning of gasoline is accurate. The balanced chemical equation is one we saw earlier.

$$2\ C_8H_{18} + 25\ O_2 \rightarrow 16\ CO_2 + 18\ H_2O$$

We'll need to know the mass of a gallon of gasoline. As you might guess, this depends on the exact composition of the gasoline in question. But since we are already basing our calculation on octane, it might be reasonable to use the density of octane, which is 0.7025 g/mL. Use that and some unit conversion factors to find the mass in grams of a gallon of octane.

$$1\ \text{gallon} \times \frac{3.7854\ \text{L}}{1\ \text{gallon}} \times \frac{1{,}000\ \text{mL}}{1\ \text{L}} \times \frac{0.7025\ \text{g}}{\text{mL}} = 2{,}659\ \text{g}$$

Now we've reduced our original question to something that looks a lot more like the previous examples: Want mass of CO_2 is produced by the complete combustion of 2,659 g of C_8H_{18}? Use molar masses and the mole ratio from the balanced equation to set up the calculation.

$$2{,}659\ \text{g}\ C_8H_{18} \times \frac{1\ \text{mol}\ C_8H_{18}}{114.21\ \text{g}\ C_8H_{18}} \times \frac{16\ \text{mol}\ CO_2}{2\ \text{mol}\ C_8H_{18}} \times \frac{44.00\ \text{g}}{1\ \text{mol}\ CO_2} = 8{,}195\ \text{g}\ CO_2$$

To put that in more familiar units, use the fact that a pound equals 454 g. Our calculation then says that burning a gallon of gasoline will produce 18 pounds of carbon dioxide. While we made a couple of simplifying assumptions, our result should certainly be a reasonable estimate.

4.3 Limiting Reactants

When we carry out chemical reactions, it is common for one reactant to be consumed prior to the other reactants. If we run out of one of the reactants the reaction stops. The reactant completely consumed has determined how far the reaction can go, limiting the quantity of product produced. We say that the reactant completely consumed in the reaction is the **limiting reactant.**

To illustrate this important concept, let's imagine that we are somehow able to do an experiment involving just a few molecules. Suppose we start with six molecules each of H_2 and O_2, and allow them to react. How much water can be formed? We know the equation for this reaction from Example Problem4.2.

$$2\ H_2(g) + O_2(g) \rightarrow 2\ H_2O(g)$$

Figure 4.4 shows what such an experiment might look like. Reaction will proceed as long as both reactants are present. But after six molecules of water have been formed, all of the H_2 will have been consumed. Even though three more molecules of oxygen remain, they cannot react. So the amount of product formed is limited by the amount of hydrogen that was available, and we say that hydrogen was the limiting reactant.

In many cases we manipulate the amounts of reactants present to make certain that one compound is the limiting reactant. If the synthesis of a particular compound or material involves one scarce or expensive reactant, then a clever process engineer would most likely make sure to make that the limiting reactant by using large excesses of other reactants. Combustion reactions like the burning of gasoline in a car engine provide a ready example. A large excess of oxygen is always available because it can simply be drawn from the air: your car may run out of gas, but it is not likely to run out of oxygen. So in any calculation we might do related to fuel consumption or exhaust emissions, it would be reasonable to assume that the gasoline would always be the limiting reactant.

In other circumstances, the identity of the limiting reactant may not be so obvious. In a laboratory experiment, we might simply mix conveniently available quantities of substances to begin to study their reaction. Determining the limiting reactant requires the comparison of the amount of each reactant present. As always, we should keep in mind that balanced chemical equations tell us the ratios in which moles of one substance react with moles of another. So we cannot directly compare masses to determine the limiting reactant; we must do the comparison using the relevant mole ratios. Example Problems 4.4 and 4.5 show how we can determine which substance is in excess and which is the limiting reactant.

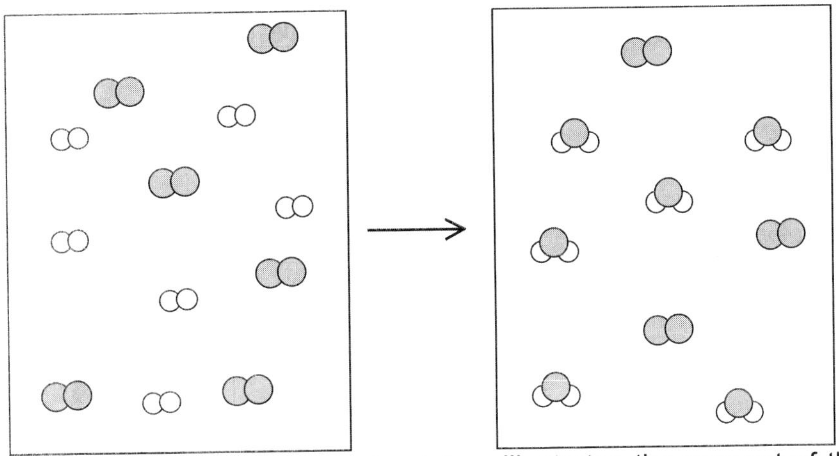

Figure 4.4 This molecular scale picture illustrates the concept of the limiting reactant. The box on the left represents the mixture of reactants in our hypothetical experiment: six H_2 molecules (shown in white) and six O_2 molecules (in gray). The right hand side shows the molecules present once the reaction has gone to completion: six H_2O molecules and three unreacted O_2 molecules. Here H_2 was the limiting reactant: since no more H_2 molecules are available, no further reaction is possible.

Example Problem 4.4

A solution of HCl(aq) contains 5.22 g of HCl. When it is allowed to react with 3.25 g of solid K_2CO_3, the products are KCl, CO_2, and H_2O. Which reactant is in excess?

Strategy

As for any reaction stoichiometry problem, we should start with a balanced equation. One way to proceed from there is to calculate the amount of one reactant that would combine with the given amount of the second reactant. Comparing that with the amount actually available will reveal the limiting reactant.

Solution

We know the reactants and products, so we can write a skeleton equation easily.

$$HCl + K_2CO_3 \rightarrow KCl + CO_2 + H_2O$$

Balancing this gives us the chemical equation we need for our stoichiometry calculation.

$$2\ HCl + K_2CO_3 \rightarrow 2\ KCl + CO_2 + H_2O$$

Let's use the quantity of HCl given, and calculate the amount of K_2CO_3 that would react completely with it.

$$5.22\text{ g HCl} \times \frac{1\text{ mol HCl}}{36.5\text{ g}} \times \frac{1\text{ mol }K_2CO_3}{2\text{ mol HCl}} \times \frac{138.1\text{ g }K_2CO_3}{1\text{ mol }K_2CO_3} = 9.87\text{ g }K_2CO_3$$

So the given quantity of HCl (5.22 g) requires 9.87 g K_2CO_3, but we have only 3.25 g of K_2CO_3 available. The reaction will stop once all of the K_2CO_3 is consumed. K_2CO_3 is the limiting reactant, and HCl is in excess.

Check Your Understanding

The important industrial reaction $N_2 + 3\ H_2 \rightarrow 2\ NH_3$ is usually carried out using an excess of hydrogen. If a reaction vessel contains 1.5×10^3 moles of N_2, how much hydrogen would need to be used to ensure that it will be present in excess?

Example Problem 4.5

In Example Problem 4.3, we used the reaction that produces P_4S_3, one of the reactants in the combustion of a match.

$$8\ P_4(s) + 3\ S_8(s) \rightarrow 8\ P_4S_3(l)$$

If 28.2 g of P_4 are allowed to react with 18.3 g of S_8, which reactant is the limiting reactant?

Strategy

We can choose either reactant and determine how much of the other reactant is required to consume it entirely. Comparing the amount calculated with the amount given in the problem will let us determine which reactant is limiting.

Solution

Let's begin with P_4:

$$28.2\ g\ P_4 \times \frac{1\ mol\ P_4}{124\ g} \times \frac{3\ mol\ S_8}{8\ mol\ P_4} \times \frac{256\ g\ S_8}{1\ mol\ S_8} = 21.8\ g\ S_8$$

So, 28.2 g of P_4 requires 21.8 g of S_8 to react completely. We have only 18.3 g of S_8, so there is not enough S_8 to react with all of the P_4. So S_8 is the limiting reactant.

Check Your Understanding

To convince yourself that you can start with either reactant, redo the example starting with the quantity of S_8 given.

The determination of the limiting reactant is typically just a piece of a larger puzzle. In most limiting reactant stoichiometry problems, the real goal is to determine how much product could be formed from a particular reactant mixture. Perhaps the first critical step in such a problem is to correctly recognize the fact that it is a limiting reactant situation. Whenever a particular amount of more than one reactant is specified in a stoichiometry problem, that fact should tell us to begin by finding the limiting reactant. Once we know which substance runs out first, we are left with an ordinary stoichiometry calculation to solve the problem. Example Problem 4.6 shows how this type of problem can be approached.

We will discuss the use of such additives in the closing section of this chapter.

Example Problem4.6

MTBE (methyl tertiary-butyl ether) is used as an additive in the gasoline sold in several urban areas. The compound is produced by the reaction of methanol and isobutene, according to the following equation.

$$\underset{\text{methanol}}{CH_3OH} + \underset{\text{isobutene}}{(CH_3)_2C{=}CH_2} \rightarrow \underset{\text{MTBE}}{(CH_3)_3COCH_3}$$

If 45.0 kg of methanol are allowed to react with 70.0 kg of isobutene, what is the maximum mass of MTBE that could be obtained?

Strategy

We are given amounts of two different reactants here (methanol and isobutene). So we should realize this is a potential limiting reactant situation. We can identify the limiting reactant as in the previous examples: choose one reactant and find the amount of the other that would be needed to react with it. Comparing that result with the given amount, we can determine which of the reactants is limiting. And once we have done that, it will be straightforward to calculate the expected amount of product.

Solution

We'll need the molar masses of all three compounds. Taking care to interpret the parentheses in the molecular formulas correctly, we find the following.

methanol (CH_3OH): 32.042 g/mol

isobutene ($(CH_3)_2C{=}CH_2$): 56.108 g/mol

MTBE ($(CH_3)_3COCH_3$): 88.150 g/mol

To identify the limiting reactant, we can calculate the mass of isobutene that would react with 45.0 kg (or 45,000 g) of methanol.

$$45{,}000 \text{ g methanol} \times \frac{1 \text{ mol methanol}}{32.042 \text{ g methanol}} \times \frac{1 \text{ mol isobutene}}{1 \text{ mol methanol}} \times \frac{56.108 \text{ g isobutene}}{1 \text{ mol isobutene}}$$

$= 7.880 \times 10^4$ g isobutene

$= 78.80$ kg isobutene

This tells us that to use 45.0 kg of methanol we would have to have at least 78.8 kg of isobutene. But we have only 70.0 kg of isobutene. So the amount of isobutene available will determine the amount of MTBE that can be formed. Isobutene is the limiting reactant and methanol is in excess.

The problem is now reduced to a simpler stoichiometry calculation starting from the mass of isobutene available.

$$70{,}000 \text{ g isobutene} \times \frac{1 \text{ mol isobutene}}{56.108 \text{ g isobutene}} \times \frac{1 \text{ mol MTBE}}{1 \text{ mol isobutene}} \times \frac{88.150 \text{ g MTBE}}{1 \text{ mol MTBE}}$$

$= 1.10 \times 10^5$ g MTBE $=$ 110 kg MTBE

Discussion

This problem combines two types of calculations we've seen in earlier examples. The first step is the identification of the limiting reactant, and the second is the actual stoichiometry calculation. Note that here isobutene was the limiting reactant even though the available mass of isobutene was greater than that of methanol. This is a reminder that reaction stoichiometry calculations should always be handled in terms of moles, and not masses.

Check Your Understanding

The compound diborane, B_2H_6, has sometimes been used as a rocket propellant. It can be produced through the following reaction.

$$3\,LiBH_4 + BF_3 \rightarrow 2\,B_2H_6 + 3\,LiF$$

If 24.6 g of $LiBH_4$ is combined with 62.4 g of BF_3, what mass of diborane could be formed?

The concept of limiting reactants has some important implications for engineering applications where chemical reactions are involved. In some situations, one reactant in a particular reaction might be very scarce or expensive compared to the other reactant(s). In this case a good design for the process would probably assure that the scarce or expensive substance would be the limiting reactant, so that none of it would go to waste. In the combustion of gasoline, for example, the gasoline is much more expensive than the oxygen. So the fuel mixture is adjusted so that sufficient oxygen is available to allow complete combustion. A rather different situation exists for the design of rocket engines. Here the total mass of the rocket is a major consideration, so the aim would likely be to use the minimum mass of fuel that could deliver the needed thrust. Many rocket engines rely on so-called two component fuel mixtures, where energy is released as two compounds are allowed to react. The optimal design for such a rocket would generally assure that the two compounds are present in a stoichiometric mixture, so that neither is in excess. The following example illustrates how such a mixture might be chosen.

Example Problem 4.7

The solid fuel booster rockets of the space shuttle are based on the following reaction between ammonium perchlorate and aluminum.

$$3\ NH_4ClO_4(s) + 3\ Al(s) \rightarrow Al_2O_3\ (s) + AlCl_3(s) + 3\ NO\ (g) + 6\ H_2O(g)$$

If either reactant is present in excess, unnecessary mass will be added to the shuttle, so a stoichiometric mixture is desired. What mass of each reactant should be used for every kilogram of the fuel mixture?

Strategy

Here we want to assure that there will be no excess of either reactant. Since we are asked for the composition "for every kilogram" of fuel, we might start by assuming the total mass of fuel will be 1,000 g. We can write that as an equation with two unknowns.

$$m_{NH_4ClO4} + m_{Al} = 1{,}000\ g$$

Since that has two unknowns, we will need a second relationship between the two masses. The balanced chemical equation shows us that the *mole* ratio between the two reactants should be 3:3, or 1:1. We can use molar masses to convert that to a ratio in terms of *masses,* and then use that ratio as a second equation to solve the problem.

Solution

The molar mass of NH_4ClO_4 is 117.49 g/mol, and that of Al is 26.98 g/mol. To assure that we have a stoichiometric mixture, we need equal molar amounts of each reactant. So if we were to use 117.49 g (1 mole) of NH_4ClO_4, we would need to use 26.98 g (also 1 mole) of Al. We can write this as an equation.

$$\frac{m_{NH_4ClO_4}}{m_{Al}} = \frac{117.49\ g}{26.98\ g} = 4.355$$

And although our actual amounts will need to be larger than those, that ratio of masses will still be right. So we now have two equations for the two unknown masses, and we can solve by doing a little bit of algebra. Our two equations are:

$$m_{NH_4ClO4} + m_{Al} = 1{,}000\ g$$

$$m_{NH_4ClO4} = 4.355\ m_{Al}$$

Substituting the right hand side of the second equation into the first equation gives:

$$4.355\ m_{Al} + m_{Al} = 1{,}000\ g$$

Simplifying that and solving we get the mass of aluminum needed for each kilogram of the fuel mixture.

$$5.355\ m_{Al} = 1{,}000\ g$$

$$m_{Al} = \frac{1{,}000 \text{ g}}{5.355} = 186.8 \text{ g Al}$$

The mass of NH_4ClO_4 is now easy to obtain.

$$m_{NH_4ClO4} = 1{,}000 \text{ g} - m_{Al}$$
$$= 1{,}000 \text{ g} - 186.8 \text{ g} = 813.2 \text{ g}$$

So each kilogram of fuel should comprise 186.8 g Al and 813.2 g NH_4ClO_4.

Discussion

The mass of fuel required for a space shuttle engine would clearly be much larger than a kilogram, but this result could easily be scaled up to the needed mass. As for most problems, there are a variety of other approaches that could be used to solve this. The particular strategy we've used is based on the general idea that if we have two unknown quantities (like the two masses here), we need to find two separate relationships between them in order to solve the problem.

Check Your Understanding

A fuel mixture comprising 600.0 g of NH_4ClO_4 and 400.0 g of Al is used in a test of a laboratory scale mock-up of a shuttle engine. When the engine stops burning, some unused fuel remains. What substance is this unburned fuel, and what mass of it should be found?

4.4 Theoretical and Percentage Yields

For chemical reactions and especially for reactions that produce a commercial product, we would generally prefer to have a reaction that is efficient, producing as much of the desired products as possible while minimizing any unwanted byproducts. Why is this important? In many industrial situations, one big reason is economics. If a business can increase the amount of product obtained from a given amount of reactants, it should become more profitable. Many factors such as temperature of the reaction, the possibility of "side" reactions or the further reaction of the product to form something else can decrease the amount of desired product obtained. Minimizing these unwanted reactions could reduce the amount of waste produced, thereby lowering environmental impacts and associated costs.

The idea of designing chemical reactions so that waste is minimized forms the basis of Green Chemistry, an increasingly important idea in the chemical industry.

We can rate the efficiency of a reaction by calculating how much product would form under perfect or ideal conditions and then comparing the actual measured result with this ideal. The ideal amount of product is called the **theoretical yield**, and it is obtained by working a stoichiometry problem. Measuring the amount of product formed gives us the **actual yield**. And from the ratio of the actual yield to the theoretical yield we can calculate the **percentage yield**.

$$\text{Percentage yield} = \left(\frac{\text{actual yield}}{\text{theoretical yield}}\right) \times 100 \tag{4.1}$$

Let's determine percentage yield for an industrial reaction.

Example Problem 4.8

One of the steps in the commercial process that produces both $NaHCO_3$ and Na_2CO_3 is the conversion of $NaHCO_3$ to Na_2CO_3 by heating:

$$2\ NaHCO_3(s) + \text{heat} \rightarrow Na_2CO_3(s) + CO_2(g) + H_2O(g)$$

In a laboratory experiment, a student heats 42.0 g of $NaHCO_3$ and determines that 22.3 g of Na_2CO_3 are formed. What is the percentage yield of this reaction?

Strategy

We know the actual yield from the experiment. In order to calculate the percentage yield we first need to find the theoretical yield. We can do that by calculating the maximum quantity of product that could form, based on the stoichiometry of the reaction. Once we have both the theoretical yield, and the actual yield, finding the percentage yield is straightforward.

Solution

Begin by working the stoichiometry problem.

$$42.0\text{ g} \times \frac{1\text{ mol } NaHCO_3}{84.0\text{ g}} \times \frac{1\text{ mol } Na_2CO_3}{2\text{ mol } NaHCO_3} \times \frac{106.0\text{ g}}{1\text{ mol } Na_2CO_3} = 26.5\text{ g } Na_2CO_3$$

Then compute the percentage yield.

$$\text{Percentage yield} = \frac{\text{actual yield}}{\text{theoretical yield}} \times 100$$

$$= \frac{22.3\text{ g}}{26.5\text{ g}} \times 100$$

$$= 84.2\%$$

Discussion

A quick estimate of the answer can tell us that we have done the stoichiometry correctly. Since 42 is half of 84, we have 0.5 mol of $NaHCO_3$. From the equation, the mole ratio of $NaHCO_3$ to Na_2CO_3 is 2:1, so we would expect to obtain 0.25 mol of Na_2CO_3. One fourth of 106 is about 26, so we know that the stoichiometry part is correct. And since the actual yield is less than the theoretical yield, an answer less than 100% is plausible. (We may sometimes see yields of greater than 100% for reactions like this, but that indicates that the product is not completely dry or contains some sort of impurity.)

Check Your Understanding

In the "Check Your Understanding" in Example Problem 4.6 you considered a reaction to produce diborane. An alternative route to this propellant is shown below.

$$3\ NaBH_4 + 4\ BF_3 \rightarrow 2\ B_2H_6 + 3\ NaBF_4$$

If 173.2 g of BF_3 react with excess $NaBH_4$ to give 28.6 g of B_2H_6, what is the percentage yield?

How does the idea of percentage yield apply to the chemistry of fuels? For a hydrocarbon like gasoline, complete combustion liberates more energy than incomplete combustion or other side reactions. And complete combustion also reduces the levels of several potentially harmful compounds, including CO and soot. So we might define the efficiency in terms of the extent of complete combustion. If we wanted to monitor this experimentally, one option would be to monitor the amount of CO_2 in the exhaust stream. Similarly, the presence of CO, CH_2O, or other carbon-containing compounds in the exhaust would allow us to monitor the extent of various side reactions. Such measurements are routinely carried out on test engines in order to determine the optimum operating conditions.

4.5 Solution Stoichiometry

In the stoichiometry problems we have discussed so far, the numbers of moles involved have been obtained by using the ratio established by the molar mass of the substances. We can now expand the range of problems that we can consider by using Equation 3.2 to determine the number of moles of substances when volumes are measured rather than masses. The heart of the stoichiometry problem, however, remains the same. The balanced chemical equation still provides the critical ratio between numbers of moles of various species involved in the reaction. Again, a flow diagram for solving a stoichiometry problem shows the manipulations we need to consider.

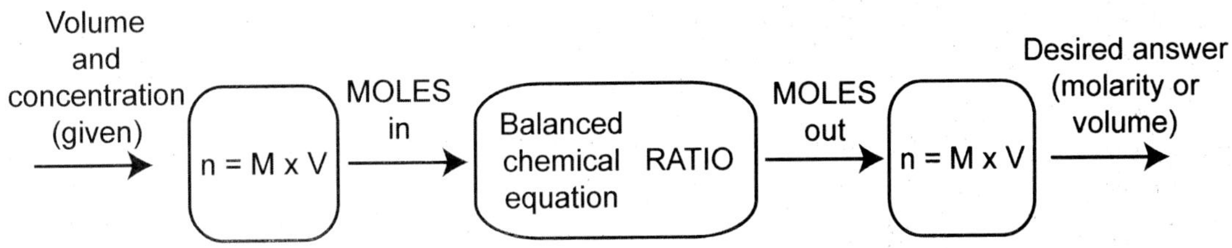

Figure 4.5 Flow diagram showing the important steps in a typical solution stoichiometry calculation.

The only complication added by considering reactions in solutions is the need to relate three variables, using Equation 3.2, rather than two variables using the molar mass ratio. Example Problem 4.9 shows how to approach these solution stoichiometry problems.

Example Problem 4.9

As we mentioned in Example Problem 3.10, the fuel hydrazine can be produced by the reaction of solutions of sodium hypochlorite and ammonia. The relevant chemical equation is

$$NaOCl(aq) + 2\ NH_3(aq) \rightarrow N_2H_4(aq) + NaCl(aq) + H_2O(l)$$

If 750 mL of 0.806 M NaClO is mixed with excess ammonia, how many moles of hydrazine could be formed? If the final volume of the resulting solution is 1.25 L, what will be the molarity of hydrazine?

Strategy

We are asked for the expected amount of a product, so this is a reaction stoichiometry problem. And since NH_3 is said to be in excess, we know that NaClO will be the limiting reactant. So we will use the given volume and concentration to find the number of moles of NaClO reacting. Then use the mole ratio from the balanced equation to find the number of moles of N_2H_4 that can be formed. Finally, we can use that number of moles and the given final volume to obtain the molarity.

Solution

Convert 750 mL to 0.750 L, and determine the number of moles of NaClO reacting.

$$n_{NaClO} = M \times V = 0.806\ mol/L \times 0.750\ L = 0.604\ mol\ NaClO$$

Use that along with the 1:1 mole ratio from the balanced equation to find the corresponding number of moles of hydrazine.

$$0.604 \text{ mol NaClO} \times \frac{1 \text{ mol } N_2H_4}{1 \text{ mol NaClO}} = 0.604 \text{ mol } N_2H_4$$

To answer the second part of the problem, we just need to use this number of moles along with the given final volume to find molarity.

$$M = \frac{n}{V} = \frac{0.604 \text{ mol } N_2H_4}{1.25 \text{ L}} = 0.484 \text{ M } N_2H_4$$

Discussion

Notice that the actual stoichiometry calculation is done using the mole ratio from the balanced equation, just as in all of our earlier examples. Only the conversions to and from numbers of moles have changed because this reaction is in solution.

Check Your Understanding

What volume (in milliliters) of 0.150 M HCl is required to completely react with 0.503 g of dry Na_2CO_3?

Indicators are also examples of weak acids or bases, as defined in Chapter 3.

Another situation that requires an understanding of solution stoichiometry is a common laboratory technique called **titration**. A solution-phase reaction is carried out under controlled conditions so that the amount of one reactant can be determined with high precision. A carefully measured quantity of one reactant is placed in a beaker or flask. A dye called an *indicator* can be added to the solution. The second reactant is added in a controlled fashion, typically by using a burette (Figure 4.6). When the reaction is complete, the indicator changes color. At the point where the indicator first changes color, we have a stoichiometric mixture of reactants. We know the number of moles of the first reactant (or the molarity and volume) and the volume of the second reactant used. So as long as we know the balanced equation for the reaction, we can find the unknown concentration of the second reactant.

Figure 4.6 The photo sequence shows some of the steps in a typical titration. One solution is poured into a burette in the left hand panel. The second solution is placed into an Erlenmeyer flask, and the burette is positioned above the flask in the second panel. The valve on the burette allows the controlled addition of solution, and the shape of the Erlenmeyer flask permits easy mixing by swirling. In the final panel, the solution in the flask changes color when the end point is reached. (Because the clear solution turns pink, the end point is hard to see in this grayscale photo.)

Example Problem 4.10

Many common titrations involve the reaction of an acid with a base. If 24.75 mL of a 0.503 M NaOH solution is used to titrate a 15.00 mL sample of sulfuric acid, H_2SO_4, what is the concentration of the acid?

Strategy

A titration problem is an applied stoichiometry problem. So we will need a balanced chemical equation. We know the molarity and volume for the NaOH solution, so we can find the number of moles reacting. The mole ratio from the balanced equation lets us go from moles of NaOH to moles of H_2SO_4. And since we know the volume of the original H_2SO_4 solution, we can find its molarity.

Solution

The acid-base reaction will produce sodium sulfate and water. The balanced equation is

$$H_2SO_4(aq) + 2\ NaOH(aq) \rightarrow Na_2SO_4(aq) + 2\ H_2O(l)$$

Determine the number of moles of NaOH from the given molarity and volume.

$$\text{mol NaOH} = 0.02475\ \text{L solution} \times \frac{0.503\ \text{mol NaOH}}{1\ \text{L solution}} = 0.0124\ \text{mol NaOH}$$

Use the balanced chemical equation to determine the number of moles of H_2SO_4.

$$0.0124\ \text{mol NaOH} \times \frac{1\ \text{mol}\ H_2SO_4}{2\ \text{mol NaOH}} = 6.22 \times 10^{-3}\ \text{mol}\ H_2SO_4$$

Determine the concentration of HNO_3.

$$M = \frac{6.22 \times 10^{-3}\ \text{mol}\ H_2SO_4}{0.01500\ \text{L solution}} = 0.415\ M\ H_2SO_4$$

Check Your Understanding

Analysis of a solution of NaOH showed that 42.67 mL of 0.485 M HNO3 was needed to titrate a 25.00 mL sample of NaOH. What is the concentration of the NaOH solution?

4.6 Insight into Alternative Fuels and Fuel Additives

In the opening insight we pointed out that gasoline is a rather complicated mixture of chemical compounds. Most of these are hydrocarbons of one sort or another, whose origins can be traced back to the oil from which the gasoline was refined. But the gasoline you buy is also likely to contain a range of **additives**. Specific reasons for using these fuel additives generally fall under three broad categories: improving engine performance, reducing undesirable engine emissions, or reducing dependence on imported petroleum products. Some additives can achieve more than one of these goals.

Let's first examine how additives can improve performance. In the gasoline industry, different grades of fuel are usually described in terms of their octane ratings, with higher octane numbers expected to deliver better engine performance. The octane number is really a measure of how highly the gas can be compressed in the engine cylinder before it will spontaneously ignite. This allows an engine to operate at higher compression without unwanted "knocking," which occurs if the fuel mixture in a cylinder ignites prematurely. Some carmakers recommend that only premium gas be used in their cars because the engines are designed to run at higher than usual compression ratios. The term "octane number" has its origins in the fact that octane itself is highly compressible; one way to achieve a higher octane number is to increase the fraction of octane present in a particular gasoline. But what is important is the compressibility of the gas, not the amount of octane present. And there are a number of ways to improve that compressibility, many of which are easier and cheaper than boosting the actual octane content.

We will discuss catalysis and catalytic converters in Chapter 11.

The first widely used fuel additive was tetraethyl lead, which was found to substantially increase compressibility. From the 1920's into the 1970's, practically all gasoline sold in the U.S. contained lead additives. But the increased performance and reduced fuel costs came at a price. Lead is toxic, and most of the lead introduced into gasoline was ultimately emitted into the atmosphere and settled into soil and water. While these health concerns were known early on, it was not until heightened environmental interests took hold in the 1970's that they received widespread attention. The final blow to leaded gas came with the introduction of the catalytic converter in the 1970's. As you may know, the catalytic converter is designed to reduce the emission of CO and other pollutants. The important chemistry of the converter takes place on a platinum metal surface. But the microscopic lead particles in the exhaust stream "poison" the platinum surface by binding to it so strongly that the catalytic reaction cannot take place. So "unleaded" gas was introduced, and gradually took over the market; since 1996 it has been illegal to sell gasoline containing lead additives for use as a motor fuel in the United States.

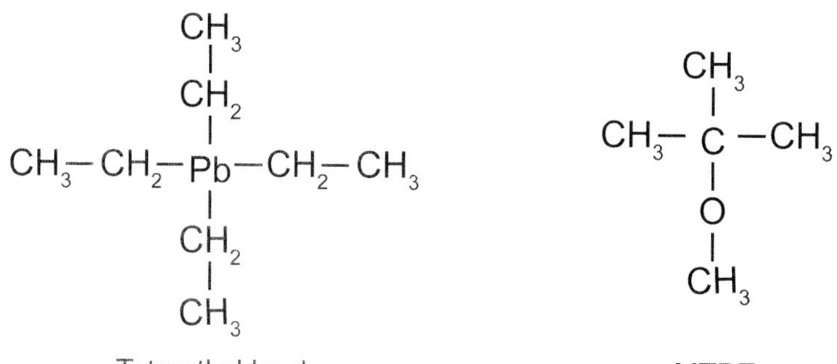

Figure 4.7 Structural formulas of tetraethyl lead and MTBE are shown.

Gasoline formulations are typically seasonal. This fact helps explain the commonly observed increase in gasoline prices in late spring because refineries are changing from winter formulations to summer formulations and demand outpaces supply during the transition.

Other additives have stepped in to take on the octane-boosting role formerly played by tetraethyl lead. The two most widely used today are probably ethanol and MTBE. Because both ethanol and MTBE contain oxygen, they are often referred to as **oxygenates**, and gasoline containing these additives is known as **oxygenated fuel**. The presence of added oxygen in the gasoline helps to assure more complete combustion, and therefore reduces the emission of carbon monoxide, hydrocarbons, and soot. Gasoline containing at least 2% oxygen by weight is known as **reformulated gasoline (RFG)**, and the use of such fuels is mandatory in some areas with severe pollution problems. Ethanol for use as a fuel additive is produced from crops such as corn, barley, and wheat. Starch from these crops can easily be converted to sugar, and the sugar is then fermented to produce ethanol. Because the feedstocks are agricultural, ethanol production does not rely on imported petroleum products. Gasoline containing up to 10% ethanol can be burned in any modern automobile, and mixtures containing up to 85% ethanol can be used in specially designed engines. When used at such high levels, ethanol would more properly be thought of as an alternative fuel rather than a gasoline additive. In the U.S., ethanol use is most common in the midwestern states.

Like ethanol, MTBE (methyl tertiary-butyl ether, $(CH_3)_3COCH_3$) increases both the octane rating and oxygen content of gasoline, and fuel with as much as 15% MTBE can be used in any modern engine. After the passage of the 1990 Clean Air Act, the use of MTBE increased dramatically; the law mandated use of oxygenated fuels, and the fuel industry saw MTBE as offering advantages over ethanol in the cost of production and transportation. More recently, possible health concerns over MTBE have called that choice into doubt. MTBE is a possible carcinogen, meaning it may cause some forms of cancer. As the usage of MTBE in gasoline increased, reports of detectable levels of the compound in groundwater also became more widespread. The source of this MTBE is generally believed to be leaks from pipelines or storage tanks. (Unlike lead, MTBE and ethanol undergo combustion along with the other components of gasoline, so no significant levels of these additives are found in vehicle exhaust streams.) While the extent of possible health impacts is still somewhat uncertain, some areas (including the state of California) have begun to move toward the elimination of MTBE from gasoline.

Focus on Problem Solving

Question

In the 1950's both the United States and the Soviet Union had programs to use decaborane, $B_{10}H_{18}$, as a rocket fuel. This chemical reacts violently with oxygen according the reaction, $B_{10}H_{18} + 12\,O_2 \rightarrow 5\,B_2O_3 + 9\,H_2O$. For maximum efficiency in rocket design an engineer devises a design that allows both fuels (in this case decaborane and liquid oxygen) to "run out" at the roughly same time. Suppose a particular rocket design allows for a total of $1.20\text{x}10^6$ kg of the two fuels. Describe how you would determine how much fuel should be decaborane and how much should be oxygen.

Strategy

This is a type of stoichiometry problem because the two components of the fuel react according to the chemical equation that is provided. There are several ways that this problem could be approached, but all would have to deal with the fact that the chemical equation relates moles of materials, and the problem must be addressed in terms of masses of materials. Essentially, we must devise a problem with two equations in two unknowns to obtain this answer.

Solution

The first equation to devise should take advantage of the fact that there are specifically two reactants involved. This means that the desired total mass must be the sum of the masses of the two components.

$$\text{mass } B_{10}H_{18} + \text{mass } O_2 = 1.20\text{x}10 \text{ kg}$$

The second equation can be determined from the reaction mole ratios and the molar masses of the chemicals involved. (Those could be looked up and calculated as usual.)

$$\text{mass } B_{10}H_{18} = \text{mass } O_2 x \frac{1}{\text{molar mass } O_2} x \frac{12 \text{ mol } O_2}{1 \text{ mol } B_{10}H_{18}} x \frac{\text{molar mass } B_{10}H_{18}}{1}$$

Assuming we have values for the needed molar masses, we can solve these two equations in two unknowns to obtain the answer.

Chapter 4
Problems & Exercises

Insight into Gasoline and Other Fuels

4.1 List at least two factors that make it difficult to accurately describe the combustion of gasoline. What assumptions can be made to address these complications?

4.2 What is an *alkane*?

4.3 Explain the difference between complete and incomplete combustion.

4.4 Automobile exhaust often contains traces of formaldehyde (CH_2O), which is another possible product of incomplete combustion. Write a balanced equation for the formation of formaldehyde during the combustion of octane. (Water will also be formed as a product.)

4.5 Methane. ethane and propane are also hydrocarbons but they are not major components of gasoline. What prevents them from being part of this mixture?

4.6 Use the web to research prices of gasoline for consumers in your area at the pump for the past year. Also find information about the price of crude oil for the same period. Discuss any correlations you observe between the two types of prices.

Fundamentals of Stoichiometry

4.7 For the following reactions, write the ratios that can be defined between molar amounts of the various compounds.

(a). $2\ H_2 + O_2 \rightarrow 2\ H_2O$

(b). $H_2O_2 \rightarrow H_2 + O_2$

(c). $P_4 + 5\ O_2 \rightarrow P_4O_{10}$

(d). $2\ KClO_3 \rightarrow 2\ KCl + 3\ O_2$

4.8 In an experiment carried out at very low pressure, 1.3×10^{15} molecules of H_2 are reacted with acetylene, C_2H_2, to form ethane, C_2H_6, on the surface of a catalyst. Write a balanced chemical equation for this reaction. How many molecules of acetylene are consumed?

4.9 Consider the complete combustion of propane, $C_3H_8 + 5\ O_2 \rightarrow 3\ CO_2 + 4\ H_2O$. If 3.04 moles of propane are burned in air, how many moles of CO_2 are produced? How many moles of O_2 are consumed?

4.10 For the reaction $H_2 + Cl_2 \rightarrow 2\ HCl$, 3.7 moles of H_2 are mixed with excess Cl_2. How many moles of HCl can be formed?

4.11 MTBE, $C_5H_{12}O$, is one of the additives that replaced tetraethyl-lead to increase performance in gasoline. How many moles of O_2 are needed to completely combust 1.5 mol of MTBE?

4.12 In petroleum refining, hydrocarbons are often manipulated by reacting them with H2(g). If hexene, C_6H_{12} is reacted with hydrogen to form hexane, C_6H_{14}, how many moles of hydrogen are needed to react with 453 moles of hexene?

4.13 For the following reactions, determine the value of "*x*."

(a). $4\ C + S_8 \rightarrow 4\ CS_2$	3.2 mol S_8 react with *x* mol CS_2
(b). $CS_2 + 3\ O_2 \rightarrow CO_2 + 2\ SO_2$	1.8 mol CS_2 react with *x* mol SO_2
(c). $N_2H_4 + 3\ O_2 \rightarrow 2\ NO_2 + 2\ H_2O$	7.3 mol O_2 react with *x* mol NO_2
(d). $SiH_4 + 2O_2 \rightarrow SiO_2 + 2\ H_2O$	1.3×10^{-3} mol SiH_4 react with *x* mol H_2O

4.14 For the reaction $S_8 + 8\ O_2 \rightarrow 8\ SO_2$, how many moles of SO_2 can be prepared from 3.23 g of sulfur and excess O_2?

4.15 How many grams of the unknown compound are formed in the following reactions? (Assume reactants for which amounts are not given are in excess.)

(a). $C_2H_4 + H_2 \rightarrow C_2H_6$	3.4 g C_2H_4 reacts to produce *x* g C_2H_6
(b). $CS_2 + 3\ Cl_2 \rightarrow CCl_4 + S_2Cl_2$	5.78 g Cl_2 reacts to produce *x* g S_2Cl_2
(c). $PCl_3 + 3\ H_2O \rightarrow H_3PO_3 + 3\ HCl$	3.1 mg PCl_3 reacts to produce *x* mg HCl
(d). $B_2H_6 + 3\ O_2 \rightarrow B_2O_3 + 3\ H_2O$	4.6 kg B_2H_6 reacts to produce *x* kg B_2O_3

4.16 When arsenic reacts with excess oxygen, O_2, tetraarsenic decaoxide is formed as the only product. Write a balanced chemical equation for this reaction. Determine the mass of product generated through the reaction of 10.4 g of As with excess oxygen.

4.17 Phosgene is a highly toxic gas that has been used as a chemical weapon at times in the past. It is now used in the plastics industry, and is produced by the reaction, $CO + Cl_2 \rightarrow COCl_2$. Given an excess of carbon monoxide, what mass of chlorine gas must be reacted to form 4.5 g of phosgene?

4.18 A gas is known to be either CH_4 or C_2H_6. A 1.00 g sample of this gas produces 1.80 g of water when combusted in excess O_2. Assuming complete combustion to CO_2 and H_2O, which gas is it?

4.19 How many metric tons of carbon are required to reduce 7.83 metric tons of Fe_2O_3 according to the following reaction?

$$2\ Fe_2O_3 + 3\ C \rightarrow 3\ CO_2 + 4\ Fe$$

How many metric tons of iron are produced?

4.20 Assuming a charcoal briquette is composed entirely of carbon, what mass of oxygen is needed to completely burn a 37.3-g briquette?

4.21 Ammonium nitrate, NH_4NO_3, will decompose explosively to form N_2 and H_2O, a fact that has been exploited in terrorist bombings. What mass of nitrogen is formed by the decomposition of 2.6 kg of ammonium nitrate?

Limiting Reactants

4.22 If 3.4 mol Al and 6.2 mol Fe_2O_3 are mixed, what is the limiting reactant?

$$2\ Al + Fe_2O_3 \rightarrow Al_2O_3 + 2\ Fe$$

4.23 If 8.4 moles of disilane, Si_2H_6, are combined with 15.1 moles of O_2 what is the limiting reactant?

$$2\ Si_2H_6 + 7\ O_2 \rightarrow 4\ SiO_2 + 6\ H_2O$$

4.24 For the reaction $Sn + O_2 \rightarrow SnO_2$, an excess of O_2 is generally needed. What is the minimum number of moles of oxygen required to oxidize 7.3 moles of tin?

4.25 In the reaction of arsenic with bromine, $AsBr_5$ will form only when excess bromine is present. Write a balanced chemical equation for this reaction. Determine the minimum number of moles of bromine that are needed if 9.6 moles of arsenic are present.

4.26 For the reaction $4\ C + S_8 \rightarrow 4\ CS_2$, if 10.0 g of carbon are mixed with 20.0 g of S_8, what is the limiting reactant?

4.27 When octane has inadequate oxygen supplies during combustion, carbon monoxide may form. If 100 g of octane is burned in 200 g of O_2, are conditions conducive to the formation of carbon monoxide?

4.28 If 23.6 g of nitrogen are mixed with 17.3 g of hydrogen, how much ammonia may form?

4.29 Copper reacts with sulfuric acid according to the equation,

$$2H_2SO_4 + Cu \rightarrow CuSO_4 + 2H_2O + SO_2.$$

How many grams of sulfur dioxide is created by this reaction is 14.2 g of copper is reacted with 18.0 g of sulfuric acid?

4.30 One of the steps in the manufacture of nitric acid is the oxidation of ammonia shown in this equation.

$$4NH_3(g) + 5O_2(g) \rightarrow 4NO(g) + 6H_2O(g)$$

If 43.0 kg of NH_3 is reacted with 35.4 kg of O_2, what mass of NO forms?

4.31 When $Al(OH)_3$ reacts with sulfuric acid the following reaction occurs.

$$2\ Al(OH)_3 + 3\ H_2SO_4 \rightarrow Al_2(SO_4)_3 + 6\ H_2O$$

If 1.7×10^3 g of $Al(OH)_3$ are combined with 680 g of H_2SO_4, how much aluminum sulfate can form?

4.32 Copper reacts with nitric acid via the following equation.

$$3\ Cu(s) + 8\ HNO_3(aq) \rightarrow 3\ Cu(NO_3)_2 + 2\ NO + 4\ H_2O$$

What mass of NO(g) can be formed when 10.0 g of Cu is reacted with 115 g HNO_3?

4.33 How much HNO_3 can be formed in the following reaction if 3.6 kg of NO_2 gas are bubbled through 2.5 kg of water ?

$$3\ NO_2(g) + H_2O(l) \rightarrow 2\ HNO_3(aq) + NO(g)$$

4.34 Hydrogen and oxygen are reacted and the water formed is collected at 25°C, where it has a density of 0.997 g/mL. If 36.8 g of H_2 and 168 g of O_2 are reacted, how many mL of water are collected?

4.35 If 43.3 mg of Al(OH)3 are reacted with 14.6 mg of HCl, how much AlCl3 is formed?

$$Al(OH)_3 + 3\ HCl \rightarrow AlCl_3 + 3\ H_2O$$

Theoretical and Percentage Yield

4.36 Define the term *theoretical yield.*

4.37 What types of events happen in chemical reactions that lead to percentage yields that are less than 100%?

4.38 Sometimes students in chemistry labs determine percentage yields greater than 100%. Assuming there is no calculation error, how could this happen?

4.39 The theoretical yield and the actual yield for various reactions are given below. Determine the corresponding percentage yields.

	Theoretical Yield	Actual Yield
Reaction 1	35.0 g	12. 8 g
Reaction 2	9.3 mg	120 μg
Reaction 3	3.7 metric tons	1250 kg
Reaction 4	40 g	41 g

4.40 A reaction that produced 4.8 mg of taxol, an anticancer drug, is reported to have a yield of 38%. What was the theoretical yield?

4.41 If a synthesis of aspartame, an artificial sweetener, occurs with 76 % yield and a reaction mixture has a theoretical yield of 530 kg, how much is the actual yield?

4.42 Copper (II) oxide reacts with carbon monoxide to form copper metal.

$$CuO + CO \rightarrow Cu + CO_2$$

If 23.5 g of CuO is reacted with excess CO, and 14.1 g of Cu is recovered, what is the percentage yield?

4.43 Urea is used in the synthesis of fertilizers and is formed by the following reaction.

$$CO_2 + 2\ NH_3 \rightarrow (NH_2)_2CO + H_2O$$

If 3.5 kg of ammonia reacts with excess CO_2 and 1.3 kg of urea forms, what is the percentage yield?

4.44 Sulfur hexafluoride is a very stable gas useful in electrical generators and switches. It is formed by direct synthesis from the elements.

$$S(s) + 3\ F_2(g) \rightarrow SF_6(g)$$

If 92 g of SF_6 is produced from the reaction of 115 g of sulfur in excess fluorine, what is the percentage yield?

4.45 Magnesium nitride forms in a side reaction when magnesium metal burns in air. This reaction may also be carried out in pure nitrogen.

$$3\ Mg(s) + N_2(g) \rightarrow Mg_3N_2(s)$$

If 18.4 g of Mg_3N_2 forms from the reaction of 20.0g of magnesium with excess nitrogen, what is the percentage yield?

4.46 Industrial production of hydrogen gas uses the reaction shown below. If 1.00 metric ton of propane reacting with excess water yields 270 kg of H_2, what is the percentage yield?

$$C_3H_8(g) + 3\ H_2O(l) \rightarrow 3\ CO(g) + 7\ H_2(g)$$

4.47 If 21 g of H_2S is mixed with 38 g of O_2 and 31 g of SO_2 forms, what is the percentage yield?

$$2\ H_2S + 3\ O_2 \rightarrow 2\ SO_2 + 2\ H_2O$$

4.48 A mixture of 10.0 g of NO and 14.0 g of NO_2 results in the production of 8.52 g of N_2O_3. What is the percentage yield?

$$NO(g) + NO_2(g) \rightarrow N_2O_3(l)$$

Solution Stoichiometry

4.49 Small quantities of hydrogen gas can be prepared by the following reaction.

$$Zn(s) + H_2SO_4(aq) \rightarrow ZnSO_4(aq) + H_2(g)$$

How many grams of H_2 can be prepared from 25.0 mL of 6.00 M H_2SO_4 and excess zinc?

4.50 Aluminum hydroxide can be dissolved in hydrochloric acid to form aluminum chloride and water. Write a balanced chemical equation for this reaction. How many mL of 3.37 M HCl are needed to react with 10.0 g of $Al(OH)_3$?

4.51 What is the role of the indicator in a titration?

4.52 The titration of 25.00 mL of 3.441 M HCl required 41.64 mL of NaOH. What was the concentration of the NaOH?

4.53 What mass of magnesium hydroxide, $Mg(OH)_2(s)$ is required to neutralize 69 mL of 0.105 M HCl(aq)?

4.54 A 0.150 g sample of benzoic acid, C_6H_5COOH is titrated with a 0.240 M solution of barium hydroxide, $Ba(OH)_2$. What volume of this solution is needed to completely react with this acid (only one hydrogen reacts).

4.55 A 10.00 mL sample of vinegar, which is dilute acetic acid, requires 14.32 mL of 0.0523 M NaOH to titrate. What is the molar concentration of the acetic acid?

$$CH_3COOH + NaOH \rightarrow NaCH_3COO + H_2O$$

Insight into Alternative fuels and additives

4.56 Why are fuel additives used?

4.57 What is actually measured by the octane ratings for different grades of gasoline?

4.58 Offer an explanation as to how oxygenated fuels might increase the extent of combustion.

4.59 Perform a web search to learn what restrictions (if any) on the use of MTBE in gasoline might exist in your local area.

4.60 Using the web, find information about the amount of lead in the environment for the past 40 years. Correlate what you observe with the presence or absence of tetraethyl lead in gasoline.

4.61 Using the web, find out how lead "poisons" the catalyst in a catalytic converter.

Additional problems

4.62 Estimate the mass of CO_2 emitted by a typical automobile in a year.

4.63 An acid-base titration was performed using HCl and $Ba(OH)_2$. (The products of this reaction include barium chloride, $BaCl_2$.) If 33.33 mL of 0.1500 M HCl solution react completely with 25.00 mL of $Ba(OH)_2$ solution, what is the molarity of the barium hydroxide?

4.64 In this chapter we used a simplified equation for the combustion of gasoline. But the products generated in the operation of real automobile engines include various oxides of nitrogen, and it is desirable to remove these compounds from the exhaust stream rather than release them into the atmosphere. Currently this is done by the "catalytic converter," but other technologies are being explored. One possible method involves the use of isocyanic acid, HNCO, which reacts with NO_2 to produce N_2, CO_2, and H_2O, as shown in the equation below.

$$8\ HNCO + 6\ NO_2 \rightarrow 7\ N_2 + 8\ CO_2 + 4\ H_2O$$

What mass of HNCO would be required to react completely with the estimated 1.7×10^{10} kg of NO_2 produced each year by US automobiles?

4.65 Phosphoric acid (H_3PO_4) is distributed commercially as a solution with a concentration of about 14.8 M. Approximately 2.1×10^9 gallons of this concentrated phosphoric acid solution are produced annually in this country alone. Assuming that all of this H_3PO_4 is produced by the reaction below, what mass of the mineral fluoroapatite ($Ca_5(PO_4)_3F$) would be required each year?

$$Ca_5(PO_4)_3F + 5\ H_2SO_4 \rightarrow 3\ H_3PO_4 + 5\ CaSO_4 + HF$$

4.66 The reaction shown below is used to destroy Freon-12 (CF_2Cl_2), preventing its release into the atmosphere. What mass of NaF will be formed if 250 kg of CF_2Cl_2 and 400 kg of $Na_2C_2O_4$ are heated and allowed to react to completion?

$$CF_2Cl_2 + 2\ Na_2C_2O_4 \rightarrow 2\ NaF + 2\ NaCl + C + 4\ CO_2$$

4.67 Blood alcohol levels are usually reported in mass percent: A level of 0.10 would mean that there were 0.10 g of alcohol (C_2H_5OH) present in 100 g of blood. This is just one way of expressing the concentration of alcohol dissolved in the blood, and other concentration units could be used. Find the molarity of alcohol in the blood of a person with a blood alcohol level of 0.12. (The density of blood is 1.2 g/mL.)

4.68 One way of determining blood alcohol levels is by performing a titration on a sample of blood. In this process, the alcohol from the blood is oxidized by dichromate ions ($Cr_2O_7^{2-}$) according to the following net ionic equation.

$$C_2H_5OH + 2\ Cr_2O_7^{2-} + 16\ H^+ \rightarrow 2\ CO_2 + 4\ Cr^{3+} + 11\ H_2O$$

A 10.00-g sample of blood was drawn from a patient, and 13.77 mL of 0.02538 M $K_2Cr_2O_7$ was required to titrate the alcohol. What was the patient's blood alcohol level? (See the previous problem for definition of blood alcohol level. $K_2Cr_2O_7$ is a strong electrolyte, so it dissociates completely in solution.)

4.69 Ammonium sulfate ($(NH_4)_2SO_4$) is a common fertilizer, and can be produced by the reaction of ammonia (NH_3) with sulfuric acid (H_2SO_4). Each year about 2 x 10^9 kg of ammonium sulfate are produced worldwide. How many kilograms of ammonia would be needed to generate this quantity of $(NH_4)_2SO_4$? (Assume the reaction goes to completion, with no other products formed.)

4.70 The pictures below show a molecular scale view of a chemical reaction between the compounds AB_2 and B_2. (Shaded circles represent B atoms and open circles are A atoms). The box on the left represents the reactants at the instant of mixing, and the box on the right shows what is left once the reaction has gone to completion.

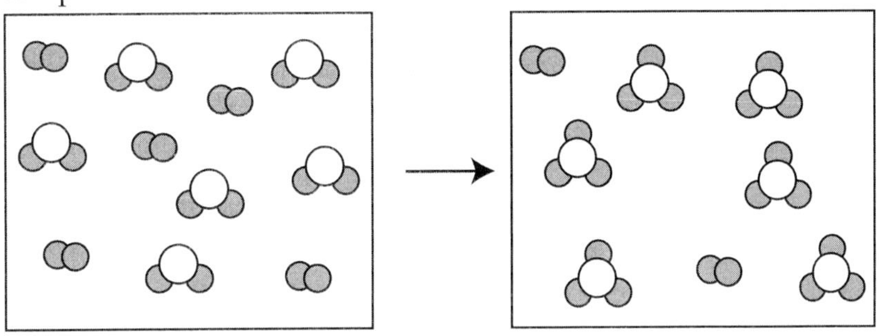

Was there a limiting reagent in this reaction? If so, what was it? Write a balanced chemical equation for this reaction. As usual, your equation should use the smallest possible whole number coefficients for all substances.

4.71 The pictures below show a molecular scale view of a chemical reaction between the elements A and B. (Open circles represent A atoms and shaded circles are B atoms). The box on the left represents the reactants at the instant of mixing, and the box on the right shows what is left once the reaction has gone to completion.

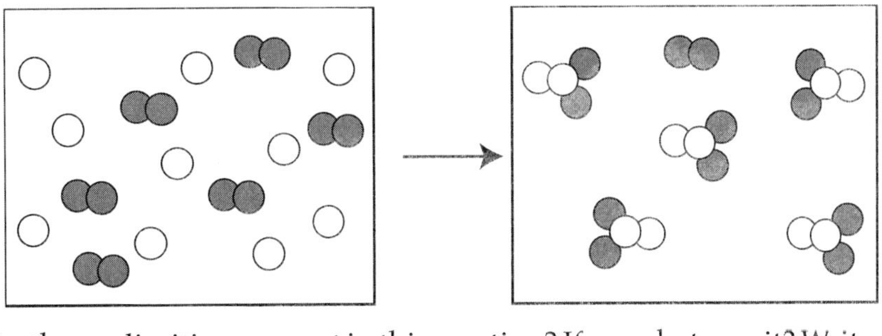

Was there a limiting reagent in this reaction? If so, what was it? Write a balanced chemical equation for this reaction. As usual, your equation should use the smallest possible whole number coefficients for all substances.

4.72 Aluminum chloride ($AlCl_3$) is used as a reagent in many industrial chemical processes. It is usually made by reacting scrap aluminum metal directly with chlorine gas (Cl_2). Suppose that 2.70 g of Al and 7.10 g of Cl_2 are mixed. What is the maximum mass of $AlCl_3$ that could be formed?

4.73 Disulfur dichloride (S_2Cl_2) is used in the production of vulcanized rubber. It can be produced by the reaction of molten sulfur ($S_8(\ell)$) with chlorine ($Cl_2(g)$). What is the maximum mass of S_2Cl_2 that could be produced by reacting 32.0 g of sulfur with 71.0 g of chlorine?

4.74 An aspirin tablet is to be analyzed for its content of the active ingredient, acetylsalicylic acid. (The remainder is filler designed to keep the tablet together.) Acetylsalicylic acid has the molecular formula, $HC_9H_7O_4$ and is titrated by a base such as NaOH according to the following reaction.

$$HC_9H_7O_4 + NaOH \rightarrow NaC_9H_7O_4 + H_2O$$

If a 0.400 g sample requires 19.3 mL of 0.121 M NaOH to reach the endpoint, what is the mass percentage of acetylsalicylic acid in the original tablet?

4.75 Cryolite (Na_3AlF_6) is used in the commercial production of aluminum from ore. Cryolite itself is produced by the following reaction.

$$6\ NaOH + Al_2O_3 + 12\ HF \rightarrow 2\ Na_3AlF_6 + 9\ H_2O$$

A mixture containing 800 kg of NaOH, 300 kg of Al_2O_3, and 600 kg of HF is heated to 950°C and reacts to completion. What is the maximum mass of Na_3AlF_6 formed?

4.76 A quality control technician needs to determine the percentage of arsenic (As) in a particular pesticide. The pesticide is dissolved and all of the arsenic present is converted to arsenate ions (AsO_4^{3-}). Then the amount of AsO_4^{3-} is determined by titrating with a solution containing silver ions (Ag^+). The silver reacts with the arsenate according to the following net ionic equation.

$$3\ Ag^+(aq) + AsO_4^{3-}(aq) \rightarrow Ag_3AsO_4(s)$$

When a 1.22 g-sample of pesticide was analyzed this way, it required 25.0 mL of 0.102 M Ag^+ solution to precipitate all of the AsO_4^{3-}. What was the mass percentage of arsenic in the pesticide?

4.77 Calcium carbonate (limestone, $CaCO_3$) dissolves in hydrochloric acid, producing water and carbon dioxide as shown in the following ***unbalanced*** net ionic equation.

$$CaCO_3(s) + H_3O^+(aq) \rightarrow H_2O(l) + CO_2(g) + Ca^{2+}(aq)$$

Suppose 5.0 g of $CaCO_3$ is added to 700 mL of 0.10 M HCl. What is the maximum mass of CO_2 that could be formed? What would the final concentration of Ca^{2+} ions be? (Assume that the final solution volume is still 700 mL.)

4.78 Iron metal can be refined from the mineral hematite (Fe_2O_3). One way of converting the mineral to iron is to react it with carbon monoxide, as shown below.

$$Fe_2O_3 + 3\ CO \rightarrow 2\ Fe + 3\ CO_2$$

Because the hematite is obtained from various ores, it is usually not in a pure form. Suppose an iron manufacturer has available 2.00×10^5 kg of ore which is 93% Fe_2O_3 by mass. (There is no iron in the remaining 7% of the ore.) How many moles of Fe_2O_3 are present in this ore? How many kg of pure iron could be obtained from this ample of ore? Assume that the process has a 100% yield and that excess CO is available.

4.79 A solution is prepared by dissolving 2.50 g of $CaCl_2$ in enough water to give 250. mL of solution. What is the molarity of $CaCl_2$ in this solution? ($CaCl_2$ and NaCl are both strong electrolytes.) If 2.00 g of NaCl are added to this solution, what will the molarity of Cl^- ions be? (Assume that the final solution volume is still 250.0 mL.)

Focus on Problem Solving Exercises

4.80 Chromium metal is found in ores as an oxide, Cr_2O_3. This ore can be converted to metallic Cr by the Goldschmidt process, $Cr_2O_3(s) + 2Al(s) \rightarrow 2Cl(l) + Al_2O_3(s)$. Suppose you have 4.5 tons of chromium(III)oxide, describe how you would determine how much aluminum metal you would need to carry out this process. List any information you would have to look up.

4.81 Hydrochloric acid reacts with many metals to form hydrogen gas, H_2. If you needed to produce hydrogen as a source of fuel in some design, how would you determine which metal, Mg, Al or Zn would produce the greatest amount of hydrogen per gram of metal? Describe the steps you would need to take in the decision making process.

4.82 When heated in air some element, X, reacts with the oxygen present to form an oxide with the chemical formula, X_2O_3. If a 0.5386 g sample of this unknown element yields an oxide with a mass of 0.7111 g, describe how you would determine the atomic mass of the element.

4.83 Chlorofluorocarbons were used for many years as refrigerants. One such molecule, $C_2Cl_2F_2$ is synthesized from the reaction of CCl_4 with HF. Suppose you have 150 tons of carbon tetrachloride as a starting point for this reaction, show how you would determine the minimum amount of HF needed to carry out the reaction.

4.84 Bismuth selenide, Bi_2Se_3, is used in semiconductor research. It can be prepared directly from it elements by the reaction, $2Bi + 3Se \rightarrow Bi_2Se_3$. Suppose you find you need 5 kg of bismuth selenide for a prototype system you are working on. You find a large bottle of bismuth metal but only 560 g of selenium. Describe how you can calculate if you have enough selenium on hand to carry out the reaction or if you need to order more.

4.85 When 2.750 g of the oxide of lead Pb_3O_4 is heated to a high temperature it decomposes to produce 0.0640 g of oxygen gas and 2.686 g of some new lead oxide compound. How can you use this data to determine the formula of the new compound?

4.86 The best way to generate small amounts of oxygen gas in a laboratory setting is by the decomposition of potassium permanganate, $2KMnO_4(s) \rightarrow K_2MnO_4 + MnO_2 + O_2$. Suppose you are testing a new fuel for automobiles and you want to generate specific amounts of oxygen with this reaction. For one test you need exactly 500 g of oxygen. Describe how you would determine the mass of potassium permanganate you need to obtain this amount of oxygen.

4.87 An ore sample with a mass of 670 kg contains 27.7% magnesium carbonate, $MgCO_3$. If all of the magnesium carbonate in this ore sample is decomposed to form carbon dioxide, describe how to determine what mass of CO_2 is evolved during the process.

4.88 Existing stockpiles of the refrigerant Freon-12, CF_2Cl_2, must be destroyed under the auspices of the Montreal Protocol because of their potential for harming the ozone layer. One method for doing this involves reaction with sodium oxalate.

$$CF_2Cl_2 + 2Na_2C_2O_4 \rightarrow 2NaF + 2NaCl + C + 4CO_2$$

If you had 150 tons of Freon-12 describe how you would know how much sodium oxalate you would need to make that conversion.

Chapter 5: Gases

In many ways the essence of chemistry is really considering things on the atomic and molecular level. And this idea applies not only when we study chemical reactions, but also when we examine the properties of matter. So as we begin to look at matter and materials from a chemist's perspective, we will always want to focus on how the properties of matter are related to the properties and behavior of its constituent molecules. Gases provide an excellent starting point for this, because in many ways the observable physical properties of gases are very direct results of the behavior of individual molecules within the gas.

Most of the earliest investigations of gas properties were carried out using air. The reasons for this are clear: air is both readily available and obviously important to our lives. And while the sophistication of the measurements we make has increased tremendously, the availability and importance of air have assured that it remains a very carefully studied gas. Nowadays much of our interest in air is actually driven by concerns over the levels and effects of various pollutants. The U.S. Environmental Protection Agency (EPA) defines pollutants as "unwanted chemicals or other materials found in the air." And although this definition does include a variety of small solid particles like dust or soot, most air pollutants are gases. Thus an exploration of some of the issues surrounding the monitoring and control of air pollution will form the backdrop for our study of gases and their properties.

5.1 Insight into Air Pollution

Air pollution is one of the unintended but unavoidable side effects of modern society. And the issues surrounding the monitoring, prevention, and remediation of air pollution involve an exceedingly complex interplay between science, business, and public policy. Our discussion here will be fairly limited, but we will try to explore some of the ways in which chemical science can help us to understand the complexity of the situation.

We should begin by looking at the chemical composition of "clean" air. As you probably know, air is a mixture rather than a pure substance. Nitrogen and oxygen account for the majority of the molecules present, and other important components include CO_2 and water. The amount of water in the air varies from place to place and day to day, as described by the humidity in a weather report. So it will be convenient to use "dry air" as a reference point in our discussion. Table 5.1 lists the composition of a 1-m^3 sample of dry air at 25°C and normal atmospheric pressure. The four gases listed account for at least 99.99% of the molecules present, with N_2 and O_2 alone making up roughly 99%.

Particulate matter includes soot, dust, and other small solid particles dispersed in the air.

In contrast to the short list in Table 5.1, if we were to collect a sample of air in a modern urban area and analyze it carefully we might detect dozens of compounds at trace levels. Some of these additional trace components can be traced to natural sources, but most are directly attributable to human activities like transportation and manufacturing. Out of the large number of chemical compounds found in urban air, the EPA has identified six principal *criteria pollutants:* CO, NO_2, O_3, SO_2, Pb, and particulate matter (PM). These six substances are commonly found throughout the country and have been shown to have a variety of negative effects on health, the environment, and property. The term "criteria pollutants" comes from the fact that the EPA has established a set of science-based standards, or criteria, for the acceptable levels of these pollutants. *Primary standards* are intended to protect our health, while *secondary standards* are intended to protect our environment and property. In many cases the allowable levels of these pollutants correspond to less than one molecule per every million molecules of air. Levels of these pollutants are monitored periodically throughout the country. If the detected concentrations exceed the primary standard in a given region, that area is designated as a *nonattainment area* for that particular pollutant.

Gas	# Moles Present
N_2	31.929
O_2	8.567
Ar	0.382
CO_2	0.013
Other trace gases	0.002
Total	40.893

Table 5.1 The composition of a one cubic meter sample of dry air at 25°C and normal atmospheric pressure is shown.

The various oxides of nitrogen are often referred to collectively as "NO_X".

As an example to illustrate the complexity of pollution issues, let's consider one of the criteria pollutants, NO_2. We might begin by asking where this pollutant comes from. Nitrogen dioxide is emitted as part of the exhaust gas from our automobiles. At the high temperatures at which engines operate, the nitrogen and oxygen gases present in air react with each other to form a variety of nitrogen oxides. Such reactions do not occur at lower temperatures, but are practically unavoidable in an engine. When engines run hotter, from standing in a traffic jam for example, they tend to produce even more nitrogen oxides. Once it is present in the air, NO_2 poses a health risk because it attacks the membranes in our lungs. The brown color of smog is also largely due to NO_2. Nitrogen dioxide molecules absorb some portion of the visible light of the Sun, leaving the colors that produce brown to pass through.

But the detrimental effects of NO_2 in the air extend well beyond this rather unappealing color. Light energy from the Sun can also initiate chemical changes through what is known as a *photochemical reaction*. In the case of nitrogen oxides, these photochemical processes lead to products that can react further to trigger the formation of ground-level ozone.

But nitrogen oxides cannot generate ozone on their own. The key additional components that are needed are molecules collectively referred to as *volatile organic chemicals*, or VOCs. Organic chemicals contain predominantly carbon and hydrogen, while volatile compounds tend to evaporate easily. So VOCs are carbon-based molecules that evaporate very readily. Many different VOCs are present in urban air. The primary source of these chemicals is fuels: the gasoline we burn in our automobiles and lawn mowers is a complex mixture of organic molecules. All combustion engines release some VOCs when they are operating, and older vehicles tend to release more. Our noses detect this phenomenon every time we are near a vehicle that produces large amounts of VOCs.

When photochemically activated nitrogen oxides react with VOCs, the resultant mixture of gases is collectively referred to as smog. Many molecules in smog are irritants to human lungs, most prominently ozone (O_3). When present at ground level, ozone poses a range of health hazards, even for healthy young adults. People with respiratory problems, such as asthma, are particularly susceptible to problems associated with ozone. Figure 5.1 shows a graph of concentrations of various chemicals associated with photochemical smog as a function of time of day in a large city. Notice that the nitrogen oxides and VOCs are both present prior to the formation of ozone. If either of these molecules is missing, the ozone will not form. As we will see, this observation provides a way to combat the formation of ozone. Many cities have adopted programs that target one or more of the key constituents in photochemical smog, hoping to avoid the formation of its most hazardous components.

We will look at the desirable effects of ozone in the upper atmosphere in Chapter 11.

Even in this brief overview, we see that a single pollutant – NO_2 in this case – can have a complex impact on the air around us. We also see that the effects of various pollutants can be coupled in unexpected ways; the simultaneous presence of NO_2 and VOC's leads to the generation of O_3, another of the criteria pollutants. Clearly the chemistry of air pollution is quite complex, so we will not be able to delve into many aspects of it in great detail. But it should be clear that to improve our understanding of these important issues, we must first explore the properties of gases, gas mixtures, and chemical reactions involving gases.

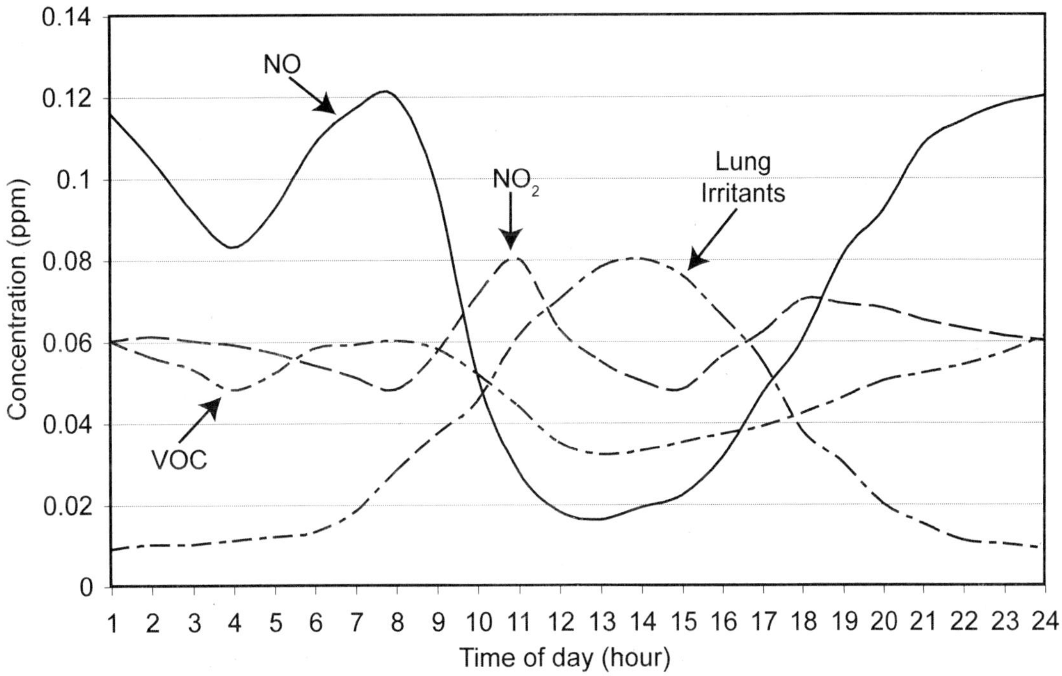

Figure 5.1 The graph shows how the levels of various air pollutants in an urban area varies over the course of a day. The levels of nitrogen oxides and VOCs rise in the morning hours, mainly due to emissions from vehicles. The presence of both NO_x and VOCs leads to increased production of lung irritants like ozone, whose levels go through a maximum during the afternoon hours.

Properties of Gases

Several observable properties are common to all gases, and distinguish them from solids and liquids.

- Gases expand to fill the volume of any container.
- Gases have much lower densities than solids or liquids.

- Gases have highly variable densities, depending on the conditions.
- Gases mix with one another readily and thoroughly.
- Gases change volume dramatically with changing temperature.

Some of these properties help explain why air pollution is difficult to combat. The fact that gases expand to fill the available volume, for example, means that exhaust gases released from a car or an industrial plant will readily spread throughout the surrounding area. Similarly, the fact that gases mix readily with one another means that once they are present, traces of gaseous pollutants will usually be difficult to separate from the air.

As we work toward developing an accurate model for gas behavior, we will want to be sure that it can account for all of these observations. We'll want our model to be based on a molecular level understanding of what a gas is like, and we'll also want it to enable us to make accurate numerical predictions about how a gas will respond to different conditions.

As it turns out, you probably already know an important quantitative part of the model we use most often to describe gases. If there is just one equation that you remember from your high school chemistry class, chances are it is the ideal gas law.

$$PV = nRT \tag{5.1}$$

R can also be expressed in other units, as discussed in Section 9.3.

P is the gas pressure, V is the volume, n is the number of moles of gas present, and T is the temperature. R is a proportionality constant, and is called the **universal gas constant** because its value - 0.08206 L atm mol^{-1} K^{-1} - is the same for any gas. This familiar equation will form the basis of much of our discussion in this chapter, so it will be important that we have a thorough understanding of the terms that appear in it. Let's start by taking a close look at the meaning and origin of gas pressure.

5.2 Pressure

From inflation guidelines for car or bicycle tires to weather reports, we regularly encounter the word *pressure*, so you undoubtedly have some notion of what it means. But to investigate the behavior of gases scientifically, we need to make our intuitive understanding of pressure more formal.

Pressure is defined as force exerted per unit area.

$$P = \frac{F}{A} \tag{5.2}$$

The obvious question arising from this definition is: "What force causes the pressure of a gas?"

For atmospheric pressure the force can be attributed to the weight of the molecules of air being attracted to the surface of the planet by gravity. The existence of atmospheric pressure may not be apparent but can readily be demonstrated by removing the air from a container. Even an apparently sturdy container collapses under the pressure of the atmosphere when the air inside is no longer supporting the container walls. Because the total weight of the molecules increases as the number of molecules increases, the atmospheric pressure changes as a function of altitude. At sea level there are 20,320 feet more atmosphere above us than at the summit of Mount McKinley, and the atmospheric pressure at sea level is much higher than at the mountaintop. The variation of pressure with altitude is shown in Figure 5.2.

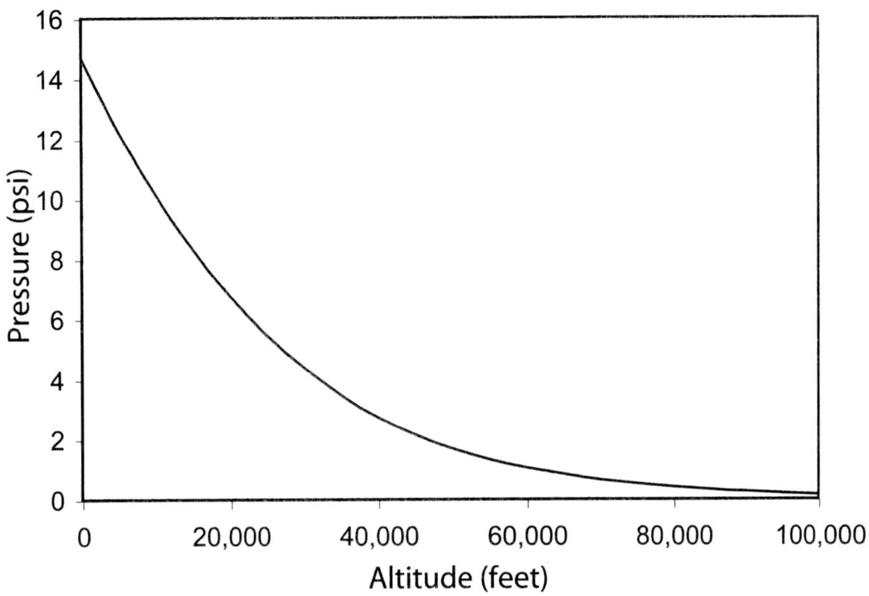

Figure 5.2 Atmospheric pressure varies as a function of altitude. Pressure is given here in units of pounds per square inch. Other units of pressure are discussed below.

But what about the gas trapped in a balloon? It exerts pressure uniformly on all the walls of the balloon, without regard to direction. And this clearly can't be due to gravity. How does a gas exert force on the walls of its container? To answer this we need to zoom in to the molecular level, as in Figure 5.3, where we will see individual gas molecules colliding with the walls and bouncing off. Each molecular collision imparts a small force to the container. And when these tiny forces are summed across the vast number of gas molecules present, they are able to produce the macroscopic force that leads to gas pressure. Later we will see how several observable gas properties can readily be explained by considering the behavior of gas molecules on the microscopic scale.

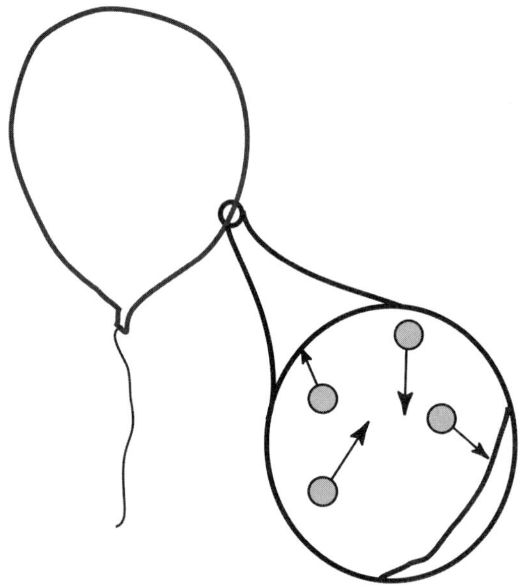

Figure 5.3 The inset in this figure shows a microscopic scale "close-up" of the gas molecules in a balloon. As molecules move through the volume of the balloon, they occasionally collide with the rubber walls. In every such collision, the gas molecule exerts a small force on the balloon wall, pushing it outward. When summed over the large number of gas molecules present, these small forces give rise to the observed pressure of the gas, and are responsible for the shape of the balloon.

Measuring Pressure

Because pressure is one of the important measurable properties of a gas, several different ways have been devised to measure it, beginning with the barometer. The essential feature of a barometer is a long tube that is closed at one end and is filled with mercury. Mercury is used because it is a dense liquid. When the mercury filled tube is inverted in a pool of mercury, some, but not all, of the liquid runs out of the tube. The closed end of the tube then is a vacuum, meaning a region of reduced pressure. If this end of the tube is somehow opened, all of the mercury pours out. What prevents the movement of mercury when the tube is closed? We can understand how the pressure of air is responsible by referring to Figure 5.4.

Mercury is chosen because it is a liquid with a large density. Barometers could be built with other liquids.

The pool of mercury is exposed to the atmosphere and thereby experiences the pressure of air weighing down on it. The column of mercury in the tube has a vacuum over it and so has no pressure weighing on it. Now consider a point at the mouth of the tube in the pool of mercury. This point experiences pressure from two places, the atmospheric pressure on the pool and the weight of the mercury from the column. When these two pressures are equal, the height of mercury suspended in the tube remains constant and provides a way to determine the atmospheric pressure. When atmospheric pressure decreases (as due to weather variations) the decrease will cause the column to shorten to maintain the equality of pressure at that point. Near sea level, the column of mercury is typically between 730 mm and 760 mm in height.

We'll look at some more modern ways of measuring pressure in the closing section of this chapter.

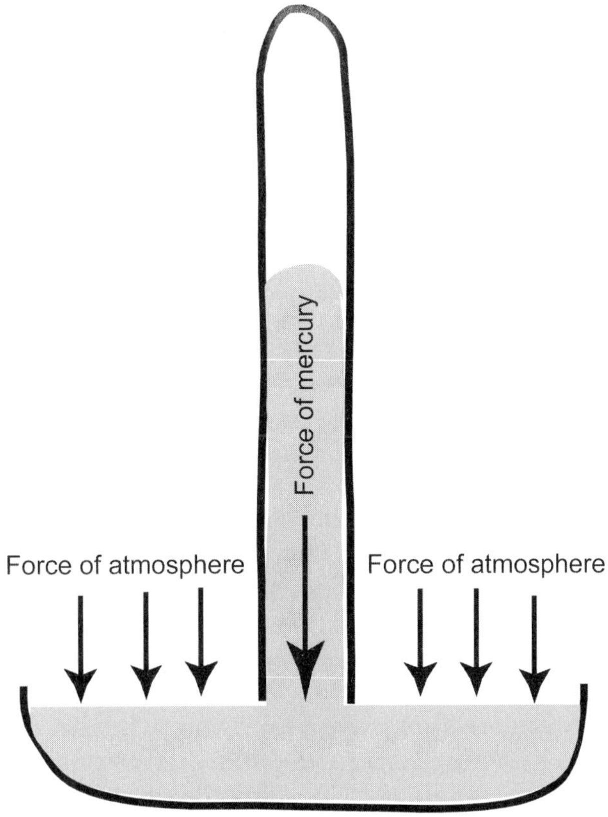

Figure 5.4 A schematic diagram of the forces at work in a mercury manometer. Atmospheric pressure exerts a force pushing down on the surface of the mercury, and tends to push the column of mercury in the barometer higher up the tube. The force due to the weight of the mercury in the column tends to push the column of mercury lower in the tube. When the two forces are equal, the column height remains constant.

Units of Pressure

The SI unit for pressure is a derived unit that reflects the definition of pressure. The SI units for force and area are the newton and the square meter, respectively. So pressure is measured in N m^{-2}. This unit has also been named the Pascal, Pa. When spread out over a square meter, a newton is not very much pressure, so the Pa is a relatively small quantity. Typical atmospheric pressures are on the order of 10^5 Pa. Thus a more practical unit is kPa, which is the unit used in weather forecasts in Canada and other "metric countries."

The construction of the barometer has influenced the units commonly used for reporting pressures. Most weather forecasts in the United States report pressure in inches (inches of mercury), while most barometers in school laboratories provide measurements in millimeters of mercury (mm Hg). Both of these units suffer from the shortcoming that they are lengths, which is confusing when we consider that pressure is a force per unit of area. The explanation of this contradiction comes directly from the barometer; the height of the mercury column in a barometer is directly proportional to the applied pressure. So it became common to report pressure directly in terms of the measured height. In spite of this confusion, the mm Hg unit is still widely used, and has been given an additional name, the torr. For our purposes here we can consider these two units equivalent.

1 torr = 1 mm Hg

The study of the atmosphere motivates yet another unit of pressure. The average pressure at sea level is close to 760 torr. The common observation of an apparent average atmospheric pressure led to the definition of a unit called an *atmosphere* (atm), which was set equal to 760 torr. As science has become more precise in its measurement this equivalence still remains.

1 atm = 760 torr (exactly),

But the definition of the unit atm has since been changed to reflect SI units. It is now defined in terms of the Pascal.

1 atm = 101,325 Pa (exactly),

Combining the last two relationships gives us a formal definition of the torr as well.

760 torr = 101,325 Pa (exactly)

The three units we will use most often in this book are the atm, torr and kPa. In some engineering fields you will also be likely to encounter other units, including pounds per square inch. Of course none of these units should cause you particular difficulty as long as you are careful that any calculations you do properly account for the units being used.

We might now ask how gas pressure plays a role in air pollution. Gases naturally flow from higher pressure to lower pressure. And pollutants are often produced in a high-pressure environment, such as in the cylinder of an auto engine or a reaction vessel in an industrial processing plant. When gases are released from these high-pressure environments, they are rapidly dispersed into the surrounding air. The pressure difference helps speed the spread of the pollutants.

5.3 History and Application of the Gas Law

Unlike solids and liquids, gases change significantly when the conditions in which they are found are altered. Changing the temperature, pressure, or volume of a gas's container results in a dramatic response from the gas. All gases respond to these physical changes in the same way despite their chemical differences, and the gas law equation allows us to calculate the impact of a particular change. Historically the study of gases preceded most of the important discoveries of modern chemistry. The relationships between physical properties of gases summarized by the ideal gas equation were initially based on empirical observations, and often still carry the name of the investigator who first characterized the relationship.

The early study of gases was important in the development of the field of chemistry and is called pneumatic chemistry.

Jaques Charles first investigated the relationship between temperature and volume in the early 19th century. He noted that when he plotted volume versus temperature for different amounts of gas, the resulting lines converged to the same temperature at zero volume. (See Figure 5.5.) This fact was used to establish the Kelvin temperature scale. Once the temperature is expressed on the Kelvin scale, the volume and temperature are directly proportional, an observation now known as Charles's Law.

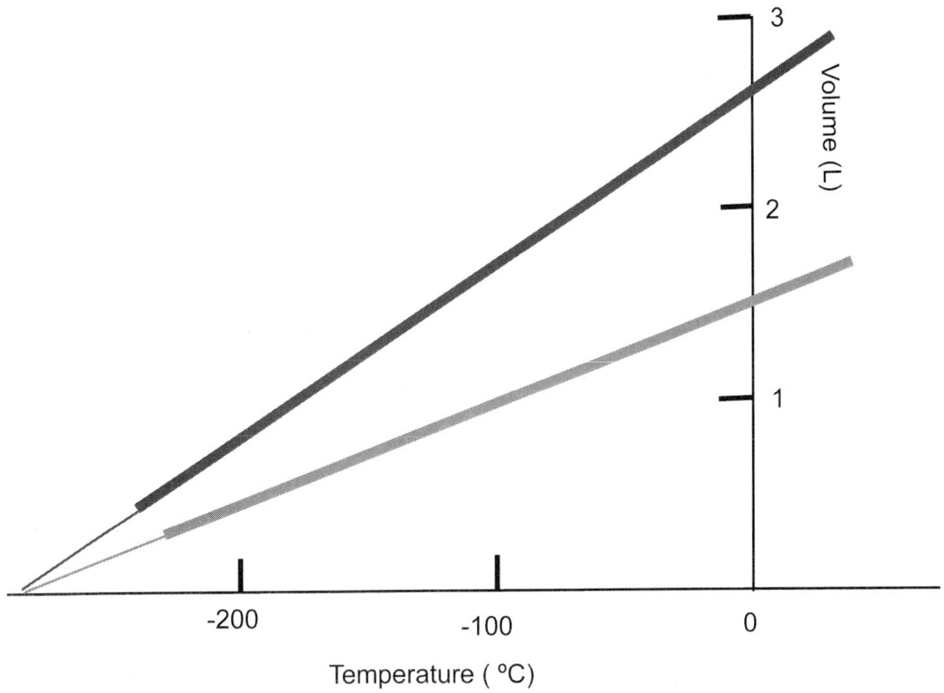

Figure 5.5 The graph shows volume as a function of temperature for two different samples of gas. The sample represented by the upper, darker line contains more gas than that represented by the lower, lighter line. Note that for both samples, the relationship between volume and temperature is clearly linear. Also note that when both lines are extrapolated back towards lower temperatures, they have the same x-intercept at -273°C. This extrapolation defines the so-called "absolute zero" temperature, or 0 K.

Other early scientists examined the connections between pairs of gas variables, and arrived at gas laws to express their observations. Robert Boyle studied the relationship between gas pressure and volume, and noted that for any gas sample, when pressure increases volume decreases or when volume increases pressure decreases. Thus Boyle's Law states that at constant temperature, pressure and volume are inversely proportional. Similarly, Avogadro's Law states that the volume of a gas is proportional to the number of molecules (or moles) of gas present.

These laws are frequently used to calculate the effect of some particular change in gas conditions. Fortunately, we will not need to memorize specific equations for all of these individual laws, because they are all incorporated into the ideal gas law. As long as we know the ideal gas equation and know how to work with it, we can easily solve problems that might also be approached using any of the "individual" gas laws. To do this, just start by identifying all of the parameters from the gas law - P, V, T, or n - that do *not* change. (And since R is a universal constant, it never changes!) Then rearrange the gas law equation to collect all of those terms onto the same side, leaving those that *do* change on the other side. If pressure and volume change while the number of moles and the temperature remain constant, for example, we'd get

$$PV = nRT = \text{constant}$$

Since nRT remains constant, we know that PV must also be constant. This tells us that the product of P times V must be the same before and after the change, so we can write

$$P_1V_1 = P_2V_2 \tag{5.3}$$

Here the subscripts '1' and '2' indicate the conditions before and after the change in conditions takes place. Now if we know any three of those four terms we can solve for the missing one, as seen in the following example.

Example Problem 5.1

In addition to ongoing sources, like car engines, pollutants can also be introduced into the air through isolated incidents like the rupture of a gas storage tank. Most gases are stored and transported at high pressures. A common laboratory cylinder of methane, for example, has a volume of 49.0 L and is filled to a pressure of 154 atm. Suppose all of the CH_4 from this cylinder is released and expands until its pressure falls to 1.0 atm. What volume would the CH_4 occupy?

Strategy

The problem involves a change in the conditions – pressure and volume – of a sample of gas. We will want to use the ideal gas law, since it is our model that relates the various properties of a gas to one another. So we begin by assuming that the gas will obey the ideal gas law under both the initial and final conditions. We will further assume that temperature is constant, since we are not given any data to suggest otherwise. And since we are working with the same sample of gas throughout, we also know that n is constant. Collect the constant terms together and work towards a solution.

Solution

In this case n and T are constant. And R is always constant. So we have:

$$PV = nRT = \text{constant}$$

And therefore

$$P_1V_1 = P_2V_2$$

Solving this for the final volume (V_2) gives

$$V_2 = \frac{P_1V_1}{P_2}$$

We know all three terms on the right: $P_1 = 154$ atm, $V_1 = 49.0$ L, and $P_2 = 1.00$ atm. Insert those above and solve.

$$V_2 = \frac{(154 \text{ mL})(49.0 \text{ L})}{(1.00 \text{ atm})} = 7{,}550 \text{ L}$$

Discussion

This is much higher than the initial volume, which makes sense since we reduced the pressure by a large factor. Because of the high initial pressure, the assumption that the gas behaves ideally may not be very accurate. We'll examine that assumption later in Example 5.8.

Check Your Understanding

A small sample of gas is generated in a laboratory. In a 125-mL vessel it exerts 115 torr of pressure. Then a valve is opened and the gas expands until its volume is 175 mL. If the temperature is constant throughout this process, what is the new pressure of the gas?

In other situations we might find that V, and T vary while n and P are constant. That would lead us to write the following relationship.

$$\frac{V}{T} = \frac{nR}{P} = \text{constant}$$

This in turn leads to Equation 5.4.

$$\frac{V_1}{T_1} = \frac{V_2}{T_2} \tag{5.4}$$

Example Problem 5.2

A balloon is filled with helium, and its volume is 2.1 L at 298 K. The balloon is then dunked into a thermos bottle containing liquid nitrogen. When the helium in the balloon has cooled to the temperature of the liquid nitrogen (77 K), what will the volume of the balloon be?

Strategy

We should start by recognizing this as a change in conditions for a gas, and realize that we will want to use the ideal gas law. So we assume that the gas will behave ideally under both the initial and final conditions. We will further assume that the pressure is constant, since we are not given any data to suggest otherwise. And since the balloon is tied off throughout, we can also assume that n is constant. Collect the constant terms together and work towards a solution.

Solution

$$\frac{V}{T} = \frac{nR}{P} = \text{constant}$$

So

$$\frac{V_1}{T_1} = \frac{V_2}{T_2}$$

Solve this for the final volume (V_2).

$$V_2 = \frac{V_1T_2}{T_1}$$

Since the three terms on the right are known, we can just insert values and solve.

$$V_2 = (2.2\text{ L})\,(77\text{ K})\,/\,(298\text{ K}) = 0.57\text{ L}$$

Check Your Understanding

The balloon in the example above will burst if its volume exceeds 2.3 L. At what temperature would you expect the balloon to burst?

For *any* change in the conditions of a gas sample, we should always be able to employ this same idea by separating those parameters that are constant from those that are changing. And since you are very unlikely to forget the ideal gas law, this approach is almost certainly more reliable than attempting to remember a list of equations covering all the possible combinations of variables.

Units and the Gas Law

Suppose we had expressed the temperature in °C rather than in kelvin in the above example? It should only take you a little work to convince yourself that the equation we used will not hold true. Changing the temperature of a gas from 10°C to 20°C, for

Although absolute zero is a fairly abstract concept, temperatures well below 1 K can routinely be achieved in the laboratory.

example, will not double its volume. Indeed, the ideal gas law itself is only valid when temperature is expressed on an *absolute* scale, where T = 0 corresponds to absolute zero temperature. To see why this must be so, notice that putting T = 0 into the gas law implies that either P or V or both must also go to zero, meaning that there must be no gas present. This simply can't be correct for T = 0°C or T = 0°F; that's obvious because the winter temperature routinely reaches these levels in many areas, and yet the air in the atmosphere obviously does not vanish. Absolute zero (0 K), on the other hand, is a very special temperature, and it is reasonable that there might be peculiar consequences if we actually cooled a gas to this level.

No degree sign is used when writing Kelvin temperatures.

In contrast to the temperature, which *must* be expressed on an absolute scale, we can use *any* convenient units for pressure and volume, so long as we then reconcile the units on those quantities with those on the gas constant, R. Why this difference? The common temperature scales are fundamentally different from other kinds of units, because interconverting among them involves not only multiplying by a conversion factor, but also adding or subtracting. This, in turn, is a direct result of the fact that different temperature scales (like K, °C, or °F) all have different zero points. In all the other quantities that we work with, the meaning of "zero" is the same no matter what units we use. If you say the mass of a sample is zero, for example, it doesn't really mater much what units you measured in - there's still nothing there. But for temperature, the difference between 0°C or 0°F or 0 K is obvious. In some engineering fields you may encounter the Rankine temperature scale (°R). This is also an absolute scale, meaning that 0°R = 0 K. But one degree on the Rankine scale is the same size as 1°F. Because they are both absolute scales, Rankine and Kelvin temperatures can be interconverted by using a multiplicative unit factor: 1°R = 1.8 K.

If you look at the table like the one in Appendix B of this book, you will find values for R in several combinations of units: 0.08206 L atm mol^{-1} K^{-1}, 8.314 J mol^{-1} K^{-1}, 62.366 L torr mol^{-1} K^{-1}. Students often ask how to decide which of these values is "the right R" to use in a given situation. The answer is that they are all really the same quantity, just expressed in different units. (If you are curious, you can do unit conversions to verify this for yourself.) If pressure is given in atm and volume in L, then it will be convenient to use R as 0.08206 L atm mol^{-1} K^{-1}, for example. If pressures are given in torr, it might be easier to insert 62.366 L torr mol^{-1} K^{-1} for R. Or, if you prefer to always use the same value of R, you could simply convert pressures from any other units into atm before doing any calculations. It really doesn't matter which set of units we choose, as long as we are careful to reconcile the units in all of our calculations.

Example Problem 5.3

A sample of CO_2 gas has a volume of 575 cm^3 at 752 torr and 72°F. What is the mass of carbon dioxide in this sample?

Strategy

We are trying to find the mass of a sample of gas. If we knew how many moles of CO_2 were in the sample, we could easily use the molar mass to find the mass of the sample. And since the problem involves a gas, we can attempt to use the ideal gas law to find the number of moles. We are given values for V, P, and T, so we will be able to solve for n. Some of the units given are unusual, so we will need to be careful to handle them properly.

Solution

We can begin by converting some of the given information into units more commonly used for gases.

$$T = (72 - 32) \times 5/9 = 22°C = 295\ K$$

$$V = 575\ cm^3 = 575\ mL = 0.575\ L$$

For P, we will use the value in torr and choose the corresponding value for R. We could also convert from torr to atm, of course.

$$n = \frac{PV}{RT} = \frac{(752\ torr)(0.575\ L)}{(62.366\ L\ torr\ mol^{-1}\ K^{-1})(295\ K)} = 0.0235\ mol$$

The gas in our sample is CO_2, with a molar mass of 44.0 g/mol, so

$$0.0235\ mol \times 44.0\ g/mol = 1.03\ g\ CO_2$$

Check Your Understanding

Find the volume in ft^3 of 1 mole of an ideal gas at -25°C and 710 torr.

5.4 Partial Pressure

By using the ideal gas law we can determine many properties of an individual gas. In many instances, however, we encounter interesting situations with more than one gas present. Our consideration of the levels of pollutants in air is an obvious example. Even if we ignore any pollutants present, "clean" air is already a mixture of gases. Aside from the possibility of chemical reactions, how do the observed properties of a gas mixture differ from those of pure gases?

The answer to this question is actually suggested by the gas law. In particular, note that none of the terms in the gas law equation depend on the identity of the gas. Consider a sample of air, and for simplicity assume that it consists entirely of N_2 and O_2. Assuming that the air behaves ideally, its pressure can be expressed using the gas law.

$$P = \frac{nRT}{V}$$

Here n is the total number of moles of air present. But this total number of moles will simply be the sum of the moles of each component.

$$n_{total} = n_{N_2} + n_{O_2}$$

We can insert this into the expression above, giving

$$P = \frac{nRT}{V} = \frac{(n_{N_2} + n_{O_2})RT}{V} = \frac{n_{N_2}RT}{V} + \frac{n_{O_2}RT}{V}$$

Each of those nRT/V terms has a special significance; they are the pressures we would see if only one of the components of the gas mixture was present. If the nitrogen was the only gas present, n_{O2} would be zero, and we would be left with

$$P = P_{N_2} = \frac{n_{N_2}RT}{V}$$

And of course if only the oxygen was present we could write a similar result for P_{O2}. When the actual mixture is present, the observed pressure is just the sum of these hypothetical **partial pressures** due to each individual gas.

John Dalton is also known for his model proposing the existance and nature of atoms.

This concept was first established by John Dalton (who also hypothesized the existence of atoms), and is sometimes referred to as **Dalton's Law of Partial Pressures**: *The pressure of a mixture of gases is the sum of the partial pressures of the component gases.* In simple terms, the partial pressure is what the pressure would be if only that component were present. As long as the component gases behave ideally and do not actually react with one another, this relationship will hold for any gas mixture.

For an arbitrary gas mixture, we can write the partial pressure of each component as

$$P_i = \frac{n_i RT}{V} \tag{5.5}$$

where the index, i, indicates which gas (for a mixture of four gases, i would vary from 1 to 4 yielding, P_1, P_2, P_3 and P_4) and n_i is the number of moles of component i . The total pressure of the mixture, P, is the sum of these partial pressures,

$$P = \sum_i P_i = \sum_i n_i \frac{RT}{V} \tag{5.6}$$

And the total number of moles is the sum of the numbers of moles of the individual components, so we have,

$$P = n_{total} \frac{RT}{V} \tag{5.7}$$

These expressions suggest another way to understand partial pressures. We might choose to express the concentration of each component gas in terms of its mole fraction, defined as

$$X_i = \frac{n_i}{n_{total}} \tag{5.8}$$

For a pure gas, $n_i = n_{tot}$ and the mole fraction is one. Any mixture will have $n_i < n_{tot}$ so the value of the mole fraction has a range from zero to one. If we take the ratio of a partial pressure to total pressure, using Equations 5.7 and 5.8 we have,

$$\frac{P_i}{P} = \frac{n_i\ (RT/V)}{n_{total}\ (RT/V)} = \frac{n_i}{n_{total}} = X_i \tag{5.9}$$

So the partial pressure is given by the mole fraction times the total pressure.

$$P_i = X_i P \tag{5.10}$$

This set of equations provides the numerical means for us to use partial pressures as shown in Example Problems 5.4 and 5.5.

Example Problem 5.4

Not all pollution is due to human activity. Natural sources, including volcanoes, also contribute to air pollution. A scientist tries to generate a mixture of gases similar to those found in a volcano by introducing 15.0 g of water vapor, 3.5 g of SO_2 and 1.0 g of CO_2 into a 40.0L vessel held at 120°C. Calculate the partial pressure of each gas and the total pressure.

Strategy

As usual, we will begin by assuming that all the gases behave ideally. Then we can treat each gas separately to determine the partial pressure it exerts. Finally, sum these partial pressures to calculate the total pressure.

Solution

For all of the gases, T = 120°C = 393 K, and V = 40.0 L.

$$H_2O: 15.0 \text{ g } H_2O \times \frac{1 \text{ mol } H_2O}{18.0 \text{ g } H_2O} = 0.833 \text{ mol}$$

$$P_{H_2O} = \frac{(0.833 \text{ mol})(0.08206 \text{ L atm mol}^{-1} \text{ K}^{-1})(393 \text{ K})}{40.0 \text{ L}} = 0.672 \text{ atm}$$

$$SO_2: 3.5 \text{ g } SO_2 \times \frac{1 \text{ mol } SO_2}{64.1 \text{ g } SO_2} = 0.055 \text{ mol}$$

$$P_{SO_2} = \frac{(0.055 \text{ mol})(0.08206 \text{ L atm mol}^{-1} \text{ K}^{-1})(393 \text{ K})}{40.0 \text{ L}} = 0.044 \text{ atm}$$

$$CO_2: 1.0 \text{ g } SO_2 \times \frac{1 \text{ mol } CO_2}{44.0 \text{ g } CO_2} = 0.023 \text{ mol}$$

$$P_{CO_2} = \frac{(0.023 \text{ mol})(0.08206 \text{ L atm mol}^{-1} \text{ K}^{-1})(393 \text{ K})}{40.0 \text{ L}} = 0.019 \text{ atm}$$

Total pressure is the sum of these:

$$P_{tot} = P_{H_2O} + P_{SO2} + P_{CO_2} = 0.672 + 0.044 + 0.019 = .735 \text{ atm}$$

Note that you could also have used the total number of moles to find P_{tot}, and then used mole fractions to find the partial pressures.

Check Your Understanding

A mixture containing 13.5 g of oxygen and 60.4 g of N_2 is found to exert a pressure of 2.13 atm at 25°C. What is the volume of the container, and what are the partial pressures of each gas?

Example Problem 5.5

An experiment is designed to determine the effect of volcanic gases on plants near hot springs. Among the variations used is a mixture that has mole fractions given in the table.

Gas	N_2	O_2	H_2O	SO_2
Mole fraction	0.75	0.15	0.08	0.02

If the desired total pressure is 750 torr, what are the partial pressures? If the gas is to be in a15-L vessel held at 30°C, how many moles of each substance are needed?

Strategy

We know the mole fractions and the desired total pressure. So we can calculate partial pressures using the relationship defined in Equation 5.10. From the total pressure and the volume we can calculate the total number of moles and thereby the number of moles of each gas.

Solution

$$P_{N_2} = (0.75)(750 \text{ torr}) = 563 \text{ torr}$$

$$P_{O_2} = (0.15)(750 \text{ torr}) = 112 \text{ torr}$$

$$P_{H_2O} = (0.08)(750 \text{ torr}) = 60 \text{ torr}$$

$$P_{SO_2} = (0.02)(750 \text{ torr}) = 15 \text{ torr}$$

The desired total pressure of 750 torr can also be expressed as 0.987 atm. So

$$n_{total} = \frac{PV}{RT} = \frac{(0.987\text{ atm})(15.0\text{ L})}{(0.08206\text{ L atm mol}^{-1}\text{ K}^{-1})(303\text{ K})} = 0.595\text{ mol}$$

$$n_{N_2} = (.75)(.595) = .45\text{ mol}$$

$$n_{O_2} = (.15)(.595) = 8.9\text{x}10\text{-}2\text{ mol}$$

$$n_{H_2O} = (.08)(.595) = 4.8\text{x}10\text{-}2\text{ mol}$$

$$n_{SO_2} = (.02)(.595) = 1.2\text{x}10\text{-}2\text{ mol}$$

Check Your Understanding

A mixture of SO_2(g) and SO_3(g) is to be prepared with a total pressure of 1.4 atm. If the mole fractions of the gases are 0.7 and 0.3 respectively, what are the partial pressures? If the mixture is to occupy 2.50 L at 27°C, what mass of each gas is needed?

5.5 Stoichiometry of Reactions Involving Gases

When we discussed quantitative aspects of chemical reactions in Chapter 4, we emphasized the importance of ratios of moles. And the ideal gas law provides a relationship between the number of moles of a gas and some easily measurable properties: pressure, volume, and temperature. So when gases are involved in a chemical reaction, the ideal gas law often provides the best way to determine the number of moles. Using the ideal gas law in a stoichiometry problem really doesn't involve any new ideas. It just combines two kinds of calculations that you've already been doing. We'll still do the stoichiometric calculation in terms of mole ratios, just like always. And we'll use the gas law to connect the number of moles of gas with its temperature, pressure, and volume.

Example Problem 5.6

When an experiment required a source of carbon dioxide, a student combined 1.4 g of sodium bicarbonate ($NaHCO_3$) with excess hydrochloric acid. If the CO_2 produced is collected at 722 torr and 17°C, what volume will the gas occupy?

Strategy

We are asked to find the volume of a gas, and we are given its pressure and temperature. We'll assume that the gas behaves ideally. So if we knew the number of moles, we could easily use the gas law to get the volume we need. Looking a little closer, we should recognize this as a reaction stoichiometry problem because it asks us how much CO_2 will be produced. The new wrinkle here is that it asks us to express the answer as a volume rather than as a mass or a number of moles. So we will first do a stoichiometric calculation to find the number of moles of CO_2 produced, then use the gas law to find the volume of that amount of gas at the indicated temperature and pressure. As in any stoichiometry problem, we'll need to start with a balanced equation for the reaction to be sure we use the correct mole ratio.

Solution

$$NaHCO_3 + HCl \rightarrow NaCl + H_2O + CO_2$$

$$1.4\text{ g NaHCO}_3 \times \frac{1\text{ mol NaHCO}_3}{84\text{ g NaHCO}_3} \times \frac{1\text{ mol CO}_2}{1\text{ mol NaHCO}_3} = 0.017\text{ mol CO}_2$$

Now use the ideal gas equation to solve for the volume. As always, we'll need to remember to convert the temperature from °C to K, and to be careful of pressure and volume units.

$$V = \frac{nRT}{P} = \frac{(0.017 \text{ mol})(0.08206 \text{ L atm mol}^{-1} \text{ K}^{-1})(290 \text{ K})}{(0.95 \text{ atm})} = 0.43 \text{ L} = 430 \text{ mL}$$

Check Your Understanding

Suppose 270 mL of CO_2 at 1.03 atm and 25°C is produced by this reaction. Assuming an excess of hydrochloric acid was present, how many grams of sodium bicarbonate were used?

STP Conditions

From our discussion so far it should be obvious that to properly describe an amount of gas we need to specify its temperature, volume, and pressure. If we want to compare various quantities of gas, it is convenient to use some standard set of conditions. In order to set a standard for an ideal gas, we might specify a temperature and a pressure; that way the volume would be directly proportional to the number of moles of gas. The conventional choice for standard temperature and pressure, or STP, is a temperature of 0°C or 273 K and 1 atm of pressure. Plugging these values into the ideal gas equation we find the molar volume (volume of one mole) of an ideal gas at STP.

$$V = \frac{nRT}{P} = \frac{(1 \text{ mol})(0.08206 \text{ L atm mol}^{-1} \text{ K}^{-1})(273 \text{ K})}{(1 \text{ atm})} = 22.4 \text{ L}$$

This number provides a conversion factor for stoichiometric problems that include gases *provided STP conditions are maintained*. Example Problem 5.7 shows this type of problem.

Example Problem 5.7

One way to reduce air pollution is to remove potential pollutant gases from an exhaust stream before they are released into the air. Carbon dioxide can be removed from a stream of gas by reacting it with CaO to form calcium carbonate. If we react 5.50 L of CO_2 at STP with excess CaO, what mass of calcium carbonate will form?

Strategy

You should recognize this as a reaction stoichiometry problem, because it is asking us how much $CaCO_3$ will be produced. As for any stoichiometry problem, we should write a balanced chemical equation for the reaction. Then convert the volume of gas into moles and proceed as usual. Since the gas volume given is at STP, we can use the molar volume we calculated above as a conversion factor.

Solution

$$CO_2(g) + CaO(s) \rightarrow CaCO_3(s)$$

$$5.50 \text{ L } CO_2 \times \frac{1 \text{ mol } CO_2}{22.4 \text{ L } CO_2} \times \frac{1 \text{ mol } CaCO_3}{1 \text{ mol } CO_2} \times \frac{100.1 \text{ g } CaCO_3}{1 \text{ mol } CaCO_3} = 24.6 \text{ g } CaCO_3$$

Discussion

Note that the conversion factor of 22.4 L = 1 mole is valid **only** when the gas in question is at STP. You should be very careful not to use that molar volume for any other conditions! We could also have worked this problem by using the gas law to find the volume of the CO_2, using the appropriate temperature and pressure for STP conditions.

Check Your Understanding
SO_2 can be generated in the laboratory by heating ZnS with O_2.) (The other product is zinc oxide.) If 14.5 L of oxygen at STP is reacted with excess zinc sulfide, what volume of SO_2 at STP is generated?

At this point we have seen how to carry out a range of calculations involving gases, all based on the ideal gas law. But while we have repeatedly said that we will "assume the gas behaves ideally," we have not yet looked at what that statement implies in terms of molecular behavior. And we have also not yet considered the many practical circumstances in which we might make large errors by assuming ideal gas behavior. We take up these challenges in the next section.

5.6 Gas Kinetic Theory and Ideal vs. Real Gases

The ability to numerically model the behavior of gases was an important achievement, and has broad applicability in many areas of science and engineering. The similarity of behavior of different gases commonly encountered in the laboratory allowed for the building of models to describe how atoms must behave. But in many important practical settings, gases do not always behave ideally, so the simple ideal gas law calculations we have been doing may not always be an accurate model of reality. Non-ideal behavior is frequently seen in gases at very high pressures, such as might be found in the cylinder of an engine or in an industrial processing plant. Can an atomic perspective developed for ideal gases help explain these real world cases? Perhaps more importantly, can an improved model define quantitative methods for dealing with real gases such as those encountered in a running engine? In this section we will develop a very useful model called the Kinetic Molecular Theory, which provides connections between the observed macroscopic properties of gases, the gas law equation, and the behavior of gas molecules on the microscopic scale. Once we understand this model, we can use it to see how gases at high pressure might deviate from ideal behavior.

Postulates of the Model

Because observation of individual gas molecules is not generally feasible, we must begin with postulates that are accepted as reasonable when we construct a model. For the Kinetic Molecular Theory of gases the following postulates form the bedrock of the model.

1. A gas is made up of a vast number of particles, and these particles are in constant random motion.
2. Particles in a gas are infinitely small; they occupy no volume.
3. Particles in a gas move in straight lines except when they collide with other molecules or the walls of the container. Collisions with each other and with the walls of the container are elastic.
4. Particles in a gas interact with each other only when collisions occur.
5. The average kinetic energy of the particles in a gas is proportional to the absolute temperature of the gas, and does not depend on the identity of the gas.

As we consider the implications of these postulates we will also comment on their validity or the conditions under which they are not particularly accurate.

Starting from these postulates, it is actually possible to derive the ideal gas law. Although that result is a great example of the synergy between scientific theory and experimental observations, it is a bit beyond our reach here. So instead we will consider the connections between the postulates of the kinetic theory and our practical experience with gases. How do these postulates help to explain the observed behavior of various gases?

The statement that a gas consists of a large number of particles is consistent with what we know about the size of individual molecules. And the postulate that the molecules are in constant random motion is consistent with the observation that gases readily expand to fill their containers. The postulates are also consistent with what we said at the outset of this chapter about the origin of gas pressure. The molecules in a gas exert force when they collide with the wall of the container. And this accounts for the relationship between pressure and volume. As the volume of the container increases, the walls move farther apart and so each molecule will strike the wall less often. That means that each molecule will exert less force on the wall in a given time interval, reducing the pressure. So the kinetic theory is consistent with the inverse relationship between P and V that we've seen in the gas law equation.

How can the kinetic theory account for the effect of changes in temperature? The situation here is a little more complicated. We need to carefully consider postulate #5: The kinetic energy of the particles in a gas is proportional to the absolute temperature of the gas, and does not depend on the identity of the gas. We know from physics that kinetic energy is related to speed and to mass: $KE = 1/2\ mv^2$. So if the kinetic energy depends on the temperature, as stated in the postulate, then the speed of the molecules must be a function of temperature. Increasing the temperature must increase the speed of the molecules. When the molecules move faster, they will collide with the walls more often, and they will also impart a greater force on the wall during each collision. And both of these effects will lead to an increase in gas pressure with increasing temperature, as predicted by the gas law.

If you were paying close attention to the statement of Postulate #5, though, you may be wondering about the inclusion of the word "average" in "average kinetic energy." That word is included because in a collection of randomly moving gas molecules, some move quickly while others are virtually standing still at any given moment. A moment later the identity of fast moving and slow moving particles may have changed due to collisions, but there will still be both kinds. So it isn't really correct to think about the molecules as having a particular speed, and we should actually say that the average molecular speed increases with increasing temperature. This situation is best approached using a *distribution function* to describe the range of speeds.

Distribution functions are common mathematical devices used not only in science, but also in economics, social sciences and education. Your percentile scores on the SAT and other standardized tests are predicated on a distribution function for the students who take the exam. If you scored in the 85th percentile, for example, you would be expected to score higher than 85% of those taking the exam. And this ranking is expected to hold independent of the number of students who have actually taken the test, so that your percentile is not revised every time another round of tests are administered to more students.

Since we do not care what direction the particles are going, our discussion here is based on speed, and not velocity. Speed is always positive.

The distribution function that describes the speeds of a collection of gas particles is known as the **Maxwell-Boltzmann Distribution of Speeds**. This function, which was originally derived from a detailed consideration of the postulates of the kinetic theory, predicts the fraction of the molecules in a gas that travel at a particular speed. It has since been verified by a variety of clever experiments that allow the speed distribution to be measured in the laboratory.

Let's examine a plot of the Maxwell-Boltzmann distribution for a sample of CO_2 gas at various temperatures, as shown in Figure 5.6. As with any unfamiliar graph, you should start by looking at what is plotted on each axis. Here the x-axis shows a fairly familiar quantity: speed in meters per second. The y-axis is a little less obvious. It shows the fraction of all the molecules in the sample that would be travelling at a particular speed under the given conditions. (The actual mathematical form of the distribution function is discussed in the accompanying MathConnections box.) We've already established that the distribution of speeds must be a function of temperature, and the figure bears this out. At high temperatures, we see that the peak in the distribution shifts toward higher speeds. This is entirely consistent with Postulate #5, which states that the kinetic energy of the molecules must increase with temperature.

Furthermore, Postulate #5 also says that the kinetic energy does not depend on the identity of the gas. But the mass of the gas molecules clearly depends on their identity. In order for the kinetic energy to remain fixed as the mass changes, the speed distribution must also be a function of mass. Figure 5.7 illustrates this by showing the speed distribution for various gases at the same temperature. You can readily see that on average lighter molecules travel faster than heavier molecules.

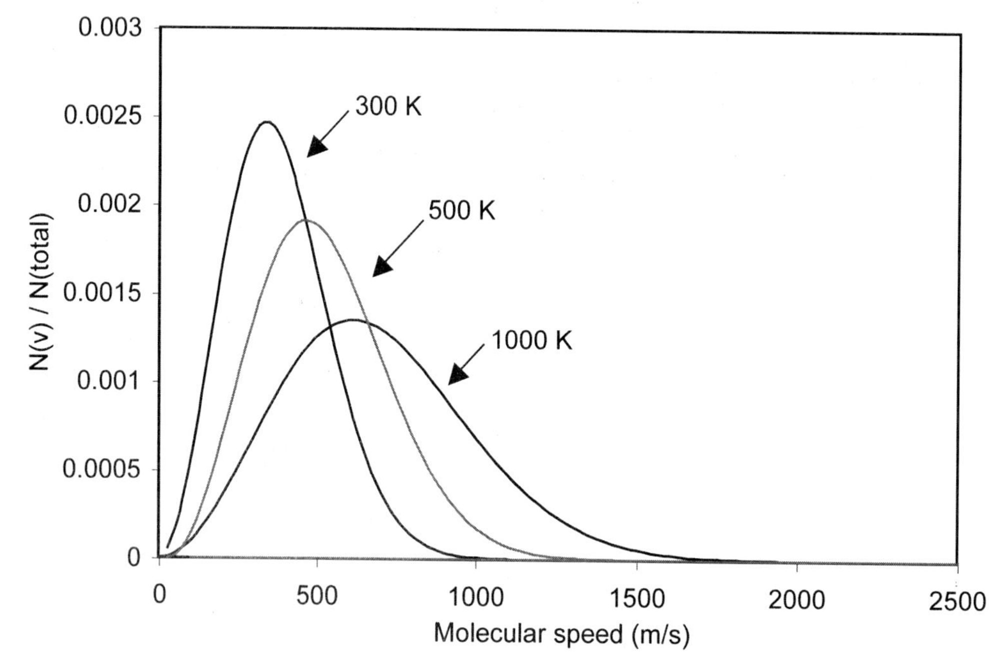

Figure 5.6 The distributions of molecular speeds for CO_2 molecules at three different temperatures are shown. The quantity on the y-axis is the fraction of all the gas molecules moving at a particular speed. Notice that as the temperature increases, the fraction of the molecules moving at higher speed increases.

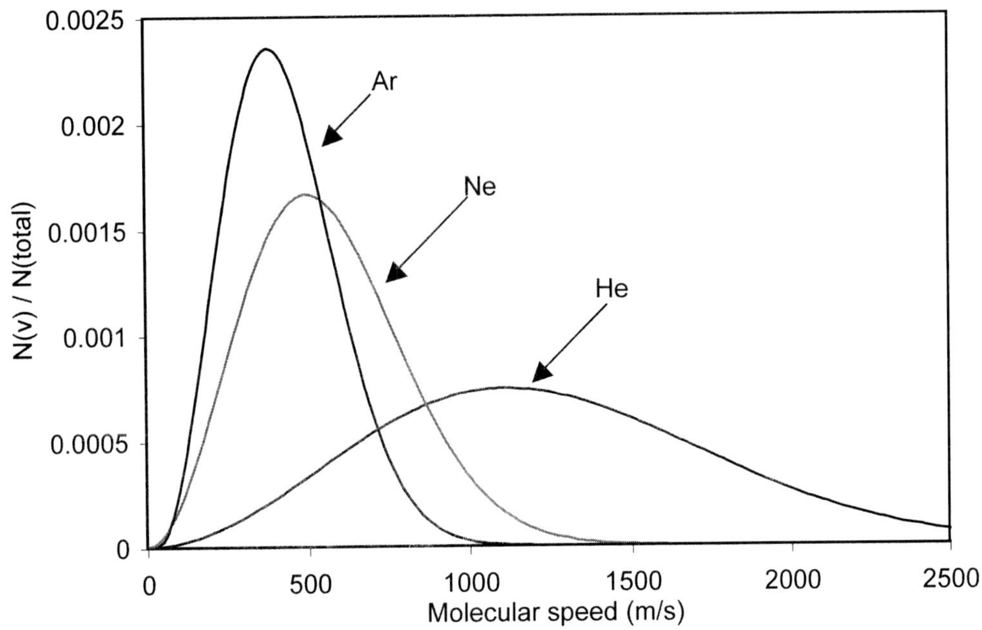

Figure 5.7 The distributions of molecular speeds for three different gases at 300 K are shown. As in Figure 5.6, the quantity on the y-axis is the fraction of all the gas molecules moving at a particular speed. Notice that as the molecular mass decreases, the fraction of the molecules moving at higher speed increases.

MathConnections

The actual equation for the Maxwell-Boltzmann distribution of speeds is somewhat complicated, but it consists entirely of terms in forms that you should recognize from your math classes.

$$\frac{N(v)}{N_{total}} = 4\pi(M / 2\pi RT)^{3/2}\, v^2\, e^{-Mv^2/2RT} \tag{5.11}$$

$N(v)$ is the number of molecules moving with speed v, and N_{total} is the total number of molecules. So the left-hand side is the fraction of the molecules having speed v, which is also the quantity plotted on the y-axis in Figures 5.6 and 5.7. N_A is Avogadro's number, M is the molar mass of the gas, R is the gas constant, and T is the temperature. (As in the gas law, this must be on an absolute scale.) Units for all of these quantities must be chosen in a consistent way, so that the right-hand side ultimately has no units. This must be true, since the fraction on the left-hand side will be unitless, and we know that the units on each side of any equation must match. Now look at Figure 5.6 or 5.7 again, and notice that the independent variable on the x-axis is the speed. So we want to think of the right-hand side of the distribution function as an elaborate function of the speed, v. We see two terms containing v. The first is just v^2, which will obviously increase as v increases. The exponential term, though, varies as $\exp(-v^2)$, which will decrease as v increases. So the overall shape of the distribution function reflects a competition between these two terms. At small values of v, the v^2 term dominates and the distribution function increases. At larger values of v, the exponential term takes over and the overall function decreases. So in between those two limits the distribution function goes through a maximum.

The speed at which the peak in the plot occurs is known as the **most probable speed**, for fairly obvious reasons: more molecules travel at this speed than at any other.

Although we will not do the derivation here, the most probable speed can be shown to be given by

$$v_{mp} = \sqrt{\frac{2RT}{M}} \qquad (5.12)$$

If you know some calculus you can do the derivation. Remember that the most probable speed is the maximum in the distribution function. And to find this maximum we would just take the derivative of Equation 5.11 and set it equal to zero. We might also want to describe our distribution by talking about the **average speed**. Because the distribution function is not symmetric, the most probable speed and the average speed will not be the same. In particular, the existence of the "tail" on the distribution curve at high speeds will pull the average to a higher speed than the most probable value. The average speed actually turns out to be 1.128 times the most probable speed. Finally, we sometimes characterize the distribution by talking about something called the **root-mean-square** speed. This value is actually the square root of the average value of v^2. At first glance it might sound like that should be the same as the average speed, but for the Boltzmann distribution the root mean square speed is actually 1.085 times the average speed. (If this sounds wrong to you, try a simple test. Pick a few numbers at random, and calculate their average. Then square all of your numbers and take the average of the squares. Finally, take the square root of that result and you'll see that it is larger than the average of the original numbers.) The root mean square speed is useful if we are interested in kinetic energy, because the average kinetic energy is given by

$$KE_{avg} = \frac{1}{2} m v_{rms}^2 \qquad (5.13)$$

Real Gases and Limitations of the Kinetic Theory

In all of our calculations so far we have begun by assuming that any gas or gas mixture of interest behaves ideally. But the postulates of the kinetic theory - and the gas law itself - are only strictly true for purely hypothetical ideal gases. How good is the assumption that gases behave ideally, and under what kinds of conditions might we want to question this assumption? The answers to these questions can help us to further develop our molecular level understanding of gases. By learning how real gases deviate from ideal behavior we will understand the nature of gas kinetic theory better. We'll start by taking a more critical look at some of the postulates of the theory.

Contrast this value to what is true for molecules in a liquid, which typically encounter another particle within one radius.

The assertion that gas molecules occupy no volume may seem at odds with reality, as we know that all matter occupies space. What is actually implied by this postulate? The essential idea is that when compared to the volume of "empty space" between particles, the volume of the particles themselves is not significant. One way to assess the validity of this assertion is to define the mean free path of the particles. The **mean free path** is the average distance a particle travels between collisions with other particles. In air at room temperature and atmospheric pressure, the mean free path is about 70 nm, a value 200 times larger than the typical radius of a small molecule like N_2 or O_2. When we consider the cubic relationship of volume to distance, the difference in volume is on the order of 200^3 or 8×10^6. So the volume of empty space at room temperature and pressure is on the order of 1 million times greater than the volume of individual molecules; the assumption in the kinetic theory that the volume of the molecules is negligible seems reasonable under these conditions.

But under what circumstances would this assumption break down? We cannot generally increase the size of molecules substantially by changing conditions, so the only way the volumes become more comparable is to "squeeze" the molecules closer together by decreasing the volume. Thus under conditions of high pressure, when gases are highly compressed, the volume of individual molecules may become significant and the assumption that the gas will behave ideally can break down. This makes sense if we realize that one way to condense a gas into a liquid is to compress it. And of course we don't expect the ideal gas law to hold for a liquid.

The postulates also assert that gas molecules always move in straight lines and interact with each other only through perfectly elastic collisions. Another way to put this is to say that the molecules neither attract nor repel one another. If this were strictly true, then the molecules could never stick together, and it would be impossible to condense a gas into a liquid or a solid. Clearly there must be some attractive forces between molecules, a topic we will explore further in Chapter 8. The ideal gas model works because under many conditions these attractive forces are negligibly small compared to the kinetic energy of the molecules themselves. And the fact that these forces lead to the formation of liquids and solids points us to the conditions under which the forces become important. We know that gases will condense if we lower the temperature sufficiently. The reason for this is that at lower temperatures the kinetic energy of the molecules decreases, and eventually the strength of the attractive forces between molecules will become comparable to the kinetic energy. At that point particles will be diverted from straight-line trajectories as shown in Figure 5.8. The overall effect of the attractive forces between molecules is to slightly increase the number of collisions between molecules above what would be predicted from the kinetic theory.

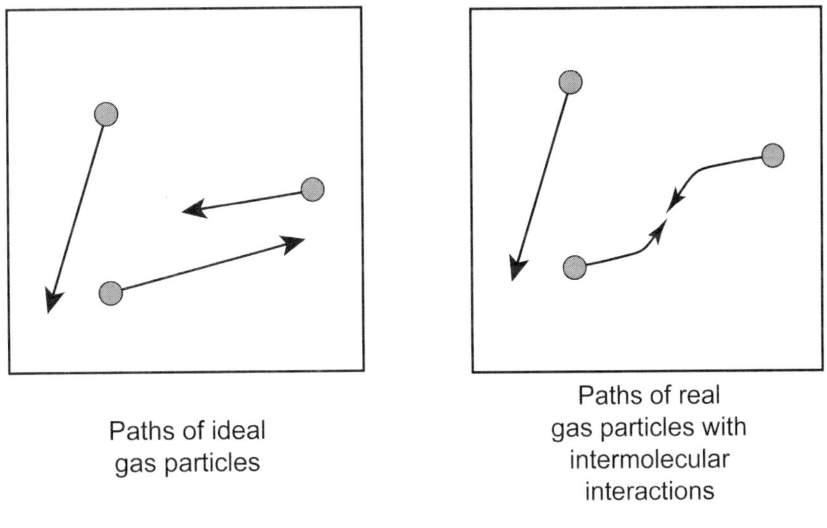

Figure 5.8 In the ideal gas model, molecules always follow straight-line trajectories, as shown in the left hand panel. In a real gas, though, intermolecular forces can distort these straight-line paths. The right hand panel shows the effect of attractive forces, which can bend the trajectories by causing two molecules to pull each other together.

Correcting the Ideal Gas Equation

When the postulates of gas kinetic theory are not valid, the observed gas will not obey the ideal gas equation. In many cases, including a variety of important engineering applications, gases need to be treated as non-ideal and empirical mathematical descriptions must be devised. There are many equations that may be used to describe the behavior of a real gas, but the most commonly used is probably the van der Waals equation.

$$\left(P + \frac{an^2}{V^2}\right)(V - nb) = nRT \tag{5.14}$$

An empirical equation contains one or more adjustable parameters that are found from a best fit to observed data rather than a theoretical model.

Here a and b are called the van der Waals constants. Unlike the universal gas constant (R), the values of these van der Waals constants must be set for each specific gas. If a gas behaves ideally, both a and b are zero and the van der Waals equation reverts to the ideal gas equation. Investigating the connection of this equation to the ideal gas equation further, we see that the term involving a accounts for the attraction between particles in the gas. The term involving b adjusts for the volume occupied by the gas particles. Table 5.2 provides the values of the van der Waals constants for some common gases. Notice that larger molecules have larger values for b, consistent with its role in accounting for the volume of the molecules themselves. The largest value for a is for HF. This means that compared to other gases, HF molecules interact with each other more and require larger corrections from ideal behavior. When we consider chemical bonding in the next few chapters we will begin to understand why this is true. Example Problem 5.8 demonstrates the use of the van der Waals equation.

Gas	a (atm L^2 mol^{-2})	b (L mol^{-1})
Ammonia, NH_3	4.170	0.03707
Argon, Ar	1.345	0.03219
Carbon dioxide, CO_2	3.592	0.04267
Helium, He	0.034	0.0237
Hydrogen, H_2	0.2444	0.02661
Hydrogen fluoride, HF	9.433	0.0739
Methane, CH_4	2.253	0.04278
Nitrogen, N_2	1.390	0.03913
Oxygen, O_2	1.360	0.03183
Sulfur dioxide, SO_2	6.714	0.05636
Water, H_2O	5.464	0.03049

Table 5.2 Van der Waals constants for several common gases.

Example Problem 5.8

In Example Problem 5.1 we considered a CH_4 storage tank with a volume of 49.0 L. When empty, the tank has a mass of 55.85 kg, and when filled, its mass is 62.07 kg. Calculate the pressure of CH_4 in the tank at an ambient temperature of 21°C using both the ideal gas equation and the van der Waals equation. What is the percentage correction achieved by using the more realistic van der Waals equation?

Strategy

We will use two different models to describe the same gas, and each model will be represented by its own equation. So we will do two independent calculations, solving for P in each case. We can find the number of moles of CH_4 from the given masses, and the van der Waals constant for CH_4 can be found in Table 5.2. To calculate the percentage difference we will divide the difference of the two numbers by the value for the ideal gas case.

Solution

$$\text{Mass of } CH_4 = \text{mass of full cylinder} - \text{mass of empty cylinder}$$

$$= 62.07 \text{ kg} - 55.85 \text{ kg} = 6.22 \text{ kg} = 6{,}220 \text{ g}$$

$$\text{Moles of gas} = 6{,}220 \text{ g } CH_4 \ x \ \frac{1 \text{ mol } CH_4}{16.04 \text{ g } CH_4} = 387.5 \text{ mol } CH_4$$

Ideal Gas:

$$P = \frac{nRT}{V} = \frac{(387.5 \text{ mol})(0.08206 \text{ L atm mol}^{-1} \text{ K}^{-1})(294 \text{ K})}{49.0 \text{ L}} = 191 \text{ atm}$$

van der Waals:

$a = 2.25 \text{ L}^2 \text{ atm mol}^{-2}$, $b = 0.0428 \text{ L mol}^{-1}$

$$P = \frac{nRT}{V - nb} - \frac{n^2a}{V^2}$$

$$= \frac{(387.5 \text{ mol})(0.08206 \text{ L atm mol}^{-1} \text{ K}^{-1})(294 \text{ K})}{49.0 \text{ L} - (387.5 \text{ mol})(0.0428 \text{ L mol}^{-1})} - \frac{(387.5 \text{ mol})^2(2.25 \text{ L}^2 \text{ atm mol}^{-2})}{(49.0 \text{ L})^2}$$

$$= 148 \text{ atm}$$

Percentage correction:

$$\frac{191 - 148}{191} \text{ x } 100\% = 22.5\%$$

Discussion

The correction here is quite significant. If you look at the calculated pressures, you should realize that they are quite high. Recall that the ideal gas law works best at low pressures and high temperatures. In this example the pressure is sufficiently high that the ideal gas model is a rather poor description of the actual behavior of the gas. Still, the simplicity of the ideal gas law makes it a good starting point for many calculations. In a practical application it would be crucial that an engineer recognized the type of conditions under which the assumption of ideal gas behavior might not be reasonable.

Check Your Understanding

Under less extreme conditions, the corrections for non-ideal behavior are smaller. Calculate the pressure of 0.500 mol of methane occupying 15.0 L at 25.7°C using both the ideal gas law and the van der Waals equation. Compare the percentage correction with that calculated in the example above.

5.7 Insight into Gas Sensors

The U.S. Environmental Protection Agency lists more than 180 substances as toxic air pollutants, and specifies acceptable levels that can be less than 1 ppm in some cases.

In order to track the effects of pollutants at such trace levels, we must first be able to measure their concentrations reliably. One way to describe the concentration of a gas, of course, is to report its pressure or partial pressure. And while a mercury barometer quite similar to one that might have been used hundreds of years ago will allow us to monitor the weather, it will not be of much help here. Instead we will need to rely on a number of more modern instruments that offer fundamental advantages over a barometer. A variety of devices now provide the ability to measure exceedingly small pressures or changes in pressure. And some of these can also give us the ability to selectively measure the partial pressure of a particular component in a gas mixture. In this section we will look at a couple of representative examples and try to show how their clever designs exploit the basic properties of gases to make incredibly sensitive measurements.

Capacitance Manometer

The term manometer refers to any pressure-measuring device.

One sensor frequently used to measure pressure in the laboratory is the capacitance manometer. This type of pressure gauge, shown schematically in Figure 5.9, directly exploits the fact that pressure is actually force per unit area.

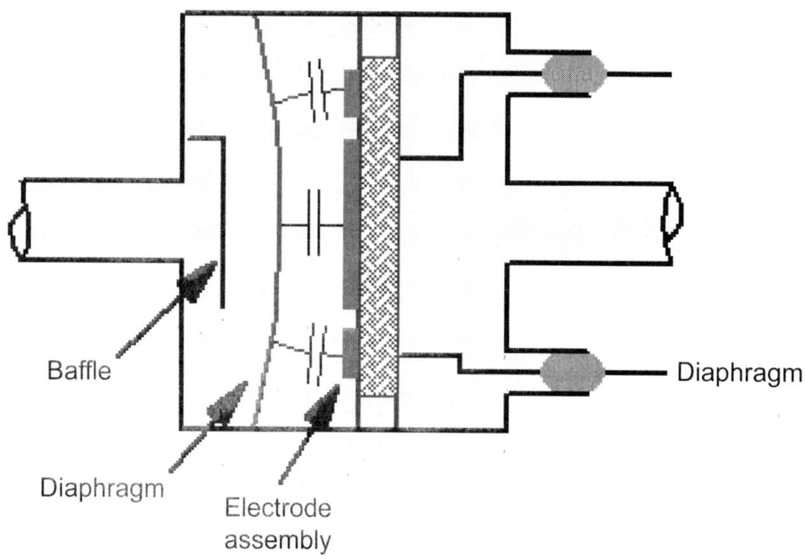

Figure 5.9 Schematic diagram of a capacitance manometer. The tube on the left is connected to the gas sample whose pressure is to be measured. As the gas molecules strike the diaphragm, the force of the collisions determines the space between the diaphragm and the fixed electrode assemblies.

A capacitor is an electronic component, as you may know from a physics class. One way to create a capacitor is to position two parallel metal plates with a space between them. In such an arrangement the distance between the plates determines the capacitance. A capacitance manometer is based on a capacitor in which one of these plates is actually a thin metal diaphragm and the other is fixed and rigid. The space between the plates is kept under vacuum. We connect the sensor so that the other side of the diaphragm is exposed to the gas whose pressure we want to measure. The pressure of the gas produces a force on the diaphragm, causing it to bend towards the second (fixed) plate of the capacitor. This, in turn, changes the capacitance of the device. If we construct an electric circuit in which the capacitor is one component, we can measure a voltage in that circuit that will be related to the pressure applied to the diaphragm. In order to make the sensor easy to use, that circuit can be designed in a way that we can measure a single voltage whose value is directly proportional to the pressure. Monitoring the voltage allows us to see any changes in pressure.

Capacitance manometers are most often used to measure pressures in the range of 0.001 – 1000 torr. This range includes moderate levels of vacuum frequently encountered in the laboratory. One advantage of this type of gauge is that it has virtually no moving parts; as long as the diaphragm is not damaged, the gauge should work reliably over very long periods of time.

Thermocouple Gauge

Both the barometer and the capacitance manometer respond fairly directly to the force exerted by gas molecules. But not all pressure sensors work that way. A useful pressure-measuring device can be based on any gas property that varies in a predictable way with pressure.

Pressures between 0.01 and about 1.0 torr, for example, are frequently measured using a thermocouple vacuum gauge. A thin filament is heated by passing a fixed electric current through it, and the temperature of the filament is then measured using a device called a thermocouple. So how does this measurement tell us anything about pressure? The filament is in contact with the gas whose pressure we want to determine. And as the molecules in that gas collide with the hot filament, they will tend to carry away some energy, reducing the filament temperature. The higher the pressure is, the more molecules will collide with the filament, and the lower its temperature will be. The thermocouple produces a voltage that is related to the filament temperature, and therefore to the gas pressure. A readout device connected to the gauge "translates" the voltage output into pressure units.

Ionization Gauge

Capacitance manometers can be effective down to pressures of about 0.001 torr. That is a very low pressure, just over a millionth of an atmosphere. But in certain types of experiments, laboratory vacuums as low as 10^{-11} torr are necessary. At such low pressures, it is no longer feasible to measure the force produced by molecular collisions, so other types of gauges are needed. The most common of these is the *ionization gauge*, shown in Figure 5.10.

Pressures on the order of 10^{-14} torr are found in outer space.

A fixed electric current flows through the filament of the ion gauge, heating it until it glows red. At these high temperatures, the metal atoms in the filament emit electrons. Some of these electrons then collide with gas molecules, and the resulting impact can knock an electron off of the molecule, producing a positive ion. Suppose some or all of the gas present is N_2. Electrons from the filament would then produce N_2^+ ions, through a process that we could write as

$$e^- + N_2 \rightarrow N_2^+ + 2e^-$$

The resulting positive ions are then collected by a grid. Measuring the current flowing into this grid will give us a measurement of how many ions were collected. And the number of ions formed will be proportional to the number of gas molecules in the region of the gauge: more gas will allow more ions to be formed. Finally, we just have to realize that the number of gas molecules present is proportional to the pressure. So the current, which is fairly easy to measure, will be proportional to pressure. In practice, the same electronics that read the current will also be calibrated so that they can display the reading directly as pressure.

Mass Spectrometer

Both the capacitance manometer and the ionization gauge measure the total pressure of a gas. But if we want to measure the level of a pollutant in air, for example, we will really want to know its partial pressure. One way to measure partial pressures is with a mass spectrometer. Like in the ion gauge, gas molecules are converted to positive ions by electron bombardment. But before the ions are collected, they are sent through a magnetic field that serves as mass filter, sorting the stream of ions by mass. In this arrangement, the current that is measured consists only of ions with a chosen mass. This lets us monitor the partial pressure of a gas by looking at the ions of the corresponding mass. Control electronics allow these instruments to continuously scan over a range of masses, and can thus tell us the partial pressure of a number of components in the same gas mixture.

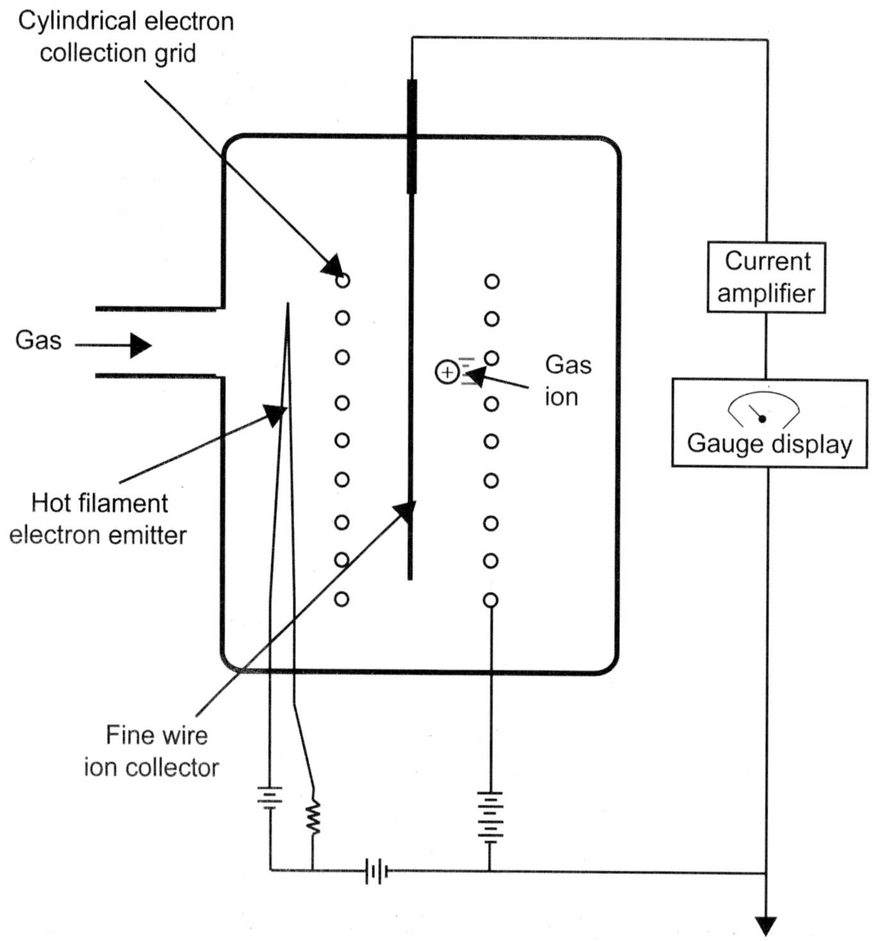

Figure 5.10 Schematic diagram of an ionization gauge. Electrons emitted from the hot filament collide with gas molecules and knock an electron free. The resulting cations are collected at the center of the gauge, and the current they produce is a measure of the gas pressure.

Focus on Problem Solving

Question

A balloon is to be used to float instruments for observation into the upper atmosphere. At sea level, this balloon occupies a volume of 95 L at a temperature of 20°C. If helium gas is used to fill the balloon, how would you determine the maximum height that this balloon and its instruments will attain?

Strategy

This is the type of question that the engineers who design an apparatus such as this must answer. The law of nature that is in play here is the fact that objects that are less dense float on those that are more dense. The balloon will float upward until it reaches a height where its density is equal to that of the atmosphere. With this understanding, we can begin to define the variables we need to understand in this problem, and note the relationships between those variables that would allow us to estimate an answer to such a problem.

Solution

First identify what is known about the variables in the problem. In this case we have a gas that will be at low pressures, so we can assume it will behave as an ideal gas. We note that we will release this balloon from sea level where the pressure can be estimated as 1 atm. We can establish, therefore, initial pressure, volume and temperature. We also know the molar mass of helium.

The key relationship is provided by the ideal gas law, which yields the ability to calculate density. To see this, think about n/V as the number of moles per unit volume. This is equal to the (mass) density divided by the molar mass. Replacing n/V with that relationship gives us a gas law in terms of density.

$$P = \frac{\rho RT}{MM}, \text{ or } \rho = \frac{MM \cdot p}{RT}$$

Looking at this set of equations, the importance of temperature and pressure is apparent.

Because the key condition is the point at which the density of the helium and the balloon is equal to the atmosphere we must look up and/or determine the density of the atmosphere as a function of altitude. The source of this information can also provide temperature and pressure of the atmosphere as a function of altitude.

Knowing T and P provides the density of helium at the various altitudes via the equation above. Having looked up the density of the atmosphere as a function of altitude we must simply find the altitude at which this calculation equals the looked up value to estimate the maximum height that can be achieved by the balloon. We must recognize, however, that the mass of the instruments must be considered if we wish an accurate answer to this type of problem.

Chapter 5
Problems & Exercises

Insight into Air Pollution

5.1 List two types of chemical compounds that must be present in the air for photochemical smog to form. What are the most common sources of these compounds?

5.2 When ozone levels in urban areas reach unhealthy levels, residents are typically urged to avoid refueling their cars during daylight hours. Explain how this might help to reduce smog formation.

5.3 In the production of urban air pollution shown in Figure 5.1, why does the concentration of NO decrease during the daylight hours?

5.4 VOC's are not criteria pollutants. Why are they monitored for the purpose of understanding urban air pollution?

5.5 Asphalt is composed of a mixture of organic chemicals. Does an asphalt parking lot contribute directly to the formation of photochemical smog? Explain your answer.

5.6 Do a web search to identify which (if any) of the EPA's criteria pollutants are prevalent in your city or state.

5.7 One observable property of gases is the variability of density based on conditions. Use this observation to explain why hot air balloons rise.

Pressure

5.8 If you are pounding a nail and miss it, the hammer may dent the wood, but it does not penetrate it. Use the concept of pressure to explain this observation.

5.9 How do gases exert atmospheric pressure?

5.10 Why do mountain climbers need to wear breathing apparatus at the tops of high mountains such as Denali in Alaska?

5.11 If you had a liquid whose density was half that of mercury, how tall would you need to build a barometer to be able to measure atmospheric pressure in a location where the record pressure recorded was 750 mmHg?

5.12 Water has a density that is 13.6 times less than that of mercury. How high would a column of water need to be to measure a pressure of 1 atm?

5.13 Water has a density that is 13.6 times less than that of mercury. If an undersea vessel descends to 1.5 km, what pressure does the water exert in atm?

5.14 Does the vacuum above the mercury in the column of a barometer affect the reading it gives? Why or why not?

5.15 If the pressure is given as "29.58 inches and rising," what is the pressure in atm?

5.16 If the atmospheric pressure is 97.4 kPa, what is it in mmHg? In atm?

5.17 Why do your ears "pop" on occasion when you swim deep underwater?

The Gas Law

5.18 When helium escapes from a balloon its volume decreases. Based on your intuition about stretching rubber, explain how this observation is consistent with the gas law.

5.19 A gas that occupies an initial volume of 2.5 L at a pressure of 4.1 kPa is compressed to a pressure of 53 kPa. What is the new volume?

5.20 A gas has an initial volume of 39 mL at an unknown pressure. If the same sample occupies 514 mL at 720 torr, what was the initial pressure?

5.21 When you buy mylar balloons in the winter months in colder places, the shopkeepers will often tell you not to worry about it loosing its shape when you take it home (outside), because it will return to shape once inside. What behavior of gases is responsible for this advice?

5.22 Why should temperature always be converted to kelvin when working with gases in problem solving?

5.23 What evidence gave rise to the establishment of the absolute temperature scale?

5.24 Calculate the missing variable in each set.

$$V_1=2.0L,\ T_1=15°C,\ V_2=?,\ T_2=34°C$$

$$V_1= ?,\ T_1=149°C,\ V_2=310\ mL,\ T_2=54°C$$

$$V_1=150L,\ T_1=180\ K,\ V_2=57\ L,\ T_2=?$$

5.25 Why is it dangerous to store compressed gas cylinders in places that could become very hot?

5.26 A balloon filled to its maximum capacity on a chilly night at a carnival pops without being touched when it is brought inside. Explain this event.

5.27 A gas bubble forms inside a vat containing a hot liquid. If the bubble is originally at 68°C and a pressure of 1.6 atm with a volume of 5.8 mL, what will its volume be if the pressure drops to 1.2 atm and the temperature drops to 31°C?

5.28 Use the fact that at STP the volume of one mole of an ideal gas is 22.4 L to verify the value of the universal gas constant.

5.29 If 3.5 moles of an ideal gas occupy 43.1 L when the pressure is 690 torr, what is the temperature of the gas?

5.30 How many moles of an ideal gas are there if the volume of the gas is 158 L at 14°C and a pressure of 89 kPa?

5.31 What is the molar mass of an ideal gas when a 5.6-g sample of it occupies 12.0 L at 25°C and a pressure of 0.95 atm?

5.32 Calculate the mass of each gas at STP. (a) 1.4 L of SO_2 (b) $3.5x10^5$ L of CO_2

5.33 What is the density of the following gases at STP? (a) CF_2Cl_2 (b) CO_2 (c) HCl

5.34 A cylinder containing 15.0 L of helium gas at a pressure of 165 atm is to be used to fill party balloons. Each balloon must be filled to a volume of 2.0 L at a pressure of 1.1 atm. What is the maximum number of balloons that can be inflated? Assume that the gas in the cylinder is at the same temperature as the inflated balloons. (HINT: The "empty" cylinder will still contain helium at 1.1 atm.)

5.35 A cylinder is filled with toxic COS gas to a pressure of 800 torr at 24°C. According to the manufacturer's specifications, the cylinder may rupture if the pressure exceeds 35 psi (pounds per square inch; 1 atm = 14.7 psi). What is the maximum temperature to which the cylinder could be heated without exceeding this pressure rating?

5.36 Cylinders of compressed gases are often labeled to show how many "SCF" or "standard cubic feet" of gas they contain. 1 SCF of gas occupies a volume of 1 ft^3 at a standard temperature and pressure of 0°C and 1 atm. A particular cylinder weighs 122 lbs when empty, and 155 lbs when filled with krypton gas at 26°C. How many SCF of Kr does this cylinder contain?

Partial Pressure

5.37 Define the term partial pressure.

5.38 Define the term mole fraction.

5.39 How does the mole fraction relate to the partial pressure?

5.40 What is the total pressure of a 4.50 L container at 15.5°C if it contains, 0.45 mol of Ar, 0.60 mol of Ne and 1.2 mol of He?

5.41 What is the total pressure of a 15.0 L container at 28.0°C if it contains 3.5 g N_2, 4.5 g O_2 and 13.0 g Cl_2?

5.42 For a gas sample whose total pressure is 740 torr, what are the partial pressures if the moles of gas present are 1.3 mole N_2, 0.33 mol O_2, 0.061 mol Ar.

5.43 A sample containing only NO_2 and SO_2 has a total pressure of 120 torr. Measurements show the partial pressure of NO_2 to be 43 torr. If the vessel has a volume of 800 mL and the temperature is 22.0°C, how many moles of each gas are present?

5.44 A sample of gas is made entirely of carbon dioxide and water, and there are 259 moles of CO_2 and 513 moles of water. If the total pressure of the sample is 21 atm, what is the partial pressure of each gas?

5.45 A sample containing nitrogen and oxygen at 23°C with a total pressure of 1.3 atm in a container with a volume of 820 mL has a partial pressure of 0.82 atm for nitrogen. How many moles of each gas are present.

5.46 Air is often modeled as "dry air," ignoring the water present. If the atmospheric pressure outside is 0.944 atm and the pressure associated with water is 0.039 atm, what are the partial pressures of nitrogen and oxygen – assuming the composition given in Table 5.1

5.47 In an experiment, a mixture of gases occupies a volume of 3.0 L at a temperature of 22.5°C. The mixture has 14.0 g of water, 11.5 g of oxygen and 37.3 g of nitrogen. Calculate the total pressure and the partial pressure of each gas.

5.48 An experiment is being carried out to test the removal of sulfur dioxide from gases being released by a power plant. The initial sample, which contains only carbon dioxide and sulfur dioxide, occupies a volume of 35 L at a temperature of 41°C and a pressure of 715 torr. After all of the SO_2 has been removed, the sample occupies a volume of 23.5 L at a temperature of 29°C and a pressure of 715 torr. Determine the partial pressures of both CO_2 and SO_2 in the initial sample.

5.49 Use the web to determine the range of partial pressures that oxygen sensors in the exhaust manifold of automobile engines must measure. Why is the amount of oxygen in an exhaust stream important for an engineer to know?

5.50 Use the web to research the types of fuel to air mixtures present in regular consumer driven automobiles. Calculate partial pressures of the gases present based on the information you find.

Stoichiometry with Gases

5.51 HCl(g) reacts with ammonia gas, NH_3(g), to form solid ammonium chloride. If a sample of ammonia occupying 250 mL at 21°C and a pressure of 140 torr is allowed to react with excess HCl, what mass of NH_4Cl will form?

5.52 Hydrogen gas is generated when acids come into contact with certain metals. When excess hydrochloric acid reacts with 2.5 g of Zn (the metal product is Zn^{2+}), what volume of hydrogen gas is collected at a pressure of 0.93 atm and a temperature of 22°C?

5.53 If you need 400 mL of hydrogen gas to conduct an experiment at 20.5°C and 748 torr how many grams of Zn should be reacted with excess HCl to obtain this much gas?

5.54 The first step in processing zinc metal from it's ore, ZnS, is to react it with O_2 according to the reaction,

$$2ZnS(s) + 3O_2(g) \rightarrow 2\,ZnO(s) + 2\,SO_2(g)$$

If 620 kg of ZnS is to be reacted, what volume of oxygen at 0.977 atm and 34.0°C is needed (at a minimum) to carry out this reaction?

5.55 What volume of oxygen at 24°C and 0.88 atm is needed to completely react via combustion with 45 g of methane gas?

5.56 N_2O_5 is an unstable gas that decomposes according to the reaction,

$$2N_2O_5(g) \rightarrow 4NO_2(g) + O_2(g)$$

what mass of N_2O_5 is present initially, if 7.52 g of NO2 are produced by this reaction?

5.57 N_2O_5 is an unstable gas that decomposes according to the reaction,

$$2N_2O_5(g) \rightarrow 4NO_2(g) + O_2(g)$$

what would be the total pressure of gases present if a 10.0 L container at 22.0°C begins with 0.40 atm of N_2O_5 and the gas completely decomposes?

5.58 One way to generate oxygen is to heat potassium chlorate, $KClO_3$ (the other product is potassium chloride). If 386 mL of oxygen at 41°C and 97.8 kPa is generated by this reaction, what is the minimum mass of $KClO_3$ used?

5.59 Ammonia is not the only possible fertilizer, sometimes it is reacted further to form urea, by the reaction, $CO_2(g) + 2\,NH_3(g) \rightarrow CO(NH_2)_2(s) + H_2O(g)$. A scientist has 75 g of dry ice to provide the carbon dioxide. If 4.50 L of ammonia at 15°C and a pressure of 1.4 atm is added, which reactant is limiting? What mass of urea will form?

5.60 Consider the following reaction.

$$6\,NiO(s) + 4\,ClF_3(g) \rightarrow 6\,NiF_2(s) + 2\,Cl_2(g) + 3\,O_2(g)$$

What mass of NiO will react with a sample of ClF_3 gas that has a pressure of 250 torr in a 2.5 L flask at 20°C?

5.61 Hydrogen peroxide breaks down to form oxygen gas and water.

$$2H_2O_2(aq) \rightarrow 2\ H_2O(l) + O_2(g)$$

This exothermic reaction can be made to occur rapidly by the addition of a catalyst. In a demonstration, 4.3 L of oxygen gas is generated at 84°C and a pressure of 722 torr. What mass of peroxide must have decomposed?

5.62 Automakers are always investigating alternative reactions for the generation of gas to inflate air bags. One system that has been considered is the oxidation of graphite by strontium nitrate ($Sr(NO_3)_2$), as shown in the equation below.

$$5\ C(s) + 2\ Sr(NO_3)_2(s) \rightarrow 2\ SrO(s) + 2\ N_2(g) + 5\ CO_2(g)$$

Suppose a system is being designed using this reaction, and the goal is to generate enough *gas* to inflate a bag to 60 L and 1.3 atm at 20°C. What is the minimum mass of graphite that should be used in the design of this system?

5.63 Nickel metal reacts with carbon monoxide to form a compound called nickel tetracarbonyl ($Ni(CO)_4$), which is a gas at temperatures above about 316 K. A 2-L flask is filled with CO gas to a pressure of 750 torr at 350 K, and then 5.00 g of Ni is added. If the reaction described above occurs and goes to completion at constant temperature, what will the final pressure in the flask be?

5.64 Ammonium dinitramide (ADN), $NH_4N(NO_2)_2$, is being considered as a possible replacement for aluminum chloride as the oxidizer in the solid fuel booster rockets used to launch the space shuttle. When detonated by a spark, ADN rapidly decomposes to produce a gaseous mixture of N_2, O_2, and H_2O. (This is NOT a combustion reaction. The ADN is the only reactant.) The reaction releases a lot of heat, so the gases are initially formed at high temperature and pressure. The thrust of the rocket results mainly from the expansion of this gas mixture. Suppose a 2.3-kg sample of ADN is detonated and decomposes completely to give N_2, O_2, and H_2O. If the resulting gas mixture expands until it reaches a temperature of 100°C and a pressure of 1.0 atm, what volume will it occupy? Is your answer consistent with the proposed use of ADN as a rocket fuel?

5.65 Clouds of hydrogen molecules have been detected deep in interstellar space. It is estimated that these clouds contain about 1×10^{10} hydrogen molecules per m^3, and have a temperature of just 25 K. Using these data, find the approximate pressure in such a cloud.

Gas Kinetic Theory and Real Gases

5.66 State the postulates of gas kinetic theory.

5.67 Under what conditions do the postulates of gas kinetic theory break down?

5.68 Which postulates of gas kinetic theory are the most likely to break down under the conditions of high pressure present inside volcanoes?

5.69 What happens to the distribution of speeds for molecules in a gas as it is cooled down?

5.70 Use gas kinetic theory to explain what happens to the pressure exerted by a gas as its temperature is increased.

5.71 The figure below shows the distribution of speeds for two samples of N_2 gas. One sample is at 300 K, and the other is at 1000 K. Which is which?

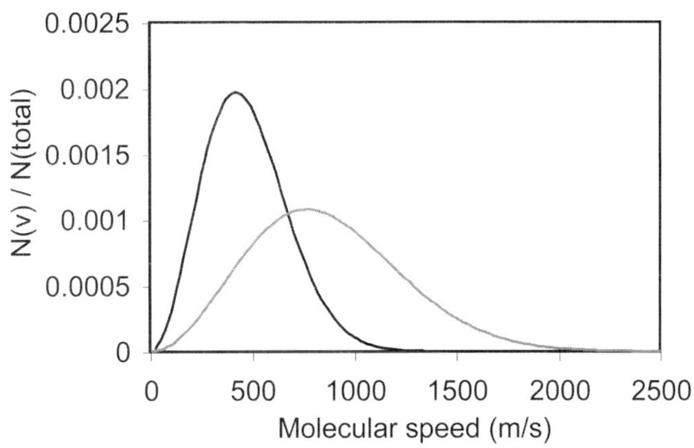

5.72 Why do heavier gases move more slowly that light gases at the same temperature?

5.73 Define the term *mean free path*.

5.74 There are two parameters in the van der Waals equation for real gases. Which parameter corrects for the volume of gas molecules? Use the equation to explain your answer.

5.75 Calculate the pressure of 15.0 g of methane gas in a 1.50 L vessel at 45.0°C using (a) the ideal gas law and (b) the van der Waals equation using constants in Table 5.2

5.76 Calculate the pressure of 11.4 g of sulfur dioxide in a 24 mL vessel at 25.0°C using (a) the ideal gas law and (b) the van der Waals equation using constants in Table 5.2

5.77 To what volume would 15.0 mol of an ideal gas at 500 K need to be compressed for the correction in pressure (relative to the ideal gas law) to be 10%?

Pressure Sensors

5.78 Scientists routinely use several different types of pressure gauges, and sometimes a single vacuum chamber may be equipped with more than one type of gauge. Why do you suppose this is?

5.79 What is actually measured in an ionization gauge pressure sensor. How is this actual measurement related to pressure?

5.80 Use the web to find out what type of sensor is used to determine the oxygen pressure in the exhaust manifold of fuel injection system engines for cars.

5.81 Do a web search to learn about at least one other type of pressure sensor not described in the chapter. Describe how the gauge operates in a way that your roommate could understand.

Putting it All Together

5.82 Consider two identical glass bulbs, each having a volume of 1.00 L. The bulbs are evacuated, and then one is filled with oxygen and the other with neon. When filled, each bulb is at a pressure of 500 torr and a temperature of 50°C.

Assume that the gases behave ideally. (Note that you should be able to answer these without doing any significant calculations.)

(a). Which bulb contains more moles of gas?

(b). In which bulb is the density of the gas higher?

(c). In which bulb is the average kinetic energy of the molecules lower

(d). In which bulb is the average molecular speed greater?

(e). If one of these bulbs were opened in this room, would the net flow of gas be into or out of the bulb?

5.83 A gas mixture contains 10.0% CH_4 and 90.0% Ar *by moles*. Find the density of this gas at 250°C and 2.5 atm.

5.84 Liquid oxygen for use as a rocket fuel can be produced by cooling dry air to -183°C, where the O_2 condenses. How many liters of dry air at 25°C and 750 torr would need to be processed to produce 150 L of *liquid* O_2 at -183°C? (The *mole fraction* of oxygen in dry air is 0.21, and the density of liquid oxygen is 1.14 g/mL.)

5.85 A number of compounds containing the heavier noble gases, and especially xenon, have been prepared. One of these is xenon hexafluoride (XeF_6), which can be prepared by heating a mixture of xenon and fluorine gases. XeF_6 is a white crystalline solid at room temperature, and melts at about 325 K.

0.0600 g of Xe and 0.0304 g of F_2 are sealed into a 100-mL bulb. (The bulb contains no air or other gases.) The bulb is heated, and the reaction above goes to completion. Then the sealed bulb is cooled back to 20°C. What will be the final pressure in the bulb, expressed in Torr?

5.86 You work in a semiconductor production plant that relies on several chlorofluorocarbons in its manufacturing process. One day you find an unlabelled gas cylinder, and you are assigned to figure out what is in the tank. First you fill a 1.0 L-flask with the gas. At a pressure of 250 torr and a temperature of 25°C, you determine that the mass of the gas in the flask is 2.2980 g. Then you send the flask to an outside lab for elemental analysis, and they report that the gas contains 14.05% C, 44.46% F, and 41.48% Cl by mass. Find the molecular formula of this gas.

5.87 When a 0.817 g sample of a copper oxide is heated with excess hydrogen gas, a reaction takes place and 0.187 g of water is formed. What is the apparent formula of the copper oxide?

5.88 A mixture of Ar (0.40 mol), O_2 (0.50 mol) and CH_4 (0.30 mol) exerts a pressure of 740 mmHg. If the methane and oxygen are ignited and complete combustion occurs, what is the final pressure of Ar, CO_2, H_2O and the remainder of the excess reactant? What is the total pressure of the system?

5.89 The complete combustion of octane can be used as a model for the burning of gasoline.

$$2\ C_8H_{18} + 25\ O_2 \rightarrow 16\ CO_2 + 18\ H_2O$$

Assuming that this equation provides a reasonable model of the actual combustion process, what volume of air at 1 atm and 25°C must be taken into an engine to burn 1 gallon of gasoline? (The partial pressure of oxygen in air is 0.21 atm, and the density of liquid octane is 0.70 g/mL.)

Focus on Problem Solving Exercises

5.90 Before the introduction of unleaded gasoline, in severe pollution episodes the amount of lead in air could be measured at levels of 2.85 μg/m^3. Describe how you could determine the number of lead atoms in the active space of your lungs (the air that is actively exchanged during a breath) if the volume of that space is 120 cm^3. What, if anything, would need to be looked up?

5.91 A soft drink can has a label that indicates that the volume of the soda it contains is 12 oz or 355 mL. There is probably some empty space at the top of the can. Describe what you can measure and how that measurement allows you to determine the actual density of the soda.

5.92 The best way to generate small amounts of oxygen gas in a laboratory setting is by the decomposition of potassium permanganate, $2KMnO_4(s) \rightarrow K_2MnO_4 + MnO_2 + O_2$. Suppose you are carrying out an experiment for which you need exactly 2.00 L of oxygen. Describe how you would determine the mass of potassium permanganate you need to obtain this amount of oxygen. What would you have to measure in your experiment in addition to this mass?

5.93 An ore sample with a mass of 670 kg contains 27.7% magnesium carbonate, $MgCO_3$. If all of the magnesium carbonate in this ore sample is decomposed to form carbon dioxide, describe how to determine what volume of CO_2 is evolved during the process. What would have to be measured to determine be able to predict the needed volume in advance?

5.94 Suppose you are going to collect H_2 gas produced from the reaction of an acid with a metal like zinc. This reaction gives off heat, so the temperature of the solution will increase as the reaction proceeds. You need to know the partial pressure of the H_2 you generate as precisely as possible. What will you need to account for in your calculation? What things would you need to measure to make accurate calculations? What information would you need to look up?

Chapter 6: The Periodic Table and Atomic Structure

From the fuels that power our planes and cars to the generation and control of air pollution, chemical reactions are central to our existence. What drives these reactions that are so much a part of who we are and what we do? How can we predict what reactions are likely to happen? To answer questions such as these we will need an understanding of what makes a chemical bond. And to understand the bond between atoms we must first refine our descriptions of the atoms themselves. Fortunately, there are tools that allow us to observe atoms and other tools to help us organize what we have observed and learned. No tool is more important to building our understanding of chemistry than the periodic table, and its structure is closely related to the nature of atoms. Our understanding of atomic structure, in turn, has influenced the development of countless modern technologies. Nowhere is this influence more important and direct than in the lighting industry.

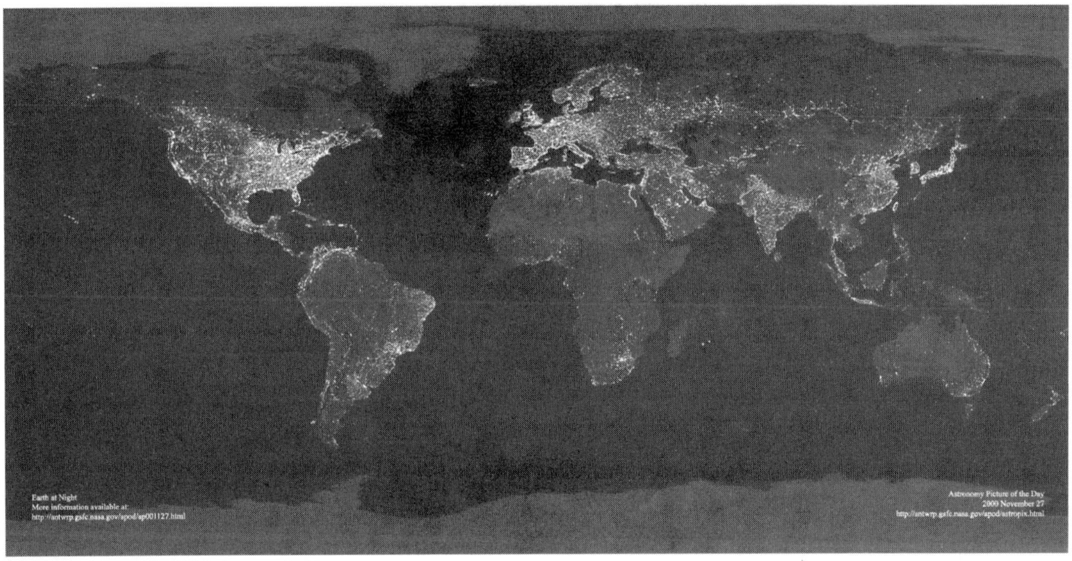

Figure 6.1 This image shows the Earth at night, as seen from orbiting satellites. The concentration of lights in industrialized nations, and even in specific cities, is very easy to see. The image is actually a composite of a large number of individual satellite photos. (Image is from http://antwrp.gsfc.nasa.gov/apod/ap001127.html)

6.1 Insight into Incandescent and Fluorescent Lights

An increasing ability to light the darkness has coincided with the development and advancement of human civilization. Prehistoric campfires gave way to torchlight, which was replaced by candlelight. Modern civilization, of course, relies on various types of electric lights: few visit Las Vegas or Times Square without being awed by the lights.

As we set out to explore how those lights operate, we'll choose a rather unlikely point of departure: a blacksmith or metalworker heating a piece of iron in a furnace. You may have seen someone playing the part of a 19th century blacksmith in an open-air museum. As the blacksmith works a piece of iron, he begins with a cold, dull gray piece of metal. After a short time in the hot fire, the iron slowly becomes a dull red color. After further heating, the metal shines bright orange, and finally it appears almost white. Why does that metal change from dull gray to orange in color as it is heated? This question may seem to have little to do with a light bulb until we consider the material that makes up the bulb's filament. It too is a metal, usually tungsten. Thus the heating of iron in a blacksmith's furnace provides some clues about the technology behind some types of light bulbs.

The filament in a typical 60-watt bulb is about 0.01 inch thick, and would be about 6.5 feet long if uncoiled. When the bulb is lit, the filament temperature reaches about 4,000°F.

Every light bulb operates on the same principle. Electricity is supplied to "excite" a substance, and that excitation somehow gives rise to light. What does it mean to "excite" a substance? This is one of the questions that we will discuss in this chapter, and the answer will lead us into a consideration of the nature of the atom itself.

Just what happens in the excitation process depends on the type of light bulb. The filament in a standard incandescent light bulb is a length of very thin tungsten wire. When an electric current is passed through the filament, the resistance of the wire causes some energy to be dissipated as heat. The tungsten is heated until it is "white hot" and it emits visible light. (Figure 6.2) Because the tungsten wire is so thin, it heats up to a high temperature quickly. But it is a metal that is behaving very similarly to the iron of the blacksmith at the museum. The specific color of light emitted by the metals is related to their temperature. The initial dull red color of the iron being heated suggests that while warm, the piece is not yet especially hot. The move towards orange as the piece remains in the fire, or white light for the tungsten filament suggests higher temperatures. With the two examples we have noted so far, the range of colors is not that wide, but the progression of color is similar to the colors of the rainbow that begin with red and move towards orange. Is there anything more than coincidence associated with this observation? We will investigate the nature of electromagnetic radiation to gain insight into this question.

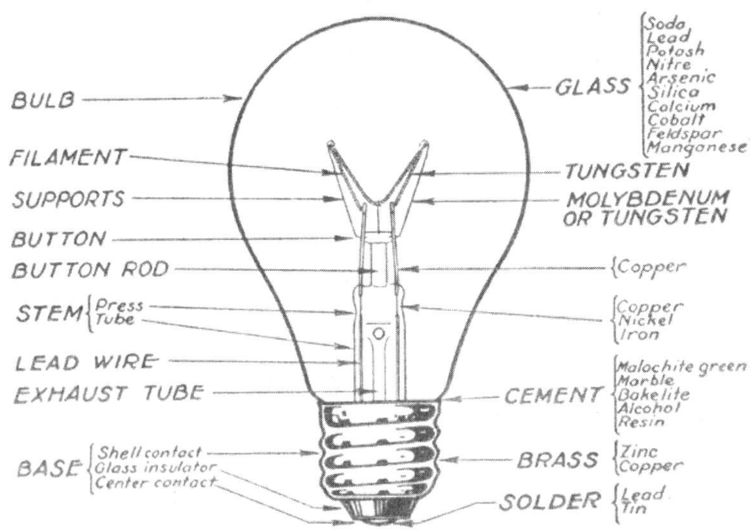

Figure 6.2 The diagram shows important components of a standard incandescent light bulb, and lists the chemical substances from which many of the components are made.

Fluorescent lights operate in a different way. Instead of heating a metal, electricity travels through a gas, as an arc. An arc occurs whenever electricity passes through material that does not conduct a current very well. The discharge of static electricity that sometimes occurs when you touch a light switch provides a painful, personal example of an electric arc. Figure 6.3 provides an illustration of the processes in a fluorescent light bulb. In such a bulb, specific atoms such as mercury interact with the electricity and become "excited," meaning they are in a state of higher than normal energy. In order to return to their normal state, these excited atoms emit light. That light interacts with a material called the phosphor, which coats the inside surface of the fluorescent tube. This phosphor then emits light of many colors, giving rise to what we perceive as white light coming from the bulb. What happens when atoms get excited or relax? Why does the bulb need to use the phosphor? Could elements other than mercury be used? All of these questions can be answered as we establish some of the foundation upon which to build a deeper understanding of chemical bonding. We proceed therefore, to learn how we view atoms in modern chemistry and physics, and how that understanding provides the insight needed to light the world.

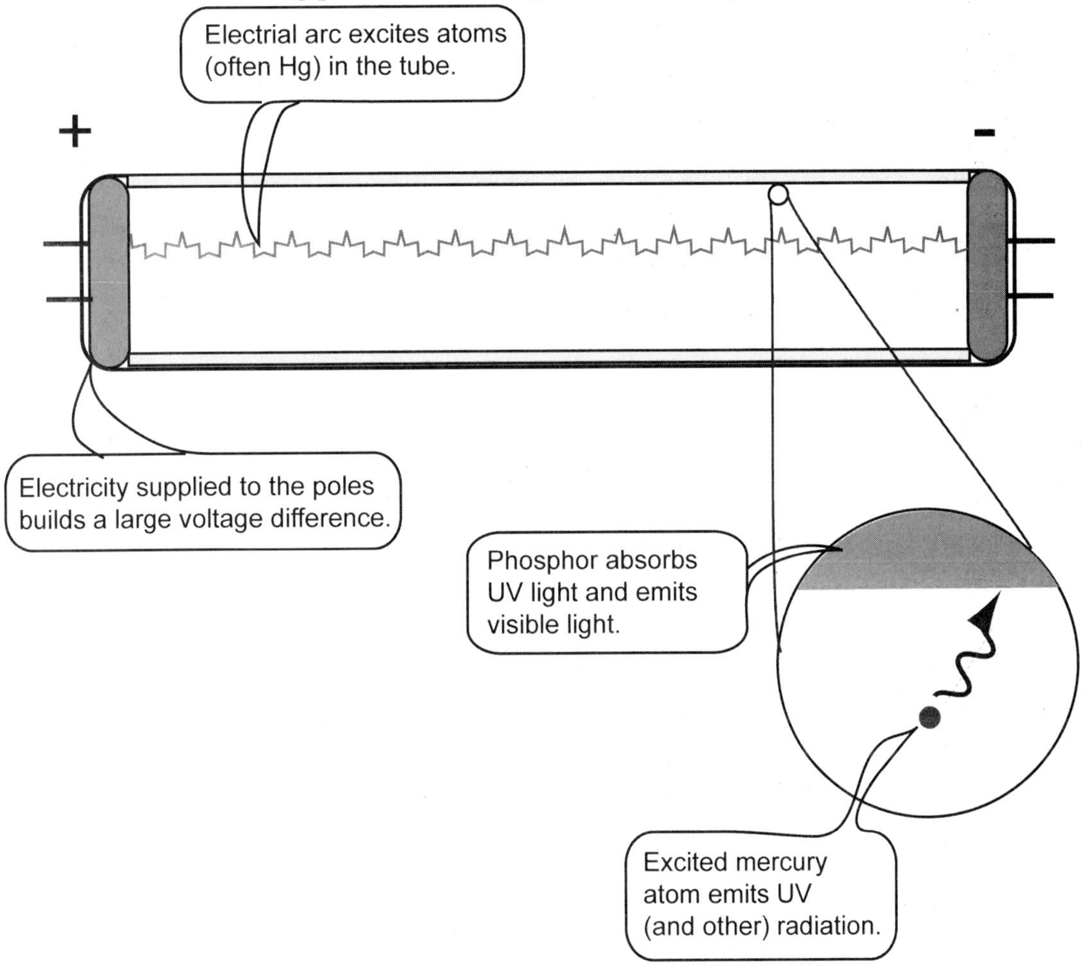

Figure 6.3 The diagram shows the important components and processes in a standard fluorescent light.

6.2 The Electromagnetic Spectrum

The engineering of a light bulb was a landmark achievement of the industrial revolution. What were the guiding principles that had to be observed when Thomas Edison and those like him were working to provide a way to generate light from electricity? Perhaps the first principle they needed was an understanding of what light, and particularly visible light, really is.

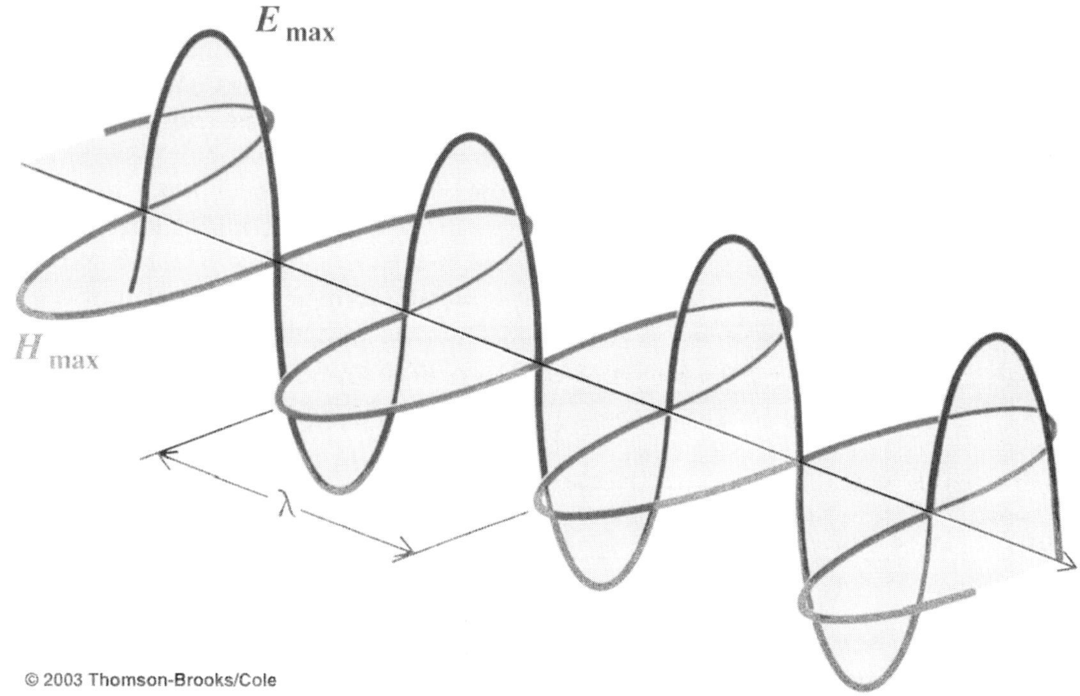

Figure 6.4 Light is one form of electromagnetic radiation, which consists of oscillating electric and magnetic fields.

The light that reacts with molecules in the cells of our retinas is what we call *visible light*. It comprises only a small portion of the possible *electromagnetic spectrum*, and accounts for only a small part of the emission of most light bulbs. Other familiar forms of electromagnetic radiation include radio waves, microwaves, and x-rays. The origin of the word "electromagnetic" lies in the nature of light. Historically, light has been described as a wave traveling through space, with one component being an electric field and another a magnetic field (Figure 6.4). To understand the wave nature of light, then, we should first be familiar with the general features of waves.

The Wave Nature of Light

Many of the features of light that we encounter can be explained as properties of waves. The central characteristics of a wave can be defined by four variables: wavelength, frequency, velocity and amplitude. Figure 6.5 illustrates the definition of these terms. The **wavelength** of any wave is the distance between corresponding points on adjacent waves. In the illustration, for example, the wavelength (designated as λ, and measured in units of length) is defined as the distance between "peaks." We could also define the wavelength as the distance between "valleys," and the value would be the same. The **amplitude** is the size or "height" of the wave. The lower wave in Figure 6.5 has larger amplitude than the upper wave. The **frequency** is the number of complete cycles of the wave passing a given point per second. Frequency is usually designated as ν and measured in units of 1/s, or hertz (Hz). Frequency and wavelength are not independent of one another. The speed of light in a vacuum (designated c and measured in units of distance/time) is a constant of nature that has been measured with notable precision.

$$c = 2.997925 \times 10^8 \text{ m s}^{-1}$$

Because the speed of light is a constant, the number of waves that will pass a certain point in time will be inversely related to the wavelength. Long wavelengths will have fewer cycles "pass by" in one second, so they will have lower frequencies (Figure 6.5). This relationship is expressed mathematically in Equation 6.1.

$$c = \lambda \times \nu \qquad (6.1)$$

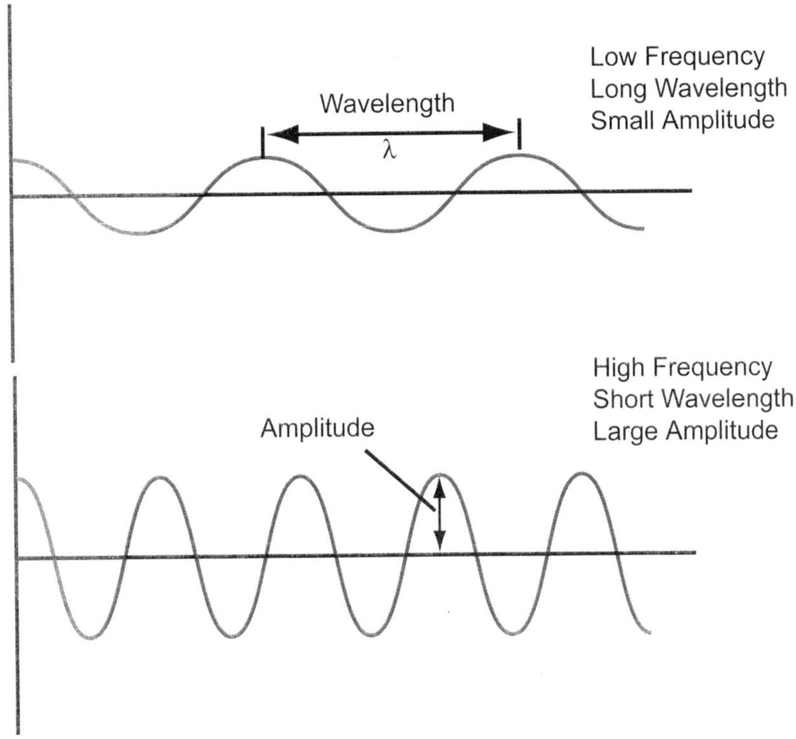

Figure 6.5 Two sine waves are shown, and the meaning of some important terms are illustrated. Because light has a fixed speed, its frequency and wavelength are always inversely related. Amplitude is independent of frequency; the choice of which of these waves would have a larger amplitude was made arbitrarily.

MathConnections

If we want to describe the detailed shape of a wave, we need to write some sort of mathematical equation for it. Let's consider a very simple example, in which the wave happens to follow the shape of a sine function. (This turns out to be true for many naturally occurring waves.) Imagine that we take a "snapshot" of the wave at a particular instant, and picture it as an x-y graph. The wave originates at $x = 0$ and travels along the x-axis. The y-values represent the height of the wave at each point. We could write a very general equation for this wave.

$$y = A \sin (bx)$$

A is fairly clearly the amplitude of the wave, but what about b? Let's assume that we want to evaluate the sine function in terms of radians. That means that the quantity bx must have no units. Since x is the distance from the origin, it would have units of length. So the b in our equation must have units of 1 / length. If we know the wavelength of the wave - which we could read off the graph - then we can also fix the value of b. The "wavelength" of the $sin(x)$ function is 2π. So suppose our wave has a wavelength of 450 nm, in the range of blue light. For our equation to match the wave, we require that

$$b \times 450 \text{ nm} = 2\pi, \text{ or } b = 2\pi / 450 \text{ nm}$$

More generally, we can see that the b in our equation must always be $2\pi / \lambda$.

If we wanted to describe a more complicated wave, we might have to use a combination of several sine (or cosine) functions, each of the same general form as we see here but with different wavelengths. Some of these component sine functions might be shifted so that their origin would not be at $x = 0$. By combining enough such functions, we could describe a waveform of arbitrary complexity. If we wanted to describe a wave as a function of time rather than position, we could proceed in a similar fashion.

Mathematically, the intensity of a light wave is actually proportional to the square of its amplitude.

Speed, wavelength, frequency, and amplitude are characteristics that can describe any wave. But since we are interested in light, it should help to compare these quantities with some familiar properties of light. If someone asks you to compare two different lights, one of the things you would probably consider is how bright each one appears. How does this relate to the wave parameters we've defined? In the wave model for light, the amplitude of the wave determines the brightness of the light: the larger the amplitude of the wave, the brighter the light will appear. Another property that you would surely consider in comparing two lights is color. And in the wave model, wavelength and frequency both correspond to the color of the light. How can these two seemingly different properties of the wave both correspond to color? The answer can be seen in Equation 6.1. Since the speed of light must remain constant, specifying *either* the frequency or the wavelength automatically fixes the value of *both* of those properties of a light wave. So we could say that the frequency and the wavelength are really just two different ways of expressing the same information.

The wave nature of light also helps to explain many of the phenomena we experience with light. *Refraction* is the bending of light when it moves between air and other more dense media such as water or glass. One common manifestation of refraction is the rainbow, in which white light is separated into its colors. In water or glass the speed of light changes and the light bends with an angle that depends on its wavelength. Thus the visible spectrum is separated into its colors, and the colors are dispersed in order of their corresponding wavelengths. Figure 6.6 illustrates this phenomenon.

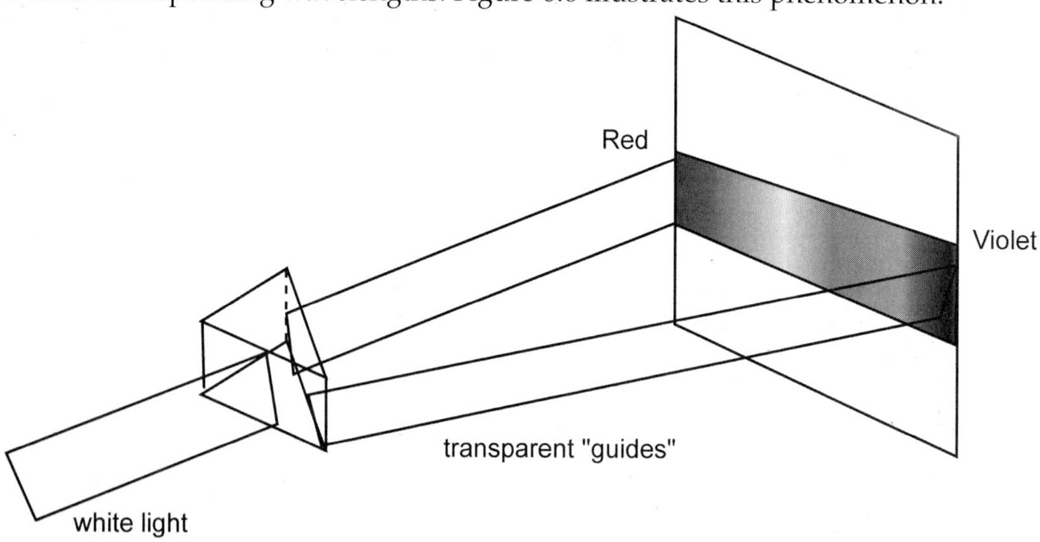

Figure 6.6 When white light passes through a prism, it is separated into its constituent colors by refraction.

Because of phenomena like refraction, it is common to categorize electromagnetic radiation in terms of its wavelength or frequency. Figure 6.7 shows the entire spectrum including information about both frequency and wavelength. In addition to visible light, the spectrum includes x-ray, ultraviolet(UV), infrared(IR), microwave and radio wave radiation, with visible falling between UV and IR. This order lists the regions of the spectrum in increasing wavelength. We will often find that we can measure or are provided with the wavelength of light, but need to express it as a frequency. Example 6.1 shows this type of conversion.

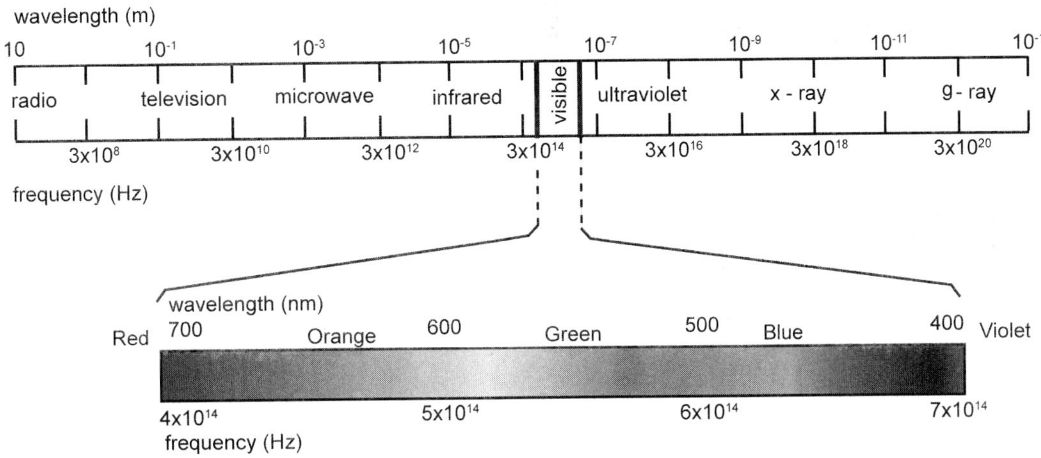

Figure 6.7 Visible light is just the small portion of the entire electromagnetic spectrum that our eyes can detect. Other regions of the spectrum are utilized in many technological applications, including remote controls, cell phones, and wireless networks.

Example Problem 6.1

Neon lights emit an orange-red colored glow. This light has a wavelength of 670 nm. What is the frequency of this light?

Strategy

We know that wavelength and frequency are related through Equation 6.1. In using this equation, we will need to be careful of units. Since the wavelength is given in nanometers, we will convert it to meters so that it is consistent with the meters per second in the speed of light.

Solution

$$670\ \text{nm} \times \frac{1\ \text{m}}{1 \times 10^{9}\ \text{nm}} = 6.7 \times 10^{-7}\ \text{m}$$

$c = \lambda \nu$ so $\nu = \dfrac{c}{\lambda}$

$$\nu = \frac{2.998 \times 10^{8}\ \text{m} \cdot \text{s}^{-1}}{6.7 \times 10^{-7}\ \text{m}} = 4.4 \times 10^{14}\ \text{s}^{-1}$$

Check Your Understanding

When trace amounts of xenon gas are added to neon lights, the color becomes blue-green. If the wavelength of this light is 480 nm, what is the frequency?

We generally think of a lamp as something emitting visible light. But for some specific applications, bulbs are designed to produce radiation in different regions of the electromagnetic spectrum. Heat lamps in fast food restaurants, for example, emit infrared radiation to keep french fries warm. Compared with ordinary incandescent bulbs, heat lamps are designed to emit a range of wavelengths shifted towards the IR region. These bulbs appear red to our eyes, because that is the visible color that is adjacent to the infrared. Moving to the opposite end of the visible spectrum, lamps designed to emit ultraviolet radiation are used both as "black lights," causing various dyes to glow, and in tanning beds. Our eyes do not detect ultraviolet light, but these lamps appear purple since the range of wavelengths they emit overlaps with the violet end of the visible spectrum. Microwave ovens, radios, televisions, and wireless telephones or computer networks all function by exploiting other specific regions of the electromagnetic spectrum.

The Particulate Nature of Light

The wave model of electromagnetic radiation is capable of explaining many observed properties of light. In the early 1900's, however, scientists acquired more sophisticated apparatus, and experiments were carried out that challenged the wave model of light. One pivotal group of experiments involved the observation of what came to be known as the photoelectric effect. Figure 6.8 is a schematic diagram of a typical photoelectric effect experiment. The heart of the experiment is light shining on a piece of metal. If the metal is enclosed in a container under vacuum so that there are very few surrounding gas molecules to interfere, then we can detect electrons being emitted from the metal under certain conditions. (The metal itself is grounded so that it does not develop a large positive charge during this process.) Since electrons carry an electric charge, it is fairly easy to measure how many electrons are ejected and even how fast they are going. Since every electron has a fixed charge, the number of electrons ejected can be determined by measuring an electric current. And the kinetic energy of the electrons can be found by applying a variable voltage to the metal plate that collects the electrons. The experimenter could control the color or intensity of the light and see how that changes the emission of electrons from the metal target. Or the identity of the metal could be changed.

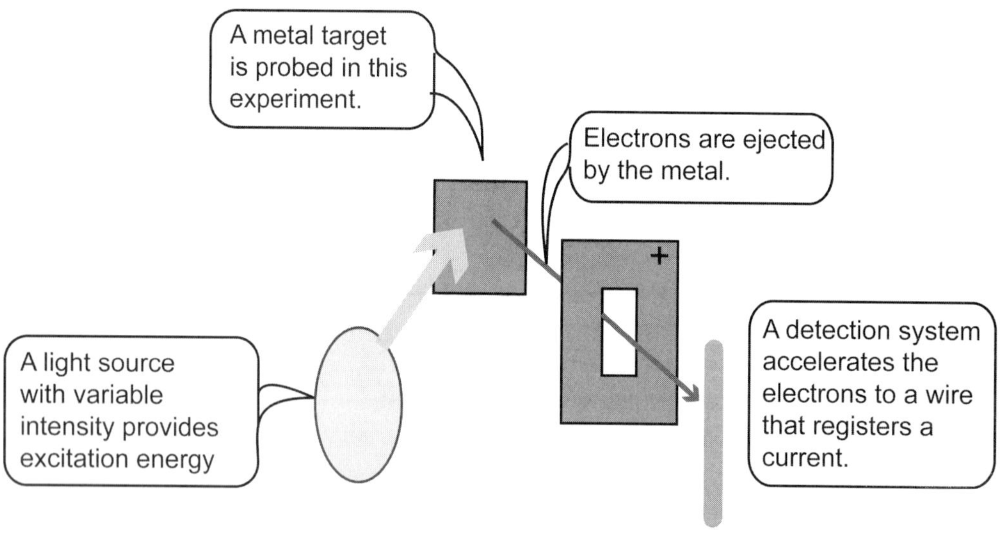

Figure 6.8 Important components of a photoelectric effect experiment are shown. Such experiments were largely responsible for the emergence of the particulate model of light, and the underlying idea is now used in various light sensors and other technologies.

How can this photoelectric effect be understood? The simplest explanation is that energy from the light is transferred to the electrons in the metal. If an electron picks up enough energy, it can "break free" from the surface of the metal. The more energy is given to the electron, the faster it will be traveling after it leaves the metal.

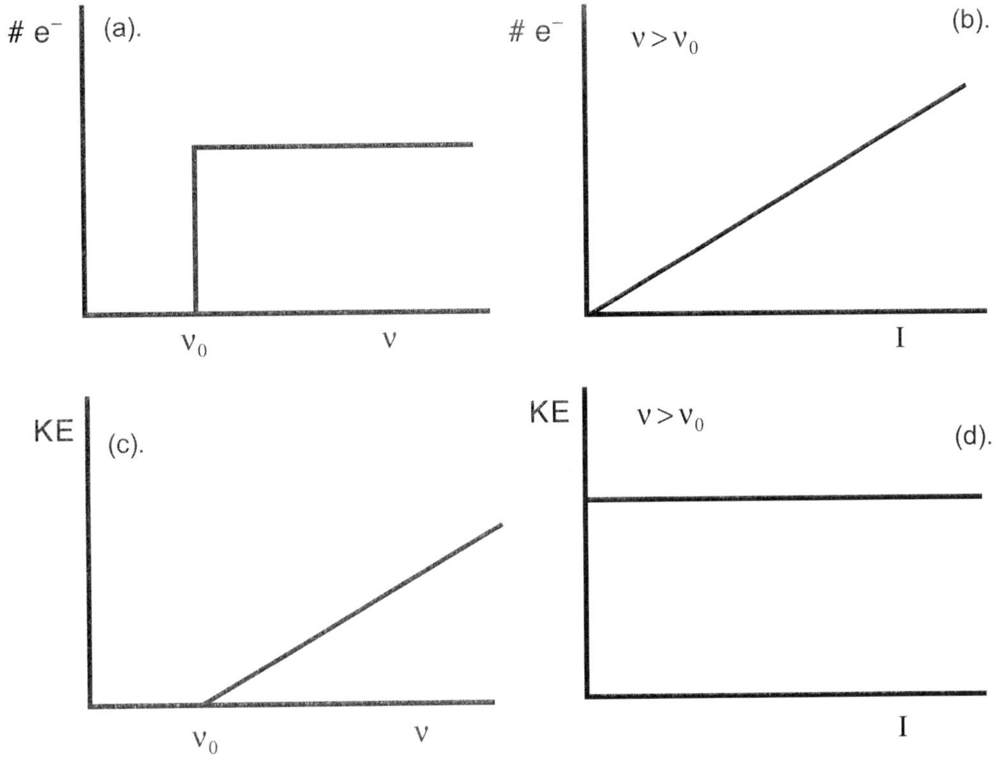

Figure 6.9 The graphs sketched here show the most important observations from experiments on the photoelectric effect. Below some threshold frequency (ν_0), no photoelectrons are emitted. For frequencies greater than ν_0, the number of electrons emitted is independent of frequency (graph (a)), but increases with increasing light intensity (graph (b)). The kinetic energy of the emitted electrons, on the other hand, increases linearly with frequency (graph (c)) but is independent of intensity (graph (d)).

Figure 6.9 summarizes the sort of results obtained in these experiments. A close look at these graphs reveals that the detailed observations are not what we would expect from the wave model for light. In a wave description of a light source, the amplitude of the wave, or the intensity of the light determines the energy. So we would expect that as the light gets brighter, the kinetic energy of the outgoing electrons should increase. But Figure 6.9(d). shows that the kinetic energy of the electrons is actually independent of the intensity of the incoming light. Only the number of electrons detected increases when the intensity goes up (graph (b)). But the kinetic energy does depend on the frequency or color of the light used (graph (c)). And this is also at odds with the wave model. So these experiments (and others) caused scientists to re-examine the nature of light.

Despite elaborate efforts to make these observations of the photoelectric effect conform to the concepts of waves, consistent explanations of light based solely on waves ultimately failed. The only way to explain all of the experimental results was to invoke the notion of *wave-particle duality*. That is, in some situations light is best described as a wave, while in other cases a particle description works better. It's important to realize that this does not mean that there are "two different kinds of light!" It just means that neither the wave model nor the particle model provides an accurate description of all of the properties of light. So in a given situation we use whichever model works best to describe the properties we are interested in. Someday, advances in our understanding of light may produce a new model that reconciles the wave-particle conflict. But until that happens, we take a pragmatic approach and use the model that's best suited to the problem at hand

The context in which we normally think of light as a particle is when it is imparting energy to another object. Albert Einstein proposed that light could behave as a collection of packets of energy, called **photons**, when it has energetic consequences. Bright light has many photons while dim light has few. The energy of a photon of light has been shown to be proportional to its frequency, leading to a simple equation.

$$E = h\nu \qquad (6.2)$$

Here E represents the photon energy, and ν is the frequency of the light. The h term is a constant, called Planck's constant, after German scientist Max Planck. Its value is

$$h = 6.626 \times 10^{-34} \text{ J s.}$$

Because frequency is related to wavelength we can substitute for ν in this equation, using Equation 6.1.

$$E = \frac{hc}{\lambda} \qquad (6.3)$$

Thus, if we know either the wavelength or the frequency of radiation, we can determine the energy of a photon of that radiation as shown in Example Problem 6.2.

Example Problem 6.2

The laser in a standard laser printer emits light with a wavelength of 780 nm. What is the energy of a photon of this light?

Strategy

The photon energy and wavelength are related by Equation 6.3. Again, care with units requires conversion from nanometers to meters.

Solution

$$780 \text{ nm} \times \frac{1 \text{ m}}{1 \times 10^{9} \text{ nm}} = 7.80 \times 10^{-7} \text{ m}$$

$$E = \frac{6.626 \times 10^{-34} \text{ J} \cdot \text{s} \times 2.998 \times 10^{8} \text{ m} \cdot \text{s}^{-1}}{7.80 \times 10^{-7} \text{ m}} = 2.55 \times 10^{-19} \text{ J}$$

Check Your Understanding

An infrared laser for use in a fiber-optic communications network emits at a wavelength of 1.2 μm. What is the energy of one photon of this radiation?

The idea that light is made up of a collection of particles, or "chunks of energy" does not seem consistent with our everyday experience. To understand why, look at the magnitude of the photon energies calculated in Example Problem 6.2. Compare those photon energies with the amount of energy emitted by an ordinary lamp. Even a modest 60-watt bulb still emits 60 J of energy each second. So the lights that you are familiar with all consist of vast numbers of photons. As the light gets brighter, the number of photons emitted per second increases. But since each photon is such a small additional amount of energy, it appears to us as if the energy coming from the lamp varies continuously rather than by discrete increments or steps.

Be careful not to confuse the photon energy with the intensity or brightness of the light. If you remember that photon energy is another way of specifying the color of the light, you will see that it is independent of intensity.

But from an atomic or molecular vantage point, the energy of a single photon can be very significant. Atoms and molecules are very small, after all, so their typical energies are small as well. Consider the results presented earlier for the photoelectric effect experiment. How can the concept of photons help us to explain those observations? Imagine that we could do the experiment by shooting a single photon at the metal. That means we would be supplying an amount of energy (equal to $h\nu$) to the metal. If this energy is enough to overcome the force that the metal atoms exert to hold onto their electrons, then when the metal absorbs the photon it will be able to eject one electron. If the energy of the photon is smaller than the *binding energy* holding the electrons to the metal, then no electron can be emitted. The fact that there is a threshold frequency (ν_0 in Figure 6.9), below which no electrons are detected, is evidence that the photoelectric process is carried out by individual photons. Below the threshold frequency, a photon does not have enough energy to overcome the binding energy and eject an electron. Why does the electron kinetic energy depend on the frequency of the light? This is really a conservation of energy question. For a particular metal, the electron binding energy is fixed. For any frequency above ν_0, a portion of the energy of the absorbed photon will be used to overcome the binding energy. And then whatever energy remains will be transferred to the electron as kinetic energy. So as the frequency increases the photon energy increases, and this leads to an increase in the observed kinetic energy of the ejected electrons.

Example Problem 6.3

In a photoelectric effect experiment, ultraviolet light with a wavelength of 337 nm was directed at the surface of a piece of potassium metal. The kinetic energy of the ejected electrons was measured as = 2.30×10^{-19} J. What is the electron binding energy for potassium?

Strategy

We know that energy must be conserved, and we can use this idea to set up an equation relating the photon energy to the kinetic energy and the binding energy. Since the photon is absorbed in this process, its energy must be accounted for. Some of that energy does work to overcome the electron binding energy, and the rest appears as the kinetic energy of the ejected electron. So we can write:

$$E_{photon} = \text{Binding E} + \text{Kinetic E}$$

We know how to find the photon energy from the wavelength, and we know the kinetic energy, so we can solve for the binding energy. As always, we must be careful with the units on all the quantities involved.

Solution

First find the photon energy.

$$337 \text{ nm} \times \frac{1 \text{ m}}{10^9 \text{ nm}} = 3.37 \times 10^{-7} \text{ m}$$

$$E_{photon} = \frac{hc}{\lambda} = \frac{(6.626 \times 10^{-34} \text{ Js})(2.998 \times 10^8 \text{ ms}^{-1})}{3.37 \times 10^{-7} \text{ m}} = 5.89 \times 10^{-19} \text{ J}$$

Then use the conservation of energy relationship to find the desired binding energy.

$$E_{photon} = \text{Binding E} + \text{Kinetic E}$$

So

$$\text{Binding E} = E_{\text{photon}} - \text{Kinetic E}$$

$$= 5.89 \times 10^{-19}\text{ J} - 2.30 \times 10^{-19}\text{ J} = 3.59 \times 10^{-19}\text{ J}$$

Check Your Understanding

The electron binding energy for chromium metal is 7.21×10^{-19} J. Find the maximum kinetic energy at which electrons can be ejected from chromium in a photoelectric effect experiment using light at 266 nm.

By combining the concepts we've just developed, we can begin to consider the issues about light bulbs raised in the insight section. We noted that when a blacksmith at a museum heats iron, it first glows red and then turns orange. Looking at the spectrum we see that red light has a longer wavelength than does orange light. So the energy of a photon of orange light is greater than that of a photon of red light. The iron changes from red to orange because the heating process has increased its energy. An object that emits orange light is at a higher temperature than one emitting red light.

Now that we have discussed the nature of light itself, we need to consider the way that light allows us to investigate atoms.

6.3 Atomic Spectra

You may have once been told to be careful to avoid breathing the gas from a broken fluorescent light bulb. That gas is dangerous because there are small amounts of mercury in the bulb, and mercury vapor can be toxic. Why is the mercury there? Like all atoms, mercury has a characteristic way of absorbing and emitting light. The particular pattern of wavelengths absorbed and emitted by mercury is called its *atomic spectrum*. The existence of such spectra has been known for over a century. What is most peculiar is that excited atoms of any particular element emit only specific frequencies. Because these frequencies tend to be well separated, the spectra are said to be *discrete*. This behavior is quite distinct from the light given off by the sun or by an incandescent bulb.

The plural of spectrum is spectra.

Investigations of the light emitted by various atoms revealed several surprising facts. All atoms display discrete spectra, with relatively few wavelengths appearing. And the wavelengths that are emitted vary from one kind of atom to the next. Discrete spectra were first observed in the late 19th century, and posed an enormous challenge for scientists to interpret. While data were gathered from many elements, including mercury, the spectrum of hydrogen played the most important role. Hydrogen gas can be excited in a glass tube by passing an electric arc through the tube. The energy from the arc breaks the H_2 molecules into individual hydrogen atoms and excites some of those atoms. The excited hydrogen atoms then emit light as they relax towards lower energy states. The light given off appears blue, and when passed through a prism it reveals four lines visible to the eye.

What conclusions can be drawn from the fact that hydrogen (or any other atom) emits only certain wavelengths of light? When an atom emits light, it is releasing energy to the surrounding world. So we should think about this situation in terms of conservation of energy. When light is emitted, the atom goes from a higher energy state to a lower energy state, and the emitted photon carries away the energy lost by the atom. The fact that only a few wavelengths of light are emitted from a particular atom is direct evidence that the atom can exist in only a few states, with very specific energies. (To see that this must be true, imagine what would happen if the atom could take on any arbitrary energy. In that case we would expect to see a continuous spectrum of wavelengths emitted, since each atom could lose an arbitrary amount of energy.) The notion that an atom could take on only certain allowed energies is extremely counterintuitive, since no ordinary macroscopic objects seem to behave in a similar way. In order to explain the observed discrete emission spectra, scientists had to develop a model for the behavior of atoms that is fundamentally different from the ways that we describe macroscopic objects.

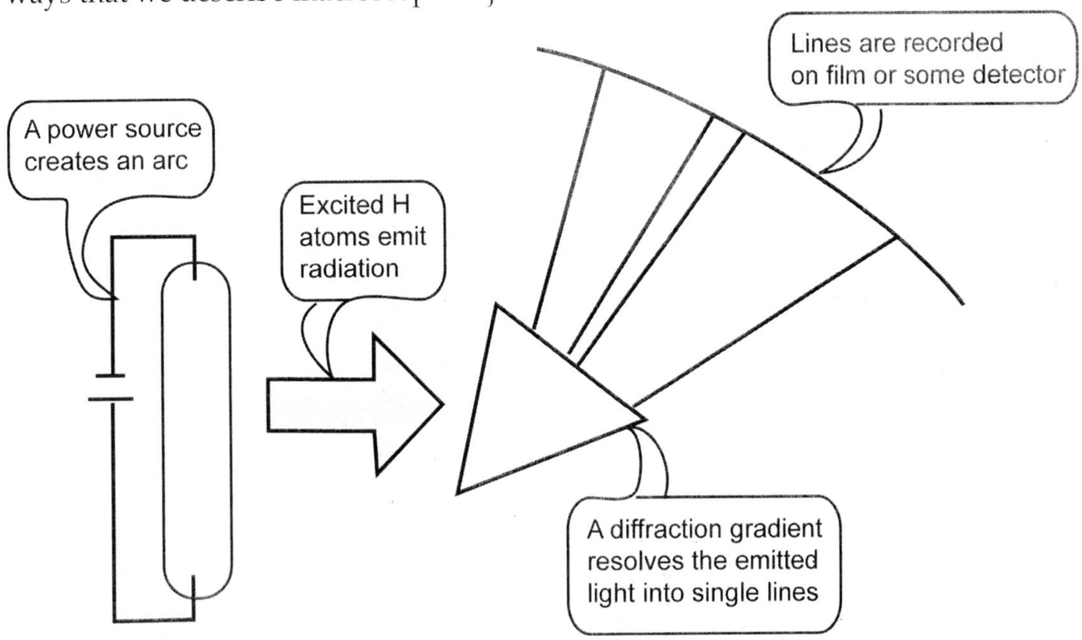

Figure 6.10 Schematic diagram of the experiment to measure an atomic emission spectrum. (Additional figure development needed. Final figure will also include a hydrogen atom spectrum.)

Example Problem 6.4

The first 4 energy levels for the hydrogen atom are shown in the diagram below. The lowest energy state is labeled as E_1, and the higher states are labeled as E_2, E_3, and E_4, respectively. When a hydrogen atom undergoes the transition from E_3 to E_1, it emits a photon with $\lambda = 102.6$ nm. Similarly, if the atom undergoes the transition from E_3 to E_2, it emits a photon with $\lambda = 656.3$ nm. Find the wavelength of light emitted by an atom making the transition from E_2 to E_1.

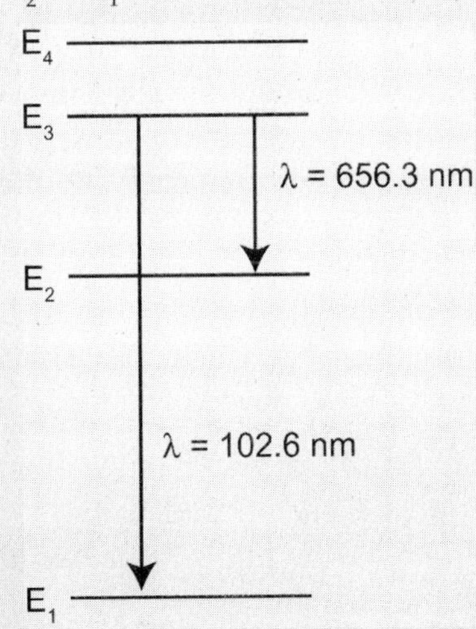

Strategy

The energy level diagram shows us the relationship between the various energy levels involved. The vertical direction in the diagram is energy. So if we represent each transition by an arrow, as shown, then the length of that arrow will be proportional to the energy lost by the atom in that transition. From the diagram, we can see that the lengths of the arrows for the E_3 to E_2 transition and the E_2 to E_1 transition must add up to the length of the arrow for the E_3 to E_1 transition. (Another way of thinking about this is that an atom going from the E_3 state to the E_1 state must lose the same amount of energy whether or not it stops at E_2 along the way.) Since we know the wavelengths for two of the transitions, we can find the corresponding photon energies. That will let us find the energy for the third transition, and then finally we can convert that to wavelength.

Solution

Converting the given wavelengths to energies gives us

$$E = \frac{hc}{\lambda} = \frac{(6.626 \times 10^{-34}\ \mathrm{J\ s})(2.998 \times 10\ \mathrm{m\ s^{-1}})}{102.6\ \mathrm{nm}} \times \frac{10^9\ \mathrm{nm}}{1\ \mathrm{m}} = 1.935 \times 10^{-18}\ \mathrm{J}$$

for the $E_3 \rightarrow E_1$ transition, and

$$E = \frac{hc}{\lambda} = \frac{(6.626 \times 10^{-34}\ \mathrm{J\ s})(2.998 \times 10\ \mathrm{m\ s^{-1}})}{656.3\ \mathrm{nm}} \times \frac{10^9\ \mathrm{nm}}{1\ \mathrm{m}} = 3.026 \times 10^{-19}\ \mathrm{J}$$

for the $E_3 \rightarrow E_2$ transition. From the diagram, we can see that

$$E_{3\rightarrow1} = E_{3\rightarrow2} + E_{2\rightarrow1}$$

so

$$E_{2\rightarrow1} = E_{3\rightarrow1} - E_{3\rightarrow2} = 1.935 \times 10^{-18}\ \mathrm{J} - 3.026 \times 10^{-19}\ \mathrm{J}$$

$$= 1.632 \times 10^{-18}\ \mathrm{J}$$

Now we just need to convert from energy to wavelength to reach the desired answer.

$$\lambda = \frac{hc}{E} = \frac{(6.626 \times 10^{-34}\ \mathrm{J\ s})(2.998 \times 10\ \mathrm{m\ s^{-1}})}{1.632 \times 10^{-18}\ \mathrm{J}} = 1.217 \times 10^{-7}\ \mathrm{m}$$

Since we know that wavelengths are most often given in nanometers, we might choose to convert that to 121.7 nm.

Discussion

Note that our result lies between the two wavelengths we were originally given, which is consistent with the spacing of the energy levels in the diagram.

Check Your Understanding

A hydrogen atom undergoing a transition from the E_4 level to the E_3 level emits a photon with a wavelength of 1.873 μm. Find the wavelength of light emitted by an atom making the transition from E_4 to E_1.

The sort of reasoning used in this example clearly allows us to do some potentially useful calculations based on data from atomic spectra. But it does not yet begin to allow us to explain or understand the very existence of such spectra. In the early part of the 20th century, many models were offered to try to explain the physics behind atomic spectra. The model proposed by Niels Bohr had the greatest influence on our subsequent understanding of the atom.

The Bohr Atom

Bohr won the Nobel Prize for physics in 1922.

As the structure of atoms was being deduced, experiments were explained by suggesting the nucleus of the atom was positively charged while the electrons around the outside have negative charge. You probably learned this model of the atom early in your time as a student, perhaps even in grade school, and so you are likely to accept it without much thought. Yet at the time of its discovery it was perplexing. In particular, the attraction between oppositely charged particles would seem to indicate that the electrons should collapse into the positively charged nucleus. Bohr addressed this concern as well as other more subtle ones in devising his model for the atom. His main creative contribution was to suggest that the electrons occupy stable orbits from which they cannot deviate without first absorbing or emitting energy in the form of light. The supposition of such orbits had never been made before, but by creating this model Bohr was able to explain observations no previous hypothesis could. The influence of the Bohr model is evident in modern society, as the atom is commonly depicted with electrons orbiting around the nucleus.

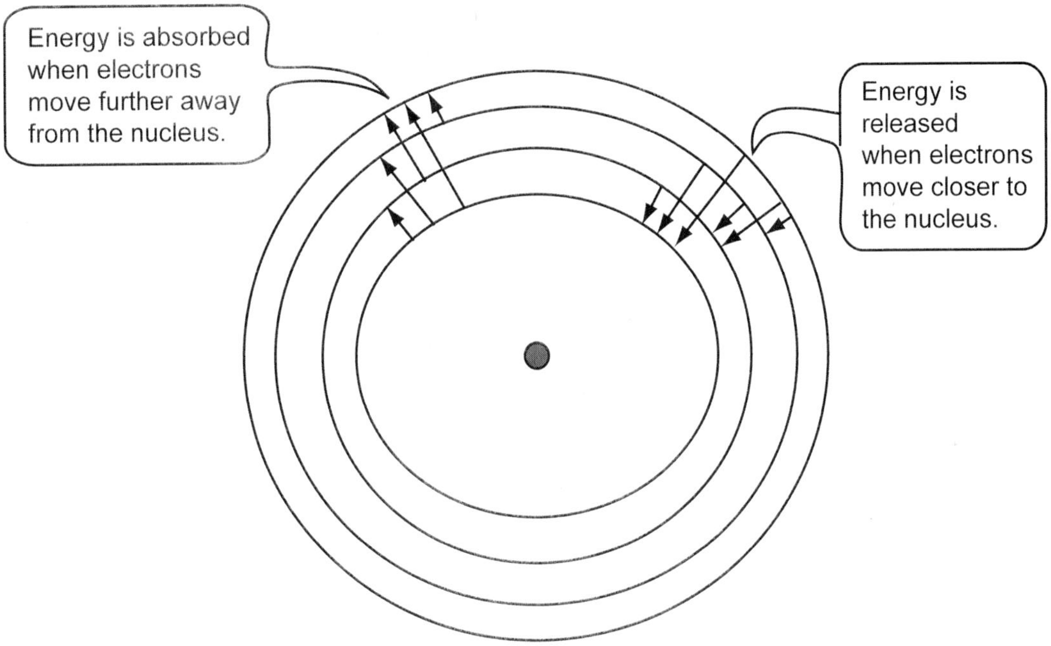

Figure 6.11 In the Bohr model, electrons move between allowed orbits when an atom absorbs or emits light. Important elements of this model, including the idea of fixed orbital paths, are not correct. But it played an essential role in the development of our understanding of atomic structure. Note the similarity between this figure and the energy level diagram we used in Example Problem 6.4; both reflect the concept that electrons in atoms are restricted to certain allowed energies.

The concept of an excited state is general in the energy levels of molecules, though we will only encounter its application for electronic energy levels.

We now know that the Bohr model for the atom is not completely correct, but it does provide useful intuition as well as a convenient way to visualize the process of "exciting" an atom. In any atom, the oppositely charged electrons and protons attract each other. Therefore, the closer to the nucleus an electron is, the stronger the attraction and the lower the energy. In order for an electron to move from an inner orbit to an outer one as depicted in Figure 6.11, energy is absorbed. The new distribution or configuration of electrons consists of a grouping that is not at the lowest possible energy and is therefore referred to as an excited state. One of the most powerful driving forces in nature is towards states of lower energy, so an excited state of an atom cannot exist indefinitely. It will return to the lowest energy or *ground state* by emitting radiation.

Both the excitation and emission processes are understandable in terms of the Bohr model of the atom. So this model has some usefulness even though it is not really an accurate representation of atomic structure.

This picture of the atom allows us to begin answering some of the questions about light bulbs we posed in the insight section. Recall that we said all lamps are based on the principle that an excited substance can emit light as it loses energy. Now we can see that the nature of the excitation largely dictates the type of bulb. In an incandescent bulb, the material being excited is a metal wire that is visible with the naked eye. This wire, like any visible material, contains many atoms - perhaps nearly a mole of them. As electric current passes through the wire, resistance causes the wire's temperature to rise. Thus the excitation process in the metal wire is affecting large numbers of atoms by heating them. Emission of energy from the macroscopic collection of excited atoms allows an incandescent bulb to provide us with the light we need.

We also noted that fluorescent lamps are usually filled with a rare gas and small amounts of mercury vapor. Remember that the molecular level characteristic of a gas is the relatively large distance between atoms. So in a fluorescent lamp the atoms in the bulb are excited individually rather than in a collective fashion. Of course many individual atoms are excited simultaneously, but the behavior of any given atom does not depend on the other atoms in the tube. In a fluorescent bulb, the mercury atoms are excited and they release the energy they gained from the electric arc by emitting light. The light emitted by the mercury atoms is in the ultraviolet region of the spectrum, so it cannot be the light that we see when we switch on the light. Instead the phosphor molecules on the glass walls absorb the emitted UV light, and another excitation and emission process takes place. In this second step the phosphor molecules emit light over a range of frequencies, providing lighting more akin to the nearly continuous light of the sun, rather than the invisible UV radiation emitted by mercury atoms. The spectrum of wavelengths emitted by the phosphor molecules is not quite the same as that emitted by the hot filament in an incandescent bulb, though, and that's why incandescent and fluorescent bulbs each have their own characteristic "looks."

6.4 The Quantum Mechanical Model of the Atom

The observation of atomic spectra and the development of the Bohr model were important advances in the understanding of atoms. But further experiments in the early part of the 20th century revealed still more unexpected properties of atoms, and important limitations of the Bohr model became apparent. So efforts to develop a more robust model of the atom continued. And while these efforts were initially driven primarily by fundamental scientific curiosity, they ultimately facilitated many important practical applications. The development of light sources such as the halogen lamp, the light emitting diode, and lasers would not have been possible without improved models of atoms. The picture that emerged to enable these technological advances is the quantum mechanical model of the atom. As we'll see in Chapter 7, this new model also allows us to describe chemical bonding more accurately, which in turn allows for materials improvements.

The essential and somewhat peculiar feature of the quantum mechanical model of the atom lies in the description of electrons as waves rather than particles. It is far more intuitive to think of electrons as finite particles, perhaps resembling tiny marbles, than to envision them as waves with poorly delineated beginning and ends. Nonetheless, experiments are now routinely performed in which the wave-like behavior of electrons is apparent. Historically the first evidence of the wave nature of electrons came through diffraction experiments. Diffraction is a well-understood phenomenon of waves, so observation of electron diffraction strongly suggests the need for a wave-based treatment of the electron. In a modern application, the scanning tunneling microscope cleverly exploits the wave behavior of electrons to make images of individual atoms or molecules on surfaces.

Mathematical tools to describe waves existed long before the observation of electron diffraction. A wide array of mathematical functions can be used to describe the periodic behavior of waves. The simplest such function may be *sin x*, which describes the movement of a pendulum reasonably well. Electron waves are notably more complicated than a pendulum, but mathematical treatments have been devised. The most famous and fundamental treatment was that provided by Erwin Schrödinger. The so-called Schrödinger equation is neatly summarized as

$$H\psi = E\,\psi \qquad (6.4)$$

but this form is deceptively simple. That's because the H term is an *operator,* which designates a complicated series of mathematical operations to be carried out. E is the energy and ψ(the Greek letter "psi") is the *wave function* of the electron. If we write out the entire operator H, we find that the Schrödinger equation is actually a second-order differential equation. (See *MathConnections* box.) It turns out that in most cases the equation cannot be solved exactly, so the study of quantum mechanics invariably involves approximation techniques. Due to the complexity of the mathematics involved, we will not make any attempt to solve this equation. But we will describe the results of its solution and look at how these results form the foundation of our modern understanding of atoms. The concept of the wave function, in particular, is central to the quantum mechanical model of the atom. The wave function contains all the information we would ever need to know about the electron, if we can determine the means by which to extract the information.

MathConnections

Written as $H\psi = E\psi$, the Schrödinger equation appears deceptively simple. If we write out the operator H, we get the more intimidating looking form below.

$$-\frac{h^2}{8\pi^2\mu}\left\{\frac{\partial^2\psi}{\partial x^2}+\frac{\partial^2\psi}{\partial y^2}+\frac{\partial^2\psi}{\partial z^2}\right\}+V(x,y,z)\psi = E\psi$$

Although this is a complicated equation, we can relate it to things you are probably familiar with. First, let's realize that the purpose of this equation is to describe the total energy of an electron as it moves through the *x,y,z*-space surrounding the atom's nucleus. We might expect that there should be terms for the kinetic energy and for the potential energy, and in fact there are. The *V(x,y,z)* term accounts for the potential energy, which is generally a function of position. And the other term on the left-hand side accounts for the kinetic energy. So in a sense the Schrödinger equation just expresses the total energy as the sum of the kinetic energy and the potential energy.

Next let's look more closely at the terms that appear. If you are familiar with calculus, you might recognize the "$\delta^2/\delta x^2$" notation as a derivative. Here it is written with deltas (δ) rather than with d's because x (or y or z) is just one of several variables on which our wave function ψ depends. And the superscript 2's tell us this is a second derivative. So for $\delta^2\psi/\delta x^2$, we would take the derivative of ψ with respect to x, and then take the derivative of the result with respect to x as well. As long as we know how to take derivatives, we could evaluate the kinetic energy term. And the potential energy term is also understandable. For an electron in an atom, the potential energy is due to the attraction between the positive charge on the nucleus and the negative charge on the electron. So the function V in the equation would just represent the way that potential varies as the position of the electron changes. From physics you may know this as the Coulomb potential, which depends on the square of the distance between the charges.

You will not actually be expected to work directly with the Schrödinger equation in a class at this level. But if we think about it step by step like this, you should begin to see the connections between this complicated equation and ideas you work with in your math and physics classes.

Potential Energy and Orbitals

The potential energy part of the wave equation plays the more significant role in our description of atomic structure. The potential energy of atoms is associated with the Coulombic attraction between the positively charged nucleus and negatively charged electrons or the repulsion between like charged electrons. The periodic nature of waves allows many solutions to the wave equation for any given potential interaction. The Schrödinger equation is often rewritten as

$$H\psi_n = E_n\psi_n \tag{6.5}$$

to emphasize the fact that multiple solutions can exist. Here n is an index that labels the different solutions. The wave functions, ψ_n, that provide solutions to this equation for a hydrogen atom have two components, called the radial and angular components. The wave functions may have positive and negative regions but the square, ψ^2, is always positive and provides the probability of finding an electron. This probability interpretation will seem odd to you, but that's just because you are more accustomed to dealing with discrete objects (or particles) rather than with waves. If you think about the standing wave that exists in a vibrating guitar string, you will realize that the wave is not at any one point along the string; the wave is "found" all along the string simultaneously. That's the same idea we need to apply to electrons in the quantum mechanical model.

The indexing system that is represented by the "n" subscript in Equation 6.5 provides important language for discussing atomic structure. The index is typically written to contain both a number and a letter, and these labels identify solutions to the wave equation. Because of their historical correspondence to the position of electrons in the orbits proposed by Bohr, each solution is referred to as an orbital. It's quite possible you have heard the names of these orbitals before: *1s, 2s, 2p, 3s, 3p, 3d, 4s, 4p, 4d, 4f,* etc. What is the origin of these "names" of orbitals, and what information do those names provide?

Quantum Numbers

While we don't need detailed treatments of quantum mechanics to understand light bulbs or most introductory chemical concepts, we will find the vocabulary established by the theory to be important. For example, the names attached to atomic orbitals come from the functions that solve the wave equations. Collectively they are referred to as *quantum numbers*.

Before examining the specific quantum numbers we need to describe an atomic orbital, let's look at a simpler problem that can help show where these numbers come from. Suppose we take a guitar string, pin down both ends of it so they can't move, and then think about what sorts of waves might possibly be produced if we pluck the string so it vibrates. (See Figure 6.12.) In the simplest case, the length of the string would just be 1/2 of the wavelength of a sine function. In mathematical terms, we could write the following equation for this wave. (A is the amplitude of the wave, x is the position along the string, and L is the length of the string.)

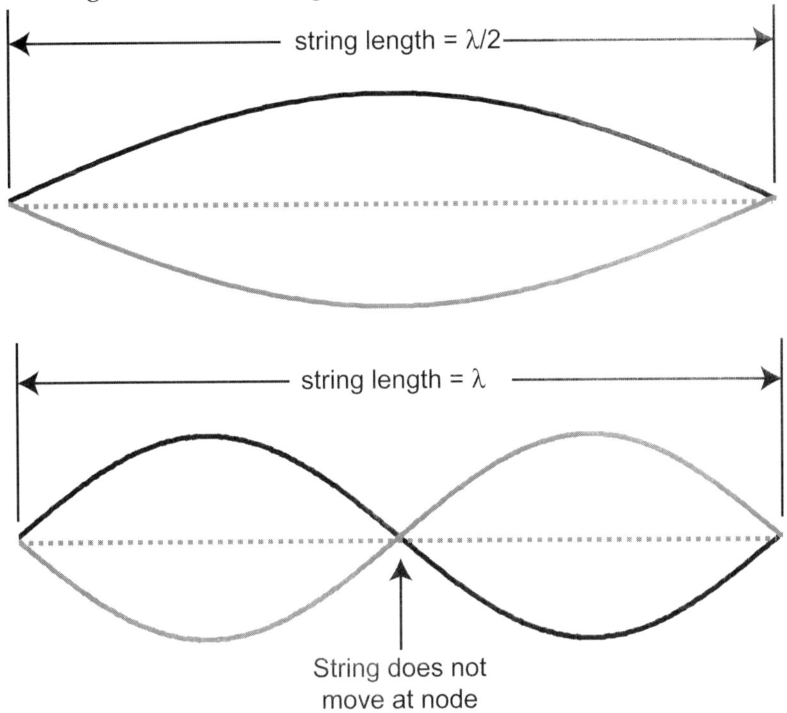

Figure 6.12 A guitar string provides an example of how multiple waves can satisfy a particular set of conditions. In the upper figure, the length of the string is equal to one-half of the wavelength of the wave. In the lower figure, the length of the string is equal to a whole wavelength. A standing wave like those shown can be formed as long as the length of the string is equal to an integer multiple of half-wavelengths.

$$\psi(x) = A \sin \frac{\pi x}{L} \tag{6.6}$$

But other waves are also possible without moving the ends of the string. If a guitarist wants to play what is called the "harmonic," a finger is placed very lightly at the midpoint of the string. The string can still vibrate on both sides of the midpoint, so now we can form a sine wave where the length of the string is one complete wavelength, as represented by the following equation.

$$\psi(x) = A \sin \frac{2\pi x}{L} \tag{6.7}$$

This wave has a higher frequency, and indeed we would hear it as a higher-pitched note. Additional waves could also be produced by pinning down appropriate positions along the string. The important restriction is that the length of the string must contain an integer multiple of half-wavelengths; this assures that the ends of the string will not move. Suppose we wanted to write a general equation for the waves that can be formed in our string. Working from the cases above, you should be able to see that the necessary form will be:

$$\psi(x) = A\sin\frac{n\pi x}{L} \tag{6.8}$$

Where n must be a positive integer. This n is a quantum number. If we substitute an allowed (positive integer) value for n into the general equation, we will get an equation for a particular wave that can exist in the string.

When we solve the Schrödinger equation for an atom, the resulting wave functions are much more complicated than these sine waves. To write the mathematical equations for the wave functions of the atomic orbitals, we need to use three different quantum numbers: the principle quantum number (designated n), the secondary quantum number (designated ℓ) and the magnetic quantum number (designated m_l). But just like in our vibrating string problem, these quantum numbers are a natural consequence of the mathematics of the wave equation. Furthermore, when we solve the wave equation for the atomic orbitals, we find that the values that the three quantum numbers are allowed to take on are all interrelated. As we'll see, this has significant consequences for what kinds of orbitals actually exist.

The principle quantum number can assume integer values from 1 to infinity, (n = *1, 2, 3, 4, 5, ...*). Orbitals are grouped according to their principle quantum numbers into shells. When n=*1* we are describing the first shell, when n=*2* the second, and so on. In hydrogen, because there is only one electron, all orbitals in the same shell have the same energy, as was predicted by the Bohr model. However, when atoms have more than one electron the negative charges on the electrons repel one another. This repulsion between electrons causes energy differences between orbitals in a shell. This difficulty was one reason the Bohr model had to be replaced.

The secondary quantum number provides a way to index energy differences between orbitals in the same shell of an atom. Like the principle quantum number, it assumes integer values, but it may be zero and is constrained to a maximum value one less than n: $\ell = 0,1,2,\ldots,n-1$. The secondary quantum number specifies *subshells*, smaller groups of orbitals within a shell. When n=*1*, the only possible value of ℓ is zero, so there is only one subshell. When n=*2*, ℓ can be either zero or one, so there are two subshells. Each shell contains as many subshells as its value of n.

The secondary quantum number also provides the key to the letter designations of orbitals. Table 6.1 provides the letter designation of the first five values of the ℓ quantum number.

ℓ-value	0	1	2	3	4
letter designation	s	p	d	f	g

Table 6.1 Letter designations for the secondary quantum number.

The origin of these letters comes from the language of spectroscopy, the technique used to observe the lines from various atoms; *s* was 'sharp,' *p* was 'principle,' and *d* was 'diffuse.' Beyond *f*, the letter designation proceeds in alphabetical order. There is no known atom that has electrons in any subshell higher than *f* when it is in its ground state, but higher orbitals may be occupied in excited states, and additional "super heavy" elements in which these orbitals are populated may yet be discovered.

For atoms under normal conditions, the energies of the orbitals are specified completely using only the n and ℓ quantum numbers. In the laboratory, it is possible to place atoms in magnetic fields and observe them spectroscopically, and when this is done some of the emission lines split into three, five or seven components. This observation points to the need for a third, magnetic quantum number. This index is labeled m_ℓ and is also an integer. It may be either positive or negative and its absolute value must be less than or equal to the value of ℓ for that orbital. Thus when ℓ is zero m_ℓ is also zero. When $\ell = 1$, m_ℓ may be -1, 0 or +1, and when $\ell = 2$, m_ℓ may be -2, -1, 0, +1 or +2. The relationships between all three of the quantum numbers are summarized in Table 6.2.

Visualizing Orbitals

Table 6.3 lists the actual wave functions for the first few orbitals of a hydrogen atom. (Don't worry - you are **not** expected to memorize these functions! They are given only to show that they are made up of several standard mathematical functions that you are already familiar with.) Since we expect the atom to be spherical, it is customary to write these functions in spherical polar coordinates, with the nucleus of the atom located at the origin. You are probably familiar with two dimensional polar coordinates, where a point in a plane is specified by a radius (r) and an angle (θ). Spherical polar coordinates are the three dimensional equivalent. A point in space is specified by a radius (r) and two angles (θ and φ), as shown in Figure 6.13. So the terms that contain r determine the size of the orbital while those containing θ or φ determine the shape.

value of n	values for ℓ (letter designation)	values for m_ℓ	number of orbitals
1	0 (s)	0	1
2	0 (s)	0	1
	1 (p)	-1,0,1	3
3	0 (s)	0	1
	1 (p)	-1,0,1	3
	2 (d)	-2,-1,0,1,2	5
4	0 (s)	0	1
	1 (p)	-1,0,1	3
	2 (d)	-2,-1,0,1,2	5
	3 (f)	-3,-2,-1,0,1,2,3	7

Table 6.2 Relationships between values of the different quantum numbers are illustrated. This table allows us to make another observation about quantum numbers. If we count the total number of orbitals in each shell, it is equal to the square of the principle quantum number, n^2.

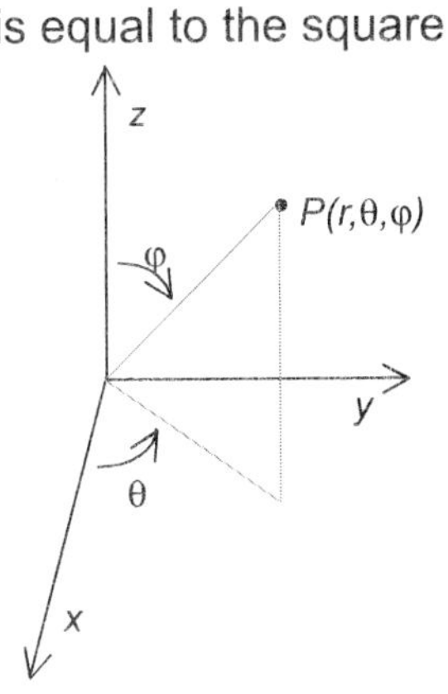

Figure 6.13 Definition of the spherical polar coordinates r, θ, and φ, and their relationship to the more familiar x, y, z coordinate system.

Orbital	Wave function
1s	$\psi_{1s} = \pi^{-1/2} a_0^{-3/2} e^{-(r/a_0)}$
2s	$\psi_{2s} = (4\pi)^{-1/2}(2a_0)^{-3/2}(2-\frac{r}{a_0})e^{-(r/2a_0)}$
$2p_x$	$\psi_{2p_x} = (4\pi)^{-1/2}(2a_0)^{-3/2}\frac{r}{a}e^{-(r/2a_0)}\sin\theta\cos\phi$
$2p_y$	$\psi_{2p_y} = (4\pi)^{-1/2}(2a_0)^{-3/2}\frac{r}{a}e^{-(r/2a_0)}\sin\theta\sin\phi$
$2p_z$	$\psi_{2p_z} = (4\pi)^{-1/2}(2a_0)^{-3/2}\frac{r}{a}e^{-(r/2a_0)}\cos\theta$

Table 6.3 Wave functions for the *n=1* and *n=2* shells of a hydrogen atom are listed. There is no need to memorize these functions – they are shown merely to point out that they are made up of ordinary mathematical terms.

It is quite common to think of orbitals as pictures rather than mathematical functions and quantum numbers. The pictorial depiction of orbitals requires an arbitrary decision. The wave function provides information about the probability of finding an orbital in a given region of space. If we wish to have 100% probability of finding an electron the amount of space needed is quite large and accordingly fails to provide insight about the tendency of that electron to be involved in chemical bonding. Conventionally, the space within which there is a 90% probability of finding an electron is chosen as a boundary for depicting orbitals. Within this convention, orbitals are most commonly shown by plotting the angular part of the wave function. Pictures of prototype *s*-, *p*- and *d*-orbitals are shown in Figure 6.14.

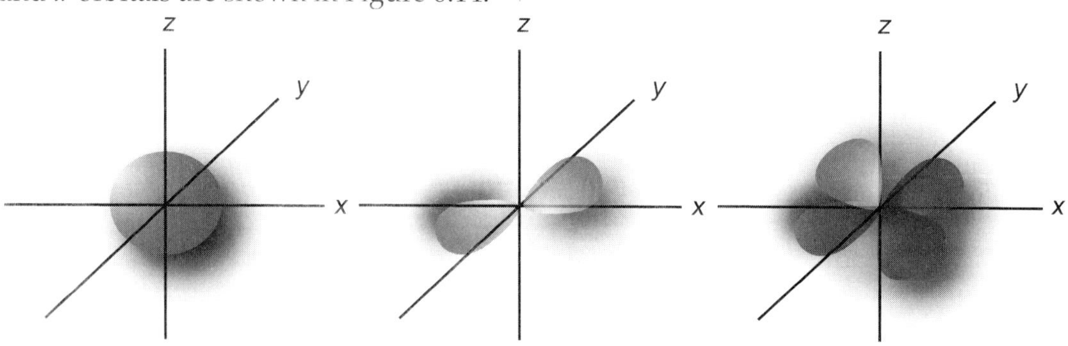

Figure 6.14 Shapes of (from left to right) *s*-, *p*-, and *d*-orbitals are shown. Although we sometimes think of figures like these as "pictures," they are actually obtained by plotting the corresponding wave functions, as described in the text.

These shapes may well be familiar to you. The *s*-orbital is spherical, while the *p*-orbital has two lobes and the *d*-orbital has four lobes. The two lobes of the *p*- are separated by a plane in which there is no probability of finding an electron. This plane is called a *node* or a *nodal plane*. The *d*-orbital has two nodes. One of the more challenging philosophical questions of science is motivated by these pictures. If the electron can be on either side of the node, but never in the nodal plane, how does it get from one side to the other? We can suggest an answer to this question in two ways. One is to remember that electrons behave as waves and even classical wave motion such as in the string of a guitar can have nodes. The second and more technical version of the answer lies in a peculiar concept of quantum mechanics called the **uncertainty principle**. This concept, stated by Werner Heisenberg, expresses the difficulty of observing an electron. Simply put, the uncertainty principle says it is impossible to determine both the position and momentum of an electron simultaneously and with complete accuracy. The uncertainty principle imposes a fundamental limit on observing the microscopic world and forces us to accept the idea that orbitals provide nothing more than probability of finding an electron.

The actual degree of uncertainty demanded by the uncertainty principle is exceedingly small, so its effects are not detectable for macroscopic objects.

In some ways, depicting the radial part of the wave function is more difficult, at least when considering the angular portion as well. Figure 6.15 shows three different *s*-orbitals corresponding to the principle quantum numbers $n = 1, 2$ and 3. Note that there are nodes in the radial portion of the wave function just as there are in many of the angular functions (*p*-, *d*-, and *f*-orbitals). There is one less node than the value of the quantum number. Thus there is one node in the *2s*-orbital, two in the *3s*-orbital and so on. Some *p*- and *d*- orbitals, such as *3p*, *4p* and *4d*, also have radial nodes. Figure 6.16 shows pictures of the orbitals in the *n=3* shell.

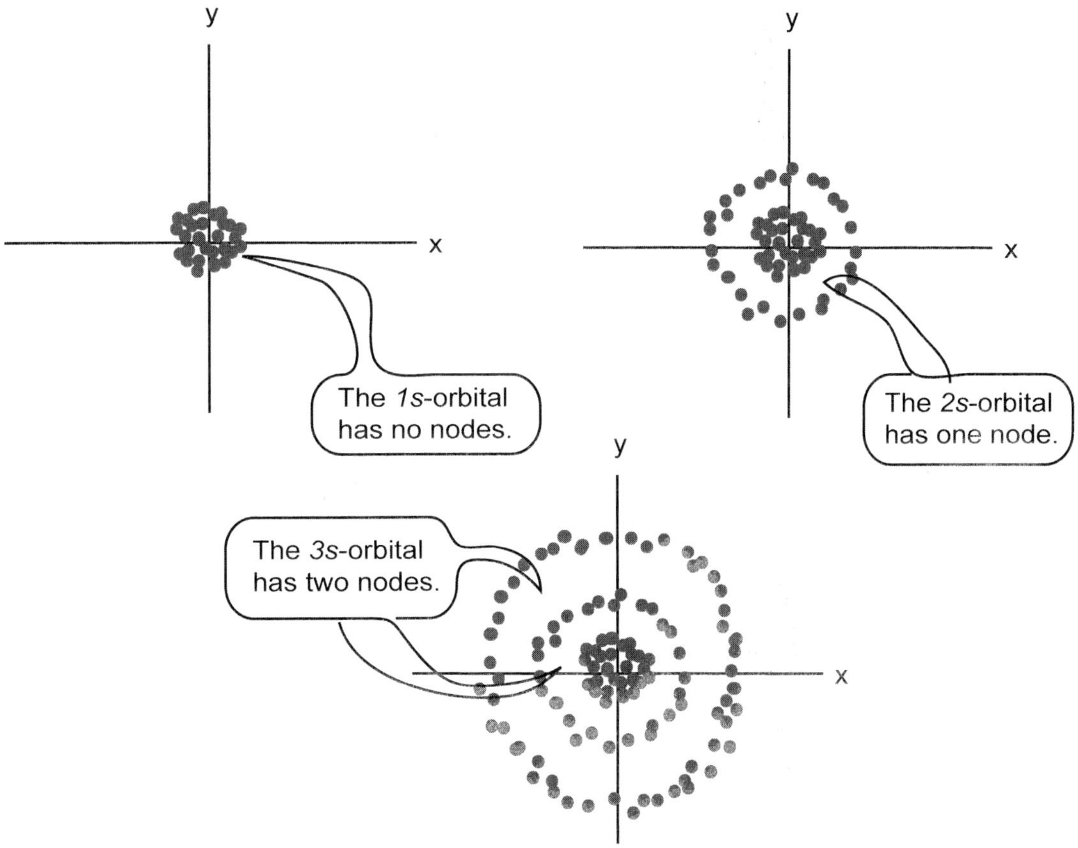

Figure 6.15 Cross-sectional views of the *1s*-, *2s*-, and *3s*-orbitals are shown. In this type of figure, the density of dots indicates the probability of finding an electron. The greatest electron density will be found in those areas that appear darkest, while the white regions seen within the *2s*- and *3s*- orbital represent nodes.

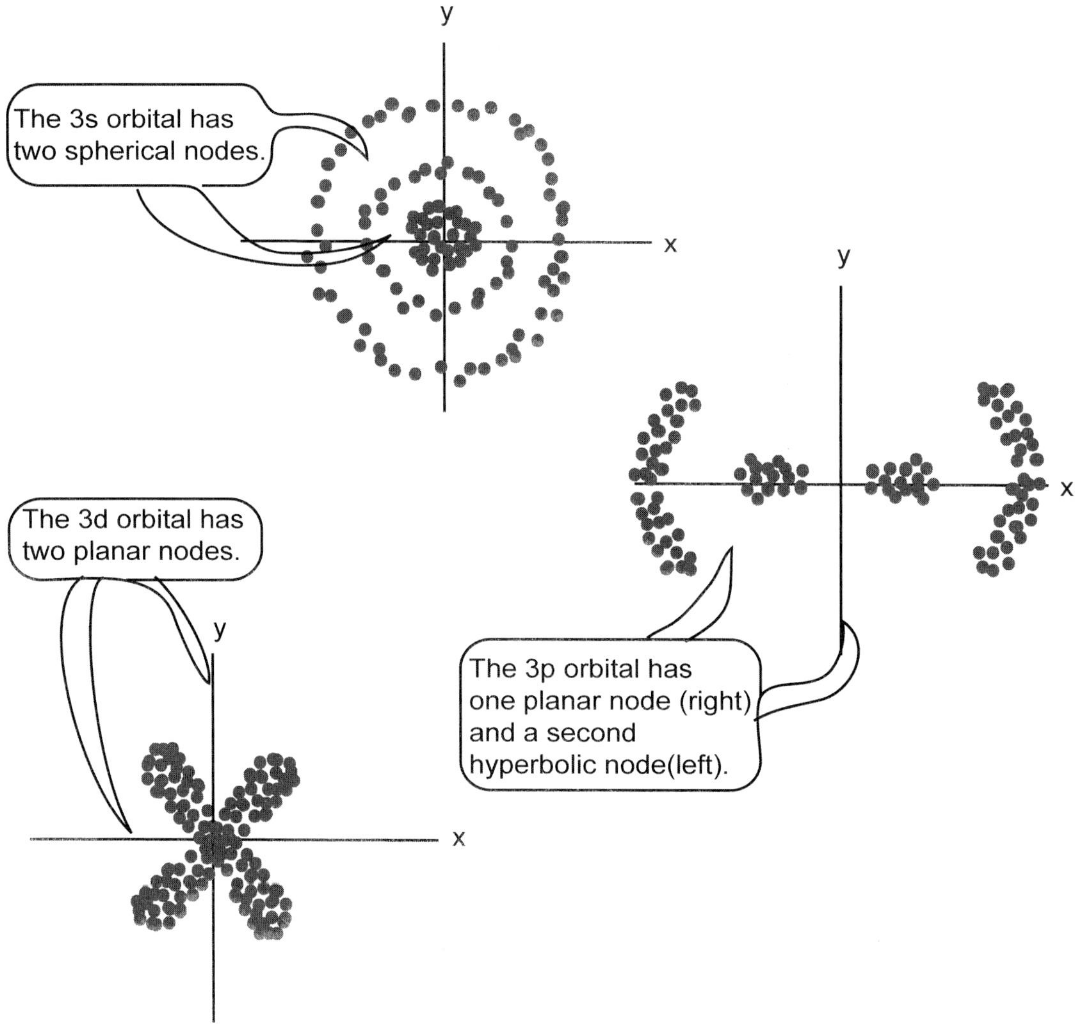

Figure 6.16 Electron density plots for *3s*-, *3p*-, and *3d*-orbitals show the nodes in each orbital. While this figure is just a sketch, accurate plots can be calculated from wavefunctions.

How did something as seemingly abstract as the quantum mechanical model of the atom help the development of modern materials and light sources? The concept of orbital shapes and energies ultimately provided ways to predict chemistry and optimize conditions in which reactions occur. Halogen lamps are predicated on the chemical reaction of halogen molecules with the tungsten wire of the lamp; the role of the halogens is depicted schematically in Figure 6.17. In a normal incandescent bulb, the choice of filament temperature involves a trade-off between the brightness and lifetime of the lamp. Higher temperatures give brighter light, but lead to shorter lifetimes due to faster evaporation of metal atoms from the filament. If halogen gas is present in the bulb, it can react with the tungsten as it vaporizes away from the high temperature filament. The resulting molecules redeposit on the wire, "recycling" the tungsten. These cyclic reactions allow higher operating temperatures and a brighter light without compromising on the lifetime of the bulb. We will not detail the orbital arguments used to understand this process, but the quantum mechanical model of halogens provides important clues. We must, therefore, know more about the way electrons occupy orbitals in atoms such as halogens.

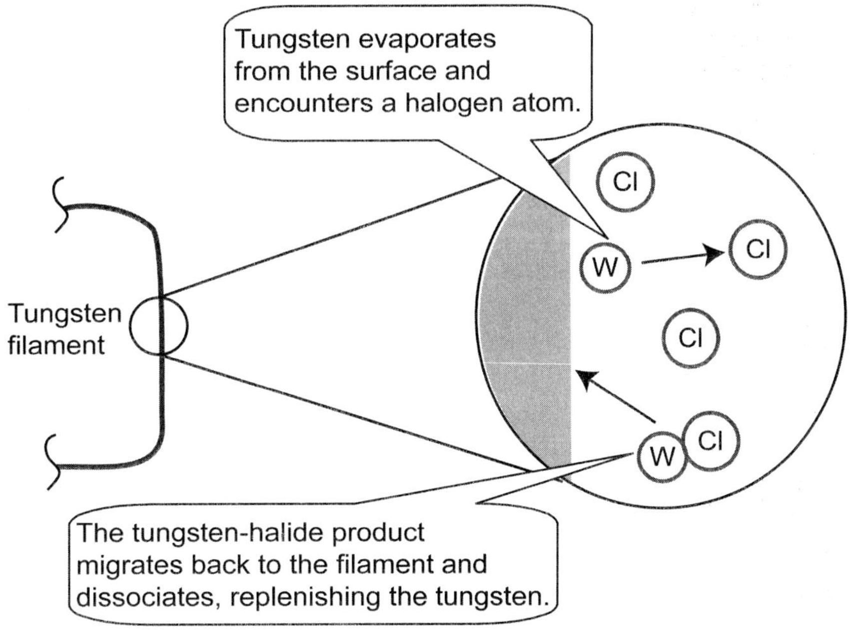

Figure 6.17 The presence of traces of halogen gas (here chlorine) allows evaporated tungsten atoms to be returned to the surface of a bulb's filament, effectively "recycling" the metal atoms. This allows the filament to operate at a higher temperature without compromising bulb lifetime, and explains why halogen lamps are brighter than regular incandescent bulbs.

6.5 The Pauli Exclusion Principle and Electron Configurations

In considering light bulbs, we have now identified four different types of atoms that play pivotal roles. Rare gases, halogens, mercury and tungsten have all been noted in one type of bulb or another. So that we may further understand the roles these elements assume, we must understand how their electrons are distributed in orbitals. Let's begin by asking the most basic question: How many electrons occupy an orbital? The answer requires the introduction of one more quantum number, the spin quantum number (designated m_s).

When placed in a strong magnetic field, like those used in Magnetic Resonance Imaging (MRI), electrons behave as though they are magnets. When electric charges move in loops they create magnetic fields, so scientists postulated that electrons (which are negatively charged) must be spinning. This postulate gave rise to the idea and the name of the spin quantum number. Since our modern model describes the electron as a delocalized wave, the spinning particle picture is no longer seen as literally correct. But the magnetic properties of electrons are consistent with the notion that they are spinning in one of two directions, and we often refer to electrons as being either "spin up" or "spin down." We use the quantum number m_s to distinguish between these two possible spins, and so it can assume only two possible values, +1/2 or - 1/2. Using this quantum number we now can specify the identity of any electron in an atom by using a set of four quantum numbers.

In MRI the "magnets" are protons in hydrogen atoms rather than electrons.

The specific values of + or – 1/2 are related to the magnetic moment of the electron.

We are constrained, however, in how we can specify the quantum numbers. We've already seen that the values of n, ℓ, and m_ℓ must conform to certain rules that determine what orbitals exist. An additional restriction was enumerated in mathematical terms by Wolfgang Pauli. In simple and practical terms, the **Pauli Exclusion Principle** states that *no two electrons in an atom may have the same set of four quantum numbers*. If the electrons have the same values for n, ℓ, and m_ℓ, then they must have different spin quantum numbers. This principle therefore dictates that any orbital may hold no more than two electrons. If there are two electrons in an orbital they must be *spin paired*; one with spin up and the other with spin down. Now that we know that orbitals can hold two electrons, how do we know what orbitals are actually occupied in a given atom?

Orbital Energies and Electron Configurations

We have already noted that electrons will occupy the lowest energy orbitals to form a ground state. But how do we know what orbitals are lower in energy, and what causes them to be lower? This question is best considered in two parts. First, recall that the closer the electron is to the nucleus, the lower its energy will be. Thus, the size of the orbital will be an important issue in determining its energy. And we've already seen that the size of orbitals tends to increase as the value of n increases. Electrons in smaller orbitals are held more tightly to the nucleus, so they have lower energies. Putting these two ideas together, we conclude that the energy of atomic orbitals will increase as the value of the n quantum number increases. Figure 6.18 confirms this trend by showing the radial probability functions for the first three shells of the hydrogen atom.

For the hydrogen atom, this is the entire story. But for multi-electron atoms we must think about this interaction further. The first difference is the size of the nuclear charge; it will be larger than for hydrogen. Larger nuclear charges exert stronger attractive forces on electrons, so the size of the orbitals will tend to decrease for larger nuclei. More importantly, the electrons in the orbitals interact with other electrons and not just the positively charged nucleus. For electrons that lie in larger orbitals therefore, the charge they "feel" is some combination of the nuclear charge and the offsetting charge of electrons located in smaller orbitals closer to the nucleus. This masking of the nuclear charge by other electrons is referred to as **shielding**. Instead of interacting with the full nuclear charge, electrons in multi-electron atoms interact with an **effective nuclear charge** that is smaller than the full charge due to shielding effects.

Referring to Figure 6.18, we can see how the concept of effective nuclear charge allows us to understand the energy ordering of the orbitals. For the shells shown, there are substantial size differences between the shells, such that *1s* is notably smaller than *2s* and *2p*, which in turn are smaller than *3s*, *3p* and *3d*. Deciding order for these shells is not complicated. When we consider subshells, however, the distinctions are subtler. Compare the *2s* and *2p* wave functions. Although the overall sizes of these orbitals are similar, the *2s* wave function shows a small local maximum close to the nucleus. And no such feature appears in the *2p* wave function. This means that a *2s* electron has a greater probability of being found close to the nucleus. And an electron closer to the nucleus will experience a larger effective nuclear charge. As a result the *2s*-orbital is lower in energy than the *2p*-subshell. In a similar way, the *3s*-orbital on average interacts with a larger charge because of the probability it will be close to the nucleus. Even though the first two lobes of the radial wave function are small, they provide the opportunity for electrons in the *3s*-orbital to "penetrate" through the shielding of other electrons. And this causes the energy of the *3s*-orbital to be lower than that of either the *3p*- or *3d*-subshells. Similar penetration in the *3p*-orbitals lowers their energy in comparison to the non-penetrating *3d*-subshell.

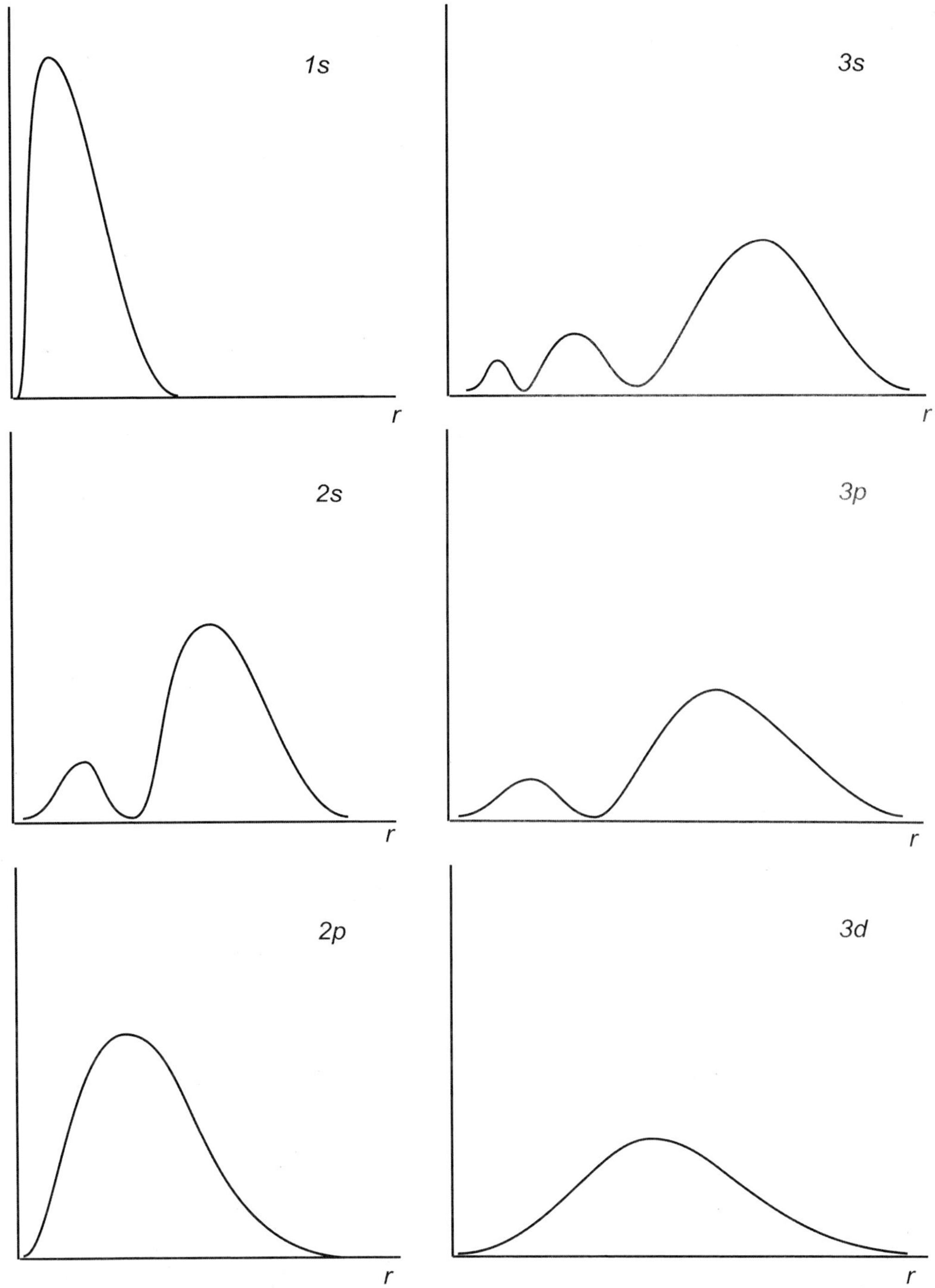

Figure 6.18 Graphs of the radial probability function are shown for hydrogen orbitals with *n* = *1*, *2*, and *3*. The increase in orbital size with increasing n is obvious. More subtle variations for orbitals with the same n value are also important in understanding the sequence of orbital energies.

As we reach the fourth shell of orbitals, the comparison of orbital size to penetrating probability becomes even more difficult. We will not attempt to carry the conceptual reasoning any further. Detailed calculations can be carried out to determine ordering. For large atoms these calculations need to include relativistic effects. The accepted order of orbital energies that arises from these calculations is: *1s, 2s, 2p, 3s, 3p, 4s, 3d, 4p, 5s, 4d, 5p, 6s, 4f, 5d, 6p, 7s, 5f, 6d, 7p*. There are no elements currently known whose ground state has electrons that occupy orbitals higher in energy than *7p*. You may have learned some device to remember this sequence of orbitals. We will reinforce this ordering in the next section by relating it to the structure of the periodic table. Now that we've established the energy sequence of the various subshells, we are ready to decide which orbitals should actually be occupied in each element. We will now consider rules for determining these *electron configurations*.

Hund's Rule and the Aufbau Principle

Aufbau comes from the German word for "build up".

The process of deciding what orbitals hold electrons in an atom requires that we "fill" orbitals starting with the lowest energy and proceeding to higher energy. In some ways this concept resembles building a building from the ground up, and is referred to as the **Aufbau Principle**. For the first few atoms this process is uncomplicated.

The hydrogen atom has one electron, and it occupies the *1s*-orbital. Helium has two electrons. Both can occupy the *1s*-orbital, so long as they are spin paired. Lithium has three electrons. Only two can occupy the *1s*-orbital, and so one electron must go into the next lowest energy orbital, the *2s*. A shorthand way to designate this orbital occupation is to write the orbital name followed by a superscript indicating the number of electrons in the orbital. Thus lithium has an electron configuration of $1s^2 2s^1$. Beryllium is the next element and its atoms contain four electrons. The *2s*-orbital can also hold two electrons, so the fourth electron is placed there to give a configuration of $1s^2 2s^2$. With five electrons, boron will also fully populate both the *1s* and *2s*-orbitals, but must account for an additional electron. The fifth electron goes into the next available orbital, a *2p*, to give $1s^2 2s^2 2p^1$.

Upon reaching carbon we are faced with a choice. There are three different *p*-orbitals, all with equal energy. So two electrons could pair and occupy a single *2p*-orbital, or they could occupy two different *2p*-orbitals, with one electron in each. The correct "choice" is motivated by the fact that electrons are all negatively charged particles. Two negatively charged electrons should repel one another, and so the most energetically favorable configuration would have them occupy separate orbitals where they can be farther apart from each other. The concept is summarized in **Hund's Rule,** which states that within a subshell, electrons occupy orbitals individually whenever possible. To depict this type of distinction in electron configuration requires a device more detailed than the representation we used for H, He, Li, Be and B. One possible choice is to use orbital diagrams with boxes representing orbitals, as shown in Figure 6.19. Another option is to identify individual orbitals with subscripts, such as p_x, p_y and p_z. Once we are familiar with electron configurations, though, the simple notation of $1s^2 2s^2 2p^2$ for carbon will suffice. Knowing Hund's Rule will allow whoever reads this configuration to immediately envision the correct orbital occupation provided by more detailed representations. Practice with this concept is provided in Example Problem 6.5.

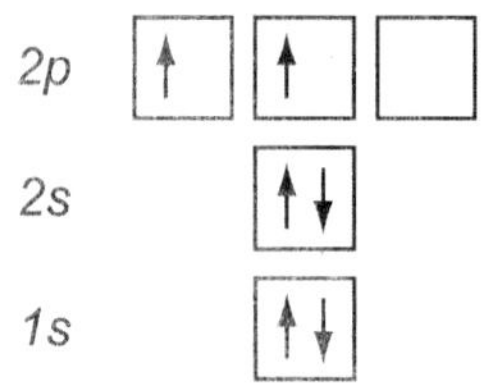

Figure 6.19 Electron configurations are sometimes depicted using diagrams of the type shown here for carbon. Such diagrams emphasize the presence of unpaired electrons, and may be helpful as you become familiar with electron configurations.

Example Problem 6.5

Sulfur may be an important element in emerging lighting technology. What is the electron configuration of a sulfur atom?

Strategy

We use the energy ordering and number of orbitals for each subshell to arrive at the overall electron configuration. Then using Hund's Rule we can predict the detailed electron configuration for an atom of sulfur.

Solution

Sulfur has 16 electrons, so we can place two in *1s*, two in *2s*, six in *2p*, two in *3s* and four in *3p* to get 16. The electron configuration is: $1s^22s^22p^63s^23p^4$.

For the detailed configuration, we must consider how to distribute the four electrons in the *3p*-subshell. We imagine placing the first three electrons each into separate *3p*-orbitals, and then filling one of those orbitals by adding the fourth electron. The resulting configuration is shown below.

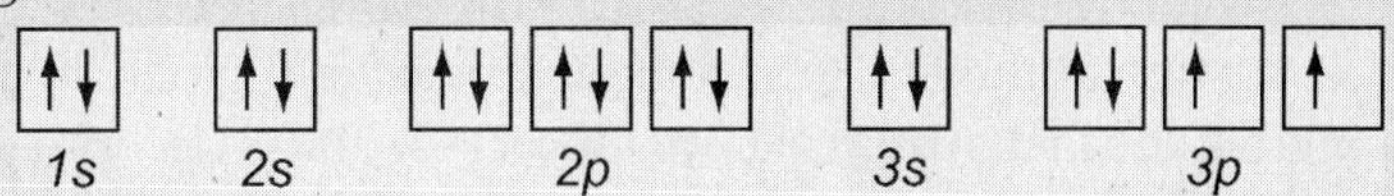

Check Your Understanding

What is the electron configuration of a silicon atom?

An observation about fluorescent lights will lead us to a way to abbreviate the writing of electron configurations. Most fluorescent bulbs are filled with argon gas, but some use krypton gas. The fact that these gases are chemically inert is important for their role in the operation of the lamp. What are the electron configurations of these two elements? Argon is $1s^22s^22p^63s^23p^6$ and krypton is $1s^22s^22p^63s^23p^64s^23d^{10}4p^6$. These two elements are chemically similar (both are essentially inert) precisely because they have similar electron configurations. In each case the "last electron" fills the *p*-subshell. This correspondence between chemical inertness and a filled *p*-subshell provides the motivation for a shortened notation for electron configurations.

This shorthand notation accomplishes two things. First, it makes it less unwieldy to write electron configurations. More importantly, it provides an introduction to a very important concept that relates electronic structure to chemical bonding. The chemical reactivity of an element is governed primarily by those electrons in its outermost occupied subshells. Electrons in filled lower shells will have little impact on reactivity. These inner electrons, which lie closer to the nucleus, are referred to as **core electrons**. They can be represented by the atomic symbol of the rare gas element that has the corresponding electron configuration. Thus, while the complete electronic configuration for potassium is $1s^22s^22p^63s^23p^64s^1$, the chemical properties of potassium are determined mainly by the outermost $4s^1$ electron. We can emphasize this fact by writing the configuration as $[Ar]4s^1$. In a similar way tin, Sn, can be written as $[Kr]5s^24d^{10}5p^2$. The electrons that are explicitly noted in this notation occupy orbitals that are further from the nucleus. While it is appropriately descriptive to designate them as outer electrons, they are typically referred to as **valence electrons**. When we take up the study of chemical bonding we will see how important it is to know the valence electron configuration of elements.

We will discuss the connections between electron configuration and chemical bonding in detail in the next chapter.

6.6 The Periodic Table and Electron Configurations

In talking about lamp technology we have mentioned two groups of elements, rare gases and halogens. Electron configurations give insight into the role these elements play in light bulbs. When we first encountered these groups, however, it was in conjunction with the periodic table (in Chapter 2). The relationship between the electron configuration predicted by the quantum mechanical model of the atom and the periodic table was vital to the acceptance of quantum mechanics as a theory. We will take advantage of this important correlation between electron configurations and the periodic table to make it relatively easy to remember configurations for the ground state of any element.

The outline of the periodic table in Figure 6.20 includes the elements that are usually sectioned off towards the bottom. The shading of regions in this figure divides the table into four blocks of elements. The elements in each block are all similar in that the electron in the highest energy orbital comes from the same subshell. Thus all the elements in the section at the left edge of the table, with the lightest shading, have their highest energy electron in an *s*-orbital. These are sometimes referred to as the s-block. The slightly darker section at the far right of the table is the *p*-block, because according to the Aufbau principle the last orbital occupied is a *p*-orbital. The transition metals are *d*-block elements (next darkest) and the lanthanides and actinides make up the *f*-block (darkest shading).

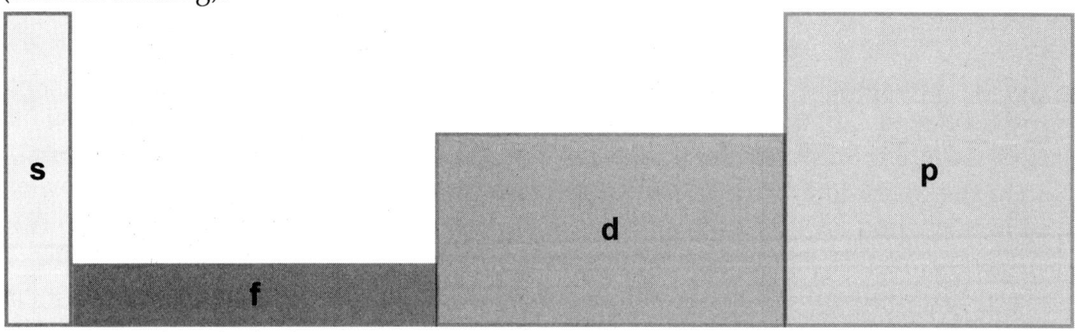

Figure 6.20 The shape of the periodic table can be broken down into blocks according to the type of orbital occupied by the highest energy electron in each element's ground state. The *f*-block elements are usually broken out of the table and displayed separately at the bottom. Here we have placed them in the body of the table to emphasize the sequence in which the various orbitals are filled.

Because of this structure, we can use the periodic table to determine electron configurations with only a few exceptions. We find the element of interest in the periodic table, and write its core electrons using the shorthand notation with the previous rare gas element. Then we determine the valence electrons by noting where the element sits within its own period in the table. Example Problem 6.6 demonstrates this procedure.

Example Problem 6.6

Use the periodic table to determine the electron configuration of tungsten (W), which is used in the filaments for most incandescent lights.

Strategy

Begin by finding tungsten in the periodic table. Use the rare gas element from the previous period to abbreviate the core electrons. Then determine the correct valence electron configuration by counting across the period in which tungsten is located.

Solution

The rare gas element preceding tungsten is Xe. When counting the valence electrons, don't forget the *f*-block elements. The correct configuration is *[Xe]* $6s^24f^{14}5d^4$.

Check Your Understanding

Mercury is used in fluorescent lamps as well as in mercury vapor lamps. Use the periodic table to determine the electron configuration of mercury, Hg.

Emphasizing the connection between the periodic table and electron configuration becomes even more important when we consider chemical reactivity. We've seen that halogen lamps are a particularly bright form of incandescent lights. All of the halogens have electron configurations ending with 'np^5,' where n is 2 for fluorine, 3 for chlorine, 4 for bromine and 5 for iodine. The similarity in electron configuration leads to similarity in chemical behavior, and explains why different halogens can be used with similar results. The choice of the specific halogen in a bulb can be adjusted to achieve desired design parameters. But all halogens behave in roughly the same way, enabling a combination of increased brightness and extended lifetime.

6.7 Periodic Trends in Atomic Properties

When we first introduced the concept of the periodic table we noted that various observations about the behavior of atoms led to its development. The reactivity of halogens and the inertness of rare gases are examples of the periodic law that influence the making of light bulbs. Using the understanding of orbitals and atomic structure we have just gained, it is possible to explain some of these periodic variations in greater detail. These atomic properties will then be used to help understand reactivity in later chapters.

Atomic Size

One important characteristic of atoms is their size. The measurement of atomic size is sometimes complicated, but the results of these measurements are summarized in Figure 6.21. There are two ways to explore trends in the periodic table, within a period and within a group. As we go down a period we observe an increase in atomic size. As we go across a period we observe a decrease. An easy way to remember this trend is to draw what look like an x,y-coordinate system axes at the bottom left of the periodic table. An increase in either the x- or y-direction on these axes corresponds to a decrease in atomic size. How do we understand this trend?

The size of the atom is largely determined by its valence electrons, since they occupy the outermost orbitals. Two factors are important: (1) the shell in which the valence electrons are found, and (2) the strength of the interaction between the nucleus and the valence electrons. As we go down a group in the table the first of these factors is at work. The size of the valence orbitals increases with n, so the size of the atom increases as we go down the period. The trend in moving across a row is less intuitive. We have already discussed the concept of effective nuclear charge. The distance between a valence electron and the nucleus will tend to be greater if there is a small effective nuclear charge. As we go across a period, we add electrons into the same subshell. These electrons do not shield each other highly, and the effective nuclear charge increases because each increment in atomic number adds a positively charged proton to the nucleus. The higher effective nuclear charge produces a stronger attraction to the nucleus. So the valence electrons are drawn closer to the nucleus and the size of the atom decreases.

Ionization Energy

Another property of atoms that plays a role in the way they interact in forming chemical bonds is the ease with which they lose electrons. This property, called the *ionization energy,* can be measured through an experiment fairly similar to the one we described earlier for the photoelectric effect. The ionization energy is defined to be the amount of energy we would have to supply to induce the following reaction.

$$X(g) \rightarrow X^{+}(g) + e^{-}$$

To be more precise, the energy required for this reaction is the first ionization energy. The energy needed to remove an additional electron from the X^+ cation, forming X^{2+}, would be the second ionization energy, and so on. We will use the term "ionization energy" for the first ionization energy.

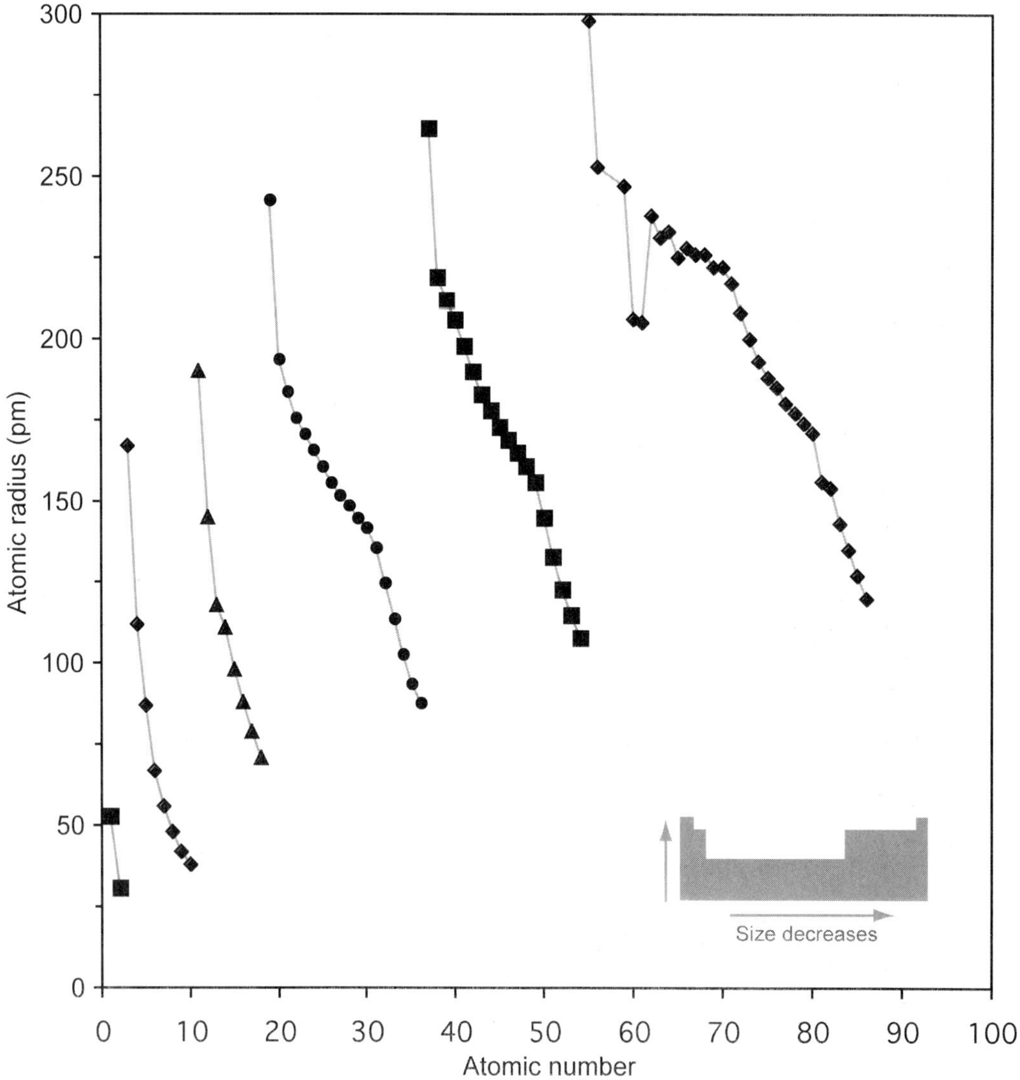

Figure 6.21 The graph shows atomic radius as a function of atomic number. The different data point symbols represent the different rows in the periodic table. The decrease in size from left to right across each period is clearly seen.

Insight into periodicity and atomic structure can be gleaned by looking at how the ionization energy varies across periods and within groups of the table. Recall from our discussion of atomic size that as we proceed across a period the effective nuclear charge increases. Not surprisingly, the ionization energy also generally increases across a period. As we move down a group, the valence electrons occupy larger and larger orbitals. As the valence electrons move farther from the nucleus, they become easier to remove. So ionization energy decreases as we move down a column in the periodic table. As we proceed up and towards the right of the periodic table the ionization energy increases, as seen in Figure 6.22.

Additional clues about atomic structure are revealed through the exceptions to the trends in ionization energy. Looking closely at Figure 6.22, we see that going from nitrogen to oxygen there is a slight decrease in ionization energy rather than the increase expected based on the periodic trend. This observation can be rationalized by considering electron pairing. Nitrogen has a half-filled *p*-subshell, with one electron in each 2*p*-orbital. To accommodate its p^4 configuration, oxygen must pair two electrons in one of the 2*p*-orbitals. Removing one electron, therefore, leaves only unpaired electrons in *p*-orbitals. Such a change leads to notably less electron-electron repulsion, enough so that the ionization energy of oxygen is lower than nitrogen. Similar exceptions to trends are seen for sulfur in the third period and selenium in the fourth.

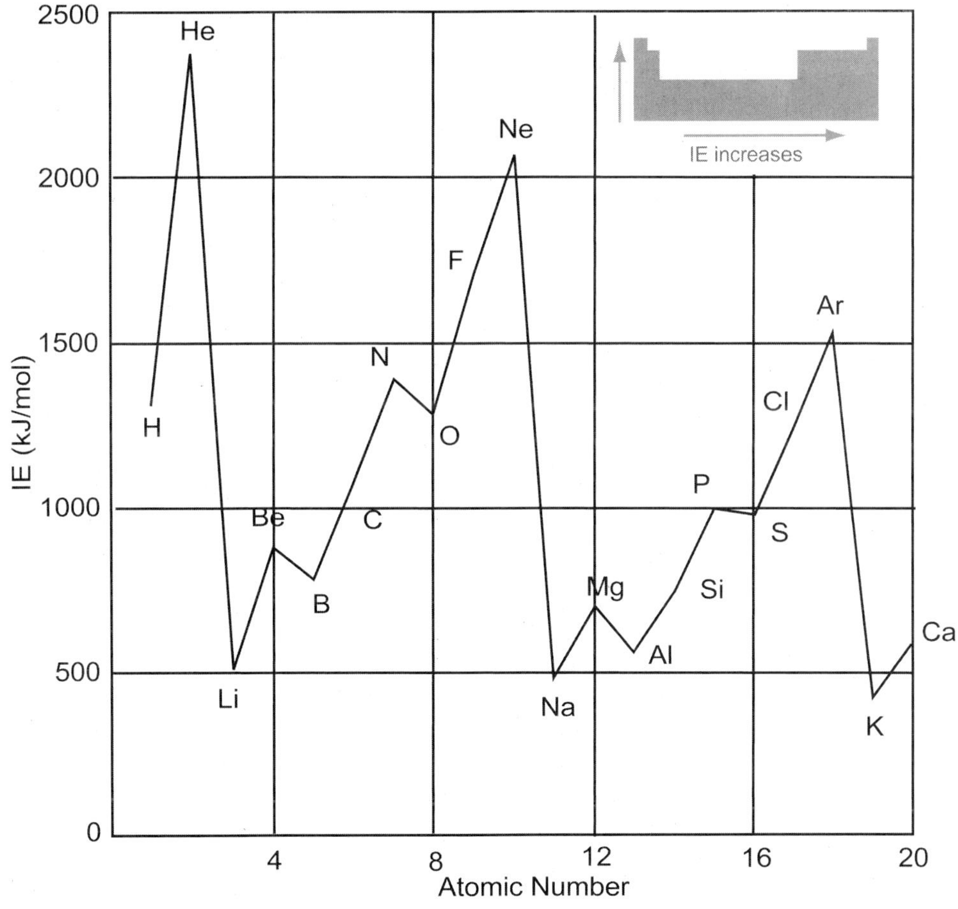

Figure 6.22 The graph shows the first ionization energy (in kJ/mol) vs. atomic number for the first 20 elements in the periodic table. The inset in the upper right emphasizes the general periodic trend: Ionization energy increases from left to right and bottom to top in the periodic table.

We can also gain insight into atomic structure by looking at successive ionization energies for the same element, as summarized in Table 6.4. For sodium, the first ionization energy is among the lowest, while the second is much larger than that for any neutral atom. For magnesium, the situation is quite different. The second ionization is still larger than the first, because the effective nuclear charge in the ion is larger than for the neutral atom. But the more dramatic change in ionization energy lies between the second and third ionizations. The point at which this large jump occurs for each element is highlighted using bold print throughout the table. In each case, the largest increase in ionization energy occurs when an electron is removed from a completely filled *p*-subshell. From this observation we can see the unusual stability of that filled *p*-subshell, and we can begin to appreciate why rare gas elements are so unreactive. Filled subshells of electrons are difficult to break up, and as we'll see this fact will help us to understand chemical bonding.

Z	Element	IE_1	IE_2	IE_3	IE_4
1	H	**1,312**	-	-	-
2	He	2,372	**5,250**	-	-
3	Li	**520.2**	7,298	11,815	-
4	Be	899.4	**1,757**	14,848	21,007
5	B	800.6	2,427	**3,660**	25,026
6	C	1,086	2,353	4,620	**6,223**
7	N	1,402	2,856	4,578	7,475
8	O	1,314	3,388	5,300	7,469
9	F	1,681	3,374	6,050	8,408
10	Ne	2,081	3,952	6,122	9,370
11	Na	**495.6**	4,562	6,912	9,544
12	Mg	737.7	**1,451**	7,733	10,540
13	Al	577.6	1,817	**2,745**	11,578
14	Si	786.4	1,577	3,232	**4,356**
15	P	1,012	1,908	2,912	4,957
16	S	999.6	2,251	3,357	4,564
17	Cl	1,251	2,297	3,822	5,158
18	Ar	1,520	2,666	3,931	5,771

Table 6.4 First four ionization energies (all in kJ/mol) for the elements of the first three periods. Those ionizations with values shown in gray cells with **bold** print involve removal of the last electron from a particular shell. Further ionization requires that an electron be removed from a more stable filled shell, and this leads to a very large increase in ionization energy.

Electron Affinity

The ionization energy is defined as the energy required to remove an electron from an atom and form a positively charged cation. But we also know that some atoms will routinely pick up an electron to form a negatively charged anion.

$$X(g) + e^- \rightarrow X^-(g)$$

We can define a property called **electron affinity** as the amount of energy we would have to supply to induce this reaction. Removing an electron from a neutral atom always requires an input of energy, since the atom will not simply eject electrons on its own. So ionization energies are always positive. But electron affinity values can be either positive or negative. If the resulting anion is stable, then energy will actually be released as the electron is added. But our definition of electron affinity says it is the amount of energy that must be *supplied* to form the anion. If energy is being released, we might say that we must "supply a negative amount of energy," so the electron affinity value will be negative. If the electron affinity is positive, that means that the anion being formed is not stable; in this case we would need to supply energy to force the atom to accept the electron. Most electron affinity values are negative.

This definition of electron affinity may seem a little twisted, since it is at odds with the normal meaning of the word "affinity." If the atom "wants" the electron, its electron affinity value is negative. But we define the signs in this way because it is consistent with the conventions usually followed when discussing energy in chemistry, physics, and engineering. Energy added to the atom is positive, and energy released by the atom is negative.

The study of energy changes in chemical reactions is called thermodynamics, and is presented in Chapter 9.

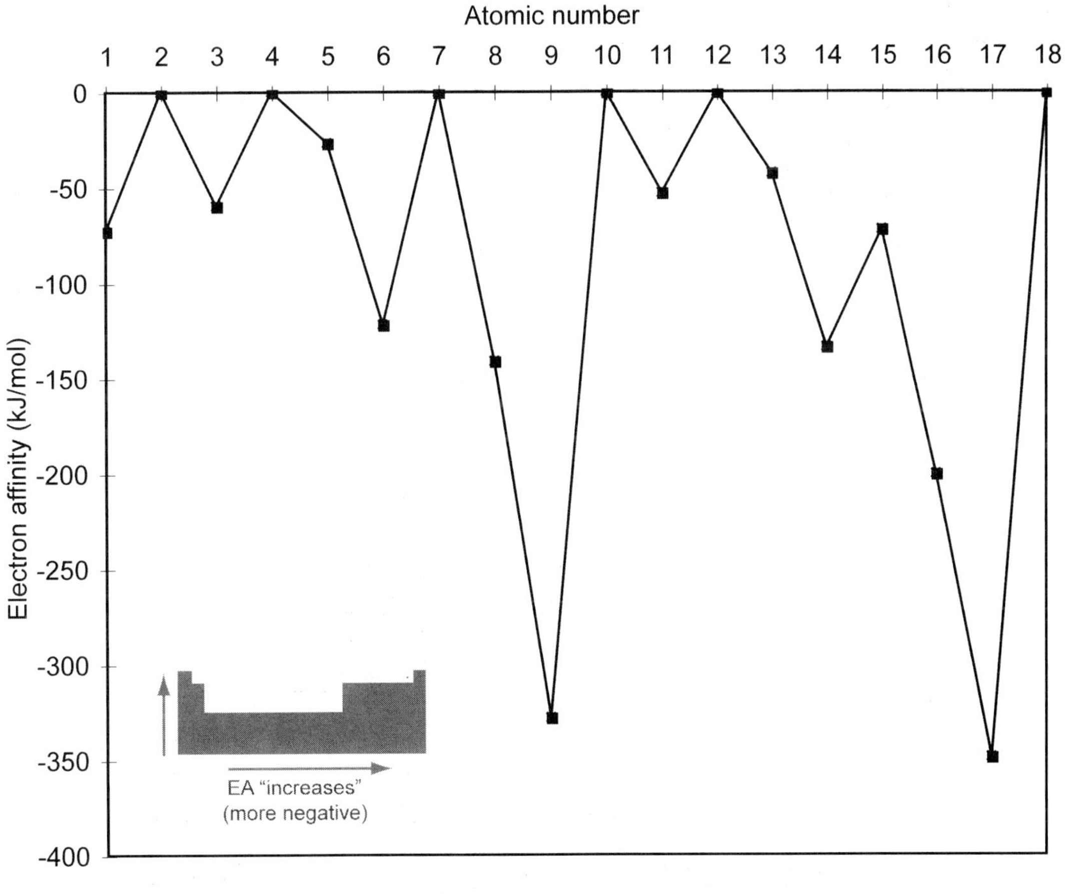

Figure 6.23 The graph shows the electron affinity (in kJ/mol) vs. atomic number for the first two rows of the periodic table. The inset at the lower left emphasizes the general periodic trend: Electron affinity "increases" - meaning the value becomes more negative - from left to right and bottom to top in the periodic table.

Electron affinity values display periodic trends based on electron configuration, much like the other properties we've been considering. The most immediate conclusion from Figure 6.23 might be that electron affinity values vary more erratically than ionization energies. Still, it should be evident from the figure that as we move from bottom to top or from left to right in the periodic table the electron affinity tends to become more negative. Thus in terms of the x, y scheme we have been using for remembering general trends, electron affinity becomes more negative with increasing x or y, much like ionization energy. It is customary to think of the trend in electron affinity in this way even though the value is actually becoming algebraically smaller as it gets more negative. That's because the more negative the electron affinity becomes, the more stable the anion will be.

A closer examination of the electron affinity values reveals that there are exceptions in the trends associated with electron pairing, just as there were for ionization energy. Thus, nitrogen does not have a negative electron affinity, because adding an electron would force the pairing of electrons in a *p*-orbital, a process that substantially increases the electron-electron repulsion.

There are other periodic trends in properties of the elements that we will not treat explicitly here. Some of those properties help dictate the use of materials for technology, including light bulbs. The choice of tungsten for the filament in incandescent bulbs is partly an economic issue, but it also depends on properties such as ductility and melting point. Because these properties are related to the bulk metal rather than isolated atoms, they are more complicated than those discussed thus far. But the electron configuration of tungsten does provide some insight. Melting a metal requires the breaking of metallic bonds. We'll see in Chapter 8 that having more unpaired electrons present in an atom tends to increase the strength of metallic bonding in solids of that element. And stronger bonding in the solid metal means the melting point will be higher. So the four unpaired *d*-electrons in tungsten help produce a solid that melts at high temperatures, allowing the filament to operate at high enough temperatures to produce a bright light output.

6.8 Insight into Modern Light Sources: LEDs and Lasers

The history of the incandescent light bulb dates back to the middle of the nineteenth century. (Although Thomas Edison is often regarded as the inventor of the electric light bulb, the first patent for an incandescent lamp was granted to Frederick de Molyens of London, in 1841.) Over the years numerous developments have improved the performance and efficiency of incandescent lamps. Advances in our understanding of chemistry have enabled several of these improvements, like the use of halogen gases. Similarly, developments in chemistry have helped identify new phosphor molecules that drove improvements in fluorescent lamps. Today's technologies rely heavily on more modern light sources, including light-emitting diodes (LEDs) and lasers. And the very existence of these twentieth century creations was made possible by improvements in our understanding of atomic structure and chemical bonding.

How does the light produced by an LED or a laser differ from that given off by an incandescent or fluorescent lamp? And what particular attributes of these devices and the light they emit make them so well suited for particular tasks?

You can almost certainly find light-emitting diodes in many of the appliances you use every day. The power indicators on everything from cell phones to TVs and the displays on many digital alarm clocks and microwave ovens are just a few examples of common uses for LEDs. As you think about some of the applications in which they are used, you may realize one of the characteristics of LEDs: each LED emits light of a particular color. Although you can buy colored incandescent bulbs, they rely on some sort of filter coated onto the surface of the bulb. But the filament still emits white light, so the colored filter greatly reduces the efficiency of the bulb. In an LED, the light emitted is usually *monochromatic*, meaning it is restricted to a single wavelength or color. So LEDs provide an easy way to produce colored light, and that's one reason why they are so widely used as indicators on electronic devices.

Some LEDs emit 2 or more wavelengths simultaneously, but usually these are so closely spaced that the human eye cannot separate them.

Another important contrast between an LED and a small incandescent bulb is in durability: light bulbs break or burn out, but LEDs last indefinitely under normal conditions. We've seen that bulbs burn out because their filaments gradually break down at the high operating temperatures needed to produce light. An LED, though, does not require a high operating temperature. It functions so long as a small electric current is flowing through it. This not only gives the LED a long useful lifetime, but also reduces the amount of power consumed and heat given off, both of which can be important design considerations in modern electronics. Because they lose very little energy as heat, LEDs can operate at very high efficiency, converting nearly all of the input energy into light. This is especially important for small portable devices, like cell phones or MP3 players, where battery lifetime is a major consideration.

LEDs are an example of what are called "solid-state devices," in which the primary component is simply a small piece of some solid material. The important functional properties of such devices are determined primarily by the composition of this solid material - and that's where chemistry comes in. In the case of an LED, the color of light that will be emitted depends on the details of the chemical bonding between the atoms in the solid. And as you might guess, that bonding depends on the identity of the atoms involved. So adjusting the composition of the solid can change the color of light that will be emitted. We'll look at some of the details behind this in Chapter 8. But the practical implication is that we can produce an LED that emits almost any color of light that we want.

Lasers share some traits with LEDs: they produce monochromatic light, and the wavelength of that light depends on the specific properties of the laser. But lasers also possess other important properties. Perhaps most obviously, the light emitted by a laser is highly directional, emerging as a well-defined beam. You've almost certainly seen laser pointers or other lasers, so you may know that the beam is generally not even visible from the side since all of the light is traveling in the same direction. Laser light is also *coherent*, which means that all of its waves are perfectly in-phase, going through maxima and minima together. Many varieties of lasers are known, ranging from small devices found in CD players, laser printers, or supermarket price scanners to very high-powered systems used in industrial machining.

The invention of the laser is a direct result of the emergence of the quantum mechanical model of the atom. Laser light is actually emitted as atoms or molecules go from a higher energy level down to a lower energy level, just like we saw in atomic spectra. So every laser's design is ultimately based on a detailed understanding of the energy levels available to the atoms or molecules in what is called the *laser medium*. The operation of a laser relies on a phenomenon called stimulated emission, which becomes important when the population of one energy level becomes greater than the population of a lower energy level. Many of the most common lasers, like the ones in your CD players, are actually solid-state devices in which this "population inversion" is established for the electrons in a solid. In other lasers, the light is emitted by atoms or molecule in a gas or in solution. The primary engineering challenges in designing a laser are to identify a medium that can emit the desired wavelength of light and to devise a way to establish and maintain the population inversion. So an understanding of the energy levels involved is clearly essential.

As new lighting technologies emerge in the twenty-first century, the connections between chemistry and engineering continue to play a major role. Recent improvements in liquid crystal displays used in flat panel computer monitors, for example, are due in part to the development of new chemical compounds whose light emitting properties are tailored for their tasks. Research into organic light emitting diodes, or OLEDs, is moving forward at a rapid pace, and holds the promise for a whole new class of paper thin and flexible color displays. Advances in all of these areas require the close cooperation of research chemists and design engineers.

Focus on Problem Solving

Question

(a) Based on your understanding of atoms and their spectra, speculate on why streetlights that use sodium lamps seem yellow/orange in color. (b) What device or instrument could be used to make a measurement that would test your hypothesis from (a)?

This question involves conceptual problem solving. We need to look at questions like this from the perspective of asking what molecular level information must be inferred to be able to answer it. Then we need to say what are the implications of the molecular level understanding to answer part (b).

Answer

(a) The key words in this question are "sodium lamp." This description implies there is something important about the lamp that is related to the element sodium. We saw in this chapter that atoms have distinctive spectra, so it is reasonable to surmise that in the case of sodium ions enough colors are missing that the remaining light appears yellow/orange. (b) We could test this hypothesis by using a spectrometer to view the light coming from the lamp. If our assertion is correct the spectrum should show lines in the yellow-orange region but not in other parts of the visible spectrum.

Chapter 6
Problems & Exercises

Insight into Incandescent and Fluorescent Lights

6.1 How is the color of molten iron (at a blacksmith museum, for example) related to a light bulb?

6.2 What is the first step in the process by which an incandescent bulb provides light?

6.3 What material is most commonly used for the filaments in an incandescent bulb? Why is this particular material chosen?

6.4 How is electricity used differently in an incandescent light bulb and a fluorescent light bulb?

6.5 Describe what is meant by the term *electric arc*.

6.6 What is the role of the electric arc in a fluorescent light bulb?

The Electromagnetic Spectrum

6.7 Explain why light is referred to as *electromagnetic radiation*.

6.8 Which of the waves depicted here has the highest frequency? Explain your answer.

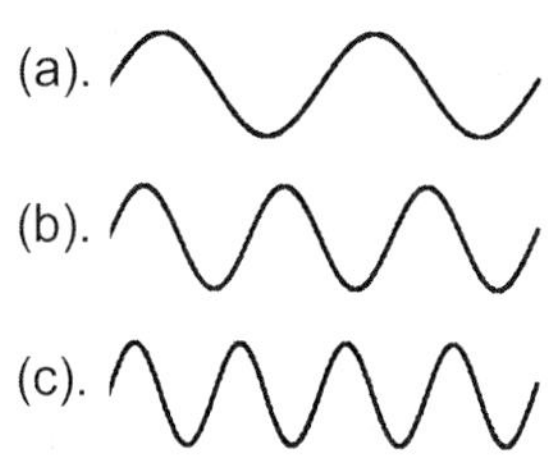

6.9 Arrange the following regions of the electromagnetic spectrum in order of increasing frequency: IR, UV, radio wave, visible.

6.10 If the microwave frequency used by a particular communications tower is 920 MHz, what is the wavelength of this radiation?

6.11 Decorative lights, such as those found on Christmas trees, often achieve their colors by painting the glass of the bulb. If a string of lights includes bulbs with wavelengths of 480 nm, 530 nm, 580 nm and 700 nm, what are the frequencies of the lights? Use Figure 6.7 to determine what colors are in the set.

6.12 Define the term *refraction*.

6.13 Define the term *photon*.

6.14 What is the relationship between the number of photons and the intensity of the light?

6.15 Find the energy of a photon with each of the following frequencies? (a). 15.3 THz (b). 1.7 EHz (see Table 1.2 if needed) (c). 6.22×10^{10} Hz

6.16 Find the energy of a photon with each of the following frequencies? (a). 4.32×10^{12} Hz (b). 106.9 MHz (c). 31.6 GHz (d). 6.11×10^{20} Hz

6.17 For photons with the following energies, calculate the wavelength and identify what region of the spectrum they are from (a). 3.5×10^{-20} J (b). 8.7×10^{-26} J (c). 7.1×10^{-17} J (d). 5.5×10^{-27} J

6.18 For photons with the following energies, calculate the wavelength and identify what region of the spectrum they are from (a). 6.0×10^{-19} J (b). 8.7×10^{-22} J (c). 3.2×10^{-24} J (d). 1.9×10^{-28} J

6.19 Optical disk drives in the Sony Playstation® 2 contain an unusual laser that can produce 2 different wavelengths of light (650 nm and 780 nm) from a single device. What is the energy of a single photon at each of these wavelengths?

6.20 The laser in most laser printers operates at a wavelength of 770 nm. What is the energy of a single photon of the 770 nm light?

6.21 Which photon has more energy, one with a wavelength of 3.52×10^{14} Hz or one with a wavelength of 115 nm? Explain your answer.

6.22 Fill in the blanks below to complete a description of the photoelectric effect experiment. (You should be able to do this with just one or two words in each blank.)

A beam of ______________strikes ______________, causing ______________to be emitted.

6.23 When ultraviolet light with a wavelength of 337 nm strikes the surface of potassium metal, electrons are ejected with a maximum kinetic energy of 2.30×10^{-19} J. What is the binding energy of electrons to the metal?

6.24 The electron binding energy for copper metal is 7.18×10^{-19} J. Find the longest wavelength of light that could eject electrons from copper in a photoelectric effect experiment.

Atomic Spectra

6.25 What is the difference between continuous and discrete spectra?

6.26 What was novel about Bohr's model of the atom?

6.27 Describe how the Bohr model of the atom accounts for the spectrum of the hydrogen atom.

6.28 According to the Bohr model of the atom, what happens when an atom absorbs energy?

6.29 Define the term *ground state*.

6.30 A neon atom emits light at many wavelengths, two of which are at 616.4 nm and 638.3 nm – both of these transitions are to the same final state. (a) what is the energy difference between the two states for each transition. If a transition between the two higher energy states could be observed, what would the frequency of the light be?

6.31 A mercury atom emits light at many wavelengths, two of which are at 435.8 nm and 546.1 nm – both of these transitions are to the same final state. (a) what is the energy difference between the two states for each transition. If a transition between the two higher energy states could be observed, what would the frequency of the light be?

6.32 Describe how excited states of atoms play a role in light bulb technology.

6.33 Why do fluorescent bulbs have phosphor on the walls of the bulb?

The Quantum Mechanical Model of the Atom

6.34 How did the observation of electron diffraction affect the development of the quantum mechanical model of the atom?

6.35 Why do we use a wave function to describe electrons?

6.36 What are the mathematical origins of quantum numbers?

6.37 What are the allowed values for the principle quantum number? For the secondary quantum number?

6.38 Use the rules of quantum numbers to determine which of the following sets of numbers are not allowed. State why they violate the rules. (a). $n=2, \ell=2, m_\ell=0$; (b). $n=0, \ell=0, m_\ell=0$; (c). $n=7, \ell=0, m_\ell=0$; (d). $n=3, \ell=1, m_\ell=-1$

6.39 When we say an electron is in a *4f*-orbital, what quantum numbers are specified?

6.40 Why are there no 2d orbitals?

6.41 What is the total number of orbitals possible with n=4? With n=5?

6.42 What is the total number of orbitals possible with n=6? With n=7?

6.43 The p and d orbitals are sometimes referred to as directional. Contrast them with s orbitals to offer an explanation for this terminology.

6.44 Define the term *nodal plane* (or *node*).

6.45 Referring to Figure 6.16, draw a 4p orbital showing all its nodes.

Pauli Exclusion Principle and Electron Configurations

6.46 Does the fact that the fourth quantum number is called the spin quantum number prove that electrons are spinning? What observations gave rise to this name?

6.47 Define the term spin paired.

6.48 On what does the Pauli Exclusion Principle place a limit?

6.49 Define the term shielding.

6.50 Why does the size of an orbital have an effect on its energy?

6.51 What is effective nuclear charge and how does it relate to the size of an orbital?

6.52 Depict two ways to place electrons in the 2p orbitals for a nitrogen atom. Which depiction is the correct one according to Hund's Rule?

6.53 How does the charge of electrons provide some rationale for Hund's rule?

6.54 Write the electron configuration for the following elements: (a). Ar (b). Hg (c). Sn (d). Fe.

6.55 Write the electron configuration for the following elements: (a). V (b). Br (c) Pt (d). W..

6.56 From the list of atoms and ion given, identify any pairs that have the same electron configurations and write that configuration. Na^{+}, S^{2-}, Ne, Ca^{2+}, Fe^{2+}, Kr, I^{-}.

Periodic Table and Electron Configurations

6.57 Halogen lamps rely on the predictable chemistry of the halogen elements. Compare the electron configurations of 3 halogens and use this comparison to justify the similarity of the chemistry of these elements.

6.58 Distinguish between the terms *core electrons* and *valence electrons*.

6.59 Describe how valence electron configurations account for some of the similarities in chemical properties among elements in a group.

6.60 Why is there no element to the immediate right of Mg in the periodic table?

6.61 If another 50 elements were discovered, how would you expect the appearance of the periodic table to change?

6.62 Which blocks of the periodic table comprise the main group elements?

6.63 What is the main difference between a halogen lamp and an ordinary incandescent lamp?

6.64 Look at the table of electron configurations in Appendix C. What elements have configurations that are exceptions to Hund's Rule? Propose a reason why these elements have these exceptions.

Periodic Trends of Atomic Properties

6.65 How does the effective nuclear charge help explain trends in atomic size across a period of the periodic table?

6.66 Use the electronic configuration of the alkali metals and your knowledge of orbitals and quantum numbers to explain the trend in atomic size of alkali metal atoms.

6.67 Which atom in each group is the largest? (a). K, Ca, Mg (b). S, Si, O (c). Ar, Ne, K

6.68 Define the term *ionization energy*. What is the difference between the definition of the first ionization energy and the second ionization energy?

6.69 At which ionization for chlorine would you expect the first large jump in ionization energy? Would this be the only large jump in energy if you continued to ionize the chlorine?

6.70 For each of the following groups, which element should have the smallest first ionization energy? (a). K, Mg, Na (b). N, O, S (c). Cl, Ar, Si

6.71 How do we explain the fact that the ionization energy of oxygen less than that of nitrogen?

6.72 Which element would you expect to have the largest second ionization energy, Na, C, or F? Why?

6.73 Why is nitrogen the only p-block element in the second period with a positive electron affinity?

6.74 Which element in each set has the largest (*i.e.*, most negative) electron affinity? (a). N, O, F (b). Al, B, Ga (c). Be, B, C

6.75 Rank the following in order of *increasing* size: F^-, Mg^{2+}, Na^+, Ne, O^{2-}

6.76 Rank the following in order of *decreasing* ionization energy: Cl, F, Ne^+, S, S^-

Insight into Modern Light Sources: LEDs and Lasers

6.77 Several excited states of the neon atom are important in the operation of a helium-neon laser. In these excited states, one electron of the neon atom is promoted from the 2p level to a higher energy orbital.

An excited neon atom with a $1s^2$, $2s^2$, $2p^5$, $5s^1$ electron configuration can emit a photon with a wavelength of 3,391 nm as it makes a transition to a lower energy state with a $1s^2$, $2s^2$, $2p^5$, $4p^1$ electron configuration. Other transitions are also possible. If an excited neon atom with a $1s^2$, $2s^2$, $2p^5$, $5s^1$ electron configuration makes a transition to a lower energy state with a $1s^2$, $2s^2$, $2p^5$, $3p^1$ electron configuration, it emits a photon with a wavelength of 632.8 nm. Find the wavelength of the photon that would be emitted in a transition from the $1s^2$, $2s^2$, $2p^5$, $4p^1$ electron configuration to the $1s^2$, $2s^2$, $2p^5$, $3p^1$ electron configuration. (It should help if you start by drawing an energy level diagram.)

6.78 Another fairly new light source, and one likely to have a larger impact on home lighting, is the compact fluorescent bulb. These bulbs combine the energy efficiency of traditional fluorescent lights with the smaller size, "warmer" light, and dimming ability of incandescent lights. Although compact fluorescent bulbs are expensive, manufacturers claim maintain that subsequent savings will more tan offset the initial costs. Analyze the relative cost of incandescent versus compact fluorescent lighting, assuming the information in the table below is accurate. Are industry claims of cost savings justified?

	Compact fluorescent	Incandescent
Initial cost	\$15.00	\$2.00
Electricity usage	18 watts	75 watts
Electricity cost	\$0.04 / kwh	\$0.04 / kwh
Lifetime	10,000 hours	2,500 hours

6.79 How much energy could be saved each year by replacing all the incandescent bulbs in a typical house with compact fluorescent bulbs? Assume there are six 75-watt lamps that are on an average of 3 hours a night, and they are each replaced with 18-watt compact fluorescent bulbs.

Additional Problems

6.80 The photoelectric effect can be used to measure the value of Planck's constant. Suppose that when a photoelectric effect experiment was carried out using light with $\nu = 7.50 \times 10^{14}\ s^{-1}$, ejected electrons were detected with a kinetic energy of 2.50×10^{-11} J. The experiment was then repeated using light with $\nu = 1.00 \times 10^{15}\ s^{-1}$ and the *same metal target*, and electrons were ejected with kinetic energy of 5.00×10^{-11} J. Use these data to find a value for Planck's constant. **HINTS:** These data are fictional, and will give a result that is quite different from the real value of Planck's constant. Be sure that you do **not** use the real value of Planck's constant in any calculations here. It may help to start by thinking about how you would calculate the metal's binding energy if you already knew Planck's constant.

6.81 A mercury atom is initially in its lowest possible (or ground state) energy level. The atom absorbs a photon with a wavelength of 185 nm, and then emits a photon with a frequency of 6.88×10^{14} Hz. At the end of this series of transitions, the atom will still be in an energy level above the ground state. Draw an energy level diagram for this process, and find the energy of this resulting excited state, assuming we assign a value of $E = 0$ to the ground state. (This choice of $E = 0$ is *not* the usual convention, but it will simplify the calculations you need to do here.)

6.82 When a photoelectric effect experiment was carried out using a metal 'M' and light at wavelength λ_1, electrons were emitted with a kinetic energy of 1.6×10^{-19} J. The wavelength was reduced to 1/2 of its original value and the experiment was repeated (still using the same metal target). This time electrons were emitted with a kinetic energy of 6.4×10^{-19} J. Find the electron binding energy for metal M. (The actual wavelengths used were not recorded, but it is still possible to find the binding energy.)

6.83 A metallic sample is known to be barium, cesium, lithium, or silver. The electron binding energies for these metals are listed below.

Metal	Electron binding energy (J)
Barium	4.30×10^{-19}
Cesium	3.11×10^{-19}
Lithium	3.94×10^{-19}
Silver	7.59×10^{-19}

One way to identify the element might be through a photoelectric effect experiment. The experiment was performed 3 times, each time using a different laser as the light source. The results are summarized below. (The kinetic energy of the ejected photoelectrons was not measured.)

Laser wavelength	Photoelectrons seen?
532 nm	No
488 nm	Yes
308 nm	Yes

Based on this information, what conclusions can be drawn as to the identity of the metal?

6.84 When a helium atom absorbs light at 58.44 nm, an electron is promoted from the1s orbital to a 2p orbital. Given that the ionization energy of (ground state) helium is 2,372 kJ/mol, find the longest wavelength of light that could eject an electron from the excited state helium atom.

Focus on Problem Solving Exercises

6.85 Suppose you are working as an engineer on a nanotechnology project where electronics components are constructed one molecule at a time. The design calls for the use of a noble gas atom to act as an insulator. You need to run a series of experiments where the size of the atom is increased for each trial. Explain the reasoning you will use to pick the smallest noble gas first and proceed to the largest.

6.86 Nanotechnology is an emerging field in which designs, such as electronic circuits, are built from molecular size components – essentially one molecule at a time. Suppose you graduate and become a nanotechnology engineer. You are working on a circuit and you need a component that will accept an electron as part of a molecular capacitor. You choose to use a halogen atom for this task. Explain your reasoning about which halogen atom will have the largest electron affinity (a trait you prefer to have in this nano-circuit design.)

6.87 In some telescopes, distant stars are studied by looking at only specific visible wavelengths. Based on the fact that stars are mostly hydrogen and knowing about the nature of the spectra of atoms, speculate on why the strategy of looking at only a small range of wavelengths is a reasonable one for constructing a telescope.

6.88 The solubility of ionic salts can be predicted by noting the relative sizes of the ions in the material. If there is a large difference in size of the ions the substance is more likely to be soluble. Lithium is the smallest ion among the alkali metals. Based on the rule noted above, which lithium halogen salt is the most soluble and which is the least soluble? Explain your reasoning.

6.89 Atomic absorption spectroscopy is based on the atomic spectra of the elements being studied. It can be used to determine the impurities in a metal sample. If an element is present, light at the appropriate wavelength is absorbed. You are working with a metal stamping company and the rolled steel you use to form panels for automobile doors is failing at an alarming rate. There is some chance that the problem is unacceptably high levels of manganese in the steel. Given that the atomic spectrum of Mn has three lines near 403 nm, how could you use a spectrometer to determine the amount of manganese in the steel?

6.90 The red color seen in fireworks is the result of having strontium containing salts in the fireworks bomb. Similarly the green/blue colors sometimes seen in fireworks arises from copper containing salts. Based on your understanding of atomic spectra and the colors seen in fireworks, describe which atom, copper or strontium, has more widely separated energy levels.

Chapter 7: Chemical Bonding and Molecular Structure

The study of atomic structure in Chapter 6 provided an interesting detour into modern physics, and there's no doubt that the paradoxes of quantum mechanics are intellectually intriguing. But quite apart from pure scientific curiosity, the concepts we developed for atomic structure offer important insights into some of the most fundamental questions in chemistry. How and why do atoms combine to form molecules? What do those molecules actually look like, and what factors determine their shapes? In our technological society, the answers to these seemingly abstract questions can have tremendous practical and economic importance. Knowing how to control or selectively exploit the formation of chemical bonds has played a crucial role in the development of many new materials and devices.

Most elements of the periodic table could provide examples of chemical bonding behavior, but here we will highlight the chemistry of fluorine. Not only does fluorine form a number of compounds that are familiar to us because of their commercial value, but the variety of fluorine compounds also allows us to introduce a range of important concepts. In this chapter we will look at the reasons atoms form bonds to one another, the different types of chemical bonds, and the connections between bonding and molecular shape. An examination of these characteristics can help us understand the properties of many substances, and is crucial to understanding how molecules interact and react with one another. We'll also improve our connection between the microscopic and symbolic views of chemistry as we become able to visualize a molecule's three-dimensional shape based on its formula.

7.1 Insight Into Familiar Fluorine Compounds

Fluorine is a highly reactive element, so it does not exist in nature in its elemental form. Many chemists have been injured by explosive reactions involving fluorine, and people who work with it need to use stringent safety precautions. Given the difficulties involved in working with elemental fluorine, it may at first seem surprising that compounds containing fluorine are quite commonplace. Examples include not only minerals like fluorite, CaF_2, but also many synthetic products, like Teflon, that are in everyday use. The explosive nature of reactions involving F_2, however, provides a clue as to why there are so many fluorine compounds. As we learned in our discussion of explosions, the products that are formed as the result of a violent reaction tend to be extremely stable under ordinary conditions.

Perhaps the most widely used fluorine compound today is Teflon, which in chemist's terms is called polytetrafluoroethylene (PTFE). As we learned in Chapter 2, PTFE consists of tetrafluoroethylene monomers joined together to form a long chain of carbon atoms, each of which is also connected to two fluorine atoms (Figure 7.1). Scientists at DuPont discovered Teflon accidentally, while trying to make new refrigerants. It is a remarkable polymer, whose properties have been exploited in a wide range of applications. Many polymers, like polyethylene, melt at low temperatures, but PTFE is stable at high temperatures. Without this property, its usefulness as a coating for cookware would be rather limited. It is also very resistant to corrosive chemicals. The only substances that react with it are elemental fluorine and molten alkali metals, all very reactive substances. Because of its resistance to chemical attack, PTFE finds use

The Manhattan Project was the US effort to develop atomic weapons during World War II.

in laboratories that employ corrosive materials. For example, PTFE was employed in the Manhattan Project to line valves that came into contact with the highly corrosive uranium hexafluoride, UF_6, used to separate isotopes of uranium. One of PTFE's most famous characteristics, of course, is its "non-stick" nature. When Teflon was first discovered, DuPont scientists were amazed at how difficult it was to find substances to which it would stick. While they saw the commercial value of this property, it also posed a number of questions. Why was the polymer resistant to sticking, and how could it be made to stick to other surfaces, such as the surface of a skillet? Why was PTFE so chemically inert? Why was it more stable at high temperatures than seemingly similar polymers like polyethylene? Why did it have such a high melting point? In this chapter we will explore concepts that will help us address some of these questions.

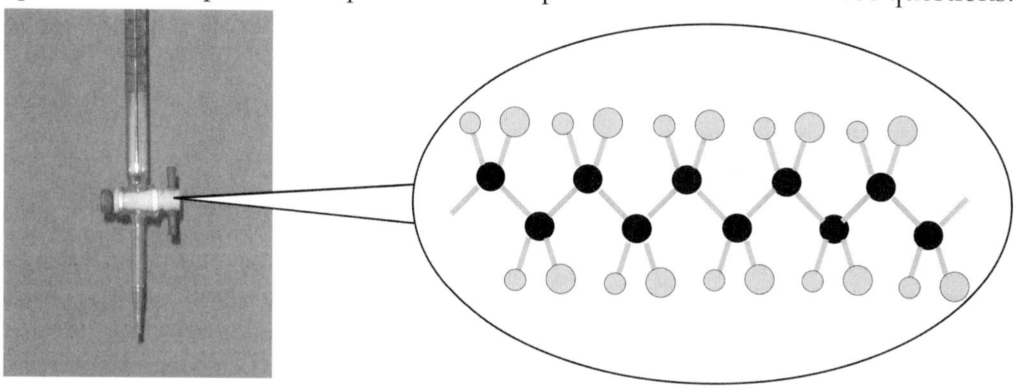

Figure 7.1 Because it is chemically inert, Teflon is often used in applications like the buret valve shown in the photo. The observable properties of Teflon can ultimately be explained in terms of its molecular structure, which is illustrated in the ball and stick model to the right of the figure.

7.2 The Ionic Bond

Fluorine forms a number of compounds that are held together by ionic bonds. An **ionic bond** is formed by the electrostatic attraction of oppositely charged ions. We have already mentioned CaF_2, composed of the Ca^{2+} and F^- ions, but there are many other compounds containing the fluoride ion. Why is fluorine so adept at forming ionic compounds? Ionic compounds always form between a metal and a nonmetal, and the more different two elements are in terms of their metallic properties, the more likely it is that a compound formed from these elements will be ionic. Sitting near the upper right hand corner of the periodic table, fluorine is among the least metallic of the elements. So there are many elements with which it can form ionic compounds. As we attempt to gain a better understanding of ionic compounds, we'll want to consider a number of fundamental questions.

- Why do metallic elements form cations, while nonmetals form anions?
- How does the energy of attraction between ions compensate for the energy costs of ion formation?
- How does ionic radius affect the strength of the bond between an anion and a cation?

Formation of Cations

We've already learned that there are energy changes associated with the formation of ions, as represented by ionization energies and electron affinities. And we also know how those energy changes vary as we move through the periodic table. Those trends correlate very well with the location of metals and nonmetals in the table, and this correlation explains the observations that metals form cations and nonmetals form anions. In each case the ions we find in compounds are those whose formation is not too costly in terms of energy.

By looking beyond the first ionization energy, we can also see why some metals form +1 cations while others show higher charges. As pointed out in our discussion of Table 6.4, there is always a large jump in ionization energy whenever the removal of an additional electron would disrupt an electron configuration ending in np^6. And this large jump in ionization energy explains which cation will actually be formed. For instance, sodium only occurs naturally as singly charged Na^+. When we look at the trend in ionization energies for sodium, this is not surprising. The first, second, and third ionization energies for sodium are 496, 4563, and 6913 kJ/mol, respectively. When we see the huge amount of energy required to take a second or third electron from a sodium atom, we can see why Nature might choose to stop at only one! Example Problem 7.1 explores the relationship between metal cations and ionization energy.

Example Problem 7.1

Using data from Table 6.4, predict the ions that magnesium and aluminum are most likely to form.

Strategy

Scan the successive ionization energies for each element to find a point at which there is a dramatic increase in the value. It is not likely that sufficient energy would be available to compensate for such a large increase in ionization energy, so the ion formed will be dictated by the number of electrons lost before the jump in ionization energy.

Solution

For magnesium, the first significant jump occurs between the second ionization energy (1451 kJ/mol) and the third (7733 kJ/mol). We would expect two electrons to be removed, and the most commonly found ion should be Mg^{2+}. For Al, the third ionization energy is 3660 kJ/mol, while the fourth is 25,000 kJ/mol. We would expect three ionizations to occur, so aluminum should form Al^{3+}.

Discussion

This type of question is designed to help you see the reasoning behind some of the facts you may already know. In many cases the way we learn chemistry is by first learning information as facts and then finding out the reason those facts are evident. In the examples from this problem, the answers are consistent with the idea that the ion formed is dictated by the number of electrons lost before the jump in ionization energy. They are also consistent with facts you may already have known – for example, that magnesium will form a +2 ion because it is an alkaline earth element, in Group 2 of the periodic table.

Check Your Understanding

Strontium is commonly found as the Sr^{2+} ion. Based on this fact, what would be your qualitative prediction for the values of the successive ionization potentials for Sr?

As our examples show, the most common cation for metals in the *s*- or *p*-blocks of the periodic table is generally the one with an np^6 configuration. But this idea does not carry over to the transition metals. As we noted in Chapter 2, many transition metals form more than one stable cation. Transition metals have partially filled *d*-subshells, but ionization usually occurs by loss of electrons from the *s*-subshell. When iron forms Fe^{2+}, for example, the electron configuration changes from $1s^2\ 2s^2\ 2p^6\ 3s^2\ 3p^6\ 4s^2\ 3d^6$ in the neutral atom to $1s^2\ 2s^2\ 2p^6\ 3s^2\ 3p^6\ 3d^6$ in the ion. Losing one additional electron from a *d*-orbital leaves Fe^{3+} with a half-filled *d*-subshell. This half-filled subshell is a fairly stable arrangement, so the Fe^{2+} and Fe^{3+} ions are both stable. With many transition metals, the situation can become quite complicated, and we will not explore this in greater detail.

If a shell is half filled, there are no electrons paired in any orbitals. Pairing of electrons is allowed (with opposite spins) but it still will increase electron-electron repulsions.

Formation of Anions

If we look at the periodic trends for electron affinity, we see that the trends are similar to those for ionization potentials. Recall that electron affinity is the energy change for the formation of an anion from the neutral atom.

$$X(g) + e^- \rightarrow X^-(g)$$

This process is often energetically favored, meaning energy is released when the anion is formed. The amount of energy released when atoms acquire additional electrons tends to increase as we go from left to right across the periodic table. Looking at this trend, those atoms that gain electrons most readily also would require multiple ionizations to produce a cation with a stable np^6 electron configuration. Thus nonmetals tend to form anions rather than cations.

For the formation of anions from nonmetals we do not find any of the complexities we encountered with the transition metals, because all of the nonmetals belong to the main groups and do not have partially filled *d*-subshells. They behave like the main group metals in that they tend to acquire an electron configuration with a filled *p*-subshell. The primary difference is that they do so by gaining, rather than losing, electrons. Thus, all the halogens form –1 ions, resulting in an np^6 electron configuration. Similarly, oxygen and sulfur form –2 ions with the same np^6 configuration.

In addition to charge, the size of an ion ultimately has an effect on the strength of an ionic bond, largely due to geometric constraints. In Section 6.7 we noted that atomic radius increases going down a group and decreases going across a period. What trend should we expect for the radii of cations and anions? The loss of an electron to form a cation decreases electron-electron repulsion, and the remaining electrons are more tightly bound to the nucleus. So we would expect positive ions to be smaller than their respective parent atoms. On the other hand, adding an electron to form an anion will lead to increased repulsion among the valence electrons. So anions should be larger than their corresponding neutral atoms. Ionic radii can be determined by measuring the spacing in crystal lattices, and the data are consistent with these expectations. With an ionic radius of 97 pm, Na^+ is significantly smaller than the sodium atom, which has an atomic radius of 190 pm. And the 133 pm ionic radius of F^- is significantly larger than the 42 pm atomic radius of neutral fluorine. Figure 7.2 shows atomic and ionic radii, and the observed trends are clear. When we compare similarly charged ions going down a group, ionic size increases in the same way that we saw for atomic size. For cations, the trend in ionic size going across a period also mimics the trend we saw for neutral atoms – decreasing as we go from left to right. For anions in the *p*-orbital block, however, we observe the opposite trend. We can explain this difference by looking at the charges of the anions. The phosphide ion, P^{3-}, has three additional electrons in its $3p$-subshell, and increased repulsions make it very large compared to the phosphorus atom. The sulfide ion, S^{2-}, is smaller than the phosphide ion; although it has the same electron configuration, it has a larger nuclear charge. Similarly, Cl^-, is slightly smaller than S^{2-}.

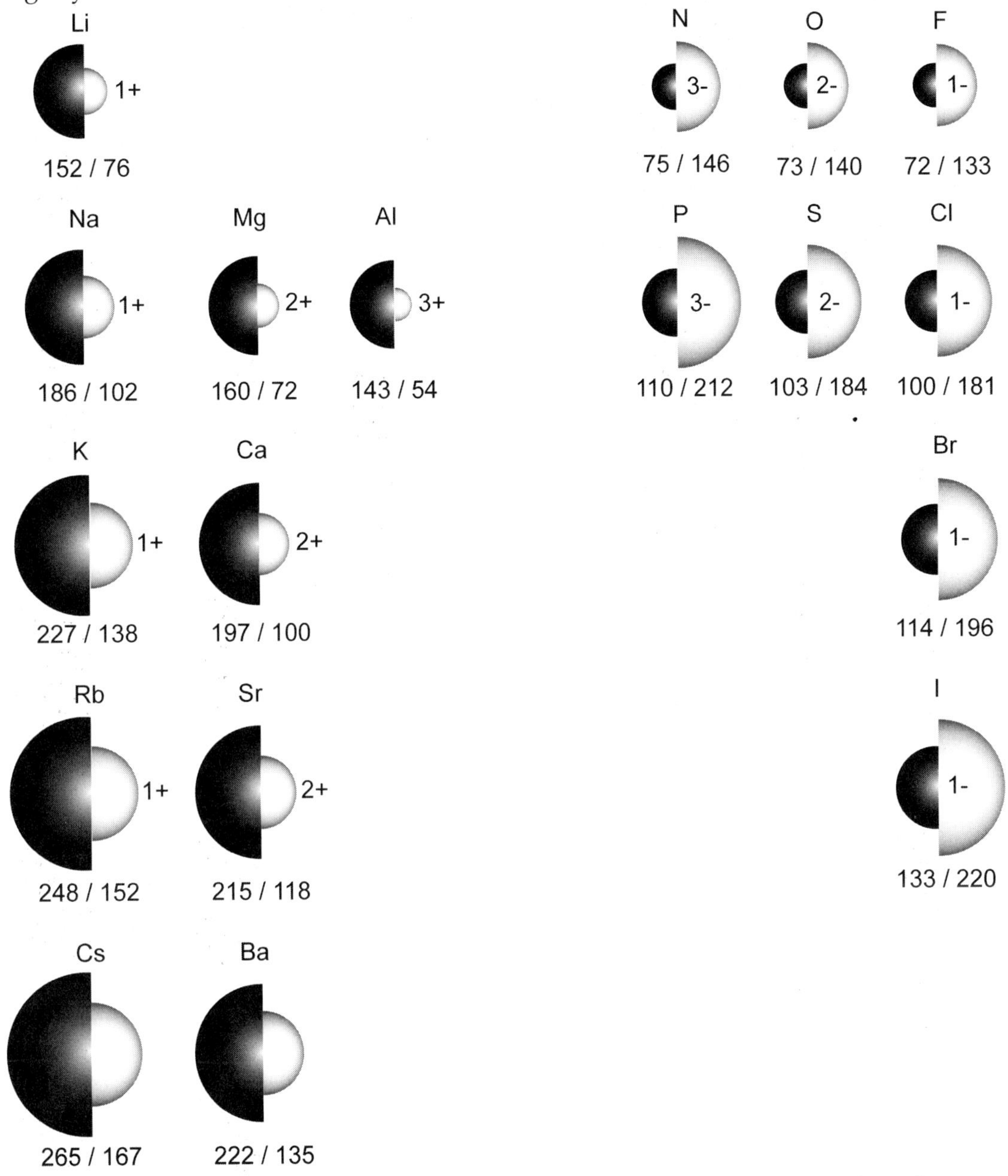

Figure 7.2 The nested spheres in the figure illustrate how the size of the indicated ions compare to the corresponding neutral atoms. Metals and their cations are shown on the left, and nonmetals and their anions on the right. All radii are given in pm.

The formation of ions alone, of course, does not constitute an ionic bond. Indeed, there is usually an energy cost associated with forming an ion pair (*i.e.*, a cation and an anion). Ionization energies are always positive, and that energy input is generally not offset by the energy released when the anion is formed. But once formed, the oppositely charged ions will attract one another, leading to a significant lowering of their overall energy. As we saw in Chapter 2, Coulomb's Law says that the *force* between two charges is proportional to the product of the magnitude of their charges divided by the square of the distance between them.

$$F \propto \frac{q_1 q_2}{r^2} \tag{7.1}$$

Here F is the force, q_1 and q_2 are the charges, and r is the distance between them (in this case, the distance between the nuclei of the two ions). It can also be shown that the *potential energy*, V, of the system is given by

$$V = k\frac{q_1 q_2}{r} \tag{7.2}$$

Here k is a proportionality constant, whose value is 1.389×10^5 kJ pm/mol if the charges are expressed in units of the electron charge.

We can now evaluate the energy change associated with the formation of a bound NaF ion pair from isolated atoms of sodium and fluorine. The ionization energy of sodium is 496 kJ/mol, and the electron affinity of fluorine is 328 kJ/mol. So we need to supply 496 kJ/mol to ionize the sodium, but 328 kJ/mol will be released in the formation of the fluoride ion. Combining these two numbers, we see that forming the individual ions will "cost" 168 kJ/mol (496 – 328 = 168). Next we need to use Equation 7.2 to determine the energy released through the Coulomb attraction between the anion and cation. The radii of Na^+ and F^- are 97 pm and 133 pm, respectively, so the internuclear distance should be 230 pm. Inserting this into Equation 7.2 and evaluating gives us the energy of attraction between the ions.

$$V = \left(1.389 \times 10^5 \ \frac{\text{kJ pm}}{\text{mol}}\right) \frac{(+1)(-1)}{(230 \text{ pm})} = -604 \text{ kJ/mol}$$

The negative sign indicates that this energy is released. Notice that the magnitude of the Coulomb energy far outweighs the energetic cost of forming the two ions, so the whole process we have described does, in fact, release energy.

Before making too much of that conclusion, though, we should consider two ways that our calculation is not realistic. First, we began with isolated atoms of sodium and fluorine. If we were to actually react elemental sodium and fluorine, though, we would not start with isolated gas phase atoms; under any reasonable conditions the sodium would be a metallic solid and the fluorine would be a diatomic gas. So there would be some added energy input needed to produce the individual atoms with which we started. On the other end of our calculation, though, we found the Coulomb energy for a single ion pair. But if we really did form sodium fluoride, we would produce a bulk crystal lattice, made up of many ions. In such an ionic lattice, ions are attracted to every other oppositely charged ion. And at the same time, ions of like charge are all repelling each other. The farther one ion is from another, the weaker the attraction or repulsion (Figure 7.3). Ultimately the lattice energy of a crystal contains contributions from all the attractions and repulsions in the lattice. In the NaF structure, each sodium ion is surrounded by six fluoride ions, while each fluoride has six neighboring sodium atoms. To get an accurate indication of the Coulomb energy associated with the lattice, we would need to account for the interaction of each ion with its many neighbors. So while starting with isolated atoms leads to an overestimation of the energy released in forming the ionic bond, using a single ion pair instead of an extended lattice greatly underestimates the energy. The actual energy released in the reaction between sodium metal and diatomic fluorine gas to produce solid sodium fluoride turns out to be 577 kJ/mol.

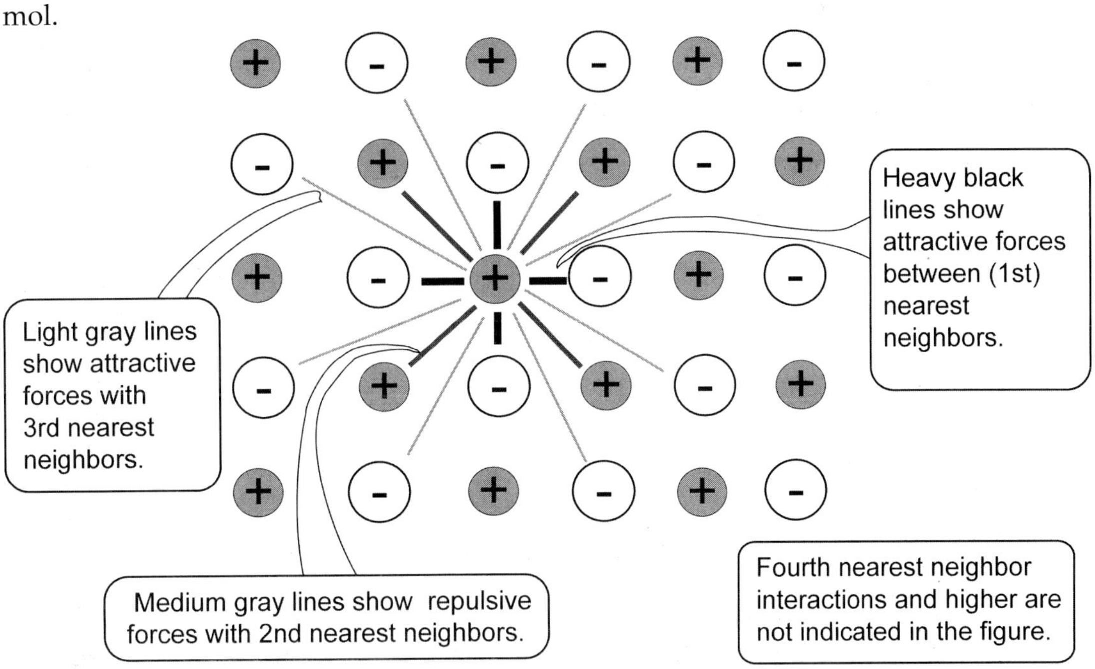

Figure 7.3 In a solid lattice any given ion has a large number of interactions with other ions, some attractive and others repulsive. This two-dimensional representation of a lattice shows the interactions for the nearest neighbors, second nearest neighbors and third nearest neighbors.

The overall results of all attractive and repulsive forces in an ionic crystal is estimated by the caclulation of a Madelung constant, a factor that is essentially a series of interactions that describes the geometry.

From the form of Equation 7.2, we can see that small, highly charged ions will tend to form ionic compounds with large lattice energies, while large ions with small charges will tend to form compounds with small lattice energies. Example Problem 7.2 explores this connection.

Example Problem 7.2

In each of the following pairs of ionic solids, the ions are arranged into similar lattices. Which one has the largest lattice energy: (a) CaO or KF? (b) NaF or CsI?

Strategy

Questions like this require us to find some basis for comparison. In this particular case the factors associated with the lattice energy have to do with the ions that make up the compound. We have two factors to consider. Smaller ions tend to have stronger Coulomb interactions because the distance between the nuclei is smaller, and ions with larger charges have stronger interactions. An examination of these two factors may allow us to predict which substance has the larger lattice energy.

Solution

(a). In CaO, calcium exists as Ca^{2+} and oxygen as the oxide ion, O^{2-}. In KF, the constituent ions are K^+ and F^-. Because they are both from elements in the same row of the periodic table, the sizes of the two cations should not be too different. The same is true for the anions, so to a first approximation, we might discount the size factor. CaO should have the larger lattice energy because the charges on its ions are larger than the charges on the ions in KF. (Note that if we wanted to look more closely at the sizes, we would expect that K^+ would be slightly larger than Ca^{2+}, while O^{2-} would be slightly larger than F^-. Since the larger cation is in KF while the larger anion is in CaO, our assumption that size would not be very important seems good.

(b). All of the ions in this problem are singly charged, so the size difference must be the key issue. Na^+ and F^- are smaller than Cs^+ and I^-, so the lattice spacing in NaF must be smaller. Therefore NaF should have the larger lattice energy.

Discussion

To get started here we needed to determine the identities of the ions involved. (One way to do this is to consider the jump in ionization energy we explored in Example Problem 7.1. However, with just a little practice it is easy to determine the likely charges for most main group elements based on their location in the periodic table.) Ionic charge and radius were taken into consideration, based on the positions of the ions in the periodic table. The use of this "periodic reasoning" points out the importance of learning trends in the periodic table. Note, however, that to be able to make the comparisons in this problem, the lattice structures of the compounds in question must be similar.

Check Your Understanding

Of the three compounds, CaF_2, KCl, RbBr, which should we predict to have the smallest lattice energy? Defend your answer.

7.3 The Covalent Bond

Consideration of ionization energy, electron affinity, and the Coulomb potential is fine for ionic compounds. But what about covalent bonds like those found in Teflon? We noted back in Chapter 2 that covalent bonds are based on the sharing of pairs of electrons between two atoms. How does covalent bonding occur, and how does it influence the behavior of compounds like Teflon?

Chemical Bonds and Energy

When chemical bonds are formed between nonmetal elements, valence electrons are shared rather than transferred. But just as for the ionic case, we must expect that the driving force behind bond formation should be a lowering of the overall energy. The energetic advantage of forming a covalent bond can be illustrated by considering how the potential energy would vary in a thought experiment in which two atoms approach each other. If the atoms are sufficiently far apart, they won't exert any discernible force on one another. (You might say that neither one knows the other is there.) But as the atoms begin to come closer together, eventually the electrons from one atom can become attracted to the positively charged nucleus of the other atom. And this added attraction lowers the potential energy of the system. Of course the two nuclei will repel each other, as do the bonding electrons, so the nuclei cannot approach too close to one another. At some point a balance between these attractions and repulsions will occur, and the energy of the system is minimized (Figure 7.4). This point of minimum energy corresponds to the formation of a covalent bond. The energy released when isolated atoms form a covalent bond is called the **bond energy**, and the distance between the nuclei of the bonded atoms is called the **bond length**.

It is important to remember that bond formation is ALWAYS an exothermic process.

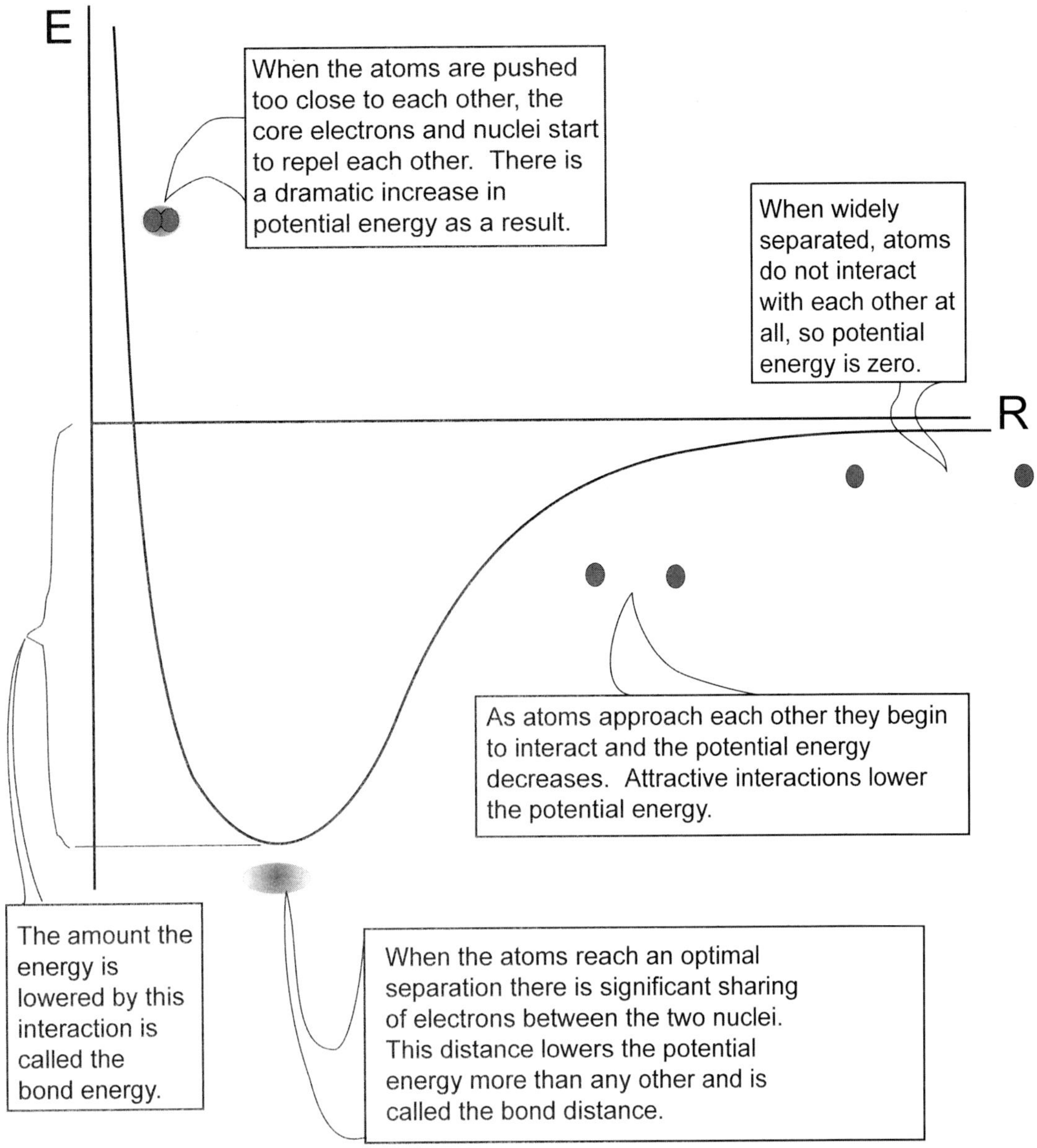

Figure 7.4 A schematic diagram shows how the potential energy of a pair of atoms might vary as the atoms draw closer together and form a covalent bond. The minimum in this "potential well" corresponds to the normal bond length and bond energy for the diatomic molecule.

When electrons are shared between two atoms, they are distributed differently than they are in the individual atoms. In Figure 7.5 we see how electron density in a molecule differs from the spherically symmetrical distribution seen in isolated atoms. In a covalently bonded molecule, there is a build-up of electron density, or a concentration of negative charge, in the region between the two bonded atoms. The increase in electron density need not be particularly large; even a slight increase of negative charge in this region lowers the potential energy of the system and can stabilize the molecule.

The formation of a chemical bond always releases energy. And once a bond is formed, that same amount of energy - the bond energy - must be supplied in order to break the bond apart. The magnitude of this bond energy depends on the atoms involved. For example, when two moles of fluorine atoms combine to form one mole of molecular fluorine, 156 kJ of energy is released. So we say that the F-F bond energy is 156 kJ per mole. Compared to many other molecules, F_2 is held together by a relatively weak covalent bond. Hydrogen fluoride, for example, has a bond energy of 565 kJ/mol, meaning its bond is more than three times stronger. Since more energy is required to break its atoms apart, we say that the bond in HF is more stable than the bond in F_2. When we want to assess the stability of a compound, some general ideas about the different types of bonds present are useful.

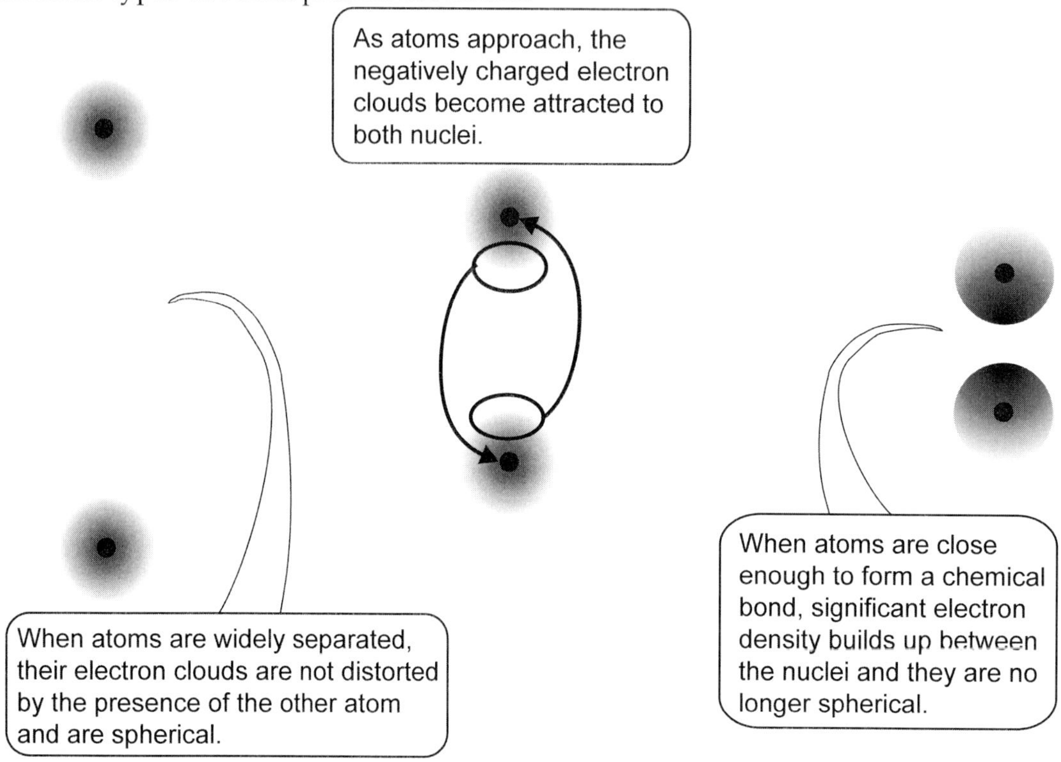

Figure 7.5 When two atoms approach one another, the negatively charged electron clouds are each attracted to the other atom's positively charged nucleus. This attractive interaction potential is largely responsible for the lowering of energy associated with the formation of a chemical bond.

Chemical Bonds and Reactions

Chemical reactions involve rearranging bonds, so that reactants are transformed into products. A reaction will be energetically favored if the energy required to break bonds is less than that released in making new bonds. The importance of Teflon in consumer products stems largely from the fact that it is not prone to react chemically. If Teflon burned easily, it would not be appropriate as a coating for frying pans, after all. So let's consider the making and breaking of bonds that would have to be involved in the combustion of Teflon. Recall that the combustion of a hydrocarbon produces CO_2 and H_2O. PTFE is a fluorocarbon, which you might think of as a hydrocarbon whose hydrogen atoms have all been replaced by fluorine. So the combustion of Teflon would have to produce CO_2 and OF_2. Table 7.1 shows the bond energies of various types of bonds that are relevant to this discussion. Note that a C-F bond is slightly stronger than a C-H bond, and thus harder to break. More importantly, notice the large difference in strength between O-H and O-F bonds. These bond energies show us that in order to burn a fluorocarbon, we would have to replace strong C-F bonds with weaker O-F bonds. Energetically this is not at all favorable, and indeed many fluorine-containing molecules are not flammable. Thus a simple look at bond energies can help explain the non-reactive nature of Teflon.

Type of Bond	Bond Energy (kJ/mol)
C-F	485
C-H	415
O-F	190
O-H	460

Table 7.1 Bond energies of some types of bonds important in the combustion of hydrocarbons and fluorocarbons.

Chemical Bonds and the Structure of Molecules

In explaining chemical bonding we must go beyond the discussion of energetics in the previous section. When we looked at ionic bonding between metals and nonmetals, we saw that the formation of cations and anions is consistent with achieving 'np^6' electron configurations that resemble those of the rare gases. For nonmetal atoms, a rare gas configuration can be achieved by gaining the electrons from metal atoms, leading to ionic bonds. When nonmetals interact with other nonmetals, however, they share electrons, effectively giving each atom access to all the shared (bonding) electrons. Thus, by sharing a suitable number of electrons, an atom can achieve an electron configuration that resembles that of a rare gas. For many molecules composed of elements from the main groups, this feature leads to eight valence electrons around each atom. This observation is the basis for what is known as the **octet rule**: *An atom will form covalent bonds in such a way as to achieve a complement of eight valence electrons.* Recall that a rare gas atom has an electron configuration in the valence shell of $ns^2\ np^6$, for a total of eight valence electrons. Because it has no *p*-orbitals in its $n=1$ valence shell, hydrogen is an exception to the octet rule and can share only two electrons. There are other exceptions as well, which we will note as we encounter them.

Lewis dot symbols of atoms help explain the valence of elements. The origincal definition of valence was derived from how many chemical bonds an element would form. Counting unpaired electrons in the Lewis symbol provides the valence of elements in the s- and p-block of the periodic table.

The basic concept of the octet rule can be conveniently depicted using a **Lewis dot symbol**, developed by the American chemist G. N. Lewis. These Lewis symbols help keep track of valence electrons, making it easier to predict the bonding in molecules, and are particularly useful for compounds of main group elements. To draw a Lewis symbol for an element we use its atomic symbol and place dots around it on four sides, to depict the number of valence electrons. As we go across a period in the main groups, the first four dots are placed singly around the symbol. Starting with the fifth dot, they are paired. Thus the elements in the second period would have the following Lewis symbols.

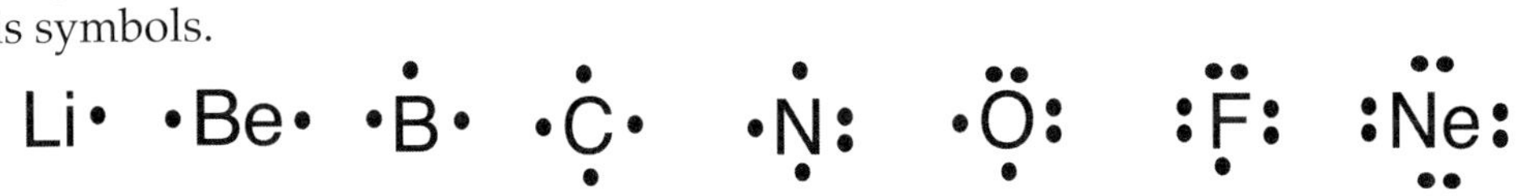

Because elements in the same group have the same number of valence electrons, the Lewis symbols for subsequent periods are the same, except for the identity of the elements. For example, the Lewis dot symbols for carbon, silicon, and germanium are as follows.

Lewis symbols for the first three rows of the periodic table are shown in Figure 7.6.

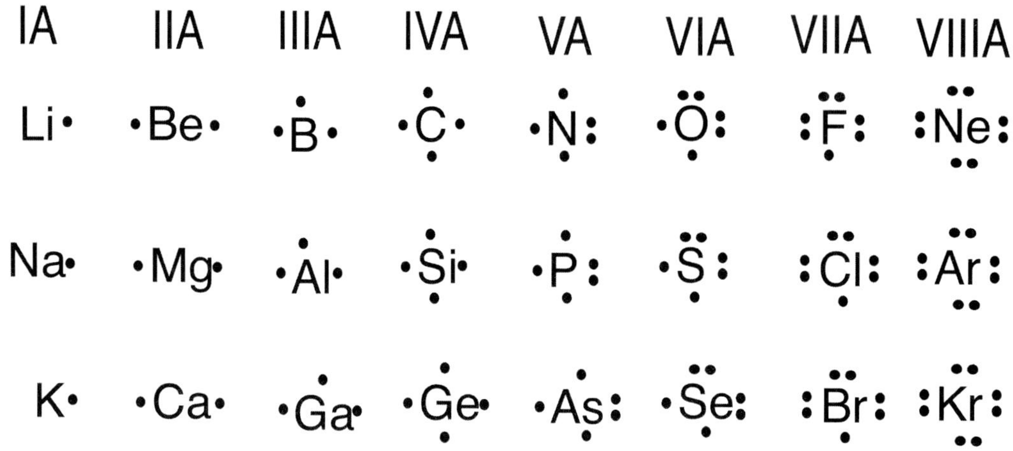

Figure 7.6 The Lewis symbol for elements in the first three periods of the periodic table are shown. Notice the similarities among elements in the same group. Lewis structures help keep track of electrons involved in chemical bonds.

These symbols for individual atoms are of fairly little utility, but similar representations for molecules are much more powerful. When we draw Lewis dot representations of molecules, we add one more level of symbolism to denote bonds between atoms. A line between two atomic symbols represents a shared pair of electrons forming a bond; thus when counting electrons one line is equivalent to two dots. Figure 7.7 shows the Lewis symbol for the hydrogen atom and the Lewis structure of the simplest molecule, H_2.

Figure 7.7 By sharing the electron from each atom, two hydrogen atoms can form a chemical bond. In doing so each atom has enough valence electrons around it to achieve a filled n=1 shell.

A **Lewis structure** shows how electrons are shared in a molecule. The Lewis structure for F_2 is shown in Figure 7.8. The sharing of electrons to form a bond effectively gives each fluorine atom a full octet of electrons. When fluorine is part of a covalently bonded system, it virtually always forms a single bond by sharing an electron pair, thereby satisfying the octet rule. The pair of electrons shared by two atoms is referred to as a **bonding pair**. The other paired electrons that are associated with a single atom are referred to as **nonbonding electrons**, or as **lone pairs**. A fluorine molecule has one bonding pair and six lone pairs.

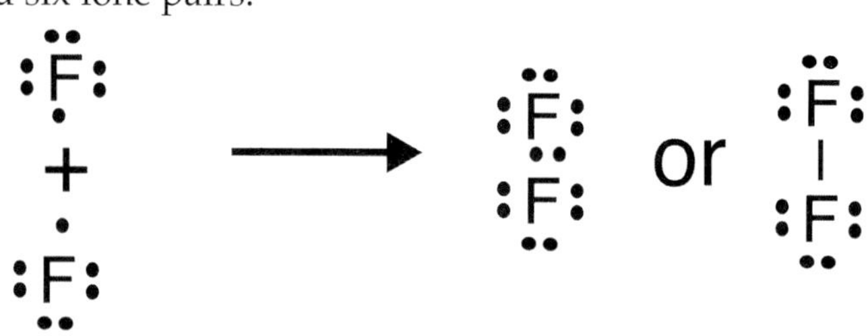

Figure 7.8 When two fluorine atoms combine, they form a stable chemical bond. The Lewis structures for the F-atoms show why just one bond forms. By sharing one pair of electrons each atom is surrounded by eight valence electrons.

Of course most molecules are more complicated than F_2, and will contain more than one pair of shared electrons. If two oxygen atoms share a single pair of electrons, for example, then neither atom has an octet of valence electrons. However, if *two pairs* of electrons are shared, then both oxygen atoms conform to the octet rule. For diatomic nitrogen, three pairs of electrons must be shared in order to satisfy the octet rule. **Multiple bonding** results from the sharing of more than one electron pair. When two pairs of electrons are shared, the result is a **double bond**, while three shared pairs result in a **triple bond**. Double bonds are stronger than single bonds, and triple bonds are stronger than double bonds. However, a double bond, while stronger than a single bond, is generally not twice as strong.

We will see in Chapter 8 that the relative strength of two single bonds versus one double bond helps understand the types of reactions that can be used to synthesize polymers,

The relative strengths of multiple bonds have important consequences. The polymerization of tetrafluoroethylene to form PTFE provides a good example. The tetrafluoroethylene monomer contains a carbon-carbon double bond. In the polymerization reaction, two of the electrons forming this double bond are used to create a bond to a second monomer unit. The end result is that the double bond in the reactant is transformed into two single bonds in the product. Because the energy released in forming two single bonds is greater than the energy needed to break a double bond, the reaction to form PTFE is an energetically favorable process. This is true for many similar reactions in which a reactant with a double bond is converted into a product with two single bonds, and explains why the common reaction of the C=C functional group is addition. The same is true for triple bonds, although those are much less common.

7.4 Electronegativity and Bond Polarity

The concept of covalent bonding provides useful insight into the chemical properties of fluorine compounds, such as PTFE. While it is certainly vital that a PTFE-coated pan is nonflammable and that the coating doesn't react with the food we cook, the property we appreciate when we are washing the dishes is the non-stick aspect of the polymer. As we will see by the end of this chapter, covalent bonding plays a role here, too.

The sharing of electrons to form covalent bonds and the more extreme transfer of electrons that occurs in ionic bonding really represent two ends of a bonding continuum. In order to describe compounds whose bonding falls somewhere in the middle of this continuum, we will introduce two additional concepts: electronegativity and bond polarity.

Electronegativity

While focusing on the chemistry of fluorine, we have indicated that this element can confer some unusual properties to molecules. What is so special about fluorine? Fluorine is an element that is decidedly "greedy" for electrons. Its high electron affinity, coupled with its small size, means that electrons in a bond between fluorine and other elements will be strongly attracted to the fluorine nucleus. The attraction of an atom for the shared electrons in a covalent bond is called **electronegativity.**

While this concept may seem simple enough, making it quantitative is actually very difficult. Unlike ionization energy or electron affinity, the electronegativity is not an energy change that can be determined experimentally. Instead, electronegativity values are established by considering a number of factors, like atomic size, electron affinity, and ionization energy, and a scale of arbitrary units has been chosen. Figure 7.9 gives the resulting values for the electronegativities of most elements. The higher the electronegativity value, the more likely an element is to attract extra electron density when it forms chemical compounds. Electronegativity tends to increase as we go up a group (decreasing atomic number) or across a period from left to right. Thus trends in electronegativity parallel those in electron affinity.

Fluorine is the most electronegative element, and is assigned a value of 4.0. Oxygen and nitrogen also have relatively high electronegativities. When these elements take part in covalent bonding, they tend to take "more than their fair share" of the bonding electron pair. What are the implications for a molecule when electrons are shared unequally?

1	2	3	4	5	6	7	8	9	10	11	12	13	14	15	16	17	18
1 H 2.1																1 H 2.1	2 He
3 Li 1.0	4 Be 1.5											5 B 2.0	6 C 2.5	7 N 3.0	8 O 3.5	9 F 4.0	10 Ne
11 Na 1.0	12 Mg 1.2											13 Al 1.5	14 Si 1.8	15 P 2.1	16 S 2.5	17 Cl 3.0	18 Ar
19 K 0.9	20 Ca 1.0	21 Sc 1.3	22 Ti 1.4	23 V 1.5	24 Cr 1.6	25 Mn 1.6	26 Fe 1.7	27 Co 1.7	28 Ni 1.8	29 Cu 1.8	30 Zn 1.6	31 Ga 1.7	32 Ge 1.9	33 As 2.1	34 Se 2.4	35 Br 2.8	36 Kr
37 Rb 0.9	38 Sr 1.0	39 Y 1.2	40 Zr 1.3	41 Nb 1.5	42 Mo 1.6	43 Tc 1.7	44 Ru 1.8	45 Rh 1.8	46 Pd 1.8	47 Ag 1.6	48 Cd 1.6	49 In 1.6	50 Sn 1.8	51 Sb 1.9	52 Te 2.1	53 I 2.5	54 Xe
55 Cs 0.8	56 Ba 1.0	57 La 1.1	72 Hf 1.3	73 Ta 1.4	74 W 1.5	75 Re 1.7	76 Os 1.9	77 Ir 1.9	78 Pt 1.8	79 Au 1.9	80 Hg 1.7	81 Tl 1.6	82 Pb 1.7	83 Bi 1.8	84 Po 1.9	85 At 2.1	86 Rn
87 Fr 0.8	88 Ra 1.0	89 Ac 1.1	104 Rf	105 Ha	106 Sg	107 Ns	108 Hs	109 Mt									

58 Ce 1.1	59 Pr 1.1	60 Nd 1.1	61 Pm 1.1	62 Sm 1.1	63 Eu 1.1	64 Gd 1.1	65 Tb 1.1	66 Dy 1.1	67 Ho 1.1	68 Er 1.1	69 Tm 1.1	70 Yb 1.0	71 Lu 1.2
90 Th 1.2	91 Pa 1.3	92 U 1.1	93 Np 1.3	94 Pu 1.3	95 Am 1.3	96 Cm 1.3	97 Bk 1.3	98 Cf 1.3	99 Es 1.3	100 Fm 1.3	101 Md 1.3	102 No 1.3	103 Lr 1.5

Figure 7.9 Electronegativity values for the elements are shown. Notice how the values increase towards the right and top of the periodic table.

Bond Polarity

Let's look in more detail at what happens when covalent bonding occurs between elements of differing electronegativities, such as hydrogen and fluorine. A convenient way to do this is to compare the charge distribution in a molecule with the charge distribution in the isolated atoms. If we observe a fluorine atom from a distance, it appears to be electrically neutral, because the nuclear and electronic charges offset each other. Since fluorine is the most electronegative element, when a fluorine atom participates in a covalent bond with another element, such as hydrogen, more than half of the electron distribution of the bonding electron pair is associated with the fluorine atom. The result is that the fluorine atom becomes a center of partial negative charge because it has more electron density than is necessary to balance its own nuclear charge. At the same time, the hydrogen atom at the opposite end of the bond takes on a partial positive charge because it has insufficient electron density to balance its nuclear charge (Figure 7.10). Thus, covalent bonding between atoms of different electronegativity results in a bond in which there is a separation of positive and negative charge centers. Because the molecule has one locus of positive charge and another of negative charge, an electric field is associated with the molecule. These two points of positive and negative charge constitute a **dipole**, and such a bond is referred to as a **polar** bond. Because the electrons are still shared rather than transferred, these bonds are often called **polar covalent bonds**.

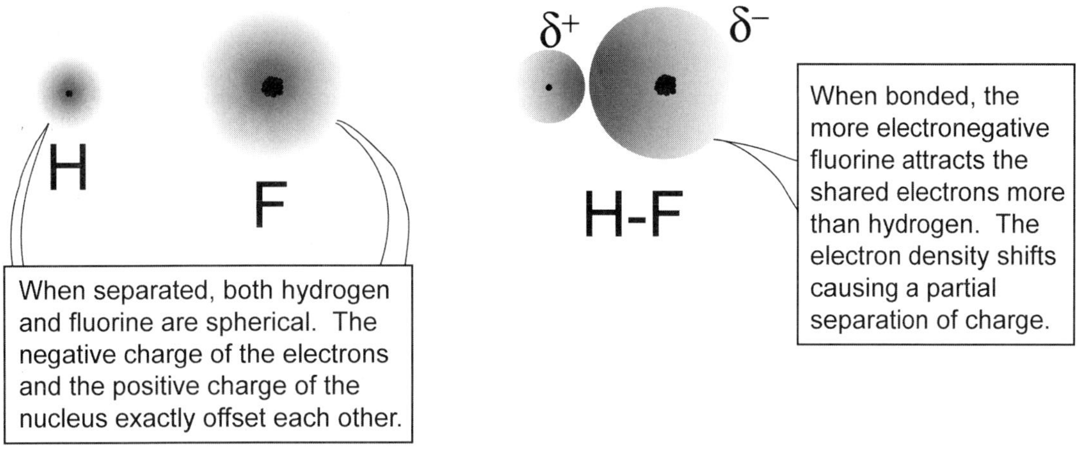

Figure 7.10 The highly electronegative fluorine atom attracts electrons more strongly than hydrogen does, so the two form a polar covalent bond.

Polar covalent bonding completes the continuum of chemical bonding. The polarity of covalent bonds varies from completely nonpolar (as in F_2 and H_2), through varying intermediate degrees of polarity (as in HF), to completely ionic (as in NaF). The type of bond formed depends on the electronegativity difference between the bonding atoms. *The greater the difference in electronegativity, the more polar the bond will be.* At some point, typically when the difference in electronegativity exceeds about 2.0, the bond is classified as ionic. Using the electronegativity values listed in Figure 7.9, we see that NaF, with an electronegativity difference of 3.1, is clearly ionic. The C-F bond in Teflon is polar covalent (electronegativity difference = 1.5) and the F-F bond is completely covalent (electronegativity difference = 0). Example Problem 7.3 explores this concept further.

The relationship between electronegativity difference and ionic character of a bond is actually continuous, but using cutoff to label a bond as ionic or polar simply makes the language more tractable.

Example Problem 7.3

Which bond is the most polar: C-H, O-H, or H-Cl? For each of these bonds, determine which atom has a partial positive charge.

Strategy

Bond polarity depends on the difference in electronegativity between the two bonded atoms. So we can look up the electronegativity values and subtract to find the difference in each case. Negatively charged electrons are drawn more strongly towards the atom with higher electronegativity, so the less electronegative atom has the partial positive charge.

The tendency of some elements to lose electron density in bonds is sometimes referred to as electropositive behavior.

Solution

From Figure 7.9, we find the following electronegativities: H = 2.1, C = 2.5, O = 3.5, Cl = 3.0. The electronegativity differences are therefore: C-H = 0.4, O-H = 1.4, H-Cl = 0.9. So the O-H bond is the most polar of the three. In the three bonds considered here, H is the less electronegative in each case, so it carries the positive charge.

Discussion

This question may appear simple, because we are using electronegativity values and differences in a straightforward way. Despite this simplicity, the results can provide important insight into material properties. Our focus at this point is on single molecules, but in laboratory observations in the real world we invariably encounter huge collections of molecules. The way these samples of molecules behave is often strongly influenced by the polarity of the bonding within the individual molecules. Therefore, understanding a bond dipole and predicting its presence using electronegativity will be useful in a number of ways as we move forward in our study of chemistry.

Check Your Understanding

Which bond should be essentially nonpolar: C-N, Sb-H, or S-Br?

Bond polarity plays a significant role in the chemistry of fluorine-containing compounds. Because fluorine has the largest electronegativity value of all the elements, it always carries a partial negative charge in all of its compounds. Bonds to fluorine tend to be highly polar, and this polarity affects the way these molecules interact with each other and with molecules of other compounds. In PTFE the partial negative charges on the fluorine atoms repel one another to such an extent that the flexible carbon backbone is forced into a shape that keeps the negative charges as far away from one another as possible (Figure 7.11). This gives PTFE molecules a much more regular shape than that of seemingly similar polymers like polyethylene. And that, in turn, means that the polymer molecules can pack into a crystalline array more efficiently, allowing Teflon to melt at a higher temperature. This is an oversimplified picture, because there are other factors that determine the stability of a crystal lattice, but it introduces us to the concept of analyzing macroscopic properties in terms of molecular bonding—a topic we will explore more fully in Chapter 8.

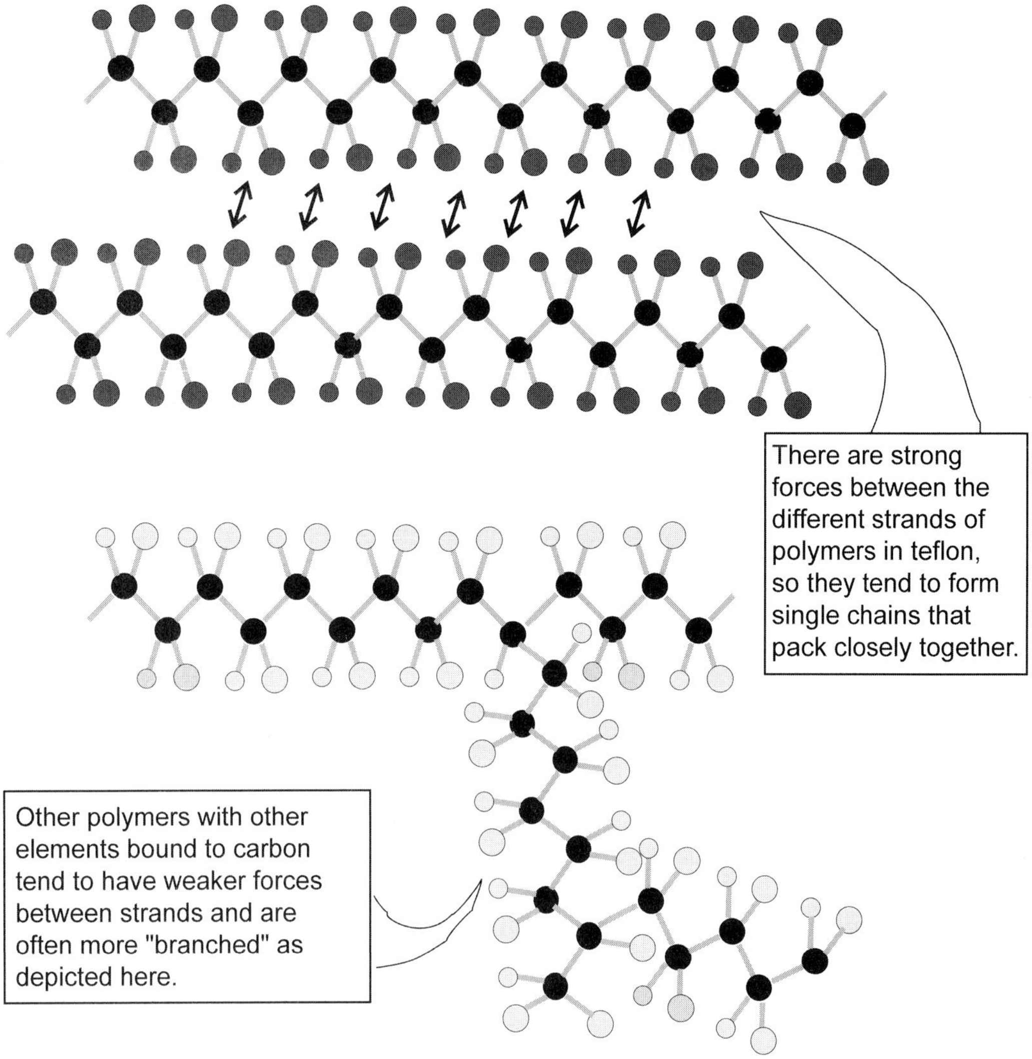

Figure 7.11 When polymers condense into a solid, the individual molecules must pack together. Because of the negative charges on the fluorine atoms in Teflon, relatively few occasions arise where the "zigzag" shape of the polymers is disrupted. This allows for fairly close packing, and helps produce the high melting point and non-stick characteristics of the material. Other polymers are generally less tightly packed.

7.5 Keeping Track of Bonding: Lewis Structures

We've now surveyed some important aspects of chemical bonding, and we've gained an understanding of a number of key concepts. But if we want to think about the connections between bonding and the properties of compounds, we often need to know not just *that* bonds have formed, but *how many* bonds have formed, and between what elements. The Lewis structures we introduced in Section 7.3 provide a powerful tool for doing this. In drawing Lewis structures for very simple molecules like H_2 and F_2, we first drew each atom separately, and then brought them together to form a bond. This approach obviously assumes that we know which atoms are to be attached to one another. The choice is obvious when we are dealing with a diatomic molecule, but can become problematic for larger molecules. If we want to be able to draw Lewis structures for more complicated molecules or ions, it is useful to use an algorithmic approach. By applying such an algorithm carefully, we can predict which bonds are likely to be formed by examining and comparing the possible structures for a molecule of interest.

As we introduce our algorithm for drawing Lewis structures, it will help if we use a simple molecule to illustrate each step. In keeping with our fluorine theme, we'll choose oxygen difluoride (OF_2) as our first example.

While it may seem routine after a little practice, it is vitally important to always carry out step 1 of this process. Using the wrong number of valence electrons is among the most common error made by students in drawing Lewis strctures.

Step 1: Count the total number of valence electrons in the molecule or ion.

We can do this by summing the numbers of valence electrons for each atom in a molecule. (If we were trying to draw the structure for a polyatomic ion, we would also need to account for the ion's charge, as we'll see in Example Problem 7.5.) For OF_2, each fluorine atom has seven valence electrons, and the oxygen atom has six. So we will have a total of 20 valence electrons in the molecule.

F	2 x 7	= 14
O	1 x 6	= 6
	total	= 20

Step 2: Draw the "skeletal structure" of the molecule.

Here we can rely on some general rules. The element written first in the formula is usually the central atom in the molecule, except when hydrogen is listed first. (Hydrogen can never be a central atom, since it can form only one bond!) Usually, the central atom is the least electronegative. Sometimes more than one possible skeleton can be drawn. In such cases, we might rely on chemical intuition to choose likely alternatives. Or we could also draw multiple possible structures and evaluate those to see which seemed more plausible. In OF_2, oxygen is written first and is the less electronegative element, so it is the central atom. We write the two fluorine atoms surrounding the oxygen. (It isn't important at this point on which two sides of the oxygen symbol we choose to place the fluorine atoms.)

Step 3: Place single bonds between all connected atoms in the structure by drawing a line between them.

We don't yet know how many bonds might be needed in our molecule. But we do know that at least one bond is needed to attach each of the outer fluorine atoms to the central oxygen atom, to keep the molecule from "falling apart." So we can go ahead and draw two single bonds.

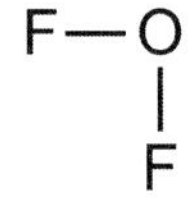

Step 4: Place the remaining valence electrons not accounted for in Step 3 on individual atoms until the octet rule is satisfied. Place electrons as lone pairs wherever possible.

It is usually best to begin placing electrons on the outer atoms first, leaving the central atom(s) for last. Remember that each bond contains two electrons, so in drawing two single bonds we have accounted for four electrons. Our molecule has 20 valence electrons, so 16 remain to be added to our structure. Placing six of these electrons around each fluorine atom brings us up to 16, and placing the remaining four on the central oxygen atom accounts for all 20 available electrons. All 16 of these electrons should be drawn as lone pairs.

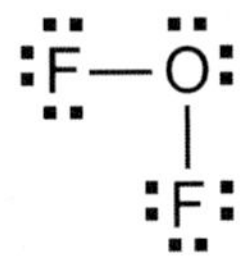

At this point we know that all of the available valence electrons appear somewhere in our structure. To decide whether this is the optimal arrangement of the valence electrons, we check to see if each atom satisfies the octet rule. In the case of OF_2, our structure now has a full octet of electrons on each atom, so this is the final structure. In other cases we might find one or more atoms having less than a full octet, in which case we would proceed to an additional step.

Step 5: For any atoms that do not have a full octet of valence electrons, create multiple bonds by shifting lone pairs into bonding positions as needed.

Our structure for OF_2 already satisfies the octet rule, so multiple bonds are not required in this molecule. We will work with this added step in future examples.

The following examples will further illustrate of our algorithm. Of course the best way to become familiar with these steps is simply to practice drawing Lewis structures yourself.

Example Problem 7.4

Chlorofluorocarbons have been used over the past 50 years as refrigerants, but their use is now being phased out worldwide due to concerns about atmospheric ozone depletion. Draw the Lewis structure of dichlorodifluoromethane, CF_2Cl_2, also known as DuPont's Freon-12.

Strategy

Use the algorithm for drawing Lewis structures.

Solution

Step 1: Count the total number of valence electrons. Carbon has four, fluorine has seven, and chlorine has seven.

C	1 x 4	= 4
F	2 x 7	= 14
Cl	2 x 7	= 14
	total	= 32

Step 2: Draw the skeletal structure of the molecule.

Carbon is listed first in the formula, and it is the least electronegative, so use it as the central atom.

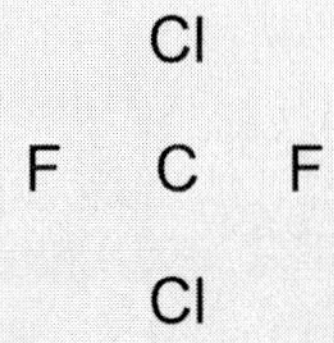

Step 3: Place single bonds between all connected atoms.

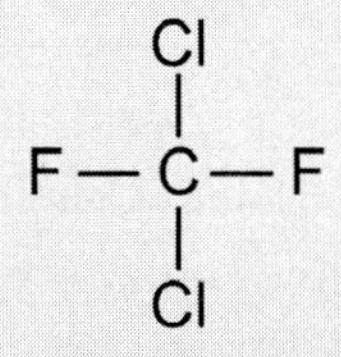

Step 4: Place the remaining valence electrons not accounted for in Step 3 on individual atoms until the octet rule is satisfied. Four single bonds have been drawn, and they require eight valence electrons. We have a total of 32 available electrons, so we are left with 32 – 8 = 24. This gives each of the four halogen atoms three lone pairs (6 electrons).

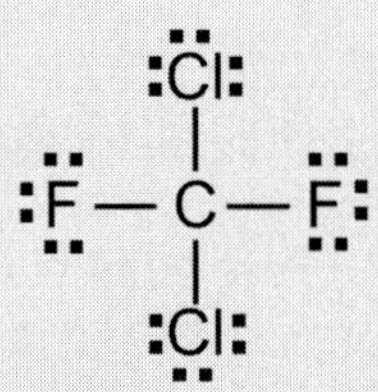

Step 5: For any atoms that do not have a full octet of valence electrons, create multiple bonds. Check to make sure each atom has an octet of eight electrons. Carbon has four bonding pairs, or eight electrons. Each halogen has one bonding pair and three lone pairs, so they also have octets. Multiple bonds are not needed, and the final structure is the one arrived at after Step 4.

Discussion

Because we are using an algorithm to draw these structures, it is important that each step be carried out correctly, as any errors may be propagated through the rest of the process. The first step is perhaps the most critical. If we lose track of how many valence electrons we need to include, we will have little chance of arriving at the correct structure. Many students have missed exam questions by adding a lone pair of electrons "they didn't have" rather than making multiple bonds.

Check Your Understanding

Draw the Lewis structure for C_2H_6.

Example Problem 7.5

Draw the Lewis structure of the sulfate ion, SO_4^{2-}.

Strategy

Follow the steps of the Lewis structure algorithm. Since sulfate is an anion, we'll need to properly account for its charge in determining the number of valence electrons.

Solution

Step 1: Count the total number of valence electrons.

The sulfur atom has six valence electrons, and each oxygen atom also has six. The ion as a whole carries a charge of –2, which means it must contain two additional electrons.

S	1 x 6	= 6
O	4 x 6	= 24
+2 extra e^-		= 2
	total	= 32

Step 2: Draw the skeletal structure of the molecule. Sulfur is the least electronegative atom, and appears first in the formula, so it should be the central atom. We'll draw square brackets around the entire ion, and indicate its charge.

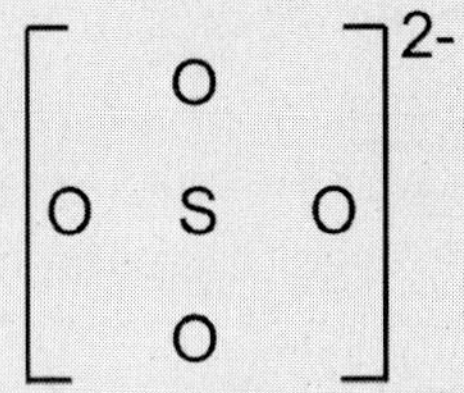

Step 3: Place single bonds between all connected atoms.

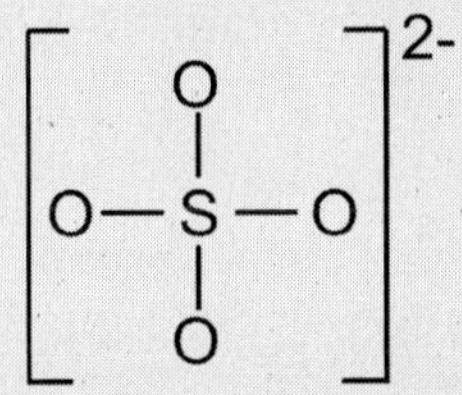

Step 4: Place the remaining valence electrons not accounted for in Step 3 on individual atoms until the octet rule is satisfied. By drawing bonds to the sulfur atoms, we have accounted for eight of the 32 electrons, and 24 are remaining to be assigned. Placing three lone pairs around each oxygen atom will account for all of the available electrons.

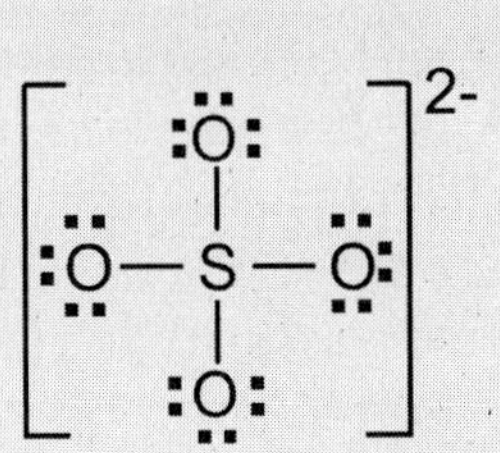

Step 5: For any atoms that do not have a full octet of valence electrons, create multiple bonds. We have fulfilled the octet rule and have accounted for all of the available electrons. Multiple bonds do not appear to be required in this ion.

Discussion

This is similar to the earlier examples, but the fact that we are working with an ion requires extra care in determining of how many valence electrons to include.

Check Your Understanding

Draw Lewis structures for the following polyatomic ions: NH_4^+, SO_3^{2-}, ClO_2^-.

Example Problem 7.6

Draw the Lewis structure of tetrafluoroethylene, C_2F_4, the monomer from which PTFE is manufactured.

Strategy

Follow the steps of the Lewis structure algorithm. Because we have two carbon atoms, we will have two "central" atoms in this case.

Solution

Step 1: Count the total number of valence electrons. Each carbon atom has four and each fluorine atom has seven, so there are 36 valence electrons.

$$\text{Number of valence electrons} = 2(4) + 4(7) = 36$$

Step 2: Draw the skeletal structure of the molecule. Carbon occurs first in the formula and it is the least electronegative, so we expect it to be the central atom. This is also consistent with our knowledge of organic chemistry, and suggests that the carbon atoms will be in the center of the molecule, and will be linked by a chemical bond. So we'll place the carbon atoms in the middle, surrounded by fluorine atoms. We have some choices as to how to arrange the fluorine atoms around the carbon atoms. *It is a good general rule to start off with the most symmetrical arrangement*, so we'll place two fluorine atoms near each carbon, rather than three near one and one near the other.

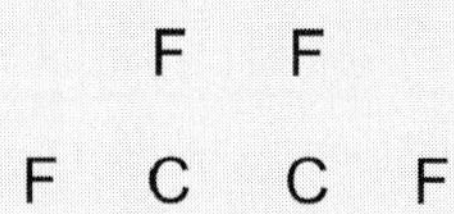

Step 3: Place single bonds between all connected atoms. In order to hold the molecule together, we need at least one bond between the two carbon atoms, and at least one bond between each carbon atom and each of the two adjacent fluorine atoms.

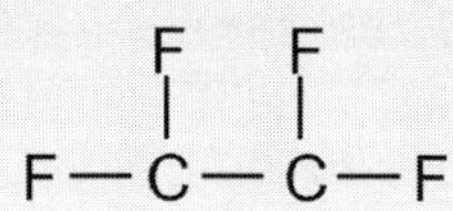

Step 4: Place the remaining valence electrons not accounted for in Step 3 on individual atoms until the octet rule is satisfied. Making five bonds requires ten electrons, so we still have 36 – 10 = 26 electrons available. Start by assigning three lone pairs to each of the outer fluorine atoms so that each has an octet. We use 24 electrons in this process, so there are two remaining valence electrons, which are placed for the moment on one of the carbon atoms. (Since the two carbon atoms look identical in the structure before we add this last pair, the choice of which carbon gets the electrons is arbitrary.)

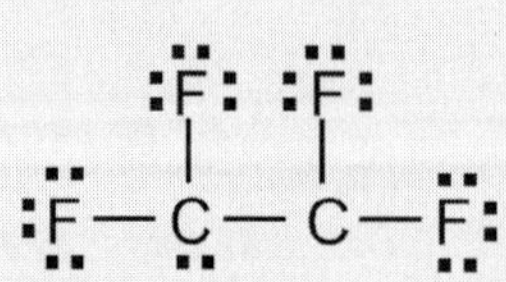

Step 5: For any atoms that do not have a full octet of valence electrons, create multiple bonds. The carbon atom that did not receive a lone pair in the last step does not have an octet. So we'll need to convert a lone pair from an adjacent atom into a bonding pair, creating a double bond. But should this double bond be formed between the two carbon atoms, or between the carbon atom and one of the fluorine atoms? At least two factors point toward the carbon-carbon choice. First, we have already seen that carbon often forms

multiple bonds, whereas fluorine does not. Second, placing the double bond between the two carbon atoms gives the molecule a much more symmetric structure, and such arrangements are usually preferable.

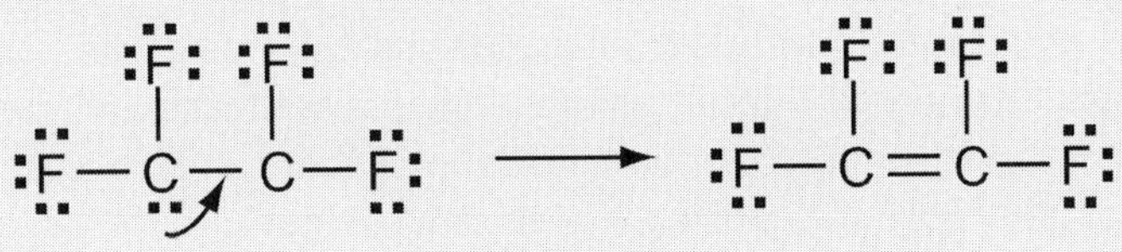

Discussion
This example shows that using our algorithm for obtaining a Lewis structure still leads to the need to make an occasional decision. As you grow more experienced in chemistry, decisions such as where to place a double bond become easier to make. Often you can look for clues, as we did in this case. An element like fluorine does not make double bonds very often. Hydrogen atoms never make double bonds. As a general rule it is not likely that an element with a valence of one (*i.e.*, an element that forms one bond with hydrogen) will form a double bond. But what happens when the choice is less obvious? Such a situation is discussed in the next subsection.

Check Your Understanding
Draw the Lewis structure for sulfur trioxide, SO_3.

Chemists are quite fond of Lewis structures, and the reason for this is clear. The structures are generally simple to draw; yet they can often provide powerful insight into the bonding within a molecule. On occasion, however, when the final step of our algorithm is carried out, the choice of where to put a double bond appears entirely arbitrary. The concept of **resonance** has been developed to help describe the bonding in such molecules within the framework of Lewis structures.

Resonance

Let's consider sulfur dioxide, one of the EPA's criteria pollutants we discussed in Chapter 5. What should its Lewis structure look like? Both sulfur and oxygen have six valence electrons, so we have a total of 18 valence electrons. Sulfur occurs first in the formula and is less electronegative, so we place it at the center of the skeletal structure.

O S O

We next place bonding pairs between the sulfur atom and each oxygen atom.

O—S—O

We still have 14 valence electrons to add, starting with the oxygen atoms. When we place all 14 electrons in the structure as lone pairs, we have the following structure.

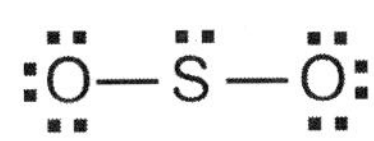

Checking for octets, we find that the oxygen atoms both have filled octets, but the sulfur has only six electrons. It needs two more, which can be obtained by forming a double bond. But on which side should we place this double bond? Unlike Example Problem 7.6, here we are confronted with two completely equivalent choices for the position of the double bond. In essence, we could draw either of two structures.

Either of these structures has the same fundamental weakness: it implies that the central sulfur atom somehow interacts more strongly with one oxygen atom than with the other. The explanation for this contradiction is that the structure of the molecule cannot be accurately represented by a single Lewis structure. Instead we say that the "real" structure corresponds to an average of all the valid Lewis structures, and we call this average a **resonance hybrid** of the contributing Lewis structures. It is very important to emphasize that the double-headed arrow does *not* mean that the structure alternates between the two possibilities; it indicates that the two structures shown for SO_2 are **resonance structures**, contributing to a resonance hybrid. *In resonance structures, the positions of all atoms are identical; only the positions of the electrons are different.*

The key point to emphasize here is that a resonance hybrid has a structure that is an average of the contributing resonance structures. To better explain this statement, let's take another look at the two possible structures of SO_2 shown above. Either resonance structure taken alone would suggest that in the SO_2 molecule there is one single bond and one double bond. If this were true, then one bond (the single bond) should be longer than the other (the double bond). But bond lengths are experimentally measurable, and the data clearly show that the two bonds are equal in length. Moreover, the bond lengths measured for SO_2 is intermediate between those typically seen for an S–O single bond and an S=O double bond. These experimental results confirm the idea that neither resonance structure alone gives a good picture of the SO_2 molecule, and reinforce the view that the real structure is best thought of as an average of the possible resonance forms. *When we can draw resonance structures, the actual structure is a hybrid, an average of the contributing structures, and NOT a mixture of them.* The bonds can be thought of as "smeared out" over the molecule. Given its simplicity, it is not really surprising that the Lewis dot model can't adequately describe every molecule we come across. The concept of resonance can be thought of as a "patch" that attempts to rationalize the experimental facts while preserving most of the simplicity of the Lewis model.

The source of the energy differences associated with resonance is the kinetic energy part of the quantum mechanical model. The details of this treatement are not necessary for an entry level chemistry course.

We have considered the structural implications of resonance, but there are also energetic consequences of resonance. This means that molecules with resonance structures behave differently than one might expect on the basis of the individual resonance structures, which show a combination of single and double bonds. Benzene is the classic example of a substance that displays resonance, and in fact the concept of resonance originated from a need to explain the chemistry of benzene. Benzene, C_6H_6, is a volatile liquid obtained from coal tar. It was accepted in the latter half of the nineteenth century that benzene was a cyclic compound, with its carbon atoms forming a six-membered ring. The only structures that could be drawn for it, however, had alternating single and double bonds (Figure 7.12). If these structures really represented the structure of benzene, the substance should have shown the addition reactions typical of compounds with C=C bonds. But benzene is actually much less reactive than alkenes, meaning that the molecule is more stable than any one Lewis structure would predict. Over time it became apparent from examining many compounds that this result is more general: when a molecule can be represented by multiple resonance structures it is more stable (*i.e.*, has a lower energy and is less reactive) than would be predicted on the basis of any one of the contributing resonance structures. There are ways to understand this energy lowering based on quantum mechanics, but we will not present them here.

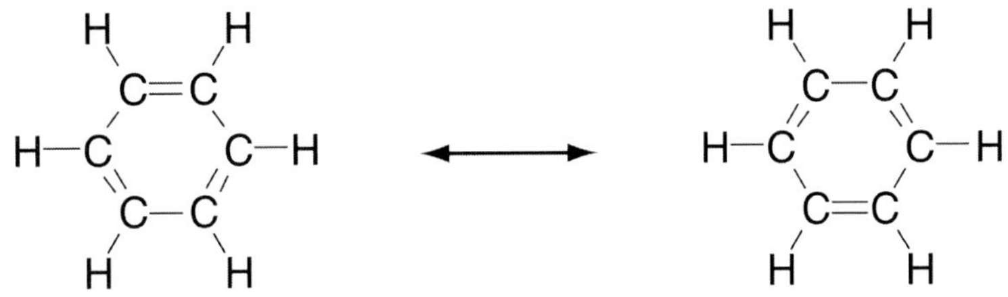

Figure 7.12 The two resonance structures of benzene show there are two ways to assign which carbon atoms are doubly bonded. Benzene does not behave chemically like a compound with three double bonds, however, because the true structure is an average between these two resonance structures.

7.6 Orbital Overlap and Chemical Bonding

Lewis structures provide us with a convenient way to gain some insight into a molecule's bonding. But they don't really tell us much about *how* chemical bonds might be formed, other than through the general notion of electron sharing. Scientific curiosity leads us to want to know more about covalent bonds, as does a sense that understanding bonding will pay off in practical applications. Since we recognize shared electrons as the "glue" that holds covalent bonds together, we should expect any deeper explanation to focus on electrons. In Chapter 6 we learned that the best current model for the atom views electrons as delocalized waves, described by orbitals. If electrons behave as waves, then to explain how electrons can form bonds between atoms we must consider the way that waves interact with one another.

To explain how two waves - or two orbitals - might interact with one another we will need to invoke the concept of interference, which provides a clear contrast between waves and particles. When two particles occupy the same space they collide; but when two waves occupy the same space they interfere. As you probably learned in a high school physics or physical science class, this interference may be either constructive or destructive. In constructive interference, two interacting waves reinforce one another to form a larger wave, while in destructive interference the two interacting waves offset or "cancel" one another to produce a smaller wave or no wave at all. Both constructive and destructive interference are illustrated in Figure 7.13.

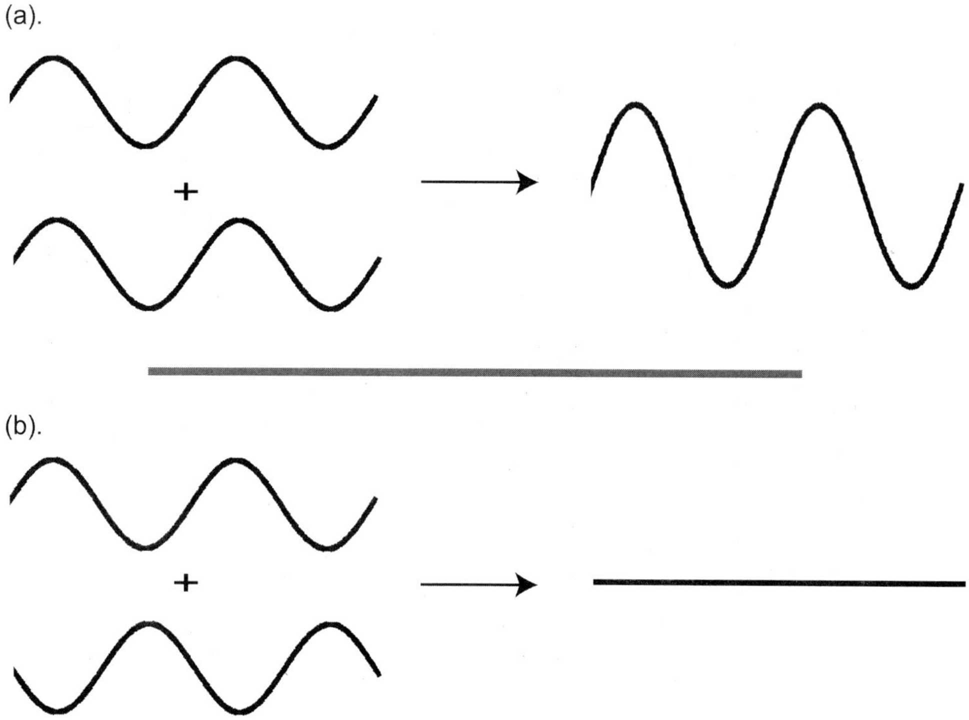

Figure 7.13 Simple sine waves illustrate the concept of interference between waves. In the upper panel (a), the two identical sine waves on the left are totally in phase with one another, meaning their peaks and valleys line up. Adding these two waves will produce constructive interference, and the resulting wave will have twice the amplitude of the original waves. In the lower panel (b), the two waves are out of phase, so that the peaks of one correspond to the valleys of the other. Combining these waves will result in destructive interference, where the two waves effectively cancel one another and the result is a flat line.

The name valence bond model arises from the fact that only valence electrons are considered.

In the quantum mechanical model, electrons are simply one type of wave, and we can picture the formation of chemical bonds as an example of constructive interference between electron waves. But in order for electron waves from different atoms to interact, they must occupy the same region of space simultaneously. Another way of saying this is that the atoms must be positioned so that the valence orbitals from one atom must overlap with those of the other atom if a bond is to form. This idea forms the basis for the **valence bond model** of chemical bonding, in which all bonds are seen as the result of overlap between atomic orbitals. Let's examine what this means by looking at the simplest possible molecule, H_2.

Hydrogen atoms have a single valence electron, in a *1s*-orbital. So to form a diatomic hydrogen molecule, the *1s*-orbitals from two different hydrogen atoms must overlap. Figure 7.14 shows what this overlap might look like and how interference between the wave functions of the two electrons produces a bond between the atoms. Note that since the *1s*-orbitals are spherical, there is no preferred direction of approach. But when *p*-orbitals are involved, their more complex shape will lead to different possible geometries for orbital interaction. And these different geometries can be used to distinguish between types of chemical bonds.

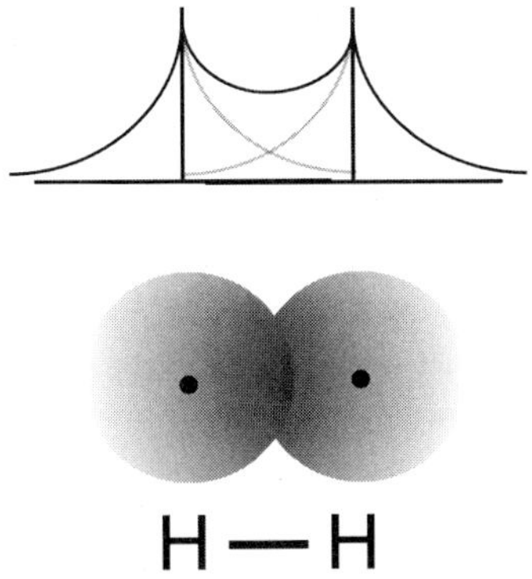

Figure 7.14 The covalent bond in molecular hydrogen can be described in terms of the overlap between the 1s-orbitals from the two H-atoms. The upper diagram in the figure shows this overlap by plotting the wavefunctions for the 1s-orbitals. Constructive interference between these wavefunctions in the region between the nuclei results in bond formation. The lower diagram uses shading to represent the same build up of electron density.

Using Lewis dot structures we can easily show that N_2 must have a triple bond between the two nitrogen atoms. While this large build up of negative charge density between the nuclei allows very strong attraction between them, it also raises a question. How do six electrons occupy the same region of space without strongly repelling each other? We can address this question by considering the shape of the orbitals involved and their arrangement as overlap occurs. The valence electron configuration for a nitrogen atom is $2s^2 2p^3$. Since the $2p$-orbitals are all singly occupied, overlap of a $2p$-orbital from one atom with a corresponding $2p$-orbital from the other atom should enable bond formation. But because of their shape, p-orbitals can overlap in different orientations.

Recall that the three $2p$-orbitals from any atom can be described as lying along the x-, y-, and z-axes. Let's assume that two nitrogen atoms approach one another along the z-axis. As seen in Figure 7.15 the $2p_z$-orbitals can overlap in an "end-to-end" fashion. This leads to a buildup of electron density along a line drawn between the two nuclei, just as we saw in the H_2 case. This type of constructive interference produces what are called **sigma bonds**. The accumulation of electron density in a sigma bond sits along the line connecting the two nuclei. This type of overlap can also occur in bonds in which an s-orbital overlaps with a p-orbital, or when two s-orbitals overlap.

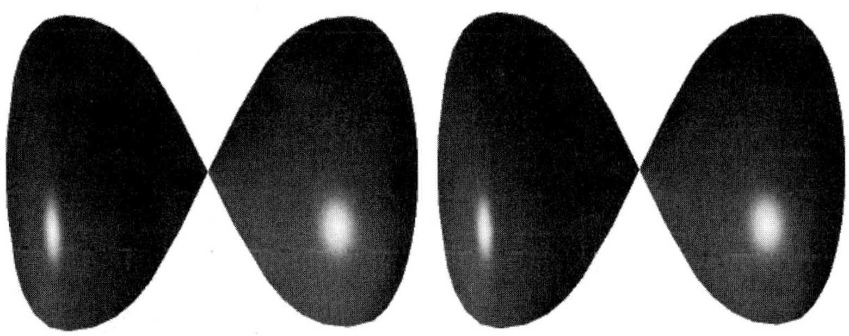

Figure 7.15 When two p-orbitals approach one another and overlap in an end-to-end arrangement, as shown here, they form a sigma bond. (This is sometimes denoted using a lower case sigma: σ.)

Now let's consider the $2p_x$- and $2p_y$-orbitals of the nitrogen atoms. As the nuclei draw closer together, these orbitals can overlap in a "side-to-side" orientation, as shown in

Figure 7.16. Once again the overlap leads to constructive interference; this allows the buildup of electron density between nuclei that is crucial to chemical bonding. When this "sideways" overlap occurs, however, the buildup of electron density will not be localized along the line between the nuclei. As shown in Figure 7.16, the electron density is localized above and below or in front and in back of this line. Such bonds are called **pi bonds**. In the N_2 molecule, one pi bond is formed through the overlap of the $2p_x$-orbitals from each atom and another pi bond is formed through the overlap of the $2p_y$-orbitals. The triple bond thus comprises one sigma bond and two pi bonds.

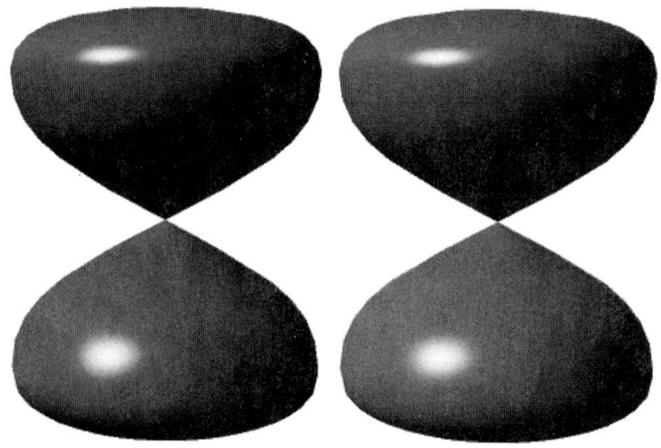

Figure 7.16 When two *p*-orbitals approach one another and overlap in a side-to-side arrangement, as shown here, they form a pi bond. (This is sometimes denoted using a lower case pi: π.) In the example shown, orbital overlap will produce a build up of electron density above and below the line defined by the two nuclei.

The idea that bonding involves the overlap and interaction of orbitals from different atoms is a fundamental building block to more detailed understanding of the chemical bond. For this reason, the valence bond model continues to play an important role in the way scientists envision bonding. We do, however, encounter serious difficulties in describing bonding through the overlap of atomic orbitals. Since *p*-orbitals form 90° angles with one another we might predict that many molecules, including H_2O and OF_2, would show 90° bond angles. But in fact such 90° angles are quite rare. Furthermore, looking at the electron configuration of carbon, we might expect compounds like CH_2 or CF_2 to be stable and commonplace. But instead we find CH_4 and CF_4, and the simple overlap of atomic orbitals is not able to help us rationalize this. So we must extend our valence bond model to allow for more extensive interactions between orbitals.

7.7 Hybrid Orbitals

Our notion of orbitals arose from the quantum mechanical treatment of individual atoms. Despite the obvious conceptual challenges that quantum mechanics poses, we accept the existence of *s*-, *p*-, *d*- and *f*-orbitals because, among other things, doing so provides such a reasonable way to understand the shape of the periodic table. When atoms combine to form molecules, however, the observations noted above suggest that those same orbitals often cannot explain the observed facts. We just noted that electrons in *p*-orbitals would be expected to lead to an H-O-H bond angle of 90° in water, for example, and this does not agree with the observed geometry of the molecule. And yet the relative simplicity of the idea that bonds result from orbital overlap remains alluring. This leads to the introduction of the concept of orbital **hybridization**, which is intended to reconcile the notion of orbital overlap with a wide range of observations on molecular shapes and structures. At its heart, the idea of hybridization acknowledges the fact that when two or more atoms approach one another closely enough to form chemical bonds, the interactions can be strong enough to reshape the orbitals on those atoms.

The simplest stable binary compound formed by carbon and fluorine is CF_4, with the Lewis structure shown below.

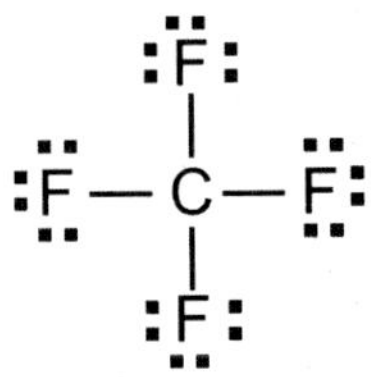

Bond lengths and strengths can be measured experimentally, and such measurements clearly show that all four C-F bonds in CF_4 are identical to one another. Furthermore, molecular shapes can also be measured by observing the rotation of molecules, and such measurements show that the F-C-F bond angles in CF_4 are all 109.5°. These data suggest that if we want to describe the bonds in CF_4 in terms of orbital overlap, we would need four identical orbitals on the carbon atom, and that these orbitals should be arranged at angles of 109.5° from one another. When we introduced the idea of orbitals, we noted that they were essentially mathematical in origin. In mathematics we can manipulate functions and transform them according to accepted rules. The wave functions for atomic orbitals can also be manipulated in the same ways. We need not be concerned with all the details of such manipulations, but one rule will be important; the number of orbitals we have must not change due to the manipulation. If we have two orbitals to begin with, we must still have two orbitals when we finish. Among other things, this assures that we can still accommodate all the needed electrons. If we take linear combinations of atomic orbitals we can devise new orbitals that can be used as the basis for a model of bonding. The resulting orbitals are called **hybrid orbitals** because they are mixtures of two or more atomic orbitals.

Orbital theory began as a model. So long as adjustments, like the invention of hybrid orbitals, do not change the premise of the model, such changes can be made - and are worth making if they provide new insight.

One way to envision this process is to use boxes to represent orbitals as we did in Chapter 6. Figure 7.17 shows this for the hybridization of a carbon atom. The *s*-orbital and the three *p*-orbitals are mixed to form four new hybrid orbitals, all having the same energy. The name given to these orbitals is sp^3, where the superscript indicates that three *p*-orbitals are included. Placing electrons in these four equivalent sp^3-orbitals means that each will be singly occupied. Thus these orbitals readily explain the ability of carbon to form four identical bonds. If we were to examine the details of the mathematics, we would also discover that each of these sp^3-orbitals is strongly oriented in a particular direction, with angles of 109.5° between orbitals. These hybrid orbitals thus make it easy to explain the observations for CF_4; overlap of an sp^3-orbital from carbon with a *p*-orbital from fluorine forms each of the four C-F bonds.

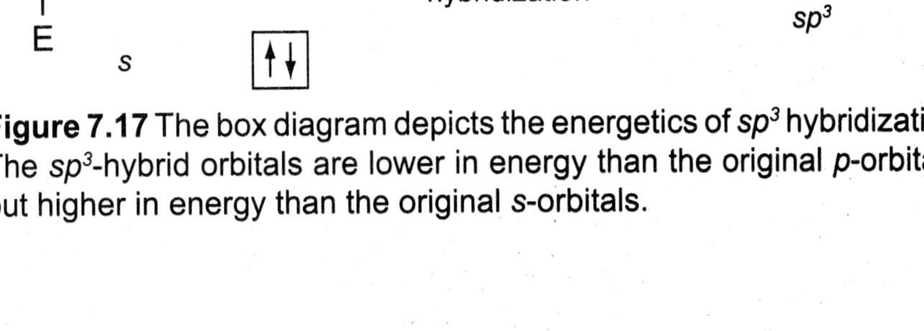

Figure 7.17 The box diagram depicts the energetics of sp^3 hybridization. The sp^3-hybrid orbitals are lower in energy than the original *p*-orbitals, but higher in energy than the original *s*-orbitals.

Hybridiztion with 5 or 6 orbitals also allows us to explain the fact that some elements form compounds with expanded octets - more than 8 electrrons around a single atom.

The concepts behind the hybrid orbital model are very flexible with regard to the number of orbitals that can be combined, as shown in Table 7.2. When only one *p*-orbital is included in the scheme we form *sp*-hybridized atoms and when two *p*-orbitals are included the hybrid orbitals are sp^2. For atoms with valence electrons in the n=3 shell or higher, hybridization can also include *d*-orbitals. Thus sp^3d-hybridization yields five orbitals, and when six orbitals are included we have sp^3d^2-hybrids. All of these schemes can be used to help explain bonding in covalent molecules.

Orbitals Combined	Hybridization	Orbital Geometry
s, p	*sp*	
s, p, p	sp^2	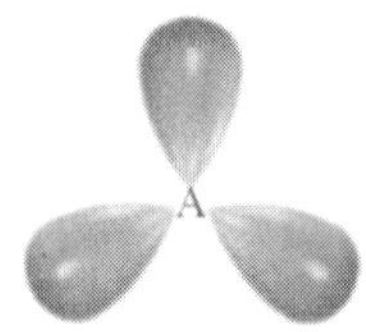
s, p, p, p	sp^3	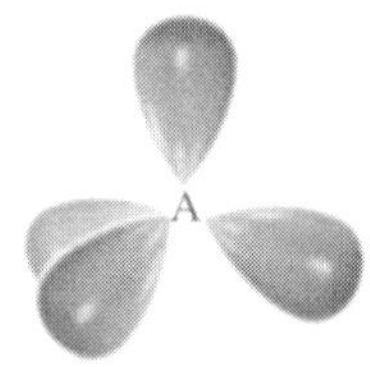
s, p, p, p, d	sp^3d	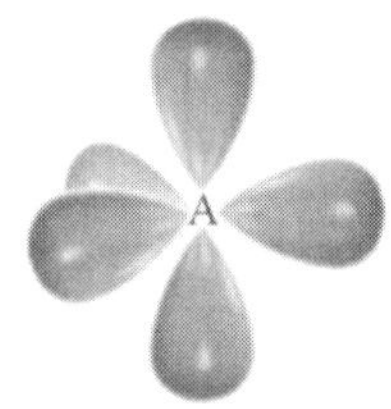
s, p, p, p, d, d	sp^3d^2	

Table 7.2 The five common orbital hybridization schemes are shown. The names for these hybrid orbitals are derived from the type and number of orbitals that combine to form them. Thus sp^2-hybrids result from the combination of an *s*-orbital with a pair of *p*-orbitals. The orbital geometries shown in the left hand column give rise to the common molecular shapes described in the next section.

7.8 Shapes of Molecules

While Lewis structures allow us to account for the number of bonds in a molecule and show us the bonding pattern (which atoms are joined together and by how many bonds), they give us an incomplete, two-dimensional picture of the molecular structure. If we extend our analysis of the molecular structure to three dimensions, we need to examine not only the order in which the atoms are joined to one another, but also how they are arranged in space—in other words, the molecular shape. In addition to being of theoretical interest, the shapes of molecules are important because they affect molecular properties, including reactivity. Being able to predict the shapes of molecules is crucial to understanding the properties and reactivity of a polymer like PTFE, the polymeric proteins and nucleic acids that are the operating systems in living organisms, the catalysts we use in industrial processes, and the molecules we synthesize for medicines and consumer goods. Indeed, one thing that sets experienced chemists apart from beginning students is the ability to read a molecular formula and visualize what the molecule would look like on the microscopic scale.

In order to arrive at an understanding of the shapes of molecules we need to analyze Lewis structures in more detail. A very useful technique for doing this relies on the **valence shell electron pair repulsion (VSEPR)** theory, which states that *molecules assume a shape that allows them to minimize the repulsions between electron pairs in the valence shell of the central atom.* Lewis structures emphasize the importance of pairs of electrons around atoms, either as bonding pairs or as lone pairs. If we consider these pairs as occupying localized regions of space, we would expect that the various regions would be as far away from each other as possible. After all, these regions are centers of negative charge, and like charges repel one another. Couple this *electron pair repulsion* with the fact that Lewis structures include only electrons in the *valence shell*, and the origin of the name VSEPR is apparent. The theory provides a very simplistic view of the electrons in molecules, but it leads to surprisingly accurate predictions of molecular geometries.

In order to use VSEPR we need to apply concepts from geometry. In particular, we want to answer this question: If the bond lengths are fixed, what geometric shapes allow the regions of negative charge represented by various electron pairs to be as far away from one another as possible? The answer depends on the number of electron pairs that surround the atom. Table 7.3 shows the basic geometry for distributing between two and six electron pairs in such a way as to minimize repulsions. Examining this table, you should realize that most of the shapes involved are fairly simple. It should be intuitively clear that if only two regions of electron density are present, for example, that a linear arrangement will maximize their separation. Of all the shapes shown, the tetrahedral and trigonal bipyramidal arrangements are probably the least familiar. You should be sure that you have a real understanding of what these different geometries look like. (Models – either physical of software-based - can help a great deal.) Most of the bonding situations we will encounter can be analyzed in terms of the five basic shapes shown here.

If you don't hsve access to molecular models, you can build your own by using small balloons and holding them together with a clip of some sort. The balloons must stay out of each others way physically, analagous to the electronic repulsions of electrons.

We can predict molecular geometries in a systematic way. Because our approach centers on the arrangement of electron pairs around the central atom, we will always start by drawing a Lewis structure. Once we've done that, it is simple to count the number of bonding and nonbonding electron pairs around the central atom. We'll start with the most straightforward examples, in which all the electrons around the central atom are in single bonds. Then we'll look at how to determine the molecular geometry when there are also nonbonding pairs or multiple bonds involved.

If the central atom is surrounded only by single bonds, the process of determining the molecular geometry is fairly simple. We count the number of single bonds emanating from the central atom, and use the information in Table 7.3 to correlate that number of electron pairs with the geometry predicted by VSEPR, and this describes the shape of the molecule.

Number of electron pairs	Geometry	Bond angles	Ball & Stick Model
2	Linear	180°	
3	Trigonal planar	120°	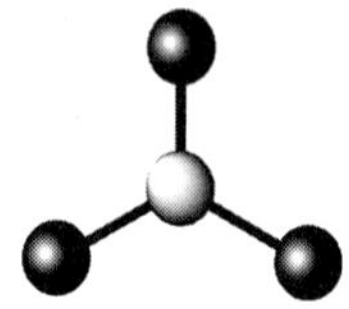
4	Tetrahedral	109.5°	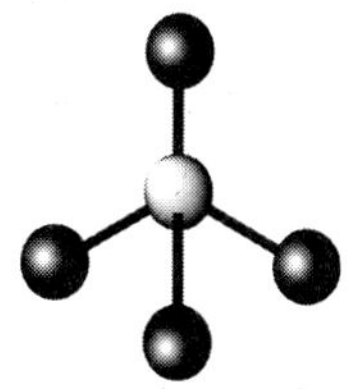
5	Trigonal bipyramidal	120°, 90°	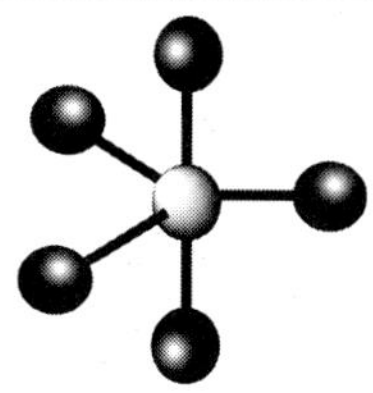
6	Octahedral	90°, 180°	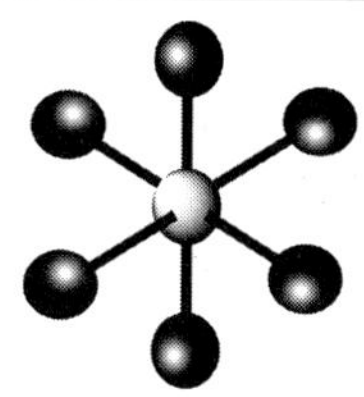

Table 7.3 Each of the geometrical arrangements shown in the table minimizes the electron pair repulsions for the indicated number of electron pairs. In order to be able to visualize the shapes of molecules, it is essential that you have a sound mental picture of what each of these geometries looks like.

Example Problem 7.7

Determine the shape of each of the following molecules. (a) CF_4 (b) PF_5

Strategy

Draw the Lewis structures. Each of these molecules has only single bonds around its central atom. So we can count the number of bonding pairs around the central atom and assign the geometry by consulting Table 7.3.

Solution

(a). The Lewis structure of CF_4 is:

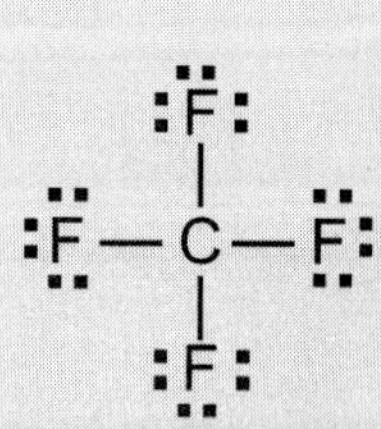

The central carbon atom has a complete octet of electrons, contributed by four bonding electron pairs. The four electron pairs have minimal interaction when they point towards the corners of a tetrahedron. The shape of the molecule is *tetrahedral* (Table 7.3).

(b). The Lewis structure of PF_5 is somewhat unusual because the central phosphorus atom has more than an octet around it. There are some molecules with this property and they serve as useful examples for structures in VSEPR.

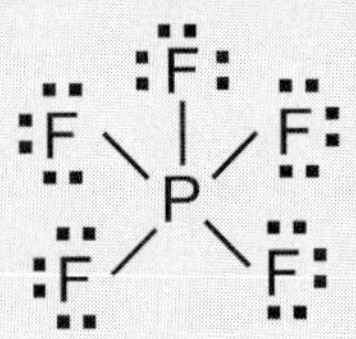

The central phosphorous atom has an expanded octet, with ten electrons arranged in five bonding pairs. The five electron pairs occupy the vertices of a *trigonal bipyramid* (Table 7.3). Note that, unlike the fluorine atoms in the tetrahedral molecule we looked at in (a), the fluorine atoms in the trigonal bipyramid are not all in equivalent positions. There is a spacing of 120° between the fluorine atoms arranged in the horizontal triangle (the **equatorial** positions), while the two atoms above and below the plane of the triangle (**axial** positions) form an angle of 90° with the plane of triangle.

Discussion

These examples can be difficult to see using only the text. Model kits or software models can help you visualize the three-dimensional nature of the shapes described here.

Check Your Understanding

Predict the shape of the SF_6 molecule.

Whenever we talk about the geometry or shape of a molecule, we refer to the positions of the atoms. This seems obvious in cases like those in Example Problem 7.7, where there are no lone pairs on the central atom. But the same idea holds even if there are lone pairs involved; it is the positions of the atoms that define the shape of a molecule. But lone pairs also contribute to the overall repulsion between electron pairs, and so they do *influence* the geometry of molecules. When the central atom is surrounded by both bonding and nonbonding electron pairs, we can choose a "base geometry" from the list in Table 7.3 by counting the total number of regions of high electron density – bonding pairs plus lone pairs. To determine the actual molecular geometry, we then visualize what that shape would look like if the positions to be occupied by lone pairs were "empty." In some cases more than one shape may seem possible. When that happens, the guiding premise of VSEPR says that the correct shape will be the one that minimizes the overall repulsion between electron pairs. In deciding which structure that is, the order of repulsive interactions is lone pair–lone pair > lone pair–bond pair > bond pair–bond pair. This order can be explained in terms of the fact that lone pairs are attracted to only one nucleus. This means that a lone pair is less tightly localized, so it occupies a larger space. Examples of the various possible geometries are shown in Table 7.4. The determination of the geometry of a molecule with both bonding and nonbonding electrons is best illustrated by looking at some examples.

Number of Electron Pairs	Number of lone pairs	Shape	Ball & Stick Model
3	0	Trigonal planar	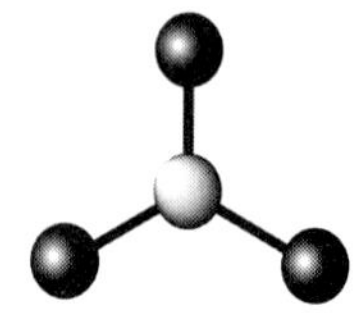
3	1	Bent (120°)	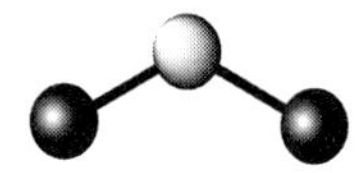
4	0	Tetrahedral	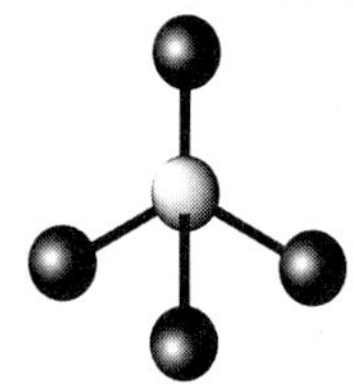
4	1	Trigonal pyramidal	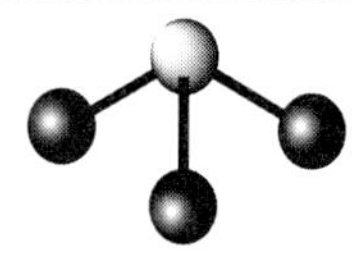
4	2	Bent (109.5°)	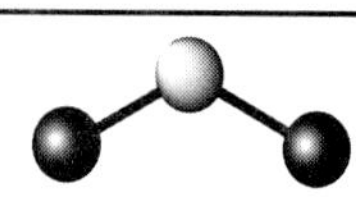
5	0	Trigonal bipyramidal	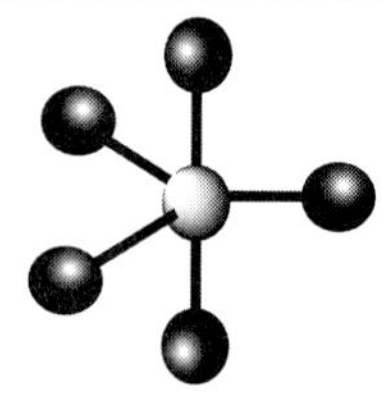
5	1	See-saw	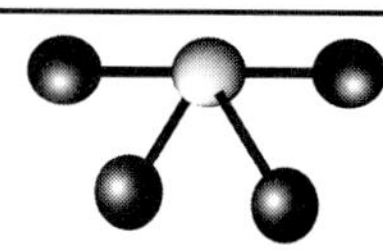
5	2	T-shape	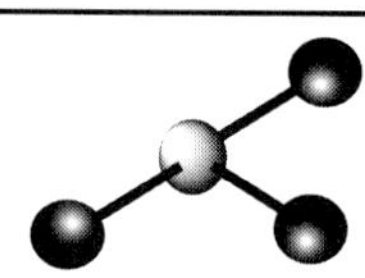
5	3	Linear	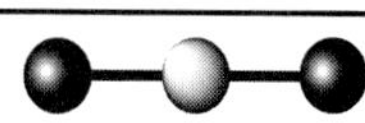
6	0	Octahedral	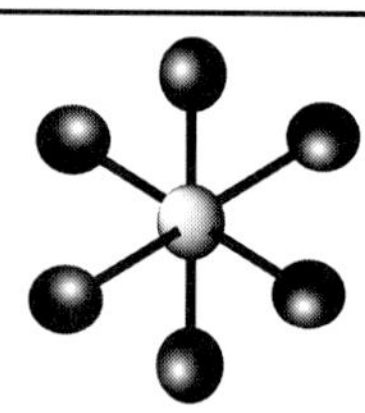
6	1	Square pyramidal	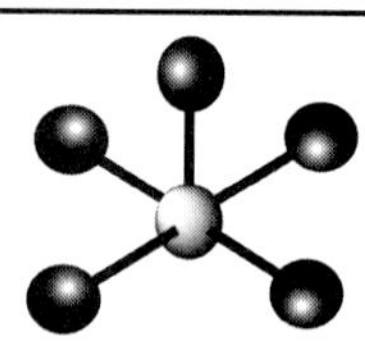

Number of Electron Pairs	Number of lone pairs	Shape	Ball & Stick Model
6	2	Square planar	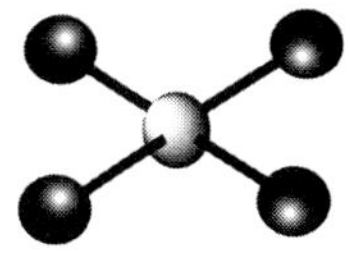
6	3	T-shape	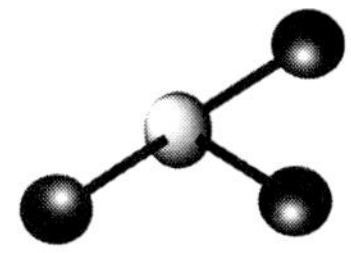
6	4	Linear	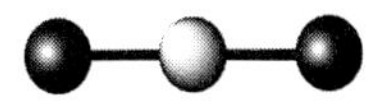

Table 7.4 The molecular shapes resulting from various combinations of the total number of electron pairs around the central atom and the number of lone pairs are shown. Note that as long as you have a good grasp of the basic geometries (as shown in Table 7.3), it is not necessary to memorize this extensive listing. Instead you should be able to start from the basic geometry for the number of electron pairs present and visualize the shape that would result from "emptying" a number of positions corresponding to the lone pairs.

Example Problem 7.8

Determine the shape of the following molecules using VSEPR theory. (a) SF_4 (b) BrF_5

Strategy

As always, we start by drawing the Lewis structures. Then count the number of electron pairs around the central atom, and determine the spatial arrangement of electrons pairs, consulting Table 7.3 as necessary. Place the lone pairs in positions where the electron repulsions are minimized, and describe the resulting geometric arrangement of the atoms.

Solution

(a) The Lewis structure of SF_4 is

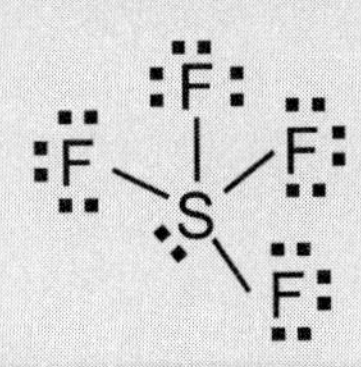

There are five pairs of electrons around sulfur, so the shape with minimum interaction is a trigonal bipyramid. Unlike the previous example, though, here the central sulfur atom has one lone pair. The five positions of a trigonal bipyramid are not equivalent, so we need to determine the best location for the lone pair. Electron pairs in the planar triangle (equatorial positions) have two near neighbors at an angle of 120° and two at 90°, whereas the axial positions have three near neighbors, each at an angle of 90°. Thus there is "more space" for electron pairs in the equatorial positions. Since lone pair-bond pair repulsion is larger than bond pair-bond pair repulsion, this means that the lone pair should occupy an equatorial position. This reduces the overall repulsive interactions (only the two closest bonding pairs are 90° away, whereas if the lone pair were in an axial position there would be three bonding pairs 90° away.)

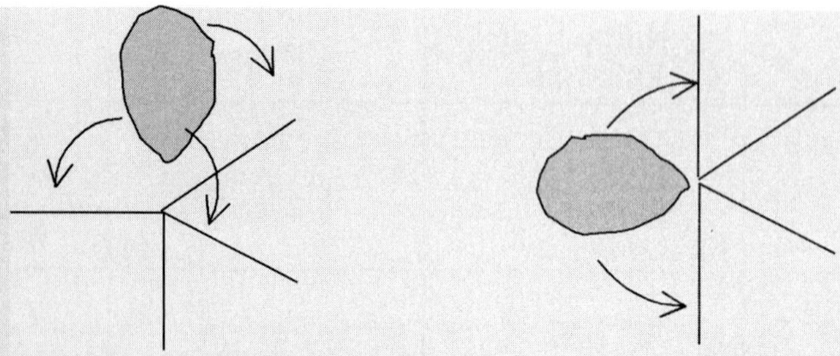

The resulting shape is usually described as a seesaw. (If you don't see why, try looking at a model.)

(b) The Lewis structure of BrF_5 is

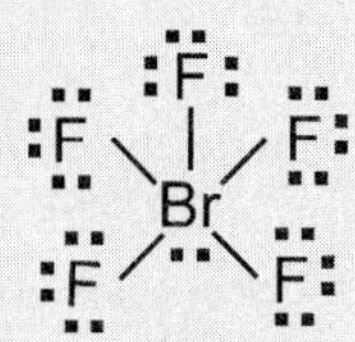

There are six electron pairs around the central bromine atom, so the distribution of electrons is based on an octahedron. One of the six positions is occupied by a lone pair. An octahedron is completely symmetric, and all of its vertices are equivalent, so the lone pair can occupy any corner. The resulting shape of the molecule is a square pyramid.

Discussion

For many students the confusing part of this type of problem lies in the appearance that lone pair electrons are considered for part of the problem but not all of it. There are two main things to remember. (1) The shape of a molecule is determined by the positions of its atoms. This is a result of the ways we measure shapes. Because nuclei are so much more massive than electrons, our experiments detect them more readily, and thus we have long considered the shape to be dictated by the nuclei alone. (2) Regardless of whether it is involved in bonding or not, every electron pair in the Lewis structure constitutes a region in space where there is a relatively high concentration of negative charge. The key concept of VSEPR theory is that these negatively charged regions will repel each other and therefore be arranged as far apart as possible. So the electrons dictate the shape of a molecule, but they are not directly measured. That's why we have a process where we appear to consider them at some stages but not at others.

Check Your Understanding

What are the shapes of the following molecules? (a) ClF_3 (b) XeF_2

In determining the geometry of molecules by VSEPR, multiple bonds are treated as a single region of electron density. This is intuitively reasonable, because the position of greatest electron density for a multiple bond lies in the region between the two bonding atoms, just as it does for a single bond. To put this in another way, it wouldn't make sense to say that the electron pairs making up a double or triple bond should point in totally different directions from one another. Thus for purposes of determining molecular shapes, any double or triple bonds behave as if they were a single electron pair.

Example Problem 7.9
Use VSEPR theory to determine the shape of the NOF molecule.

Strategy
Once again, we start by drawing the Lewis structures. Then count the number of regions of electron density around the central atom, remembering to count any double or triple bonds as a single region. Determine the spatial arrangement of electrons pairs, consulting Tables 7.2 as needed. Place any lone pairs in positions where the electron repulsions are minimized, and describe the resulting geometric arrangement of the atoms.

Solution
The Lewis structure of NOF is

There are 3 regions of electron density around the oxygen atom, so the shape with minimum interaction is a trigonal planar. The central oxygen atom has one lone pair. The three positions of a trigonal planar geometry are equivalent, so it does not matter which one the lone pair occupies. The molecule will be bent, with an N-O-F bond angle of approximately 120°. Since lone pairs provide stronger repulsion, the actual bond angle should be slightly less than 120°.

Discussion
Much like the situation we saw for lone pairs, double bonds can cause confusion for many students. When counting the number of valence electrons to draw the Lewis structure, we must be sure to count the double bond as four electrons. (And a triple bond would count as six electrons, if one were present.) But in determining the shape, we treat the double (or triple) bond as a single region of electron density. It may help if you realize that both of the bonds in the double bond must "point in the same direction" since they connect the same pair of atoms.

Check Your Understanding
What are the shapes of the following molecules? (a) COF_2 (b) COS

Applying the same principles as for smaller molecules, VSEPR theory can also be used to predict the geometry of larger molecules. In predicting the structure for a molecule with more than one "central" atom, we look at each atom within the backbone of the molecule and predict the bonding about that atom, using the same procedures that we have just outlined. For example, let's see what happens when we apply VSEPR theory to the structure of tetrafluoroethylene (C_2F_4, the monomer from which PTFE is produced). The Lewis structure we drew in Example Problem 7.6 is a good place to start.

Which order we use to look at the different atoms is unimportant.

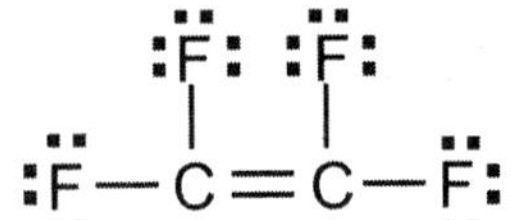

Each of the two carbon atoms is bound to three other atoms and has no lone pairs, implying a trigonal planar geometry. Next we need to ask whether those two trigonal planar ends of the molecule are coplanar or not. Here it helps to invoke hybridization and think about the orbitals involved in forming the double bond. Each can be described as using sp^2-hybrid orbitals; two of those hybrid orbitals will overlap with *p*-orbitals from fluorine atoms, while the third overlaps with an sp^2-orbital from the other carbon. This gives us a single sigma bond between the carbon atoms so far. But each carbon also still has one remaining unhybridized *p*-orbital. And in each case that *p*-orbital is perpendicular to the plane of the sp^2-hybrids, as shown in Figure 7.18. If these *p*-orbitals are aligned parallel to one another, they will be able to overlap side to side, forming a pi bond. This not only explains the existence of the double bond, but also tells us that the entire molecule must be planar. If one end of the molecule were to rotate with respect to the other, the *p*-orbitals would no longer be in the right orientation to form the pi bond.

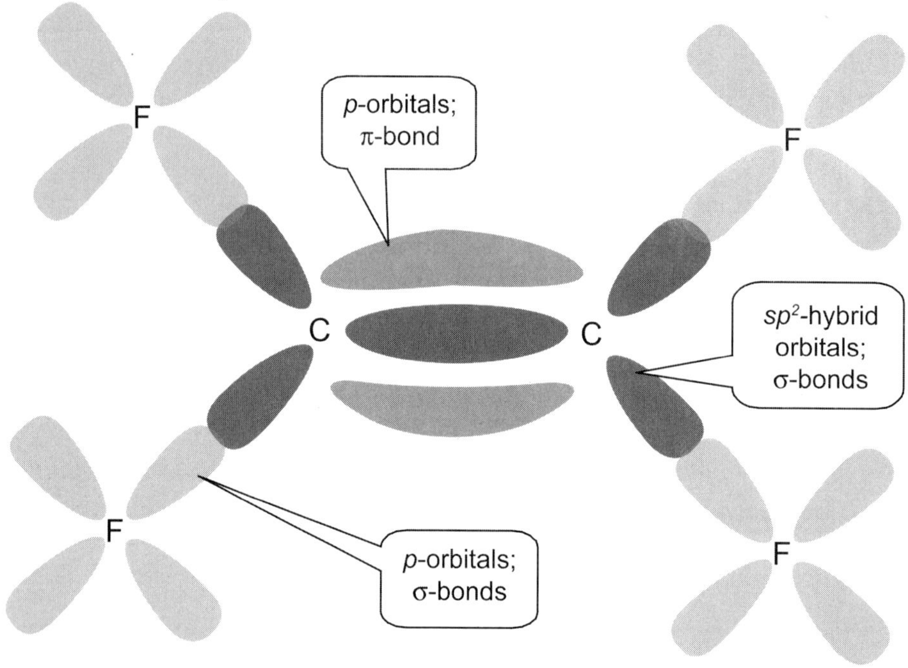

Figure 7.18 The orbitals involved in the bonding in C_2F_4 are shown. All six atoms lie in the plane of the paper. The *s*-bonding framework results from overlap of sp^2-hybrid orbitals on the carbon atoms and *p*-orbitals on the fluorine atoms. The π-bond is due to overlap of the remaining unhybridized p-orbitals from the carbon atoms. Thus the electron density in the π-bond would be above and below the plane of the paper. (The third *p*-orbital for each fluorine atom, which would be perpendicular to the plane of the paper, is not involved in the bonding and is not shown here.)

7.9 Insight into Chlorofluorocarbons

Throughout this chapter, we've learned quite a bit about the connection between the properties of PTFE and the chemical bonds that hold the polymer together. And while the discovery of PTFE was accidental, it is widely recognized as an extremely useful and beneficial material. But another group of fluorine-containing compounds, the chlorofluorocarbons, or CFCs, has a much different history. We'll close this chapter by looking at some of the uses and unintended consequences of CFCs.

The term chlorofluorocarbon refers to any of a group of volatile compounds that have been used as refrigerants, propellants, and degreasers. More specifically, these are usually derivatives of methane or ethane in which all of the hydrogen atoms have been replaced by either fluorine or chlorine atoms. These compounds were first produced in the 1930's as a safer alternative to refrigerants of the day. Since the refrigerants in use included ammonia and sulfur dioxide, the desirability of safer options was clear. Over time, the coolant tends to leak from a refrigerator and find its way into the atmosphere. So many of the problems surrounding the use of something like ammonia as a refrigerant stem from chemical reactivity and toxicity. Thus if you had been an engineer at DuPont assigned to develop new refrigerants, it is almost certain that you would have put things like "unreactive," "chemically inert," and "non-toxic" on the list of desirable properties in your first brainstorming session. We've noted how the bonding in PTFE makes it essentially inert, and the same arguments hold for CFCs. It turns out that the energy changes associated with the vaporization and condensation of CFCs are also extremely desirable for use as refrigerants. And this combination of factors made the CFCs an irresistibly attractive choice. By almost any measure, the performance of CFCs as refrigerants is unmatched by any other class of compounds.

You are probably aware that CFCs eventually sparked considerable environmental debate, and that their use has now been almost completely phased out. You may also know that much of the criticism of CFCs stems from the fact that they have been implicated in the destruction of ozone in the stratosphere. If these molecules are almost completely inert, how can they cause such havoc with the chemistry of the atmosphere? Paradoxically, their inertness under ordinary conditions leads directly to their presence in the ozone layer, and hence to their negative environmental impact.

We will look at the issue of stratospheric depletion of ozone in more detail in Chapter 11.

As we noted above, refrigerants will eventually escape, either through leaks or when a refrigerator or air conditioner is destroyed. And the fact that CFCs were also used as propellants in aerosol spray cans greatly accelerated their buildup in the atmosphere. Once released into the atmosphere, the CFC molecules are so inert that they will survive for a long time without undergoing chemical reactions. During that time, the CFCs drift higher and higher into the upper atmosphere. At sufficiently high altitudes, though, these molecules begin to encounter high-energy ultraviolet photons coming from the sun. These photons carry sufficient energy to break chemical bonds, producing individual halogen atoms. And unlike the original CFCs, these free halogen atoms are extremely reactive, and play a role in the breakdown of the protective ozone layer surrounding the earth.

From an engineer's perspective, the major lesson to be drawn from this story is probably the dangers of any unforeseen consequences of design decisions. The choice of CFCs as refrigerants was completely reasonable in terms of their performance, and the fact that these compounds are entirely non-toxic seemed to alleviate many concerns surrounding other options. And yet the chemical inertness that seemed so appealing ultimately proved to be the downfall of CFC coolants. This is not to say that this outcome should necessarily have been foreseen back in the 1930's, but rather to point out the need to consider as many possible ramifications as possible when making design choices.

Focus on Problem Solving

Question

A solid is placed in water to form a clear colorless solution that is observed to conduct electricity. A second solution is added and you observe the formation of a precipitate. After the precipitate settles the liquid above it does not conduct electricity perceptibly. What does this experiment tell you about the bonding in the initial solid and in the precipitate, and how does it tell you that?

This question is another example of conceptual problem solving. We have seen three types of chemical bonding in this chapter, and from the observable behavior of this experiment we need to infer what one is present. The other piece of information we need to include involves what allows a solution to conduct electricity, something that was discussed in Chapter 3.

Answer

In order for something to conduct electricity, it must allow charge to move. In solution the way that charges are present is in the form of ions. This strongly suggests that the solids use ionic bonding. How does this explain the observations? When the first ionic solid is dissolved the ions present are dispersed in solution and it then conducts electricity. The addition of the second solution forms a precipitate. Because it was formed by combining ions from the two initial solutions, this precipitate must also have ionic bonding. Note that since the solution o longer conducts electricity, it must not contain significant concentrations of any ions. So it is possible that two different precipitates may have formed, removing all of the ions. Another possibility is that the second product could be water, formed from the reaction between an acid and a base.

Chapter 7
Problems & Exercises

Insight into Familiar Fluorine Compounds

7.1 What is the chemical name of Teflon?

7.2 List some of the properties that make Teflon a useful coating material. Why are these properties useful?

7.3 What similarities exist between Teflon and CFCs?

7.4 Elemental fluorine is highly reactive and yet many fluorine-containing compounds are quite stable. Explain how these two observations are connected to each other.

7.5 Several examples of familiar fluorine containing molecules were introduced throughout this chapter. Having completed the chapter, list several of those familiar compounds and indicate the type of chemical bonding they display – ionic, polar covalent or covalent.

7.6 Both ionic sodium fluoride and polymeric Teflon have extended, repeating structures. Draw a picture of the two types of molecules and discuss the differences in bonding that are depicted by your pictures.

The Ionic Bond

7.7 Why is the Na^{2+} ion not found in nature?

7.8 Why do non-metals tend to form anions rather than cations?

7.9 How does ionization energy play an important role in the formation of ionic compounds?

7.10 Based on electron configurations, predict the two most stable ions of copper.

7.11 Why do transition metals often form more than one cation?

7.12 Which pair will form a compound with the larger lattice energy: Na and F or Mg and F? Why?

7.13 In a lattice, a positive ion is often surrounded by eight negative ions. We might reason, therefore, that the lattice energy should be related to eight times the potential of interaction between these oppositely charged particles. Why is this reasoning too simple?

7.14 Use the concept of lattice energy to rationalize why sodium fluoride dissolves in water while calcium fluoride does not. Extending this reasoning, would magnesium fluoride be soluble?

The Covalent Bond

7.15 Describe the difference between a covalent and an ionic bond.

7.16 Considering the potential energy of the interacting particles, how does covalent bonding lower the energy of the system?

7.17 Sketch a graph of the potential energy of two atoms as a function of the distance between them. On your graph label how bond energy and bond distance are defined.

7.18 When a covalent bond forms is energy absorbed or released? Explain how your answer is related to the graph you sketched in the previous problem.

7.19 Coulomb forces are often used to explain ionic bonding. Are Coulomb forces involved in covalent bonding as well? Explain.

7.20 In terms of the strengths of the covalent bonds involved, why are combustion reactions exothermic?

7.21 If the formation of chemical bonds is always exothermic, why don't all elements form dozens of bond to each atom?

7.22 Draw the Lewis symbol for each of the following atoms. (a) boron (b) fluorine (c) selenium (d) indium

7.23 Researchers are currently trying to synthesize super-heavy elements. There is some theoretical evidence that certain elements may be stable long enough to carry out chemical reactions if they could be made. Suppose element 114 has been synthesized and given the atomic symbol Bh. What would the Lewis dot symbol be for this new element?

7.24 Use Lewis symbols to explain why chlorine bonds with only one hydrogen atom.

7.25 Define the term *lone pair.*

7.26 Under what circumstances do multiple covalent bonds form?

7.27 How does the bond energy of a double bond compare to that of two single bonds between the same elements? How does this relationship explain the types of reactions compounds with double bonds undergo.

Electronegativity and Bond Polarity

7.28 How is electronegativity defined?

7.29 Distinguish between electron affinity and electronegativity.

7.30 Certain elements in the periodic table shown in Figure 7.9 were noted as having no electronegativity value defined. Based on the identity of these elements, hypothesize as to why they have no electronegativity value.

7.31 When two atoms that have different electronegativities form a covalent bond, what does the electron distribution in the bond look like?

7.32 When an HF bond is formed it is said to be polar. For this to be true the hydrogen atom must have less than one electron around it – yet the Lewis dot structure of HF attributes two electrons to hydrogen. Draw a picture of the electron density distribution for HF and use it to describe how the hydrogen atom can carry a partial positive charge. How can these two models of the HF bond (the electron density and the Lewis structure) seem so different and yet describe the same thing?

7.33 Why is a bond between two atoms with different electronegativities called a polar bond?

7.34 Sometimes great confusion arises for people who do not study science because scientists use "common" words in ways that are different than their "everyday" usage (referred to as vernacular usage). What does the word polar imply in every day usage? Based on the origin of this "everyday" word how are the scientific and vernacular usages of the word polar similar?

7.35 In each group of three bonds, which bond is likely to be the most polar? Which will be the least polar? (a) C-H, O-H, S-H (b) C-Cl ,Cl-Cl, H-Cl (c) F-F, O-F, C-F (d) N-H, N-O, N-Cl

7.36 Considering the trends in electronegativity shown in Figure 7.9, explain why alloys that form between two or more transition metals are not ionic substances.

7.37 Can fluorine atoms ever carry a partial positive charge in a molecule? Why or why not?

7.38 Hydrochlorofluorocarbons are interim replacements for CFCs as refrigerants. An example would be $CHCl_2F$. In this molecule, what bond(s) are the most polar? What bond(s) are the least polar?

7.39 Using only halogen atoms to form a diatomic molecule, what would be the most polar molecule that could be formed? Explain your reasoning.

7.40 Suppose you wanted to make a diatomic molecule using two different halogen atoms. What combination of halogens would be the least polar?

Keeping Track of Bonding: Lewis Structures

7.41 Draw the Lewis structure for each of the following molecules. (a) CO (b) H_2S (c) SF_6 (d) NCl_3

7.42 Draw the Lewis structure for each of the following molecules. (a) IBr (b) $Si(OH)_4$ (c) OF_2 (d) CCl_4

7.43 Draw the Lewis structure of the following hydrochlorofluorocarbons (a) $CHClF_2$ (b) $C_2H_2Cl_2F_2$ (The two carbons are central.) Is there more than one possible way to draw the skeletal structure?

7.44 Why is it impossible for hydrogen to be the central atom in the Lewis structure of a polyatomic molecule?

7.45 In the context of Lewis structures, what is resonance?

7.46 How many important resonance structures can be drawn for SO_3?

7.47 How does the actual structure of a molecule that exhibits resonance justify the usage of the term resonance hybrid?

7.48 Draw the Lewis structure of two resonance forms for ozone, O_3. What prediction can you make about the relative lengths of the O-O bonds?

7.49 Why does the existence of resonance structures for a molecule such as benzene explain its unusually unreactive chemical behavior?

7.50 Which of the molecules listed has a Lewis Structure with only one lone pair of electrons? CO_2, H_2S, H_3O^+, HCl

7.51 Which of the molecules listed has a Lewis Structure with only one lone pair of electrons? F_2, CO_3^{2-}, CH_4, PH_3

7.52 Can the formal charge of an atom in a compound be observed experimentally?

7.53 Draw the Lewis structure for the compound NH_3BCl_3. (There is a central N-B bond.) What are the formal charges of each of the species in this molecule?

7.54 In the text we discussed the case of N_2O and the use of formal charges. If one nitrogen atom in this compound were replaced by a phosphorous to form NPO, how many different structures could be drawn? Using formal charges, which would likely be most stable? (Ignore the possibility of an expanded octet in this problem.)

Orbital Overlap and Chemical Bonding

7.55 Explain the concept of wave interference in your own words.

7.56 Distinguish between constructive and destructive interference.

7.57 How is the concept of orbital overlap related to the wave nature of electrons?

7.58 How does overlap explain the build-up of electron density between nuclei in a chemical bond?

7.59 How do sigma and pi bonds differ? How are they similar?

7.60 CO, CO_2, CH_3OH, and CO_3^{2-} all contain carbon-oxygen bonds. Draw Lewis structures for these molecules and ions. Given that double bonds are stronger than single bonds, and triple bonds are stronger than double bonds, rank the four species in order of increasing C–O bond strength.

7.61 Draw the Lewis dot structure of the following compounds and identify the number of pi bonds in each. (a) CS_2 (b) CH_3Cl (c) NO_2 (N atom central) (d) SO_2

7.62 Draw the Lewis dot structure of the following compounds and identify the number of pi bonds in each. (a) Cl_2O (b) H_2CCH_2 (c) HCCCN (d) SiO_2

Hybrid Orbitals

7.63 What observation about molecules compels us to consider the existence of hybridization of atomic orbitals?

7.64 The number of hybrid orbitals formed must always be equal to the number of atomic orbitals combined. Suggest a reason why this must be true.

7.65 What type of hybrid orbital is generated by combining the *s*-orbital and all the *p*-orbitals of an atom? How many hybrid orbitals result?

7.66 Considering only *s*- and *p*-atomic orbitals, list all the possible types of hybrid orbitals that can be used in the formation of (a) single bonds, (b) double bonds, and (c) triple bonds.

7.67 Atoms that are sp^3 hybridized generally do not form double or triple bonds. Suggest an explanation for this observation.

7.68 What type of hybridization would you expect for the central carbon atom in each of the following species? (a). CO (b). CO_2 (c). H_2CO (d). CH_2F_2

Shapes of Molecules

7.69 What physical concept forms the premise of VSEPR theory?

7.70 Draw a trigonal bipyramid shape, as expected for PF_5. Distinguish between the axial and equatorial fluorine atoms.

7.71 Predict the shape of each of the following molecules or ions. (a) IF_3 (b) ClO_3^- (c) TeF_4 (d) XeO_4

7.72 Predict the shape of each of the following molecules or ions. (a) PH_4^+ (b) OSF_4 (c) ClO_2^- (d) I_3^-

7.73 Which of these molecules would be linear? Which have lone pairs around the central atom? (a) XeF_2 (b) CO_2 (c) BeF_2 (d) OF_2

7.74 When multiple bonds are encountered during the application of VSEPR theory how are they considered?

7.75 Propene has the chemical formula of $H_2C{=}C{-}CH_3$. Describe the shape of the molecule by considering the geometry around each carbon atom.

7.76 Describe why a central atom with four bonding pairs and one lone pair will have the lone pair in an equatorial location rather than an axial one.

7.77 Describe what happens to the shape about the carbon atoms when a C=C double bond undergoes an addition reaction in which it is converted into a C-C single bond.

Insight into Chlorofluorocarbons

7.78 List some properties of chlorofluorocarbons that made them attractive candidates for use as refrigerants.

7.79 How is the strength of C-F and C-Cl chemical bonds related to the fact that CFCs are relatively inert?

7.80 How do CFCs ultimately break down in the atmosphere to produce species that contribute to the degradation of ozone?

7.81 Do a web search to identify some compounds that have been introduced as replacements for CFCs. Compare the chemical bonds that are present in these molecules to the CFCs. Does this comparison imply that the newer refrigerants should not have the same detrimental effect on ozone as the CFCs?

Additional Problems

7.82 In 1999 a group of chemists working for the Air Force Research Laboratory reported that they had produced a compound containing an unusual ion with the formula N_5^+. This ion had not been expected to exist, and its instability makes it both difficult and dangerous to prepare.

Consider the possible Lewis structure below. Indicate the hybridization expected for each nitrogen atom and the expected bond angles.

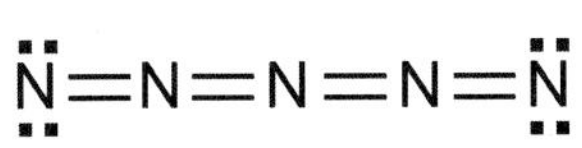

Assuming the structure shown above is correct, how many of the 5 nitrogen atoms would ALWAYS lie in the same plane?

7.83 The following molecules have similar formulas, but each has a different shape. Draw Lewis structures for these compounds, then determine the expected hybridization of the central atom and the molecular shape. (a). SiF_4 (b). KF_4 (c). SeF_4

7.84 Consider the Lewis structure below. Is this an ion? If so, what is its charge?

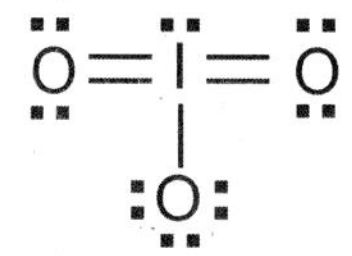

7.85 A Lewis structure for the oxalate ion is shown below. (One or more other resonance forms are also possible.)

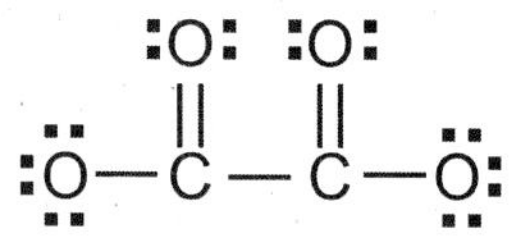

What is the correct charge on the oxalate ion? What type of orbital hybridization is expected for each of the carbon atoms in this structure? How many sigma bonds and how many pi bonds does the structure contain?

7.86 Methyl cyanoacrylate, $C_5H_5O_2N$, is the compound commonly sold as "super glue." The glue works through a process called polymerization, in which molecules of methyl cyanoacrylate form strong chemical bonds to one another and to the surfaces being glued. This process is initiated by traces of water on the surfaces.

The skeletal arrangement of the atoms in methyl cyanoacrylate is shown below. Complete the Lewis structure of the compound by adding lone pairs of electrons or multiple bonds where appropriate.

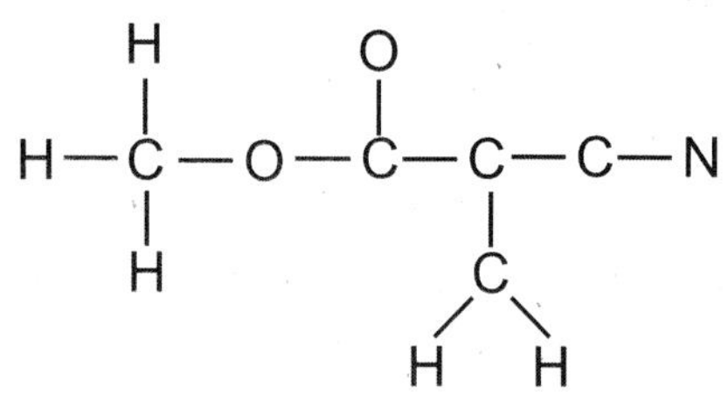

Using your Lewis structure, predict the hybridization for the six central atoms (the five carbon atoms and the leftmost oxygen atom). How many sigma bonds and how many pi bonds does your structure contain?

7.87 An unknown metal M forms a chloride with the formula MCl_3. This chloride compound was examined and found to have a trigonal pyramidal shape. Draw a Lewis structure for MCl_3 that is consistent with this molecular geometry. Use your structure to propose an identity for the metal M, and explain how you have made your choice.

Focus on Problem Solving Exercises

7.88 Ozone, O_3, and oxygen, O_2, both absorb UV light in the upper atmosphere. The ozone, however, absorbs slightly longer wavelengths of UV light – wavelengths that would otherwise reach the surface of the planet. (a) How can Lewis structures help explain the bonding difference between ozone and oxygen that gives rise to the difference in wavelength of absorption? (b) What relationship or relationships between light and energy do you need to know in order to understand this photochemical application of Lewis structures?

7.89 Nitrogen is the primary component of our atmosphere. It is also used as an inert reagent to fill containers of chemicals that might react with regular air. Draw a Lewis structure of nitrogen and use this drawing to help explain why nitrogen does not react readily with other molecules.

7.90 Hydrogen azide, HN_3, is a liquid that explodes violently when subjected to shock. In the HN_3 molecule one nitrogen – nitrogen bond length has been measured to be 112 pm and the other 124 pm. Draw a Lewis structure that accounts for these observations. Explain how your structure reflects these experimental facts.

7.91 If leads are attached to opposite sides of a crystal of sodium chloride it does not conduct electricity. Distilled water does not conduct electricity either. Yet, if the sodium chloride crystal is dissolved in the distilled water the resulting solution does conduct electricity. Use pictures at the molecular level of detail to describe why this observation occurs.

Chapter 8: Molecules and Materials

The concept of the chemical bond is among the most pervasive ideas in chemistry. For as long as they have known that atoms join together to form molecules, scientists have continually refined their models of the bond. Much of modern chemical research involves probing ways to study and manipulate chemical bonds in highly selective ways. In some sense, chemical bonding is not only the "glue" that holds together molecules, but also the concept that holds together the study of chemistry.

The concepts of chemical bonding we have introduced so far may seem rather detached from most of the materials that are used in engineering designs. In order to bridge the gap between isolated molecules and those materials, our vantage point must be extended. In this chapter we will look at materials such as polymers and metals, and see how ideas related to chemical bonding can be applied to understand their important engineering properties. We begin by considering the various forms of carbon, an element that plays a role in many of these materials.

8.1 Insight into Carbon

When you think about the elements in the periodic table, it's common to assume that most things are known about them. For decades, textbooks of chemistry said there were two forms of the element carbon, graphite and diamond. In 1985 that picture changed overnight. A team of chemists at Rice University discovered a new form of carbon, whose 60 atoms formed a framework that looks like a tiny soccer ball. Because the structure resembled the "geodesic domes" popularized by the architect Buckminster Fuller, it was given the whimsical name "buckminsterfullerene." The discovery of C_{60} and other related molecules, now collectively known as the fullerenes, helped usher in not only a new form of carbon, but also a new branch of science – nanotechnology. New carbon structures are certainly not the only area where nanotechnology research is flourishing, but the story of C_{60} provides an excellent example of how the study of chemistry affects the development of new materials. More traditional forms of carbon have long been used as materials in important applications as well.

The 1996 Nobel Prize in Chemistry was awarded to Professors Curl, Kroto and Smalley for the discovery of C_{60}.

Diamond is one form of carbon whose physical properties have led to a range of uses. The impressive appearance of cut diamonds is responsible for the most familiar use of diamond in jewelry. From an engineering perspective, however, the hardness of the diamond probably represents a more significant property. Diamond drill bits find uses in many areas, including the drilling of oil wells in some locations. (The choice of drill bits depends on the nature of the rock that must be traversed.) The hardness of diamonds is sufficiently useful that industrial processes to convert graphite into small diamonds were sought for many years. And while a number of processes for making synthetic diamonds for certain applications are now viable, their use remains rather limited. Considering the cost of naturally occurring diamonds, this immediately suggests that making diamonds must be both difficult and expensive.

Figure 8.1 shows us an indication of why it is so hard to make a diamond. The figure shows what is known as a **phase diagram** for carbon, and is essentially a map showing what state of the element will be most stable at a given combination of temperature and pressure. If you pay close attention to the units on the vertical scale, you will see that diamond becomes the preferred form of the element only for extraordinarily high pressures – greater than 200,000 atm! Fortunately the process by which diamond would be converted to graphite is so slow as to be essentially nonexistent, so at least the diagram does not imply that our existing diamonds will soon be turned to graphite. But it does tell us that in order to convert bulk graphite into diamond we will need to achieve very high pressures. (As it turns out, high temperatures are also needed to speed up the conversion process to a useful rate.) Generating and maintaining these high temperatures and pressures is difficult, expensive, and dangerous, and that's why we don't see synthetic diamonds in every jewelry store. As we explore bonding in solids, we will want to look into some fundamental questions about why these extreme conditions might be needed to form diamonds. How is the form of an element related to variables like pressure and temperature? How do graphite and diamond differ in the first place so that an observable transformation can be affected from one to the other?

Of course, graphite is not just a raw material for the production of diamonds; it has important applications of its own. For example, if you happen to be writing notes in the margin of this book using a pencil, you are using graphite. What are the properties of graphite that allow it to be useful in a pencil lead, and how does the bonding in graphite give rise to those properties? Graphite also finds applications as a lubricant. How do lubricants work and what is their impact on engineering designs?

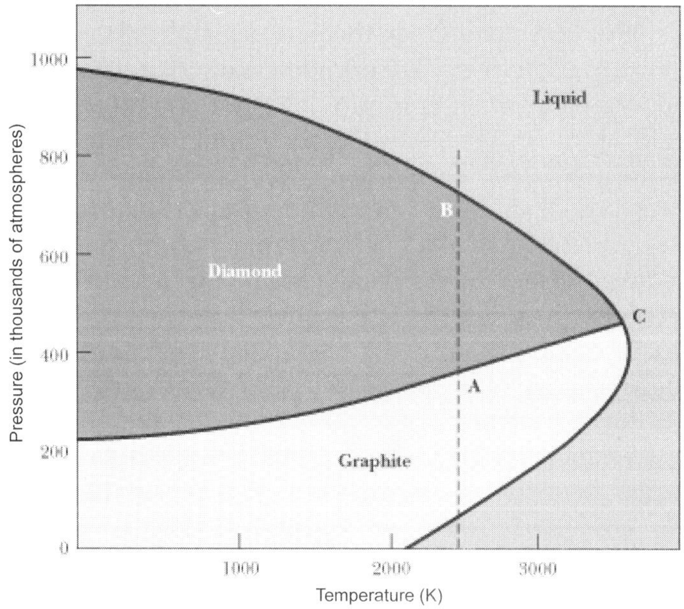

Figure 8.1 This phase diagram for carbon shows which form of the element is most stable as a function of temperature and pressure. Note that pressure is shown in units of thousands of atmospheres. It is clear that graphite is the favored form at all ordinary combinations of T and P, and that the formation of diamonds requires extremely high pressures.

Other more recent developments in materials also use the graphite form of carbon as a key, if not exclusive, component. Carbon fiber materials offer an appealing combination of high strength and low weight, and are increasingly popular in sports equipment. Graphite shafts on golf clubs or carbon fibers in snow skis suggest that durability and flexibility are associated with this material. It's difficult to imagine making a golf club out of a bundle of pencil leads, so what type of manipulation is required to convert this material into one that can be used in this application? When we have a composite material such as carbon fibers, what decisions must be made in order to control the properties?

Moving from the practical to the possible, what do the new discoveries associated with C_{60} hold for materials that engineers might use in the 21st century? While the small sphere of C_{60} may not have any immediate uses, manipulation of experimental conditions allows the growth of tubes of carbon, called nanotubes, as shown in Figure 8.2. These materials, while not yet viable for large-scale design projects, have remarkable properties. They have significantly greater tensile strength than many commonly used materials such as steel, and futurists have already begun to contemplate their potential for things as speculative as space elevators. If carbon nanotubes have another, metallic element enclosed within them, they might represent a way to build wires that are essentially one molecule wide. The prospect for miniaturization of electronic devices is one of the most immediate and exciting opportunities promised by new nanotechnologies. What are the challenges of nanotechnology? For example, how does the inclusion of metal atoms within a nanotube provide the properties that might lead to a molecular wire? This question points toward the need to understand the nature of bonding in bulk samples of metals as well.

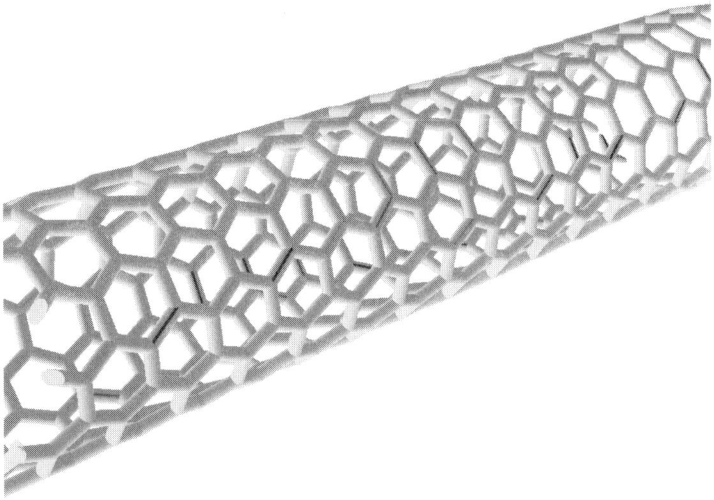

Image provided by Prof. R.B. Weisman of Rice University.

Figure 8.2 Nanotubes like the one pictured here are the most recently discovered form of the element carbon. The structure of the tubes has similarities to both graphite and buckminsterfullerene. Wide ranging applications of these tubes are already beginning to emerge.

The connections between the molecular level behavior of molecules and the macroscopic properties of materials are not always easy to discern. Current research efforts in areas like material science and computer modeling are advancing our ability to make these connections, however. In this chapter we will introduce the ideas that can be used to infer the molecular scale explanations for why materials behave the way they do. Along the way, we will be able to answer at least some of the questions we have raised about carbon-based materials and the emerging field of nanotechnology.

8.2 Condensed Phases - Solids

The various forms of carbon that we have just described share one important physical property – they are all solids under normal laboratory conditions of temperature and pressure. Indeed, the vast majority of elements are solids under such conditions. What factors contribute to the stability of condensed phases? Forces between atoms and molecules certainly play a role in the answer to this question, but the basic structure of condensed phases also contributes to an understanding of their stability. We will begin our treatment with this structural perspective.

Back in Chapter 1, in our first description of the microscopic view of matter, we noted that the atoms or molecules in a solid are generally packed together much more closely than they would be in a liquid or gas. When we consider how those atoms or molecules are arranged into a solid, there are two broad categories. Many substances assume regular, repeating geometric arrangements and are referred to as **crystalline** structures. Other substances solidify into essentially random arrangements, known as **amorphous** structures. While both categories include important examples, much more is known about crystalline materials. This is simply because their regular periodic arrays can be studied much more easily with available experimental tools. We will, therefore, focus on crystalline solids in our description here.

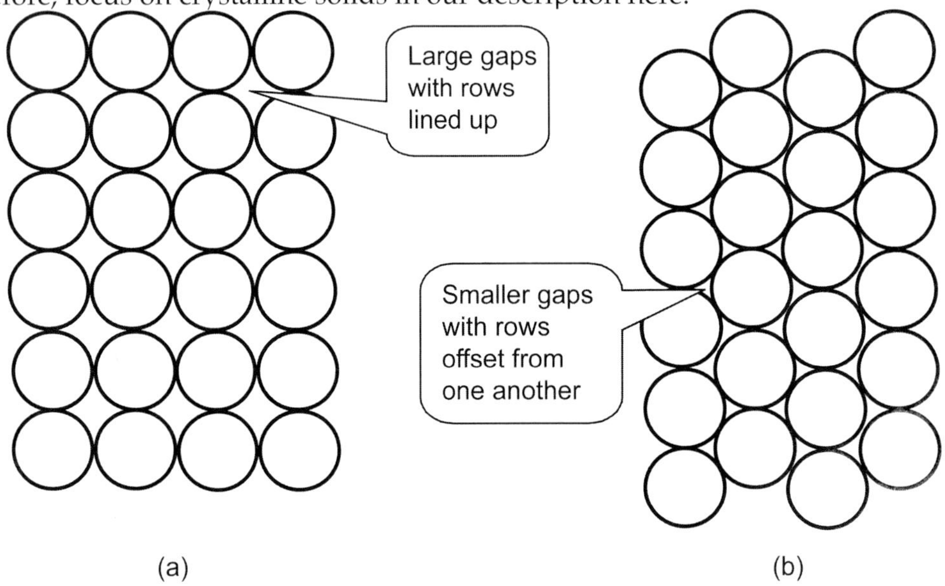

Figure 8.3 The circles in the picture on the left are arranged in a very orderly pattern, but significant gaps remain. If our aim were to get as many circles as possible into a given area, we would do better by staggering the rows, as shown in the right hand panel. Both pictures show 24 circles, but the overall "footprint" is smaller in the arrangement on the right.

Non-spherical molecules can also be packed into crystals, but we will limit our discussion to spherical atoms for the sake of simplicity.

Since atoms can be thought of as spherical, arranging atoms into a crystalline solid is loosely analogous to stacking a collection of balls into a box. It is intuitively clear that no matter how we arrange the balls in the box, there will always be some gaps between them; the spherical balls can never completely fill the volume of the box. But it should also be clear that the amount of empty space between the balls could be minimized by a careful arrangement of the balls. We can illustrate the differences in how the balls – or atoms – may be packed using a two-dimensional model as shown in Figure 8.3. For simplicity, the figure uses circles rather than spheres. But it is easy to see that the gaps can be reduced if the rows of balls are offset from one another. This idea can be quantified by thinking about the **packing efficiency** of a structure, which represents the percentage of space that is actually occupied in a given arrangement. We'll save the calculation of this result for a problem at the end of the chapter, but the offset structure has a packing efficiency of 91% while the structure in which the rows all line up has a packing efficiency of 76%. Now let's shift our thinking from balls to atoms. The packing efficiency of the atoms will clearly be related to the density of a material, since increasing the packing efficiency will put more atoms into the same volume. Experiments on crystal structures show that many elements condense into a solid that fills space as completely as possible. Such arrangements are called close-packed structures, and are the three-dimensional equivalent of the staggered rows of balls in Figure 8.3.

In moving from circles in two dimensions to spheres in three dimensions, though, we'll have some additional choices to make. Continuing with the analogy of stacking balls into a box, suppose we have completed a single layer of balls in the close-packed arrangement of Figure 8.3(b). When a second layer of balls is added (Figure 8.4), we could place them directly above the balls in the first layer, or we could again offset them so that each ball rests in a gap between balls in the lower layer. It is probably clear that this offset choice will again let us achieve a higher packing efficiency and fit more balls into the box. When the second layer of balls is complete and we contemplate where to place a third layer, we face an additional choice. The third layer balls could be placed directly above those in the first layer. Or they could be placed above hollows between the first layer balls. If the third layer is directly over the first layer the structure is said to have hexagonal close-packing (hcp), whereas the offset third layer results in cubic close-packing (ccp). Both of these structures occupy roughly 74% of available space.

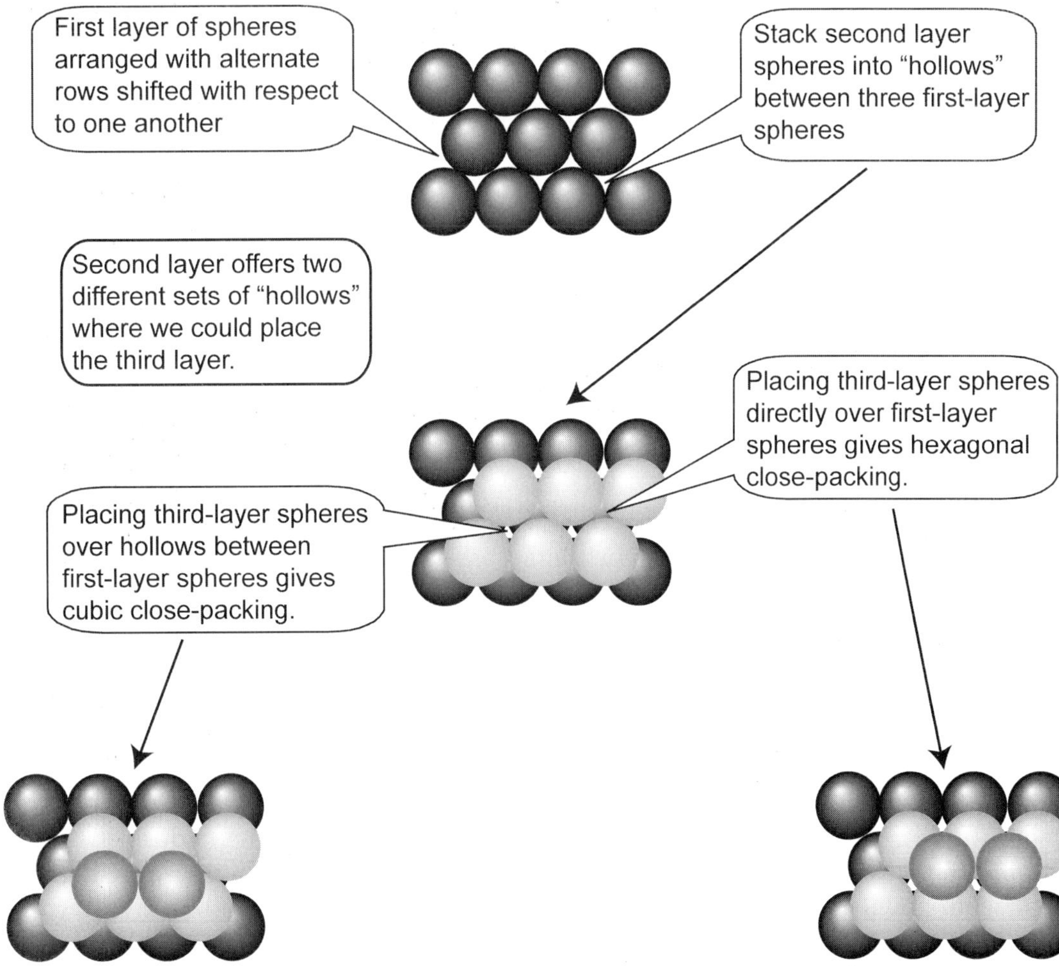

You can "build" this figure using any coins you might have if you want to have additional insight into the difference between the two packing arrangements.

Figure 8.4 Stacking spheres offers two different ways to achieve maximum packing density. The structure on the left is called cubic close-packing, while that on the right is called hexagonal close packing. The two arrangements differ in the way that the layers line up with one another. The cubic structure is said to feature an '*a,b,c,a,b,c...*' stacking pattern, while the hexagonal structure shows an '*a,b,a,b,...*' pattern.

The cubic close-packed structure is often referred to as "face-centered cubic" (fcc), for reasons that are apparent when we look at the structure as shown on the right in Figure 8.5. One atom is located at each corner of a cube, and additional atoms are placed at the center of each of the six faces of the cube. (It is admittedly difficult to see that this is the same arrangement as we saw in Figure 8.4. The best way to see that is probably to build models with a set of marbles or small balls.) The atoms at the four corners of a face all "touch" the atom at the center of that face. In order to "fit" an atom into the center of the face, the edge length of the cube must be larger than the diameter of the atoms, so the atoms at the corners do not touch one another. The single cube we have shown in Figure 8.5 is known as the **unit cell** for the face-centered cubic structure. It is the smallest collection of atoms that displays all of the features of this structure. To move from this unit cell to the structure of an actual sample of an fcc metal, we would simply need to extend the crystal lattice by repeating the unit cell in all directions.

As you will likely discover in a subsequent Materials Science course, not all unit cells are cubic, but we will restrict our discussion to them.

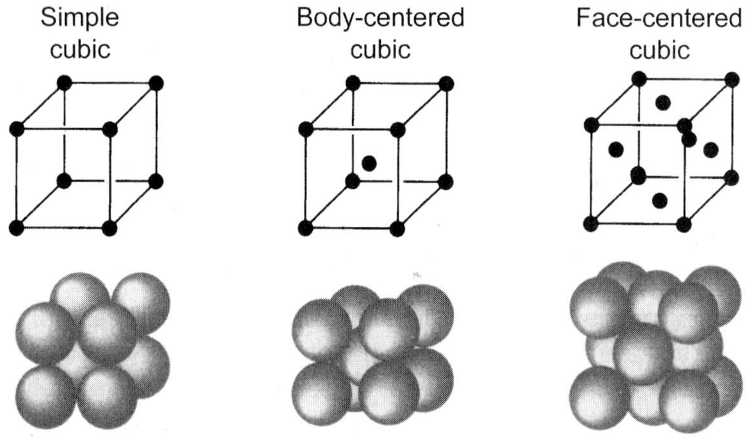

Figure 8.5 The three cubic crystal lattices are shown. In a simple cubic crystal, atoms are located at each of the corners of a cube. In a body-centered cubic crystal, an additional atom sits at the center of the cube. And in a face-centered cubic crystal, atoms are found at the center of each face of the cube. In each case these arrangements repeat throughout the crystal.

Some substances do not occupy space with the level of efficiency afforded by close-packed structures. For example, some metals assume a body-centered cubic structure (designated bcc) as shown in Figure 8.5. Here we again have atoms at the corners of a cube, but now with an atom placed at the center of the cube instead of at the center of each face. The corner atoms touch the center atom, but not each other, similarly to the case for the face-centered structure. The packing efficiency in this structure is lower, with the atoms occupying only about 68% of the space. In rather rare cases, the central atom in this bcc structure is missing; this means that the edge length of the cube becomes equal to the diameter of one atom, allowing the corner atoms to touch one another. The resulting structure is called simple cubic (designated sc). And although it is the simplest looking cubic arrangement, it is also the least efficient way to fill space, with a packing efficiency of only 52%.

Example Problem 8.1

Show that the packing efficiency of the face-centered cubic structure is actually 74%, as we have claimed above.

Strategy

This is essentially a geometry problem, and it requires us to understand the location of and spacing between the atoms in the fcc lattice. In order to find the packing efficiency, we'll need to compare the volume of the atoms in the fcc unit cell and the volume of the cell itself. It should help if we start with a good diagram of the structure. We'll rely in part on Figure 8.5, but will also use a simpler diagram of a single face of the cube, since three-dimensional crystal structures can be challenging to draw and interpret.

Solution

Before we can calculate the volume of the atoms in a unit cell, we have one very basic question to answer: How many atoms are contained in the unit cell? To answer that, we need to realize that an individual atom is frequently "shared" between multiple unit cells. The atoms on the faces, for example, will be in the unit cell we are looking at as well as the adjacent one. And the corner atoms will each appear in a total of eight unit cells, since corners of eight cubes would meet if we were to stack the unit cells up like blocks. That gives us the following result.

$$\text{\# atoms per unit cell} = 1/2\ (\text{\# face-center atoms}) + 1/8\ (\text{\# of corner atoms})$$

A cube has eight corners and six faces, giving us 6 face-center atoms and 8 corner atoms.

$$\text{\# atoms per unit cell} = 1/2\ (6) + 1/8\ (8) = 3 + 1 = 4$$

So each unit cell will contain 4 complete atoms. If we let r be the radius of an atom, then the volume of each atom will be $4/3\ \pi r^3$. So the total volume occupied by the atoms themselves will be

$$4 \text{ atoms}\left(\frac{4}{3}\pi r^3\right) = \frac{16}{3}\pi r^3$$

Now we just need to find the volume of the unit cell in terms of the radius of the atoms. Here a drawing of a single face of the cube will come in handy.

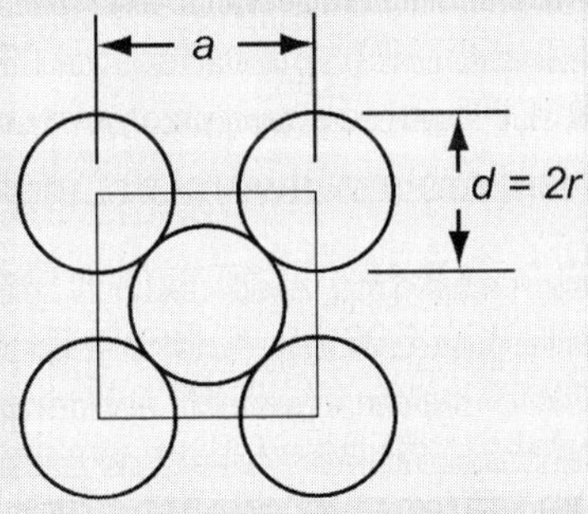

The circles represent the atoms, while the square shows the face of the unit cell cube. The diameter of an atom (d) is $2r$, and the edge length of the cube is a. From the picture, we can see that a diagonal line through the center of a face of the cube will have a length of $2d$, as it includes the full diameter of the center atom and the radius of two corner atoms. Since the face of the cube is a square, we know that the diagonal line makes a 45° angle with the sides. So recalling a little bit of trigonometry, we can write the following relationship.

$$\sin(45°) = \frac{a}{2d} = \frac{1}{\sqrt{2}} = 0.707$$

So

$$a = 1.414\ d = 2.828\ r$$

The volume of the cube is just a^3, and we can now express this in terms of r.

$$V_{\text{unit cell}} = a^3 = (2.828\ r)^3 = 22.63\ r^3$$

Combine this with the volume filled by atoms from above.

$$\text{packing efficiency} = \frac{\text{volume of 4 atoms}}{\text{volume of unit cell}} \times 100\%$$

$$= \frac{\frac{16}{3}\pi r^3}{22.63\ r^3} \times 100\% = \frac{16.76}{22.63} \times 100\% = 74.0\%$$

This verifies the number cited in the text.

Discussion

This simple calculation helps to point out the importance of being able to visualize crystal structures when working with solids. Once we constructed an accurate diagram, the calculation here only required high school geometry.

Check Your Understanding

Calculate the packing efficiencies for the simple cubic and body-centered cubic structures. Verify your answers by comparing with the numbers given above.

Packing efficiency is not the only important factor associated with solid structures. From the perspective of bonding, an equally important concept is the **coordination number** – the number of atoms immediately adjacent to any given atom. Looking at the structures we have introduced thus far, we see that a simple cubic structure has a coordination number of six, a body-centered cubic structure has a coordination number of eight and the close-packed structures (both ccp and hcp) have coordination numbers of twelve. This suggests a reason why the close-packed structures are much more common: by increasing the coordination numbers, they allow each atom in the lattice to interact with more nearest neighbor atoms. And it is the attractive forces of those interactions that are ultimately responsible for holding the crystal together.

The periodic table in Figure 8.6 shows the preferred crystal lattice for each element. Scanning this table, we see that metals typically assume one of three structures: bcc, ccp or hcp. There is no discernable pattern as to what structure is most likely for any particular metal. Nonetheless, knowledge about what structure a metal assumes can be very important in the development of new materials. Steel, for example, is predominantly iron, but by adding relatively small amounts of other elements its properties can be fine-tuned. Deciding what atoms might usefully combine within the solid framework of iron requires that the materials engineer know what that structure is and how it might be altered by exchanging iron atoms for those of some other element.

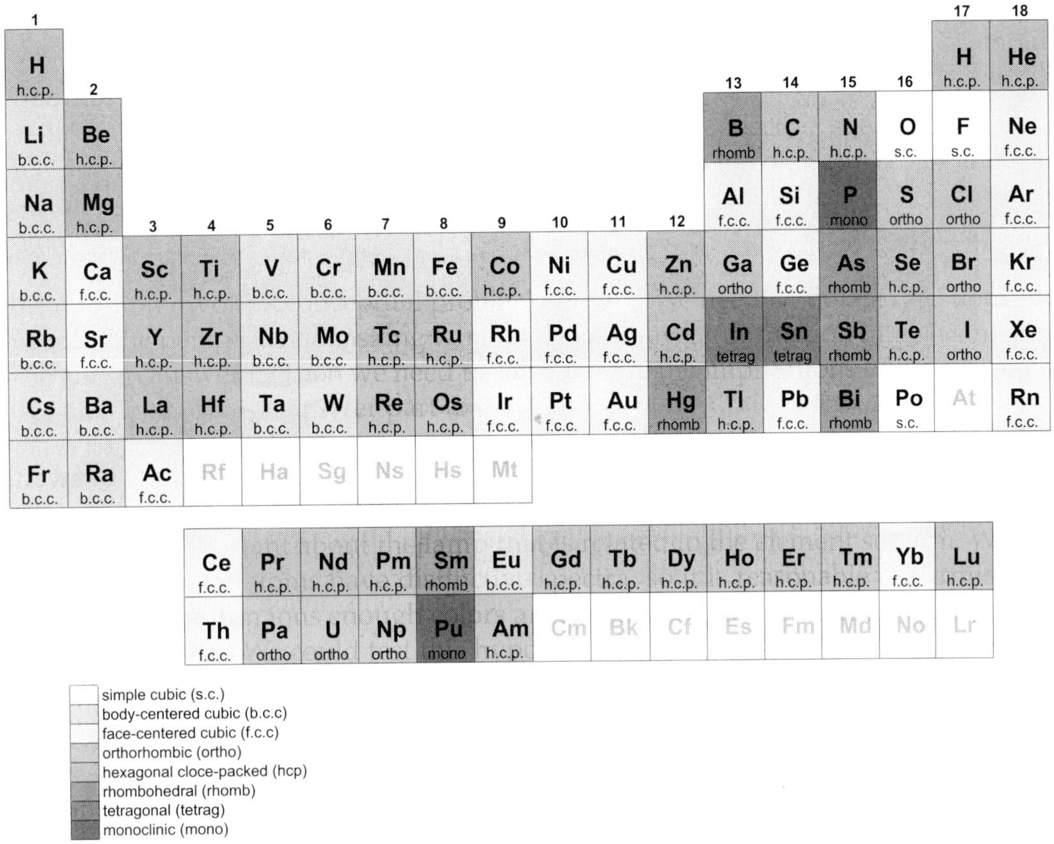

Figure 8.6 The shading in this periodic table indicates the most stable crystal structure for the solid state of the elements. Most of the elements display one of the structures we have discussed, but a few less common crystal arrangements are also known.

Some elements, including carbon, form more than one solid phase. (In such cases the structure shown in Figure 8.6 represents the most stable form.) The familiar diamond and graphite forms of carbon have dramatically different physical properties, and those properties are very directly related to differences in crystal structure. The crystal structures of diamond and graphite are both rather unusual, in large part because they are based to one degree or another on covalent bonding. The diamond structure is illustrated in Figure 8.7, and is consistent with what we already know about bonding in carbon compounds. As we saw in Chapter 7, a carbon atom with four single bonds will have a tetrahedral structure. If this geometry is extended in all directions, the result is the diamond structure, in which each carbon atom is bound to four neighboring atoms. Figure 8.7 emphasizes the local tetrahedral geometry for one carbon atom. The fact that each carbon atom is held in place by four strong C-C bonds explains why diamond is such a hard material. In order to break a diamond, we would need to break many covalent bonds, which would require a prohibitively large amount of energy. Silicon also displays the diamond structure.

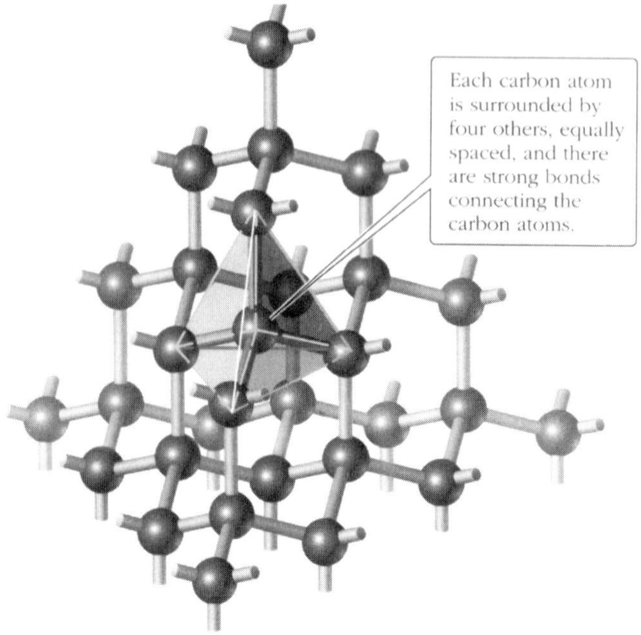

Figure 8.7 The carbon atoms in diamond are joined by covalent bonds, and each atom displays the same tetrahedral geometry we saw in Chapter 7 for molecules like methane. Fracturing a diamond crystal requires breaking many covalent bonds, and this explains the inordinately high strength of diamond as a material.

As we've already noted in the opening "Insight" section, though, diamond is not the only form of solid carbon, or even the most common. And graphite obviously doesn't share the hardness of diamond, so its structure must be very different. The layered structure of graphite (Figure 8.8) is somewhat more complicated than that of diamond. In graphite, the distance between carbon atoms within a single plane is relatively short – on the same scale as the bond length for carbon-carbon covalent bonds. Between the planes (or layers), however, the distances are much larger – more than double that seen within the layer. We will see in Section 8.4 that this distinction arises from fundamental differences in the forces operating within a layer versus those between layers, and it helps explain the unusual and useful physical properties of graphite.

Structural information about solids is a rich topic of study. We are only able to scratch the surface of this field in this introduction, but as engineering students many of you will later take whole courses exploring the structures and properties of materials. For our present purposes, we want to introduce the structural concepts associated with solids so that we can extend our knowledge of chemical bonding and use it to better understand solids.

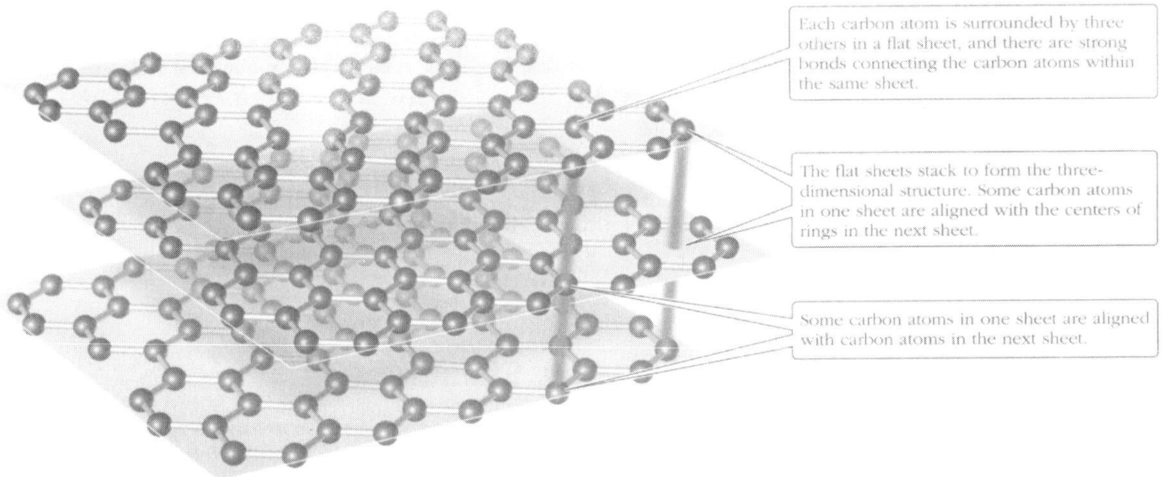

Figure 8.8 Graphite has a layered structure. The carbon atoms within each planar sheet are held together by covalent bonds, but much weaker forces hold the different sheets together. As we'll see in Section 8.4, many of the useful properties of graphite arise from the fact that the sheets can slide across one another.

8.3 Metals and Metallic Bonding

The physical properties of metals that lead to their widespread use in countless designs can readily be listed. And indeed, you are probably familiar with the basic properties of metals from science classes dating back at least to middle school. Metals are **malleable**, meaning they can be pounded into useful shapes or foils. They are **ductile**, meaning they can be pulled into wires. Metals are good conductors of electricity and heat. How does the bonding in metals help explain these properties? Can the inclusion of metal atoms inside a nanotube provide the same properties? By looking at a model of metallic bonding, we can gain significant insight into these questions.

With many useful properties and unusual bonding, metals have been described with different models through the years. We introduce only two, very successful models/theories.

Models of Metallic Bonding

In considering what we know about metallic elements, we might choose to start from the periodic trends noted in Chapter 6. Looking at metals in general, and the transition metals in particular, there is little difference in electronegativity from element to element – so metals and alloys are not likely to undergo ionic bonding. They are also a long way from having filled subshells, so they would require a tremendous number of covalent bonds to achieve a complete octet. And the experience we've gained so far suggests that individual metal atoms are not likely to form such large numbers of covalent bonds. If neither ionic nor covalent bonding can be expected for a metal, what models can account for the fact that metal atoms do bind to one another?

The first and simplest explanation of metallic bonding is often referred to as the "**sea of electrons**" model. The essential feature of this model is that the valence electrons associated with metal atoms move freely throughout the solid rather than being tied to any specific atom, as depicted in Figure 8.9. This model accounts for several observed properties of metals. To understand malleability, consider what would happen at the atomic level if a metal were struck with a hammer. At least some of the atoms would respond to this force by moving relative to others. If the electrons are floating freely, however, as postulated in the sea of electrons model, they can adjust to the new position of the atoms and the bonding is relatively unaffected. In a similar way, pulling a metal into a wire adjusts the positions of the atoms, but freely floating electrons can adjust and maintain the metallic bonding. Finally, freely floating electrons are able to move according to the influence of an electrical field, so the conductivity of metals can also be understood with the model.

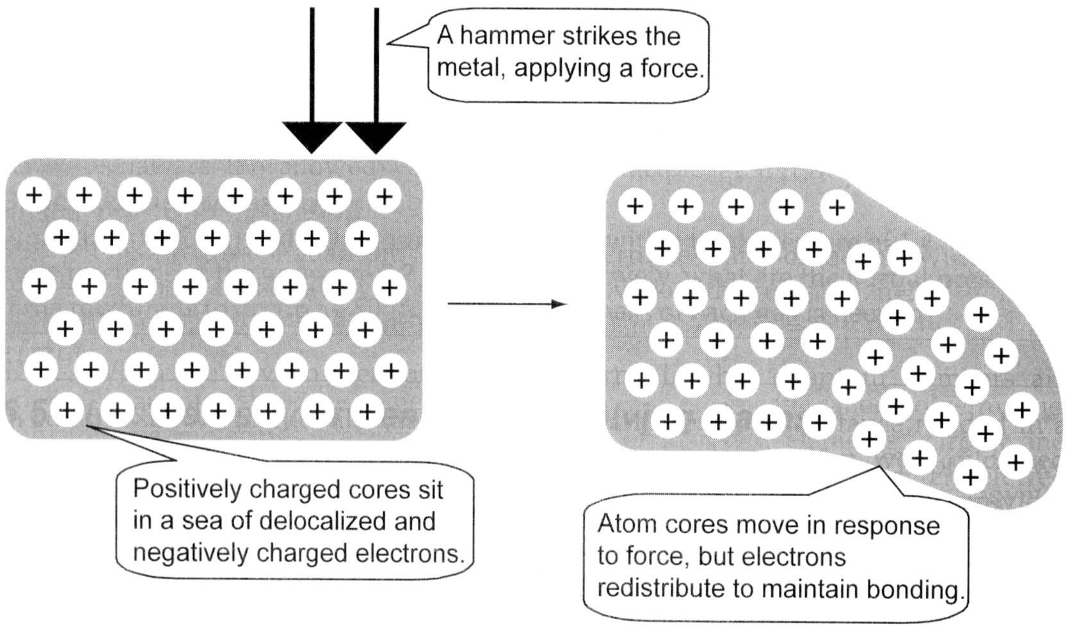

Figure 8.9 The "electron sea" model of metallic bonding is somewhat crude, but does account for the malleability and ductility of metals, as well as their electrical conductivity. The gray shaded area represents the delocalized (and negatively charged) valence electrons, spread throughout a piece of metal. The white circles depict the positively charged cores of the metal atoms. In the left hand picture, the atoms are arranged in a regular array. A force is applied as indicated by the large arrows – perhaps by striking the metal with a hammer. The cores of the metal atoms move in response to this force, and the sample is deformed as shown in the right hand picture. But the free flow of electrons maintains the bonding throughout this process. A similar argument could be made for the process of drawing a metal through a die to make a wire.

The sea of electrons model provides a qualitative understanding, but quantitative models for metallic bonding also exist. The most important such model is band theory, and we can use our example of putting metal atoms into a carbon nanotube to explain the origins of this theory.

If you could put lithium atoms inside a suitably sized nanotube, the metal atoms could be made to line up into a "wire" just one atom wide. (We use lithium for the simplicity of the s-orbital overlap, not because this it the most practical choice.) Since the valence electron configuration of lithium is $2s^1$, any bonding interactions between the metal atoms will have to be based on the 2s-orbitals. Those orbitals, of course, are waves, and so we can use the concept of wave interference to think about how they will interact with one another. We'll start with the simplest case: two lithium atoms placed close enough together for their 2s-orbitals to overlap with one another and interact. Now recall that waves can interfere either constructively or destructively. If the pair of 2s-orbitals interact in phase with one another, the interference will be constructive. The resulting wave, which will have increased amplitude (or electron density) between the nuclei, is often referred to as a **bonding molecular orbital**. If, on the other hand, the pair of 2s-orbitals interact out of phase with one another, the resulting wave will have a node between the nuclei. This is known as an **antibonding molecular orbital**. Rather than being associated with either atom individually, these new molecular orbitals can be thought of as belonging to the pair of lithium atoms together. Figure 8.10 shows how the energies of these molecular orbitals are related to that of the original 2s-orbitals. Looking at Figure 8.10 we can think through what would happen as more and more lithium atoms were loaded into the nanotube. With two atoms there would be one bonding and one antibonding orbital. With four lithium atoms there would be two bonding and two antibonding orbitals, as shown in the center panel of Figure 8.10. (With an odd number of atoms, one non-bonding orbital would be formed, keeping the number of bonding and antibonding orbitals equal to one another.) As we proceed further to very large numbers of atoms, the number of molecular orbitals formed becomes so large that there is essentially no energy difference from one orbital to the next. At this point the orbitals have merged into a band of allowed energy levels, and this is the origin of the term **band theory**. While our presentation here is entirely qualitative, it is important to note that a more thorough development of these ideas leads to a quantitative model.

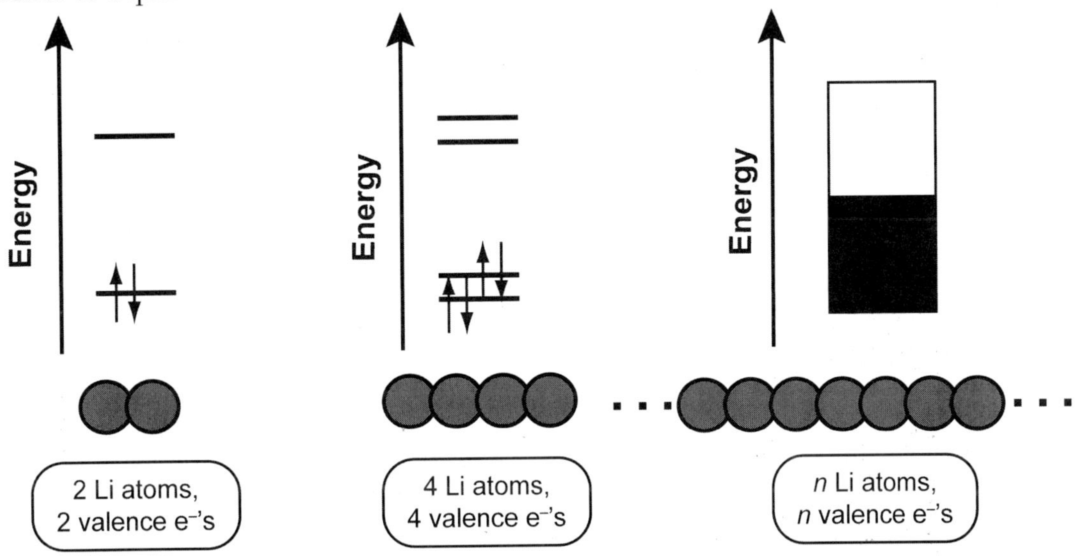

Figure 8.10 Here we examine the bonding in a 1-dimensional row of lithium atoms, as might be assembled in a nanotube. In the left hand panel we have just two lithium atoms, each of which contributes one valence electron. With four atoms, as in the center panel, we have four electrons. As the number of atoms becomes very large, the available orbitals stack up very close together in energy. With one electron from each atom, the lower half of the available orbitals will be filled, creating the valence band. The higher energy states in the conduction band remain empty.

Our one dimensional depiction of band energies would also have to be replaced for detailed descriptions of metals.

In this development we used lithium atoms within a nanotube because doing so provided a plausible one-dimensional system. In a three-dimensional bulk metal the situation is slightly more complicated and the way the orbitals will combine has some dependence on the structure of the solid. For metals with valence electrons beyond the *s*-subshell, the orbitals that ultimately form bands will include *p*- or *d*-orbitals. These will form additional bands, and the resulting *s*-bands, *p*-bands or *d*-bands may overlap with one another in energy. To understand properties such as conductivity, the band structure of the material provides a very powerful model.

Band Theory and Conductivity

The band structure of a material plays much the same role as the atomic orbital energy levels did for atoms in Chapter 6. Just as the Aufbau principle dictates that electrons occupy the lowest energy orbitals, in bulk materials the electrons fill the bands starting at the lowest energy. The highest energy with electron occupation is called the Fermi level. The location of the Fermi level, and the energy gap between the Fermi level and the next available band provide a means for understanding the electrical conductivity of the material. Electrons carry a current by moving, and this motion can be thought of as electrons moving from one orbital to another. So in order to have mobile electrons we will need to have an unfilled portion of a band that is close in energy to a filled portion of a band. An instructive analogy to this idea can be found by thinking about how difficult it is to move in a very crowded room. Similarly, electrons in a filled band cannot move to conduct electricity. The band populated by the valence electrons is referred to as the **valence band**, while the unoccupied band above it is known as the **conduction band**.

Figure 8.11 depicts some general features of the bands for three different classes of materials: metals, semiconductors and nonmetals. Looking at these depictions, we note that for the metal, which is conductive, the Fermi level falls within a band. What this means is that the first unoccupied energy level is infinitesimally close to the Fermi level, so it takes essentially no energy to promote an electron to the next highest level. Electrons can move easily, and so metals are good conductors of electricity. On the other hand, looking at the nonmetal, it has its Fermi level at the very top of a band. There is a large energy gap between this band and the next, so it takes a significant amount of energy to reach that empty band where electron movement would be allowed. Under any ordinary circumstances, electrons will not have enough energy available to overcome this large **band gap**, so nonmetals do not conduct and are referred to as *insulators*. The diamond form of carbon is an excellent example of a non-conductive material.

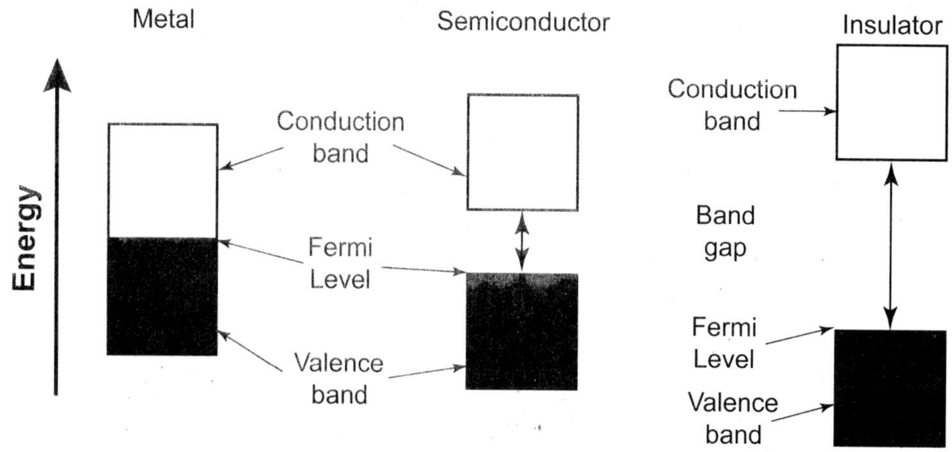

Figure 8.11 These band energy diagrams point out the key differences between metals, insulators, and semiconductors. The dark shading shows the distribution of electrons: In all three examples here, the conduction band is full and the valence band is empty. The differences in conductivity can be explained in terms of the band gap.

The intermediate case is a semiconductor, shown in the center panel of the figure. The Fermi level is at the top of the band, but the next band is only modestly higher in energy. For semiconductors, some electrons have enough thermal energy to reach the upper band even at room temperature, and therefore they conduct electricity somewhat. If the temperature is increased, more electrons occupy the upper band and the conductivity of semiconductors is observed to increase at higher temperatures. Silicon, from the same group as carbon, is the prototypical example of a semiconductor.

Semiconductors

As an engineering student, you are undoubtedly aware of the pervasive importance of semiconductors in our twenty-first century lives. Semiconductor technology is at the heart of the electronics industry. And with microprocessors finding their way into our cars and appliances, the scope of that industry has never been broader. The rise of the semiconductor provides an excellent example of the interplay between the atomic scale view of chemistry and the macroscopic scale thinking of engineering. For it is fair to say that the semiconductor revolution could not have taken place without the ingenious development of a host of new materials.

Pure silicon provides an interesting textbook example of a semiconductor, but is not really all that useful in most applications. In order to generate significant conductivity in pure silicon, we must somehow promote electrons from the valence band into the conduction band. (This is the origin of the term "conduction band.") This means we must add energy, and the two most likely ways to do that are by exposing the silicon to significant heating or allowing it to absorb light with sufficient photon energy to excite the electrons across the band gap. Neither of those would be easy to achieve in most electronic devices. Looking elsewhere in the periodic table doesn't really offer many better options, though. There are relatively few elements that can be classified as semiconductors, and none of them is suitable for widespread use in pure form. Instead, the entire semiconductor industry is founded on the idea that by cleverly combining elements we can produce materials with the specific semiconducting properties that might be desired for a particular application. This is typically accomplished through a process called "doping," in which carefully controlled trace amounts of another element are incorporated into a semiconductor like silicon.

Doping a material to derive useful properties is not limited to semiconductors. Many alloys including specialty steels are created by intentional dopiing.

To understand how this doping works, let's start with the band diagram of silicon, as shown on the left hand side of Figure 8.12. Suppose we wanted to somehow increase the conductivity of pure silicon slightly. From what we have learned about the conductivity of metals, we might conclude that a way to do this might be to put some electrons into the conduction band. So if we had some way to just "inject" an extra electron, it would go into the upper band and then it would be able to move around in that band. And if we could add enough electrons, then we might get something that could be a decent conductor. So how might we go about adding electrons to silicon? Each silicon atom has four valence electrons. So in order to increase the number of valence electrons, we might try adding a small amount of an element with more than four. Suppose we took a sample of pure solid silicon, and somehow replaced just one silicon atom with a phosphorous atom. Since we only changed one atom, we could assume that the overall structure of the crystal and its band diagram shouldn't really be disrupted. But we do have one extra electron now. And since the valence band is full, that additional electron must go into the conduction band, where it will be free to move about. So in this thought experiment, adding one phosphorous atom into our silicon crystal will boost its conductivity ever so slightly.

It is not actually possible to carry out such a single atom replacement, of course. And even if it were, the resulting change in conductivity would be immeasurably small. But the same idea will hold true if we add more phosphorous atoms. The fifth valence electron from each phosphorous atom must go into the conduction band, leading to a band diagram like that shown on the right hand side of Figure 8.12. The "stripe" of electron density in the conduction band conveys added conductivity, and the extent of that conductivity can be adjusted by controlling the amount of phosphorous present. This makes what is known as an "n-type" semiconductor. The name comes from the fact that we have added additional electrons, each of which has a negative charge. (Note that the overall lattice remains neutral, because each phosphorous atom also has an extra proton when compared to the silicon atom it replaces.) Those negatively charged electrons allow the semiconductor to carry a current.

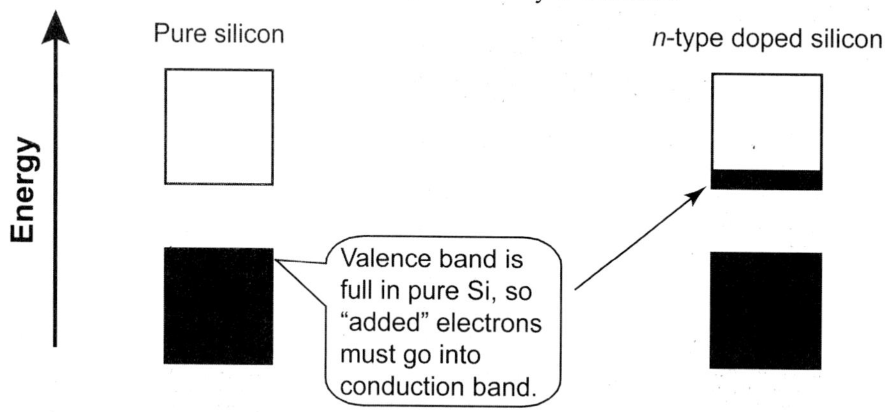

Figure 8.12 Imagine that we add electrons into pure silicon. Since the valence band is already full, the added electrons must go into the conduction band, leading to an *n*-type semiconductor with the band diagram on the right. The electrons in the conduction band do not need to cross the band gap in order to be promoted into unfilled levels, so they boost conductivity. The number of electrons in the conduction band can be controlled by adjusting the extent of doping.

For a macroscopic analogy of the importance of size, think of a pile of oranges at a grocery store. If you were to replace an orange near the bottom and still maintain the pile, which would be easier to use - a tangerine (about the same size) or a large grapefruit?

A typical n-type semiconductor might contain on the order of 0.00001% phosphorous added to a silicon crystal. (Note that in order to be able to do this, the silicon itself must be incredibly pure.) We say that he silicon is doped with phosphorous, and this doping usually takes place as the material is grown. As long as the dopant atoms are randomly dispersed throughout the solid, the overall effect on the crystal structure will be minimal. Other elements with more than four valence electrons could also be used. But phosphorous is a convenient choice because it has an atomic radius close to that of silicon, so it won't disrupt the crystal lattice appreciably.

The second common category of doped semiconductors is called "p-type." The idea in this case is to decrease the number of electrons in the valence band, so that conductivity can be achieved without needing to promote electrons across the band gap. The strategy is similar to what we described for the n-type case, except that now we want to use a dopant with *fewer* than four valence electrons. The most common choice in this case is aluminum, and Figure 8.13 shows how the band diagram will change as the material is doped. Each aluminum atom that we add will produce a single vacancy, or "hole" in the valence band. These holes behave as if they had a positive charge, so electrical engineers refer to this as a p-type material. The extent of doping controls the level of conductivity achieved, just as for the n-type case.

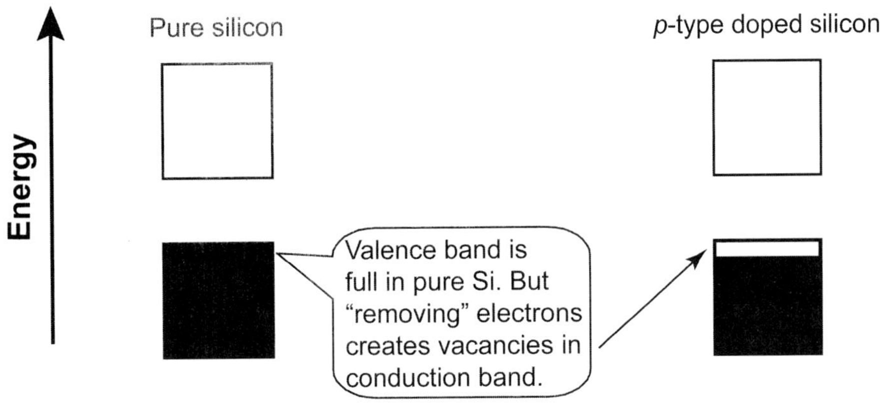

Figure 8.13 We can envision a *p*-type semiconductor as having some electrons "missing" from the valence band. The "missing" electrons are often referred to as "holes."

Example Problem 8.2

Although much less common than silicon devices, germanium-based semiconductors can also be fabricated. What kind of material (*n*- or *p*-type) would result if pure germanium were doped with (a). gallium, (b). arsenic, (c). phosphorus.

Strategy

Like silicon, germanium has four valence electrons per atom. So a dopant with fewer than four valence electrons will produce a *p*-type material, while a dopant with more than four valence electrons will give an *n*-type material. We can use the periodic table as needed to find the number of valence electrons in each element.

Solution

(a). Gallium is in Group 13, so it has three valence electrons. Germanium doped with gallium will be a *p*-type semiconductor.

(b). Arsenic is in Group 15, so it has five valence electrons. Germanium doped with arsenic will be an *n*-type semiconductor.

(c). Phosphorus is also Group 15, so germanium doped with phosphorus will be an *n*-type semiconductor, just as in (b).

Check Your Understanding

A materials engineer needs to have an *n*-type semiconductor made from silicon. What atoms might be used as a dopant? Sketch a band diagram for the resulting material.

From what we've said so far it probably isn't clear why we need both types of materials or how we might make any useful devices out of these. In practice all the important circuit functions in a chip take place at the junctions between different types of materials. The point at which a piece of p-type material meets a piece of n-type material is called a **p-n junction**, and the flow of electrons across this junction can be regulated easily by applying voltage from a battery or other power supply. Clusters of these junctions can be made to function as switches and gates of varying degrees of complexity. We don't have time to explain much of this here, but can try to describe in qualitative terms how a simple p-n junction works.

Figure 8.14 shows a diagram of what the junction might look like. The p-type material on the left is shown with some "positive holes" drawn. Remember these are just "missing" electrons. And the n-type material on the right has its "extra" electrons shown. Now imagine that we connect a battery across the ends of the joined materials, as shown in the second panel of the figure. The electrons from the p-type material will drift towards the positive side of the battery, while the holes will move toward the negative side. (Holes act like they have positive charges.) This means that current flows across the junction when the battery is connected in this way. But suppose we connect the voltage in the other direction, as in the third panel of the figure. Now the electrons and holes are both pulled away from the junction, so no current flows across the boundary. Our junction can serve as a voltage-controlled switch: we can turn the current across the junction on or off by applying the voltage in the right direction. And this simple idea can be applied to develop circuits of great complexity. That takes a lot of junctions of various sorts, and a lot of clever design, but it can be done. Many of you will go on to take a materials properties course in engineering, where you will probably see these ideas from a different perspective. But for our purposes here, the main aim is to show how important bulk properties can be manipulated by adjusting the atomic composition of a solid.

For a common type of device - the Metal Oxide Semiconductor Field Effect Transistor (MOFSET) the size of voltage needed is in the range of 1-3 V.

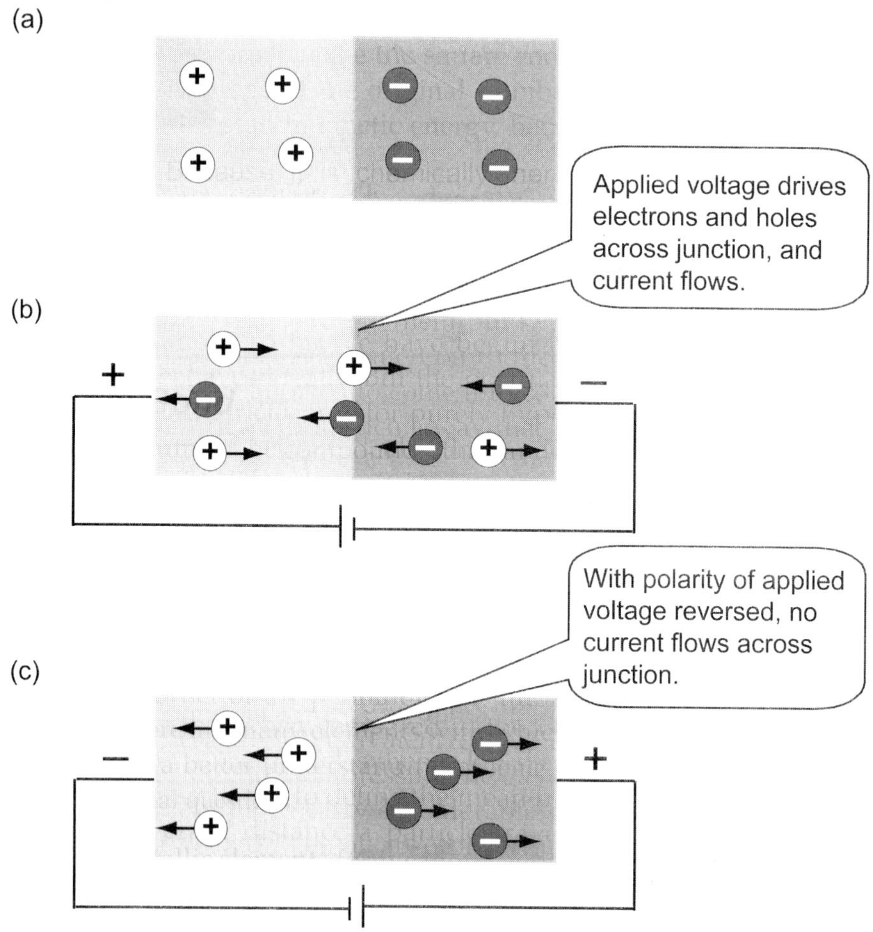

Figure 8.14 A p-n junction can serve as a simple switch, allowing current to pass when a voltage is applied with the correct polarity. This simple functionality can be exploited to build a variety of more complex circuit functions.

The features of metallic conductors, semiconductors and insulators that we've outlined here apply to more than just elemental forms of substances. Alloys are combinations of metallic elements (with an occasional minor component that is a non-metal), and their bonding resembles what we have just described for metals. Semiconductors are often made from combinations of elements, too, with gallium arsenide being a particularly important example. Insulators, too, are often made from combinations of elements, and many ceramic materials are insulators.

Not all materials have chemical bonds that are responsible for holding a solid together. We will now look at how other forces play a role in condensed matter.

8.4 Intermolecular Forces

We saw in Chapter 7 that the formation of chemical bonds is governed by the minimization of the energy of the atoms involved. As we turn our attention to the structures of solids, we'll see a very similar circumstance. The structure of any solid material, whether crystalline or amorphous, is determined by the balance of attractive and repulsive forces between the atoms or molecules involved. In this section we will consider the nature of **intermolecular forces** – the forces between molecules.

Forces Between Molecules

The bonding interactions between the atoms in a molecule that we studied in Chapter 7 constitute an extremely important facet of the science of chemistry. But understanding chemical bonding still leaves us a bit short of being able to understand many aspects of the structure of matter. The forces between molecules are also important. While typically much weaker than the forces in actual chemical bonds, these **intermolecular forces** are largely responsible for determining the structure and properties of condensed phases. So they are especially important from an engineering perspective. Here we will look at a variety of types of intermolecular forces separately.

Intermolecular forces are also critically important in biochemical molecules such as protiens and DNA. Biomedical engineers need to consider this type of interaction at least as often as materials engineers.

Dispersion Forces

Dispersion forces, sometimes called London forces, are common to all molecules. They are also referred to as **instantaneous dipole - induced dipole forces**. This rather awkward sounding term points us to the origin of these forces, so let's consider each part of the name separately. To begin with, recall that a dipole exists whenever we have two oppositely charged objects separated by some distance.

The quantum mechanical picture of the atom, and hence our pictures of various orbitals, arises from ideas of probability and average position. In order to understand the notion of an instantaneous dipole, though, we'll need to step away from that viewpoint. Instead, we'll need to consider electron positions or distributions at a single instant in time. For an electron in an *s*-orbital, for example, the ideas presented in Chapter 6 tell us that the probability of finding that electron at some distance from the nucleus is spherically symmetric. But now imagine that we could take a single instantaneous "snapshot" of the electron. You'll probably begin to think that the electron might be more likely to be found on one "side" of the nucleus. (A similar snapshot a moment later might show the electron on the other side, so the average position will still be consistent with the spherical shape we ascribe to the orbital.) This dynamic fluctuation in the distribution of electrons holds in molecules as well. Thus, at any given instant, we may find more electrons on one end of a molecule than on the other, even if on average they are evenly dispersed throughout. Whenever such a fluctuation occurs, there will be a momentary asymmetry of charge; one end of the molecule will have a small, transient negative charge while the other will have a small, transient positive charge. This immediately implies a transient charge separation, which gives rise to a dipole moment. This dipole differs from those described earlier in our consideration of polar bonds, since it lasts only for that instant that the electrons happen to be unevenly distributed. Accordingly it is referred to as an instantaneous dipole. The lifetime of this instantaneous dipole is fleeting and it soon disappears only to form again as the momentary distribution of electrons again becomes asymmetrical.

This dynamic movement of the electron is not the only way to achieve an uneven distribution of electron density. Since electrons have very little mass and are charged, they can also be influenced by electric fields external to the molecule, such as those from a dipole in a neighboring molecule. When a molecule encounters an external electric field, the negatively charged electrons will tend to migrate away from the negative pole of the applied field. This response establishes a new dipole within the molecule itself. Thus the external field has forced a dipole to exist where it had not existed previously; it has *induced* the dipole. This process is shown schematically in Figure 8.15.

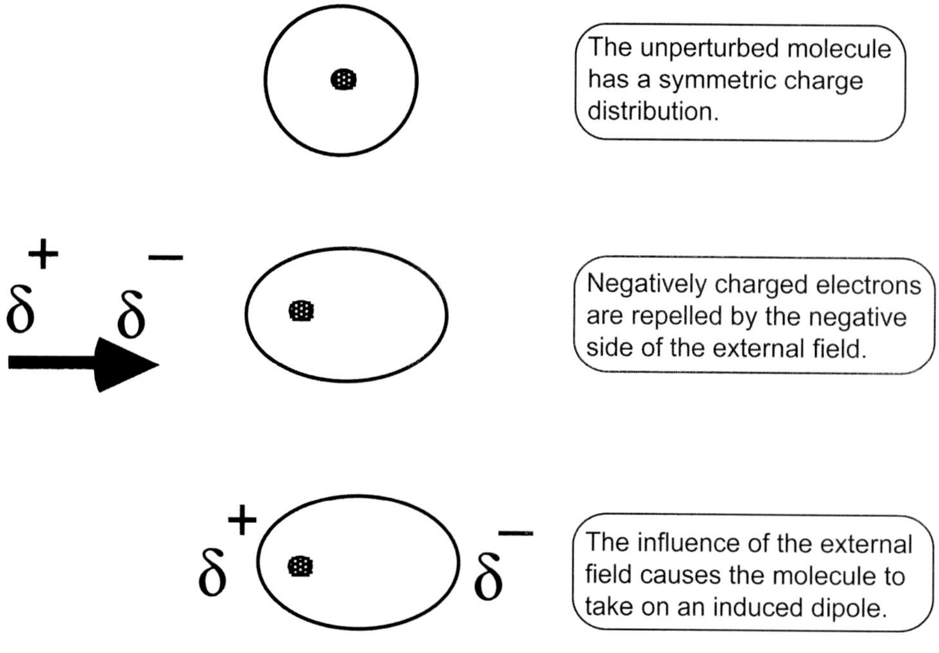

Figure 8.15 An external electric field distorts the shape of the electron cloud around a molecule, creating a transient or induced dipole. The distorted molecule can then act as an external field on a nearby molecule, propagating the effect. Dispersion forces result from the mutual attractions among such induced dipoles in a collection of molecules.

Now we combine the ideas of the instantaneous and induced dipoles. Consider two molecules that are separated by a small distance. One molecule experiences a fluctuation in its electron density, creating an instantaneous dipole. The second molecule will then see that dipole as an external electric field; this will give rise to an induced dipole in the second molecule. This combination is the instantaneous dipole - induced dipole interaction, or dispersion. Because the dipoles paired in this way are aligned with the positive end of one near the negative end of the other, dispersion forces are attractive. You may be thinking that such fleeting forces between pairs of molecules must be incredibly weak – and if so, you are right. But since macroscopic samples contain on the order of Avogadro's number of molecules, the sum of all these weak interactions can be a significant amount of energy. It is this energy that holds many liquids and some solids together.

We have described the concept of dispersion forces using atoms for simplicity. The basic concept is readily transferred to molecules as well.

How does one estimate the relative size of dispersion forces for different materials? The key concept needed to answer this question is **polarizability**. This term refers to how susceptible a molecule is to having its charge distribution perturbed by an external field. This feature of molecules can be quantified experimentally by measuring strengths of interactions between molecules, and there are some general conclusions that we can draw. Perhaps the most important of these is that large molecules tend to be more polarizable than small ones. Since the interaction between positively charged nuclei and negatively charged electrons decreases as the distance increases it makes sense that large molecules are more polarizable. The electrons are less tightly held and thus more easily influenced by the external field.

Dipole – Dipole Forces

When molecules that have permanent dipoles interact with each other, there is a tendency for those dipoles to align. This alignment, particularly in a liquid, is not rigid. But the positive region of one molecule prefers to be near the negative end of other molecules, and *vice versa*. The charge at these poles, however, is seldom more than a fraction of the charge of an electron. Thus the extent of attraction for opposite charges or repulsion for the same charges, while more than what one sees in dispersion forces, is quite modest when compared to the strength of chemical bonds. With only small attractions and repulsions operating, the alignment of the dipoles will not be absolute. At any given time there will be pairs or groups of molecules in a sample that experience repulsive interactions, but there will be more attractive ones. Figure 8.16 shows this situation for a snapshot of 50 molecules in a sample. Dark lines indicate attractive interactions and dashed lines show attractive interactions. There are clearly more attractive situations than repulsive, and this picture shows only a small number of molecules. If this snapshot were taken moments later, the location of the various interactions would change but the number of attractive and repulsive ones would remain similar on average. Because they are based on permanent rather than transient dipoles, these dipole-dipole interactions are typically stronger than dispersion forces.

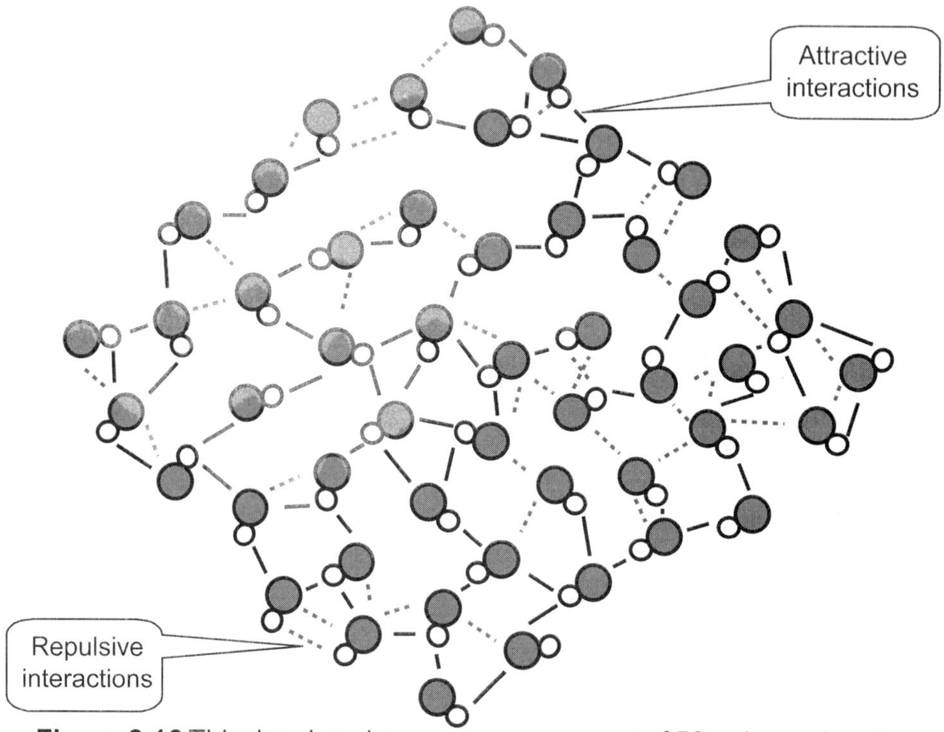

Figure 8.16 This drawing shows an arrangement of 50 polar molecules, as might be found in an instantaneous "snapshot" of a liquid. The solid lines indicate attractive interactions between opposite ends of the dipoles of pairs of molecules, and the dashed lines indicate repulsive interactions. The number and strength of the attractions outweigh those of the repulsions, so there is a net force holding the molecules together.

Hydrogen Bonding

The remaining intermolecular interaction is among the most poorly labeled terms in all of science. Although they are frequently called "hydrogen bonds," hydrogen bonding interactions are not really bonds at all. That is, they do not involve a chemical bond in the same formal sense that we used that term in Chapter 7. They are, rather, an especially strong form of dipole-dipole interactions that are observed when certain combinations of elements are present in compounds. Hydrogen bonding is typically the strongest of all intermolecular interactions. And because hydrogen bonds play a prominent role in biological chemistry, these interactions have a profound effect on life as we know it. There are some specific criteria that must be met in order for hydrogen bonding interactions to form.

Among the most famous roles for hydrogen bonding, perhaps the most important is the fact that the two strands of DNA are held together by them .

The "hydrogen" in hydrogen bonding is always a hydrogen atom bound to a highly electronegative atom, typically N, O, or F. The highly electronegative atom will tend to attract the electron pair it shares with hydrogen, and so the hydrogen atom will take on a partial positive charge. Since hydrogen has no other valence electrons, this positive charge will be especially "exposed" to other molecules. If an atom with a partial negative charge and an available lone pair of electrons approaches this exposed positive charge, a strong dipole-based attraction will result. And the list of atoms best able to provide this source of electron density is the same as we just mentioned above: N, O, and F. In virtually all of their compounds, these elements tend to have one or more lone pairs of electrons available, and these electrons can be at least partially shared with a nearby hydrogen atom from another molecule.

In the simplest cases of hydrogen bonding, the hydrogen atom and the highly electronegative atom with which it will interact are found in the same compound. Thus in hydrogen fluoride the H-atom of one molecule can be attracted towards the F-atom of a second molecule. A similar situation exists for water, with the oxygen atoms providing the lone pairs to which the hydrogen atoms will be attracted. It is crucial to remember that these hydrogen bonding interactions are much weaker than the actual chemical bonds within a single molecule. The hydrogen bonds in water, for example, are responsible for holding the molecules to one another in the solid or liquid phases. When liquid water is heated, the hydrogen bonds break down, but the chemical bonds do not. Thus as the water boils, the vapor driven off is still made up of H_2O molecules. In more complicated cases, including many that arise in biological molecules, hydrogen bonding can occur between different molecules or even between different pieces of large molecules like proteins. The common thread in all of these cases is the presence of hydrogen atoms bound to highly electronegative atoms.

Example Problem 8.3

What type(s) of intermolecular forces would need to be overcome in order to vaporize each of the following liquids? (a). CH_4 (b). CH_3F (c). CH_3OH

Strategy

Dispersion forces will exist for any substance, but will only be the most important forces in the absence of stronger dipole-dipole or hydrogen bonding interactions. In order to see if those stronger forces are possible, we need to look at the structure of each compound. So we might start by using Lewis structures to establish the molecular geometries. In this case we are dealing with methane, whose structure we know well, and two closely related compounds.

Solution

(a). We are quite familiar with the structure of CH_4 from Chapter 7: it is a tetrahedral molecule with four identical C-H bonds. Since the structure is totally symmetric, no dipole-dipole forces are possible. And since the H-atoms in CH_4 are not bound to N, O, or F, no hydrogen bonding is possible. Only dispersion forces need to be overcome to vaporize liquid methane.

(b). Methyl fluoride also has a tetrahedral geometry, but one of the hydrogen atoms from methane has been replaced by fluorine. With only one fluorine atom present, the molecule is no longer symmetric. The C-F bond is highly polar, and will give the molecule a dipole moment. So to vaporize CH_3F, dipole-dipole forces must be overcome.

(c). Methanol's structure is also closely related to methane's. Here we replace one of the hydrogen atoms from CH_4 with an –OH group. As in CH_3F, the molecule will have a dipole moment. But in this case the dipole arises from the –OH group. And that same functional group means that hydrogen bonding will be possible here. The hydrogen atom in one molecule's –OH group can be strongly attracted to the oxygen atom in another molecule. So to vaporize methanol, we would need to overcome hydrogen bonding interactions.

Discussion

Identifying the forces involved lets us predict the order of the boiling points for these liquids. With only weak dispersion forces involved, CH_4 should have a very low boiling point. Dipole-dipole interactions should make it more difficult to vaporize CH_3F, giving it a higher boiling point. (Dispersion forces would also be slightly stronger than for CH_4, due to the increased mass.) Finally, the hydrogen bonding in CH_3OH should give it the highest boiling point among this group. These predictions are correct, as the boiling points are -162°C (CH_4), -78°C (CH_3F), and 65°C (CH_3OH).

Check Your Understanding

Rank the following in order of increasing boiling points by considering the important intermolecular forces in each case: Ne, CO, CH_4, NH_3.

The forces we've reviewed here – dispersion, dipole-dipole interactions, and hydrogen bonding – give rise to many of the interactions that can occur between molecules and allow us to understand both the existence and properties of condensed phases. None of these forces is as strong as even fairly weak chemical bonds. Hydrogen bonding interactions are typically the strongest of the intermolecular interactions, but even they are an order of magnitude weaker than the average covalent bond. Despite this apparent weakness, when considering a macroscopic sample the shear number of such interactions can make them a key factor in determining bulk properties.

The comparison between intermolecular and intramolecular forces can be illustrated very clearly by looking more closely at graphite. (Recall the general features of the structure from Figure 8.8.) Within a plane, the carbon atoms are bound by strong, covalent bonds, as shown in the left hand panel of Figure 8.17. A large number of resonance structures could be drawn, but it is reasonably clear that on average each carbon-carbon bond will be something between a single bond and a double bond. Between the sheets, however, the interactions are due to dispersion forces, making them dramatically weaker. So the carbon atoms within a given sheet are held together by relatively strong chemical bonds, while much weaker dispersion forces hold pairs of adjacent sheets to one another. The result is that the sheets of graphite are capable of sliding past each other, as shown in the figure. This helps explain why graphite is useful as a lubricant. Lubricants must be capable of withstanding large forces, and the strong bonds within a sheet provide that strength. They must also be capable of viscous flow, and the movement of one sheet relative to the next makes that possible.

Though fairly simple, the case of graphite points to the importance of both chemical bonding and intermolecular forces in determining the properties of a solid. Additional models are needed to gain a more complete understanding of this type of bonding, so we will consider these models now.

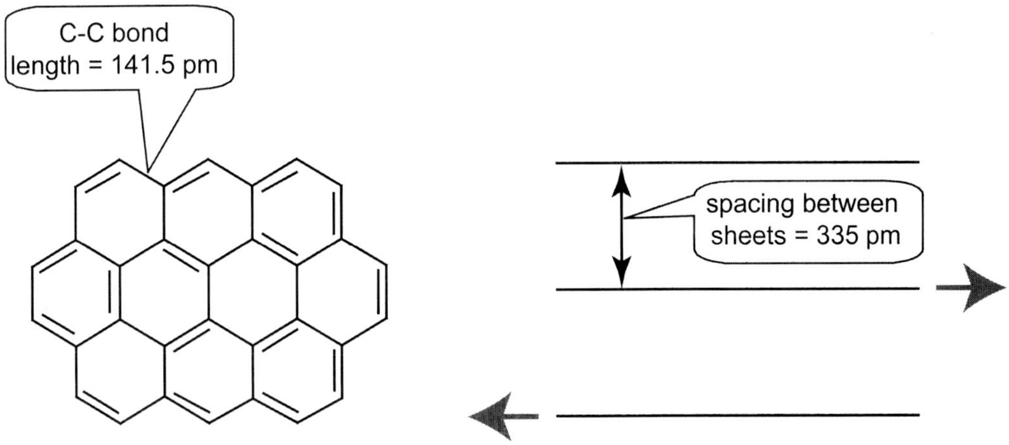

"Top view" of a graphite sheet "Side view" of a stack of three graphite sheets

Figure 8.17 Carbon atoms within each sheet in the graphite structure are joined by strong covalent bonds, with a fairly short C-C distance. Since each carbon atom forms four bonds within its own sheet, intermolecular forces must be responsible for holding the sheets together. This is apparent from the much larger spacing between layers. Because the forces holding the layers to one another are weak, sliding of one layer relative to the next is fairly easy, and this explains the softness and lubricating properties of graphite.

8.5 Condensed Phases – Liquids

Solids are not the only condensed form of matter. Intermolecular forces of the type we just described often result in the formation of liquids as well as solids. From a structural perspective, liquids are not as "neat" as solids, but we can look into some important characteristics that they possess.

The key difference between liquids and solids lies in the relative mobility of the atoms or molecules that are present in a liquid. For a solid, the average position of each particle remains essentially unchanged; even though there may be significant vibration about that average position, the atoms remain in a fixed arrangement over time. By contrast, particles in a liquid are free to move with respect to one another, and they do so constantly. This constant motion plays an important role in the fact that liquids often serve as useful solvents capable of having other chemicals dissolved in them. The dissolved particles, like those of the liquid itself, also become capable of movement. This allows dissolved particles to encounter other dissolved materials (or solids in contact with the liquid) and such encounters may allow chemical reactions to take place.

Confining our interests for the moment to looking at pure liquids, though, what can we learn about their properties from looking at intermolecular forces? We have already seen how such considerations allow us to understand the lubricating characteristics of graphite. Let's consider some physical properties that might be important in understanding the properties of liquids.

Vapor Pressure

After a springtime rain shower the sidewalks are covered with puddles. Those puddles disappear, but clearly the water never boils. The evaporation that takes place from these puddles is indicative of a phenomenon known as vapor pressure that is actually characteristic of both liquids and solids. The vapor pressure is a characteristic property of a particular substance at a particular temperature. A puddle of the same size will evaporate faster on a hot day than a cool one. And if the liquid on the sidewalks were acetone rather than water, the puddles would disappear much more quickly.

We can reach a qualitative description for the idea of vapor pressure by considering intermolecular interactions. Recall from the previous section that such interactions are attractive in nature. Thus in a solid or a liquid the attractive interactions that are holding the molecules together are due to those intermolecular forces. From a molecular perspective the system consists of a large number of molecules, constantly moving but interacting with one another. In any molecular system the velocities of the individual particles varies, but there are always some particles with a high kinetic energy. (The Boltzmann distribution discussed in Chapter 5 for gas molecules is an example of this. Similar ideas apply to molecules in liquids, too, although their exact form is more complicated.) If an especially energetic molecule happens to find its way to the surface of the liquid, it may have enough energy to overcome the attractive tugs of its neighboring molecules and escape into the vapor phase. And that is the molecular origin of evaporation.

This molecular viewpoint lets us predict the behavior of vapor pressure with temperature changes. As temperature increases, the number of molecules with high kinetic energy also increases, so more molecules will be capable of escaping the attractive forces of intermolecular interaction. Thus vapor pressure should increase as temperature increases, and this is seen experimentally. Our molecular description also allows us to understand the relative vapor pressures of different molecules. In order to escape from the liquid, the molecules must have kinetic energies sufficiently high to overcome the intermolecular interactions in the liquid. So the strength of those intermolecular forces sets the bar that evaporating molecules must overcome in order to escape. If a system has strong intermolecular interactions, molecules will need particularly large amounts of kinetic energy. There are simply fewer molecules capable of escaping when intermolecular interactions are strong, and so the vapor pressure will be low. Liquids with high vapor pressures evaporate rapidly and are said to be **volatile**. Many engineering designs – such as those associated with gasoline powered engines – must account for volatility of some component, such as the fuel. Volatility is sometimes an asset, as when the fuel is being injected into a cylinder for combustion. In other cases it can be a liability, as when trying to minimize the release of unburned fuel from a storage tank into the atmosphere.

Measurements are made in evacuated vessels so the only source of pressure is from the vapor.

This gives us a reasonably useful qualitative view of vapor pressure. But how can this be quantified. Water has a lower vapor pressure than acetone, but how much lower? To measure the vapor pressure of a liquid or a solid requires that the system reach equilibrium. In other words the molecules that are escaping from the liquid must be replaced by other molecules that are being captured. This situation is shown in Figure 8.18. Once this equilibrium has been established, the pressure of the molecules in the gas phase will remain constant and can be measured fairly easily. This situation, in which the net state of the system is unchanged but there are still processes occurring (molecules escaping and being captured by the liquid), is called **dynamic equilibrium**. We'll study this idea in detail in Chapter 12, because it is important in many more cases than just vapor pressure. The values of the vapor pressure for a number of liquids near room temperature are provided in Table 8.1.

Substance	Vapor Pressure (mmHg)	Normal boiling point (°C)
Br_2	184	58.8
$CClF_3$	24,940	-81.1
CCl_2F_2	4,448	-29.8
CCl_3F	717	23.8
CCl_4	99.0	76.54
HCN	657	26
formaldehyde	3,525	-21
methanol	108	64.96
n-pentane	549	36.07
neopentane	1,365	9.5
isobutane	2,393	-11
n-butane	1,658	-0.5
propane	6,586	-42.07
ethane	29,380	-88.63

Table 8.1 Vapor pressures at 295 K and normal boiling point temperatures are shown for a variety of substances. Note the correlation between these two closely related properties.

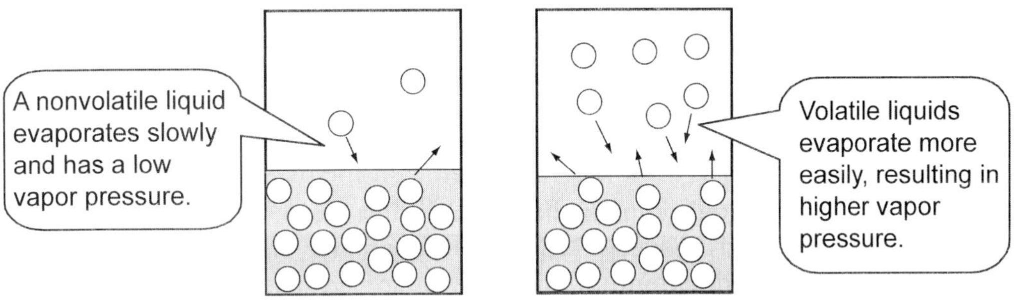

Figure 8.18 When a liquid is held in a closed container, it will establish an equilibrium with its vapor phase. The vapor pressure provides a measure of the ease with which a particular liquid evaporates.

Boiling Point

The boiling point of a liquid is also dependent on the strengths of the intermolecular interactions and is closely related to the vapor pressure. As was just noted, the vapor pressure of a material increases as the temperature increases. This phenomenon is true over all temperature ranges so that eventually as temperature continues to increase, the vapor pressure of the liquid equals the external pressure. When this point is reached, small bubbles of vapor form within the liquid and the system boils. The **normal boiling point** of a liquid is defined as the temperature at which its vapor pressure is equal to one atmosphere. The well-known normal boiling point of water is 100°C. If the external pressure is below one atmosphere, the temperature at which boiling will occur is lower. This idea gives rise to the special cooking instructions for people who live in the mountains, where atmospheric pressure is reduced. If one wishes to

Formation of gas bubbles in a liquid is not automatic, so it is possible to "superheat" a liquid above it's boiling point under certain circumstances.

boil an egg at high altitude, the time it takes will be longer since the water is boiling at a lower temperature. The vapor pressure of water as a function of temperature is shown in Figure 8.19. The difference in temperature of boiling between sea level and the mountains is a direct result of the temperature dependence of the vapor pressure. At higher altitude atmospheric pressure is lower, and boiling occurs whenever the vapor pressure is equal to the external pressure. Note also the shape of the curve, with the pressure increasing rapidly as temperatures exceed 100°C, where water typically boils.

Normal boiling points of several liquids are listed in Table 8.1. Note that those liquids with low vapor pressures have high boiling points. It is possible therefore to rate the strength of intermolecular interactions by observing the normal boiling point of the liquids. It is also possible to separate molecules of different intermolecular interaction strengths by distillation, which is an important step in the refining of oil.

Note that vapor pressure continues to increase past the boiling point. Familiar things like popcorn or pressure cookers rely on this fact.

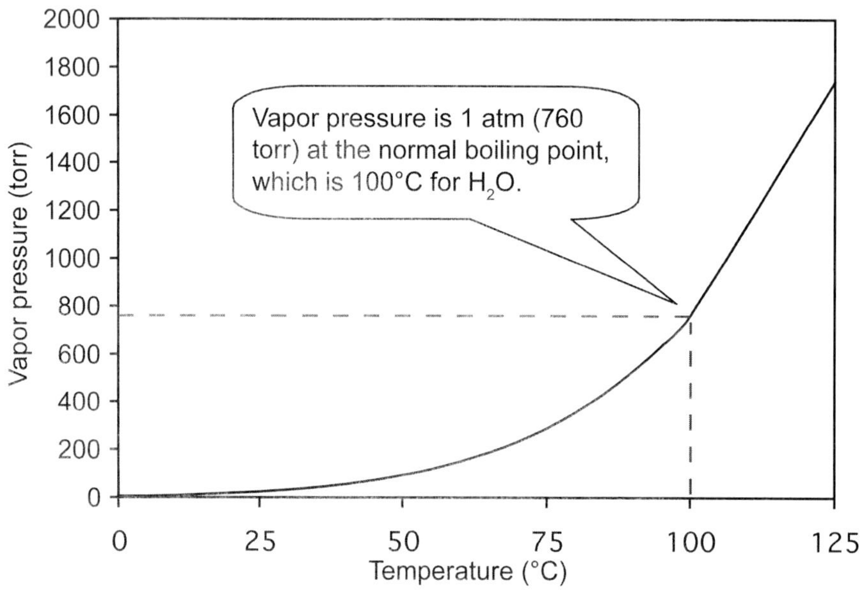

Figure 8.19 The vapor pressure of water varies dramatically as a function of temperature.

Surface Tension

The attractive nature of intermolecular interactions means that a liquid will, in general, behave in a way that will maximize the number of interactions. One effect of this is to minimize the number of molecules at the liquid's surface. Such surface molecules will necessarily encounter a different environment than those in the bulk of the liquid. If the surface of the liquid is in contact with a gas, there will be fewer molecules available to interact with these surface molecules, a situation shown in Figure 8.20. Since the typical intermolecular interactions are attractive, these surface molecules find themselves in a state of higher energy than bulk molecules. The system will naturally minimize the number of molecules in this higher energy environment, giving rise to a phenomenon that is called **surface tension**. Something under tension is in a higher energy state; in this case the origin of the "high energy state" is the existence of the surface and the fact that molecules at the surface have fewer nearest neighbors than do molecules in the bulk of the liquid.

Surface tension explains a variety of observable phenomena, including the fact that water tends to "bead up" into spherical (or nearly spherical) drops. The energy of any given sample of liquid will be reduced if fewer molecules sit at the surface. The shape that encloses the greatest volume with the least amount of surface is a sphere. The fact that water experiences strong intermolecular interactions due to hydrogen bonding means that its surface tension is relatively high, and provides a strong incentive to form spherical drops. This effect is best seen on a waxed surface because the polar water molecules do not interact strongly with the nonpolar wax. On other surfaces, including a dirty car hood, strong interactions between the water molecules and substances on the surface can overcome the surface tension, producing a more uniform film of water.

In engineering applications where surface tension can create problems, often molecules called surfactants are added to decrease the surface tension. Surfactants also rely on an understanding of intermolecular forces in their deisgn.

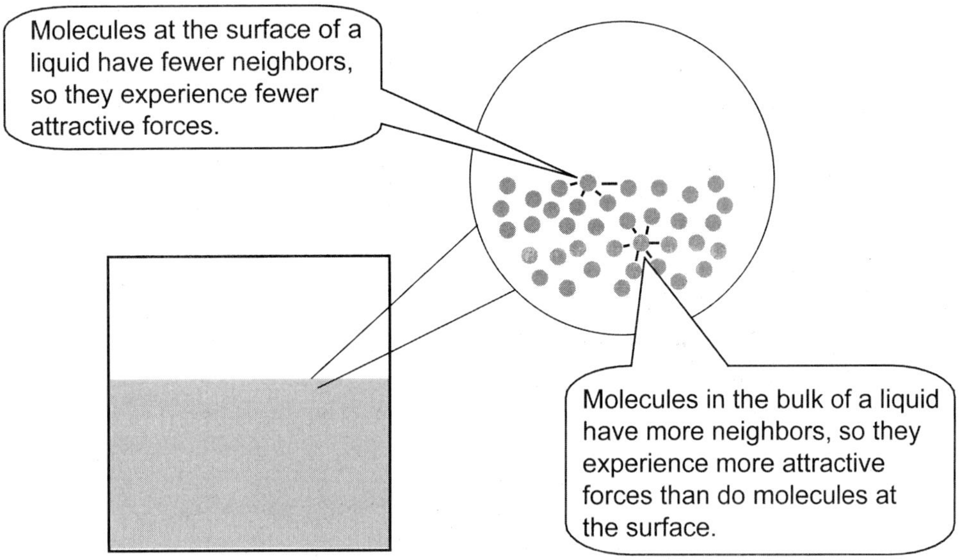

Figure 8.20 Molecules at the surface of a liquid have fewer nearby neighbors than do molecules in the bulk of the liquid. Surface tension arises this difference in molecular scale environments.

When the surface of a liquid is interacting with a solid rather than a gas, there are competing interactions to consider. The liquid molecules interact with each other but they also interact with the molecules in the solid. The liquid-liquid interactions are referred to as **cohesion** and the liquid-solid interactions are called **adhesion**. The relative strengths of these forces will dictate the shape of the meniscus that forms for the liquid. Figure 8.21 shows a comparison of two different liquids in a glass tube. Water (on the top) has strong interactions with the glass, which has numerous polar bonds on its surface. So the adhesion forces are larger than cohesion and the meniscus is concave. The liquid in the bottom of the tube is mercury, which does not interact strongly with the polar glass surface. So in that case adhesion is small relative to the cohesion and a convex meniscus results. A liquid for which the adhesive and cohesive forces were equally strong would be expected to show a horizontal surface, with virtually no meniscus.

Figure 8.21 The graduated cylinder shown in this photograph contains a layer of water on top of a layer of mercury. The meniscus at the top of the mercury column is concave down, meaning the mercury is higher in the center of the cylinder than at the walls. The meniscus at the surface of the water is concave up, with the water higher along the walls than in the center of the cylinder. These shapes result from the differences in strength of the adhesive and cohesive interactions between the two liquids and the glass of the graduated cylinder.

The carbon structures we noted in the opening insight for this chapter do not readily exist in liquid phases. To melt diamond or graphite requires temperatures that are high enough that there are no practical consequences of the liquid phases. There are, however, carbon-based materials that show important properties of both solids and liquids, and they comprise a sizable fraction of the materials used in modern engineering designs. Before we can move on past our study of materials, we need to take another look at polymers – particularly carbon-based polymers.

8.6 Polymers

The importance of carbon in the overall landscape of chemistry and materials science is due not to the elemental forms we have noted thus far in this chapter, but rather the myriad chemical compounds it is capable of forming. While there are more than enough organic compounds to provide a complete course devoted to their properties and reactions, we will limit our consideration here to carbon-based polymers. As we saw back in Chapter 2, polymers are giants among molecules, constructed by the sequential stringing together of smaller molecules called monomers. They constitute the largest component of the materials we call plastics – though plastics are actually defined by their ability to be molded, and usually include additives.

The utility of polymers and plastics in engineering applications arises from the fact that materials chemists and engineers are able to control the physical properties they display. The main things that can be empirically adjusted to modify polymers are the monomers used, the type of reactions needed to generate the polymer, and the catalysts that are employed to speed the reactions. A careful choice of these factors can ultimately control the physical properties of the resulting polymer.

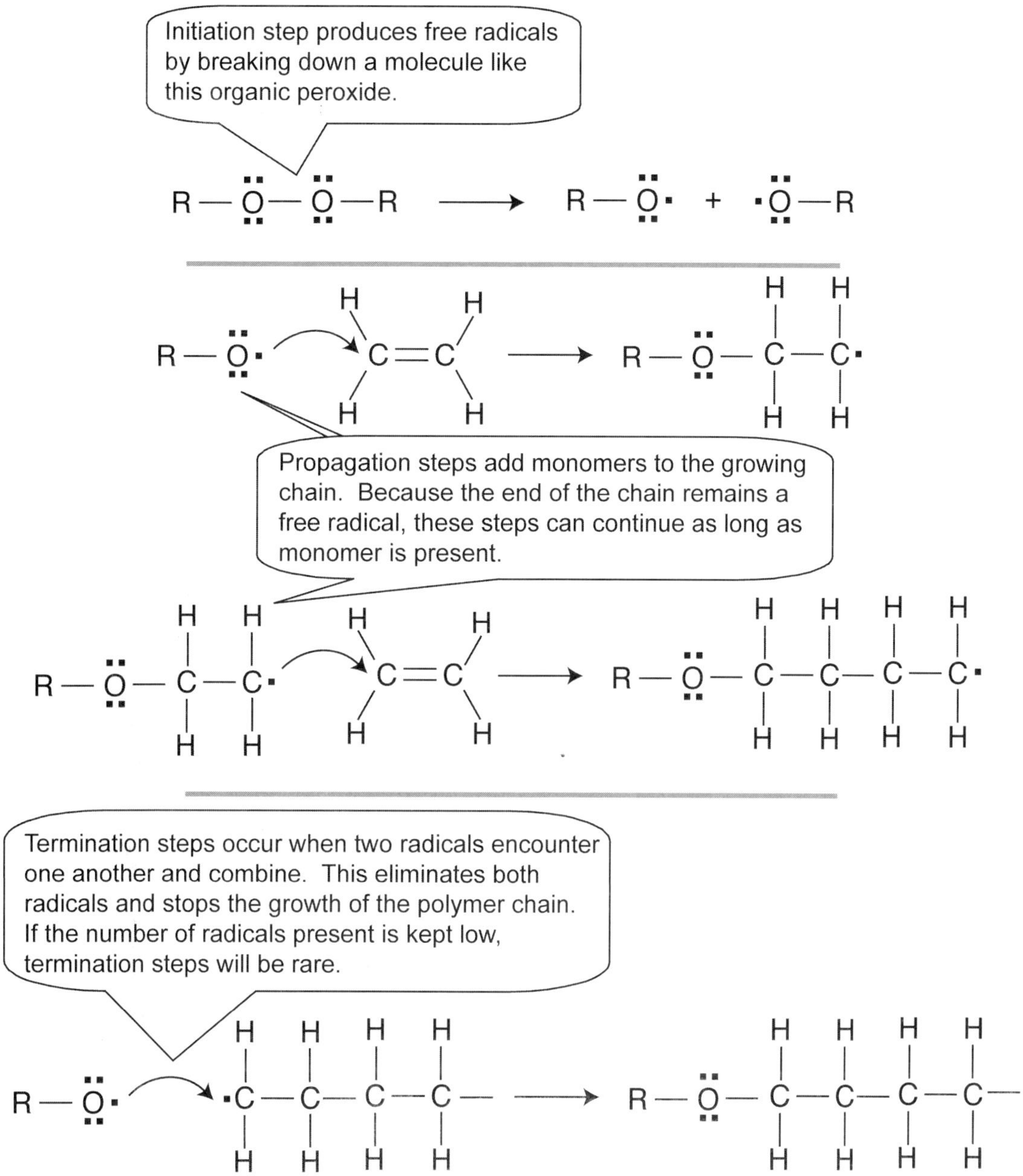

Figure 8.22 Addition polymerization often proceeds by a free radical mechanism, featuring initiation, propagation, and termination steps. Conditions are chosen so that the initiation and termination steps occur infrequently. Each propagation step adds one monomer unit to the growing polymer chain.

Addition Polymers

Now that we understand a bit about chemical bonds and chemical reactions, it should not seem surprising that the manner in which polymers are generated depends strongly on the characteristics of the monomers involved. In the closing "Insight" section of Chapter 2, we showed how monomers containing one or more double bonds undergo free radical addition reactions to form polymers. As shown in Figure 8.22 an **addition polymerization** begins with the generation of a free radical. (This step is normally accomplished by heating a molecule that falls apart into two radicals.) The resulting radical attacks the double bond of another monomer, breaking it and forming an additional single bond. Importantly, the product is also a radical that can attack yet another monomer, and with enough of these steps the polymer can be formed. Just how many times this reaction repeats itself controls the length of the polymer chain. Eventually a radical will be eliminated, either by encountering another radical or by some other reaction. This stops the polymer molecule from growing longer, and is referred to as "chain termination." Because the occurrence of these termination steps is somewhat random, not all of the polymer chains will grow to the same length. Instead, there is normally a range of lengths and hence a distribution of molecular masses. This distribution can be roughly described in terms of the **degree of polymerization**, which is the average number of repeating units in the polymer. This is usually calculated in terms of mass.

$$\text{Degree of polymerization} = \frac{\text{molecular mass polymer}}{\text{molecular mass monomer}}$$

One reason that polymers are such common engineering materials is that their properties can be tailored in a number of ways. Perhaps the most obvious is the choice of the monomer unit, since we know that different polymers can offer vastly different characteristics. But it is also possible to have distinctly different properties for polymers derived from the same monomer. One variable is the degree of polymerization, which can be controlled by varying parameters used in the synthesis reaction. Another approach is to control the way in which the monomers are linked together. Perhaps the most important example of this is the formation of both low density polyethylene (LDPE) and high density polyethylene (HDPE), which we explored in the closing section of Chapter 2.

Polystyrene is an example of a commercial polymer whose properties depend on whether it is syndiotactic or atactic.

When the monomer is less symmetric than ethylene, there is additional variability that might result from the polymerization process. For example, propylene has a structure similar to that of ethylene, but with a methyl (CH_3) group replacing one hydrogen atom. This change means that the resulting polymer may have methyl groups attached to the carbon backbone in three different ways, resulting in the various geometric arrangements shown in Figure 8.23. It is important to realize that the geometry at each carbon atom in the chain is tetrahedral, so the methyl groups are either in front of or behind the plane of the page in this illustration. If the methyl groups are all pointing forward (as in Figure 8.24(a)) or backward, then the polymer is said to be **isotactic**. If the position of the methyl groups systematically alternates between forward and backward, the configuration is called **syndiotactic**. In the third case the arrangement of the methyl groups is random, sticking "into" and "out of" the page with no regular pattern, and the polymer is said to be **atactic**.

The physical properties of a polymer can be strongly influenced by this type of variability. Because the structure of atactic polymers is not very well defined, their physical properties can vary considerably from one sample to another. The regular arrangement of the monomers in isotactic and syndiotactic polymers generally leads to more predictable and controllable properties. So the ability to selectively produce polymers in one of these arrangements offers a powerful tool to the materials scientist or engineer. This ability to control the specific configuration of a polymer was first achieved by Karl Ziegler and Giulio Natta. These scientists discovered new catalysts for the addition polymerization reaction that not only increased the rate of reaction, but also controlled the structure. The catalysts came to be known as Ziegler-Natta catalysts. Their discovery revitalized the study of polymers, and they were awarded the Nobel Prize in chemistry for their efforts.

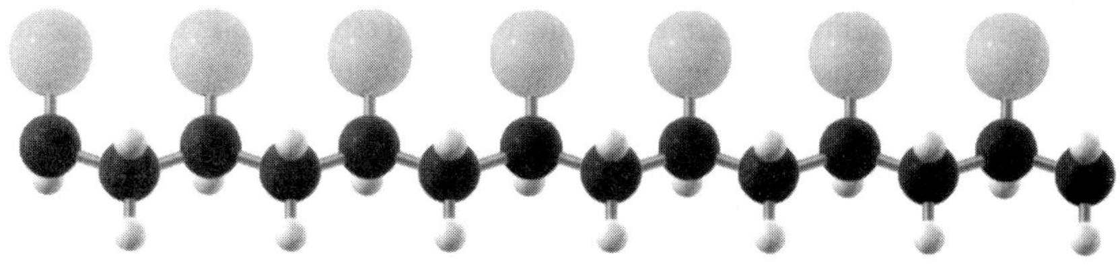

(a).

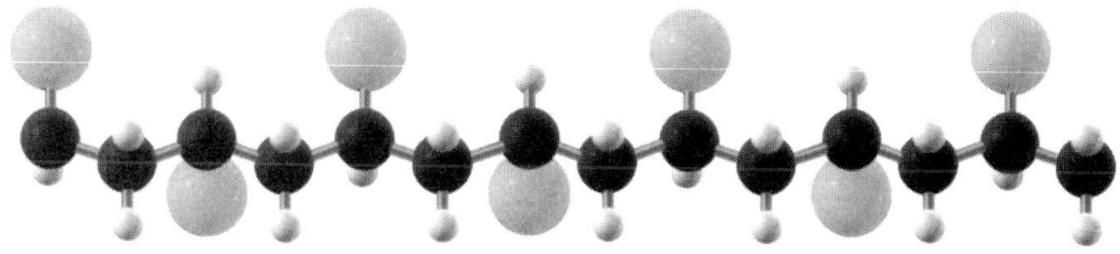

(b).

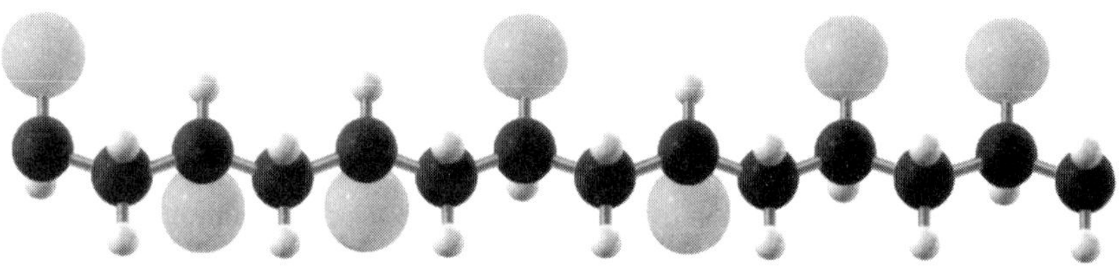

(c).

Figure 8.23 For monomers with more complicated structures, different ways of linking them together become possible. Here we see polypropylene, which differs from polyethylene in that every other carbon atom has a methyl group in place of one hydrogen atom. The large gray balls here represent those methyl groups. In (a), the methyl groups are all arranged on the same side of the polymer chain, giving isotactic polypropylene. In (b) the methyl groups are on alternate sides of the chain, making syndiotactic polypropylene. Finally, the random arrangement of the methyl groups in (c) is atactic polypropylene.

Condensation Polymers

The second commonly observed type of reaction for the formation of polymers is called condensation. In this reaction, there are two functional groups on the monomer, which together can react to form small molecules; those small molecules then split off, and the remainder of the two monomers are joined together. (The term "condensation polymer" originates from the fact that the small molecule being eliminated is often water.) Figure 8.24 shows the condensation reactions that produce two polymers that are important in fiber manufacturing, Nylon and Dacron. Nylon is referred to as a polyamide, while Dacron is a polyester. These names refer to the characteristic amide and ester linkages found in the molecules, as highlighted in Figure 8.24. Looking at these two examples, we can see the functional groups that must be present in monomer for condensation polymerization to occur.

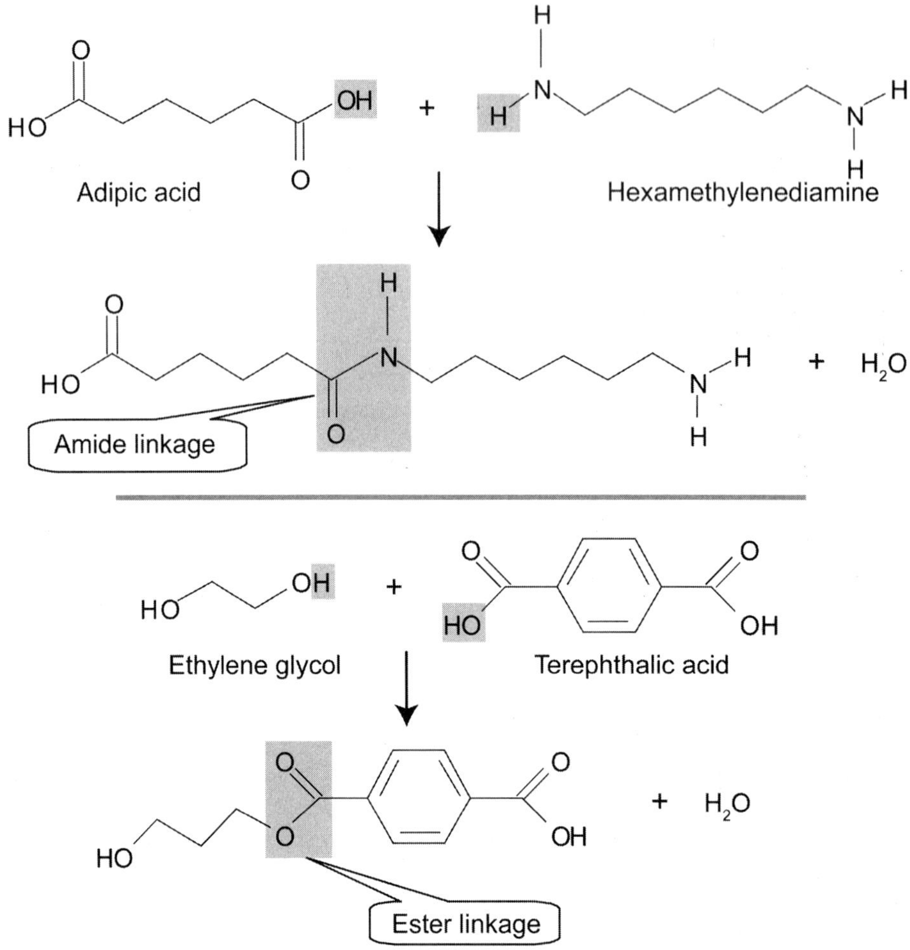

Figure 8.24 In condensation polymerization, a molecule of water is eliminated as each monomer is added to the chain. Here we show the first steps in the formation of Nylon (a) and Dacron (b). Both are copolymers, with two different monomer molecules combining to form the polymer. Since each monomer contributes its own functional group to the condensation step, a regular alternation of monomers is assured in these cases. Other condensation polymerizations can involve a single monomer with different functional groups on each end.

Example Problem 8.4

Each of the following monomers or pairs of monomers can undergo condensation polymerization reactions. Draw the structures showing the repeat units and linkages in each of the resulting polymers.

(a). glycine:

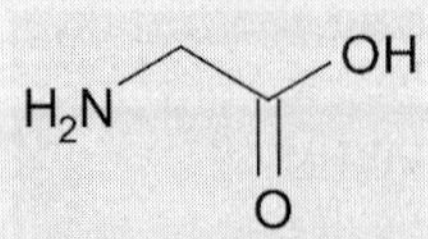

(b). 6-aminocaproic acid:

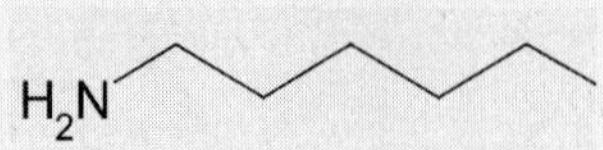

(c). *p*-phenylenediamine and terephthalic acid:

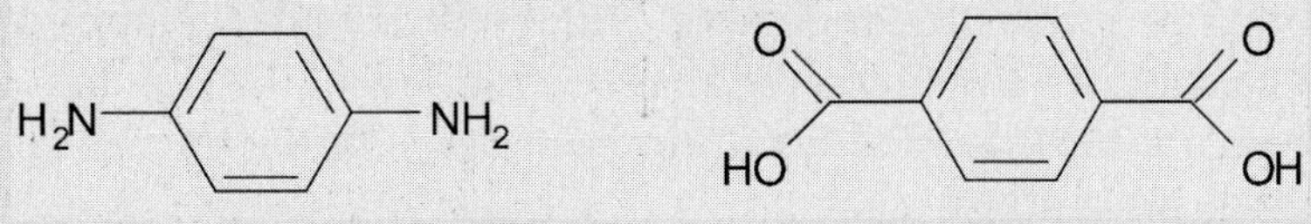

Strategy

Condensation polymerization involves the elimination of a small molecule – usually water – as each monomer is added to the growing chain. We can look for functional groups on the ends of the monomers that can undergo such a reaction.

Solution

(a). We have an amine group (-NH_2) on one end of the monomer and a carboxyl group (-COOH) on the other. Elimination of H_2O from these groups will produce an amide linkage. The polymer will have the following structure, with the monomers connected through amide linkages.

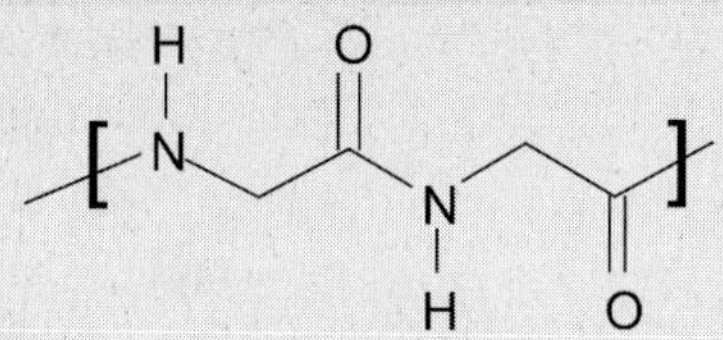

(b). This is very similar to (a).

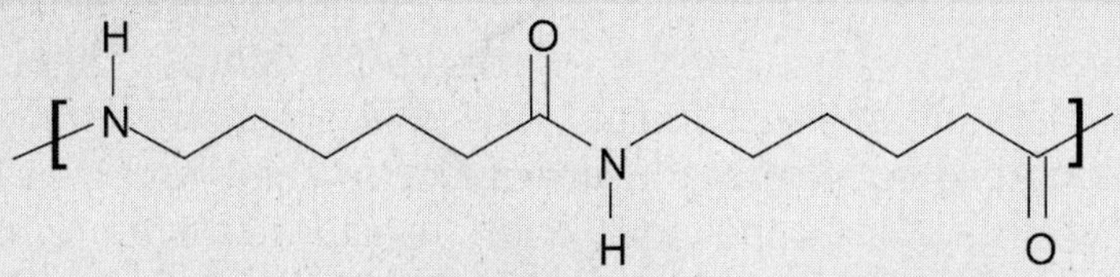

(c). Here we have amide groups on one monomer and carboxyl groups on the second. We will still form amide linkages, here with the different monomers alternating. (The polymer formed here is Kevlar.)

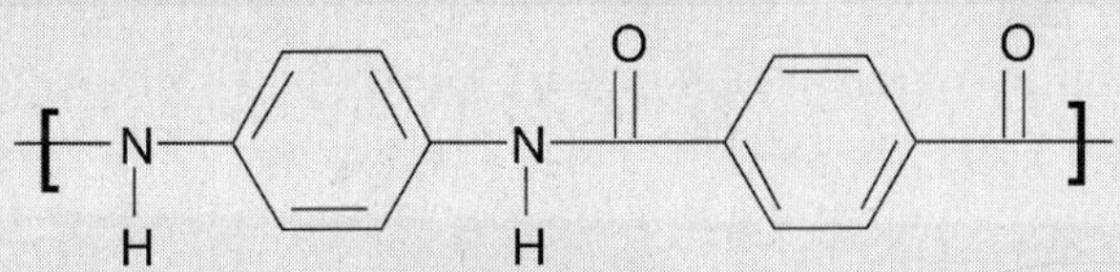

Check Your Understanding

The polymer whose structure is shown below is produced through a condensation reaction in which water is eliminated. Draw the structure of the corresponding monomer.

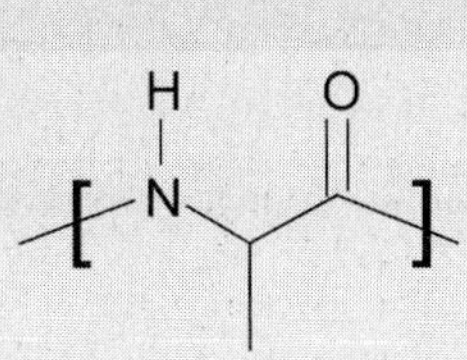

Copolymers

Nylon and Dacron are just two examples of **copolymers**, made up of more than one type of monomer. In these cases, because the monomers are arranged in a regular, alternating series they can be further classified as *alternating copolymers*. There are other important types of copolymers as well. One of the more creative ways to engineer materials is to design block copolymers. In a *block copolymer* there are regions of the material where a single monomer unit is repeated, interspersed with other regions have a different monomer unit as the repeating unit. Spandex is an example of a material that is a block copolymer. In this material, some regions of the polymer are relatively stiff while others are flexible. The resulting material is both strong and flexible, and finds widespread use in sports gear.

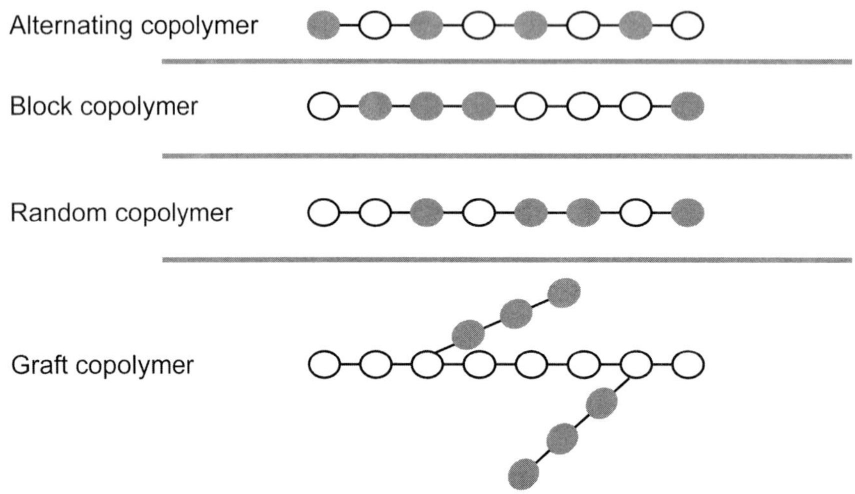

Figure 8.25 Copolymers can be grown with a variety of different arrangements of the monomers. Here the gray and white ovals represent two different monomers. For the most part the names of these types of copolymers are self-explanatory. The different monomers alternate in an alternating copolymer, or are randomly combined in a random copolymer. In a block copolymer, each type of monomer tends to be clumped together. Graft copolymers involve side chains of one polymer attached onto a backbone of a different polymer.

Another engineering material that provides an example of a copolymer is commonly called ABS because the monomer units involved are acrylonitrile, butadiene and styrene. This material is a *graft copolymer,* in which both butadiene and acrylonitrile are attached to a polystyrene backbone. Polystyrene provides ease of processing, glossiness, and rigidity. Acrylonitrile adds chemical resistance and hardness, while butadiene provides impact resistance. The specific composition can be tailored to a particular application. ABS is used in the strong plastic used in the cases of computers, televisions, and other household electronic goods.

Physical Properties

There are levels of polymer structure, such as the formation of ordered regions with the polymer foleded back onto itself called lamellar regions that we are not considering here.

There are two key reasons for the importance of polymers as engineering materials. First and most obviously, polymers offer physical properties like strength or elasticity that can be desirable in a wide range of uses. Second, those properties can be controlled or tailored to a greater degree than is usually seen in metals or other classes of materials. Through the choice of a particular polymer and the details of its synthesis and processing, materials engineers can choose the properties they need with a striking degree of specificity. In this section we will look at a few examples of the types of choices that can be made.

We begin by considering the thermal properties of polymers. In choosing a polymer for a specific design application, you would surely need to make sure that your finished object will be suitable for the temperature range at which it will be used. But an experienced engineer will also recognize the need to think about how the object will be manufactured; many plastic components are formed at high temperatures, and different polymers respond in dramatically different ways to heat. Polymers are often divided into two broad categories: thermoplastic and thermosetting. Thermoplastic polymers melt or deform on heating. This may seem like a weakness, since it clearly means that they are not suitable for use in high temperature applications. But a great many plastic objects, including children's toys and bottles of many sorts, are generally used at ambient temperatures. So the fact that they would melt if heated appreciably is not a major drawback. And when we think about how to go about making these objects, which can come in rather complicated shapes, the ability to melt the polymer at reasonable temperatures becomes a major advantage. Depending on their complexity, products made from thermoplastic polymers are typically extruded or formed in molds or presses. The fact that the material softens or melts when heated allows it to be shaped into the desired form, and once cooled the polymer solidifies and regains its structural properties.

If we are trying to design an object that might need to be used at higher temperatures, though, the fact that thermoplastics soften and melt will make such materials a poor choice. Instead we might turn to thermosetting polymers, which can maintain their shape and strength when heated. The name "thermosetting" comes from the fact that these polymers must be heated in order to set or "lock in" their structure. But once this has been done, the materials offer increased strength and do not lose their shape upon further heating. Rather than being extruded, most thermosetting polymers are molded. The molecular origin of the difference between thermoplastic and thermosetting polymers is shown in Figure 8.26. The initial heating and setting of a thermosetting polymer produces a number of links between sites on the carbon backbone of different molecular chains. These links are referred to as crosslinks, because they cross between and link the individual molecular strands of the polymer. Chemically, these crosslinks are really additional covalent bonds joining the polymer chains to one another. Like most covalent bonds, they are strong enough that they do not readily fail upon heating. So the crosslinked polymer keeps its shape.

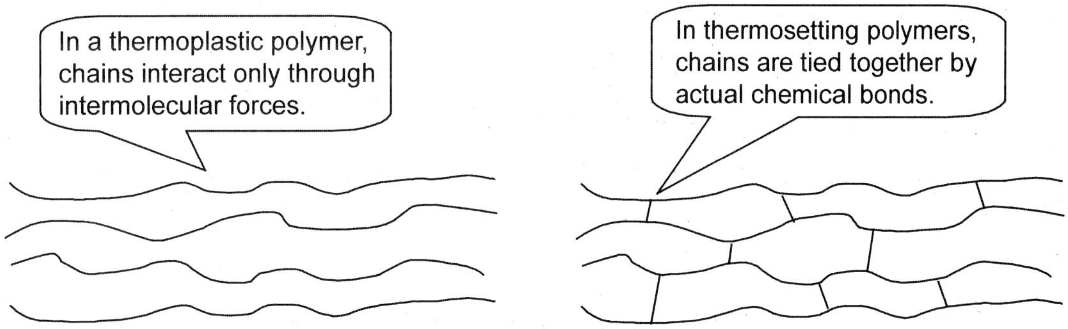

Figure 8.26 The different properties of thermoplastic and thermosetting polymers are due to the ways in which the polymer chains interact with one another.

An important example of the engineering importance of crosslinking and of American industrial history is the discovery of vulcanization. In vulcanization, natural rubber is heated in the presence of sulfur. This produces crosslinking and leads to a harder material that is markedly more resistant to heating. Until vulcanization was discovered, natural rubber was difficult to use in applications such as automobile tires because it would become sticky when heated. Natural rubber is no longer widely used, having been replaced by synthetic forms. But the large-scale development of vulcanized tires and the design freedoms they afforded automotive engineers was an important component of the growth of the U.S. automobile and tire industry.

Another critical physical property associated with polymers is elasticity. The ability of many polymeric materials to be stressed and deformed but return to their original shape is a characteristic that is often valuable to engineers. Fibers, in particular, need elasticity. Polymers that are particularly flexible and elastic are sometimes categorized as elastomers. Looking at the structure of elastomers, we find that they tend to share some common traits. In particular, the geometry of the carbon backbone is often such that the elastomer forms an amorphous solid rather than a crystalline one. The forces between polymer molecules in an amorphous solid are not as strong as those in a more crystalline system, so they can be deformed and restored with less force.

Polymers and Additives

Despite their wide range of useful properties, polymers alone often do not have the needed characteristics for a particular design. Fortunately, additives can be included in the material to further control polymer. Some additives have only modest mechanical affects. Pigments, for example, can be added with the sole intent of changing the color of the material, and don't really influence the underlying molecular structure of a polymer. Other additives play critical roles in the performance of the materials. PVC alone, for example, is too brittle for many applications. To rectify this, relatively small molecules called plasticizers are added to improve the flexibility of the materials. A plasticizer has to have several critical features. It must be capable of being incorporated into the solid polymer, so its structure should resemble the polymer being used. And it must be non-volatile, or it will escape from the solid too quickly and no longer impart the desired flexibility. Plasticizers do escape slowly, however, and often the smell associated with plastic materials arises from the release of trace amounts of these additives. Other additives that find regular usage include anti-static agents, fillers, fire retardants and stabilizers.

There are many more features associated with polymers that many of you will undoubtedly learn about as you proceed in your engineering curriculum. From the perspective of chemistry, they serve as a tremendous example of the impressive structural variability associated with carbon-containing molecules. While other elements are capable of forming polymeric materials, it is the chemical characteristics of carbon that make the modern plastics industry possible.

8.7 Insight into the Invention of New Materials

Few components of research in the chemical sciences have a more profound impact on the everyday life of humans than the discoveries associated with new materials. There are, of course, many ways that new materials are found, including good fortune or serendipity.

The discovery of buckminsterfullerene, for example, shows several of the components of how research in chemistry proceeds. The researchers at Rice University who discovered C_{60}, led by Richard Smalley and Robert Curl, had already established the ability to study other elemental clusters. In doing so, they constructed sophisticated and unique instruments to generate small, gas-phase clusters as well as to detect their existence and measure the number of atoms in each cluster. When Harry Kroto visited the Smalley laboratories and suggested they study carbon, there was no expectation that they would discover an entirely new field. Then they noticed that the experimental peak associated with C_{60} was unusually large. They then decided to optimize the production of this species and attempted to explain its apparent unusual stability. Once they announced the discovery, other scientists repeated and extended the studies and five years later, physicists at the University of Arizona discovered a means to make large quantities of C_{60} and other fullerenes. With this discovery the world of nanotechnology was opened up and the road to potentially amazing materials was available for further research.

Other materials have been discovered by using well-established techniques more than an accidental discovery. Traditional synthesis methods for new materials involve the mixing of solid precursors in ratios that are predicted to provide a desired stoichiometry. The mixture is then heated to allow the atoms in the separate solids to diffuse into each other. This process can be quite slow, so heating times are often long and temperatures are sometimes quite high. In many cases the process has to be repeated several times to achieve the synthesis. Normally the intermediate products must be ground into small particles, and hence this process has sometimes been dubbed the "heat-and-beat" method of synthesis.

Sometimes, even established techniques like heat-and-beat lead to unexpected results. Superconductivity arises in some materials when all resistance to electricity flow vanishes. Because the resistance to electrical flow represents a loss process in designs that use electricity, superconductors are very attractive materials. The ability to use them in electrical designs would improve the efficiency of many designs. Traditional superconductive materials required very low temperatures, but in 1987 Paul Chu and his research team at the University of Houston discovered a new form of superconductivity that lost all resistance at much higher temperatures. Here "much higher" is a relative term, as the needed temperatures are still far below room temperature. One of the key goals of ongoing research in materials science is motivated by the desire to build room temperature superconducting materials.

The concept of careful doping that we considered in semiconductors can also be used in the design of superconducting materials.

The initial discovery of a group of materials like the fullerenes most often occurs in the context of "pure" chemistry, carried out by researchers who are not necessarily striving towards specific practical applications. Identifying and developing practical uses for scientific knowledge is at the heart of engineering, of course. So it is quite common to see engineers become interested in newly discovered materials. A good example of this can be seen in the development of electrically conducting polymers, a field that currently holds great promise in many emerging applications. By now you should closely associate the word polymer with plastics. And when you think of plastics, you clearly expect them to be electrical insulators. After all, many plastics are routinely used for the insulation on wires. So how might a polymer be made into an electrical conductor?

We learned in this chapter that electrical conductivity in metals and semiconductors is closely linked to the idea of electron mobility or delocalization. And that turns out to be the key to achieving conductivity in polymers, too. When we explored chemical bonding in Chapter 7, we saw that the concept of resonance in Lewis structures was invoked to allow us to draw structures for molecules where electrons were not specifically localized in a particular chemical bond. If we create a polymer for which we can draw extensive resonance structures, then, there might be a possibility of achieving enough delocalization of electrons to lead to conductivity. Thinking back to Chapter 7, you may realize that one way to do this would be to set up an alternating sequence of double and single bonds down the backbone of the polymer. A prototypical example would be poly(acetylene), for which a portion of a chain would have the structure shown below.

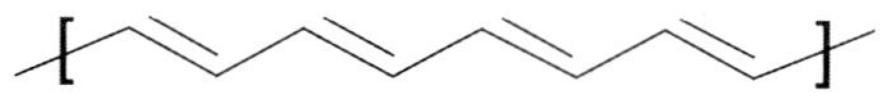

It should be apparent that a second resonance structure could be drawn in which the positions of all of the C-C and C=C bonds are reversed. This opens up the possibility of delocalizing significant numbers of electrons within a single polymer molecule – a big step towards conductivity. In fact a few specific examples of conducting organic materials have been known since the early 1900's. But they were generally regarded as curiosities until our understanding of both polymer structure and electrical conductivity evolved over the past century. That progress now makes it possible not only to explain earlier observations, but also to begin to rationally design new polymers with specific electrical properties. The production of these conducting polymers involves a form of doping that is somewhat analogous to what we saw for inorganic semiconductors, and the fact that this doping can be carefully controlled holds great potential for tailoring the properties of individual polymers. In many ways the possibilities are even greater for the case of polymers. Only a handful of elements can be used as the underlying material for semiconductors, but the range of available monomers and polymers is much broader.

Applications of conducting polymers are currently being pursued in several areas, perhaps most notably as organic light-emitting diodes, or OLED's. Different preparations can produce polymers that will display different colors. And because the polymers can be dissolved into solution, they can be manipulated in intriguing ways. A polymer solution can be deposited through a process like ink-jet printing, and adding contacts can produce a highly customized display that can be applied to almost any surface.

The development of the field of conducting polymers has seen important contributions from both chemists and engineers, as is usually the case. The interplay between the discoveries of the scientist, driven largely by curiosity, and the advances of the engineer seeking a solution to a specific problem is synergistic. Progress on either side fuels the further thinking of both groups and spurs the next generation of advances.

Focus on Problem Solving

Question

The metals in groups 11 and 12 all have face-centered cubic structures. They are also among the least reactive metals. Aside from looking up the density itself, what could you look up to determine which metal was the most dense and how would that information help you determine the density?

This question has both a conceptual component and computational component. First we need to understand how the lattice is related to the density – and then we need to provide the computational pathway that relates structures at the atomic level to densities at the macroscopic level.

Answer

Knowing that the structure is face-center cubic tells us about the geometry of the atoms – how they are connected. If we can look up the length of the cubic edge we can calculate the number of atoms in a given volume (related to the volume of the unit cell). Then we would need to look up the atomic mass to convert between number of atoms in the given volume and the mass of those atoms. Density is mass per unit of volume so these steps would provide a density estimate that could be used to estimate which metal would have the greatest density.

Chapter 8
Problems & Exercises

Insight into Carbon

8.1 How many forms of elemental carbon are known?

8.2 Use the web and find out why the C_{60} form of carbon is called buckminsterfullerene.

8.3 What property of diamond leads to the most engineering applications? What types of applications would benefit from this property?

8.4 What variables must be controlled and in what manner to produce diamonds from graphite?

8.5 What is the relationship between the structure of buckminsterfullerene and carbon nanotubes?

8.6 Use the web to look up information on nanotubes. Distinguish between single walled and double walled nanotubes.

8.7 Use the web to look up the experimental conditions required to synthesize buckminsterfullerene in an electrical arc. Would running an arc in air work?

Condensed Phases – Solids

8.8 What is the difference between an amorphous and a crystalline solid?

8.9 Define packing efficiency.

8.10 Using circles, draw a regular two-dimensional arrangement that demonstrates a low packing efficiency and one that illustrates a high packing efficiency.

8.11 Using pentagons, draw an arrangement that demonstrates a low packing efficiency and one that illustrates a higher packing efficiency.

8.12 Using circles, show that a cubic structure (squares in 2D) has a lower coordination number than a hexagonal structure.

8.13 Calculate the packing efficiencies for the two 2-dimenional arrangements shown in Figure 8.3.

8.14 Explain why hcp and ccp structures have the same coordination number.

8.15 What is the coordination number of atoms in the diamond structure?

8.16 Explain why graphite is described as having a layered structure.

8.17 Polonium is the only metal that forms a simple cubic crystal structure. Use the fact that the density of polonium is 9.32 g/cm^3 to calculate its atomic radius.

8.18 Iridium forms a face-centered cubic lattice, and an iridium atom is 271.4 pm in diameter. Calculate the density of iridium.

8.19 Europium forms a body-centered cubic unit cell and has a density of 4.68 g/cm^3. From this information, determine the length of the edge of the cubic cell.

8.20 Manganese has a body-centered cubic unit cell and has a density of 7.88 g/cm^3. From this information, determine the length of the edge of the cubic cell.

8.21 The spacing in a crystal lattice can be measured very accurately by x-ray diffraction, and this provides one way to determine Avogadro's number. One form of iron has a body-centered cubic lattice, and each side of the unit cell is 286.65 pm long. The density of this crystal at 25°C is 7.874 g/cm³. Use these data to determine Avogadro's number.

Metals and Metallic Bonding

8.22 How many electrons per atom are part of the sea of electron model for the following metals (a) iron (b) vanadium (c) silver

8.23 One important physical property of metals is their malleability. How does the electron sea model account for this property?

8.24 The electron sea model is not generally used for quantitative predictions of properties. What factors are left out of this model that prevent quantitative precision?

8.25 How does the electron sea model of metallic bonding explain that metals are good conductors of electricity?

8.26 What is the key difference between metallic bonding (in the electron sea model) and ionic bonding (as described in Chapter 7) that explains why metals conduct electricity and ionic solids do not?

8.27 What is the difference between a bonding and antibonding orbital?

8.28 Describe how the combination of atomic orbitals gives rise to bands in the limit of large numbers of atoms.

8.29 Draw a depiction of the band structure of a metal. Label the valence band, conduction band and Fermi level.

8.30 How do conductors, insulators and semiconductors differ according to band theory?

8.31 Semiconductors are found to increase in conductivity as temperature is increased. Using band theory, explain this observation.

8.32 Use the web to find out what a III-V semiconductor is. How is this type of semiconductor related to silicon?

8.33 Explain how doping affects the electronic properties of a semiconductor.

8.34 What type of atom is needed as a dopant in an n-type semiconductor? Why is it called n-type?

8.35 What type of atom is needed as a dopant in a p-type semiconductor? Why is it called p-type?

8.36 Is an n-type semiconductor actually negatively charged?

8.37 How does a p-n junction serve as voltage-gated switch?

8.38 Suppose a device is using a 15.0-mg sample of silicon that is doped with $1 \times 10^{-5}\%$ phosphorous. How many phosphorous atoms are in the sample?

Intermolecular Forces

8.39 What is an instantaneous dipole?

8.40 Why are dispersion forces attractive?

8.41 If a molecule is not very polarizable, how will it respond to an external electric field?

8.42 What is the relationship between polarizability and dispersion forces?

8.43 Most gaseous compounds consist of small molecules, while polymers are never gaseous at room temperature. Explain this observation based on intermolecular forces.

8.44 Why are dipole-dipole forces typically stronger than dispersion forces?

8.45 Under what circumstances are ion-dipole forces important?

8.46 Is hydrogen bonding a chemical bond? Explain your answer.

8.47 What is the unique feature of N, O, and F that causes them to play a role in hydrogen bonding?

8.48 List all the intermolecular interactions that would be found in a sample of the following molecules. (a) H_2S (b) CCl_4 (c) CH_3OH (d) NF_3 (e) F_2 (f) NH_2OH

8.49 Arrange the following molecules in the order of increasing intermolecular forces: F_2, CH_3OH, H_2S.

8.50 Arrange the following molecules in the order of increasing intermolecular forces: C_2H_5OH, C_3H_7OH, H_2O

8.51 Carbon tetrachloride (CCl_4) is a liquid at room temperature and pressure, while ammonia (NH_3) is a gas. How can these observations be rationalized in terms of intermolecular forces?

8.52 Explain from a molecular perspective why graphite has properties that are useful for lubrication.

Condensed Phases – Liquids

8.53 Describe how interactions between molecules affect the vapor pressure of a liquid.

8.54 What makes a chemical compound volatile?

8.55 Describe how a ski hill in operation provides an analogy for dynamic equilibrium.

8.56 Why does the vapor pressure of a substance require a dynamic equilibrium in order to be measured?

8.57 Which substance has the highest boiling point: CCl_4, BF_3, or NF_3?

8.58 Which substance has the lowest vapor pressure: NF_3, PF_3, or BF_3?

8.59 A substance is observed to have a high surface tension. What predictions can you make about its vapor pressure and boiling point?

8.60 Suppose you have three unknown pure substances that are liquids at room temperature. You make vapor pressure measurements and find that substance Q has a pressure of 110 torr, substance R has a pressure of 42 torr and substance S has a pressure of 330 torr. If you slowly increase the temperature, what substance will boil first and which will boil last?

8.61 Suppose you have three unknown pure substances that are liquids at room temperature. You determine the boiling point of substance A to be 53°C, substance B to be 117°C and substance C to be 77°C. Based on this information, predict the vapor pressure of the three substances at room temperature.

8.62 Rank the following hydrocarbons in order of increasing vapor pressure: C_2H_6, $C_{10}H_{22}$, CH_4, C_7H_{16}, $C_{22}H_{46}$.

8.63 Draw the meniscus of a fluid in a container where the interactions among the liquid molecules are weaker than those between the liquid and the container molecules.

8.64 When water is in a tube such as a buret, in what direction does the meniscus curve? What does this observation say about the relative magnitude of adhesion and cohesion in that system?

Polymers

8.65 If shown structures of monomers that might polymerize, what would you look for to determine if the type of polymerization would be addition?

8.66 Why is temperature increased to start most addition polymerization reactions?

8.67 In a polymerization reaction, what is meant by the term *chain termination*?

8.68 The monomer of polyvinylchloride is $H_2C{=}CHCl$. Draw an example of an isotactic PVC polymer.

8.69 Why is there no isotactic or syndiotactic form of polyethylene?

8.70 Why are isotactic or syndiotactic polymers often more attractive for materials development?

8.71 What are the products in a condensation polymerization reaction?

8.72 Use the web to find the structure of two types of nylon and describe how they differ from each other.

8.73 Distinguish between a block copolymer and a graft copolymer.

8.74 Use the web to look up the polymers used for six synthetic fibers. Classify the polymers as being (a) addition or condensation polymers and (b) as alternating, block or graft copolymers when they are copolymers.

8.75 What is the key physical property that characterizes a thermoplastic polymer?

8.76 What conditions would prohibit the use of a thermoplastic polymer?

8.77 What happens molecularly in a thermosetting polymer when it is heated?

8.78 What structural characteristics are needed for additives like plasticizers?

8.79 Use the web to find out what pencil erasers are made out of. Would it be possible to form a new shape of an eraser by melting it an pouring it into a mold?

8.80 Use the web to find an application where the presence of isotactic, syndiotactic or atactic polymers has an effect on the physical properties important for an application of a polymer.

Invention of New Materials

8.81 Use the web to investigate the story of the discovery of buckminsterfullerene. What was the initial reaction to the announcement of this material?

8.82 What was the observation that caused Smalley and coworkers to investigate further and ultimately discover C_{60}?

8.83 In the "heat-and-beat" method of materials synthesis, why is heat needed?

8.84 Many experts feel that applications of OLED technology will soon become common. Do a web search to identify emerging uses for OLED's. Can you find any available commercial products using OLED displays?

8.85 Use the web to find information about new materials being considered in potential battery technology for the future. Describe the characteristics that are described as helpful for applications in batteries.

8.86 There is a new material sometimes referred to as memory metal. Use the web to find information about memory metal. What is this material? Why does the word memory appear in it's name?

Focus on Problem Solving

8.87 If you know the density of a material and the length of the edge of its cubic lattice how would you determine if it was face-centered cubic, body-centered cubic or simple cubic? Would any information have to be looked up?

8.88 If a pin is run across a magnet a number of times it will become magnetized. How could this phenomenon be used as an analogy to describe polarization in atoms and molecules.

8.89 You go into the laboratory to look for a squirt bottle containing acetone. You find two unlabeled bottles with different colored tops suggesting they are different liquids. Unfortunately, you have a terrible cold and cannot tell by smell which one might be acetone. What simple test could you use to determine which liquid is acetone and which is water? How does this test tell you this information?

8.90 A material engineer, with an eye towards cost, has an interest in obtaining a material where the degree of polymerization is high. What types of measurements would need to be made in the laboratory to be able to determine if the degree of polymerization was acceptable?

Chapter 9: Energy and Chemistry

No aspect of chemistry has more widespread impact than its role in energy and energy transformation. Arguably the first use of chemistry as humanity evolved was the taming of fire, chemical combustion. We'll begin by focusing on the types of energy production and usage in North America, a component of life in the developed world that we'll refer to as the **energy economy**.

9.1 Insight into Energy Use and the World Economy

Few factors reveal the wealth of a country or region more vividly than its energy usage. The central role of the United States in the world economy is reflected in the fact that the nation accounts for 24% of world energy usage. Canada, Japan and Western Europe are also large users of energy. Energy use is closely connected to a broad array of societal, economic, and political factors. For example, in the early 1990's, Pacific Rim countries such as Thailand and South Korea experienced growing economies and saw substantial increases in their energy usage. Yet during this time energy usage worldwide was stagnant, as shrinking economies in Eastern Europe offset the increases elsewhere in the world.

Figure 9.1 illustrates the link between economic growth (as measured by the Gross Domestic Product) and energy usage. Data are plotted for several countries as energy consumption per dollar of GDP from 1990-2000. It is clear from the figure that the lines shown are all nearly horizontal. This confirms the correlation between economic performance and energy consumption, since a horizontal line in this plot implies that energy usage is proportional to GDP.

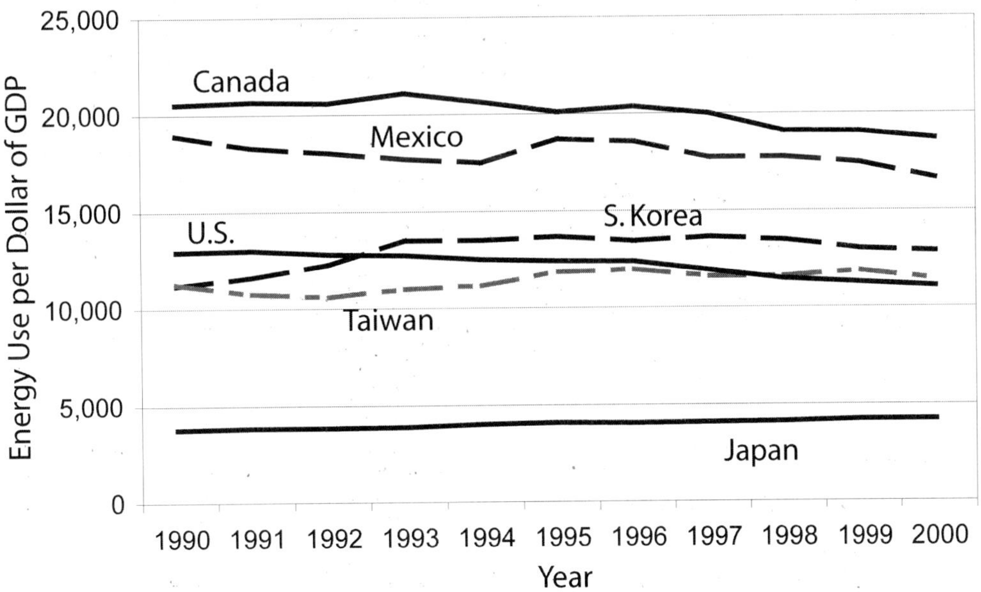

Figure 9.1 The graph demonstrates the link between energy consumption and gross domestic product (GDP). GDP is a common indicator of economic strength. The data are plotted in units of Btu per U.S. dollar, with all data converted to constant (1995) dollar values. The fact that these lines are all essentially horizontal confirms the connection between energy use and economic strength. (Data are from the Energy Information Administration, http://www.eia.doe.gov.)

Energy Usage in the United States

We can learn more about the pivotal role of energy in an economic system by examining the sources and uses of energy. While there are differences between countries, the United States is representative of energy production and usage for an industrialized country.

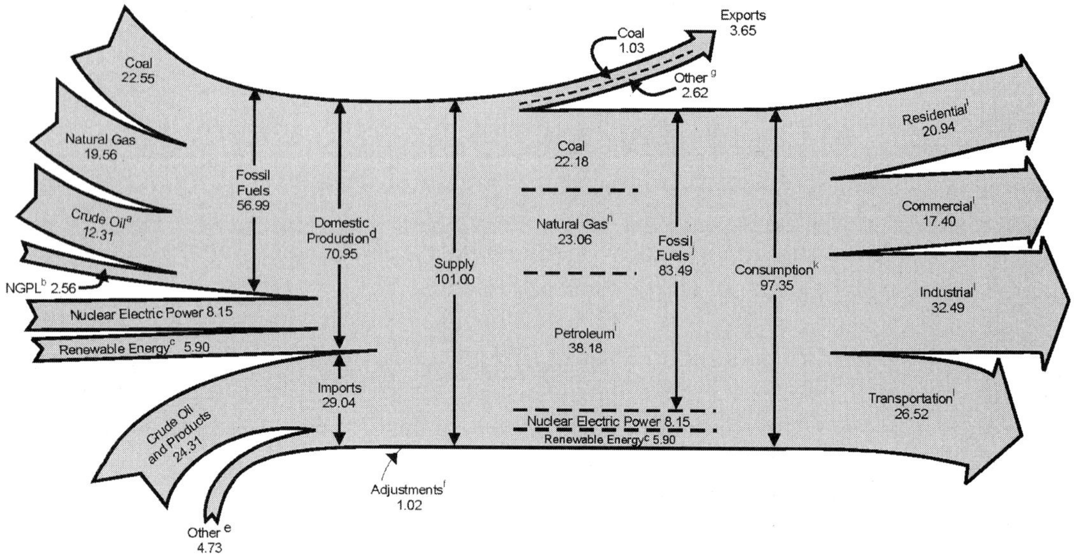

Figure 9.2 Energy production and consumption in the United States during the year 2002. The discussion in the text explains how to read this complex figure, which contains an enormous amount of information about the energy economy. (Figure is from the Energy Information Administration's Annual Energy Review. The most recent report is available at http://www.eia.doe.gov/emeu/aer/.)

Figure 9.2 is a visual representation of the pattern of energy production and usage for the U.S. in 2002. At first glance this figure is somewhat complicated, but it can be analyzed to reveal a great deal of information. Start by looking at each end: the left side represents sources of energy, while the right side shows consumption. Thus the whole figure represents the flow of resources through the U.S. energy economy. All numbers on the figure are given in units of quadrillion Btu. (The Btu is a unit of energy. The letters stand for "British thermal unit," and 1 Btu = 1054.35 J.) Starting in the upper left, we can examine the sources of energy production. Coal accounted for 22.55 quadrillion Btu, or 31.8% of the total domestic production of 70.95 quadrillion Btu. Natural gas and crude oil provided 27.6% and 17.4% of the total, respectively. Smaller contributions came from nuclear energy and renewable energy sources (including hydroelectric power, wood burning, solar energy, and wind). Energy imports appear at the bottom left of the figure, where we see that the main imports are crude oil and related petroleum products. Combining the domestic production and the imports gives the total supply of energy, which was 101.00 quadrillion Btu for 2002. Moving to the right, the figure divides this supply up to show the ways in which the energy was consumed. The four main components of usage are classified as (1) residential, (2) commercial, (3) industrial and (4) transportation, accounting for roughly 21%, 18%, 33% and 27% of total energy usage, respectively.

Certainly some assumptions are built into the division of uses into these four categories - so absolute numbers are not as important as comparative ones.

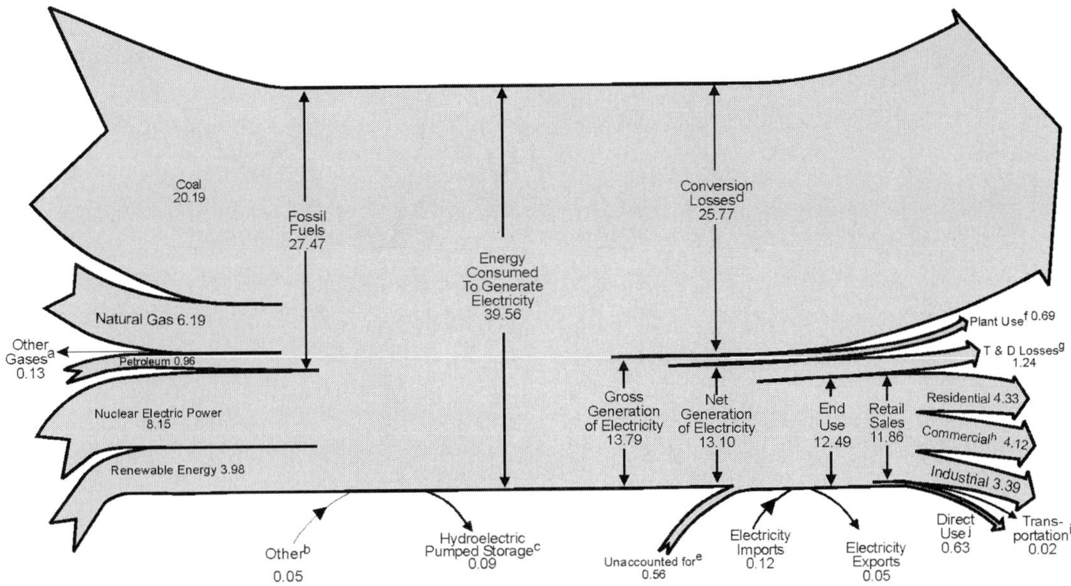

Figure 9.3 Visual summary of the generation and consumption of electricity in the United States during the year 2002, presented in a manner similar to Figure 9.2. (Figure is again from the Energy Information Administration's Annual Energy Review.)

Nearly half of all domestic energy usage goes toward the production of electricity. The generation and consumption of electricity is detailed in Figure 9.3. (The form of this diagram is similar to that of Figure 9.2, so you should examine it in the same way.) This figure shows that "conversion losses" account for nearly 2/3 of the energy consumed to generate electricity. Indeed these "conversion losses" in electricity generation mean that slightly more than 20% of the total domestic energy consumption is actually "lost" from the economy. This raises some important questions: How is energy from other sources converted to electricity, and why is this conversion necessary? Why is so much energy lost or wasted in this process, and how can we minimize or eliminate this loss?

Other questions arise when considering the relative roles of various fossil fuels in the energy economy. Figure 9.4 shows the history of US domestic consumption of energy from various sources. It is clear that our consumption of fossil fuels has grown steadily over the entire the history of our country. But since the early 1970's that growth has been sporadic, and there have been periods in which consumption of petroleum and natural gas have actually decreased substantially. A variety of factors contribute to these trends in consumption, notably the availability of raw material and the price and availability of imported fuels. The decrease in petroleum consumption in the 1970's, for example, coincides with the "energy crisis" of that era, a period in which international politics created dramatically higher prices for imported oil. This, in turn, triggered an emphasis on energy conservation. Since our interest here is chemistry, though, we will want to ask whether there are more fundamental differences between these fuels in terms of energy production and transformation. What is the role of chemistry in deciding which of these fossil fuels is most attractive in a given set of circumstances? To address all of the questions raised here we must consider some aspects of **thermochemistry** - the study of the energetic consequences of chemistry.

Ulitmately, energy is a commodity in an economic sense, so when supplies are low or demand is high, prices will tend to increase.

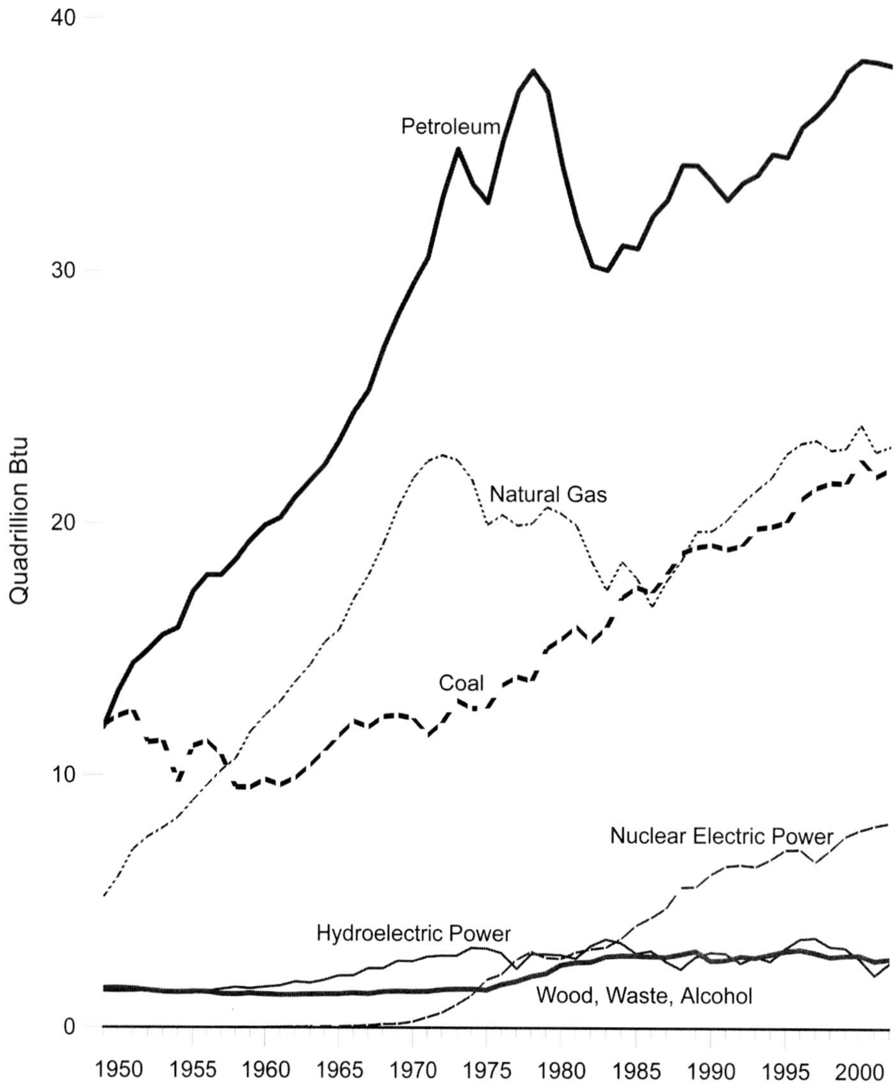

Figure 9.4 US consumption of energy from various sources has changed markedly over the last 50 years. (Figure is again from the Energy Information Administration's Annual Energy Review.)

9.2 Defining Energy

The questions raised in Section 9.1 lead us into a study of energy. Most people have an intuitive feel for the concept of energy, and we see the word every day. Breakfast cereals advertise their ability to provide energy in the morning. Substitute players are described as boosting their team's energy in sports. The "energy crisis" brought on by oil embargoes in the 1970's is a critical component of the history of that decade, and energy issues continue to influence political issues as we enter the 21st century. This widespread usage of the concept of energy underscores its importance. It also suggests that we'll need to be very careful in defining energy and related terms, because it is apparent that in casual use the word "energy" conveys many related but subtly different meanings.

Forms of Energy

Most of the energy we encounter can be placed into two broad categories: potential energy and kinetic energy. **Potential energy** is associated with the relative position of an object. For example, a roller coaster gains potential energy as it is pulled up the initial slope because it acquires a higher position relative to the ground and gravity is attracting it downward. But gravity is not the only force that gives rise to objects having energy associated with their positions. The attraction and repulsion between electrical charges also lead to potential energy, as we've seen in our consideration of the structure of atoms and molecules. **Kinetic energy** is associated with motion. When the roller coaster heads down the first hill it transforms the potential energy into kinetic energy. You probably know from physics that kinetic energy is described mathematically by Equation 9.1

$$KE = 1/2\ m\ v^2 \tag{9.1}$$

where KE stands for kinetic energy, m is mass and v is velocity.

The roller coaster example shows kinetic and potential energy on a macroscopic scale. But on a microscopic level all substances and objects, including fuels or the paper on which these words are printed, have these same forms of energy, because they are composed of atoms and molecules. These atoms and molecules have kinetic energy associated with their constant motion, and they have potential energy due to the various forces they exert on one another. The combined kinetic and potential energies of the atoms and molecules that make up an object constitute its **internal energy**. Thus the roller coaster cars have three basic forms of energy: kinetic (from their motion), potential (from their position relative to the ground), and internal (from the molecules that compose the materials from which they are made).

Much of the internal energy of an object, be it a roller coaster car or a piece of coal, is associated with the potential energy arising from the relative positions of the atoms making up the object. As we saw in Chapter 8, the making or breaking of chemical bonds changes this potential energy. In most chemical reactions, bonds are broken in the reactants and new bonds are formed in the products. If the energy liberated in bond making is greater than that consumed in bond breaking, the overall process releases energy. This energy release is often referred to as **chemical energy**, and the harnessing of chemical energy is an important aspect of the overlap between chemistry and engineering.

As we saw in Chapter 3, the liberation of chemical energy in an explosion can have dramatic effects on kinetic energy of particles - but chemical energy itself is just a form of potential energy.

These categories for energy encompass a broad spectrum of phenomena, from roller coasters to petroleum to the molecules in a piece of paper. Occasionally we will find utility in having more specific terms for various forms of energy as well. *Radiant energy* is associated with light or electromagnetic radiation. The radiant energy of the sun is ultimately responsible for the majority of energy resources on this planet. *Mechanical energy* is associated with the motion of macroscopic objects. *Thermal energy* arises from the temperature of an object. We can associate thermal energy with the molecular level motion of the atoms and molecules making up any object. *Electrical energy* results from moving charge (usually electrons in a metal). *Nuclear energy,* which can be released in nuclear fusion and fission processes, is a form of potential energy associated with the arrangement of protons and neutrons into atomic nuclei.

Heat and Work

While we can use a wide variety of classifications for types of energy, all energy flow is either heat or work. We need to understand both of these forms of energy transfer to assess the energy economy of the world and the role chemistry plays in that economy. **Heat** is the flow of energy between two objects, from the warmer one to the cooler one, because of a difference in their temperatures. Thus if we are speaking carefully, heat is a process and not a quantity. Although we routinely hear statements such as "turn up the heat," heat is not an entity we can pump into a room or a cup of coffee. An object does not possess heat. In a strictly scientific sense, a furnace does not produce heat but rather a body of warm air or hot water that has a higher temperature than the cool air in a room. What emerges from the vent on the floor is not "heat," but warm air. While these distinctions are essentially semantic, they can be very important in many cases.

Work is the second form of energy transfer. **Work** is a transfer of energy accomplished by a force moving a mass some distance against resistance. Lifting a set of roller coaster cars up the initial hill against the pull of gravity is an example of work. When we consider examples of bulk materials such as this, we are typically viewing work in terms of mechanical energy. Work, however, encompasses a wider range of phenomena than just mechanical movement of macroscopic objects. The most common type of work we will encounter in chemical processes is **pressure-volume work (PV-work)**. When a gas expands it can do work. If an inflated balloon is released before it is "tied off," it flies around as the gas inside the balloon expands into the large volume of the room. Since the flying balloon has mass, it is easy to see that the expanding gas is doing pressure-volume work on the balloon. A more productive example is the work done by the expansion of hot gases in the piston of a car engine after the spark plug fires to ignite the fuel-oxygen mixture present. This PV-work is ultimately responsible for the motion of your car.

Fuels and Combustion

From a practical point of view, chemical energy is usually tapped for a variety of uses through the reaction of molecules that are collectively referred to as fuels. Fossil fuels comprise the bedrock on which the energy economy is built. The origin and nature of fossil fuels provide a useful example of the role of chemistry in the production of energy.

Recall from earlier that hydrocarbons are chemical compounds composed of only carbon and hydrogen.

Fossil fuels are mostly composed of mixtures of hydrocarbons. There are a vast array of hydrocarbon molecules, but for an example we can focus on small alkane molecules. This class of hydrocarbons has the general formula, C_nH_{2n+2}, and the names and formulas of alkanes containing 1-12 carbon atoms are provided in Table 9.1. Of this group several are prominent fuels, including methane, propane and octane. When hydrocarbons are burned the products are carbon dioxide and water vapor.

$$\text{Hydrocarbon} + O_2(g) \rightarrow CO_2(g) + H_2O(g)$$

Chemists refer to this burning process as **combustion**. Looking at the general combustion equation above, we can see one critical advantage of hydrocarbons as fuels: the products are common gases and therefore easily disposed. And in light of our introduction to PV-work above, we might already imagine that the production of gaseous products will make it easier to convert any energy released in these reactions into useful work.

Compound	Formula
Methane	CH_4
Ethane	C_2H_6
Propane	C_3H_8
Butane	C_4H_{10}
Pentane	C_5H_{12}
Hexane	C_6H_{14}
Heptane	C_7H_{16}
Octane	C_8H_{18}
Nonane	C_9H_{20}
Decane	$C_{10}H_{22}$
Undecane	$C_{11}H_{24}$
Dodecane	$C_{12}H_{26}$

Table 9.1 Names and formulas of some small alkanes. Note that different isomers are possible for alkanes containing four or more carbon atoms.

Energy Units

Like many quantities, energy is often measured and discussed in a wide variety of units. And various disciplines of science and engineering tend to favor particular choices of units.

In the SI system, energy is measured in **joules**, J. In terms of more fundamental units, the joule is equal to 1 kg m^2 / sec^2. Realizing that work should have units of energy allows us to make sense of this unit. Work is force multiplied by distance, and force is mass multiplied by acceleration. Thus, work is

$$(\text{mass}) \times (\text{acceleration}) \times (\text{distance})$$

or

$$(\text{kg}) \times (\text{m/sec}^2) \times (\text{m}) = \text{kg m}^2 / \text{sec}^2.$$

It's often handy to develop an intuition for the magnitude of a particular unit. A joule is roughly the amount of energy it takes to raise a 1 kg (about 2 lb.) book 10 cm (4 in.) above a table-top (against the force of gravity). If we think about chemistry on a molecular level, a joule is a very large amount of energy; breaking a single chemical bond typically requires only about 10^{-18} J. But we seldom break only one bond at a time. If we look at a reaction from the macroscopic viewpoint we might consider breaking a mole of bonds. In that case the joule seems like too small a measure of energy, and we might use kJ (or kJ/mole) instead. Breaking one mole of C-H bonds, for example, requires about 410 kJ.

Many older units for energy relied on some readily observable property for their original definition. Thus the British Thermal Unit (**Btu**) was first defined as the amount of energy needed to raise the temperature of one pound of water by one degree

Fahrenheit. The Btu still finds widespread use in several engineering disciplines. Another classic unit for energy is the **calorie**, which was originally defined as the amount of energy required to raise 1 gram of water from 14.5°C to 15.5°C .

The joule is now the most widely accepted unit for energy, and so these other units are now defined in terms of it. A calorie is defined to be 4.184 J and a Btu is defined to be 1055 J. A frequent complication arising from units of energy is that the Calorie reported for foods is actually a kilocalorie. Thus, 1 Calorie is actually equivalent to 4,184 J or 4.184 kJ. (The food Calorie should be written with an upper case 'C.') The amount of energy present in foods is surprisingly large.

You may have seen a popular chemistry demonstration in which the energy present in a small piece of "gummy bear" candy is rapidly released by reaction with molten potassium chlorate.

The energy amounts that are typical of large economies are enormous, and are usually reported in quadrillion (10^{15}) Btu, as we saw when analyzing Figures 9.2 and 9.3. Thus the numbers shown in Figure 9.2 are for the United States in 2000, when the total energy input was 102.60×10^{15} Btu.

9.3 Energy Transformation and Conservation of Energy

We have considered several ways to categorize energy. But these multiple forms of energy that we have specified are not all of equal utility, so in many cases it is desirable to transform energy from one form into another. For example, the lighting in your room is provided by electricity, but that electricity was probably generated by the release of chemical energy through the combustion of coal. Unless you want to try to light your room by burning a chunk of coal, you need a way to harness the chemical energy released as the coal burns and then convert it to electrical energy. That electrical energy must then be conveyed to your room. We have already noted the reality of energy waste in these processes in Section 9.1. Now we begin to consider the laws of Nature that apply when one form of energy is converted into another.

The issue of defining the system and its surroundings is also important in many engineering problems.

The first and perhaps most prominent constraint on energy transformation is that total energy must be conserved. If we properly account for all energy conversion and energy transfer processes, the total amount of energy present must remain constant. In order to assure that this constraint is followed, we will need to define a number of terms quite carefully. First, we must specify precisely what is being studied. The part of the universe that is being studied is defined to be the **system**. The remainder of the universe is considered the **surroundings**, even though it is not generally necessary to consider everything else in the actual universe. These definitions assure that the system plus the surroundings must equal the universe. The system and the surroundings are separated by a **boundary**.

These definitions may seem obvious, but they are not always so. For example, if the system to be studied is "the atmosphere of the earth," definition of the boundary is somewhat challenging. The atmosphere simply gets gradually less dense and it is arbitrary to suggest precisely where it ends. In many cases like this, the main requirement is that we be consistent. The same choice of system and surroundings *must* be used throughout a particular problem, even if the choice is made somewhat arbitrarily. (This notion of a system and its surroundings is also commonly used in many engineering problems, too.)

Presuming that an appropriate choice of a system has been made, the concept of conservation of energy immediately becomes useful. Since we said that heat and work are the only possible forms of energy transfer, we can say that the overall change in energy, E, of a system is attributable to these two components. Heat is commonly designated as "q" and work as "w," so we can write:

$$\Delta E = q + w \tag{9.2}$$

The symbol Δ (delta) is introduced here as a notation meaning "the change in." This symbol, which will be used frequently in our study of thermodynamics, is always defined as the difference between the final state and the initial state.

$$\Delta E = E_{final} - E_{initial} \tag{9.3}$$

Equation 9.2 is deceptively simple because it includes arbitrary choices for the meaning of the signs for the quantities heat and work. Convention dictates that energy transferred into the system is always designated positive and energy flowing out of the system carries a negative sign. Thus, when heat flows into the system from the surroundings the value of q is positive and when work is done on the system the value of w is positive. Conversely, when heat flows out from the system or work is done by the system on the surroundings q and w will both be negative. Example Problem 9.1 demonstrates the use of this equation.

Note that everyday usage of the word heat can sometimes imply that heat is a object of some sort. It's important to remember in thermodynamics that heat is a process - nothing possesses heat.

Example Problem 3.1

515 J of heat is added to a gas that does 218 J of work as a result. What is the change in energy of the system?

Strategy

Energy flow occurs as either heat or work, so we must identify how much of each process has occurred and in what direction. Then we can use Equation 9.3 to calculate the result. This type of problem points to the importance of considering the sign associated with changes in energy.

Solution

Heat added TO the system: $q > 0$, $q = +515$ J

Work done BY the system: $w < 0$, $w = -218$ J

$$\Delta E = q + w = 515 \text{ J} + (-218 \text{ J}) = +297 \text{ J}$$

Note that in most cases when a value is positive the sign "+" is not placed in front of the number.

Check Your Understanding

408 J of work is done on a system that releases 185 J of heat. What is the energy change of the system?

In the example above, we see that ΔE is not zero. How can this be consistent with the idea that energy must be conserved? By following our sign conventions for q and w, we considered the processes of heat and work from the perspective of the system (the unspecified gas in this case). So the value of ΔE we calculated is also the change in internal energy for the system. This change in the internal energy of the system is precisely offset by a change in the surroundings: $\Delta E_{surroundings} = -\Delta E_{system}$. (We usually just write ΔE rather than ΔE_{system}, since by definition the system is what we are most interested in.) Thus the energy of the universe remains constant.

$$\Delta E_{universe} = \Delta E_{surroundings} + \Delta E_{system} = 0$$

Energy can be transformed from one form to another, but cannot be created or destroyed. This observation of Nature is known as the **first law of thermodynamics**.

Waste Energy

The conversion of energy from one form to another has already been noted in a number of places in this chapter. Anyone who sits in front of a campfire to fight back the evening chill can argue the virtues of heat, but in most cases civilization prefers its energy in the form of work. The combustion of gasoline is not inherently useful, but when the heat released is harnessed in the engine of an automobile, the work that results gets us where we need to go. The **second law of thermodynamics**, however, observes that it is impossible to convert heat completely to work. (We'll examine the second law in detail Chapter 10.) The car's engine gets hot when it runs. The heat that warms the engine is energy that does not propel the car towards its destination. So a portion of the energy released by the combustion of gasoline does not contribute to the desired work of moving the car. This energy can be considered "wasted" in terms of the energy economy.

Of course on a cold winter day, the waste heat from the engine is partially used to warm the passenger cabin, so clever engineering can put "waste" heat to use.

Device	Energy Conversion	Typical Efficiency, %
Electric heater	Electricity/Thermal	~100
Hair drier	Electricity/Thermal	~100
Electric generator	Mechanical/Electricity	95
Electric motor (large)	Electricity/Mechanical	90
Battery	Chemical/Electricity	90
Steam boiler (power plant)	Chemical/Thermal	85
Home gas furnace	Chemical/Thermal	85
Home oil furnace	Chemical/Thermal	65
Electric motor (small)	Electricity/Mechanical	65
Home coal furnace	Chemical/Thermal	55
Steam turbine	Thermal/Mechanical	45
Gas turbine (aircraft)	Chemical/Mechanical	35
Gas turbine (industrial)	Chemical/Mechanical	30
Automobile engine	Chemical/Mechanical	25
Fluorescent lamp	Electricity/Light	20
Silicon solar cell	Solar/Electricity	15
Steam locomotive	Chemical/Mechanical	10
Incandescent lamp	Electricity/Light	5

Table 9.2 Typical efficiencies of common energy conversion devices.

One common way to obtain work from a system is to supply it with heat. So heat flows into the system and it does work. But the second law asserts that the amount of heat flow will necessarily exceed the amount of useful work achieved. This excess heat is essentially "lost" from the energy economy and may contribute to thermal pollution. (*Thermal pollution* is the raising or lowering of water temperature in streams, lakes, or oceans above or below normal seasonal ranges as the result of discharge of hot or cold waste streams into the water.) The efficiency of conversion from heat to work can be expressed as a percentage.

Typical efficiencies for some common conversion processes are shown in Table 9.2. The prospect of improving the efficiencies of appliances is often discussed as an important means of conserving energy in the future. The possible energy savings from this proposed action is shown in Figure 9.5. Even with savings in energy associated with efficient appliances, the generation of electricity will still always result in energy being lost as heat. Energy output in the form of electrical energy is always less than the chemical or nuclear energy input at the generating plant.

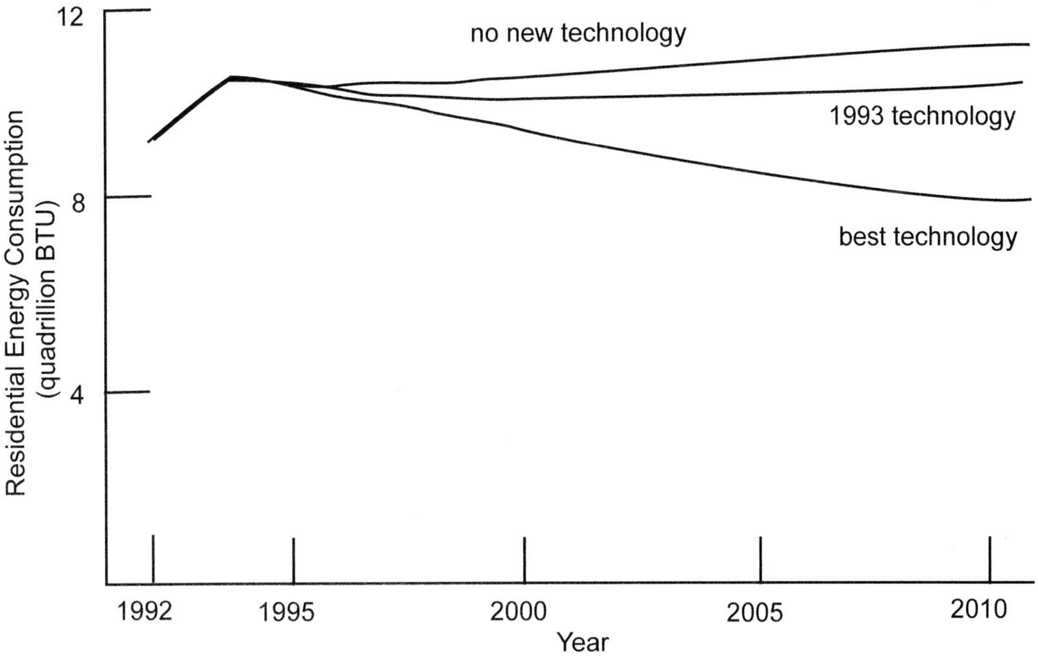

Figure 9.5 The graph shows the predicted impact on energy consumption as a result of the implementation of more efficient technologies for common household appliances.

Graphs such as this one use models that assume "average" consumer behavior for the replacement of appliances with new, energy efficient ones.

9.4 Heat Capacity and Calorimetry

General statements about nationwide or worldwide energy resources and consequences may seem abstract in many respects, if only because the numbers are so large. Making the connection between these global systems and systems we can observe in a laboratory requires a systematic way to measure energy flow. We can do this by observing the production or consumption of heat by a system through a set of techniques collectively called **calorimetry**.

Heat Capacity and Specific Heat

Suppose we wish to raise the temperature of two different systems or objects. In general, the different systems will absorb different quantities of energy based on three main factors: amount of material, type of material, and temperature change. One way to note the importance of the amount of material is to compare the behavior of a glass of water and an ocean of water. On a hot summer day at the beach, a glass of cold water will quickly become warm, while the ocean water temperature does not change noticeably over the same time. The small amount of water in the glass behaves differently than the large amount in the ocean. The type of material is also important and the summer beach provides some insight here, too. In most locations the sand at the beach is predominantly silicon dioxide, and it heats up much more quickly than even the shallow water. Both materials are exposed to similar amounts of energy from the sunlight, but they behave differently. Finally the amount of energy supplied and the temperature change are also related. On a cloudy day at the beach when less energy is supplied to the sand (because the clouds absorb some energy from the sun) it does not warm up as much. Summarizing these observations, if we want to be able to calculate the heat associated with a given temperature change, we'll need to account for the amount and identity of the material being heated as well as the extent of the temperature change.

This idea can easily be expressed as an equation, and we will use two slightly different versions of this relationship. We have two options for expressing the amount of material: by mass or by moles. Either choice leads to a usable equation. You may already be familiar with the case where we choose to use mass, since that is commonly done in physics. In that case the identity of the substance is included through a term called the **specific heat capacity** (c, sometimes simply called the **specific heat**), leading to Equation 9.4.

$$q = mc\Delta T \qquad (9.4)$$

The specific heat is a physical property of a material that measures how much heat is required to raise the temperature of one gram of that material by one degree (°C).

Similarly, the **molar heat capacity** is a physical property that describes how much heat is required to raise the temperature of one mole of a substance by one degree (°C) . So if we choose to express the amount of material in terms of moles rather than mass, our equation changes only slightly. (The subscript p on C_p indicates that this is the heat capacity at constant pressure. Under other conditions, such as constant volume, the value of the heat capacity may differ slightly.)

$$q = nC_p\Delta T \qquad (9.5)$$

Either equation provides the same information about the heat needed to produce a given change in temperature, and as long as we know the molar mass of the substance it should be simple to convert between the specific heat and the molar heat capacity. Table 9.3 provides a list giving specific heats and molar heat capacities for a few materials. A more extensive table may be found in Appendix D.

Note that water has a much higher heat capacity than the metals shown in this table. This observation is generally true for the comparison of metals to water.

Substance	Specific heat, c ($J\ g^{-1}\ K^{-1}$)	Molar heat capacity, C_p ($J\ mol^{-1}\ K^{-1}$)
Al (s)	0.900	24.3
Cu (s)	0.385	24.5
H_2O (s)	2.09	37.7
H_2O (l)	4.18	75.3
H_2O (g)	2.03	36.4

Table 9.3 Specific heat and molar heat capacities for some common substances.

Example Problems 9.2 and 9.3 show applications of these equations. As you work through these examples, notice that the heat capacity will have the same numerical value whether the temperature is expressed in K or in °C. While this may seem odd at first, it makes sense when you realize that the heat depends only on the temperature *difference*, not on the temperature itself. Verify for yourself that if the temperature of a piece of metal rises from 20°C to 30°C , ΔT can correctly be written as either 10°C or 10 K.

Example Problem 9.2

Heating a 24.0-g aluminum can raises its temperature by 15.0°C. How much heat flowed into to the can?

Strategy

Heat flow depends on the heat capacity of the material being heated, the size of the sample and the change in temperature. Since the information given is in mass, it will be easiest to use the specific heat rather than the molar heat capacity. The value of the specific heat of aluminum must be found in a table.

Solution

$$\begin{aligned} q &= mc\Delta T \\ &= (24.0\text{ g}) \times (0.901\text{ J / g °C}) \times (15.0\text{°C}) \\ &= 324\text{ J} \end{aligned}$$

Check Your Understanding

1.50 kJ of heat is supplied to a block of iron weighing 207. g. What is the change in temperature of the iron?

Example Problem 9.3

The molar heat capacity of liquid water is 75.3 J / mol K. If 37.5 g of water is cooled from 42°C to 7°C what is the heat flow out of the system?

Strategy

The heat flow is proportional to the amount of water and the temperature change. As before, we can do the calculation with the amount of water expressed in either grams or moles. Molar heat capacity is given but the amount of water is provided in grams. So we will use the molar mass to convert the amount from grams to moles. (We could also leave the amount in grams and convert the molar heat capacity into the specific heat.) Care must also be taken in defining the change in temperature. Such changes are always taken to be the final state minus the initial state. Since the water is cooling here, ΔT will be negative.

Solution

$$\begin{aligned} q &= nC_p\Delta T \\ &= (37.5\text{ g} \times (1\text{ mol water / }18.0\text{ g water})) \times (75.3\text{ J / mol °C}) \times (-35\text{ °C}) \\ &= -5.49 \times 10^3\text{ J} = -5.49\text{ kJ} \end{aligned}$$

The negative value indicates that the system (in this case the water) has lost energy and the surroundings have gained that energy. Notice that as long as we correctly express ΔT as $T_{final} - T_{initial}$, the correct sign for q will result automatically.

Check Your Understanding

If 47.0 kg of copper increases in temperature by 12.5°C when 226 kJ of heat is added, what is the molar heat capacity of copper?

In these examples, we have simply stated that a given amount of heat was added to or removed from our system, without considering the source of that heat. While this lets us familiarize ourselves with the equations, it isn't a very realistic situation. In a more common application, we can use the same type of equations to determine heat flow between two objects, as shown in Example Problem 9.4.

Example Problem 9.4

A glass contains 250.0 g of warm water at 78.0°C. A piece of gold at 2.30°C is placed in the water. The final temperature reached by this system is 76.9°C °C. What was the mass of gold? The specific heat of water is 4.184 J / g °C, and that of gold is 0.129 J / g °C.

Strategy

We will need to assume that heat flows only between the gold and the water, with no heat lost to or gained from the surroundings. In that case the heats must balance; the heat gained by the gold was lost by the water. Each of those can be calculated using Equation 9.3, and the two must be equal and opposite in sign.

Solution

$$q(\text{gold}) = -\, q(\text{water})$$

$$m_{gold} \times c_{gold} \times \Delta T_{gold} = -\, m_{water} \times c_{water} \times \Delta T_{water}$$

$$m_{gold} \times (0.129 \text{ J / g °C}) \times (74.6\text{°C}) = -\,(250 \text{ g}) \times (4.184 \text{ J / g °C}) \times (-1.1\text{°C})$$

$$m_{gold} = 120.\text{ g}$$

Check Your Understanding

A 125-g sample of cold water and a 283-g sample of hot water are mixed in an insulated thermos bottle and allowed to equilibrate. If the initial temperature of the cold water is 3.0°C, and the initial temperature of the hot water is 91.0°C, what will the final temperature be?

These examples provide a means to calculate heat flow or related information. But we have yet to look at how we can actually make laboratory measurements of heat flow.

Calorimetry

Calorimetry is the term used to describe the measurement of heat flow. Experiments are carried out in devices called *calorimeters* that are based on the concept established in Example Problem 9.4. The heat evolved or absorbed in the system of interest is determined by measuring the temperature change in its surroundings. In the case of the calorimeter, every effort is made to thermally isolate (prevent heat flow between) the immediate surroundings and the rest of the universe. Figure 9.6 shows a typical instrument for calorimetry. If the instrument is thermally isolated from the rest of the universe, the only heat flow that needs to be considered is that between the system being studied and the immediate surroundings, whose temperature can be measured.

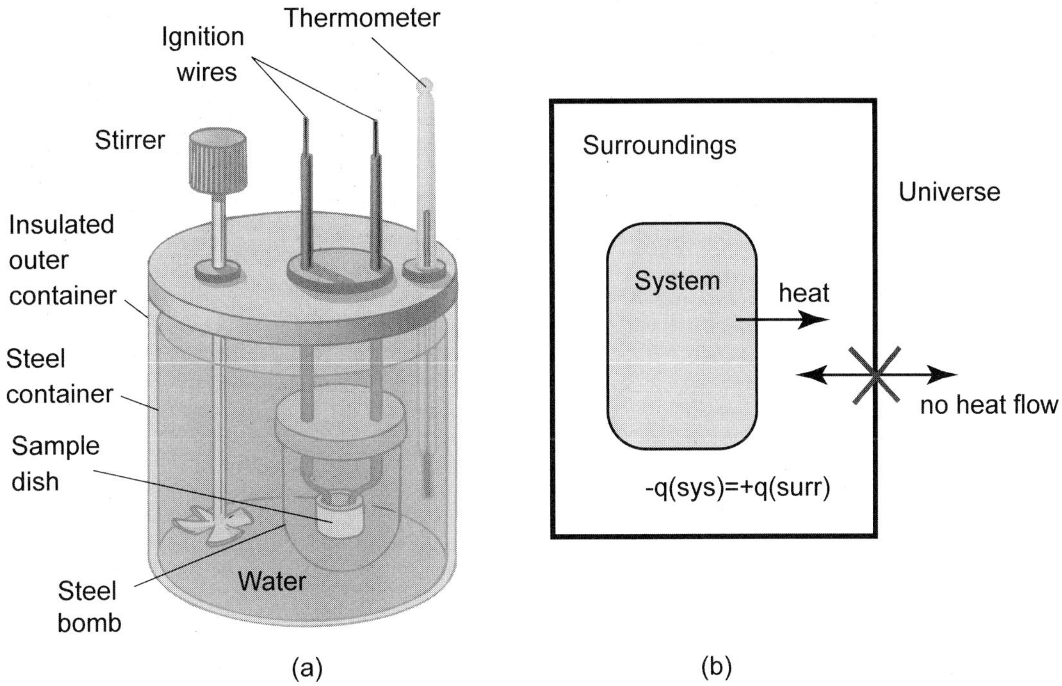

Figure 9.6 A bomb calorimeter is a fairly complicated piece of equipment, as shown in the diagram in (a). But the general premise of the device is simply to carry out a reaction at constant volume and with no heat flow between the calorimeter and the outside world. The diagram in (b) shows the standard choice of system and surroundings for a bomb calorimetry experiment. The system consists of the contents of the bomb itself. The surroundings include the bomb and the water bath surrounding it. We assume that no heat is exchanged with the rest of the universe, outside the insulated walls of the apparatus.

A two-step process is used to make the calorimetric measurement. The first step is a calibration, and involves generating a known amount of heat in the apparatus. The second step is the actual measurement, in which we determine the amount of heat absorbed or released in the reaction of a given amount of material. Two methods are common for performing the calibration: (1) burning a known amount of a well-characterized material or (2) resistive heating, in which a known amount of current is passed through a wire that heats due to its electrical resistance. By measuring the change in temperature of the surroundings resulting from a known heat input, the heat capacity of the entire calorimeter may be obtained.

There are many types of specialized calorimetry, but the general process of standardizing the measurement with a known item and then making the experimental measurement is generally applicaable.

$$\text{known amount of heat} = \text{heat capacity (calorimeter constant)} \times \Delta T \quad (9.6)$$

Note that in contrast to our earlier equations relating q and ΔT, there is no mass or number of moles term here for the quantity of material. The calorimeter constant is the heat capacity of a particular object (or set of objects) rather than of a material. It may help to think of it as "per calorimeter," and then realize that we have just one calorimeter. For someone who routinely uses a calorimeter this approach is much simpler than the alternative, which would be to keep track of the masses of steel, water, and other materials in the calorimeter. In the case of a bomb calorimeter, for example, the calorimeter constant is largely attributable to the water that surrounds the bomb, but also includes the thermometer, the stirring system, and the bomb itself.

Once the calorimeter constant is known, we are ready to use the calorimeter for our actual measurement. We place known amounts of reactant(s) into the calorimeter, initiate reaction, and then measure the temperature change that results in the calorimeter. The calorimeter constant allows us to determine the amount of heat released or absorbed in the reaction. Example Problem 9.5 shows how this process quantitatively.

Example Problem 9.5

A calorimeter is going to be used to compare the energy content of some fuels. The system is calibrated by using an electrical resistance heater to supply 100.0 J of heat. When this heat is added, the observed temperature rise is 0.850°C. When 0.245 g of a particular fuel is burned in this same calorimeter, the temperature rise is 5.23°C. What is the energy density of the fuel in kJ/g?

Strategy

The calibration step is used to determine the value of the calorimeter constant. Once this value is obtained the amount of heat evolved from the fuel can be determined by Equation 9.5 and the energy density is calculated by dividing this heat by the mass that generated it.

Solution

Step 1: Calibration

$$q = \text{C(calorimeter)} \times \Delta\text{T}$$

so

$$\text{C(calorimeter)} = q / \Delta\text{T}$$
$$= 100.0\ \text{J} / 0.850°\text{C}$$
$$\text{C(calorimeter)} = 118\ \text{J} / °\text{C}$$

Step 2: Determination of heat evolved by fuel

$$q = \text{C(calorimeter)} \times \Delta\text{T}$$
$$= 118\ \text{J} / °\text{C} \times 5.23°\text{C}$$
$$= 617\ \text{J}$$

Step 3: Calculation of the energy density

$$\text{energy density} = q / \text{m}$$
$$= 617\ \text{J} / 0.245\ \text{g}$$
$$= 2{,}520\ \text{J} / \text{g} = 2.52\ \text{kJ/g}$$

Check Your Understanding

Naphthalene ($C_{10}H_8$) is often used to calibrate calorimeters *via* its combustion reaction, which releases 5150.1 kJ/mole. A 1.05-g sample of naphthalene is burned in a calorimeter whose temperature rises 3.86°C. This system is then used to measure the energy density of a sample of coal. A 1.83-g sample causes a temperature change of 4.90°C. What is the energy density of the coal?

9.5 Enthalpy

The concept of heat flow is reasonably intuitive. Cool objects warm up at the expense of hotter objects. From a scientific standpoint, however, this natural process requires more careful examination. The conditions under which the heat flow occurs ultimately have an impact on the measurements that are made. For example, when one mole of octane, C_8H_{18}, burns to form $CO_2(g)$ and $H_2O(l)$ under constant volume conditions, 5.45 x10^3 kJ of heat is released. The same reaction under constant pressure conditions releases 5.48x10^3 kJ. As a percentage, the difference between the two conditions is small, but consider the engineering consequences for designing safety systems for large volume storage containers for octane. Such vessels may hold in excess of 1,000 moles of octane. Each mole releases 30 kJ more energy under constant pressure combustion situations versus constant volume. When thousands of moles burn, the difference in total energy release may be significant. Materials may be predicted to not melt for heat releases calculated assuming constant volume, whereas they may melt in the constant pressure situation. Why do the two cases release different heats, and how can we properly account for these differences?

Defining Enthalpy

Since there are differences associated with the reaction conditions, it has proven to be an advantage to define two mathematical functions for energy. Internal energy has already been introduced. By using the definition of internal energy we can show that under constant volume conditions the internal energy equals the heat flow. Start with

$$\Delta E = q + w$$

In chemical reactions we can usually consider only PV-work. When a gas expands, it does an amount of work equal to $P\Delta V$ on its surroundings. But if the expanding gas is our system, we want w to be the work done *on* the gas, and that will be $-P\Delta V$. So we can replace w in the equation above by $-P\Delta V$.

$$\Delta E = q - P\Delta V \qquad (9.7)$$

This equation now allows us to see how internal energy is related to heat. If the volume is held constant, ΔV is zero, so the second term is zero. All that remains is

$$\Delta E = q_v, \qquad (9.8)$$

where the subscript "v" is used to denote that the equation is correct *only* under constant volume conditions. Equation 9.8 tells us that if we do a calorimetry experiment at constant volume, we will get a direct determination of ΔE.

But what if we do the experiment at constant pressure? It would be useful to have a function that we could show to be equal to heat flow under constant pressure conditions. This function is **enthalpy**, and is defined as

$$H = E + PV \qquad (9.9)$$

Working from this definition, we can show that the change in enthalpy (ΔH) will be equal to the heat flow under constant pressure conditions. From the above definition, the change in enthalpy (ΔH) must be

$$\Delta H = \Delta E + \Delta(PV)$$

We can expand this by substituting for ΔE using Equation 9.7.

$$\Delta H = (q - P\Delta V) + \Delta(PV)$$

If the pressure is held constant, then the $\Delta(PV)$ term will simply become $P\Delta V$, giving

$$\Delta H = q - P\Delta V + P\Delta V$$

The second and third terms clearly cancel leaving the desired result.

$$\Delta H = q_p \tag{9.10}$$

The enthalpy change therefore equals the heat flow under constant pressure conditions. (This is denoted with a subscript "*p*," just as we did above for the constant volume case.)

Now we have two ways to define the heat flow into the system, corresponding to two different sets of conditions. For a process that takes place at constant volume, the measurable heat flow is equal to ΔE, the change in internal energy. And for a process occurring at constant pressure, the measurable *q* is equal to the change in enthalpy, ΔH. In many ways enthalpy is the more useful term, because constant pressure conditions are more common. A reaction carried out in a beaker in the general chemistry laboratory, for instance, occurs under constant pressure conditions (or very nearly so). Thus when we refer to the "heat of" a process, we are typically referring to a change in enthalpy, ΔH. As in previous definitions, ΔH refers to $H_{final} - H_{initial}$.

When heat is evolved from a system, the process is **exothermic** and the value of ΔH is less than zero. Exothermic processes "feel" hot; if you pick up the beaker in which an exothermic reaction is taking place, heat will flow from the reacting system into your hand. Conversely, when heat is absorbed by the system, the process is **endothermic** and the value of ΔH is greater than zero. Endothermic processes "feel" cold because they draw heat from their surroundings.

ΔH of Phase Changes

Heat flow into a substance does not always raise its temperature. If heat flows into an ice cube at 0°C, for example, the ice will melt to form liquid water at 0°C. (If heat continues to flow into the resulting water, its temperature will, of course, begin to rise.) How can the temperature remain constant in spite of the influx of heat? To understand this we need to remember that intermolecular forces are more extensive in the solid than in the liquid. So as the ice cube melts, energy must be expended to break some of the intermolecular attractions. The internal energy of the liquid water is higher than that of the solid ice, even though both are at the same temperature.

For similar reasons, there will be a heat flow associated with any phase change. Names of the phase changes between solids, liquids or gases are summarized in Figure 9.7. Since these phase changes will generally take place at constant pressure, the heat associated with them should be viewed as a change in enthalpy. Some phase changes are so common that their enthalpy changes have specific names and symbols assigned to them. The heat required to melt a substance is the heat of fusion, designated ΔH_{fus}. The enthalpy change for converting a liquid to a gas is known as the heat of vaporization, ΔH_{vap}. We know that to go from a liquid to a gas the molecules in a sample must all overcome whatever intermolecular forces are present. That means that energy must flow into the substance to vaporize it, so the heat of vaporization will always be positive. It follows that the reverse process, condensation, will release heat. The values for enthalpy changes in either direction have the same numeric value and differ only by the sign. Since the strength of the intermolecular forces varies from one substance to another, the magnitude of the enthalpy change for any phase transition will also depend on the substance. Values can be found in many standard tables, including one in Appendix D. Data for water are provided in Table 9.4 as an example.

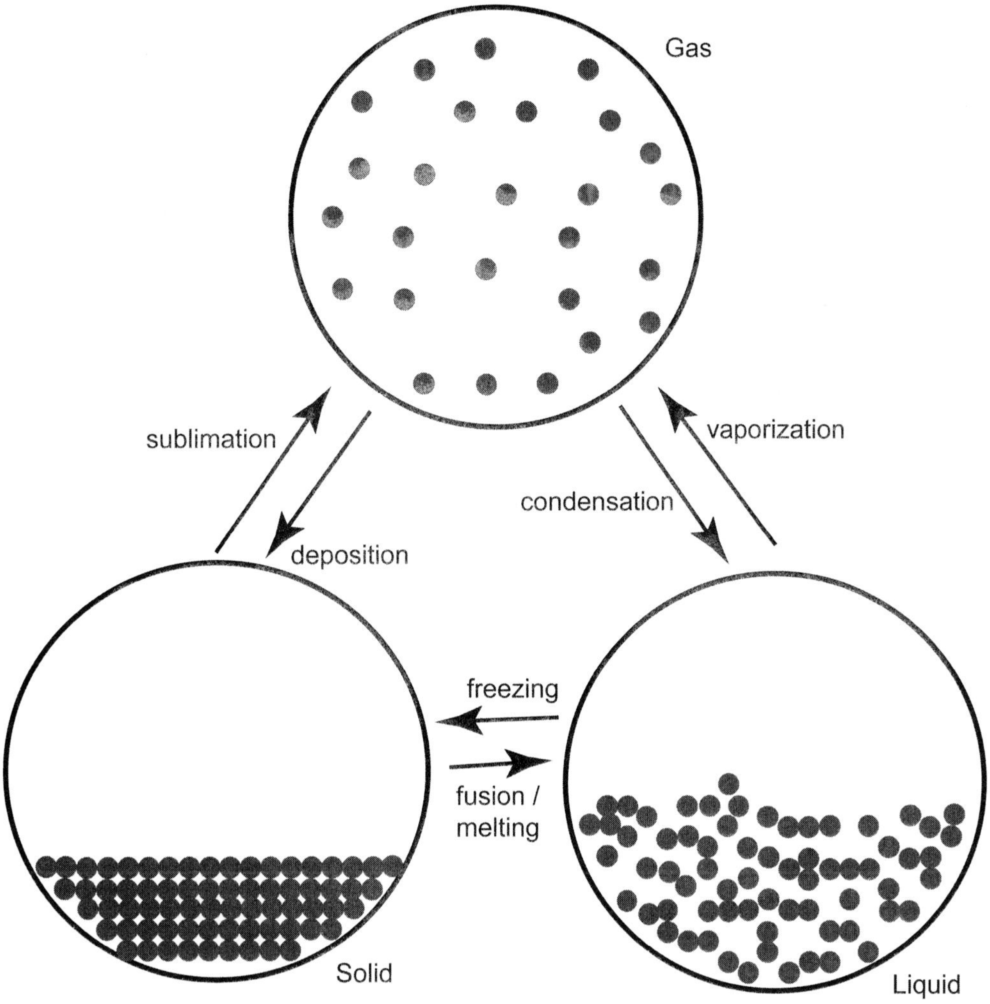

Figure 9.7 Microscopic level views of the solid, liquid, and gas phases are shown, along with the terms commonly used for the possible phase transitions.

Note that the numerical value of the enthalpy associated with the phase change between a solid and a liquid is always smaller than for a the change for a liquid to a gas.

Phase change	**Fusion**	**Freezing**	**Vaporization**	**Condensation**
Transition temperature	0°C	0°C	100°C	100°C
ΔH (J/mol)	$\Delta H_{fus} = 6009.5$	$\Delta H_{freeze} = -6009.5$	$\Delta H_{vap} = 4.07\text{x}10^4$	$\Delta H_{cond} = -4.07\text{x}10^4$

Table 9.4 Standard molar enthalpies and temperatures for phase changes of water.

These data would allow us to calculate the heat required for a phase transition involving any particular sample of water. Just as we saw when we considered temperature changes, the heat will depend on the amount of substance undergoing the transition: it will take more heat to melt a large block of ice than a small cube. The values shown in Table 9.4 are given in units of J/mol, so we might want to specify the amount of water in terms of moles, leading to the following equation.

$$\Delta H = n \ x \ \Delta H_{phase\ change} \qquad (9.11)$$

Here *n* is the number of moles, as usual. Notice that this relationship does not include ΔT, in contrast to Equations 9.4 and 9.5. This should make sense if we keep in mind that phase changes occur at a constant temperature, so there is no ΔT associated with the transition.

Example Problem 9.6

Calculate the enthalpy change when 240. g of ice freezes.

Strategy

The ΔH_{fus} value in table 9.4 is in J/mol, so the amount of ice must be converted into moles. Multiplying the number of moles by the ΔH_{fus} will provide the desired quantity.

Solution

$$240 \text{ g } H_2O \times (1 \text{ mol} / 18.0 \text{ g}) = 13.3 \text{ mole } H_2O$$

$$\Delta H = n \times \Delta H_{fus}$$

$$= 13.3 \text{ mole} \times 6009.5 \text{ J/mol}$$

$$= 8.01 \times 10^4 \text{ J}$$

Check Your Understanding

Calculate the enthalpy change when 14.5 g of water vapor condenses to form liquid water.

Just as for the case of a temperature change, there is no fundamental reason why we could not express the quantity of material in terms of mass rather than moles. But it is essential that we be consistent in the choice of units: if the quantity is in mass, then we will need an enthalpy change value that is expressed in terms of mass as well. Since you know the molar mass of water, you should easily be able to convert the values in Table 9.4 from J/mol to J/g. And if you repeat the calculations from the last example that way you should get the same result. In most engineering fields it is more common to use mass rather than moles to specify the amount of a substance. But the underlying concept is the same no matter which units we use.

At this point we can combine what we've learned about heat flow for both phase changes and temperature changes to obtain the enthalpy change for converting water from ice to liquid and then steam. A plot of the temperature versus heat flow for such a process is shown in Figure 9.8. The heat transfers corresponding to sections (1), (3) and (5) are for changes of temperature, while sections (2) and (4) represent phase changes.

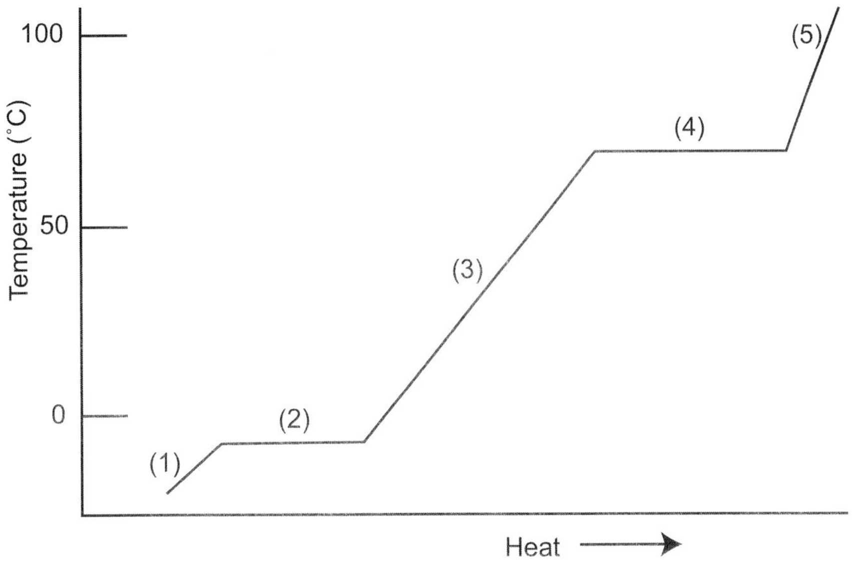

Figure 9.8 The figure shows how the temperature of a sample of H_2O varies as it absorbs heat. Here the sample is originally ice at -20°C. In the segment of the graph marked as (1), the temperature of the ice rises until it reaches the melting point at 0°C. In region (2), ice melts at a constant temperature of 0°C until it is all converted into liquid. In region (3), only liquid water is present, and its temperature rises until it reaches the boiling point at 100°C. In region (4), the water boils at a constant temperature of 100°C until it is all converted into steam. Finally, in region (5), the resulting steam continues to absorb heat, and its temperature rises.

Vaporization and Electricity Production

The amount of energy required to convert water from a liquid to a gas, through the series of processes shown in Figure 9.8, is put to good use in the conversion of chemical energy into electricity. The basic features of a fossil fuel powered electricity plant are shown schematically in Figure 9.9.

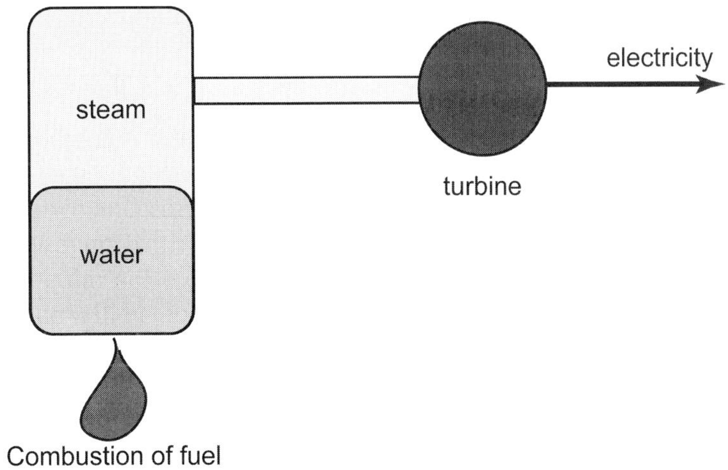

Figure 9.9 The important elements of a standard electric power plant are shown in this schematic diagram. The process exploits the large heat of vaporization of water.

You might like to speculate what would have to happen to the design of power plants, etc., if something with a much smaller heat of vaporization than water was used.

Chemical energy is initially released as heat through the combustion of the fuel, typically coal or natural gas. The goal of the power plant is to convert as much of this energy as possible into electricity. The critical step in this process is to "trap" heat given off in the combustion reaction. Water is the chosen material for this process because it has a large heat of vaporization. You'll recall from Chapter 8 that intermolecular forces are unusually strong in water because of hydrogen bonding. These hydrogen bonds between water molecules in the liquid are responsible for the large value of ΔH_{vap}. If a material with a small heat of vaporization were substituted for water, much more of it would be required to absorb the same amount of heat. The large heat of vaporization coupled with the relative abundance of water has led to its widespread use in the energy sector of the economy.

Heat of Reaction

So far we have considered enthalpy changes for simple physical processes like temperature changes and phase transitions. But the importance of chemistry to the energy economy arises from the fact that there are enthalpy changes associated with chemical reactions as well. This enthalpy change is commonly referred to as the **heat of reaction**. Since many reactions are carried out under constant pressure conditions, this term is sensible, if slightly imprecise.

Bonds and Energy

Chemical reactions involve energy changes because chemical bonds are broken and formed when reactants are transformed into products. Consider a fairly simple reaction - the combustion of methane.

$$CH_4(g) + 2O_2(g) \rightarrow CO_2(g) + 2H_2O(l)$$

We are familiar with all of these molecules, so we can easily write Lewis structures for them.

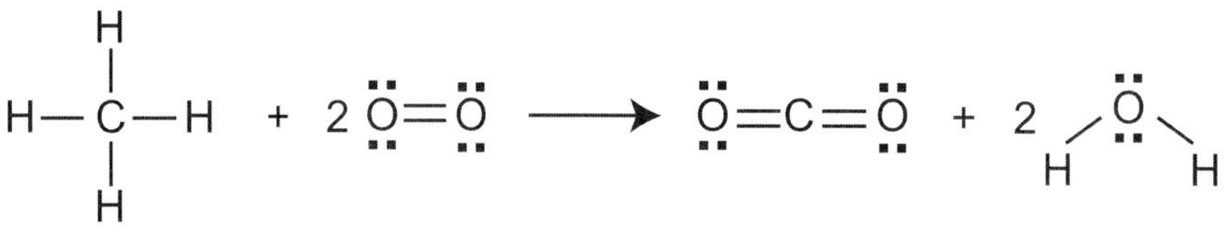

On the reactant side of the equation, we have four C-H single bonds and two O=O double bonds. On the product side we have two C=O double bonds and four O-H single bonds. Over the course of the reaction, the bonds in the reactants must all be broken. And that will require an input of energy. On the other hand the bonds in the products must all be formed, and that will release energy. (Recall that bond formation is always exothermic while bond breaking is always endothermic.) If the energy released in forming new bonds is greater than the energy required to break the original bonds, than the overall reaction should be exothermic. Conversely if the energy needed for bond breaking is greater than that released in bond making, than the reaction will be endothermic. Since methane is the principle component of natural gas, we know that burning it must release energy. So we should anticipate that this particular reaction must be exothermic. And calorimetric measurements bear this out; ΔH for the combustion of one mole of methane is -890.4 kJ.

We might attempt to use estimates for the individual bond energies to calculate an approximate value of ΔH for a reaction. And indeed this is sometimes done to obtain an estimate for reactions involving compounds where little or no thermochemical data is available. The accuracy of this sort of approach is not very good, however. And as we will see in the following section, a simple and accurate alternative exists in most cases.

The overall energetics of a chemical reaction can be summarized in a **thermochemical equation**. For the combustion of methane, the thermochemical equation could be written as

$$CH_4(g) + 2O_2(g) \rightarrow CO_2(g) + 2H_2O(l) \qquad \Delta H = -890.4 \text{ kJ}$$

This equation tells us two important things. (1) The value of the heat of reaction is for the equation as written; if more fuel is burned, more heat will be released. So if the stoichiometric coefficients are multiplied by some factor, the heat of reaction must also be multiplied by that factor. Thus another thermochemical equation that deals with the combustion of methane is

$$2CH_4(g) + 4O_2(g) \rightarrow 2CO_2(g) + 4H_2O(l) \qquad \Delta H = -1780.8 \text{ kJ}$$

(2) The thermochemical equation provides information about whether a reaction is exothermic or endothermic upon inspection, because ΔH carries a sign. We know that the combustion of methane must be exothermic, and the negative sign on ΔH confirms this.

Heats of Reaction for Some Specific Reactions

Some classes of chemical reactions are sufficiently common or useful that they have been assigned their own label for heats of reaction. The example reaction for methane we have already used falls into one such category, combustion. Since combustion is a common part of the energy economy, enthalpy changes for combustion reactions are used to compare various fuels. These heats of combustion are usually designated as ΔH_{comb}. Similarly, the neutralization reactions between acids and bases have heats of reaction that are called heats of neutralization, ΔH_{neut}.

Another class of reactions, formation reactions, has specially designated heats of reaction known as heats of formation, ΔH_f°, even though many of them are never actually carried out in practice. Their importance is due to the computational utility derived from the following definition. A **formation reaction** is the chemical reaction by which 1 mole of a compound is formed from its elements in their standard states. The "standard state" is the most stable form of the element at room temperature (25°C) and pressure (1 atm). The formation reaction for carbon monoxide is

The science of thermodynamics often defines things for the advantage of mathematical ease. We will soon see how formation reactions are an example of this strategy.

$$C(s) + 1/2\, O_2\,(g) \rightarrow CO(g) \qquad \Delta H^\circ = \Delta H_f^\circ(CO)$$

Because we require that there be just one mole of CO on the right hand side, we must use a fractional coefficient on the O_2 in order to balance the equation, even though this may look odd. Two very common errors when writing formation reactions are to have elements that are not in their standard form or to have more than one mole of product compound. For carbon monoxide these mistakes might produce either of the following *incorrect* "formation" reactions.

$$C(s) + O(g) \rightarrow CO(g) \qquad \Delta H^\circ \neq \Delta H_f^\circ(CO)$$

$$2C(s) + O_2(g) \rightarrow 2CO(g) \qquad \Delta H^\circ \neq \Delta H_f^\circ(CO)$$

Although these are both valid chemical equations, they do not describe the formation reaction for CO. The first reaction is not correct because oxygen is not shown as a diatomic molecule, its standard state. The second reaction is not correct because two moles of carbon monoxide are formed. Formation reactions are quite useful for determining heats of reaction, as we will see in the following section.

9.6 Hess's Law and Heats of Reaction

The enthalpy changes associated with chemical reactions play an important role in a wide range of situations. Comparisons of possible new fuels or explosives are obvious examples. In many cases it might be desirable to determine the enthalpy change for a reaction without necessarily carrying out the reaction itself. The synthesis of a new explosive might be very difficult and dangerous, for example, so it would help to know in advance if its potential usefulness might justify the effort. In other cases it may simply be very difficult to perform a direct calorimetric determination of a heat of reaction. For a variety of reasons, we often need to obtain heats of reaction indirectly.

Hess's Law

To obtain information about heats of reaction in an indirect manner, we use what is usually called **Hess's Law**: the enthalpy change for any process is independent of the particular way the process is carried out. The underlying concept upon which this idea is built is that enthalpy is a **state function.** A state function is a variable whose value depends only on the state of the system and not on its history. When you drive your car, your location is a state function but the distance you have traveled to get there is not. For a chemical reaction, the concept of state functions can be very important. We rarely if ever know the microscopic details of how reactant molecules are actually converted into product molecules. But it is relatively easy to determine what the reactants and the products were. Since enthalpy is a state function, if we can find a way to determine the change in enthalpy for *any* particular path that leads from the reactants to the desired products, we will know that the result will apply for *whatever* actual path the reaction may take. The block diagram in Figure 9.10 illustrates the use of this concept.

You could use your book for another analogy of a state function. If you drop the book to the floor from a height of one meter, it doesn't matter how you first achieved that height. If you lift it over your head first and then bring it to one meter the "sound" it acheives when you drop it will be - on average - the same as if you lift it directly to that height.

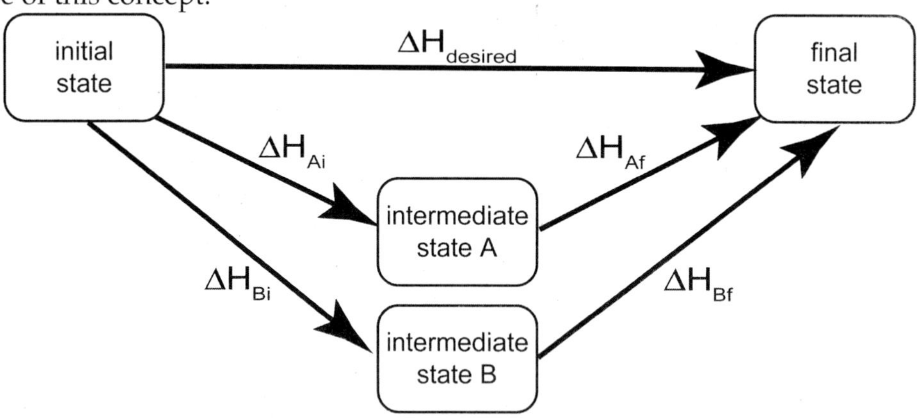

Figure 9.10 Conceptual diagram representing Hess's Law. Because enthalpy is state function, we can choose any convenient path from the initial state to the final state, and use that path to calculate the enthalpy change.

In this case, the desired enthalpy change can be obtained via either of two alternative pathways.

$$\Delta H_{desired} = \Delta H_{Ai} + \Delta H_{Af}$$

$$\Delta H_{desired} = \Delta H_{Bi} + \Delta H_{Bf}$$

We can further develop this idea by including an energy axis in the picture. Such a figure is called an **enthalpy diagram**. Figure 9.11 shows an enthalpy diagram for the combustion of methane. This combustion may either be complete, producing carbon dioxide directly, or incomplete producing carbon monoxide. (The production of carbon monoxide from burning of methane presents a danger for natural gas burning furnaces that are improperly vented.)

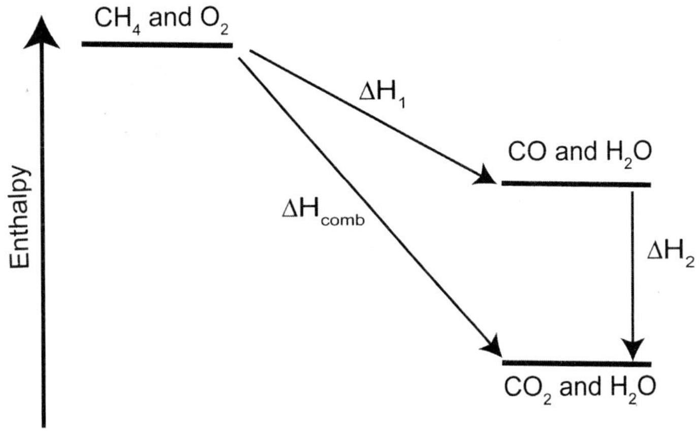

Figure 9.11 Enthalpy diagram for the combustion of methane. If we know the values of ΔH_1 and ΔH_2, we can use them to calculate ΔH_3.

Practically speaking, any time we can write a chemical reaction as a group of sequential reactions whose net product is the same as the desired product, we may evaluate the enthalpy of the desired reaction by using the sequential ones. Example Problem 9.7 shows how such a series of reactions is used.

Example Problem 9.7

Sulfur trioxide reacts with water to form sulfuric acid, a major contributor to "acid rain." One origin of SO_3 is the combustion of sulfur that is present in small quantities in coal, according to the following equation.

$$S(s) + 3/2\ O_2(g) \rightarrow SO_3(g)$$

Given the thermochemical information below, determine the heat of reaction for this reaction.

$$S(s) + O_2(g) \rightarrow SO_2(g) \qquad \Delta H° = -296.8\text{ kJ}$$

$$2SO_2(g) + O_2(g) \rightarrow 2\ SO_3(g) \qquad \Delta H° = -197.0\text{ kJ}$$

Strategy

We must construct a pathway that results in the desired reaction using reactions whose energetics are known. In this case we can see that $SO_2$is formed in the first reaction we are given, and then consumed in the second. It is critical to pay attention to the precise stoichiometry of the given and desired reactions. The reaction of interest produces just one mole of SO_3, while the second reaction we are given produces two moles.

Solution

We start with the frst of the two given reactions.

$$S(s) + O_2(g) \rightarrow SO_2(g) \qquad \Delta H° = -296.8\text{ kJ}$$

Next we multiply the second given reaction by 1/2. The 1/2 is needed because the desired reaction has only one mole of SO_3.

$$1/2 \times (2\ SO_2(g) + O_2(g) \rightarrow 2\ SO_3(g)\ \Delta H° = -197.0\text{ kJ})$$

This gives an equation in which 1 mole of SO_2 is consumed, along with its enthalpy change.

$$SO_2(g) + 1/2\ O_2(g) \rightarrow SO_3(g) \qquad \Delta H° = -98.5\ kJ$$

Adding this to the first reaction above gives the desired amount of SO_3 and the heat of reaction.

$$S(s) + O_2(g) \rightarrow SO_2(g) \qquad \Delta H° = -296.8\ kJ$$
$$SO_2(g) + SO_2(g) \rightarrow SO_3(g) \qquad \Delta H° = -98.5\ kJ$$
$$S(s) + 3/2\ O_2(g) \rightarrow SO_3(g) \qquad \Delta H° = -395.3\ kJ$$

Check Your Understanding

Use the following thermochemical equations as needed to find the heat of formation of diamond.

$$C\ (diamond) + O_2\ (g) \rightarrow CO_2\ (g) \qquad \Delta H° = -395.4\ kJ$$
$$2\ CO_2\ (g) \rightarrow 2\ CO\ (g) + O_2\ (g) \qquad \Delta H° = 566.0\ kJ$$
$$C\ (graphite) + O_2\ (g) \rightarrow CO_2\ (g) \qquad \Delta H° = -393.5\ kJ$$
$$2\ CO\ (g) \rightarrow C\ (graphite) + CO_2\ (g) \qquad \Delta H° = -172.5\ kJ$$

Formation Reactions and Hess's Law

Scientific reference books like the Handbook of Chemistry and Physics contain much more tabulated data of heats of formation and other thermodynamic variables.

The type of calculation shown in the previous example has occasional application in chemistry. But Hess's Law is also useful in another more common approach. Heats of formation for many substances are widely tabulated (see Appendix E). And Hess's Law allows us to use these tabulated values to calculate the enthalpy change for virtually any chemical reaction. The block diagram shown in Figure 9.12 shows how this useful scheme arises.

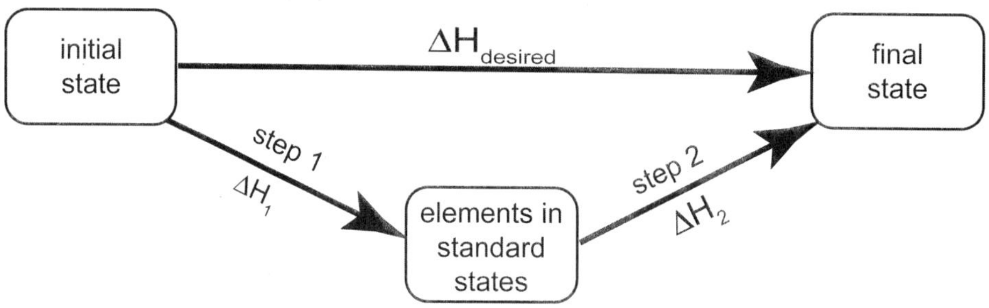

Figure 9.12 This conceptual diagram shows how to use tabulated enthalpies of formation to calculate the enthalpy change for a chemical reaction. We imagine that first the reactants are converted to elements in their standard states, and then those elements recombine to form the products. Because enthalpy is a state function, we do not need to know anything about the actual pathway the reaction follows.

Since Step 1 is the opposite of the formation reaction of the reactants, the enthalpy change of the process is $-\Delta H_f°$(reactants). Similarly, Step 2 has an enthalpy change of $\Delta H_f°$(products). Remember, however, that the formation reaction is defined for the generation of *one mole* of the compound. Consequently, to use tabulated heats of formation we must multiply by the stoichiometric coefficient in the reaction to account for the number of moles of reactants used or products generated. Taking these factors into account provides one of the more useful equations in thermochemistry.

$$\Delta H^\circ = \sum_i \nu_i \Delta H_f^\circ(\text{product})_i - \sum_j \nu_j \Delta H_f^\circ(\text{reactant})_j \tag{9.12}$$

In this equation we have designated the stoichiometric coefficients with the Greek letter ν. The first summation is over all of the reactants while the second is over all of the products. Two example problems are provided to show how heats of formation are used in understanding the thermochemistry of reactions useful in the production of energy.

Example Problem 9.8

What is the heat of combustion of one mole of propane, C_3H_8, to form gaseous carbon dioxide and liquid water?

Strategy

We can use tabulated value for the heats of formation of the reactants and products to determine the desired value. First we must write the balanced chemical equation for the process. Then we can use Equation 9.12 to calculate the heat of the reaction (in this case the heat of combustion) by looking up heats of formation in the tables in Appendix E. The needed stoichiometric coefficients will be obtained from the balanced equation. Note that the heat of formation of an element in its standard state – like the O_2 in this equation - will always be zero.

Solution

$$C_3H_8(g) + 5\ O_2(g) \rightarrow 3\ CO_2(g) + 4\ H_2O(l)$$

$$\begin{aligned}\Delta H^\circ &= 3\text{ mol }\Delta H_f^\circ(CO_2) + 4\text{ mol }\Delta H_f^\circ(H_2O) - 1\text{ mol }\Delta H_f^\circ(C_3H_8) - 5\text{ mol }(0)\\ &= 3\text{ mol }(-393.51\text{ kJ/mol}) + 4\text{ mol }(-285.83\text{ kJ/mol}) - 1\text{ mol }(-103.85\text{ kJ/mol})\\ &= -2220\text{ kJ}\end{aligned}$$

Discussion

We use kJ/mol as units for the heat of formation of a substance. But in writing the enthalpy change of a chemical reaction, we will use kJ as our preferred unit, and not kJ/mol. The reaction in this example illustrates why we do this. The value we calculated, ΔH° = -2220 kJ, is for a reaction in which one mole of propane reacts with five moles of oxygen to form three moles of carbon dioxide and four moles of water. So if we were to say "-2220kJ/mol," we would need to carefully specify per mole of what substance. We choose to write the ΔH value in kJ, with the understanding that it refers to the reaction as written. This is also dimensionally consistent with Equation 9.12, provided that we treat the stoichiometric coefficients as carrying units of moles. You may see other texts that refer to values as being "per mole of reaction."

Check Your Understanding

Use heat of formation data from Appendix E to calculate ΔH° for the following reaction.

$ClO_2\ (g) + O\ (g) \rightarrow ClO\ (g) + O_2\ (g)$

Example Problem 9.9

Ethanol, C_2H_5OH, is used in some blends of gasoline. It has a heat of combustion of 1366.8 kJ/mol. What is the heat of formation of ethanol?

Strategy

We know the relationship between the heat of combustion for a reaction and the heats of formation of the substances involved. We will need a balanced chemical equation for the combustion, and we must recognize that the combustion of ethanol yields the same products as hydrocarbon combustion. We can write the balanced equation and then use Equation 9.12 to determine the desired quantity.

Solution

$$C_2H_5OH(l) + 3O_2(g) \rightarrow 2CO_2(g) + 3H_2O(l) \qquad \Delta H° = -1366.8 \text{ kJ}$$

Using Equation 9.12:

$$\Delta H° = 2\,\Delta H_f°(CO_2(g)) + 3\,\Delta H_f°(H_2O(l)) - \Delta H_f°(C_2H_5OH(l)) - 3\,\Delta H_f°(O_2(g))$$

$$\Delta H° = 2\,\Delta H_f°(CO_2(g) + 3\,\Delta H_f°(H_2O(l)) - \Delta H_f°(C_2H_5OH(l)) - 3\,(0)$$

$$-1366.8 \text{ kJ} = 2 \text{ mol}(-393.51 \text{ kJ/mol}) + 3 \text{ mol}(-285.83 \text{ kJ/mol}) - \Delta H_f(C_2H_5OH(l))$$

$$\Delta H_f°(C_2H_5OH(l)) = -277.7 \text{ kJ/mol}$$

Check Your Understanding

Incomplete combustion of hydrocarbons leads to the generation of carbon monoxide rather than carbon dioxide. Improperly vented furnaces can poison people who live in an affected building due to the toxicity of CO. Calculate the heat of reaction for the incomplete combustion of methane, CH_4, to yield liquid water and CO(g).

9.7 Energy and Stoichiometry

The ability to predict the energetic consequences of a chemical reaction is an important skill in chemistry, with many practical applications. Writing a thermochemical equation allows for a treatment of energy that is similar in many ways to the stoichiometry problem solving we learned in Chapter 4. For an exothermic reaction, we can treat energy as a product, while in an endothermic reaction energy can be thought of as a reactant. We must keep in mind that the stated value of ΔH corresponds to the reaction taking place exactly as written, with the indicated numbers of moles of each substance reacting.

The emphasis on the importance of the mole as the heart of the stoichiometry problem remains when we solve problems involving the energy of reactions. So if we wanted to calculate the amount of energy released by burning a given mass or volume of methane, we would start by converting that quantity into moles. Then we could use the thermochemical equation to relate the amount of energy to the number of moles of methane actually burned. In stoichiometry problems we have used the balanced chemical equation to effectively convert from the number of moles of one compound to the number of moles of another. Now we can use the thermochemical equation to provide a means to convert between number of moles of a reactant or product and the amount of energy released or absorbed. Schematically we can view this method of addressing the problem as shown in Figure 9.13. The conversion factors needed to obtain the number of moles are the same as earlier: molar mass, density, gas pressure or volume, etc.

In stoichiometry problems we devise ratios from the balanced equation and we sometimes simplify those ratios. For thermochemical equation, we must remember that the enthalpy is related to the specific stoichiometry of the balanced equation.

Consider the reaction between nitrogen gas and oxygen to form nitric oxide. This occurs regularly as a side reaction when hydrocarbons are burned as fuel. The exothermic combustion reaction that powers an automobile uses air rather than pure oxygen, so there is always a large amount of nitrogen present. At the high temperatures produced in a running engine, some of the nitrogen reacts with oxygen to form NO.

$$N_2(g) + O_2(g) \rightarrow 2NO(g) \qquad \Delta H° = 180.5 \text{ kJ}$$

The nitric oxide gas formed in this way is an important species in several pollution pathways and is itself an irritant molecule. For example, NO(g) reacts further with oxygen to form nitrogen dioxide, which is brown and is largely responsible for the dark haze typically associated with urban smog. Example Problem 9.10 shows how this type of equation can be used to determine heats associated with specific reactions.

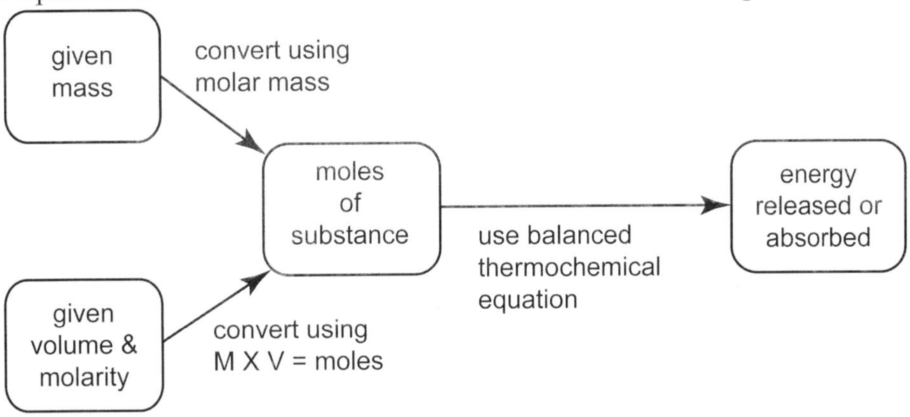

Figure 9.13 This flow chart details the sequence of steps needed to calculate the amount of energy released or absorbed when a chemical reaction is carried out using a given amount of material.

Example Problem 9.10

An engine generates 15.7 g of nitric oxide gas during a laboratory test. How much heat was absorbed in producing this NO?

Strategy

The thermochemical equation for this reaction is shown above. This equation provides a link to convert between the amount of NO formed and the energy absorbed. As in any other stoichiometry problem we will work with moles of substance and not directly with the mass (as is given in the problem). Note that the stated ΔH value is for the production of two moles of NO, since two moles appear in the equation.

Solution

$$15.7 \text{ g NO} \times (1 \text{ mole NO} / 30.0 \text{ g NO}) = 0.523 \text{ mole NO}$$

$$0.523 \text{ mole NO} \times (180.5 \text{ kJ} / 2 \text{ mole NO}) = 47.2 \text{ kJ absorbed}$$

Check Your Understanding

If 124 kJ of heat is absorbed by a reaction that forms NO from nitrogen and oxygen, what mass of NO(g) must have been produced? What mass of nitrogen was consumed?

This type of manipulation can provide insight into the relative merits of various fuels as well.

Energy Density and Fuels

When considering the economic merits of a particular fuel, there are several factors that should be considered. Among the characteristics that are typical of a useful fuel are items such as the availability of technology for extraction, the amount of pollution released upon combustion, and the relative safety of the fuel. (Since fuels are burned there is always some danger of unintended or uncontrolled combustion.) From an economic point of view the ease of transportation of the fuel is a key factor in this consideration. Fuels that are expensive or dangerous to get to the consumer are less attractive than those that can be delivered at more modest cost. Transportation costs for commodities are dictated largely by the mass that is transported. Because of this important cost, one key feature of a fuel is its **energy density**, the amount of energy that can be released per gram of fuel burned.

Fuel	Energy Density (MJ/kg)
Hydrogen	142.00
Methane	55.50
Octane	47.90
Propane	50.30
Aviation gasoline	43.1
Coal (anthracite)	31.4
Diesel fuel	45.3
Oil, crude (petroleum)	41.9
Oil, heating	42.5
Automotive Gasoline	45.8
Kerosene	46.3
Wood, oven dry	20.0

Table 9.5 Energy densities for a few possible fuels.

The energy densities of some fossil fuels, as well as some alternative fuels, are shown in Table 9.5. From this information we can see why petroleum is such a prominent source of energy. Consider the positive factors for petroleum use. (1) It is a liquid, which makes it easy to transport and deliver to the customer. (2) It is relatively safe. Explosions can occur, but they are preventable using designs that have a modest cost. (3) The products of combustion are gases. Where a liquid fuel is easy to transport, gaseous combustion products are easy to dispose of. (4) It has a high energy density. In addition to the issues of transportation cost already discussed, this means that automobile engines can be designed to produce relatively large amounts of energy without having the fuel increase the overall weight of the vehicle by very much. Imagine the complications if a tank of gas for your car weighed as much as the car itself. You would use a significant portion of your gas just to carry its own weight. And vehicle safety would be significantly compromised if the mass of the car could increase by a large amount just because the gas tank was full.

9.8 Insight into Batteries

Throughout this chapter we have concentrated on reactions that absorb or release energy in the form of heat. Although this is the most common manifestation of energy in chemical reactions, there are also many reactions that interconvert between chemical energy and other forms of energy. Some reactions, like those occurring in glow sticks or fireflies, release energy in the form of light rather than heat. And you exploit reactions that convert chemical energy into electrical energy every time you turn on a battery-powered device. At the heart of any battery is a chemical reaction that releases energy. How can chemical energy be converted directly into electrical energy? And why do some batteries simply "die" while others can be recharged more or less indefinitely? We'll take a closer look at the chemistry of batteries – called electrochemistry - in Chapter 13, but for now we can use ideas from thermochemistry to get a glimpse at the answers.

Electrical energy arises from moving charge. In practice, in order to power any conventional electrical device we need to be able to move electrons through a wire. So if a chemical reaction is going to directly produce useful electrical energy, it will need to function as a source of moving electrons. Since electric charge must always be conserved, these electrons must move through a closed cycle. Electrons "released" at one terminal of the battery must be "recaptured" at the opposite terminal in order to complete this circuit. What sort of chemical reactions would make good candidates for such a process? **Oxidation-reduction** (or **redox**) reactions involve the transfer of electrons from one species to another. A battery, then, is really a cleverly engineered application of a redox reaction. Successful battery design requires that the electrons can be made to flow through an external circuit as they are being transferred from one reacting species to another.

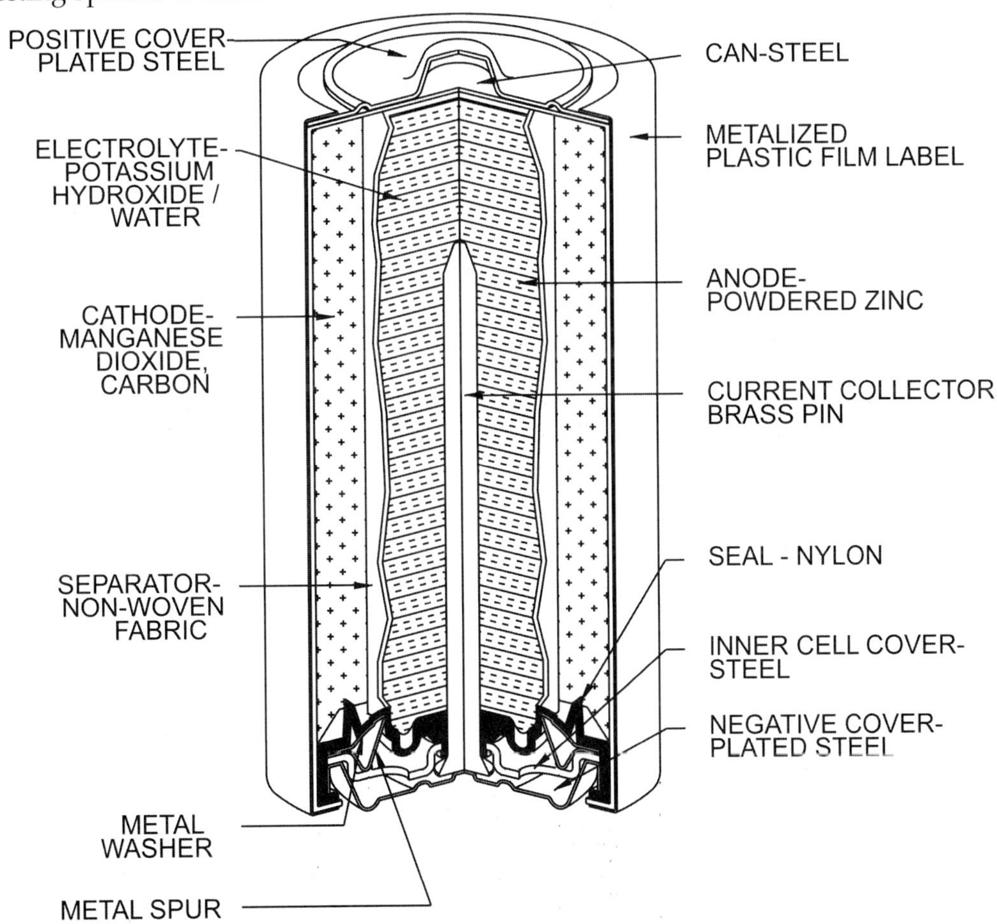

Figure 9.14 Construction of a typical alkaline battery is illustrated. (Figure courtesy of Eveready Battery Company.)

In a standard "alkaline" battery, the important chemical reaction is

$$Zn(s) + MnO_2(s) + H_2O(l) \rightarrow ZnO(s) + Mn(OH)_2(s)$$

How does this reaction involve the transfer of electrons? Look at the zinc and manganese on each side of the equation. Zinc metal reacts to form ZnO, where it appears in the +2 oxidation state. So each atom of zinc that reacts "loses" two electrons as it is converted from the 0 oxidation state to the +2 oxidation state. Manganese goes from the +4 oxidation state in MnO_2 to the +2 state in $Mn(OH)_2$, so each atom "gains" two electrons. Thus the net transfer of electrons is from zinc to manganese. The design of the battery (as shown in Figure 9.14) physically separates the Zn and MnO_2 from one another, assuring that the transferred electrons must flow through the external circuit in order for the reaction to proceed.

The redox reaction in a battery has its own characteristic thermochemistry, just like the combustion reactions we have already looked at. The energy change associated with the chemical reaction fixes the voltage of an electrochemical cell. Some batteries, like the one that starts your car, are actually combinations of multiple cells, arranged so that the effective voltage of the battery is increased.

Batteries are sometimes classified as either "primary" or "secondary." A primary cell like the typical alkaline battery becomes useless once the underlying chemical reaction has run its course. The lifetime of the battery is determined by the amounts of reactants present, so a relatively large D-cell would last longer than a small AA-cell in the same application. The battery "dies" when the reactants are completely converted into products, bringing the reaction to a halt. In practice, the voltage output of a battery usually begins to decrease near the end of its lifetime, and the cell will generally fail before the reactants are completely consumed.

A secondary battery is one that can be recharged, allowing for a much longer life cycle. In order to make a rechargeable battery, we must be able to reverse the redox reaction, converting the products back into reactants. Since we know that the cell reaction must be exothermic to supply energy, we can also see that the reverse reaction must be endothermic. So some external energy source will be needed to "push" the reaction back towards the reactants. This is the role of a battery charger, which typically uses electrical energy from some other source to drive the redox reaction in the energetically "uphill" direction. The question of whether or not a given reaction can be made reversible in this way determines whether a particular type of battery can easily be recharged.

Battery research for applications such as electric cars remains an important place for future energy (and economic) development.

Many new types of batteries have been introduced in recent years, as demand for portable electronic devices like cell phones, electronic organizers, and laptop computers has soared. What factors influence the choice of battery types for a particular application? One of the most important considerations is energy density, just as we discussed earlier for fuels. Table 9.6 summarizes typical characteristics and uses of several types of batteries.

Battery manufacturers usually report energy densities in units of W h/kg. Since a watt is 1 J/s, and an hour is 3600 s, 1 W h = 3600 J. Cost also plays a key role in the selection of a battery type. While some people might be willing to pay hundreds of dollars for a good battery for an expensive laptop computer, you are not likely to want that same battery for your MP3 player. Battery lifetime is also a factor. Digital cameras tend to require more power than traditional film cameras, for example. So consumers are much more willing to pay extra for rechargeable batteries for digital cameras, anticipating cost savings over time.

Attribute	Zinc-Air	Alkaline	Lithium
Energy Density	High	Medium	High
Energy Storage	High	Medium	Medium
Cost	Low	Low	High
Safety	High	High	Medium
Environment	High	High	Medium

Primary Battery Comparison

Attribute	Nickel Cadmium	Nickel Metal Hydride	Lithium Ion
Energy Density	Low	Medium	High
Energy Storage	Low	Medium	Medium
Cycle Life	High	High	High
Cost	Low	Medium	High
Safety	High	High	Medium
Environment	Low	Medium	Medium

Rechargeable Battery Comparison

Table 9.6 The preferred type of battery for a given application depends on a number of factors. This table offers a qualitative comparison of some of the important characteristics of common primary and rechargeable battery types. Many of these parameters can be measured quantitatively, so an engineer could make a careful comparison before specifying a choice of battery type.

Focus on Problem Solving

Question

During the nuclear power reactor accident at Three Mile Island an unknown amount (x kg) of fuel pellets melted, and molten droplets fell into water at the bottom of the reactor. Assume the pellets are pure UO_2 and were initially at 900°C and the water was initially at 8°C. Further assume that sensors indicated that the final temperature of the water was 85°C.

(a) What information would you have to look up in order to determine the mass of fuel pellets that melted?

(b) What would you have to know in order to use this information to determine the percentage of the fuel that melted in the accident? How would you calculate the percentage?

Answer

(a) This question would use the heat relationship from Equation 9.4 and realize that it applies to both the water and the fuel pellets. Thus

$$q_{pellet} = -q_{water}$$

and

$$(mc\Delta T)_{pellet} = -(mc\Delta T)_{water}$$

In this equation we can see there are six possible variables. Which do we know and which can be looked up?

ΔT: For both cases, this value is known. The initial temperature of both the water and the fuel pellets was given in the problem and both share the final temperature of 85°C.

c: The specific heat is not given for either the water or the fuel pellets. This is information that must be looked up, and that should be available from standard reference sources.

m: This is the key unknown for the fuel pellets – and clearly the goal of the exercise is to obtain it. For water, we also do not have the information given. We must find some way to look up this value, or else we must make a reasonable estimate.

Answer for (a): We would have to look up the heat capacity of water and fuel pellets and the mass of water in the reactor (initially at 8°C) to be able to answer this questions.

(b) This part of the question assumes we can accomplish the tasks in part (a) and then take it one step further. In this case we need to identify any other unknowns in the problem and ways to address them in order to answer this question.

The first question that must be addressed is, "What are the sources of heat that raised the temperature of the water?" Were the fuel pellets the only thing that melted? If they were not, we must account for other materials in our handling of this problem. If we assume everything that melted was fuel pellets, and some was not, we would over estimate the percentage of the fuel pellets that melted in the accident.

After we have subtracted the contribution from other materials and determined the mass of the fuel pellets that melted, to obtain a percentage, we must know the total mass of fuel pellets present. This number must be looked up or estimated based on some reliable information.

Answer for (b): The percentage melted can be calculated using the equation,

$$\text{Percent melted} = \frac{m_{melted}}{m_{total}} \times 100\%$$

Here we are assuming we have been able to calculate both the mass melted and the total mass present for the fuel pellets.

Chapter 9
Problems & Exercises

Energy and the World Economy

9.1 List reasons you can envision as to why there is a connection between the amount of energy used by a country and its economic development.

9.2 What are the advantages of electricity as a type of energy that make it worth generating despite the sizable losses that occur during this process?

9.3 What differences (if any) would you expect to find between the energy usage patterns of the United States shown in Figures 9.2 and 9.3 with usage in (a) European countries? (b) underdeveloped countries?

9.4 The total energy consumption for the United States in the year 2000, as shown in Figure 9.2, is 96.50×10^{15} Btu. (a). What percentage of that energy is obtained from imports? (b). What percentage of is used to generate electricity?

9.5 Using the value you calculate from (a) in the previous problem and the fact that an average barrel of petroleum has an energy equivalent of 5.85 million Btu, how many barrels of oil, on average, must be imported each day?

Defining Energy

9.6 Distinguish between kinetic and potential energy.

9.7 Define the term internal energy.

9.8 A motorcycle that weighs 450 kg is traveling at 87.4 m/sec. What is the kinetic energy of the vehicle in J?

9.9 What is the kinetic energy of a single molecule of oxygen if it is traveling at 1.5×10^3 m/sec?

9.10 The kinetic energy of molecules is often used to induce chemical reactions. The bond energy in an O_2 molecule is 8.22×10^{-19} J. Can an O_2 molecule traveling at 780 m/sec provide enough energy (assuming the kinetic energy can somehow be used to do so) to break the O=O bond? What is the minimum velocity of an O_2 molecule that would give a kinetic energy capable of breaking the bond if it is converted with 100% efficiency?

9.11 Analyze the units of the quantity (pressure x volume) and show that they are energy units, consistent with the idea of PV-work.

9.12 When a furnace is running in the winter what is produced? Why is it not completely accurate to call this product "heat?"

9.13 Define the term hydrocarbon.

9.14 What are the products of the complete combustion of a hydrocarbon?

9.15 Carry out the following conversions of energy units. (a) 14.3 Btu → cal (b) 1.4×10^5 cal → J (c) 31.6 mJ → Btu

9.16 Total energy input in the United States in 1994 was 88.45×10^{15} Btu. Express this value in J and in cal.

Energy Transformation

9.17 If a machine does 4.8×10^3 kJ of work after an input of 7.31×10^4 kJ of heat, what is the change in internal energy for the machine?

9.18 How much work is done on a system that has a change in internal energy of 405 J when 162 J of heat is added?

9.19 If the algebraic sign of ΔE is negative in which direction has energy flowed?

9.20 State the first law of thermodynamics briefly in your own words.

9.21 Which type of energy, heat or work, is "valued" more by society? What evidence supports your judgment?

9.22 PV work occurs when volume changes and pressure remains constant. If volume is held constant can PV work be done? What happens to equation 9.2 when volume is held constant?

9.23 Which system does more work? (a) ΔE=-436 J, q=400 J or (b) ΔE=317 J, q = 347 J

9.24 In which case is heat added to the system? (a) ΔE=-43 J, w=40 J or (b) ΔE=31 J, w = 34 J

9.25 Figure 9.5 shows projections for residential energy usage decreasing through the next 10 years. What factors allow such a trend to be predicted? What factors would apply pressure for continued increased demand for household energy uses?

9.26 Gas furnaces have achieved impressive efficiency levels largely through the addition of a second "heat exchanger" that condenses water vapor that would otherwise escape out the exhaust system attached to the furnace. How does this process improve efficiency?

9.27 When energy conservation programs are promoted, they sometimes include strategies associated with such things as turning out lights. Use the web to determine the average percentage of US household energy usage that is attributable to lighting and use the information to comment on how effective the "turn out the lights" strategy might be.

9.28 Draw a rough picture of a refrigerator and show everywhere heat flow is occurring. Which location has the largest heat flow?

Heat Capacity and Calorimetry

9.29 Define the term calorimetry.

9.30 For the example of shallow water and sandy beaches, which material has a larger heat capacity or specific heat? How does a hot day at the beach provide evidence for your answer?

9.31 A metal radiator is made from 26.0 kg of iron. The specific heat of iron is 0.449 J/g °C. How much heat must be supplied to the radiator to raise its temperature from 25.0 °C to 55.0 °C?

9.32 The material typically used to heat metal radiators is water. If the boiler generates water at 79.5 °C, what mass of water was needed to provide the heat required in the previous problem? Water has a specific heat of 4.184 J/g °C.

9.33 Copper wires used to transport electrical current heat up due to the resistance in the wire. If a wire weighing 140 g attains 280 J of heat through this mechanism, what is the change in temperature in the wire? Copper has a specific heat of 0.384 J/g °C.

9.34 If a copper nail and an iron nail of the same mass and initially at the same room temperature are put into a vessel containing boiling water, which one would you expect to reach 100°C first?

9.35 A copper penny weighing 7.3 g has a temperature of 18.8°C. This penny is placed in 10.0 g of water having a temperature of 54.0°C. What is the final temperature of the penny plus water, assuming no heat is lost to the surroundings?

9.36 Define the term calibration.

9.37 Resistive heating is used to calibrate a calorimeter. When 300.0 J of heat is added to the system, the temperature change is 2.45°C. What is the calorimeter constant for this apparatus?

9.38 The energy densities of various types of coal are listed below.

Anthracite 35 kJ/g	Sub-bituminous 31 kJ/g
Bituminous 28 kJ/g	Lignite 26 kJ/g

An unknown sample of coal is burned in an apparatus with a calorimeter constant of 1.3 kJ / °C. When a 0.367-g sample is used the temperature change is 8.75°C. What type of coal is the sample?

9.39 Science fiction films sometimes invent creatures composed of metals or ceramics that have a specific heat much lower than water. Suppose one of these creatures was made of copper, with a specific heat of 0.384 J/g °C. What type of climate would you expect on the planet where this creature was found?

9.40 How does the specific heat of water explain why cities on the coast of large bodies of water tend to be cooler in the summer than cities several miles inland?

Enthalpy

9.41 Under what conditions does the enthalpy equal the heat of a process?

9.42 Why is enthalpy generally more useful than internal energy in thermodynamics of real world systems?

9.43 Define the terms exothermic and endothermic.

9.44 List at least two phase changes that are exothermic processes.

9.45 What happens to the temperature of a material as it undergoes an endothermic phase change? If heat is added, how can the temperature behave in this manner?

9.46 The ΔH_{fus} of pure silicon is 43.4 kJ/mol. How much energy would be needed to melt a 5.24 g sample of silicon at its melting point of 1693 K?

9.47 If 14.8 kJ of heat are given off when 1.6 g of HCl are condensed from a vapor, what is ΔH_{cond} for this material?

9.48 Calculate the energy required to convert 1.70 g of H_2O from ice at -12.0°C into steam at 105°C.

9.49 Acetone has ΔH_{vap} = 31.3 kJ/mol. If 1.40 kg of water is vaporized to steam in a boiler, how much acetone (in kg) would need to be vaporized to utilize the same amount of heat from the boiler?

9.50 Distinguish between a chemical equation and a thermochemical equation.

9.51 Define the term formation reaction.

9.52 Write the formation reaction for each of the following substances. (a) $CH_4(g)$ (b) $C_3H_8(l)$ (c) $HCl(g)$ (d) $C_6H_{12}O_6$ (s) (e) NaF(s)

9.53 Explain why each of the following chemical equations is *not* a correct formation reaction.

(a). $4Al(s) + 3O_2(g) \rightarrow 2Al_2O_3(s)$

(b). $N_2(g) + 3/2\ H_2(g) \rightarrow NH_3(g)$

(c). $2Na(s) + O(g) \rightarrow Na_2O(s)$

Hess's Law and Heats of Reaction

9.54 What mathematical concept about enthalpy is the basis of Hess's Law?

9.55 State Hess's Law in words. Why does it hold?

9.56 The phase change between graphite and diamond is difficult to observe directly. Both substances, however, can be burned. From these equations, calculate the ΔH for the conversion of diamond into graphite.

$C(s, graphite) + O_2(g) \rightarrow CO_2(g)$ $\Delta H°$=-393.51 kJ

$C(s, diamond) + O_2(g) \rightarrow CO_2(g)$ $\Delta H°$=-395.94 kJ

9.57 Hydrogen gas will react with either acetylene or ethylene gas. The thermochemical equations for these reactions are provided below. Write the thermochemical equation for the conversion of acetylene into ethylene by hydrogen gas.

$C_2H_2(g) + 2H_2(g) \rightarrow C_2H_6$ $\Delta H°$ = -311 kJ

$C_2H_4(g) + H_2(g) \rightarrow C_2H_6$ $\Delta H°$ = - 136 kJ

9.58 Using heats of formation tabulated in Appendix E, calculate the heats of reaction for the following,

(a) $C_2H_2(g) + 5/2\ O_2(g) \rightarrow 2CO_2(g) + H_2O(l)$

(b) $PCl_3(g) + Cl_2(g) \rightarrow PCl_5(g)$

(c) $C_2H_4(g) + H_2O(g) \rightarrow C_2H_5OH(g)$

(d) $Fe_2O_3(s) + 2Al(s) \rightarrow Al_2O_3(s) + 2Fe(l)$

9.59 The heat of combustion of butane is -2877 kJ/mol. Use this value to find the heat of formation of butane. (You may also need to use additional thermochemical data found in Appendix E.)

9.60 When a chemical bond is broken, is energy added or released?

9.61 When a reaction is exothermic, is the sum of bond energies of products or reactants greater?

Energy and Stoichiometry

9.62 For the reaction $C_2H_2(g) + 2\ H_2(g) \rightarrow C_2H_6$, $\Delta H° = -136$ kJ. What are the ratios that can be defined between moles of substances and energy?

9.63 For the reaction $N_2(g) + O_2(g) \rightarrow 2NO(g)$, $\Delta H° = 180.5$ kJ. How much energy is needed to generate 35 mol of NO(g)?

9.64 For the reaction $2NO(g) + NO_2(g) + Na_2O(s) \rightarrow 2NaNO_2(s)$, $\Delta H° = -427.14$ kJ. How much energy is produced when 4.3 mol of NO is consumed?

9.65 Silane, SiH_4, burns according to the reaction, $SiH_4 + O_2 \rightarrow SiO_2 + 2H_2O$, with $\Delta H° = -1429$ kJ. How much energy is released if 15.7 g of silane is burned.

9.66 Suflur trioxide can be removed from the exhaust gases of power plants by reaction according to the equation, $CaO(s) + SO_3(g) \rightarrow CaSO_4(s)$ $\Delta H° = -886$ kJ. If 240 kg of SO_3 is to be removed, how much heat is released?

9.67 Reactions of hydrocarbons are often studied in the petroleum industry. One such reaction is, $2C_3H_8(g) \rightarrow C_6H_6(l) + 5H_2(g)$ $\Delta Hv = 698$ kJ. If 35 L of propane at 25°C and 0.97 atm is to be reacted, how much heat must be used?

9.68 In principle, ozone could be consumed in a reaction with lead and carbon with the thermochemical equation, $Pb(s) + C(s) + O_3(g) \rightarrow PbCO_3(s)$ $\Delta H° = -841.0$ kJ. How much energy would be released if 110 g of ozone reacts with excess lead and carbon?

9.69 0.0157 g of a compound with a heat of combustion of 37.6 kJ/mole is burned in a calorimeter, releasing 18.5 kJ of heat. What is the molar mass of the compound?

9.70 Define the term energy density.

9.71 Why is energy density such an important consideration when considering transportation?

9.72 What are some features of petroleum that make it such an attractive fuel?

Batteries

9.73 What type of chemical reaction takes place in a battery?

9.74 A small AAA battery and a much larger D battery both supply the same voltage. How does the underlying chemistry of the alkaline battery explain this? What factors would lead an engineer to specify D batteries in the design of a particular device rather than the smaller AAA cells?

9.75 Explain the meaning of the terms "primary" and "secondary" in the context of batteries. Give examples of both types of batteries.

9.76 Consumers face a variety of choices in selecting batteries. A typical AA alkaline battery might have a rated lifetime of about 2800 mA hours and cost around \$0.50. New high-capacity alkaline cells are good for 3100 mA hours, but typically cost about \$1.00 each. And rechargeable nickel metal hydride (NiMH) cells offer yet another option, with a lifetime of 2100 mA hours (per charge cycle) and a cost around \$3.50 each. What kinds of considerations might lead you to choose each of these options? What type of applications do you think would be best suited for each of these battery types?

Additional Problems

9.77 The figure below shows a cross-section of a "self-cooling" beverage can. The can is equipped with an outer jacket containing sodium carbonate (Na_2CO_3), which dissolves in water rapidly and endothermicly.

$$Na_2CO_3\ (s) \rightarrow 2\ Na^+\ (aq) + CO_3^{2-}\ (aq) \qquad \Delta H° = 67.7\ kJ$$

The user adds water to the outer jacket, and the heat absorbed in the chemical reaction chills the drink. The can contains 200 g of drink, the jacket contains 55 g of Na_2CO_3, and 100 g of water are to be added. If the initial temperature of the can and the water is 32°C on a summer day, what is the coldest temperature that the drink can reach? The can itself has a heat capacity of 40 J/°C. Assume that the Na_2CO_3 solution and the drink both have the same heat capacity as pure water, 4.184 J g^{-1} $°C^{-1}$. (**HINT:** Treat this like a calorimetry problem.)

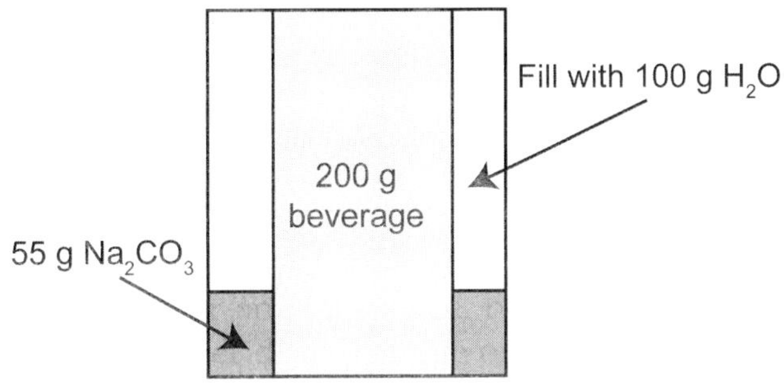

9.78 You make some iced tea by dropping 134 g of ice into 500.0 mL of warm tea in an insulated pitcher. If the tea is initially at 20.0°C and the ice cubes are initially at 0.0°C, how many grams of ice will still be present when the contents of the pitcher reach a final temperature? The tea is mostly water, so assume that it has the same density (1.0 g/mL), molar mass, heat capacity (75.3 J/K/mol), and heat of fusion (6.0 kJ/mol) as pure water. The heat capacity of ice is 37.7 J/K/mol.

9.79 A student performing a calorimetry experiment combined 100.0 mL of 0.50 M HCl and 100.0 mL of 0.50 M NaOH in a Styrofoam cup calorimeter. Both solutions were initially at 20.0 °C, but when the two were mixed the temperature rose to 23.2°C.

(a). Suppose the experiment is repeated in the same calorimeter, but this time using 200 mL of 0.50 M HCl and 200.0 mL of 0.50 M NaOH. Will the ΔT observed be greater than, less than, or equal to that seen in the first experiment, and why?

(b). Suppose the experiment is repeated once again in the same calorimeter, this time using 100 mL of 1.00 M HCl and 100.0 mL of 1.00 M NaOH. Will the ΔT observed be greater than, less than, or equal to that seen in the first experiment, and why?

9.80 The specific heat of gold is 0.13 J g^{-1} K^{-1}, and that of copper is 0.39 J g^{-1} K^{-1}. Suppose we take a 25-g sample of gold and a 25-g sample of copper, heat them both to 80°C, and then drop them each into identical beakers containing 100 mL of cold water at 10°C.

Once the 2 beakers reach thermal equilibrium, which of the following will be true, and why? You should not need to do any calculations here.

(i). Both beakers will be at the same temperature.

(ii). The beaker with the copper sample in it will be at a higher temperature.

(iii). The beaker with the gold sample in it will be at a higher temperature.

Focus on Problem Solving

9.81 Suppose the working fluid inside an industrial refrigerator absorbs 680 J of energy for every gram of material that is evaporated in the evaporator. The refrigerator unit uses this energy flow as part of a cyclic system to keep foods cold. A new pallet of fruit with a mass of 500 kg is placed in the refrigerator. Assume that the specific heat of the fruit is the same as that of pure water, because the fruit is mostly water? Describe how you would determine the mass of the working fluid that would have to be evaporated to lower the temperature of the fruit by 15°C. List any information you would have to measure or look up.

9.82 Hydrogen combines with oxygen in fuel cells according to the thermochemical equation

$$2H_2(g) + O_2(g) \rightarrow 2\,H_2O(g) \qquad \Delta H° = -571.7\text{ kJ}$$

Suppose you are working with a firm that is using hydrogen fuel cells to power satellites. The satellite requires $4.0\text{x}10^5$ kJ of energy over its useful lifetime to stabilize its orbit. Describe how you would determine the mass of hydrogen you would need in your fuel cells for this particular satellite.

9.83 The chemical reaction, $BBr_3(g) + BCl_3(g) \rightarrow BBr_2Cl(g) + BCl_2Br(g)$ has an enthalpy change, ΔH°, that is very close to zero. Using Lewis structures of the molecules, all of which have a central boron atom, provide a molecular level description of why there is so little energy change in this reaction.

9.84 Two baking sheets (such as those used for making cookies) differ by the type of metal they use. You purchase one of each and bake a dozen cookies on each sheet at the same time in your oven. You observe that after nine minutes the cookies on one sheet are slightly burned on the bottom while on the other sheet they are fine. (You are curious and you vary the conditions so you know the observation is not because of something about the oven) (a) How can you use this observation to infer something about the specific heat of the materials in the baking sheets? (b) What is the mathematical reasoning (equation) that you need to support your conclusion?

9.85 Silicon nitride, $Si_3N_4(s)$ has physical, chemical and mechanical properties that make it a useful industrial material. It is crucial for some project that you know the heat of formation of this material. A clever experiment allows the direct determination of the ΔH° of the following reaction,

$$3CO_2(g) + Si_3N_4(s) \rightarrow 3\,SiO_2(s) + 2\,N_2(g) + 3C(s)$$

Based on the fact that you know the heat of this reaction, state what additional data must be either looked up or measured in order to use this reaction to determine the $\Delta H_f°$ for silicon nitride.

9.86 A runner generates 418 kJ of energy per kilometer from the cellular oxidation of food. The runner's body must dissipate this heat or the body will overheat. Suppose that sweat evaporation is the only cooling mechanism used by the body of this runner. If you estimate the enthalpy of evaporation of water is 44 kJ/mol and that sweat can be treated as entirely water, describe how you would estimate the volume of sweat that would have to be evaporated if the runner runs a 10 km race.

9.87 The energy density of a fuel is an important value to know because to move a vehicle one must also move the as yet unburned fuel. Octane is a major component of gasoline. It burns according to the reaction,

$$2\ C_8H_{18}(l) + O_2(g) \rightarrow 16\ CO_2(g) + 18\ H_2O(g)\quad \Delta H° = -1.10\ x10^4\ kJ$$

Based on the fact that you know the heat of this reaction, describe how you would determine the energy density – in kJ/g – for octane. Be sure to indicate what you would need to calculate or look up to complete this problem.

9.88 An engineer is using sodium metal as a cooling agent in a design because it has useful thermal properties. Looking up a value for heat capacity the engineer finds a value of 28.2 J / mol °C. Inadvertently the number is taken to be a specific heat. (a) What would be the difference between these two values? (b) Would the engineer over estimate the ability of sodium to remove heat from the system or under estimate it because of this error? Be sure to explain you reasoning.

9.89 Which material would work better in an implementation of passive solar heating (where the goal is to absorb heat from the sun during the day and release it during the night) one with a high heat capacity or one with a low heat capacity? Explain your reasoning.

9.90 A 1.0-kg sample of stainless steel is heated to a temperature of 400°C. Suppose you take this heated sample and drop it into an insulated bucket that contains water at an initial temperature. Assuming there is no difficulty in the transfer of heat from the steel to the water, describe how you can determine the maximum mass of water that could be made to boil with only the heat given off by this sample of steel. Be sure to list any quantities you would need to look up to solve this problem.

Debate Questions

D9-1 The idea of taxing energy resources is occasionally broached in the political arena. There are several possible ways to implement such a tax including,

(a). taxes on gasoline (per gallon basis)

(b). taxes on natural gas (per cubic foot basis)

(c). taxes on electricity usage

(d). taxes based on how much carbon is released as carbon dioxide in the generation of the useful energy.

Choose one of these types of taxes and argue either for or against it in a legislative debate. Highlight when you use science as part of your argument.

D9-2 There are member of environmental groups who suggest the automobile fleet of the United States should be converted to electrically powered vehicles, in order to reduce emissions. Analyze the factors about energy production that would need to be considered to offer an informed opinion of such proposals.

D9-3 Energy production and usage is correlated with the strength of an economy. What issues need to be considered when developing countries attempt to expand their power generating resources? As a case study consider a proposal by a developing country to build a hydroelectric dam. The flooding of land behind the dam will endanger important archeological ruins and wildlife habitats. Argue either for or against the building of this dam. How does science measure as a consideration in this debate? If the proposed dam were being built in the U.S. or Canada would the situation require different consideration?

Chapter 10: Entropy and the Second Law of Thermodynamics

In our discussions of chemical bonding we introduced the idea that bonds form because doing so reduces the overall energy of the collection of atoms involved. And we've seen many examples of chemical reactions, like combustion and explosions, that also reduce the energy of the atoms and molecules involved. But if you think for just a moment, you will also be able to think of many common chemical and physical processes where the energy of the system under consideration clearly increases. Ice cubes melt. The batteries in your laptop or MP3 player are recharged. Clearly these endothermic chemical reactions occur regularly. As each of these cases shows, the energy of a system does not always decrease, despite our intuitive sense of nature's preference. So how can we understand and predict which changes nature will actually favor? We will need to extend our understanding of thermodynamics, and especially of the second law of thermodynamics. While applications abound in virtually all fields of science and engineering, the impact of thermodynamics on our understanding of chemical reactions has been especially profound. We'll explore the implications of thermodynamics by looking into the chemistry of recycling plastics.

10.1 Insight into Recycling of Plastics

One way to begin to think about the issues surrounding recycling is to consider the "life cycle" of a plastic bottle. Let's choose a soft drink bottle made from poly(ethylene terephthalate), or PET. For simplicity, let's assume that our bottle is originally made entirely out of new or "virgin" PET and contains no recycled material. Where does this virgin plastic actually come from? We already know that polymers like PET contain long chain molecules, and that these chains are made up primarily of carbon atoms. We also know that these chains are generally produced through the sequential addition of small molecules, or monomers. So to produce polymers we need a large supply of organic molecules. As we saw in the previous chapter, petroleum provides just such a source of hydrocarbons. And indeed most polymers have their origins in petroleum. The refining process allows individual compounds to be obtained in appropriate purity, and subsequent chemical reactions lead first to the needed monomers and then to the polymers.

In the industrial scale synthesis of PET, the usual starting materials are dimethyl terephthalate and ethylene glycol. These compounds react to form bis-(2-hydroxyethyl) terephthalate (BHET) and methanol. The methanol boils off at the reaction temperature (typically around 210°C), leaving fairly pure BHET. The BHET is then heated further to around 270°C and undergoes a condensation reaction to form the PET polymer.

Condensation reactions were introduced in Chapter 8.

Ethylene glycol is formed as a byproduct in this second step, and can thus be reused within the plant to produce more BHET.

The resulting polymer can then be melted, blown, and molded into bottles of the desired shape. These bottles are then filled, capped, shipped, and sold. You pick up your soda, perhaps from a vending machine on your way to class. Once the bottle is empty, you toss it into the recycling bin, and feel good that you've done your part to help protect the environment. Chances are you may never have thought much about what happens to the bottle from there.

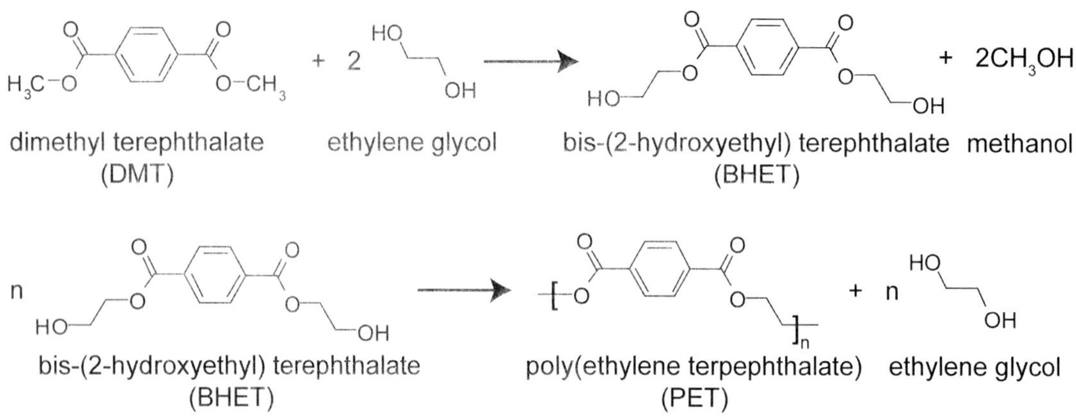

Figure 10.1 Steps in the industrial synthesis of PET are illustrated. Typical values for n in the polymer formula are 130 – 150, giving a molar mass of around 25,000 for the polymer.

Typically the contents of the recycling bin are sold to a reclaimer – a business specializing in processing plastics. The material from the bin must usually be sorted into different types of plastics, and any other materials that may have been thrown into the bin are discarded. Some of this sorting is done by hand, and some takes advantage of differences in density among the various polymers that might be present. The plastics are then crushed to reduce their volume before being shipped for further processing. The next step is called reclamation, in which the sorted and compressed plastics are processed into a useable form. In most reclamation processes, the plastic is first chopped into small, uniform-sized flakes. The flakes are washed and dried, then melted and extruded into spaghetti-like strands. These are then cut into smaller pellets, which are sold to manufacturers for use in new products. The most important uses for recycled PET include fiberfill for sleeping bags and coats, fleece fabrics for outdoor wear, carpeting, and industrial strapping.

You may have noticed one thing that does not appear on that list of uses: new drink bottles. Although such bottles are the dominant source of PET for recycling, only very limited amounts of recycled PET are used to make new bottles. Thus the recycling of PET is far from being a "closed loop" process; large amounts of virgin plastic continue to be used in bottling despite increased collection of used bottles at the consumer level. Why is this? The simplest and shortest answer is economics: bottles can be made from virgin plastic at a lower overall cost. Several factors contribute to this. In many cases there are legal restrictions on the use of recycled materials for food and beverage containers, due to concerns over possible contamination. Satisfying these regulations adds cost to the overall equation. Degradation of the plastic over repeated recycling processes is another concern. The average chain length of the polymer molecules will generally be somewhat lower after recycling, and so if bottles were made from 100%

recycled PET they might have to be thicker and heavier. Although progress is being made on increasing the recycled content of drink bottles, most U.S. bottles still contain at least 90% virgin plastic.

One possible way to achieve a closed loop, in which plastic bottles could be recycled back into plastic bottles, might be to convert the polymer into monomers and then re-polymerize it to produce new plastic. Under what circumstances might such a scheme be feasible? To explore that type of question, though, we will first need to learn more about thermodynamics.

10.2 Spontaneity

Nature's Arrow

The idea of time travel frequently drives the plot in science fiction stories. Whether one moves forward or backward in time, the prospect of existing in some other era appeals to our imagination in a way that provides fertile ground for authors. But our actual experience is that time marches inexorably from the past towards the future, and that this direction is not to be subverted. In a sense, time is an arrow that points in the direction that nature is headed. We've seen that large hydrocarbon molecules, like those found in gasoline, can react readily with oxygen to produce carbon dioxide and water. But your experience also tells you that the reverse reaction doesn't happen; water vapor and carbon dioxide are always present in the air, but they never react to produce gasoline. Nature clearly "knows" the correct direction for this process. This sense of the direction of life and our experience of the universe is an important intuition to carry into this chapter. But what gives Nature this direction? And how can we convert our intuition into a useful quantitative model? We'll try to answer these questions by imparting a bit of mathematical rigor to our observations. But we will not be invoking anything that is fundamentally new to our everyday experience.

Spontaneous Processes

A more formal way of expressing the directionality of Nature is to note that our intuition is predicated on the fact that some things "just happen" while others do not. Some processes occur without any outside intervention, and we say that such a process is *spontaneous*. From a thermodynamic perspective, then, a **spontaneous** process is one that takes place without continuous intervention. The concept of a spontaneous reaction is sometimes obvious and other times more complicated.

Students often misinterpret the word *spontaneous* as indicating that a process or reactions will take place quickly. But note that our definition does not actually refer in any way to the speed of the process. Some spontaneous processes are very fast, but others occur on extremely long time scales. We understand that the chemical compounds in some waste materials, like paper, may spontaneously react to decay over time. (This process can be more complicated than a simple chemical reaction, though, because of the involvement of bacteria). But some spontaneous reactions are so slow that we have a hard time observing them. Other reactions occur quickly once they start, but they don't just start on their own. Gasoline, for example, can sit more or less indefinitely in a can in the garage, in contact with oxygen in the air. Nonetheless, no reaction is observed. Yet, upon being mixed with air in the cylinder of your car and ignited by the spark plug, the reaction proceeds until virtually all the gasoline is burned. Is this reaction spontaneous? The answer is yes. Despite needing a flame or spark to initiate

Some spontaneous processes take place over geological time scales - the formation of petroleum used for plastics feedstocks, for example.

it, once it begins, the reaction continues without any further intervention. This example emphasizes the importance of the phrase "continuous intervention" in our definition. A useful analogy to this chemical reaction is to think of a rock perched precariously on a cliff. If it is nudged over the edge, it proceeds to the bottom. It does not stop midway down, unless, of course, it's a prop in a Roadrunner cartoon!

The reactions used to produce many polymers behave in a way much like the combustion of gasoline: once initiated, the reaction is usually spontaneous and can proceed with no need for "intervention." The production of poly(methyl methacrylate) – Plexiglas, or PMMA – provides a good example.

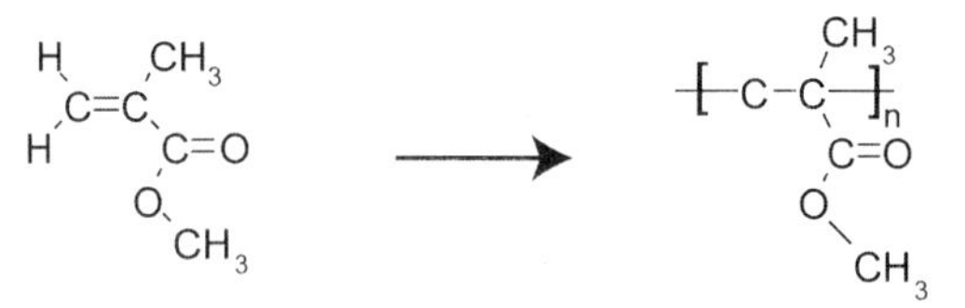

methyl methacrylate monomer — poly(methyl methacrylate)

This reaction proceeds by a free radical process, like the one we described back in Chapter 2 for polyethylene. A small trace of an initiator is needed to get the reaction started, and then it proceeds until virtually all of the available monomer has been converted into polymer.

But suppose we wanted to convert the polymer back into monomer. In that case the necessary reaction is the reverse of the polymerization, and it is not a thermodynamically spontaneous process at ordinary temperatures. We could still drive the reaction "backwards" to produce the methyl methacrylate monomer. But to do so we would need to maintain a high temperature, providing enough energy to allow the molecules to go against Nature's preferred direction. So what is the role of energy in the directionality of Nature?

Enthalpy and Spontaneity

Recall from Chapter 9 that the enthalpy change for a chemical reaction is equal to the heat flow at constant pressure.

$$\Delta H = q_p$$

When ΔH is negative, the reaction is exothermic, while a positive value of ΔH points to an endothermic reaction. What can we say about a reaction's spontaneity based on its enthalpy change? If we were to stop and list spontaneous processes that we observe around us, and then determine whether those process were exothermic or endothermic, chances are that a majority would be exothermic. This implies that there is some relationship between enthalpy and spontaneity. The relationship is not exclusive, however. If you think for a moment you should be able to point out some endothermic reactions that obviously do occur spontaneously. The melting of an ice cube at room temperature is one simple example. So at this point we might conclude that exothermic reactions seem to be preferred in some way. But there clearly must be things other than energy or enthalpy considerations at work in determining whether or not a process is spontaneous. In order to develop a way to predict the spontaneity of a reaction, we must first introduce an additional thermodynamic state function: entropy.

10.3 Entropy

The flow of energy in the form of heat, as we have just seen, is not capable of predicting whether or not a process will happen spontaneously. So we must also consider another thermodynamic state function, called **entropy**. Historically, entropy was first introduced in considering the efficiency of steam engines. Figure 10.2 illustrates the Carnot cycle, which utilizes a combination of adiabatic processes in which no heat is exchanged and isothermal, or constant temperature, processes. The Carnot cycle demonstrated that a previously unknown state function existed, because the sum around the closed path of q/T (heat divided by temperature) was shown to equal zero. But more importantly, the changes in entropy of a system and its surroundings allow us to predict whether or not a process is spontaneous. What is entropy and how can it help us understand the production or recycling of polymers?

Because a state function does not depend on it's history, any time the system is returned to the same state, the value of the state function must be the same. So there can be no change in the state function.

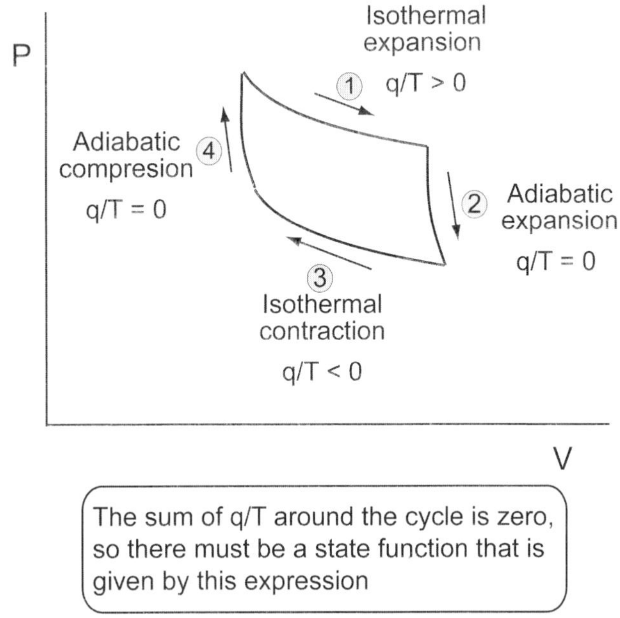

Figure 10.2 In the Carnot cycle, an ideal gas undergoes a series of four processes. Two of these (labeled as 1 & 3 in the figure) are isothermal, which means they occur at constant temperature. The other two steps (2 & 4) are adiabatic, meaning that q = 0 for those parts of the cycle. Carnot showed that the sum of the quantity q/T for the entire cycle is equal to zero. Since the cycle begins and ends with the system in the same state, this means that there must be a state function equal to q/T. We call this state function entropy.

Probability and Spontaneous Change

We can observe a pattern in many changes that occur in everyday life that will provide an analogy into events at a molecular level. For a familiar example, let's think about autumn. While the falling leaves may be welcome as a sign of cooler temperatures, they also mean an added chore – raking the leaves into piles. Why can't the leaves simply fall in a pile to begin with? Such an event goes against our intuition, because it is so unlikely to happen that we know we'll never see it happen. This macroscopic example with perhaps thousands of leaves serves as a reasonable analogy for molecular systems with Avogadro's number of particles. Let's look at the concept of mathematical probability to solidify our understanding.

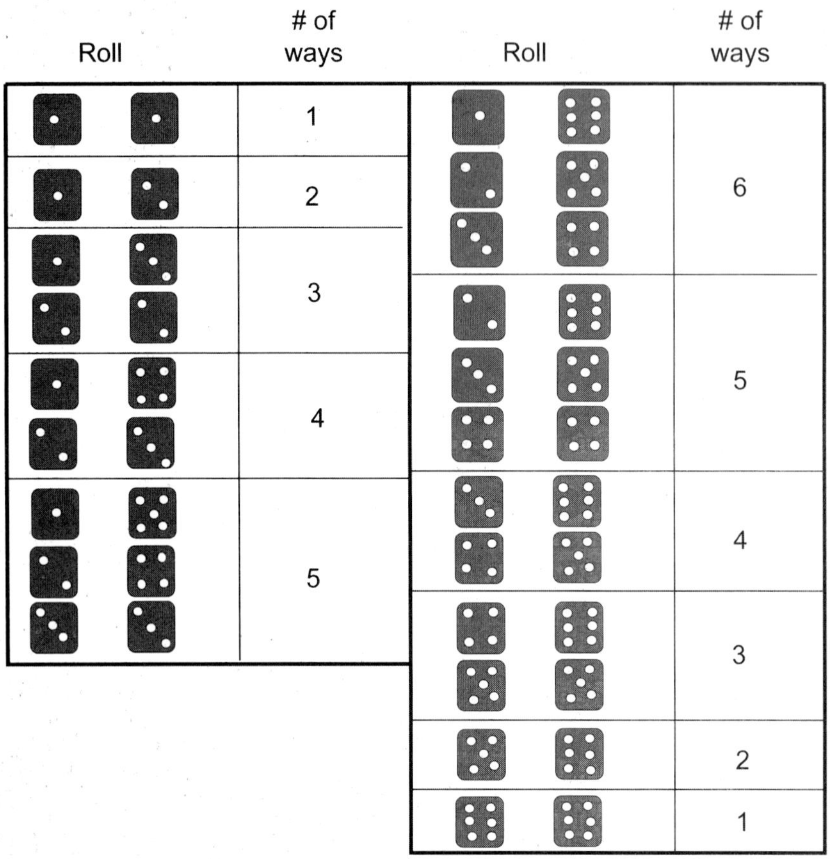

Figure 10.3 The probability of rolling a given total value on a pair of dice depends on the number of different combinations that produce that total. The least likely rolls are 2 and 12, for example, because there is only one possible combination that gives each of those totals. The most likely total is 7, because there are 6 different rolls that all add up to that number. (Note that for rolls in which the two individual dice show different values, two possibilities exist. For a total roll of 3, for example, the two combinations would be 1,2 and 2,1.)

The example of leaves not falling in a pile, though perhaps obvious, is somewhat challenging to describe in mathematical terms. To establish a foundation in ideas of probability, let's think instead about rolling dice. If you take just one die and roll it, what is the chance that the roll will be a four? With six possible outcomes the chance is one in six. For two dice, what is the chance the roll will be a pair of fours? This time the counting is a bit more involved, but we can quickly see from Figure 10.3 that the chances are one in 36. If a third die is added, the chances of rolling three fours in one throw are one in 216. We can quickly see the relationship that is developing for rolling all fours. There is only one way to achieve it, and the number of possible combinations grows quickly according to the following relationship.

$$\text{probability} = \left(\frac{1}{6}\right)^N, \text{ where } N \text{ is the number of dice being thrown}$$

We should note that this relationship is not general. The factor of one in the numerator is present because we are looking for a single specific roll (of a number four) and the six in the denominator is there because there are six possible rolls for each die. With this relationship, however, we could easily predict that the chances of rolling the same number with five dice in one roll are one in 1296. (Note that the chances of rolling a *specified* number on all five dice – say all fours – are 1 in 7,776. But if we do not specify in advance which of the six possible numbers we want on all five dice, then there will be six possible outcomes instead of just one.) Our common experience with rolling dice is that we expect to have some random assortment of numbers present when five dice are rolled. Why? There are very many ways to obtain a "random" roll. Such a roll occurs far more often precisely because it is more probable.

The mathematical relationship of probability we noted above has two important features. First, it shows that to obtain the probability of a collection of events based on the probability of an individual event we must multiply. This observation becomes important when we consider just how many molecules are involved when we observe something in nature or in the laboratory. Second, the number of ways to make an ordered observation (like all dice turning up "four") is smaller than the number of ways to make a random observation (no particular pattern present in the dice). When we transpose these observations to a collection of molecules with $\sim 10^{23}$ particles present, the chances for highly specified arrangements become phenomenally small. If we rolled Avogadro's number of dice (to start addressing numbers with more chemical relevance) the chances of all of them coming up four is 1 in $10^{4.7\times10^{23}}$. The chance of that occurring is so small that if we used all the zeros after the decimal point in that number to replace *all* the letters in *all* the books on the planet with zeros, we would still have zeros left over!

Definition of Entropy

For large numbers of particles, then, probability favors random arrangements. Using this insight, we can tentatively define **entropy** as a measure of the randomness or disorder of a system. However, we still have to establish a definition that can be used in quantitative ways and from a molecular perspective. To do this, we turn to a branch of physical chemistry called **statistical mechanics**, or **statistical thermodynamics**, where we find a subtle addition to the definition: The probability of events that must be counted is not the number of ways particles can be arranged physically, but rather the number of ways those particles can achieve the same energy.

These two probabilites are often correlated with one another, however.

If we recall the Maxwell-Boltzmann distribution of molecular speeds (see Section 5.6) we know that in any sample we observe at room temperature some particles must move slowly and others quite rapidly. We cannot, however, say precisely which particle is moving very fast or which particle is moving more slowly (Figure 10.4). There are a large number of different ways, with different particles assuming the various required speeds, that the sample can have the same energy content (and hence the same temperature). In statistical mechanics the way the collection of particles assumes a given energy is associated with a concept called a **microstate**. The number of microstates for a given energy is commonly designated by the upper case Greek letter omega (Ω), and entropy of a system is related to the number of microstates by the equation,

$$S = k_B \ln \Omega \qquad (10.1)$$

Here k_B is a numerical constant called the Boltzmann constant. It is not easy to have an intuition about the number of microstates of a system, so this equation is hard to use directly at this stage of your study of chemistry. And we'll soon see that we won't need to use it. It is important, however, to realize that as a system becomes "more random," the value of Ω will increase. So, the entropy of a system increases as the system moves towards more random distributions of the particles it contains because such randomness increases the number of microstates.

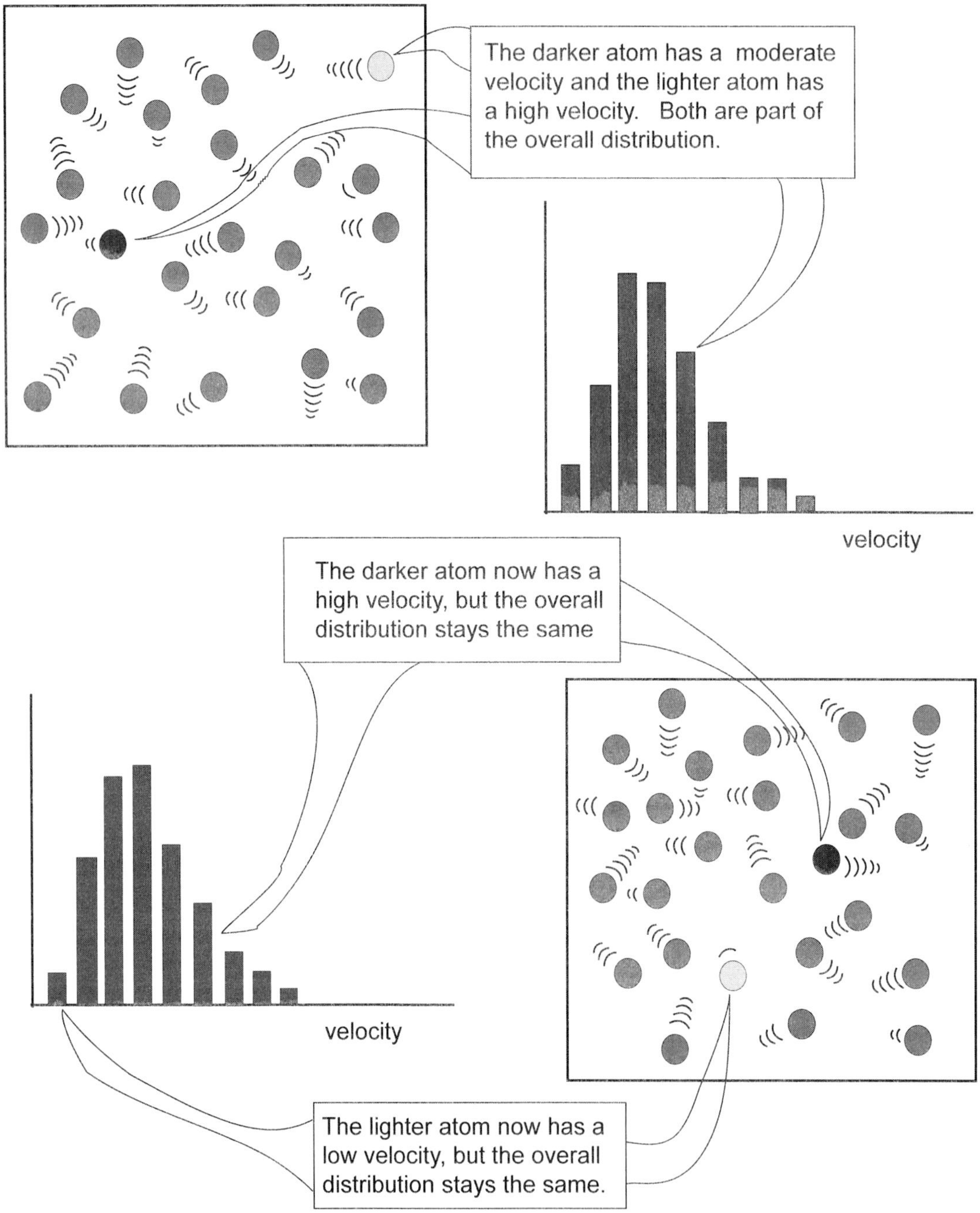

Figure 10.4 The Maxwell Boltzmann distribution tells us the overall collection of molecular speeds, but does not specify the speed of any individual particle. Energy exchange during molecular collisions can change the speed of individual molecules without disrupting the overall distribution.

Judging Entropy Changes for Processes

While the concept of a microstate is abstract, we can still assert that certain types of changes will lead to an increase in entropy (because there are more available microstates). Let's see why this is so. First consider the melting of a solid to form a liquid. As a solid, the particles are held rigidly in place, so the number of ways they can have a specific energy is limited. When the liquid forms, the movement of particles relative to each other presents a much wider array of ways to achieve a specific energy, so the number of microstates increases and so does the entropy. Similar reasoning can be applied to boiling, when molecules originally confined near each other in a liquid become much more randomly distributed in the gas phase. The increase in random molecular motion corresponds to more microstates, so entropy increases. Another possible way to increase the entropy of a system is to increase the number of particles present. Thus, a chemical reaction that generates two moles of gas where only one was present initially will generally increase the entropy.

Another very important process to understand is the change in entropy associated with heating. We can think of a very low temperature limit of the Boltzmann distribution to reason our way through this change. In a distribution of molecules, if the temperature were quite low it would be very unlikely to have molecules moving at high speeds because they would account for too high a percentage of the available energy. So the usual Boltzmann distribution could not be obtained. However, if the system were heated to a higher temperature, a small selection of the molecules could then move at high speeds, because there is greater total energy available. We have only considered a small portion of the distribution of speeds of the molecules, but we can already see that adding heat provides more ways to achieve the energy of the system. This type of reasoning extends to the whole distribution of speeds, and the important result is that *an addition of heat to a system results in an increase of entropy.*

Recall that heat is a process, so the "addition of heat" to a system doesn't imply the transfer of any tangible particles. Rather, it simply indicates that the system has come into thermal contact with something at a high temperature and energy has flowed into the system in the fform of heat.

What are the implications of entropy for polymer synthesis or recycling? When a polymer is formed, a large number of monomers are converted into a single giant molecule. This will generally lead to a decrease in the entropy of the system, since there are more possible ways to arrange the individual monomers. (Note that in many polymerization reactions other small molecules, like water, may be formed as byproducts. In such cases the sign of the entropy change may not always be obvious.) The fact that polymerization reactions are still spontaneous under appropriate conditions tells us that entropy alone is not the only important consideration. Other factors, like energy, must favor the formation of the polymer. What about the role of entropy in recycling? As plastics are recycled, there is a possibility that the long polymer chains may be broken. From an entropy point of view this should be a favorable process. Breaking the chains gives a smaller average molecular size, and the same system of atoms will have more microstates available if more individual molecules are present. Reducing the polymer chain length tends to weaken the mechanical properties of the plastic, though, because the shorter chains do not interact with one another as strongly. So entropy provides a challenge to the recycling process. To recycle polymers without a steady loss in the quality of the material, we need to overcome Nature's preference for increasing entropy. And as we'll see by the end of this chapter, this leads very directly to real economic obstacles to recycling.

10.4 The Second Law of Thermodynamics

In Chapter 9 we stressed the importance of thermodynamics in terms of the way human society uses energy. At that time we noticed that whenever there is an attempt to convert energy from one form into another, some energy is "lost" or "wasted." In other words, not all of the energy potentially available gets directed into the desired process. How does this fact arise from thermodynamics? Entropy provides the key we need to understand why the loss of "useful" energy is inevitable.

The Second Law

There are several ways to state the Second Law, but all would submit to the same reasoning we are using here to relate them to processes on the molecular scale.

The second law of thermodynamics has already helped us provide an explanation of the energy that is lost in the energy economy. In Chapter 9 we noted that the second law could be expressed in the following way: *It is impossible to convert heat completely to work with no other result.* Now we can begin to understand why this is true. First consider heat. Heat flows due to random collisions of molecules, and an increase of heat increases the random motions of molecules. Work, by contrast, requires moving a mass some distance. To yield a net movement, there must be a direction associated with a motion, and that direction implies that there is an order to the motion. Converting heat into work, therefore, is a process that moves from random motions towards more ordered ones. We have just seen how this type of a change goes against Nature's tendency to favor a more probable state (the more random one). How can we consider this law of thermodynamics in the context of entropy?

To make this connection, we must be careful to realize that changes in the universe involve both the system and its surroundings. Indeed, if we focus on the system alone we cannot understand how order is created at all. And yet the synthesis of polymers shows that it does happen, as do everyday situations like the growth of plants, animals, and people. To express the second law of thermodynamics in terms of entropy we must focus on the total change in entropy for the universe, ΔS_u.

$$\Delta S_u = \Delta S_{sys} + \Delta S_{surr} \tag{10.2}$$

Because Nature always tends to proceed towards a more probable state, we can assert an equivalent form of the second law of thermodynamics: *In any spontaneous process, the total entropy change of the universe is positive, ($\Delta S_u > 0$).* That this statement of the second law is equivalent to our original version is certainly not obvious. But remember, energy that is not converted into work (a process that would decrease entropy) is transferred to the surroundings as heat. Thus the entropy of the surroundings increases, and the total entropy change of the system and surroundings is positive.

Implications and Applications

The implications of this expression of the second law are far-reaching in terms of being able to calculate and predict the outcome of chemical reactions and other processes we might wish to study. We can focus on these implications in a qualitative way first, by considering a polymerization reaction from the perspective of thermodynamics. Later, we will develop a more quantitative approach to these issues.

Let's return to the polymerization of methyl methacrylate to form PMMA. The monomer and polymer structures are shown in Section 10.2. Looking at both structures, we see that most of the chemical bonds are unchanged during this reaction. The only exception is the C=C double bond in the monomer, which is converted into two C-C single bonds in the polymer. From our knowledge of bond energies, then, we can say that this reaction must be exothermic: two C-C single bonds are stronger than a C=C double bond. Since the reaction converts a large number of monomer molecules into a single polymer molecule, we can also predict that the entropy change for the system must be negative. So why is the reaction spontaneous? The fact that the reaction is exothermic means that heat must be released from the system. And that same heat must therefore flow into the surroundings. This will lead to an increase in the entropy of the surroundings. And as long as that increase in entropy is larger than the decrease in the entropy of the system, the overall change in entropy for the entire process can still be positive.

We can take this reasoning a little further to begin to understand the role of temperature in determining the spontaneity of a process. The surroundings will absorb an amount of heat equal to -ΔH. But the surroundings represent a very large reservoir, so this heat will not really produce a measurable temperature change. This means that the entropy change for the surroundings can be shown to be

$$\Delta S_{surr} = -\Delta H/T$$

The entropy change for the system is just ΔS, and while we don't know its value, we do know that it will be negative. The criterion for having the polymerization be spontaneous is that

$$\Delta S_u = \Delta S + \Delta S_{surr} > 0,$$

and this will be true so long as the magnitude of ΔS_{surr} is greater than that of ΔS. (Remember that ΔS is negative and ΔS_{surr} is positive.) The magnitudes of ΔS and ΔH are essentially independent of the temperature. But the magnitude of ΔS_{surr} will decrease as the temperature increases. So at some sufficiently high temperature, ΔS_u will no longer be positive and the reaction will cease to be spontaneous.

This same argument points out a possible route to depolymerization, which might be useful for recycling. Suppose we raise the temperature high enough that ΔS_u for the polymerization reaction becomes negative. That must mean that ΔS_u for the reverse reaction, in which polymer is converted back into monomer, must become positive. So if we heat the polymer above some threshold temperature, we should be able to regenerate methyl methacrylate monomer. And indeed, when PMMA is heated above a temperature of about 400°C it is converted into monomer with a very high efficiency. This process, called thermolysis, is one example of what is often called advanced recycling or feedstock recycling. Since the recovered monomer can be purified by distillation or other means, it can then be repolymerized to produce PMMA that is indistinguishable from virgin material. Thermolysis is not practical for most plastics, though, because the monomers themselves often break down or undergo other undesirable reactions at the high temperatures that would be required. In the particular case of PMMA, thermolysis is used mainly within manufacturing plants to reclaim scraps that are left behind in the production of items like automobile taillight lenses.

In this case purifying implies separating the monomer from other chemicals present. Distillation is a common method for separation of chemical mixtures.

10.5 The Third law of Thermodynamics

In our previous example we did not obtain any numbers for entropy changes. Instead, we took a purely qualitative approach. In order to move towards a quantitative view, though, all we really need is to define some reference point with a fixed value of entropy. Then as long as we are able to calculate entropy changes, we should be able to obtain values for any cases of interest. When we seek to calculate entropy changes for chemical reactions, the most convenient method comes from the **Third Law of Thermodynamics**, which says that *the entropy of a perfect crystal of any pure substance approaches zero as the temperature approaches absolute zero*. An additional implication of the third law is that it is impossible to attain a temperature of absolute zero, but scientists have come very close to that value. Because all substances can be cooled to temperatures near zero (at least in principle), it is possible to evaluate the entropy of one mole of any given chemical substance under standard conditions, by determining the change in entropy from 0 K to 298 K at a pressure of 1 atm. This process yields the **standard molar entropy**, S°. Table 10.1 provides a partial listing of S° for some substances, and a more complete table is provided in Appendix E.

Because entropy is a state function and because the third law allows us to obtain a value for the standard molar entropy of any substance, we can derive a useful equation for entropy change in a reaction. Figure 10.5 shows how the entropy change of a reaction may be determined by a method that is reminiscent of the heats of formation version of Hess's Law from Chapter 9.

$$\Delta S^\circ = \sum_i \nu_i S^\circ(\text{product})_i - \sum_j \nu_j S^\circ(\text{reactant})_j \tag{10.3}$$

Here we designate the stoichiometric coefficient with the Greek letter, ν, and the subscripts i and j refer to individual product and reactant species. In practice this equation is used much like Hess's Law, as shown in the following example. One subtle difference between this and Hess's Law is that here we use *absolute* entropy values (S°) in contrast to the enthalpy *changes* for formation reactions (ΔH_f°) we used in Equation 9.12.

Note that there are chemicals for which the standard entropy is negative. In particular, ions in solution cannot be generated alone (without an oppositely charged counter ion) so to assign an entropy in such cases a standard must be chosen and entropies determined relative to that standard - meaning that negative values can be found for ions in aqueous solutions.

Compound	S° J mol^{-1} K^{-1}	Compound	S° J mol^{-1} K^{-1}
$H_2(g)$	130.6	$CO_2(g)$	213.6
$O_2(g)$	205.0	$C_4H_{10}(g)$	310.03
$H_2O(l)$	69.91	$CH_4(g)$	186.2
$H_2O(g)$	188.7	$C_2H_4(g)$	219.5
$NH_3(g)$	192.3	$C_3H_3N(l)$	178.91

Table 10.1 Standard molar entropies (S°) for selected substances are shown. A much larger listing appears in Appendix E. Values for many compounds can also be found online in the NIST Chemistry WebBook at http://webbook.nist.gov/chemistry.

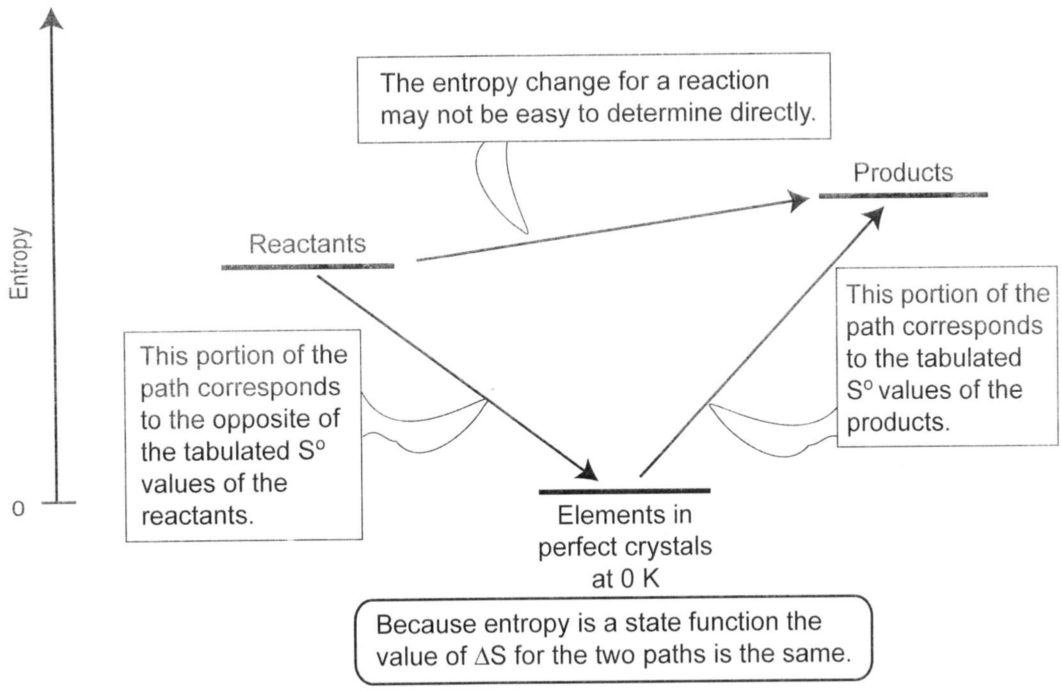

Figure 10.5 Entropy is a state function, so the value of ΔS must be independent of the path taken from reactants to products. When working with tabulated standard molar entropies, we implicitly choose a path in which reactants are converted to pure, perfect crystals of elements at 0 K, followed by the reaction of those elements to form the desired products. Such a path is fairly obviously not really feasible. Nevertheless, it allows us to obtain accurate values for entropy changes.

Example Problem 10.1

Polymerization reactions are complicated somewhat by the fact that they involve very large numbers of molecules. But we can demonstrate the general features of the thermodynamics of polymerization by considering a much smaller model system. Instead of considering the formation of polyethylene, for example, we can begin with the following reaction, in which two ethylene molecules combine to form butane.

$$2\,C_2H_4(g) + H_2(g) \rightarrow C_4H_{10}(g)$$

Use data from Table 10.1 to calculate $\Delta S°$ for this reaction.

Strategy

Any time we are asked to calculate the standard entropy change for a reaction, our first thought should be to try looking up values for standard molar entropy and using them in Equation 10.3. The two main things we need to be careful about are (1) to watch the state of the substances (in this case all are gases), and (2) to make sure we don't forget to include the stoichiometric coefficients in our calculations. Unlike heats of formation, the standard molar entropy of an element in its standard state is *not* zero, so we need to be sure to include everything appearing in the equation.

Solution

$$\Delta S° = S°(C_4H_{10}(g)) - 2S°(C_2H_4(g)) - S°(H_2(g))$$

$$= (310.03\ \mathrm{J\ K^{-1}}) - 2(219.5\ \mathrm{J\ K^{-1}}) - (130.6\ \mathrm{J\ K^{-1}}) = -259.6\ \mathrm{J\ K^{-1}}$$

Discussion

Whenever we carry out a calculation like this we should take time to decide if the result makes sense, considering the qualitative reasoning we have already developed. Does it make sense that this reaction has a negative change of entropy? The answer is "yes," because we have decreased the amount of gas present as a result of this reaction. Three moles of gaseous reactants are consumed, and only one mole of gaseous product is formed. Can this reaction be spontaneous? Yes, as it turns out. The explanation lies in the change in entropy of the surroundings. Because it forms a number of strong C–H bonds, this reaction is exothermic. And the release of heat will increase the entropy of the surroundings. This factor must be large enough to compensate for the decrease in entropy of the system itself.

Check Your Understanding

Acrylonitrile (C_3H_3N) is an important monomer for the production of many acrylic fibers. It can be produced from propene and ammonia according to the following reaction.

$$2\ C_3H_6(g) + 2\ NH_3(g) + 3\ O_2(g) \rightarrow 2\ C_3H_3N(l) + 6\ H_2O(g)$$

Given that $\Delta S°$ for this reaction is -43.22 J/mol, use data from Table 10.1 to calculate the standard molar entropy (S°) of $C_3H_6(g)$.

The case in the example problem provides a good example of one of the dangers of attempting to use ΔS to predict spontaneity. The need to consider both the system *and* surroundings not only makes for additional work, it also provides an opportunity to draw an incorrect conclusion if we forget to include the surroundings. As you might guess, having to account for the surroundings is rarely convenient. After all, the system is by definition what we are really interested in! Ideally we would be able to use a state function that predicts spontaneity with a single calculation for the system alone. Fortunately, such a state function exists, and so we now proceed to introduce the concept of free energy.

10.6 Gibbs Free Energy

In many ways, thermodynamics is one of the more powerful examples of the application of mathematics in science. If you continue in engineering, there is a good chance that you will eventually take a course in thermodynamics, in which you will develop a much more rigorous mathematical perspective. For now, we will simply introduce the fruits of the mathematical efforts of one of the scientists who invented thermodynamics, J. Willard Gibbs. Gibbs, motivated by an interest in predicting spontaneous processes, defined a new function that was ultimately called the **Gibbs free energy**, G.

Another way to define free energy called the Helmholz Free Energy is useful for constant volume conditions. Because such conditions are less common for engineering applications, we will not develop any of the mathematics for this definition of free energy.

$$G = H - TS$$

He realized that changes in this function could predict whether or not a process was spontaneous under conditions of constant pressure and temperature. This constraint is not that demanding, since many laboratory processes occur under these conditions (or approximately so). Now, if we focus on the change in the Gibbs free energy for a process at constant temperature, we have the following result.

$$\Delta G = \Delta H - T\Delta S \qquad (10.4)$$

From a practical point of view, this is perhaps the most important equation in this chapter.

Free energy and spontaneous change

How is this change in free energy connected to the spontaneity of a process? We have already established that the total entropy of the system plus its surroundings must increase for a spontaneous process. That total entropy change can be written as follows.

$$\Delta S_u = \Delta S_{sys} + \Delta S_{surr}$$

We usually refer to the entropy change for the system as simply ΔS, so we'll drop the "sys" subscript.

$$\Delta S_u = \Delta S + \Delta S_{surr}$$

The entropy change of the surroundings is due to the flow of heat, and since T is constant, it can be shown that ΔS_{surr} is just q_{surr}/T. And q_{surr} is $-q_{sys}$. For a process at constant pressure, q_{sys} is ΔH. Putting these ideas together, we have now established that

$$\Delta S_{surr} = -\frac{\Delta H}{T}$$

We can insert that into our equation above for ΔS_u.

$$\Delta S_u = \Delta S - \frac{\Delta H}{T}$$

If we multiply both sides of this by T, we should begin to see a connection to ΔG.

$$T\Delta S_u = T\Delta S_u - \Delta H$$

Comparing this with

$$\Delta G = \Delta H - T\Delta S,$$

we can see that

$$\Delta G = -T\Delta S_u.$$

In all of these equations, T is the absolute temperature, so its value is always positive. That means that the last equation assures us that ΔG will always have the opposite sign from ΔS_u. So if ΔS_u is positive – as it must be for any spontaneous process - ΔG must be negative.

All of this shows why chemists find ΔG to be such a useful thermodynamic quantity. It is a state function of the system, so it can be calculated fairly easily. And the sign of ΔG is sufficient to tell us whether or not a process will be spontaneous.

We can also use Equation 10.4 to help us formalize our understanding of the roles of ΔH and ΔS in determining the spontaneity of a given reaction. We have already argued that exothermic reactions, with $\Delta H < 0$, seem to be preferred over endothermic ones, and also that reactions where ΔS is positive seem to be preferred. Equation 10.4 shows us that if ΔH is negative and ΔS is positive, then ΔG will *always* be negative. But not all spontaneous processes have this specific set of circumstances. Table 10.2 shows the four possible combinations of signs for ΔH and ΔS. In the case in which the process is exothermic, but entropy decreases, we see that the sign on ΔG is dependent on the temperature. As T increases, the relative importance of the $-T\Delta S$ term also increases, so such processes will be spontaneous only at lower temperatures where the ΔH term is dominant. Processes that occur spontaneously only at lower temperatures are sometimes said to be **enthalpy driven**, because the enthalpy term is responsible for the negative value of ΔG. For endothermic processes, if the entropy of the system decreases, the sign on ΔG will always be positive, and the process is never spontaneous. By contrast, an endothermic process that increases the entropy may be spontaneous at high temperatures, where the –TΔS term becomes larger than the ΔH term of Equation 10.4. Reasoning similar to that above results in these processes being referred to as **entropy driven**. The reasoning associated with Table 10.2 can be used to understand the nature of phase changes as noted in Example Problem 10.2.

Sign of ΔH	Sign of ΔS	Implications for Spontaneity
–	+	Spontaneous at all temperatures
+	–	Never spontaneous
–	–	Spontaneous only at low temperatures
+	+	Spontaneous only at high temperatures

Table 10.2 The signs of ΔH and ΔS determine the way in which temperature influences the spontaneity of a chemical reaction. The four possible combinations of signs are summarized here.

Example Problem 10.2

Use the reasoning summarized in Table 10.2 to explain why ice spontaneously melts at room temperature, but not outside on a freezing winter's day.

Strategy

This problem requires us to understand and use the free energy equation qualitatively. We must be able to judge whether the process is endothermic or exothermic and how the process changes the entropy. Then, by considering Equation 10.4, we can look at the relative sizes of the two terms to explain the behavior described.

Solution

Melting is an endothermic process (ΔH > 0), because we must add heat to the system to effect the change. It is, however, also a process that increases the entropy, as molecules formerly held in place in a solid access greater space in the liquid and are therefore less ordered. Thus both ΔH and ΔS have positive values, and ΔG will be negative at high temperatures, where the TΔS term is larger. So, melting tends to occur at warmer temperatures, such as room temperature. At cold temperatures, the ΔH term is more important, so on a freezing day the sign of ΔG is positive; melting is not spontaneous and therefore is not observed. At the freezing point, ΔG is equal to zero and ice and water can coexist in any proportions.

Discussion

The result that ice melts if it is warm but not if it is cold is intuitively obvious. But this problem emphasizes the importance of gaining a qualitative understanding of entropy changes. Without that type of reasoning, we cannot hope to answer questions such as these.

Check Your Understanding

At what temperatures would you expect gas to condense into a liquid? Explain why the free energy change is negative for this process at these temperatures.

If the actual values of ΔH and ΔS are known, then the same type of argument used in Example Problem 10.2 can also be extended to obtain quantitative information.

Example Problem 10.3

The heat of fusion for crystalline polyethylene is approximately 7.7 kJ/mol, and the corresponding entropy change for melting is 19 J/mol K. Use these data to estimate the melting point of polyethylene.

Strategy

Since ΔH and ΔS are both positive, we know that ΔG must be positive at low temperatures and negative at higher temperatures. The melting point marks the dividing line between "low" and "high" temperatures, so at the melting point itself, ΔG must be equal to zero. We can therefore start with Equation 10.4 and set $\Delta G = 0$. That leaves T as the only unknown, so we can solve for the desired melting point.

Solution

We begin with Equation 10.4, and set $\Delta G = 0$. (Adding a subscript 'm' to the temperature will help remind us that this equation is only valid at the melting temperature.)

$$\Delta G = \Delta H - T_m \Delta S = 0$$

Rearranging this gives us

$$\Delta H = T_m \Delta S$$

Since ΔH and ΔS are known, solving for T is simple. (We do need to take care in handling the units, though, since ΔH is in kJ while ΔS is in J.)

$$T_m = \frac{\Delta H}{\Delta S} = \frac{7{,}700 \text{ J mol}^{-1}}{19 \text{ J mol}^{-1} \text{ K}^{-1}} = 405 \text{ K}$$

Discussion

The result seems reasonable: 405 K is 132°C, or 270°F. You may recall from our earlier discussions of polymers that the molecules in a given sample of polymer usually do not all have the same chain length. Thus there is a bit of difficulty in specifying quantities "per mole." The values used here were actually measured per mole of monomer unit.

Check Your Understanding

Poly(tetrafluoroethylene) melts at approximately 327°C. If the heat of fusion is ΔH_{fus} = 5.28 kJ mol^{-1}, what is the molar entropy change of fusion (ΔS_{fus})?

Free Energy and Work

By now you are probably wondering about the name: Why is it called *free energy*? The change in Gibbs free energy can be shown to be equal to the maximum useful work that can be done by the system,

$$\Delta G = -w_{max} \tag{10.5}$$

We must include a minus sign in this expression to be consistent with the convention that *w* is the work done *on the system*. This relationship suggests that ΔG tells how much energy is "free" or available to do something useful. For an engineer interested in making practical use of a chemical reaction, the implications should be clear.

Keep in mind that work is not a state function. So the maximum work will only be realized if we carry out the process by a specific path. In this case the requirement is that the change is carried out along a reversible path. This means that the system is near equilibrium, and a small incremental change in a variable will bring the system back to its initial state. Maximum work is possible only for reversible processes. Systems that are far from equilibrium usually undergo irreversible changes. In an **irreversible** change, a small incremental change of any variable does not restore the initial state. The amount of work available in an irreversible change is always less than the maximum work. So while the free energy change can be used to establish an upper bound to the amount of work that might be obtained from a given process, the actual work produced in any real application may be considerably less. Reactant mixtures like those for combustion reactions are generally very far from equilibrium. Systems that are far from equilibrium often change rapidly, and rapid changes tend to be irreversible.

Because irreversible changes are challenging to calculate, we will not attempt any such numerical exercises in our treatment of thermodynamics.

To make quantitative use of these free energy concepts, of course, we still need a simple method for obtaining accurate values for free energy changes.

10.7 Free Energy and Chemical Reactions

Free energy is a state function, and so the value of the free energy of a system depends on specific variables such as concentrations or pressures. As usual, to provide for consistent comparison we define a *standard state* to be one atmosphere of pressure and concentrations of solutions to be 1 M. The free energy change under these conditions is the **standard Gibbs free energy** change, ΔG^o. While it is feasible to calculate this value from ΔH° and S^o at a given temperature using Equation 10.4, the most convenient means to calculate the change in free energy for many reactions is to use a formulation analogous to Hess's Law for enthalpy changes.

Hess's Law for enthalpy in terms of heats of formation was given in Equation 9.12.

$$\Delta G^\circ = \sum_i \nu_i \Delta G_f^\circ(\text{product})_i - \sum_j \nu_j \Delta G_f^\circ(\text{reactant})_j \tag{10.6}$$

Compound	ΔG°_f kJ mol^{-1}	Compound	ΔG°_f kJ mol^{-1}
$H_2(g)$	0	$CO_2(g)$	-394.4
$O_2(s)$	0	$C_4H_{10}(g)$	-15.71
$H_2O(l)$	-237.2	$CH_4(g)$	-50.75
$H_2O(g)$	-228.6	$C_2H_4(g)$	68.12
$NH_3(g)$	-16.5	$C_3H_6(g)$	62.75

Table 10.3 Values of the free energy change of formation, ΔG°_f, are shown for selected compounds. A much larger listing appears in Appendix E.

In Equation 10.6 we are again utilizing the concept of the formation reaction from Chapter 9. Tabulated values of free energies of formation are provided in Table 10.3 for a few molecules, with a more extensive list in Appendix E. Note that elements in their standard states have ΔG_f° of zero, for the same reason their heats of formation are zero. Because the formation reaction uses elements in their standard states to define the reactants, a formation reaction of an element would have the same chemical species as both reactant and product, and this process would clearly have no change in any thermodynamic state function. We can use Equation 10.6 to calculate standard free energy changes, as demonstrated in Example Problem 10.4.

Example Problem 10.4

In Example Problem 10.1 we considered an addition reaction between two ethylene molecules and found that the entropy change was negative. We suggested at the time that this reaction would still be spontaneous because it is strongly exothermic. Confirm this by calculating the free energy change for the same reaction using values from Table 10.3.

$$2\ C_2H_4(g) + H_2(g) \rightarrow C_4H_{10}(g)$$

Strategy

Any problem that requests a calculation of a state function using values from a table generally means we'll be using a formulation like Equation 10.6. Free energy is the third such state function we have encountered that can be treated in this way. Again, if we're careful to note the states of the molecules present and watch the stoichiometric coefficients, this type of problem can readily be solved.

Solution

$$\Delta G^\circ = \Delta G_f^\circ\,(C_4H_{10}(g)) - 2\Delta G_f^\circ\,(C_2H_4(g)) - \Delta G_f^\circ\,(H_2(g))$$
$$= (-15.71\text{ kJ}) - 2(68.12\text{ kJ}) - (0) = -151.95\text{ kJ}$$

Discussion

This value is negative, indicating that this reaction would indeed be spontaneous under standard conditions. It's worth pointing out again, though, that this does not mean the reaction will occur readily if we mix the reactants. Simply because a reaction has a negative value for ΔG° does not mean we will observe the spontaneous conversion. In some cases the kinetics of the reaction are so slow that despite the thermodynamic indication of spontaneity, the reaction is not observed. Thermodynamics provides us with no information about how rapidly a spontaneous process will take place.

Check Your Understanding

For the reaction shown below, ΔG° = -1,092.3 kJ. Find ΔG_f° for liquid acrylonitrile (C_3H_3N).

$$2\ C_3H_6(g) + 2\ NH_3(g) + 3\ O_2(g) \rightarrow 2\ C_3H_3N(l) + 6\ H_2O(g)$$

Implications of ΔG° for a Reaction

Now that we have a method for calculating the standard change of free energy, what does knowing this value tell us? We've just seen that one answer is that the free energy change tells us the maximum useful work that could be obtained from a reaction. But the free energy change also has very important implications for the many chemical reactions that can be made to run in either direction. Earlier in this chapter we considered the thermal depolymerization (or thermolysis) of PMMA as a possible alternative to mechanical recycling. We can now use free energy changes to explore that idea more quantitatively. The polymerization of methyl methacrylate has ΔH° = -56 kJ and ΔS° = 117 J/K. We can use these values in Equation 10.4 to find ΔG° = -21 kJ. (We use the standard temperature of 298 K, which gives us the standard free energy change.) The negative value tells us that the formation of the polymer is spontaneous at 298 K. This, of course, means that the reverse reaction, in which PMMA is broken down into monomers, cannot be spontaneous at this temperature. Since the depolymerization amounts to running the polymerization reaction backwards, we can conclude that it must have $\Delta G^\circ = +21$ kJ.

Polymerization: methyl methacrylate monomer → PMMA polymer $\Delta G^\circ = -21$ kJ

Depolymerization: PMMA polymer → methyl methacrylate monomer $\Delta G^\circ = +21$ kJ

The fact that the numerical values of these free energy changes are relatively small hints at the fact that a relatively small shift in temperature might be able to flip the signs, making the depolymerization the thermodynamically preferred process. We can quantify this idea, too. Using the same sort of calculation as we did in Example Problem 10.3, we can find the temperature at which ΔG° will equal zero.

$$T = \frac{\Delta H}{\Delta S} = \frac{-56 \text{ kJ}}{0.117 \text{ kJ K}^{-1}} = 479 \text{ K}$$

Above this temperature, entropy wins out over enthalpy and the reaction will be driven "backwards," towards the monomer. The fact that this temperature threshold is reasonably low is one reason why depolymerization is a viable recycling strategy for PMMA. Most polymerization reactions have ΔS° values around 100 J/K, but most are more strongly exothermic than what we see here for PMMA. Looking at the equation above, we can see that this leads to a higher temperature requirement for depolymerization. So thermolysis is less feasible for two reasons. First, the need for higher temperatures implies a higher cost. Second, the higher temperatures increase the likelihood that additional chemical reactions, like the breakdown of monomers into other compounds, would compete with the thermolysis.

10.8 Insight into the Economics of Recycling

According to the Container Recycling Institute, almost 50% of all aluminum beverage cans sold in the U.S. in 2001 were recycled. But data from the American Plastics Council indicate that in the same year only about 22% of PET bottles were recycled. It seems fair to assume that individual consumers are no more anxious to recycle aluminum cans than plastic bottles. So there must be a real cause for the difference between those two numbers. And that cause lies in economics. When it comes right down to it, the business of recycling is all about trying to sell your trash. And as you might guess, selling trash can be a difficult business to be in. In this section we'll take a look at some of the factors that make aluminum recycling so much more attractive than plastic recycling.

We'll begin by looking at the recycling of aluminum, which now accounts for nearly all beverage cans sold in the U.S. The chemistry of aluminum makes it an excellent target for recycling. Recall that in Chapter 1 we pointed out how difficult it is to obtain pure aluminum from ores like bauxite. Aluminum reacts readily with oxygen, forming strong chemical bonds that are not easily disrupted. Extracting aluminum from its ores requires the use of very high temperatures, and therefore consumes a great deal of energy. When aluminum cans are recycled, however, the paint and other coatings used can be removed relatively easily, allowing the underlying aluminum to be melted down and reprocessed into a new can. Current industry estimates are that making four new cans from recycled aluminum uses the same amount of energy as producing one can from raw aluminum ore. This provides strong economic incentives throughout the entire process. First, the cost savings mean that the beverage industry has a strong motivation to encourage recycling and to actively seek recycled aluminum to produce new cans. This, in turn, means that communities and private companies in the recycling business are assured of finding a good market for any aluminum cans they can collect – an important concern since there will always be costs connected with the collection of the recyclables. Other positive factors also exist in the recycling equation for aluminum. The fact that essentially all beverage cans in use are made of aluminum means there is no need to separate collected cans into different types. And because aluminum cans are so easily crushed, the collected material can be compressed so that storage and transportation are easier and less expensive. All in all, aluminum is an ideal candidate for recycling.

Now let's contrast that with the case for plastics. We can begin by thinking about the costs associated with the production of virgin polymers, since these costs will set the standard against which the price of recycled plastics will be measured. Feedstocks for virtually all commercial polymer productions have their roots in petroleum. So the cost of the raw materials for the synthesis of most plastics is linked to the price of oil. Oil is a complex mixture of compounds, and our new understanding of entropy should make clear that such a mixture is unlikely to spontaneously separate into its various components. Various separation and purification methods are used to obtain the needed monomer molecules. Most of these schemes involve cracking and distillation, which requires heating the crude oil until various components boil off and can be reclaimed from the vapor phase. This need for heating means that there is also an energy cost associated with the production of the feedstock for polymerization. Once we have a supply of the appropriate monomer (and any other reagents that may be needed), we will have to pay to transport them to the plant where the polymer will be produced. That cost can be minimized, of course, if the plant where the polymer will be made is located close to a refinery where the oil is processed.

Many polymerization reactions are spontaneous under ordinary conditions. But as we've learned, that doesn't necessarily mean that those reactions are fast. And if we are in the business of making plastics, we will probably not be satisfied if our polymer production requires days or even years. Most reactions run faster when heated, though, as we'll see when we examine chemical kinetics in Chapter 11. So we will probably want to carry out our polymerization at higher temperatures. This will add a further energy cost to the bill for producing our polymer.

So how will this compare to the cost of recycling plastics? Just as for aluminum cans, there will be some costs associated with the collection of the bottles to be recycled. But the fact that plastic bottles are made from a wide array of different polymers offers some added complications here. The various types of polymers must usually be separated before they can be processed. Consumers can be encouraged to do some of this separation based on the recycling codes commonly found on bottles. (See Table 10.4.) But in some cases even different colors of the same type of plastic may be incompatible, and there will inevitably be some mixing of bottle types in the collection bins. So the recycler must expect to have to sort the materials before they can be sold. This is most often done by hand, although sometimes flotation and separation based on density may be possible. No matter the method, though, this separation adds cost to the overall recycling effort. Once bottles of a particular type have been separated out, they are ready to be processed; this would typically include crushing, washing, and melting the plastic before forming it into pellets to be resold. Again, each of those steps adds some cost.

Symbol	Polymer	Repeat unit	Sources	Recycled products
1 PETE	polyethylene terephthalate	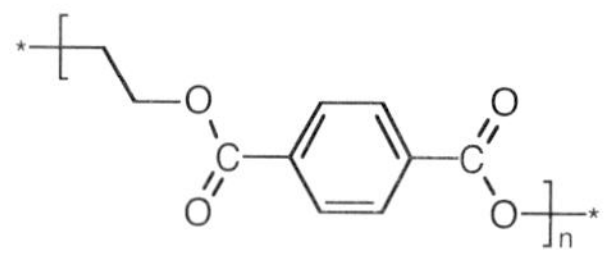	soda bottles, peanut butter jars, vegetable oil bottles	fiber, tote bags, clothing, film and sheet, food and beverage containers, carpet, fleece wear
2 HDPE	high density polyethylene	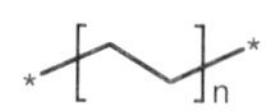	milk and water jugs, juice and bleach bottles	bottles for laundry detergent, shampoo, and motor oil; pipe, buckets, crates, flower pots, garden edging, film and sheet, recycling bins, benches, dog houses, plastic lumber
3 PVC	polyvinyl chloride	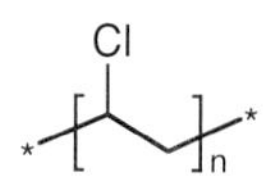	detergent / cleanser bottles, pipes	packaging, loose-leaf binders, decking, paneling, gutters, mud flaps, film and sheet, floor tiles and mats, resilient flooring, electrical boxes, cables, traffic cones, garden hose
4 LDPE	low density polyethylene	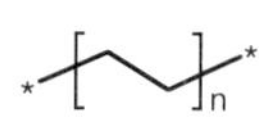	6-pack rings, bread bags, sandwich bag	shipping envelopes, garbage can liners, floor tile, furniture, film and sheet, compost bins, paneling, trash cans, landscape timber, lumber
5 PP	polypropylene	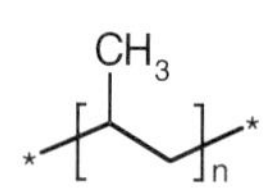	margarine tubs, straws, screw-on lids	automobile battery cases, signal lights, battery cables, brooms, brushes, ice scrapers, oil funnels, bicycle racks, rakes, bins, pallets, sheeting, trays.
6 PS	polystyrene	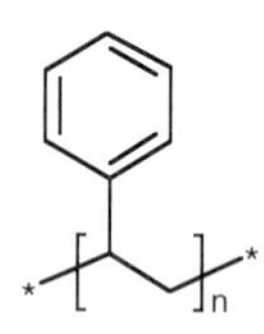	styrofoam, packing peanuts, egg-cartons, foam cups	thermometers, light switch plates, thermal insulation, egg cartons, vents, desk trays, rulers, license plate frames, foam packing, foam plates, cups, utensils
7 OTHER	miscellaneous and multi-layer plastics	N/A	squeezable ketchup and syrup bottles	bottles, plastic lumber

Table 10.4 – The symbols shown in the left hand column are used to identify common household plastics so that consumers can sort them for recycling. Many communities provide curbside pick-up of PETE and HDPE to be recycled, but recycling of other plastics is much less common.

While consumers may toss a bottle into a recycling bin out of a sense of "doing a good deed," a company that might be interested in buying the recycled plastic pellets is more likely to have an eye on the bottom line. So in order for the resulting recycled pellets to be attractive, they will have to be priced competitively with virgin polymer. And here things are not nearly as favorable as for aluminum. The production of most plastics from raw feedstocks is generally less expensive than the processing of aluminum from its ore. So in economic terms the standard to which recycled plastic will be compared is more demanding. There is also an important difference in the ability of recycling to produce high quality materials. Once the aluminum has been reclaimed from a recycled can and melted down, it is indistinguishable from "new" aluminum freshly extracted from ore. But for plastics, the recycling process leads to degradation of the polymer. This is really just an effect of the Second Law: the entropy of a polymer molecule will generally increase if the long chain is broken into shorter pieces. Chain lengths in recycled plastics are invariably shorter than in virgin materials, and this can raise concerns as to whether the recycled material offers sufficient strength to satisfy design requirements. On balance, then, many businesses find it less expensive to use virgin plastics rather than recycled polymers.

Recall that for practical samples, there is a distribution of polymer sizes. The degradation assocaited with recycling tends to shift this distribution towards smaller polymers.

Are there circumstances that would be likely to shift this balance? That would require either that recycling become less expensive or that the production of virgin polymers become more expensive (or both). The cost of virgin materials has its roots in the price of the petroleum from which the needed feedstocks are obtained. So a significant rise in oil prices would tend to drive up the cost of nearly all plastics, too. But since many of the costs associated with recycling are energy related, they are also tied in one way or another to oil prices. So it is not immediately clear that plastic recycling can be made more economically attractive in the near term.

If recycled plastics cannot compete effectively with virgin materials in applications like drink bottles, then might there be other outlets in which the material can be put to use? One increasingly popular idea is to use recycled plastic in applications typically calling for more traditional (i.e., non-plastic) materials. The rapidly growig market for "lumber" made from recycled plastic is perhaps the most prominent example. Recycled plastic is now fairly widely used in building decks, picnic tables, playground equipment, and other items that we generally picture as being made of wood. While this may not appeal to traditionalists, the plastic alternatives offer better weather resistance and durability. Thus they can be a cost-saving choice, especially if one considers long term savings on maintenance costs. Other areas in which recycled plastic now competes with non-plastic materials include filling for winter coats and sleeping bags.

Focus on Problem Solving

Question

Thermodynamics provides a way to interpret everyday occurrences. If you live in northern climates, one common experience is that during early winter snow falls, but melts when it hits the ground. Both the formation and the melting happen spontaneously. How can thermodynamics explain both of these seemingly opposite events?

This is a conceptual question that requires that we interpret events in terms of Gibbs free energy. To carry out this analysis we need to think of the two components (or terms) of the Gibbs energy and think about the role of temperature in determining which of them dominates.

Answer

When snow forms, liquid water becomes a solid, which decreases the entropy of the system and is an exothermic process. Because the entropy contributes according to the term $-T\Delta S$, a decrease in entropy provides a positive value and the enthalpy term is negative. This combination will be negative (and therefore a process is spontaneous) at lower temperatures. Thus snow forms in the cold air of clouds. The opposite process, melting, is endothermic but increases the entropy. In this case, the Gibbs free energy will be negative when temperature is higher (i.e., where the $-T\Delta S$ term dominates). When the snow hits the somewhat warmer ground, the higher temperature there makes the melting process spontaneous even though the formation process was also spontaneous.

Chapter 10
Problems & Exercises

Recycling of Plastics

10.1 "Reduce, reuse, recycle" is a common slogan among environmentalists, with the order of the three words indicating the perceived relative benefits. List some reasons why recycling is the least desirable of these three approaches to waste reduction.

10.2 Is the recycling or most plastics primarily a chemical process of a physical one? Explain and defend your choice.

10.3 List some consumer products that tend to be made from recycled PET.

10.4 Why is recycled PET only rarely used to make new soft drink bottles?

10.5 Use the web to find a company that specializes in the recycling of plastics. Does the site emphasize environmental, other scientific or economic benefits? Based on the science you can find on the site write a brief essay explaining how the science fits into one of these categories.

10.6 Use the web to research how many pounds of plastics are recycled in your area. How has this value changed over the past decade?

Spontaneity

10.7 The ability to violate the direction of spontaneous change in nature is a common theme in movies. Describe a scene from a movie or television show that is not true to the common experience of nature and the direction it always progresses.

10.8 In the thermodynamic definition of a spontaneous process, why is it important that the phrase "continuous intervention" be included rather than just "intervention?"

10.9 If butane undergoes a combustion reaction that is considered spontaneous, how can a butane lighter be carried safely in your pocket or purse?

10.10 For each of the processes listed, identify it as spontaneous or non-spontaneous. For each non-spontaneous process, describe the "opposite direction" spontaneous process.

(a). A group of cheerleaders builds a human pyramid.
(b). Table salt dissolves in water.
(c). A cup of cold coffee in a room becomes steaming hot.
(d). Water molecules in the air generate hydrogen gas and oxygen gas.
(e). A person peels an orange, and you smell it from across the room.

10.11 For each of the processes listed, identify it as spontaneous or non-spontaneous. For each non-spontaneous process, describe the "opposite direction" spontaneous process.

(a). Oxygen molecules split to form oxygen atoms.
(b). A tray of water is placed in the sun on a warm day and freezes.
(c). A solution of salt water forms a layer of acid on top of a layer of basic solution.

(d). Silver nitrate is added to a solution of sodium chloride and a precipitate forms.
(e). Sulfuric acid sitting in a beaker turns into water by giving off $SO_3(g)$.

10.12 Athletic trainers have "ice packs" that are not cold until the trainer snaps an inner container in the pack, causing the pack to become cold. What is the nature of heat flow for this spontaneous process?

10.13 Are any of the following exothermic processes not spontaneous under any circumstances?

(a). Snow forms from liquid water.
(b). Liquid water condenses from water vapor.
(c). Fossil fuels burn to form carbon dioxide and water.
(d). Monomers react to form a polymer.

10.14 Enthalpy changes often help predict whether or not a process will be spontaneous. What type of reaction is more likely to be spontaneous, an exothermic or an endothermic one? Provide two examples that support your assertion and one counterexample.

10.15 When a fossil fuel burns, is the system the fossil fuel? Explain your answer.

10.16 Murphy's Law is a whimsical rule that says that anything that can go wrong will go wrong. But in an article in the *Journal of Chemical Education,* Frank Lambert writes: "Murphy's Law is a fraud." He also writes: "The second law of thermodynamics is time's arrow, but chemical kinetics is its clock." Read Lambert's article (*J. Chem. Ed.*, 74(8), 1997, p. 947), and write an essay in which you explain in the context of the latter quote why Lambert claims Murphy's Law is a fraud. (For more of Professor Lambert's unique insights into thermodynamics, see his web site at http://www.secondlaw.com/)

10.17

Humpty Dumpty sat on a wall,

Humpty Dumpty had a great fall.

All the King's horses and all the King's men

Couldn't put Humpty together again.

In Lewis Carroll's *Through the Looking Glass*, Alice encounters Humpty Dumpty, a human-sized egg sitting on a wall. Alice, who is familiar with the nursery rhyme, asks anxiously, "Don't you think you'd be safer on the ground? That wall is so narrow." Humpty, an egg with an attitude, growls, "Of course I don't think so. Why if I ever *did* fall off—which there's no chance of—but if I did... the King has promised me—with his very own mouth—[that] they'd pick me up again in a minute, they would!"

Write a paragraph in the voice of seven-and-one-half-year-old Alice, explaining to Humpty in the context of this section (a) the probability that Humpty will fall off, and (b) the probability that the King's horses and men will be able to put him back together again.

10.18

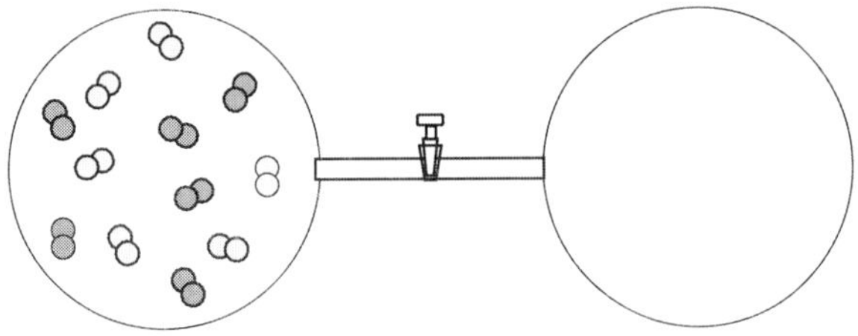

The vessel on the left contains a mixture of oxygen and nitrogen at atmospheric pressure. The vessel on the right is evacuated.

(a). Describe what will happen when the stopcock is opened.
(b). If you could see the molecules, what would you observe after a period of time has passed?
(c). Explain your answers to (a) and (b) in terms of probabilities.
(d). What is the probability at any one moment that all the oxygen molecules will be in one vessel and the nitrogen molecules in the other? Explain.

Entropy

10.19 When Carnot originally proposed that a new state function must exist, what observation about the Carnot engine caused him to assert that idea?

10.20 Some games include dice with shapes other than a cube. If you roll two dice with eight equivalent sides, numbered 1 – 8, what is the probability of rolling two 8's? What is the most probable roll?

10.21 How does probability relate to spontaneity?

10.22 Define the concept of a microstate. How is this concept related to the order or disorder of a system?

10.23 Using the Boltzmann distribution, explain how adding heat to a system will increase the number of microstates, and therefore the entropy.

10.24 When ice melts, its volume decreases. Despite this fact, the entropy of the system increases. Explain (a) why the entropy increases and (b) why under most circumstances a decrease in volume results in an entropy decrease.

10.25 If a sample of air is separated into nitrogen and oxygen (ignoring other gases present), what is the sign of ΔS for this process? Explain your answer.

10.26 (a). List three chemical reactions that increase the entropy of the system. (Ignore any changes on the surroundings.)

(b). List three chemical reactions that decrease the entropy of the system. (Again, ignore the surroundings.)

10.27 Many metals form solid carbonates, such as $CaCO_3$. When carbonates are heated, CO_2 can be driven off, leaving behind a metal oxide. What is the sign of the entropy change of the system for this type of chemical reaction? Explain your answer.

10.28 For the following chemical reactions, predict the sign of ΔS for the system. (Note that this should not require any detailed calculations.)

(a). $Fe(s) + 2HCl(g) \rightarrow FeCl_2(s) + H_2(g)$

(b). $3NO_2(g) + H_2O(l) \rightarrow 2HNO_3(l) + NO(g)$

(c). $2K(s) + Cl_2(g) \rightarrow 2KCl(s)$

(d). $Cl_2(g) + 2NO(g) \rightarrow 2ClNO(g)$

(e). $SiCl_4(g) \rightarrow Si(s) + 2Cl_2(g)$

10.29 When we consider the fact that a leaf is an example of exquisite order, how can it form spontaneously in nature? What natural process shows that the order found in a leaf is essentially temporary?

The next four questions relate to the following paragraph (Frank L. Lambert, *Journal of Chemical Education*, 76(10), 1999, 1385).

The movement of macro objects from one location to another by an external agent involves no change in the objects' physical (thermodynamic) entropy. The agent of movement undergoes a thermodynamic entropy increase in the process.

10.30 A student opens a stack of new playing cards and shuffles them. In light of the paragraph above, have the cards increased in entropy? Explain your answer in terms of thermodynamics. Explain why the agent (the shuffler) undergoes an increase in entropy.

10.31 An explosion brings down an old building. A pile of rubble is all that remains. Where did the thermodynamic entropy increase occur? Write a paragraph to explain your reasoning.

10.32 Write two examples of your own that illustrate the concept in the paragraph above.

10.33 According to Lambert, leaves lying in the yard and playing cards that are in disarray on a table have not undergone an increase in their thermodynamic entropy. Suggest another reason why leaves and playing cards may not be a good analogy for the entropy of a system containing, for example, only H_2O molecules or only O_2 molecules.

10.34 A researcher heats a sample of water in a closed vessel until it boils.

(a). Does the entropy of the system increase?
(b). Has the randomness of the molecules increased? (In other words, are there more physical positions the molecules can occupy?)
(c). What else has increased that affects the entropy of the system?

The researcher now heats the water vapor from 400 K to 500 K, keeping the volume constant.

(d). Does the entropy of the system increase?
(e). Has the randomness of the molecules increased? (In other words, are there more physical positions the molecules can occupy?)
(f). Why has an increase in temperature of the gas at constant volume caused an increase in entropy?

(To delve more deeply into the concept of entropy, read John P. Lowe's article "Entropy: Conceptual Disorder" in *Journal of Chemical Education*, 65(5), 1988, 403-406.)

The Second Law of Thermodynamics

10.35 What happens to the entropy of the universe during a spontaneous process?

10.36 Why do we need to consider the surroundings of a system when applying the second law of thermodynamics?

10.37 How does the statement of the second law of thermodynamics that heat cannot be turned completely into work relate to the statement of the second law from this chapter that is cast in terms of entropy?

10.38 According to the second law of thermodynamics, how does the sign of ΔS_u relate to the concept that some energy is "wasted" or lost to the surroundings when we attempt to convert heat into work?

10.39 How does the second law of thermodynamics explain a spontaneous change in a system to become more ordered when that process is exothermic?

10.40 When we consider chemical reactions, what is the most common way that entropy changes occur in the surroundings?

10.41 When a reaction is exothermic, how does that influence ΔS of the system? Of the surroundings?

10.42 Which reaction occurs with the greater increase in entropy? Explain your reasoning.

(a.) $2H_2O(l) \rightarrow 2H_2(g) + O_2(g)$

(b). $C(s) + O_2(g) \rightarrow CO_2(g)$

10.43 Which reaction occurs with the greater increase in entropy? Explain your reasoning.

(a). $2NO(g) \rightarrow N_2(g) + O_2(g)$

(b). $Br_2(g) + Cl_2(g) \rightarrow 2BrCl(g)$

10.44 The entropy change for the combustion of methanol (used in race cars) is negative. We know that once methanol is ignited the reaction is spontaneous. Why doesn't this reaction violate the second law of thermodynamics?

10.45 Limestone is predominantly $CaCO_3$, which can undergo the reaction, $CaCO_3(s) \rightarrow CaO(s) + CO_2(g)$. We know from common experience that limestone does not spontaneously undergo this reaction, yet the ΔS for the reaction is positive. How can the second law of thermodynamics explain that this reaction is NOT spontaneous?

The Third Law of Thermodynamics

10.46 Suppose you find out that a system has an absolute entropy of zero. Based on this fact, what else do you know about that system?

10.47 Use data from the tables in Appendix E to calculate the entropy change of each of the reactions listed below.

(a). $Fe(s) + 2HCl(g) \rightarrow FeCl_2(s) + H_2(g)$

(b). $3NO_2(g) + H_2O(l) \rightarrow 2HNO_3(l) + NO(g)$

(c). $2K(s) + Cl_2(g) \rightarrow 2\,KCl(s)$

(d). $Cl_2(g) + 2NO(g) \rightarrow 2ClNO(g)$

(e). $SiCl_4(g) \rightarrow Si(s) + 2\,Cl_2(g)$

10.48 If you scan the values for S° in the Appendix E, you will see that several aqueous ions have values that are less tan zero. The third law of thermodynamics states that for a pure substance the entropy goes only to zero at 0 K. Use your understanding of water solvation to explain how a negative value of S° must arise.

10.49 Calculate ΔS for the process by which magnesium chloride dissolves, $MgCl_2(s) \rightarrow Mg^{2+}(aq) + 2\ Cl^-(aq)$. Use your understanding of how water solvates ions at the molecular level to explain the sign of ΔS.

10.50 Calculate the standard entropy change for the reaction, $CO_2(g) + 2H_2O(l) \rightarrow CH_4(g) + 2O_2(g)$. What does the sign of ΔS° say about the spontaneity of this reaction?

10.51 Through photosynthesis, plants build molecules of sugar containing several carbon atoms from carbon dioxide, which contains only one carbon atom. In the process, entropy is decreased. The reaction of CO_2 with formic acid to form oxalic acid provides a simple example of a reaction in which the number of carbon atoms in the compound increases:

$$CO_2(aq) + HCOOH(aq) \rightarrow H_2C_2O_4(aq)$$

(a). Calculate the entropy change for this reaction and discuss the sign of ΔS°.
(b). How do plants carry out reactions that increase the number of carbon atoms in a sugar, given the changes in entropy for reactions like this?

10.52 Find sites on the World Wide Web that describe two different attempts to reach the coldest temperature on record. What features do these experiments have in common?

10.53 Look up the value of the standard entropy for the following molecules. After ranking them from 1st to 3rd, explain why this ranking makes sense: $CH_4(g)$, $C_2H_5OH(l)$, $CO_2(s)$

10.54 Look up the value of the standard entropy for the following molecules. After ranking them from 1st to 3rd, explain why this ranking makes sense: $SiO_2(s)$, $NH_3(g)$, $C_2H_6(g)$

10.55 A beaker of water at 40°C (on the left in the drawing) and a beaker of ice water at 0°C are placed side by side in an insulated container. After some time has passed, the temperature of the water in the beaker on the left is 30°C and the temperature of the ice water is still 0°C. Describe what is happening in each beaker: (a). on the molecular level, and (b). in terms of the second law of thermodynamics.

Gibbs Free Energy

10.56 Describe why it is easier to use the calculation of ΔG for determining the spontaneity of a process than a calculation of ΔS.

10.57 Under what conditions does ΔG allow us to predict whether a process is spontaneous?

10.58 There is another free energy state function, Helmhotz free energy, that is defined as F = E – TS, where internal energy has replaced enthalpy in the definition. Under what conditions would this free energy tell us whether or not a process is spontaneous?

10.59 Based on Equation 10.4, explain why the sign on ΔG for a spontaneous process must be negative.

10.60 For an endothermic process that increases disorder in the system, explain why high temperatures are needed to have a spontaneous change.

10.61 For an exothermic process that decreases disorder in the system, explain why low temperatures are needed to have a spontaneous change.

10.62 Why is the free energy change of a system equal to the *maximum* work rather than just the work?

10.63 Distinguish between a reversible and an irreversible process.

10.64 For the reaction, $NO(g) + NO_2(g) \rightarrow N_2O_3(g)$, use data in Appendix E to calculate $\Delta H°$, $\Delta S°$, and $\Delta G°$ using Equation 10.4.

(a). Is this reaction spontaneous at 25°C? Explain your answer.
(b). Based on your results for enthalpy and entropy, predict whether you need higher or lower temperatures to have a spontaneous reaction as written.
(c). To show that your prediction is accurate, choose a temperature that corresponds to your prediction in part (b), and calculate ΔG. (Assume that both enthalpy and entropy are independent of temperature.)

10.65 The equation for the combustion of acetylene is

$$C_2H_2\,(g) + 5/2\,O_2\,(g) \rightarrow 2CO_2\,(g) + H_2O(g) \quad \Delta H° = -1255.5\text{ kJ}$$

(a). Use the data in Appendix E to calculate $\Delta S°$ for this reaction.
(b). Calculate $\Delta G°$ and show that the reaction is spontaneous at 25°C.
(c). Does a temperature range exist where this reaction is not spontaneous?
(d). Do you think you could reliably use Equation 10.4 to calculate such a temperature change? Explain your answer.

10.66 Natural gas (methane) is being used in experimental vehicles as a clean-burning fuel.

(a). Write the equation for the combustion of $CH_4(g)$, assuming all reactants and products are in the gaseous phase.
(b). Use the data in Appendix E to calculate $\Delta S°$ for this reaction.
(c). Calculate $\Delta G°$ and show that the reaction is spontaneous at 25°C.

10.67 Silane, SiH_4, is the first compound in a series that is analogous to the alkanes and has the general formula Si_nH_{2n+2}. Silane is used in the electronics industry to produce thin ultra pure silicon films, but it is somewhat difficult to work with, because $SiH_4(g)$ is *pyrophoric* at room temperature (meaning that it bursts into flame spontaneously.)

(a). Write the equation for the combustion of $SiH_4(g)$. SiO_2 is a solid at 25°C.
(b). Use the data in Appendix E to calculate $\Delta S°$ for this reaction.
(c). Calculate $\Delta G°$ and show that the reaction is spontaneous at 25°C.
(d). Compare $\Delta G°$ for this reaction to that of methane. (See the previous problem.) Are the reactions in these two exercises *enthalpy* or *entropy* driven? Explain.

Free Energy and Chemical Reactions

10.68 Explain why $\Delta G°_f$ of $O_2(g)$ is zero.

10.69 Using data from Appendix E, calculate $\Delta G°$ for these reactions.

(a). $Fe(s) + 2HCl(g) \rightarrow FeCl_2(s) + H_2(g)$

(b). $3NO_2(g) + H_2O(l) \rightarrow 2HNO_3(l) + NO(g)$

(c). $2K(s) + Cl_2(g) \rightarrow 2\ KCl(s)$

(d). $Cl_2(g) + 2NO(g) \rightarrow 2ClNO(g)$

(e). $SiCl_4(g) \rightarrow Si(s) + 2\ Cl_2(g)$

10.70 Using data from Appendix E, calculate ΔG° for these reactions.

(a). $Mg_3N_2(s) + 6H_2O(l) \rightarrow 2NH_3(g) + 3Mg(OH)_2(s)$

(b). $4CH_3NH_2(g) + 9O_2\ (g) \rightarrow 4CO_2(g) + 10\ H_2O(l) + 2N_2(g)$

(c). $Fe_3O_4(s) + 4CO(g) \rightarrow 3Fe(s) + 4CO_2(g)$

(d). $P_4O6(s) + 6H_2O(l) \rightarrow 4H_3PO_4(aq)$

10.71 Calculate ΔG° for the dissolving of both sodium chloride and silver chloride using data from Appendix E. Explain how the values you obtain relate to the solubility rules for these substances.

10.72 Consider the phase transition between white and red phosphorous as a reaction, P(s, white) → P(s, red). Use the data in Appendix E to determine which form of phosphorous is more stable at 25°C. Is this form of the solid more stable at all temperatures? If not, what temperatures are needed to make the other form more stable?

10.73 How could you use Equation 10.4 and the information in Appendix E to estimate the temperature at which white and red phosphorous are at phase equilibrium under standard conditions?

10.74 Calculate ΔG° for the complete combustion of one mole of the fossil fuels, methane (CH_4), ethane (C_2H_6), propane (C_3H_8) and n-butane (C_4H_{10}). Identify any trends that are apparent based on these calculations.

10.75 Imagine you are looking through old chemicals in a storeroom and you come across an old partially empty bottle of concentrated nitric acid. Rather than looking colorless it appears to be yellow/brown in color. Calculate ΔG° for the reaction, $4HNO_3(g) \rightarrow 4NO_2(g) + 2H_2O(l) + O_2(g)$, and use this result plus your knowledge about nitrogen dioxide to explain why the solution in the old bottle of nitric acid is no longer colorless.

10.76 Incomplete combustion of hydrocarbon fossil fuels occurs when too little oxygen is present and results in the production of carbon monoxide rather than carbon dioxide. Water is the other product in each case.

(a). Write the balanced chemical equation for complete and incomplete combustion of propane.

(b). Based on these reactions predict which will have the larger change in entropy.

(c). Use data from Appendix E to calculate ΔG° for each reaction.

(d). Based on these results predict the sign and value of ΔG° for the combustion of carbon monoxide to form carbon dioxide.

Economics of Recycling

10.77 During polymerization, the system gains order as monomers link together. Could an endothermic polymerization reaction ever occur spontaneously? Explain.

10.78 If a particular polymer formation reaction happened to be endothermic, how could it ever take place? Where would the increase in entropy of the universe have to arise?

10.79 When polymers are recycled, the ends of the long chain polymer molecules tend to break off, and this process eventually results in a loss of physical properties that render the recycled polymer unusable. Explain why the breaking off of the ends of the polymer molecules is favorable from the standpoint of the entropy of the system.

10.80 The recycling of polymers represents only one industrial process that allows the creation of order in one location by creating greater disorder at some other location, often at a power plant. List three other industrial processes that must create disorder in the surroundings in order to generate the desired material.

Additional Problems

10.81 Diethyl ether is a liquid at normal temperature and pressure, and boils at 35°C. Given that ΔH for the vaporization of diethyl ether is 26.0 kJ/mol, find the molar entropy change for the vaporization of diethyl ether.

10.82 Compare the enthalpy change for melting ice and for boiling water. Use molecular level reasoning to explain the relative size of these two numbers.

10.83 Gallium metal has a melting point of 29.8°C. Use the information below to calculate the boiling point of gallium.

Substance	$\Delta H_f°$ (kJ mol^{-1})	$\Delta G_f°$ (kJ mol^{-1})	S° (J K^{-1}mol^{-1})
Ga (s)	0	0	40.83
Ga (l)	5.578	0.0888	59.25
Ga (g)	271.96	233.76	169.03

10.84 Methane can be produced from CO and H_2. The process might be carried out in two steps, as shown below, with each step carried out in a separate reaction vessel within the production plant.

Reaction #1 $CO(g) + 2 H_2(g) \rightarrow CH_3OH(l)$ ΔS° = -332 J/K

Reaction #2 $CH_3OH(l) \rightarrow CH_4(g) + 1/2\, O_2(g)$ ΔS° = +162 J/K

Substance	$\Delta H_f°$ (kJ mol^{-1})	$\Delta G_f°$ (kJ mol^{-1})	S° (J K^{-1}mol^{-1})
CO(g)	-110.5		197.674
$CH_3OH(l)$	-238.7	-166.4	126.8
$CH_4(g)$	-74.8		186.2

NOTE: You should be able to work this problem without consulting the table in Appendix E for additional data.

(a). Calculate ΔH° for reaction #1.

(b). Calculate $\Delta G_f°$ for CO(g).

(c). Calculate S° for $O_2(g)$.

(d). At what temperatures is reaction #1 spontaneous?

(e). Suggest a reason why these two steps would need to be carried out separately.

10.85 Iodine is not very soluble in water, but dissolves readily in a solution containing iodide ions due to the reaction below.

$$I_2\,(aq) + I^-\,(aq) \rightarrow I_3^-\,(aq)$$

The temperature dependence of $\Delta G°$ for this reaction has been studied, and the following graph depicts the results. Notice that the quantity on the y-axis is $\Delta G°/T$, not just $\Delta G°$. Additional data relevant to this reaction are also given in the table below the graph. The solid line is a "best fit" to the actual data points.

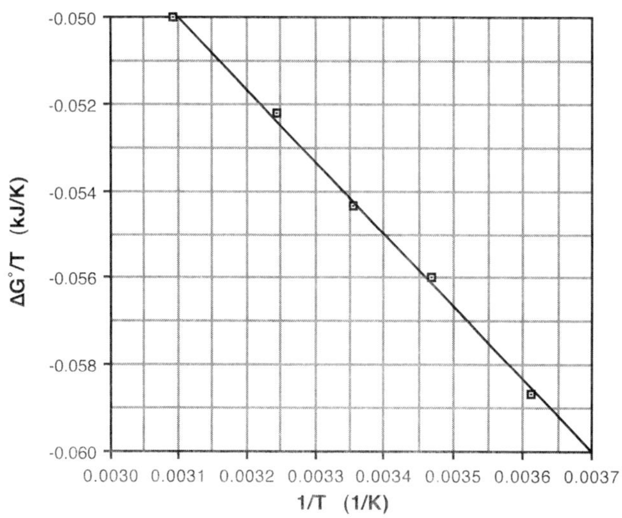

Substance	$\Delta G_f°$ (kJ mol^{-1})
I_2 (aq)	16.37
I^- (aq)	-51.57
I_3^- (aq)	-51.4

(a). Calculate $\Delta G°$ for this reaction at 298 K. (DO NOT read this value off the graph. Use the data given to calculate a more accurate value.)

(b). Determine $\Delta H°$ for this reaction. You should assume that $\Delta H°$ is independent of T.

(HINT: You will need to use the graph provided in order to find $\Delta H°$. It may help if you realize that the graph is a straight line and try to write an equation for the line.)

Focus on Problem Solving

10.86 Suppose you need to know the heat of formation of 2-pentene, C_5H_{10}, but the tables you have do not provide the value. You have a sample of the chemical. What could you do to determine the heat of formation?

10.87 Suppose you need to know the heat of formation of cyclohexane, C_6H_{12}, but the tables you have do not provide the value. You have a sample of the chemical. What could you do to determine the heat of formation?

10.88 You have a table of thermodynamic variables that includes heats of formation and free energies of formation. You need to know the standard entropy of a molecule that is not in your tables. How can you use the information you have to estimate the entropy value you need.

10.89 You have a table of thermodynamic variables that includes standard entropies and free energies of formation. You need to know the heat of formation of a molecule that is not in your tables. How can you use the information you have to estimate the entropy value you need.

10.90 Looking at a series of hydrocarbons with only single carbon-carbon bonds how could you identify a trend for the heat of formation for successive carbon-carbon bond formation?

Chapter 11: Chemical Kinetics

When we considered thermodynamics, you may have noticed that we concentrated on what should *eventually* happen to a chemical system. But we have not yet given any thought to how long it might take to reach this eventual outcome, or to exactly *how* a collection of reactant molecules might be transformed into products. Since chemistry is fundamentally about change, if we really want to understand chemistry we will need to examine both the process of change and the speed with which it occurs. On the Fourth of July, it is exhilarating to watch the explosive oxidation reactions of fireworks in the night sky, but a holiday event to watch iron rust would probably not draw a crowd! How do these chemical reactions occur? What factors determine whether a reaction proceeds with explosive speed or at a seemingly imperceptible rate? And could we learn to manipulate these factors to force a reaction to occur at the speed we choose? The area of chemistry dealing with these questions of "How?" or "How fast?" is called **kinetics**, and we begin with a glimpse into stratospheric ozone and its depletion.

11.1 Insight into Ozone Depletion

Walk outside right after a summer thunderstorm and you will probably notice a pleasantly fresh smell. The source of that enjoyable scent is ozone (O_3), the less common allotrope of oxygen. Though its pungent odor can be pleasing in very low concentration, it becomes unpleasant and even toxic in concentrations greater than about one part per million, causing headache and bronchial discomfort. In a thunderstorm, energy from lightning helps to drive reactions producing ozone from ordinary oxygen. But ozone can also be formed through reactions involving various compounds present in the exhaust gases from automobiles and industrial processes. Weather forecasts for many urban areas now include "ozone alerts" on days when the ozone level in the air is expected to rise above about 0.1 ppm. Ozone is a major air pollutant in the troposphere (the atmosphere at ground level), and yet a major topic in the news is global concern about the depletion of the stratospheric ozone layer. Why is ozone a problem in the troposphere but a necessity in the stratosphere? What is the chemistry behind these issues?

Ozone is an example of a molecule whose structure is best represented using resonance.

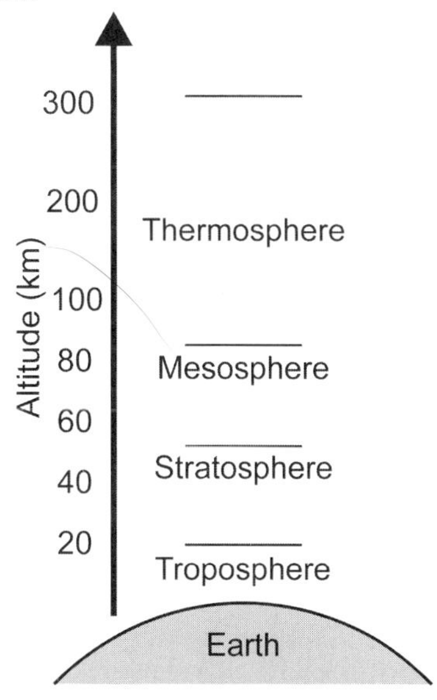

Figure 11.1 Several layers of the atmosphere are identified in the figure. The ozone layer is in the stratosphere, at an altitude of about 30 km.

The smell of ozone after a thunderstorm disappears quickly, partly due to its diffusion through the air (and to the dulling of your olfactory nerves), but also due to the fact that ozone is unstable and decomposes to form ordinary oxygen. The decomposition reaction has a deceptively simple overall equation: $2O_3 \rightarrow 3O_2$. That ozone is reactive and cannot exist for long at the Earth's surface suggests two important facts. First, O_2 is the more stable of the two allotropes of the element oxygen. So thermodynamics must favor the decomposition of ozone. Second, in order for the ozone layer to exist, the upper atmosphere must have some conditions that allow for the formation of the less stable allotrope, O_3, in significant concentration. The accumulation of ozone in the stratosphere is governed by the way these conditions influence the rates of chemical reactions, and therefore provides our initial insight into the role of kinetics in chemistry.

Sydney Chapman was a British mathematician. He devised the idea of the Chapman cycle while trying to develop a mathematical theory of gases.

The British scientist Sydney Chapman first explained the chemistry of the formation of ozone in the upper atmosphere in 1903. And while Chapman's original proposal of the cycle was quite speculative, modern measurements of a wide array of quantities support his hypothesis. The Chapman cycle begins and ends with oxygen, the more stable allotrope, as shown in Figure 11.2. The first step is the photochemical dissociation of O_2 to form oxygen atoms, which may then react with O_2 molecules to form ozone, O_3. Ozone decomposition mechanisms also exist, so that even in the stratosphere ozone is not stable. So why does the ozone layer exist? To answer this we need to consider the rates at which ozone is produced and consumed. If ozone is produced faster than it is consumed, its concentration will increase. But if the destruction of ozone is faster than its formation, the ozone level will decrease. The existence of the ozone layer, then, depends on this balance between the rates at which ozone is produced and destroyed. To understand ozone depletion, we will need to understand reaction rates.

The ozone layer itself would probably not make headlines were it not at risk. Over the past few decades, scientists have been measuring a decrease in ozone concentration that returns during the early spring months, particularly over Antarctica, and (less dramatically) over North America. The evidence of this effect is presented graphically in Figure 11.3. The decrease in stratospheric ozone concentration from 30 years ago to that of today is what is commonly referred to as the "ozone hole."

How does the ozone hole arise? We must consider the role of species other than the allotropes of oxygen in the Chapman cycle. There is evidence that chlorine and bromine, when found in the stratosphere, lead to a decrease in the amount of ozone present. How does this occur? What factors influence whether or not the decrease occurs? And what might be done to combat the loss of ozone? These questions all point to aspects of chemical kinetics that we will address in this chapter. To begin this study we focus first on the concepts of rate of reaction and the ways in which reaction rate can be measured.

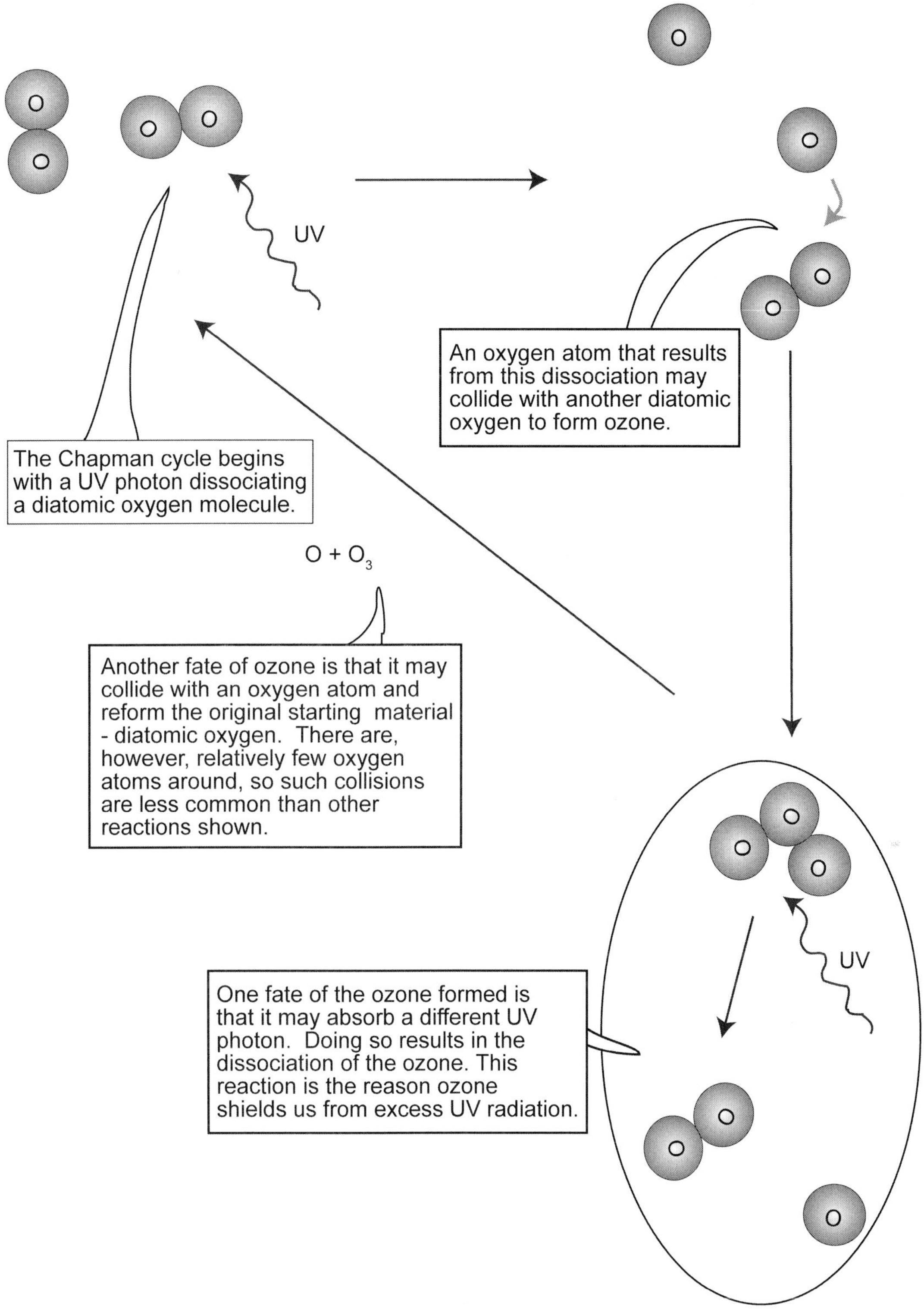

Figure 11.2 The Chapman cycle for the formation and destruction of ozone in the stratosphere.

Note that total ozone change is not the only measurement that implies a seasonal depletion of ozone over polar areas of the planet.

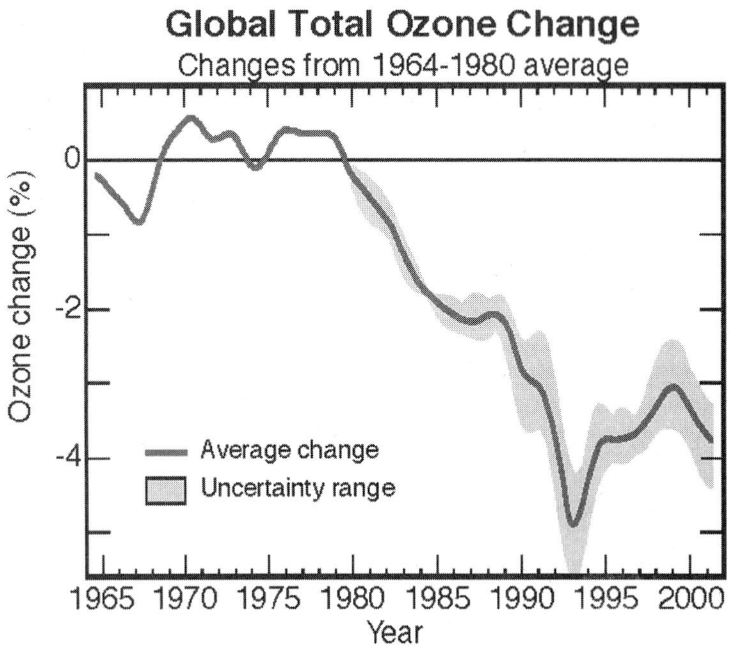

Figure 11.3 The data in this graph clearly show a decrease in the overall levels of atmospheric ozone over the past two or three decades. (Source: WMO (World Meteorological Organization) *Scientific Assessment of Ozone Depletion: 2002*, Global Ozone Research and Monitoring Project - Report No. 47, 498pp., Geneva, 2003.)

11.2 Rates of Chemical Reactions

We suggested that the rates of different chemical reactions will govern the ozone levels in the atmosphere, and this seems sensible. If ozone - or anything else - is produced faster than it is consumed, it will accumulate. To discuss these ideas in a quantitative way we must first raise two fundamental issues in chemical kinetics: 1) How do we define the rate of the reaction, and 2) How can we measure it?

Concept of Rate and Rates of Reaction

Aside from our interest in the rate of destruction of ozone, we have some general experience with rates. If we travel 55 miles from home and it takes us an hour to go that distance, we say that our average speed is 55 miles per hour.

$$\text{average speed} = \frac{\text{distance traveled}}{\text{time elapsed}}$$

In this ratio, there is an indication of progress (distance traveled) related to the time required to make that progress. Since the distance measured was from zero miles (at home) to 55 miles from home, the *change in distance* from home can be symbolized by Δd, where Δ is the symbol for change. Time changed from zero time (at home) to one hour later at our destination. So the time elapsed can be represented by Δt.

$$\text{average speed} = \frac{\Delta d}{\Delta t}$$

How do we translate this ratio into one that is meaningful in the context of chemical reactions?

We need a ratio of the progress of the reaction to the time it takes to make that progress. In a chemical reaction there is a change in the identity of the chemicals, rather than a distance traveled. So to define the chemical rate we must measure the change in chemical content that occurs during the reaction. Usually we do this by measuring chemical concentrations. Thus the rate of a chemical reaction is the ratio of the change in concentration to the elapsed time.

$$\text{rate} = \frac{\text{change in concentration}}{\text{elapsed time}} \qquad (11.1)$$

In chemical kinetics the elapsed time may vary over wide time ranges from picoseconds or shorter to thousands of years.

Notice the units of this ratio. Concentration is generally measured in molarity or mol L^{-1} and time in seconds, so rate units are usually, mol L^{-1} s^{-1}. We can express this ratio in a mathematically compact form by using square brackets, [ ], to designate a concentration or molarity.

$$\text{rate} = \frac{\Delta[\text{substance}]}{\Delta t}$$

Using our example of ozone decomposition, then, we can now begin to write an expression for the rate of reaction.

$$\text{rate} = \frac{\Delta[O_3]}{\Delta t}$$

Stoichiometry and Rate

The overall reaction for the destruction of ozone was mentioned in the opening "Insight" section.

$$2O_3 \rightarrow 3O_2$$

As we saw above, the rate of this reaction can be given by the rate at which ozone decomposes.

$$\text{rate} = \frac{\Delta[O_3]}{\Delta t}$$

But suppose we had monitored the increase in O_2 concentration over time, instead of the decrease in ozone concentration. The change in concentration we'd measure would be different, yet it must also be determined by the rate of the same reaction. As the reaction proceeds, oxygen concentration increases more quickly than ozone concentration decreases, because three oxygen molecules form for every two ozone molecules destroyed. Which rate should we measure, the rate of O_2 production or the rate of disappearance of O_3? The answer is "either one", but to do so in a consistent way we must take into account the stoichiometry of the reaction.

If we are measuring the increase in the concentration of the product, the rate of a reaction is naturally a positive number. When we observe a *decrease* in the concentration of a reactant, we must include a minus sign in the rate statement in order to obtain a positive value for the rate. Thus rate might be defined as either

$$\text{rate} = \frac{\Delta[\text{product}]}{\Delta t} \quad \text{or} \quad \text{rate} = -\frac{\Delta[\text{reactant}]}{\Delta t}$$

But this still does not account for the fact that the number of O_2 molecules produced in a given time period is greater than the number of O_3 molecules consumed. To account for this, we include the stoichiometric coefficient in the denominator of the rate expression:

$$\text{rate} = \frac{\Delta[\text{product}]}{n_{\text{prod}} \Delta t} \tag{11.2a}$$

or

$$\text{rate} = -\frac{\Delta[\text{reactant}]}{n_{\text{react}} \Delta t} \tag{11.2b}$$

where n_{prod} is the stoichiometric coefficient of the product being measured, and n_{react} is the coefficient of the reactant. Accounting for stoichiometry in this way assures that *the same reaction rate is obtained* using either the rate of disappearance of reactants or rate of appearance of products.

Example Problem 11.1

The conversion of ozone to oxygen, $2O_3 \rightarrow 3O_2$, was studied in an experiment, and the rate of ozone consumption was measured as 2.5×10^{-5} mol L^{-1} s^{-1}. What was the rate of O_2 production in this same experiment?

Strategy

Use the stoichiometry of the reaction to relate the respective rate expressions. Because the rate of the reaction is the same whichever quantity is measured, we can set the expressions for the rate of oxygen appearing and ozone disappearing equal to each other and solve for the unknown quantity.

Solution

Begin by writing the expressions for the rate of reaction in terms of both O_2 and O_3 concentration.

$$\text{rate} = -\frac{\Delta[O_3]}{2\Delta t} = \frac{\Delta[O_2]}{3\Delta t}$$

Remember that the change in concentration of the reactant is a *negative* quantity and substitute -2.5×10^{-5} mol L^{-1} s^{-1} for $\dfrac{\Delta[O_3]}{\Delta t}$.

$$\frac{2.5 \times 10^{-5} \text{ mol L}^{-1} \text{ s}^{-1}}{2} = \frac{\Delta[O_2]}{3\Delta t}$$

Solve for the rate of O_2 formation.

$$\frac{\Delta[O_2]}{\Delta t} = \frac{3\,(2.5 \times 10^{-5} \text{ mol L}^{-1} \text{ s}^{-1})}{2}$$

$$= 3.8 \times 10^{-5} \text{ mol L}^{-1} \text{ s}^{-1}$$

Discussion

For some students the most troubling part of this mathematical description of rate is the minus sign when reactant concentrations are used. By recalling our standard definition for Δ, however, we can see the need for this sign more explicitly. $\Delta[O_3]$ in this case is $[O_3]_{later} - [O_3]_{initial}$. The initial concentration of a reactant is larger than the concentration at some later time, so $\Delta[O_3]$ itself is negative. We need the minus sign in the equation so that the rate is not a negative number.

> ***Check Your Understanding***
> Like ozone, dinitrogen pentoxide will decompose to form O_2 (along with nitrogen dioxide) by the reaction $2N_2O_5 \rightarrow 4NO_2 + O_2$. If the rate of disappearance of N_2O_5 is 4.0×10^{-6} mol L^{-1} s^{-1}, what is the rate of appearance of each product?

Average Rate and Instantaneous Rate

Let's think about a simple experiment, shown in Figure 11.4. Take a glass and place a candle inside. Light the candle, let it burn for a few moments, and cover the glass with a saucer. Observe the flame. What happens? The brightly burning flame gradually decreases in size until it goes out, because the concentration of one reactant, oxygen, has decreased. Over the course of this experiment, the rate of the combustion reaction decreases until the reaction stops. What is the rate of the reaction? Is it the rate at which oxygen is consumed in the brightly burning flame or in the nearly extinguished one? This simple experiment shows that the rate we observe depends on when or how we make our observations.

Figure 11.4 When a candle burns in a closed container, the flame will diminish and eventually go out. As the amount of oxygen present decreases, the rate of combustion will also decrease. Eventually, the rate of combustion is no longer sufficient to sustain the flame, even though there is still some oxygen present in the vessel.

Figure 11.5 shows a graph of the concentration of ozone as a function of time for a laboratory experiment. Looking closely at this figure, we can see the slight difference between the average reaction rate and the instantaneous reaction rate. The essential difference between the two methods of defining rate lies in the amount of time needed to make the observation. For the **average rate**, two concentrations are measured at times separated by a finite difference, and the rate is given by the slope of the line between them. The **instantaneous rate** refers to the rate at one instant in time, and it is given by the slope of a line tangent to the curve defined by the change in concentration versus time. As Figure 11.5 shows, these two slopes can differ. (Returning to the speed analogy, your instantaneous speed at any moment during a long drive might be very different than your average speed for the same trip. You may accelerate to pass another car, or come to a stop in traffic.) In most cases we prefer to work with instantaneous rates in kinetics. One commonly measured rate is the **initial rate**: the instantaneous rate of the reaction just as it begins. In Figure 11.5, the tangent line and the concentration curve coincide for at least a brief period at the beginning of the reaction, so the initial instantaneous rate is often relatively easy to measure.

If you have taken calculus, you likely realize that the instantaneous rate is actually a derivative of the concentration versus time curve.

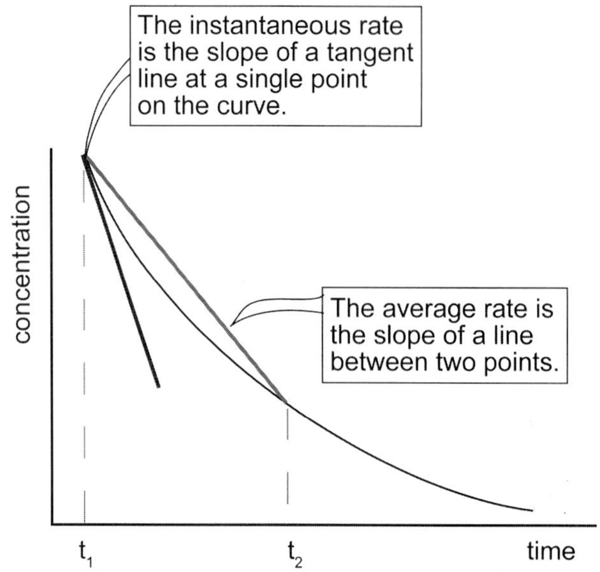

Figure 11.5 The distinction between the instantaneous rate of reaction and the average rate measured over some period of time is illustrated here. The slope of the darker gray line to the left gives the instantaneous rate at time t_1. The slope of the lighter gray line to the right gives the average rate for the period from time t_1 to time t_2.

To determine rates of reaction in complicated systems like the stratosphere, we typically try to measure the rates of the same reactions independently in the laboratory. Thus to consider the reactions of the Chapman cycle, a chemist would have to devise an experiment that measures the rate of reaction of some very reactive species, such as oxygen atoms. The reactive species is first created (by use of a pulse of laser light, for example, which can break the bond in O_2) and then its rapid consumption is monitored. This can be a difficult measurement, since the reaction may proceed extremely quickly. Reactive species like oxygen atoms will react with most chemicals present in the reaction container, so great care must be taken to isolate the reacting species to assure that only the reaction of interest is occurring. Once concentrations of reacting species are measured, an equation can be formulated to describe how the rate depends on these concentrations.

11.3 Rate Laws and the Concentration Dependence of Rates

We saw in Figure 11.4 that as O_2 was consumed, the rate of combustion decreased. The rate of a chemical reaction depends on a number of factors. One of these factors, the concentration of the reacting species, can help us understand why there is an O_3 layer in the upper atmosphere, despite the fact that O_2 is the more stable of the two oxygen allotropes. To approach this issue for the stratosphere, let's first investigate how reaction rates measured in a laboratory depend on concentration.

The Rate Law

Observations made of many chemical reactions show that the dependence of reaction rate on concentration often follows relatively simple mathematical relationships. This behavior can be summarized in a mathematical equation referred to as the **rate law**. Actually, there are two forms of the rate law, and we begin with the **differential rate law** (the name and the relationship are derived from calculus). For a reaction between substances X and Y, the rate of reaction can usually be described by an equation of the form

$$\text{rate} = k\,[X]^m[Y]^n \ldots \qquad (11.3)$$

where k is a constant called the **rate constant**, [X] and [Y] represent the reactant concentrations, and m and n are typically either integers or half integers. *The actual values of the exponents m and n must be measured experimentally.* (In some cases these values can be related to the stoichiometric coefficients of the reaction, but not always.)

The experimentally determined values of the exponents, in this case m and n, are referred to as the **order of the reaction**. For example, if $m = 1$ the reaction is first order with respect to reactant X; if $m = 2$, the reaction is second order with respect to X. It is unusual for any exponent in the rate law to have a value greater than two. When a rate law depends on more than one reactant concentration, we can distinguish between the **overall order of the reaction** and the order with respect to individual reactants. Thus a rate law such as

$$\text{rate} = k\,[A][B]$$

is second order overall, but first order with respect to reactant A and first order with respect to reactant B. (Each exponent is understood to be one.)

Example Problem 11.2

In the following rate laws, determine the orders with respect to each substance, and the overall order of the reaction. (a) rate = $k\,[A]^2[B]$ (b) rate = $k\,[A][B]^{1/2}$

Strategy

The order with respect to each individual species is the exponent, and the reaction order is the sum of the individual orders.

Solution

(a). The order with respect to A is 2 and with respect to B is 1. The overall order is 3.
(b). The order with respect to A is 1 and with respect to B is 1/2. The overall order is 3/2.

Discussion

It is important to remember that the order of the reaction is determined experimentally. While this problem focuses on the mathematical detail of the definition of reaction order, it is imperative that we keep in mind that, generally, reaction order and stoichiometry are not directly related to each other.

Check Your Understanding

The following reactions were studied in the laboratory, and their respective rate laws determined. Find the reaction orders with respect to each reactant and the overall order in each of the following:

(a). $H_2(g) + Br_2(g) \rightarrow 2HBr$; rate = $k\,[H_2][Br_2]$
(b). $2N_2O_5 \rightarrow 4NO_2 + O_2$; rate = $k\,[N_2O_5]$

Note that the rate constant depends on the temperature of the reaction as we will discuss in more detail in Section 11.5.

The rate constant gives us valuable information about the kinetics of a chemical reaction. Perhaps most importantly, the magnitude of the rate constant tells us whether a reaction proceeds quickly or not. If the rate constant is small, the reaction is likely to proceed slowly. By contrast, a large rate constant indicates a rapid reaction. Indeed, observed rate constants for chemical reactions range over many orders of magnitude. We should also point out that the value of the rate constant depends on the temperature of the reaction. (We will consider temperature dependence further in Section 11.5, but it may take some time to get used to the idea of a "constant" that is a function of a variable.) The temperature dependence of the rate of a reaction is described in terms of the rate constant.

The units of a rate constant depend on the overall order of the reaction. We have already established that rate has the units of mol L^{-1} s^{-1}, when concentrations are given in mol L^{-1}. The units of the rate constant must correctly balance units for the rate law. You should be able to verify that the units for the rate constant of a first order reaction must be s^{-1}, while for second order reactions the rate constant has the units L $mol^{-1}s^{-1}$.

Not all rate laws are as simple as those we have mentioned thus far, all of which have included concentrations of reactants only. One reaction for the destruction of stratospheric O_3, for example, has a concentration dependence on the product, O_2. Increased O_2 concentration, however, slows the reaction rather than increasing it. This observation requires a negative exponent for oxygen in the rate law. Thus the rate law for $2O_3 \rightarrow 3O_2$ is given by

$$\text{rate} = k\frac{[O_3]^2}{[O_2]}, \text{ or rate} = k[O_3]^2[O_2]^{-1}$$

It is also not uncommon to find orders of 1/2 or -1/2 in rate laws for some reactions.

Determination of the Rate Law

As noted earlier, the rate law must be determined experimentally. There are two common ways to make this determination. One is graphically, and that method is emphasized in Section 11.4. The other is to measure the initial rate of the reaction while adjusting the concentrations of the various reactants.

We can see the essential logic needed to determine the rate law for a particular reaction by manipulating the mathematical expression,

$$\text{rate} = k[A]^n$$

Keep in mind that the order of reaction is usually an integer and is rarely greater than two. So if the concentration of A is doubled, the rate changes in simple and predictable ways.

1. If $n = 0$, doubling the concentration of A does not change the rate at all, because any quantity raised to the 0 power is 1;
2. If $n = 1$, doubling the concentration of A *doubles* the rate;
3. If $n = 2$, doubling the concentration of A increases the rate by a *factor of four* (because $2^2=4$).

Other orders of reaction are also possible, and can give more complicated relationships. But these three cases cover the majority of the reactions we will encounter.

Let's determine the rate law for a simple case where there is only one reactant.

Example Problem 11.3

Earlier in the chapter we mentioned the decomposition of N_2O_5:

$$2\,N_2O_5(g) \rightarrow 4NO_2(g) + O_2(g)$$

Consider the following data for the kinetics of this reaction.

Experiment	Initial $[N_2O_5]$ (mol L^{-1})	Initial Rate of Reaction (mol L^{-1} s^{-1})
1	$3.0x10^{-3}$	$9.0x10^{-7}$
2	$9.0x10^{-3}$	$2.7x10^{-6}$

Determine the rate law and rate constant for the reaction at the temperature of these experiments.

Strategy

To establish the rate law, determine the order with respect to the reactant by noticing how the rate changes with changing concentration. Write the rate law, and calculate *k* by substituting the concentration and rate from one of the experiments into the rate law.

Solution

Looking at the data, we see the initial concentration triples from the first to the second experiment.

$$3\,(3.0 \text{ x } 10^{-3}) = 9.0 \text{ x } 10^{-3} \text{ mol L}^{-1}$$

And, the rate also triples:

$$3\,(9.0 \text{ x } 10^{-7}) = 2.7 \text{ x } 10^{-6} \text{ mol L}^{-1} \text{ s}^{-1}.$$

Because tripling the concentration triples the rate, the reaction must be first order with respect to N_2O_5.

$$\text{rate} = k\,[N_2O_5]$$

Now use the values from either experiment to determine *k*. Using Experiment #1:

$$9.0 \text{ x } 10^{-7} \text{ mol L}^{-1} \text{ s}^{-1} = k\,(3.0 \text{ x } 10^{-3} \text{ mol L}^{-1})$$

Therefore, $k = 3.0 \text{ x } 10^{-4} \text{ s}^{-1}$.

Discussion

The absolute size of the initial concentration does not matter when we carry out this type of experiment. Neither does the choice to double or triple that initial concentration. We could equally well have chosen to cut the initial concentration in half and then measured the new rate. The logic used to determine the rate law remains the same. If we were actually doing the experiment, we would probably choose to change the concentration by a factor of two or three rather than by some arbitrary amount, but even that is not really necessary.

Check Your Understanding

Would the result change if we used Experiment 2 to determine the rate constant?

Now let's confront the situation in which two reactants are used. We will need to determine the order with respect to each of them. In this case our rate law will have three unknowns: the rate constant and the two reaction orders. In order to determine values for these three parameters, we must carry out at least three experiments. (This is analogous to the idea that at least three equations are needed to solve for three unknowns.) When working with a system that depends on several variables, it is nearly always a good idea to try to separate the influence of one variable from the others. In this case, we can do that by holding one reactant concentration constant while changing the other to determine its effect on the rate.

Because rates are experimentally determined, the change in rate may not be exactly an integer multiple of the initial rate. Between Experiments 1 & 2 the rate change that is measured is slightly more than twoce the initial value.

Example Problem 11.4

The study of the kinetics of real systems can be complicated. For example, there are several ways O_3 can be converted to O_2. One such reaction is

$$NO_2 + O_3 \rightarrow NO_3 + O_2$$

Three experiments were run, and the following data obtained:

Experiment	Initial $[NO_2]$ (mol L^{-1})	Initial $[O_3]$ (mol L^{-1})	Initial Rate of Reaction (mol L^{-1} s^{-1})
1	2.3×10^{-5}	3.0×10^{-5}	1.0×10^{-5}
2	4.6×10^{-5}	3.0×10^{-5}	2.1×10^{-5}
3	4.6×10^{-5}	6.0×10^{-5}	4.2×10^{-5}

Determine the rate law and rate constant for this reaction.

Strategy

To determine the order with respect to each of the reactants, look for pairs of experiments that differ only by the concentration change of one reactant and compare their rates. Any difference in rate *must* be due to the effect of the reactant whose concentration changed. Write the rate law, and use it to determine *k*.

Solution

Order with respect to NO_2:

From experiments 1 and 2 we see that, if we double $[NO_2]$ while holding $[O_3]$ constant, we double the rate (within experimental error). Thus the order in NO_2 is 1.

Order with respect to O_3:

From experiments 2 and 3 we see that if we double $[O_3]$ while holding $[NO_2]$ constant, we double the rate. Thus the order in O_3 must also be 1.

The rate law is:

$$\text{rate} = k[NO_2][O_3]$$

Evaluate *k*, using data from any of the three experiments. Here we will use experiment 2.

$$\text{rate} = k[NO_2][O_3]$$

$$2.1 \times 10^{-5} \text{ mol L}^{-1} \text{ s}^{-1} = k\,(4.6 \times 10^{-5} \text{ mol L}^{-1})(3.0 \times 10^{-5} \text{ mol L}^{-1})$$

$$\text{so, } k = 1.5 \times 10^{4} \text{ L mol}^{-1} \text{ s}^{-1}$$

Discussion

The key to solving this type of problem lies in recognizing which concentrations are changed and which stay the same. Sometimes when rates are reported using scientific notation it is easy to miss a change in the exponent as well. These types of organizational details are the only additional complication of this type of problem relative to Example Problem 11.3.

Check Your Understanding

Why are the units on *k* in this example different from the units in Example Problem 11.3?

11.4 Integrated Rate Laws

The rate law determined by the method described in the previous section allows us to predict the rate of the reaction for a given set of concentrations. But since the concentrations of the reacting substances will change with time, the rate law does not easily let us predict the concentrations or rate that we might find at some later time. To do that we need to formulate an equation that tells us explicitly how the concentrations will change as a function of time. This new equation, called the **integrated rate law**, can be derived from the rate law itself. (See Math Connections box below.) The reasoning behind the integrated rate law is this: If we know all of the concentrations in a reacting system at any given time, and we know how the reaction rate changes with time, then we should be able to predict what the concentrations will be at some later time. This is similar in concept to a physics problem where you might calculate the position of a projectile from its initial position and velocity.

As is often the case, the choice of what mathematical technique to use is dictated by what is desired to be known. The integrated rate law is used when you need to consider concentrations are a function of time.

The form of the integrated rate law depends on the order of reaction. And if we know the integrated rate laws for a few common reaction orders, we can use them as models for comparison with data. As we'll see below, this provides a useful alternative way of determining the rate law for a reaction system.

MathConnections

You may have already learned about the integral in calculus. If so you could probably guess from the name that the integrated rate law is found by integrating the rate law. Let's take a look at how that's done. (If you have not learned about integrals yet, don't worry. You can work with the integrated rate law without knowing how to derive it.) The simplest case to start with is a zero order reaction with a single reactant we'll call A. For such a case the rate law is just

$$\text{rate} = k[\text{A}]^0 = k$$

Now remember that the rate is the change in concentration over the change in time. We've been writing this in terms of macroscopic changes, using the 'Δ' notation.

$$\text{rate} = -\frac{\Delta[A]}{\Delta t} = k$$

But if we simply let the changes become infinitesimal we can convert from deltas to derivatives.

$$-\frac{d[A]}{dt} = k$$

We can rearrange this to put one of the differential terms on each side.

$$d[\text{A}] = -kdt$$

And now we simply take the integral of both sides to eliminate the differentials. We will start at the beginning of the reaction, when t=0 and $[A] = [A]_0$, and integrate from there to some later time when the corresponding values are just t and $[A]_t$.

$$\int_{[A]_0}^{[A]_t} d[A] = -k\int_0^t dt$$

Both of these integrals are simple to evaluate, and we get the following result.

$$[A]_t - [A]_0 = -kt$$

And this is the integrated rate law for the zero order case.

If you are familiar with integration, you should be able to follow the same sort of procedure to derive the integrated rate laws for first and second order kinetics.

Zeroth Order Integrated Rate Law

In a few cases the order of a reaction is zero.

$$\text{rate} = k[A]^0 = k$$

In such instances the rate of the reaction will not change as the reactants are consumed by the reaction. The integrated rate law for this type of kinetics is given by

$$[A]_t = [A]_0 - kt \qquad (11.4a)$$

Rearrange this equation slightly, and compare it to the equation for a straight line with the form $y = mx+b$.

$$[A]_t = -kt + [A]_0 \qquad (11.4b)$$

$$y \;=\; mx + \; b$$

Zeroth order rate laws are relatively uncommon in chemical reactions.

Since the forms of these expressions match, we can see that if we plot [A] (on the y-axis) as a function of t (on the x-axis), we will get a straight line. And we can also see that the slope of that line *(m)* must be equal to $-k$ and the y-intercept *(b)* must be equal to $[A]_0$, the initial [A] value. Equation 11.3 provides us a model of the behavior expected for a system obeying a zero-order rate law. To test this model we simply need to compare it with data for a particular reaction. So we could measure the concentration of the reactant A as a function of time, and then plot [A] vs. t. If the plot is linear, we could conclude that we were studying a zero-order reaction. The catalytic destruction of N_2O in the presence of gold is an example of this type of kinetics, and a graphical analysis of the reaction is shown in Figure 11.6. (Catalysis and other catalytic reactions are discussed in Section 11.7.)

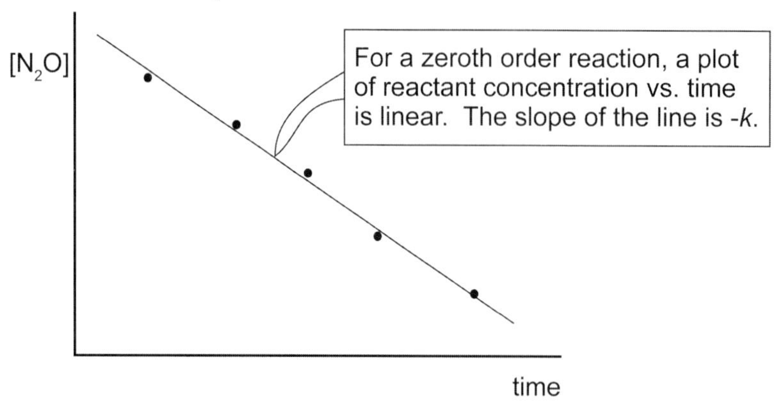

Figure 11.6 The decomposition of N_2O over gold is a zero-order process, for which the concentration decreases linearly as a function of time.

First Order Integrated Rate Law

For a first order reaction of some reactant, X, the integrated rate law is given by the equations,

$$\ln\frac{[X]_t}{[X]_0} = -kt \qquad (11.5)$$

or

$$[X]_t = [X]_0 e^{-kt} \qquad (11.6)$$

$[X]_0$ represents the initial concentration of the reactant. With these equations, if we know the rate constant and the initial concentration of a reaction, we can predict the concentration at any subsequent time, as shown in Example Problem 11.5.

Example Problem 11.5

The photodissociation of ozone by UV light in the upper atmosphereis a first order reaction with a rate constant of 1×10^{-5} s^{-1} at 10 km above the planet's surface.

$$O_3 + h\nu \rightarrow O + O_2$$

Consider a laboratory experiment in which a vessel of ozone is exposed to UV radiation at an intensity chosen to mimic the conditions at that altitude. If the initial O_3 concentration is 5.0 m*M*, what will the concentration be after 1.0 day?

Strategy

Convert the time given to seconds to match the units on the rate constant, and then use the first-order integrated rate law to find $[O_3]$ at the desired time.

Solution

$$1.0\text{ day}\left(\frac{24\text{ h}}{1\text{ day}}\right)\left(\frac{60\text{ min}}{1\text{ h}}\right)\left(\frac{60\text{ s}}{1\text{ min}}\right) = 8.6 \times 10^4\text{ s}$$

$$\begin{aligned}[O_3] &= [O_3]_0 e^{-kt} \\ &= 5.0\text{ m}M\,(e^{-(1.0 \times 10^{-5})(8.6 \times 10^4)}) \\ &= 2.1\text{ m}M\end{aligned}$$

Discussion

This problem helps point out the importance of units. The argument of an exponent must be unitless, so when we plug in values for *k* and *t* they must have the same base time unit (seconds or minutes, for example).

Check Your Understanding

In the upper stratosphere, at an altitude of about 40 km, this same reaction proceeds with a rate constant of 1×10^{-3} s^{-1}. (The difference is caused by the higher intensity of UV radiation.) If an experiment were to simulate these conditions, how long would it take for an initial ozone concentration of 5.0 m*M* to fall to 2.1 m*M*?

Figure 11.7 shows a plot of $[O_3]$ versus time for the experiment described in Example Problem 11.5.

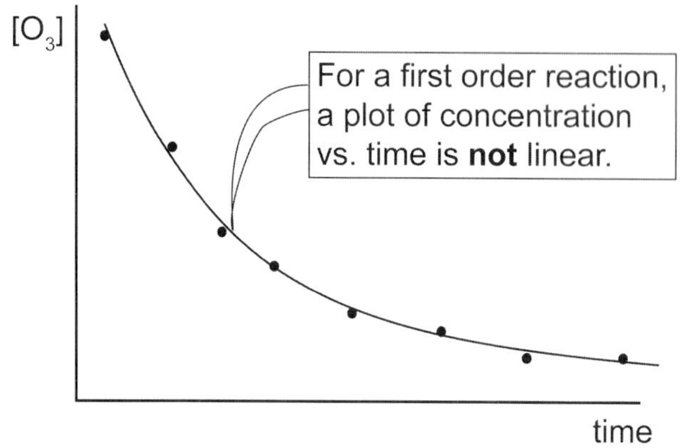

Figure 11.7 This plot of $[O_3]$ vs. time shows that the dissociation of ozone does not follow a zero-order rate law, because the graph is clearly nonlinear.

A quick glance at Figures 11.6 and 11.7 shows a glaring difference between the two plots of concentration versus time: one is linear and the other is clearly curving. The curve shown for the first-order reaction in Figure 11.7 is an example of what is known as an exponential decay. You are probably familiar with this type of curve from math classes, and you will see many other physical phenomena that show this kind of behavior as you continue your engineering studies.

Manipulating data to obtain linear plots often provides a powerful way to analyze experimental data.

Suppose we had measured the data shown in Figure 11.7, but we did not actually know the order of the reaction. The fact that the plot of concentration vs. time is not linear tells us that the reaction is *not* zero order; it doesn't fit the model for that rate law. Next we might ask if the reaction is first order. We can answer this question by comparing the data to the model for first order behavior. One way to do this is to attempt to fit an exponential decay (Eq. 11.6) to the data. This might be done using a graphing calculator, a spreadsheet, or some other tool that can fit nonlinear functions. If we are examining the data visually, though, a curved plot is not as easy to interpret as a straight line, so we might want to manipulate the data so that a straight line results. To do so in this case requires only a little algebra, to arrange the integrated rate law into a linear form.

$$\ln[X] = \ln[X]_0 - kt$$

Now we have an equation that provides a straight line when we plot ln[X] versus time. Figure 11.8 shows the same data as Figure 11.7, manipulated to give such a straight-line plot.

At this point we can see that (1) if a plot of [X] vs. time is linear, the reaction is zeroth order and (2) if a plot of ln[X] vs. time is linear, the reaction is first order. From the form of the integrated rate law equations for these cases, you should also be able to see that the slope of the linear plot must be equal to $-k$. So we can find the rate constant along with the reaction order. Next we will examine the correct model for second order kinetics.

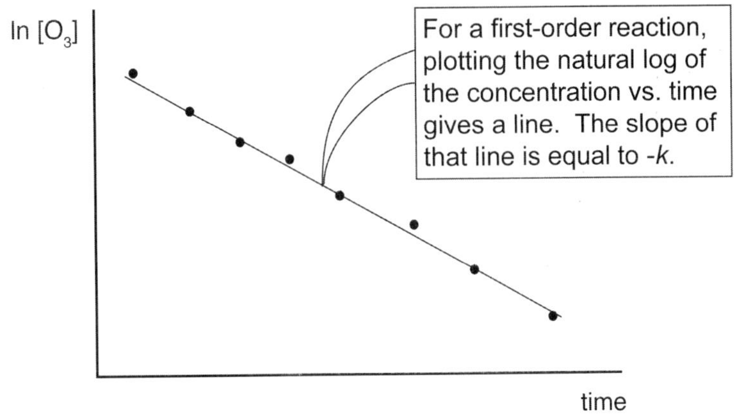

Figure 11.8 The first order integrated rate law predicts that a plot of the natural logarithm of reactant concentration vs. time should be linear. Here we see that this is indeed the case for the dissociation of ozone.

Second Order Integrated Rate Law

For second order reactions, the mathematical form of the integrated rate law is quite different from the one we used for first order reactions. For the consumption of reactant X, the integrated rate law is

$$\frac{1}{[\mathrm{X}]_t} - \frac{1}{[\mathrm{X}]_0} = kt \tag{11.7}$$

As usual, $[\mathrm{X}]_t$ is the concentration of X at time t and $[\mathrm{X}]_0$ is the initial concentration. If we know the value of k we can use this equation and the same strategies used in Example Problem 11.5 to consider second order reactions. This equation provides us with a mathematical model for the progress of a second order reaction as a function of time. As we've seen in the earlier cases, a good way to compare data to such a model is to rely on a straight-line plot. The form of the integrated rate law for second order kinetics immediately suggests the nature of a linear plot; in this case we must plot $1/[\mathrm{X}]$ vs. time. If such a plot is linear, then the reaction is second order. Example Problem 11.6 provides practice at using graphical techniques to determine the order of reactions.

Example Problem 11.6

Among the possible fates of NO_2 in atmospheric chemistry is the decomposition reaction to form NO and O_2. This reaction was studied at 370°C by a student who obtained the following data.

Time (s)	$[NO_2]$ (mol L^{-1})
0.0	0.300
5	0.0197
10	0.0100
15	0.0070
20	0.0052
25	0.0041
30	0.0035

Based on these data, determine the order of this reaction and the rate constant.

Strategy

The available data are the reactant concentration as a function of time for a single experiment, so we will need to use graphical techniques to determine the order of the reaction. There are three possibilities we can explore using the integrated rate laws we've examined: The reaction could be zero order, first order or second order with respect to the concentration of NO_2. We will need to manipulate and plot the data in various ways to determine whether there is a good fit with any of these models. (Other orders are also possible, so we should be aware that all three tests could conceivably fail.) With a spreadsheet or a programmable calculator, such manipulation of data is easy to accomplish. For this example, we will calculate all of the possible manipulations of the data first and then plot the appropriate numbers to find the linear relationship and determine the rate law.

Solution

The three versions of concentration we need to plot versus time are $[NO_2]$ (zero order model), $\ln[NO_2]$ (first order model) and $1/[NO_2]$ (second order model). The table below gives the data we will need to plot.

Time	$[NO_2]$	$\ln[NO_2]$	$1/[NO_2]$
0	0.300	-1.20	3.33
5	0.0197	-3.93	50.8
10	0.0100	-4.61	100.
15	0.00700	-4.96	143.
20	0.00520	-5.26	192.
25	0.00410	-5.50	243.
30	0.00350	-5.65	286.

First, plot concentration versus time to determine if the reaction is zero order.

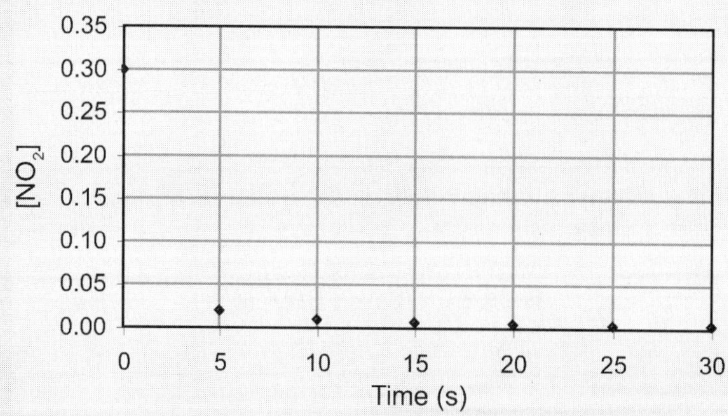

This plot is clearly not linear, so we must proceed to check out the other possibilities. To check for first order kinetics we need to plot $\ln[NO_2]$ versus time.

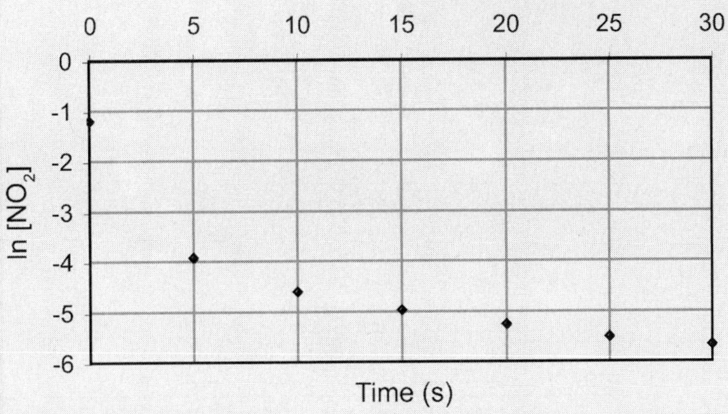

Just like the plot for zero order kinetics, this plot is not linear so the reaction is not first order. Our final option in this case is to plot $1/[NO_2]$ versus time to see if the reaction is second order.

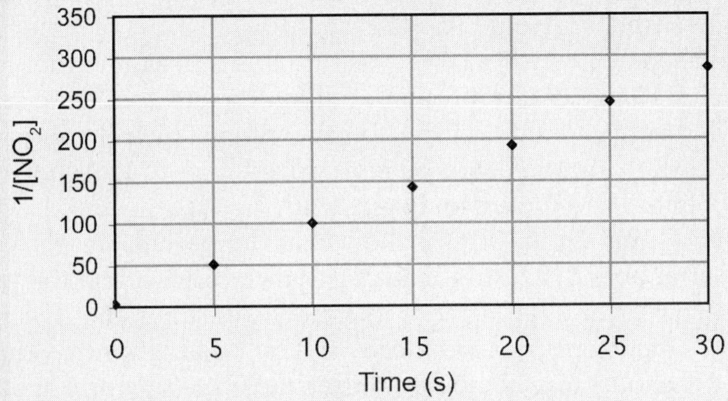

In this case the plot is linear, so the reaction must be second order. To determine the rate constant we must estimate the slope of the line. Estimating from the graph the line crosses 50 at 5 seconds and 200 at 21 seconds. Calculating the slope (rise/run) from those numbers, we have 150/16, which is 9.4 L mol^{-1} s^{-1}. (The units are also given by the rise/run concept from the graph.)

Discussion

Because this is an introductory text, we have chosen our example problems so that they will fit one of these three simple models. In some cases, however, reaction kinetics can be quite complicated and the appropriate model might be none of the three we tried here. If, after constructing these three plots, we had observed no linear graphs, we would have had to conclude that the reaction was neither zero, first, nor second order. Similar integrated rate law models can be derived for other cases, though.

Check Your Understanding

N_2O_5(g) can decompose to form NO_2 and O_2. The following data were obtained for this decomposition at 300°C. Determine the order of the reaction and the rate constant.

Time (s)	$[N_2O_5]$ (mol L^{-1})
0	0.1500
300	0.0729
600	0.0316
900	0.0203
1200	0.0108
1500	0.0051
1800	0.0020

Half-Life

It may be counter intuitive to have more than one half-life, but think about starting a new experiment once a half-life is reached. If you didn't know the first part of the experiment happened, it would make sense to measure the half-way point in the "new" system.

The **half-life** of a given reactant is the time it takes for its concentration to fall to half of its original value. While this quantity can be defined for any reaction, it is particularly meaningful for first order reactions. To see why, let's return to the system we considered in Example Problem 11.5: The photodissociation of ozone by UV light in the upper atmosphere, $O_3 + h\nu \rightarrow O + O_2$. We said this was a first order reaction with a rate constant of $1 \times 10^{-5}\ s^{-1}$ at an altitude of ten kilometers above the surface of the Earth. Suppose we designed a laboratory experiment in which the temperature and light levels found at that altitude could be reproduced. Figure 11.9 shows the type of results we might obtain. The experiment begins with $P_{O3} = 1$ atm, and drops off as time goes by. **The dashed lines on the figure** show that the partial pressure reaches half of its initial value after about 19 hours. The reaction continues, and the pressure falls further, reaching 0.25 atm, or one-fourth of its initial value after about 38 hours. In the first 19-hour period, the ozone pressure decreased by one-half of its value, from 1 to 0.5. Between about 19 hours and 38 hours on the graph, the pressure again decreased by one-half, this time falling from 0.5 to 0.25. For the particular conditions of this experiment, the ozone pressure decreases by a factor of two for every 19 hours. So the half-life is constant, no matter how much ozone we start with.

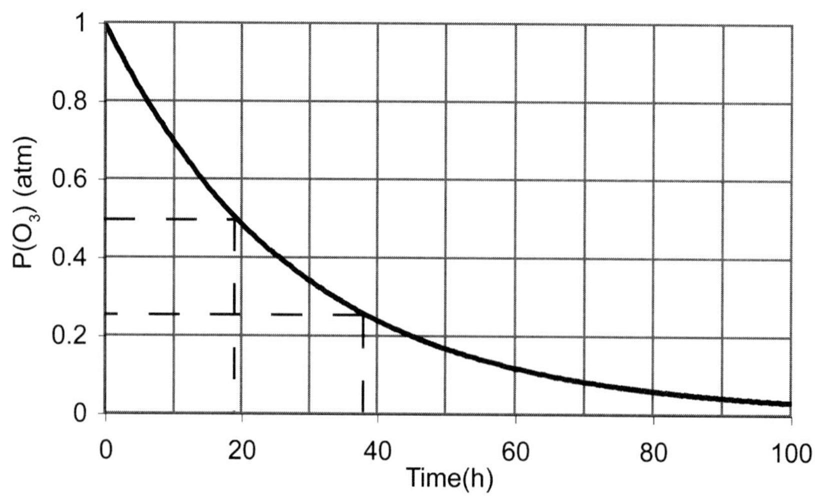

Figure 11.9 Ozone pressure as a function of time in an experiment designed to model the destruction of ozone in the stratosphere. The ozone pressure falls by a factor of one half every 19 hours.

We can obtain the mathematical expression for the half-life of a first order reaction by substituting into the integrated rate law (Equation 11.6). By definition, when the reaction has been proceeding for a time of one half-life ($t_{1/2}$), the concentration of the reactant must be $[X] = 1/2[X]_0$. Thus we have

$$\ln\left(\frac{[X]_0}{2[X]_0}\right) = -kt_{1/2}$$

which simplifies to

$$\ln\frac{1}{2} = -kt_{1/2}$$

$$\ln 1 - \ln 2 = -kt_{1/2}$$

or

$$\ln 2 = kt_{1/2}$$

Solving this equation for the half-life gives

$$t_{1/2} = \frac{0.693}{k} \qquad (11.8)$$

The half-life of *any first order reaction* is related to its rate constant by Equation 11.8. Since k does not depend on the amount of substance present, neither does $t_{1/2}$.

The half-life is most often used to describe the kinetics of nuclear decay. All radioactive decay processes follow first order kinetics, and people working with nuclear processes typically use half-lives rather than rate constants to discuss these systems. According to Equation 11.8, we can easily convert the half-life to the rate constant (or *vice versa*), so knowing either of these values gives us the same amount of information about the reaction kinetics. (For nuclear decay, the half-life is also independent of temperature.)

Example Problem 11.7

The rate of the photodissociation of ozone in the example shown in Figure 11.9 may seem slow. But it is actually tremendously faster than what we would see in the absence of ultraviolet light. The rate constant, k, for the *thermal* decomposition of ozone in the dark at 25°C is just 3×10^{-26} s^{-1}. What is the half-life of ozone under these conditions?

Strategy

We can see from the units on the rate constant that this is a first order reaction. Use the equation for first order half-life.

Solution

$$t_{1/2} = \frac{0.693}{k} = \frac{0.693}{3 \times 10^{-26}\ \text{s}^{-1}} = 2 \times 10^{25}\ \text{s}$$

Discussion

Comparing this result with the half-life of 19 hours (or about 68,000 s) we saw above shows the importance of the UV photons in the photodissociation reaction. In the absence of light, the rate constant is smaller by more than 20 orders of magnitude, making the half-life tremendously long.

Check Your Understanding

The half-life of the photochemical decomposition of N_2O_5 in the stratosphere (at 40 km) is roughly 43,000 sec. What is the rate constant for the reaction?

11.5 Temperature and Kinetics

We first looked at CFC's in Chapter 7.

The indicted culprits for ozone depletion in the stratosphere are chemicals many of us use in our refrigerators—chlorofluorocarbons (CFCs). Freon-12 is the registered trademark for a particular chlorofluorocarbon that has been used in refrigerators and air conditioners.

Have you ever asked yourself why you need a refrigerator? Aside from keeping your favorite beverage refreshingly cold, what advantages does the cooled storage space provide? The answer is in the temperature dependence of chemical reactions. Without a refrigerator, food would spoil more quickly. Maintaining healthful foodstuffs is, in fact, a critically important use for refrigeration. It works, because at cooler temperatures chemical reactions - including those that allow the growth of dangerous bacteria that can proliferate in food - are slower. You can measure this temperature effect with a simple experiment. Set up one sample of milk at room temperature and another sample in the refrigerator. Determine the time it takes the milk to smell sour. Why does temperature affect reaction kinetics, and how can we quantify that effect?

Temperature Effects and Molecules that React

We can investigate the role of temperature on kinetics by referring to the most useful model for describing the motion of molecules—the kinetic molecular theory. Let's consider the events that lead to a chemical reaction between two gas-phase molecules. (The same ideas apply to reactions involving solids and liquids, too. The gas-phase case is just easier for us to discuss.)

Based on the kinetic molecular theory, before two molecules can react they must collide. But we should not expect that all collisions between molecules result in chemical reactions. Our experience suggests that the air surrounding us is stable even though the N_2 and O_2 molecules are constantly undergoing collisions. So when nitrogen and oxygen collide, the probability of reaction must be quite low. If collisions alone cannot explain reactivity, what additional variables come to mind? One possibility is the kinetic energy of the colliding molecules. Slow moving molecules collide with gentle bumps that do not usually result in reaction, while fast moving molecules are more likely to react.

Recall that the Boltzmann distribution describes the speeds of molecules in the kinetic molecular theory. Figure 11.10 shows the speed distributions for gas molecules at two temperatures. At the higher temperature a larger proportion of the molecules move with high speeds. Thus at higher temperatures reactions tend to proceed more quickly, because more of the collisions that occur between reactants have the high energies that may be needed to lead to reaction.

Why do reactions require high-energy collisions? Chemical reactions involve the breaking and forming of chemical bonds. From thermodynamics, we know that breaking bonds requires an energy input, while forming bonds liberates energy. Intuition suggests that before new bonds can form, old ones must break, or at least begin to break. Thus, for many chemical reactions there is a need to put in energy at the outset to start breaking bonds that exist in the reactants. Figure 11.11 provides one way to view this requirement graphically. The energy inherent in either reactants or products allows us to predict whether a reaction is exothermic or endothermic, as you saw in Chapter 9. But to get from reactants to products, the reaction must first overcome an energy threshold, called the **activation energy** or **activation barrier**.

Collisions of fast-moving particles provide sufficient kinetic energy to overcome the activation barrier. The larger this barrier is, the more kinetic energy will be needed. So the activation energy is what dictates how energetic a collision must be in order to lead to reaction.

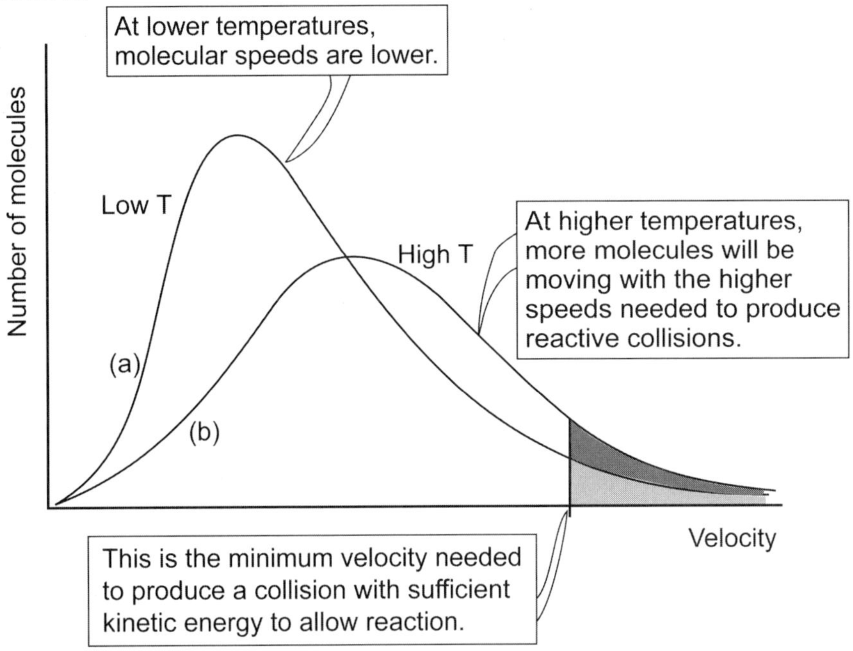

Figure 11.10 The two curves show the Maxwell-Boltzmann distribution of molecular speeds for two temperatures. The shaded areas represent the number of molecules travelling fast enough to collide with sufficient energy to react. Note that there are many more fast moving molecules at higher temperature (b).

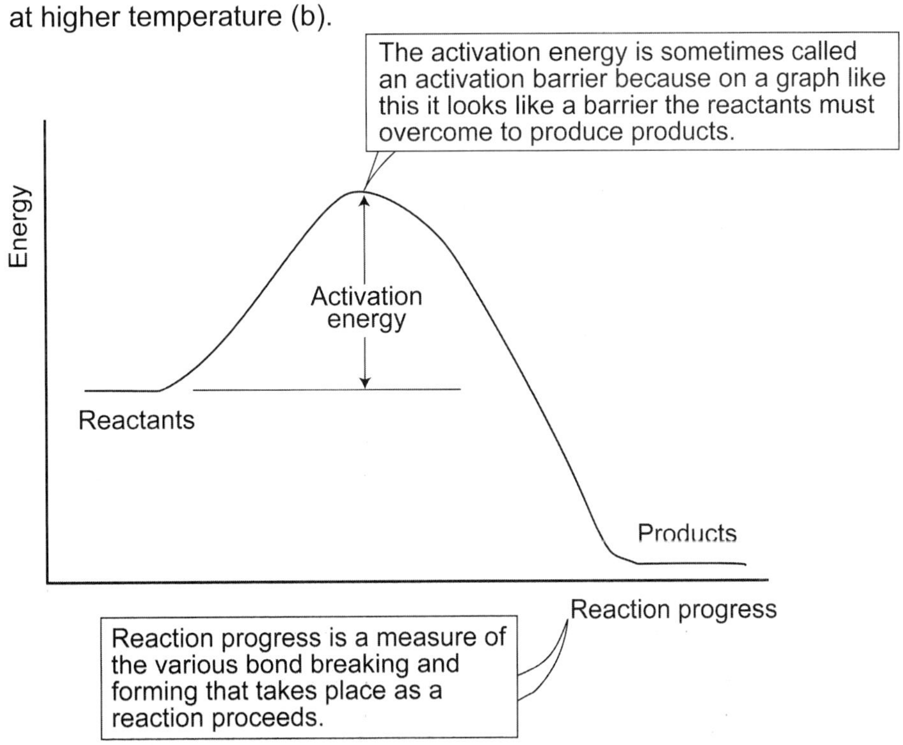

Figure 11.11 Potential energy plot for an exothermic chemical reaction. In order to progress from reactants to products, the molecules must collide with enough energy to pass over the "activation barrier."

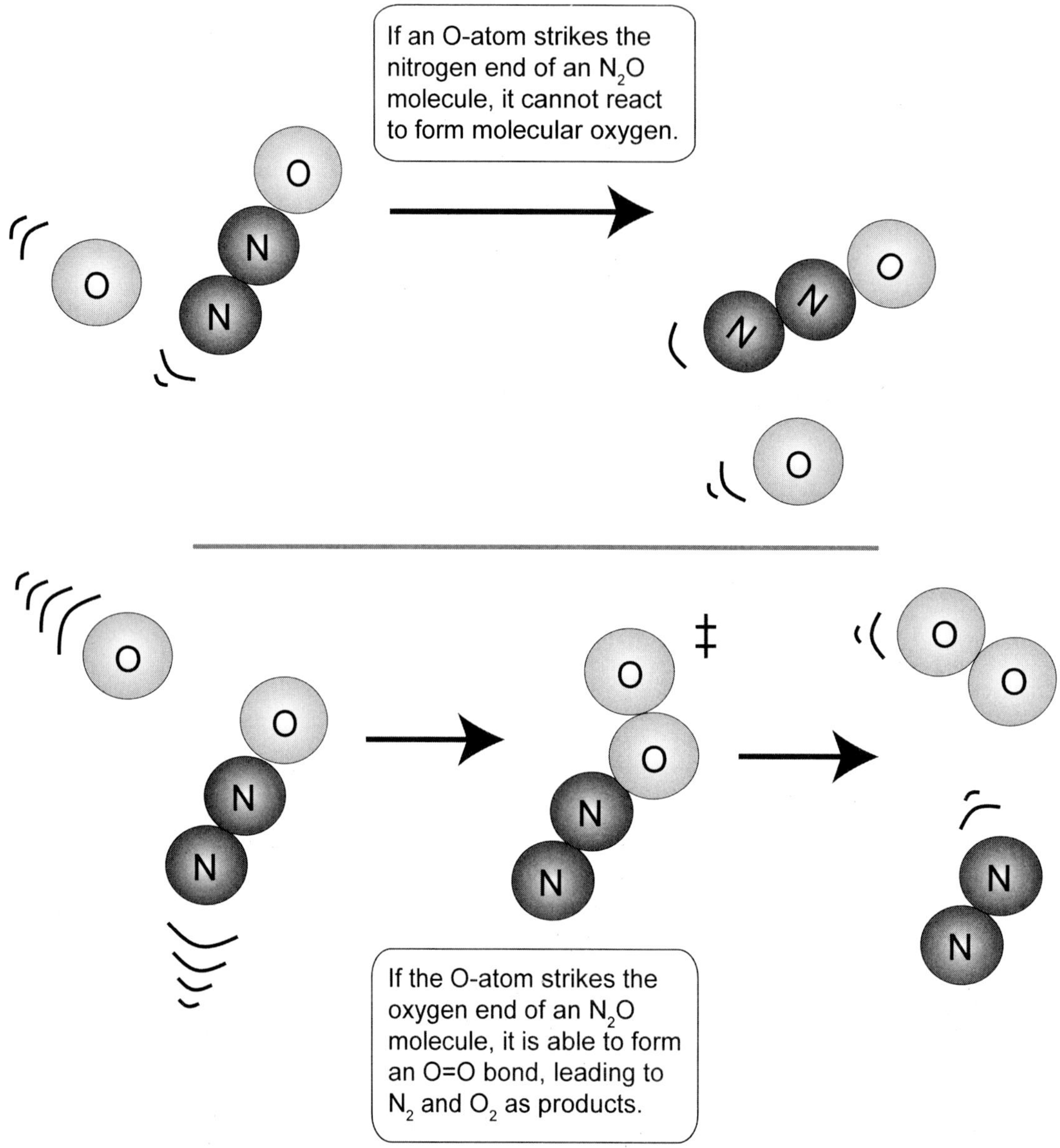

Figure 11.12 Geometric factors can also be important in determining whether a molecular collision is reactive. In order for the oxygen atom to react with N_2O, it must strike the oxygen end of the molecule. The importance of collision geometry depends on the shapes of the molecules involved in the reaction.

Another way that chemists look at activated complexes is to calculate a specific structure called a transition state.

Collision geometry also plays a role in whether a collision between fast-moving molecules is effective. As you can see in Figure 11.12, a collision between a molecule of $N_2O(g)$ and an oxygen atom may or may not lead to a reaction, depending on the geometry of the collision. At the moment of an effective collision, both bond breaking and bond formation are occurring. For an instant, as bond rearrangement is occurring, an unstable intermediate particle, called an **activated complex**, exists (denoted ‡ in Figure 11.12) in the reaction mixture. The activated complex represents the highest energy point along the route from reactants to products. Because it is highly unstable, the activated complex may have a lifetime as short as 10^{-15} s. Formation of an activated complex, and its subsequent decomposition into the reaction products, depends on both the energy and the geometry of the collision.

Arrhenius Behavior

The temperature dependence of chemical kinetics was observed before the detailed molecular picture of modern chemistry was highly developed. Thus, an empirical relationship, called the Arrhenius equation, was established to describe the temperature dependence of the rate constant, k.

$$k = Ae^{-E_a/RT} \tag{11.9}$$

Here E_a is the activation energy, R is the universal gas constant, T is the temperature (in kelvin), and A is a proportionality constant called the **frequency factor** or **pre-exponential factor**. We noted earlier that a large activation energy should hinder reaction. Equation 11.9 shows this effect: As E_a increases, k will be smaller, and smaller rate constants correspond to slower reactions. Note that since the temperature and activation energy appear in the exponent, the rate constant will be very sensitive to these parameters. That's why fairly small changes in temperature can have drastic effects on the rate of a reaction.

Svante Arrhenius also studied acids and bases and the most common definition of acids (that they form H^+ in water) and bases (that they form OH^- in water) is often attributed to him.

The Arrhenius equation can be used to determine activation energy experimentally. Temperature is a parameter that we can usually control in an experiment, so it may help to get that out of the exponent. We can do this by taking the natural log of both sides of Equation 11.9.

$$\ln k = \frac{-E_a}{RT} + \ln A$$

Rearranging, we get

$$\ln k = \frac{-E_a}{R}\left(\frac{1}{T}\right) + \ln A$$

This shows us how we can determine values for E_a or A. We will need to measure k for at least two temperatures, but we already know ways to do that. The rearranged equation is in the familiar form of a line, $y=mx+b$, so a plot of ln k versus $1/T$ should be a straight line. The line's slope is - E_a/R, where R is a known constant. So we can get an estimate of the activation energy from the slope, as shown in Figure 11.13. This type of problem is easily attacked with a graphing calculator or a spreadsheet, and there are several such problems at the end of the chapter.

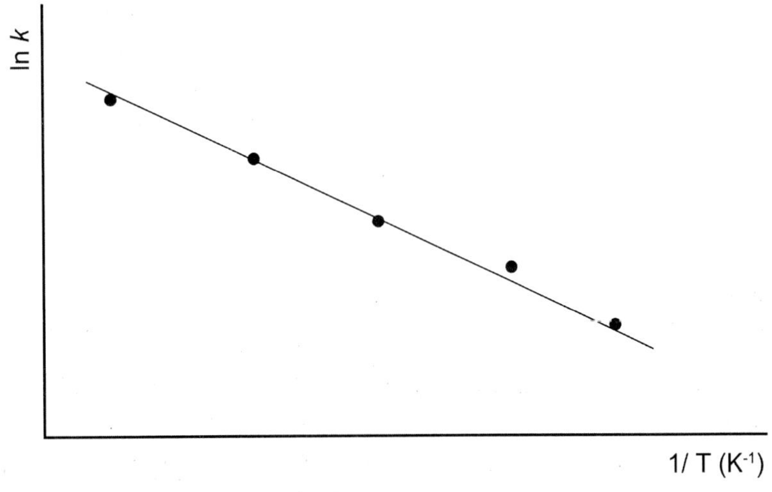

Figure 11.13 According to the Arrhenius equation, a plot of ln k vs. 1/T should give a straight line, with a slope equal to $-E_a/R$. Such a plot is frequently used to determine the activation energy of a reaction.

Example Problem 11.8

In the troposphere, ozone can be converted to O_2 by the following reaction with hydrogen oxide radicals.

$$HO(g) + O_3(g) \rightarrow HO_2(g) + O_2(g)$$

The following values for the rate constant, *k*, for this reaction were measured in experiments at various temperatures.

***k*, L mol^{-1} s^{-1}**	**Temperature, *K***
1.0×10^7	220
5.1×10^7	340
1.1×10^8	450

(a) Does this reaction display Arrhenius behavior? (b) Estimate the activation energy from these data.

Strategy

To determine if the reaction displays Arrhenius behavior, we will need to compare the data to the model represented by the Arrhenius equation. To do this we plot ln *k* vs. $1/T$. If the plotted data form a straight line, the slope of the line is $-E_a/R$. So from that slope we can determine activation energy.

Solution

First, expand the table to include the data we need to plot.

***k*, L mol^{-1} s^{-1}**	**ln k**	**Temperature, *K***	**1/T (K^{-1})**
1.0×10^7	16.1	220	4.5×10^{-3}
5.1×10^7	17.7	340	2.9×10^{-3}
1.1×10^8	18.5	450	2.2×10^{-3}

Now plot ln k (vertical axis) vs. 1/T (horizontal axis)

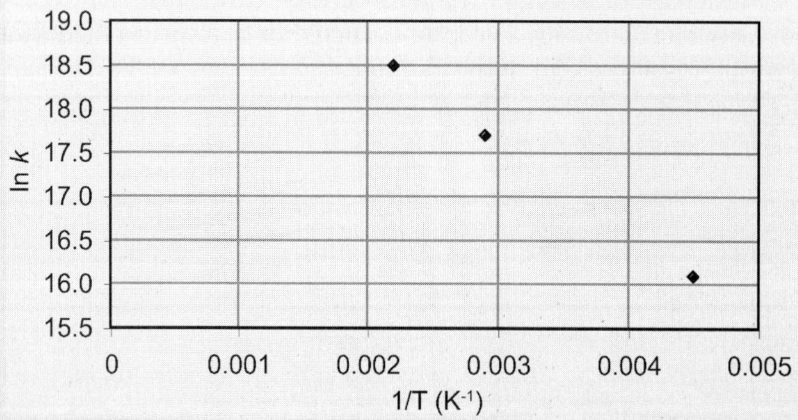

The data do appear to fall on a straight line. So our answer to question (a) is that yes, the reaction does appear to exhibit Arrhenius behavior. (Ideally we should have more than three data points on which to base this conclusion.) To determine the slope of the straight line, we could use a curve fitting routine on a graphing calculator. Or we can determine the slope from two points on the line. If our straight line for these data includes the first and last point, which visual estimation suggests is a reasonable proposition, we can use data from our expanded table directly.

$$slope = \frac{rise}{run} = \frac{\Delta \ln k}{\Delta\left(\frac{1}{T}\right)} = \frac{18.5 - 16.1}{(2.2\times10^{-3} - 4.5\times10^{-3})\ \mathrm{K}^{-1}}$$

$$= -1.04\times10^{3}\ \mathrm{K} = -\frac{E_a}{R}$$

We can insert R = 8.314 J K^{-1} mol^{-1} and solve for E_a.

$$E_a = (1.04x10^3\ \mathrm{K})(8.314\ \mathrm{J\ K^{-1}\ mol^{-1}})$$
$$= 8.7x10^3\ \mathrm{J\ mol^{-1}} = 8.7\ \mathrm{kJ\ mol^{-1}}$$

Discussion

Often the best-fit line for experimental data does not explicitly cross through any of the individual data points. Graphing calculators or spreadsheet programs provide the best and simplest way to determine the slope of such a line. But it is important to keep in mind that the slope calculation is only meaningful if the straight line model provides a reasonable fit to the data. It is always a good idea to look at the data and the best-fit line on the same graph to verify that you are satisfied with the quality of the fit. If the data do not suggest a linear relationship, then we should not put our confidence in the meaning of a slope or other parameter taken from a linear fit.

Check Your Understanding

Once generated, HO_2 may react with carbon monoxide to reform the hydrogen oxide radical and carbon dioxide. This reaction plays a role in ozone pollution in urban centers. Assume that an experiment to determine the rate constant of this reaction provides the data tabulated here. What is the activation energy of this reaction?

***k*, L mol^{-1} s^{-1}**	**Temperature, *K***
5.9	300
4.7 x 10^4	650

Making Mental Connections

Perhaps the most challenging mental connections that are forged in general chemistry are those between bulk macroscopic properties and molecular level origins of those bulk properties. The Arrhenius equation provides us with a good opportunity to focus on this type of mental connection.

$$k = \mathrm{A}e^{-E_a/RT} \tag{11.9}$$

In this equation there are two important factors – the argument of the exponent and the frequency factor, A. The accompanying figure (MM11.1) looks at how these two components of the equation arise from molecular level considerations. The figure shows a reaction in which two molecules of ClO are converted to Cl_2 and O_2,

$$2\ \mathrm{ClO} \rightarrow \mathrm{Cl_2} + \mathrm{O_2},$$

but the same arguments can be applied to any other reaction, too.

The effect of temperature on the Boltzmann distribution is perhaps familiar by now, but we show it for two separate cases – low temperature (2) and higher temperature (3). We have noted previously that Boltzmann distributions imply that more molecules are able to react at higher temperatures. In the final three illustrations we depict the connection between the molecular velocities incorporated in (2) and (3) and reactivity.

Drawing (4) shows that the slow moving molecules have insufficient energy to react; the collision energy does not exceed the activation energy. In (5), however, the high velocity molecules have large collision energies – in excess of the activation energy – and reaction occurs. We could just as easily have drawn this effective collision originating from the higher temperature Boltzmann distribution, where even more molecules will undergo these high-energy collisions.

While the illustration shows the importance of the collision energy exceeding the activation barrier, how does this concept get expressed in the Arrhenius equation? The argument of the exponential function is essentially the ratio of these two values. Recall from Chapter 5 where we introduced gas kinetic theory that the average velocity (the root mean square velocity, to be more precise) is given by the equation,

$$v_{rms} = \sqrt{\frac{3RT}{MM}}$$

Looking at the equation above, if we square both sides we can isolate RT,

$$RT = \frac{1}{3} MMv_{rms}^2$$

And noting that the molar mass is the mass of a molecule times Avogadro's number (N_A) we can see that we have,

$$RT = \frac{N_A}{3} mv_{rms}^2$$

Because kinetic energy is 1/2 mv^2, RT is directly related to the average kinetic energy.

$$RT = \frac{2N_A}{3}\left(\tfrac{1}{2} mv_{rms}^2\right) = \frac{2N_A}{3} KE_{rms}$$

Thus, the macroscopic Arrhenius equation is based on average behavior of the kinetic energy of the molecules in the sample. The Boltzmann distribution can be evaluated to provide that average behavior, and the connection between macroscopic and microscopic is made. Essentially the Arrhenius equation contains a ratio of

$$\frac{\text{activation energy}}{\text{kinetic energy}}$$

When we change the kinetic energy part of this ratio by changing the temperature, the activation energy remains the same.

We have yet to discuss the origin of the frequency factor, A, however. Illustration (6) helps show the origin of this part of the Arrhenius equation. Even at high temperatures and high collision energies, not all collisions result in reactions. In this case the collision geometry was not correct for allowing reaction. An effective collision might require that the oxygen ends of the two ClO molecules actually collide with one another, for example, to allow the O-O bond to begin to form. Thus, things such as collision frequency and orientation of colliding molecules are taken into account in the Arrhenius equation through the frequency factor.

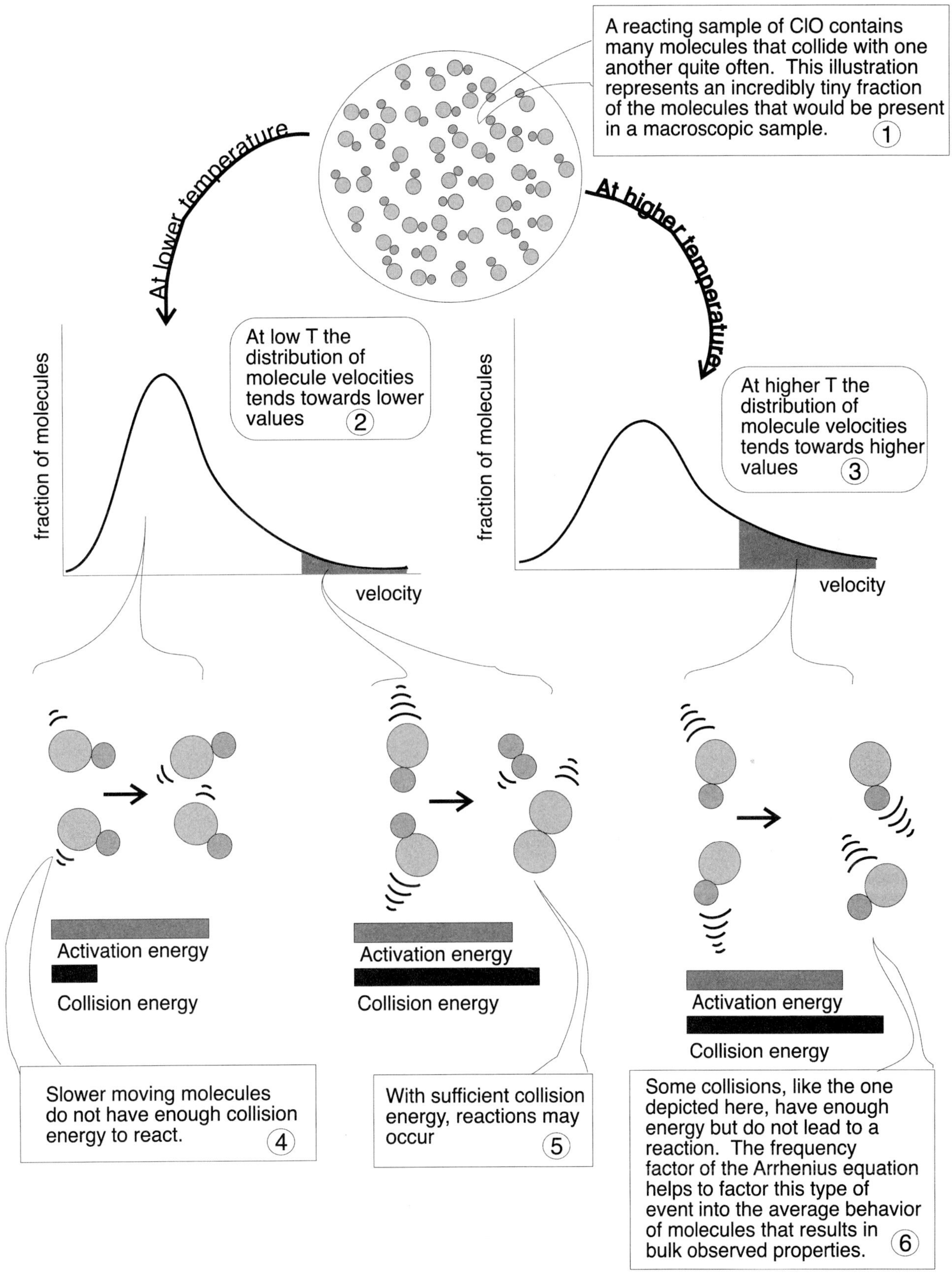

Figure MM11.1 The connection between temperature, collision energy, and reaction rate is developed here. The activation energy establishes a minimum threshold that must be overcome before a collision can lead to reaction. The number of energetic molecules increases with temperature, according to the Boltzmann distribution.

11.6 Reaction Mechanisms

The study of reaction mechanisms on a molecular level is also referred to as chemical dynamics.

The net reaction of the Chapman cycle was noted earlier: $3\ O_2 \rightarrow 3\ O_2$. In some ways it appears pointless to talk about the kinetics of a reaction that results in no net change. But as we have already seen, the importance of the Chapman cycle lies in the details. Without the Chapman cycle, the upper atmosphere would not have an ozone layer, and the quantity of harmful UV radiation reaching the surface of the planet would be substantially greater. This natural cycle points to the importance of the mechanism of the reaction. A **reaction mechanism** is a collection of one or more molecular steps that account for how reactants become products. If the overall equation is like a parts list and a finished product, then the mechanism is like the assembly instructions. In many cases, such as the Chapman cycle, to truly understand the kinetics of the process we must consider the reaction mechanism. What is the stepwise process of bond breaking and bond formation behind the overall reaction?

Elementary Steps and Reaction Mechanisms

We have already reasoned that chemical reactions occur when reactants collide. Based on the kinetic molecular theory, collisions between two particles should be rather common. Three-particle collisions are much less probable, since they require three randomly moving species to arrive at the same place at the same time. And four-particle collisions are virtually unheard of. However, when we look at the stoichiometric coefficients of chemical equations, they seem to imply that collisions occur among many particles: balanced equations often involve stoichiometric coefficients far greater than two or three. To understand reaction mechanisms, we need to distinguish between the overall stoichiometry of a reaction and the steps in the reaction mechanism.

The individual steps in a reaction mechanism are called **elementary steps**. Unlike the overall stoichiometric equation, the coefficients for reactants in an elementary step *do* provide the exponents in the rate law for that step. According to our reasoning above, only three types of elementary steps are likely to occur, those involving one, two or three molecules. Steps with one reactant are called **unimolecular**, while those with two and three reactants are called **bimolecular** and **termolecular**, respectively. The molecularity tells us the overall order of the rate law for the elementary step, as summarized in Table 11.1.

Type of elementary reaction	**Molecularity**
A → products	Unimolecular
A + B → products 2 A → products	Bimolecular
A + B + C → products 2 A + B → products	Termolecular
A + B + C + D → products	Not observed

Table 11.1 Summary of molecularity of elementary reactions

Let's consider a particular reaction mechanism in order to define some other terms. The destruction of ozone by chlorine radicals is expressed by the net reaction, $2O_3 \rightarrow 3O_2$. The mechanism of this reaction is

$$Cl + O_3 \rightarrow ClO + O_2$$
$$ClO + O_3 \rightarrow Cl + 2O_2$$
$$\text{Net:} \quad 2O_3 \rightarrow 3O_2$$

Notice that the equations for elementary steps look just like those for any other chemical reaction. So it is important to remember that when we write a reaction mechanism, each equation is always an elementary step. In the mechanism above, you can see two important characteristics of reaction mechanisms

1. The ClO generated in the first reaction is consumed in the second reaction. A chemical species that is generated in one step and consumed in a later step is called an **intermediate** or a **reactive intermediate**. Many mechanisms involve one or more intermediates.
2. The steps of the mechanism, when properly summed, give the observed stoichiometry of the overall reaction.

The product of an elementary step can be an intermediate in the overall reaction, but it is the product of that step.

Because of the enormous number of collisions occurring at any moment in a reaction in the gas phase and the rapid formation and decomposition of reaction intermediates, reaction mechanisms cannot be known with certainty. However, by observing reaction rates and comparing theoretical mechanisms with experimental rates and with the overall stoichiometry, we can propose a mechanism that fits available data. If more than one mechanism seems plausible, we may be able to design additional experiments to try to choose between them. Example Problem 11.9 provides some practice identifying valid mechanisms and the terms associated with them.

Example Problem 11.9

The decomposition of N_2O_5 is given by the equation

$$2\,N_2O_5(g) \rightarrow 4NO_2(g) + O_2(g)$$

The following mechanism is proposed for this reaction.

$$N_2O_5 \rightarrow NO_2 + NO_3$$
$$NO_2 + NO_3 \rightarrow NO_2 + NO + O_2$$
$$NO + NO_3 \rightarrow 2NO_2$$

(a) Does this mechanism as written provide the correct stoichiometry? If not how does it need adjustment? (b) Identify all intermediates in the mechanism. (c) Identify the molecularity of each step in the mechanism.

Strategy

Consider the definitions for reaction mechanism, and see how this example of a mechanism illustrates those definitions.

Solution

(a) A simple sum of the mechanism yields

$$N_2O_5 + NO_3 \rightarrow + O_2 + 3\,NO_2$$

which is not the desired stoichiometry. If, however, we multiply the first step by 2 we have;

$$2\,N_2O_5 \rightarrow 2\,NO_2 + 2\,NO_3$$

Adding it to steps 2 and 3,

$$2\,N_2O_5 \rightarrow 2\,NO_2 + 2\,NO_3$$
$$NO_2 + NO_3 \rightarrow NO_2 + NO + O_2$$

$$NO + NO_3 \rightarrow 2NO_2$$

yields

$$2\,N_2O_5 \rightarrow + O_2 + 4\,NO_2$$

so the mechanism *does* provide the correct stoichiometry.

(b) The intermediates in this case are NO_3 and NO.

(c) The first step is unimolecular, while steps two and three are bimolecular.

Discussion

This type of problem solving is really performed "in reverse." We knew what the target reaction had to be (the end of the problem) and had to go back to other data given to solve the problem. Reaction mechanisms are by no means the only time we use this problem solving strategy in general chemistry. Stop for a moment and try to think of another type of chemistry problem where working the problem "in reverse" makes the most sense.

Check Your Understanding

The following mechanism was once proposed for the decomposition of N_2O_5:

$$N_2O_5 \rightarrow N_2O_3 + O_2$$

$$N_2O_3 \rightarrow NO_2 + NO$$

$$NO + N_2O_5 \rightarrow 3NO_2$$

(a) Is this mechanism plausible? (b) Identify the intermediates in the reaction. (c) Determine the molecularity of each step.

Mechanisms and Rate: The Rate-Determining Step

Once we have proposed a mechanism, it is easy for us to write the rate law for each elementary step in that mechanism. What we can actually measure experimentally, however, is the rate law for the overall reaction. How can we relate the observed overall rate law to the rate laws for the individual elementary steps in a mechanism? The answer lies in the fact that most mechanisms include one step that is much slower than all the others. This slow step is called the **rate-determining step** and it dictates the rate law. Why does the slowest step assume this role?

Consider this analogy. Suppose you eat lunch at a very popular cafeteria where the food is great and the prices are low. The only problem is that the service is a little slow: as the line moves along, it takes each person three minutes to get their food, and then another 30 seconds to pay. If we watch the cash register in this cafeteria, we can expect to see one customer walking away with food every three minutes - the rate of the slow step in the "mechanism" for getting lunch. We can take this analogy one step further. Suppose the owner is concerned that his line is too long and is scaring away customers. In an effort to speed things up, he buys a new cash register that cuts the paying time down to just 15 seconds per customer. Much to his surprise, he finds that one customer still walks away from the register every three minutes! Speeding up the fast part of the process has no impact on the overall rate, because that rate is set by the slow step. And that's why we call the slow step in a reaction mechanism "rate determining." In a real chemical reaction, the fast steps may be several orders of magnitude faster than the slow step.

In the formation of ozone through the Chapman cycle, the slow step is the last one: the reaction of ozone with atomic oxygen.

$$O + O_3 \rightarrow 2\,O_2$$

It is because this step is rate determining for the cyclic reaction that there is a protective ozone layer in the stratosphere. Once oxygen atoms combine to form ozone, the O_3 survives long enough to absorb UV radiation, and that absorption prevents the UV from reaching the surface of the planet. If the last step were not slow, the ozone created earlier in the cycle would be destroyed chemically before it encountered a photon of UV light. And without the ozone to absorb it, that light would pass through the stratosphere to harm living organisms in the troposphere.

11.7 Catalysis

We have noted from the outset of this chapter that environmental concerns about ozone depletion arise from the increase in the rate of destruction of ozone in the stratosphere, a process apparently caused by chemicals produced by human activity. When reaction kinetics yield an enhanced rate due to chemicals that are neither reactants nor products, the process is called **catalysis**. A **catalyst** is a substance that increases the rate of the reaction but is not part of the stoichiometry. What is the catalyst in ozone depletion, and how does it work?

Homogeneous and Heterogeneous Catalysts

Catalysts can be divided into two broad categories. **Homogeneous catalysts** are those that are in the same phase as the reacting chemicals, while **heterogeneous catalysts** are in a different phase from the reacting species. For gas-phase reactions, the heterogeneous catalyst is typically a solid surface. Both types of catalysis are important in atmospheric processes.

The catalytic destruction of ozone in the stratosphere involves reactions between gases present there, so it is an example of homogeneous catalysis, in this case by chlorine. The initial source of much of the chlorine present in the stratosphere is chlorofluorocarbon molecules (CFCs) that were released in the troposphere and slowly migrated to the stratosphere. (Because they are very unreactive at ground level, nearly all CFC molecules that are released into the atmosphere eventually find their way to the stratosphere.) Upon absorption of UV light, they initiate a catalyzed reaction path:

$$\text{Step 1:}\quad CF_2Cl_2 + h\nu \rightarrow CF_2Cl + Cl$$

$$\text{Step 2:}\quad Cl + O_3 \rightarrow ClO + O_2$$

$$\text{Step 3:}\quad ClO + O_3 \rightarrow Cl + 2O_2$$

It is generally held that these reactions occur on the surface of ice particles based on work by Dr. Susan Solomon, a scientist at the National Oceanic and Atmospheric Administration in Boulder,Colorado.

The second and third reactions are ones we encountered earlier in our discussion of mechanisms. The chlorine atoms serve as a catalyst for ozone decomposition because they (a) are not part of the reaction stoichiometry, (b) are not consumed by the reaction, and (c) speed up the net reaction. Because the chlorine atom is regenerated in Step 3, it can catalyze another reaction. An individual chlorine atom may cycle through this mechanism often enough to destroy 100,000 or more ozone molecules. For this reason, in 1978 the U.S. forbade the use of CFCs as propellants in aerosol cans. An international treaty, the Montreal Protocol, calls for an end to CFC production. Non-CFC refrigerants and propellants are now in use in new equipment, although old refrigerators and air conditioners still contain CFCs that may one day be released to the environment.

Heterogeneous catalysis finds a role in the atmospheric chemistry of ozone in the troposphere as well. Catalytic converters on automobiles are filled with a porous ceramic material, which provides a surface that catalyzes the removal of CO and NO_x (nitrogen oxides) from exhaust. (Nitrogen oxides initiate the formation of ozone and other lung irritants in photochemical smog. We will examine this process in detail in Section 11.8.) The process by which catalytic converters operate is shown in Figure 11.14. The steps shown there are often found in heterogeneous catalysis.

1. The reactive species adsorb, or stick, onto the surface of the catalyst. (When a molecule is *adsorbed*, it sticks to the surface. When a molecule or substance is *absorbed*, it penetrates into the bulk of a material, like water into a sponge. Desorption is the opposite of adsorption.)
2. Species migrate on the surface until they encounter each other.
3. The reaction takes place on the surface.
4. The products desorb from the catalyst surface.

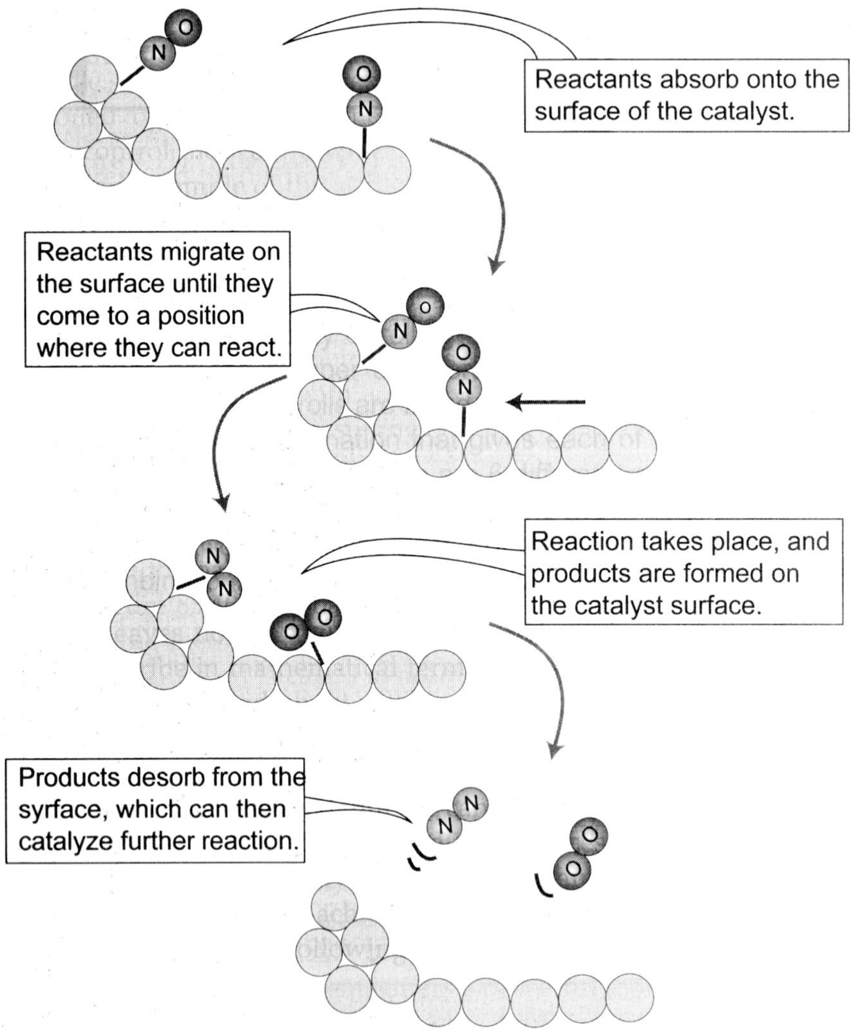

Figure 11.14 A schematic model of the steps involved in the catalytic removal of NO from exhaust gas.

Molecular Perspective of Catalysis

We noted earlier that the activation energy of a chemical reaction plays a critical role in its kinetics. Catalysts increase the rate of reaction by providing a new reaction mechanism that lowers the activation energy. Figure 11.15 displays this concept graphically for the conversion of two ozone molecules to three oxygen molecules. Despite the fact that the addition of chlorine atoms requires an extra step relative to direct collisions of two ozone molecules, the net energy required for the reaction to proceed is lower.

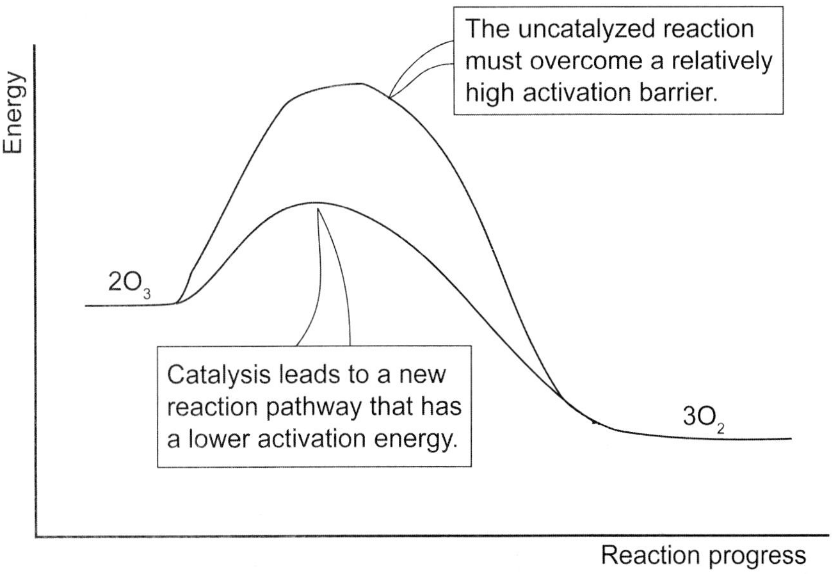

Figure 11.15 Catalysts increase the rate of reaction by providing an alternative reaction pathway with lower activation energy.

We saw in Section 11.5 that the energy needed to overcome the activation barrier is normally acquired from collisions of fast moving particles. When a catalyst is introduced, the amount of energy needed decreases, and a larger fraction of the molecules have velocities that are high enough to provide the needed energy. So, while the direct effect of a catalyst at the molecular level is to lower the activation energy, the net result is an increased rate of reaction.

Catalysis and Process Engineering

One of the most prominent challenges of modern chemistry and materials science is the rational design of catalysts for industrial processes. Successful catalysts in industries such as petroleum refining may result in billions of dollars of products, so the incentives for research in these areas are great. As in any design process, we can begin by listing some desirable features for a catalyst. It should have sufficient longevity or durability to last for many reaction cycles. While the definition of a catalyst might seem to imply that the catalyst should last indefinitely, this is not usually true in practice. Many catalysts suffer from chemical or mechanical degradation, and eventually become useless. A good catalyst should also have a high "turnover number," meaning it can generate large amounts of products in short times. (*Turnover number* is defined as the number of molecules that can react per catalyst binding site per unit time.) Another crucial property of catalysts is selectivity. We often want the catalyst to speed up one, and only one, reaction in order to minimize the formation of any unwanted byproducts. A highly selective catalyst will accomplish this goal. Historically catalysts have often been chosen by trial and error. But as our understanding of reaction mechanisms, molecular structure, and materials properties has increased, many researchers feel it is now becoming possible to apply standard engineering design principles to improve catalysts.

For an industrial process, a high turnover number is advantageous because it improves the throughput of the process and enhances the overall efficiency of production.

11.8 Insight into Tropospheric Ozone

Ozone in the stratosphere helps make life possible by blocking much of the harmful ultraviolet radiation. In the troposphere, however, it is a major lung irritant. It causes localized problems when it is formed from O_2 by an electrical discharge (such as lightning, or the electrostaticly charged mechanism in a photocopy machine) or by ultraviolet light. As the major constituent of photochemical smog, ozone contributes to poor health in urban areas. Recent regulations passed by the Environmental Protection Agency suggest that O_3 is quite prominent in the troposphere, with almost 2,000 violations of air quality standards reported during typical summer months in North America.

We first encountered photochemical smog in Chapter 5, now we can provide additional detail based on kinetic reasoning.

The kinetics of ozone formation at the surface of the planet are substantially different from those seen in the stratosphere. The formation of ozone in smog requires several steps. The key compound is NO_2, formed in the high temperature environment of automobile engines. When NO_2 absorbs sunlight, it can dissociate:

$$NO_2 + h\nu \rightarrow NO + O \qquad (11.10.a)$$

It is the involvement of light in this initiating reaction that provides the term **photochemical smog**. It is plausible to expect that the oxygen atoms generated by this reaction can then form ozone by the reaction

$$O + O_2 + M \rightarrow O_3 + M \qquad (11.10.b)$$

where we include an unidentified third species, M. This molecule must be present to remove excess energy present in the ozone molecule once it forms, or else that excess energy will cause the newly formed molecule to fall apart. Even if this reaction occurs, the ozone that forms could quickly be consumed by reaction with .

$$O_3 + NO \rightarrow NO_2 + O_2 \qquad (11.10.c)$$

Just like in the stratosphere, the balance between the rate at which it is formed and the rate at which it is destroyed will determine the ozone level. A glance at the rate constants given in Table 11.2 shows that reaction 11.10.c is extremely fast when compared with 11.10.b. Thus, if there are no other chemicals present to react with NO, ozone is not likely to form in the troposphere.

Reaction (at 300 K unless indicated)	Rate Constant
$O + O_2 + M \rightarrow O_3 + M$	5.0×10^5 L^2 mol^{-2} s^{-1}
$O_3 + NO \rightarrow NO_2 + O_2$	1.0×10^7 L mol^{-1} s^{-1}
$O_3 + NO_2 \rightarrow NO_3 + O_2$	3.0×10^4 L mol^{-1} s^{-1}
$O_3 + OH \rightarrow OOH + O_2$	4.8×10^7 L mol^{-1} s^{-1}
$O_3 + Cl \rightarrow OCl + O_2$	1.1×10^{10} L mol^{-1} s^{-1}
$O + N_2O \rightarrow N_2 + O_2$ (1200 K)	8.1×10^5 L mol^{-1} s^{-1}
$NO + CH_3O_2 \rightarrow CH_3O + NO_2$	3.8×10^8 L mol^{-1} s^{-1}
$ClO + ClO \rightarrow Cl_2 + O_2$	1.4×10^7 L mol^{-1} s^{-1}

Table 11.2 Rate constants for some reactions in atmospheric chemistry (Data adapted from NBS technical Note 866, "Chemical Kinetic and Photochemical Data for Modeling Atmospheric Chemistry", U.S. Department of Commerce, National Bureau of Standards, 1975.)

The chemicals that most often give rise to the consumption of NO are volatile organic compounds (VOCs). Common VOCs are industrial solvents like pentane, hexane, and benzene (Table 11.3). As a group, these are known as **petroleum distillates**, because they are obtained from crude oil by a process known as distillation. We can designate this class of chemicals as RH, where R can be any organic radical, such as CH_3—, C_2H_5—, etc. The role of these RH species in smog formation begins when they react with OH radicals.

$$RH + {\cdot}OH \rightarrow R{\cdot} + H_2O \qquad (11.10.d)$$

In this case we have designated those species that are radicals by a dot representing an unpaired electron. The product radical, R·, can react with oxygen in a fast reaction.

$$R{\cdot} + O_2 + M \rightarrow RO_2{\cdot} + M \qquad (11.10.e)$$

The species $RO_2{\cdot}$ is called *an alkylperoxy radical*, a species that removes NO and thereby increases the lifetime of any ozone that is formed. The reaction produces NO_2 and another radical, RO·.

$$RO_2{\cdot} + NO \rightarrow RO{\cdot} + NO_2 \qquad (11.10.f)$$

Representative rate constants for these reactions are included in Table 11.2. Note the wide range of rate constant values. We can see that the reactions to remove NO are roughly 40 times faster than the reaction of NO with ozone. The result is that if NO_2 (the initiator) and VOCs are both present, with adequate sunlight, smog may form and ozone can form within that smog.

These few reactions do not entirely describe the chemical processes that form smog. Additional reactions not shown here produce other chemicals that are strong lung irritants. But even with the few reactions we have examined here, we can already begin to appreciate how complicated the study of atmospheric chemistry is. The pollutants are present at concentrations of parts per million or less, and their concentrations may vary with the seasons, the weather, or even the time of day. And yet they result in real threats to human health.

If you read the label on a can of paint thinner you are lkely to see the term "petroleum distillates" rather than actual chemical names such as hexane.

VOC	Source
Petroleum distillates (pentane, hexane, benzene)	Gasoline spills and evaporation
Terpenes	Emitted from live plants (e.g., smells of trees)
Alcohols and aldehydes	Solvents – for paint thinners, etc
Plasticizers	Outgassing from new rugs, upholstery, electronic equipment

Table 11.3 Examples of volatile organic chemicals and their sources

How can atmospheric chemists hope to unravel such a complex problem? The usual approach is to construct a model that incorporates as many of the actual reactions as possible. Scientists then use computers to solve the rate law equations for the connected processes. One commonly used model, known as the "Urban Airshed Model," incorporates 86 chemical reactions among 36 different species. The role of individual reactions can be explored by adjusting their rate constants or even by removing them from the model entirely. Several of the species involved are those we have discussed in this chapter, and the major pollutant that results is ground-level ozone. Among all of the reactions included in the mechanism, ozone is produced in only one of them, and yet it is the key product of the entire mechanism. Of course uncertainty in the values of any of the rate constants used in this model can shift the expected concentrations of many substances, sometimes dramatically. Our brief introduction to chemical kinetics cannot provide us with the tools needed to investigate such a complicated process any further. But it does provide some insight into the roles of both chemistry and mathematical modeling in understanding and attacking an important societal problem.

Focus on Problem Solving

Question

At room temperature, it takes about three days for a container of milk to turn sour. An identical container stored in a refrigerator turns sour only after 11 days. How would you use this information to estimate the activation energy of the chemical reaction that turns milk sour? What information would have to be looked up?

Strategy

This question is an application of the kind of kinetics experiments we addressed with the Arrhenius equation. In this case we have to identify a few items before we can apply the equation, but the essential issues are similar to example problems we have already seen. The time it takes for the milk to sour is related to the inverse of the rate constant. And the crucial relationship between rate constant, temperature, and activation energy is found in the Arrhenius equation.

$$\ln k = \frac{-E_a}{R}\left(\frac{1}{T}\right) + \ln A$$

Solution

The key to this type of question is to identify the temperature conditions with some precision. Room temperature is typically taken to be in the range of 20°C to 25°C. If we accept this range for our temperature, we would have to report a range of possible activation energies for this reaction. The temperature of the refrigerator provides a similar problem. A typical refrigerator keeps the temperature in a range from 3°C to 7°C. Once we establish these temperature ranges and determine a rate of the souring reaction we can plot the rate constant versus 1/T and use the slope of this line to estimate the activation energy.

Chapter 11
Problems & Exercises

Insight into Ozone Depletion

11.1 What is an allotrope?

11.2 Describe what would happen to the ozone molecules if a small amount of ozone were generated and released in a room.

11.3 In what region of the atmosphere is ozone considered a pollutant? In what region is it considered beneficial?

11.4 What are the steps of the Chapman cycle? Explain its importance.

11.5 What is the net chemical reaction associated with the Chapman cycle?

11.6 At what points in the Chapman cycle do photochemical reactions take place?

11.7 Based on the concepts of the energy of light we learned in Chapter 6, which photochemical reaction in the Chapman cycle has the highest energy input via light?

11.8 Describe what is meant by the term "ozone hole." Would you say that the average, non-scientist pictures this issue correctly based on the phrase?

11.9 Is the ozone hole permanent?

Rates of Reaction

11.10 For each of the following, suggest appropriate rate units.
(a). driving from one place to another
(b). drying dishes by hand
(c). the beating wings of a mosquito
(d). eyes blinking

11.11 For each of the following, suggest an appropriate rate unit.
(a). heart beating
(b). video tape rewinding
(c). automobile wheels rotating
(d). gas evolving in a very fast chemical reaction

11.12 Write the following in order of increasing reaction rate
(a). dynamite exploding
(b). iron rusting
(c). paper burning

11.13 Distinguish between *instantaneous* rate and *average* rate. In each of the following situations, is the rate measured the instantaneous rate or the average rate?

In a hot dog eating contest, it took the winner only 4 minutes to eat 20 hot dogs, so he ate 5 hot dogs per minute.

At Minute 1.0, the winner was eating 6 hot dogs per minute, but at Minute 3.0, he was down to only 3 hot dogs per minute.

11.14 Candle wax is a mixture of hydrocarbons. In the reaction of oxygen with candle wax in Figure 11.4, the rate of consumption of oxygen decreased with time after the vessel was covered, and the flame went out. From the perspective of the kinetic molecular theory, describe what is happening in the vessel.

11.15 In the description of the candle in Figure 11.4, we mentioned the consumption of oxygen. Assuming the candle wax is a mixture of hydrocarbons with the general formula C_nH_{2n+2}, what other substances could be measured besides the concentration of oxygen to determine the rate of the reaction?

11.16 The reaction for the Haber process, the industrial production of ammonia, is

$$N_2(g) + 3\ H_2(g) \rightarrow 2\ NH_3(g)$$

Assume that under certain laboratory conditions ammonia is produced at the rate of 6.29×10^{-5} mol L^{-1} s^{-1}. At what rate is nitrogen consumed? At what rate is hydrogen consumed?

11.17 Ammonia can react with oxygen to produce nitric oxide and water.

$$4\ NH_3(g) + 5\ O_2(g) \rightarrow 4\ NO(g) + 6\ H_2O(g)$$

If the rate at which ammonia is consumed in a laboratory experiment is 4.23×10^{-4} mol L^{-1} s^{-1}, at what rate is oxygen consumed? At what rate is NO produced? At what rate is water vapor produced?

11.18 The following data were obtained in the decomposition of $H_2O_2(aq)$ to $O_2(g)$ and $H_2O(l)$. The rate at which oxygen gas was produced was measured.

Time, min	Volume of O_2, mL
3.3	10.4
6.9	17.3

(a). Calculate the average rate in mL/min for the first 3.3 minutes.
(b). Calculate the average rate in mL/min for the first 6.9 minutes.

11.19 A gas, AB, decomposes, and the volume of B_2 produced is measured as a function of time, *t*. The data obtained are as follows.

***t*, min**	0	8.3	15.4	19.0
Volume, L	0	4.2	8.6	11.5

What is the average rate of production of B_2 for the first 8.3 min? For the first 19 min?

11.20 The gas AB_2 decomposes into A and B. The rate of decomposition was measured by determining the volume of AB_2 as a function of time, *t*, using a buret.

***t*, min**	0	10	10	30	40	50
Volume, L	50.0	40.3	31.9	26.2	21.1	17.3

Plot the data and determine the instantaneous rate of decomposition of AB_2 at $t = 35$ min.

11.21 Azomethane, CH_3NNCH_3, is not a stable compound, and once generated it decomposes. The rate of decomposition was measured by monitoring the partial pressure of azomethane, in torr:

***t*, min**	0	15	30	48	75
volume, L	36.2	30.0	24.9	19.3	13.1

Plot the data and determine the instantaneous rate of decomposition of azomethane at $t = 20$ min.

Rate Laws and the Concentration Dependence on Rates

11.22 Which of the following would likely have a very large rate constant? Which would have a very small rate constant?

(a). dynamite exploding
(b). iron rusting
(c). paper burning

11.23 Second order rate constants used in modeling atmospheric chemistry are commonly reported in units of cm^3 $molecule^{-1}$ s^{-1}. Show that this unit is appropriate for a second order rate law.

11.24 Second order rate constants used in modeling atmospheric chemistry are commonly reported in units of cm^3 $molecule^{-1}$ s^{-1}. Convert the following rate constants to L mol^{-1} s^{-1}.

(a). 3.5×10^{-14} cm^3 $molecule^{-1}$ s^{-1}
(b). 7.1×10^{-18} cm^3 $molecule^{-1}$ s^{-1}
(c). 6.1×10^{-30} cm^3 $molecule^{-1}$ s^{-1}

11.25 Give the order with respect to each reactant and the overall order for each of the following hypothetical reactions.

(a). $2\text{ A} + \text{B} \rightarrow \text{C}$; rate = $k[\text{A}][\text{B}]^2$

(b). $\text{A} + 2\text{B} + \text{C} \rightarrow 2\text{ D}$; rate = $k[\text{B}][\text{C}]^{1/2}$

(c). $2\text{ A} + \text{B} \rightarrow 3\text{ C}$; rate = $k\dfrac{[A]^2}{[C]^{1/2}}$

11.26 Show that the unit of the rate constant in the following first order reaction is s^{-1} if the units of rate are L mol^{-1} s^{-1}.

$$2N_2O_5 \rightarrow 4NO_2 + O_2;\ \text{rate} = k\,[N_2O_5]$$

11.27 Show that the unit of the rate constant in the following second order reaction is L mol^{-1} s^{-1}, if the units of rate are mol L^{-1} s^{-1}.

$$H_2(g) + Br_2(g) \rightarrow 2HBr \qquad \text{rate} = k\,[H_2][Br_2]$$

11.28 In the stratosphere, one of the reactions that destroys O_3 molecules is

$$NO + O_3 \rightarrow NO_2 + O_2$$

When this reaction was studied in the laboratory, its rate constant was determined to be 1.9×10^4 L mol^{-1} s^{-1}, and the reaction was found to be first order with respect to both NO and O_3. When the [NO] = 1.2×10^{-5} mol L^{-1} and $[O_3] = 2.0 \times 10^{-5}$, what was the rate of the reaction?

11.29 The hypothetical reaction, $\text{A} + \text{B} \rightarrow \text{C}$, has the rate law:

$$\text{rate} = k[\text{A}]^x[\text{B}]^y$$

If the rate doubles when the concentration of A is doubled and B is held constant, but the rate increases fourfold when the concentration of B is doubled and A is held constant, what are the values of x and y?

11.30 The rate of the decomposition of hydrogen peroxide, H_2O_2, depends on the concentration of iodide ion present. The rate of decomposition was measured for various concentrations of H_2O_2 and of KI (at constant temperature and pressure). The data appear below. Determine the order in each substance, write the rate law, and evaluate the rate constant.

Rate, mL min^{-1}	$[H_2O_2]$	[KI]
0.090	0.15	0.033
1.7	0.30	0.033
1.9	0.15	0.066

11.31 Give the order with respect to each reactant and the overall order for the hypothetical reaction

$$A + B + C \rightarrow D + E,$$

which obeys the rate law: rate = $k[A][B]^2$.

11.32 The following experimental data were obtained for the reaction

$$2A + 3B \rightarrow C + 2D$$

[A], mol L^{-1}	[B], mol L^{-1}	Rate = $\Delta[C] / \Delta t$, mol L^{-1} s^{-1}
0.090	0.15	0.033
1.7	0.30	0.033
1.9	0.15	0.066

Determine the reaction order in each reactant and the value of the rate constant.

11.33 The following experimental data were obtained for the reaction in acid solution:

$$NH_4^+(aq) + NO_2^-(aq) \rightarrow N_2(aq) + 2\ H_2O(l)$$

$[NH_4^+]$, mol L^{-1}	$[NO_2^-]$, mol L^{-1}	Rate = $\Delta[N_2] / \Delta t$, mol L^{-1} s^{-1}
0.0092	0.098	34.9×10^{-8}
0.0092	0.049	16.6×10^{-8}
0.0488	0.196	315×10^{-8}
0.0249	0.196	156×10^{-8}

Determine the rate law for this reaction, and calculate the rate constant.

11.34 For the reaction A + 2 B → C, the following experimental data were obtained. Derive a rate law for the reaction, and determine the value of the rate constant.

[A], mol L^{-1}	[B], mol L^{-1}	Rate = $\Delta[C] / \Delta t$, mol L^{-1} s^{-1}
0.40	0.32	0.012
0.80	0.32	0.024
0.80	0.64	0.096

11.35 For the reaction

$$2\ NO(g) + 2\ H_2(g) \rightarrow N_2(g) + 2\ H_2O(l)$$

at 1100 °C, the following data have been obtained:

[NO], mol L^{-1}	[H_2], mol L^{-1}	Rate = Δ[NO]/Δt, mol L^{-1} s^{-1}
5.0×10^{-3}	0.32	0.012
15×10^{-3}	0.32	0.024
15×10^{-3}	0.64	0.096

Derive a rate law for the reaction, and determine the value of the rate constant.

11.36 The reaction,

$$NO(g) + O_2(g) \rightarrow NO_2(g) + O(g)$$

plays a role in the formation of nitrogen dioxide in automobile emissions. Suppose a series of experiments measured the rate of this reaction at 500 K according to the data below:

[NO], mol L^{-1}	[O_2], mol L^{-1}	Rate = Δ [NO] / Δt, mol L^{-1} s^{-1}
0.002	0.005	8.0×10^{-17}
0.002	0.010	1.6×10^{-16}
0.006	0.005	2.4×10^{-16}

Derive a rate law for the reaction, and determine the value of the rate constant.

11.37 In a heterogeneous system, like wood burning in oxygen, the surface area of the solid can be a factor in the rate of the reaction, since increased surface area means increased collisions with oxygen molecules. To understand how surface increases on dividing a solid, do the following:

(a). Draw a cube that is 2 cm on a side. What is its total surface area?

(b). Divide each face of the cube into four equal sections. If you were able to take the cube apart on the lines you have just drawn, how many cubes would you now have?

(c). What are the dimensions of the new cubes?

(d). What is the total surface area of all the cubes now? What will happen if you divide the cubes again?

(e). Explain how the surface area of a log versus that of all the toothpicks that could be made from the log has an effect on the rate that each burns. Would you rather try to start a campfire with a log or with toothpicks?

11.38 In wheat-growing areas, such as the plains regions of the central U.S. and Canada, harvested wheat is stored in tall grain elevators that are visible for miles in the flat prairie. The wheat is dumped from trucks into a lower area of the elevator and then moved up the elevators into storage areas. The combustible dust produced as the wheat grains rub together has caused grain elevator explosions. Use the cube analogy in 11.38 above to explain why wheat dust dispersed in the air in a grain elevator can react explosively if sparked, whereas wheat grains do not.

Integrated Rate Laws

11.39 The decomposition of N_2O_5 in solution in carbon tetrachloride is a first order reaction.

$$2\ N_2O_5 \rightarrow 4\ NO_2 + O_2$$

The rate constant at a given temperature is found to be 5.25 x 10^{-4} s^{-1}. If the initial concentration of N_2O_5 is 0.200 *M*, what is its concentration after 10 minutes have passed?

11.40 In Exercise 11.39 above, if the initial concentration of N_2O_5 is 0.100 *M*, how long will it take for the concentration to drop to 0.0100 times its original value?

11.41 For a drug to be effective in treating an illness, its levels in the blood stream must be maintained for a period of time. One way to measure the level of a drug in the body is to measure its rate of appearance in the urine. The rate of excretion of penicillin is first order, with a half-life of about 30 min. If a person receives an injection of 25 mg of penicillin at $t = 0$, how much penicillin remains in the body after 3 hours?

11.42 Amoxycillin® is an antibiotic packaged as a powder. When it is used to treat babies and small animals, the pharmacist or veterinarian must suspend it in water, so that it can be administered orally with a medicine dropper. The label says to dispose of unused suspension after 14 days. It also points out that refrigeration is useful. In the context of this chapter, what is implied in the latter two statements?

11.43 As with any drug, aspirin (acetylsalicylic acid) must remain in the blood stream long enough to be effective. Assume that the removal of aspirin from the blood stream into the urine is a first-order reaction, and the half-life of aspirin is about 3 hours. The instructions on an aspirin bottle say to take 1 or 2 tablets every 4 hours. If a person takes 2 aspirin tablets, how much aspirin is remaining in the blood stream when it is time for the second dose? (One aspirin tablet contains 325 mg of aspirin.)

11.44 A possible reaction for the degradation of the pesticide DDT to a less harmful compound was simulated in the laboratory. The reaction was found to be first-order, with $k = 4.0 \times 10^{-8}\ s^{-1}$ at 25°C. What is the half-life for the degradation of DDT in this experiment, in years?

11.45 The half-life for the decomposition of N_2O_5 at 30 °C is 10.0 min. What fraction of a given sample of N_2O_5 will decompose in 30.0 min?

11.46 A sample of substance undergoes a first-order decomposition. After 40.0 min at 500°C, only 12.5 percent of a sample remains. What is the half-life of the sample? If the original sample weighed 243 g, how much would remain after 2.00 hr?

11.47 Show that the half-life of a second order reaction is given by

$$t_{1/2} = \frac{1}{k\,[X]_0}$$

In what fundamental way does the half-life of a second order reaction differ from that of a first order reaction?

11.48 The following data were collected in the decomposition of N_2O_5.

Time, *t*, min	$[N_2O_5]$, mol L^{-1}
0	0.200
5	0.171
10	0.146
15	0.125
20	0.106
25	0.0909
30	0.0777
35	0.0664
40	0.0570

Plot $\ln[N_2O_5]$ versus *t*, and determine the rate constant for this first order reaction

11.49 The rate of photodecomposition of an herbicide, picloram, in aqueous systems was determined by exposing solutions containing picloram to sunlight for a number of days. One such experiment produced the following data. (Hedlund, R.T., Youngson, CR. "The Rates of Photodecomposition of Picloram in Aqueous Systems", *Fate of Organic Pesticides in the Aquatic Environment,* Advances in Chemistry Series, #111, American Chemical Society, (1972),159-172.)

Exposure Time, *t* days	[Picloram], mol L^{-1}
0	4.14×10^{-6}
7	3.70×10^{-6}
14	3.31×10^{-6}
21	2.94×10^{-6}
28	2.61×10^{-6}
35	2.30×10^{-6}
42	2.05×10^{-6}
49	1.82×10^{-6}
56	1.635×10^{-6}

Determine the order of reaction, the rate constant, and the half-life for the photodecomposition of picloram.

11.50 The rate of decomposition of SO_2Cl_2 according to the reaction

$$SO_2Cl_2\,(g) \rightarrow SO_2\,(g) + Cl_2\,(g)$$

can be followed by monitoring the total pressure in the reaction vessel. Consider the following data.

Time, *t* seconds	**Total Pressure *torr***
185.3	527.3
242.8	549.5
304.5	571.6
362.7	590.0
429.5	607.6
509.7	625.6
606.3	644.1

What must you do to convert these total pressures into changes in the pressure of the SO_2Cl_2? Once you manipulate the data, use a graphing calculator or spreadsheet to plot the data and determine the order of reaction and the rate constant.

11.51 Peroxyacetyl nitrate (PAN) has the chemical formula $C_2H_3NO_4$ and is an important lung irritant in photochemical smog. An experiment was run to determine the decomposition kinetics of PAN. The data below were determined in this experiment. Determine the order of the reaction for this decomposition and calculate the rate constant for the reaction.

Time, *t* (minutes)	**Pressure PAN (torr)**
0	2.00×10^{-3}
10	1.61×10^{-3}
20	1.04×10^{-3}
30	5.4×10^{-4}
40	2.3×10^{-4}
50	7.7×10^{-5}
60	2.1×10^{-5}

11.52 The following series of pictures represents the progress of a reaction in which A_2 molecules dissociate into atoms: $A_2 \rightarrow 2A$. Each picture represents a "snapshot" of the reaction mixture at the indicated time. Find the rate law for this reaction, including the numerical value of the rate constant.

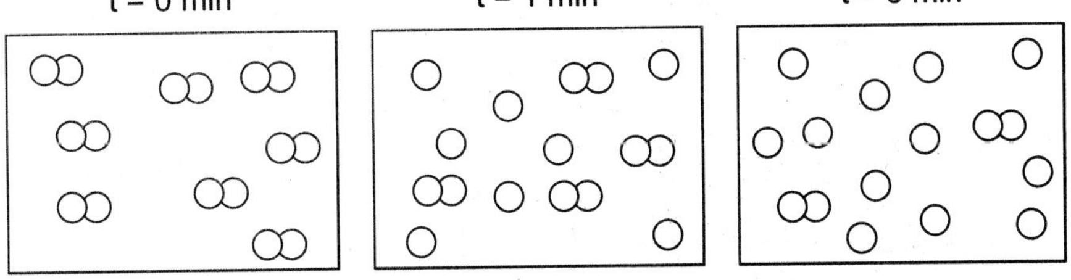

11.53 Hydrogen peroxide (H_2O_2) decomposes to water and oxygen.

$$H_2O_2\ (aq) \rightarrow H_2O\ (l) + O_2\ (g)$$

Ordinarily this reaction proceeds rather slowly, but in the presence of some iodide ions (I^-) the decomposition is fairly fast. The decomposition in the presence of iodide was studied at 20°C, and the data were plotted in various ways. Use the graphs below, where concentrations are in moles/liter and time is in seconds, to answer the questions that follow.

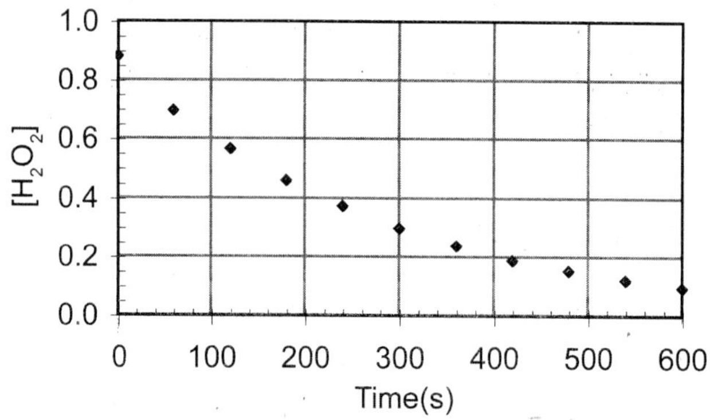

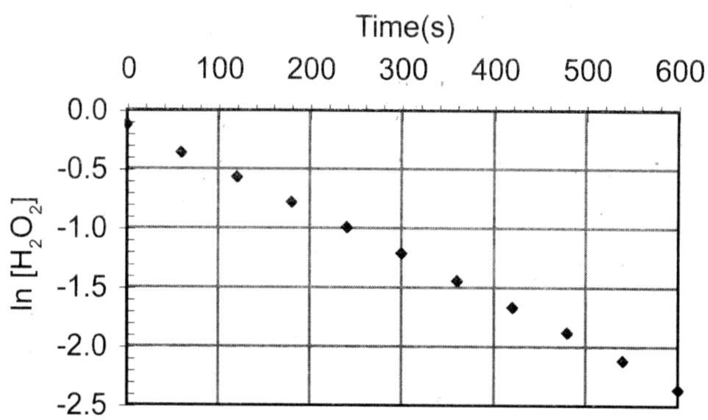

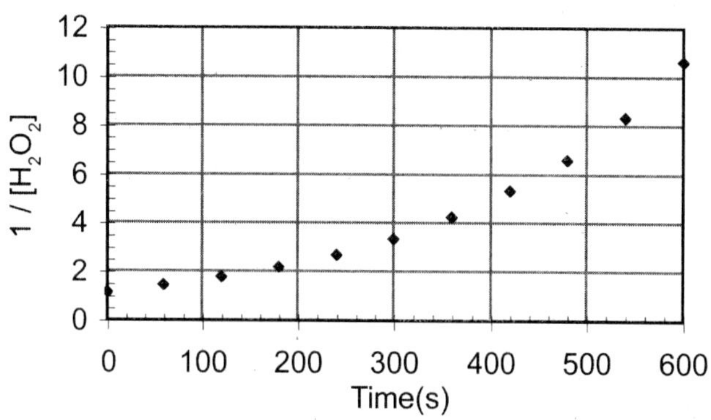

(a). What is the order of reaction for the decomposition of hydrogen peroxide?

(b). Find the numerical value of the rate constant at 20°C, including the correct units.

(c). Obtain an estimate of the initial rate of reaction in the experiment that produced the graphs (*i.e.*, the rate at time = zero in the graphs).

Temperature and Kinetics

11.54 Using the kinetic molecular theory, explain why an increase in temperature increases reaction rate.

11.55 The activation energy for the reaction in which CO_2 decomposes to CO and a free radical oxygen atom, O, has an activation energy of 460 kJ mol^{-1}. The frequency factor is 2×10^{11} s^{-1}. What is the rate constant for this reaction at 298 K?

11.56 The labels on most pharmaceuticals state that the medicine should be stored in a cool, dark place. In the context of this chapter, explain why this is sound advice.

11.57 The following rate constants were obtained in an experiment in which the decomposition of gaseous N_2O_5 was studied as a function of temperature. The products were NO_2 and NO_3.

k, s^{-1}	Temperature, K
3.5×10^{-5}	298
2.2×10^{-4}	308
6.8×10^{-4}	318
1.6×10^{-3}	328

Determine E_a for this reaction in kJ mol^{-1}.

11.58 The table below presents measured rate constants for the reaction of NO with ozone (one step that is possibly responsible for the destruction of some stratospheric ozone) at three temperatures. From these data, determine the activation energy of the reaction in kJ mol^{-1}.

k, L mol^{-1} s^{-1}	Temperature, K
1.3×10^{6}	200
4.4×10^{6}	250
9.9×10^{6}	300

11.59 The table below presents rate constants measured at three temperatures for the following reaction, which is involved in the production of nitrogen oxides in internal combustion engines.

$$O(g) + N_2(g) \rightarrow NO(g) + N(g)$$

From this data determine the activation energy of the reaction in kJ mol^{-1}.

k, L mol^{-1} s^{-1}	Temperature, K
4.4×10^{2}	2000
2.5×10^{5}	3000
5.9×10^{6}	4000

11.60 Many reactions occur in the formation of photochemical smog, including the reaction of ozone with various volatile organic chemicals, VOC's. The table below shows the Arrhenius expression for four such reactions.

Volatile Organic Chemical	Arrhenius Expression
Allene	$9.6 \times 10^5 \exp(-2750/T)$
Propane	$6.6 \times 10^6 \exp(-1970/T)$
1-butene	$1.7 \times 10^6 \exp(-1696/T)$
cis-2-butene	$1.9 \times 10^6 \exp(-956/T)$

(a). Which reaction is likely to be the fastest at 310 K?
(b). Which reaction has the lowest activation energy?
(c). Which has the highest activation energy?

11.61 Bacteria cause milk to go sour by generating lactic acid. Devise an experiment that would measure the activation energy for the production of lactic acid by bacteria in milk. Describe how your experiment will provide the information you need to determine this value. What assumptions must be made about this reaction?

Reaction Mechanisms

11.62 Can a reaction mechanism ever be proven correct? Can it be proven incorrect?

11.63 Define the term *intermediate* in the context of reaction mechanisms.

11.64 Why is a termolecular reaction step uncommon?

11.65 Describe how the Chapman cycle is a reaction mechanism. What is the molecularity of each reaction in the Chapman cycle?

11.66 What are the reaction intermediates in the Chapman cycle?

11.67 Propane is commonly used as a household fuel where natural gas is not available. The combustion of propane is given by the equation

$$C_3H_8 + 5\,O_2 \rightarrow 3\,CO_2 + 4\,H_2O$$

Does this equation represent the probable mechanism of the reaction? Defend your answer.

11.68 Five people are responsible for cleaning up after a very large banquet. Two are clearing the tables, while one washes the dishes and two dry. Which step is likely to be rate-determining?

11.69 The following mechanism is proposed for a reaction:

$$NO + Br_2 \rightarrow NOBr_2 \qquad \text{(slow)}$$

$$NOBr_2 + NO \rightarrow 2\,NOBr \qquad \text{(fast)}$$

(a). Write the overall equation for the reaction.
(b). What is the rate determining step?
(c). What is the intermediate in this reaction?
(d). What is the molecularity of each step of the reaction?
(e). Write the rate expression for each step.

11.70 HBr is oxidized in the following reaction:

$$4HBr(g) + O_2(g) \rightarrow 2H_2O(g) + 2Br_2(g)$$

A proposed mechanism is

$$HBr + O_2 \rightarrow HOOBr \qquad \text{(slow)}$$

$$HOOBr + HBr \rightarrow 2HOBr \qquad \text{(fast)}$$

$$HOBr + HBr \rightarrow H_2O + Br_2 \qquad \text{(fast)}$$

(a). Show that, after adjusting one step, the mechanism provides the correct stoichiometry.
(b). Identify all intermediates in this mechanism.
(c). What is the molecularity of each elementary step?
(d). Write the rate expression for each elementary step.
(e). Identify the rate determining step.

11.71 Suppose some reaction $A_2 + BC_2 \rightarrow A_2B + C_2$ has the following proposed mechanism.

$$BC_2 \rightarrow B + C_2 \qquad \text{(fast)}$$

$$A_2 + B \rightarrow A_2B \qquad \text{(slow)}$$

Would the rate law determined by this mechanism differ from than of a one-step mechanism that is the same as the overall reaction?

Catalysis

11.72 If a textbook defined a catalyst as "a substance that increases the rate of a reaction," would that definition be adequate?

11.73 The word catalyst has been adopted in areas other than science, such as a reserve who enters the game and acts as a catalyst for a comeback of the sports team on which she plays. Discuss how this use of the word is similar to the scientific use of the word. How is it different than the scientific use of the word?

11.74 In the process by which CFC's contribute to the depletion of stratospheric ozone, is light a catalyst? Why or why not?

11.75 What distinguishes homogeneous and heterogeneous catalysis?

11.76 Define the terms adsorbed and desorbed. Do the processes to which these terms refer only occur in heterogeneous catalysis?

11.77 Based on the kinetic theory of matter, what would the action of a catalyst do to a reaction that is the reverse of some reaction we say is catalyzed?

11.78 Draw a hypothetical activation energy diagram for an uncatalyzed reaction, and on the same set of coordinates, one for the catalyzed reaction. Why does reaction rate increase with the use of the catalyst?

11.79 The catalytic converter of your automobile likely contains two different catalysts. Based on the fact that the role of this device is to convert NO_x to N_2 and O_2 as well as to convert CO to CO_2, discuss why it makes sense that there are two catalysts. (Hint: consider the oxidation-reduction nature of the reactions involved)

11.80 The label on a bottle of 3% (*v*/*v*) hydrogen peroxide, H_2O_2, purchased at a grocery store, states that the solution should be stored in a cool, dark place. H_2O_2 decomposes slowly over time, and the rate of decomposition increases with an increase in temperature and in the presence of light. However, the rate of decomposition increases remarkably if a small amount of powdered MnO_2 is added to the solution. The decomposition products are H_2O and O_2. MnO_2 is not consumed in the reaction.

(a). Write the equation for the decomposition of H_2O_2.
(b). What role does MnO_2 play?
(c). In the chemistry lab, a student substituted a chunk of MnO_2 for the powdered compound. The reaction rate was not appreciably increased. What is one possible explanation for this observation?
(d). Is MnO_2 part of the stoichiometry of the decomposition of H_2O_2?

Insight into Tropospheric Ozone

11.81 Why is smog appropriately labeled *photochemical* smog?

11.82 When oxygen molecules, O_2, react with oxygen atoms to form ozone, a third molecule must be present.

(a). What is the molecularity of this elementary reaction?
(b). Would you expect this reaction to be fast or slow? Explain your reasoning.
(c). Why does the third molecule have to be there?

11.83 What is a VOC? What role do VOCs play in photochemical smog?

11.84 Why does the formation of ozone in smog depend on having *less* NO(g) present?

11.85 Many urban centers in North America do not meet air quality standards for ozone levels. What factors associated with living in a city contribute most strongly to the formation of potentially dangerous levels of ozone?

11.86 Ozone is not stable at pressures typically found in the troposphere. How might this contribute to the variation of ozone levels over the course of each day?

11.87 Why do ozone concentrations "lag" in time relative to other pollutants in photochemical smog?

11.88 What strategies could be proposed to lessen the formation of ozone as part of photochemical smog:

(a). at the chemical level - *i.e.*, what chemicals would need to be accentuated or eliminated?
(b). at the practical level - *i.e.*, what changes would the average person need to make to accomplish the changes you suggest at the chemical level?

11.89 Find a web site that tells you if you live in an area that has unhealthy levels of ozone.

(a). What is the URL of the site? Does this site strike you as a reliable source of information?
(b). Does your area have unhealthy air?

11.90 Use the web to find out what it means if the Environmental Protection Agency lists a city as a non-attainment area.

11.91 Northern Michigan can experience higher than normal ozone levels in the summer because prevailing southwest winds carry industrial pollution in the form of VOCs from Milwaukee and Chicago. (For maps, visit the EPA Region 5 web site at http://www.epa.gov/ARD-R5/home.htm.)

Rewrite equations 11.10(d). – 11.10(f)., using hexane, C_6H_{14}, as the VOC pollutant. Explain how the NO_2 produced in (a). can result in tropospheric ozone production.

11.92 Along the Front Range of Colorado, the greatest contributor to Denver's "Brown Cloud" is vehicle traffic. Motor vehicles emit VOCs while they stir up the dust on the highways, increasing particulates. (Visit the Regional Air Quality Control's web site at http://www.raqc.org/.)

In 1998, Denver exceeded the EPA's 8-hour ozone standard of 0.08 ppm. Imagine that you are responsible for helping Denver (or your community) decrease ozone concentrations. Using the information in Table 11.3, list four things you would tell Denver residents to do around the house that would decrease emissions of VOCs and thus decrease ozone formation.

One of the suggestions given to residents of the Denver area to help decrease ozone formation is to fill their car's gas tank after dark. Explain why this could be effective.

Putting it All Together

11.93 According to an article published in 1966 in the *Journal of Chemical Education*, xenon reacts photochemically with fluorine at room temperature (298 K) to produce XeF_2. A portion of a table showing ratios of gases in the reaction vessel for four experiments is reproduced below. (Assume atmospheric pressure is 1 atm.) The reaction is as shown.

$$Xe(g) + F_2(g) \rightarrow XeF_2(s)$$

(Data from John H. Holloway, *J. Chem. Ed*, **43**, 203 (1966).)

Experiment	1	2	3	4
Volume of Xe, in mL at atmospheric pressure	300	350	400	450
Volume of F_2, in mL at atmospheric pressure	271	260	262	183

(a). Draw the Lewis structure of XeF_2, and predict its shape.
(b). Calculate the partial pressure of each gas in Experiment 3.
(c). Calculate the theoretical yield of XeF_2 in Experiment 3.
(d). After the reactants were introduced into the reaction bulb, the bulb was placed in sunlight. XeF_2 began crystallizing on the inner surface of the bulb about 24 hours later. Compare the rate of this reaction to the rate of precipitation of CaF_2 from solutions of $Ca^{2-}(aq)$ and $F^{-}(aq)$ in the laboratory.

11.94 In the previous problem, the reaction to produce XeF_2 occurs photochemically at room temperature. However, to produce $XeF_4(s)$ a higher temperature and pressure are required. Production of XeF_6 requires even higher pressures. Discuss the differences in reaction conditions for these three reactions in terms of activation energy and collision theory.

11.95 On a particular day, the ozone level in Milwaukee exceeded the EPA's 1-hour standard of 0.12 ppm by 10 ppb. How many ozone molecules were there in 1 liter of air at the detection site?

11.96 The EPA's 8-hour ozone standard for communities is 0.08 ppm. On a certain day, the concentration of ozone in the atmosphere in Denver was 1.5×10^{-3} mol/L. Was Denver in compliance with EPA regulations that day?

11.97 The following is a thought experiment. Put a little water in a test tube and add some NaF crystals. Immediately after you add NaF, you observe that the crystals begin dissolving. The quantity of solid NaF decreases, but before long, it appears that no more NaF is dissolving. The solution is saturated.

(a). As NaF dissolves, what do you think happens to the rate of solution? Describe what is occurring on the molecular level.

(b). The equation for the dissolution of NaF in water is $NaF(s) \rightarrow Na^+(aq) + F^-(aq)$.

(c). Assume that the reverse reaction, $Na^+(aq) + F^-(aq) \rightarrow NaF(s)$, also occurs as the crystal dissolves. In other words, both dissolution and precipitation are taking place. When it appears that there is no more change in the quantity of NaF dissolving (the solution is saturated), what has happened to the rates of the forward and reverse reactions? Explain your answer.

Focus on Problem Solving

11.98 If you have a reaction for which you do not know the order, explain how you can obtain the information you need with only a single experiment.

11.99 You take a measurement of an archeological specimen and find the counts from C-14 are x% of background. How does this information allow you to establish the age of the artifact? What would you have to know or look up?

11.100 You are told that an artifact is 3200 years old. How would you use this information to predict the number of radiation counts from C-14 you would measure?

11.101 You are working on a self-heating food package. The chemicals you are using seem to be expended too quickly. What experiments could be run to try to devise conditions that would allow the heating pack to last longer.

Chapter 12: Chemical Equilibrium

In our study of kinetics we saw that the rate of a reaction changes over time, invariably slowing down as the reaction progresses. If we were to observe many reactions, this behavior would be evident each time. What does this imply? Eventually, for any chemical reaction, the net rate approaches zero. When a zero rate has been attained we have reached a state of equilibrium with respect to that reaction. This concept of chemical equilibrium is critically important in a number of ways, and we will take a brief look at several aspects of equilibrium in this chapter. Let's begin by considering an industrial chemical that plays a role in the manufacturing of several important materials: hydrogen cyanide.

12.1 Insight into Hydrogen Cyanide

If you've ever thought about hydrogen cyanide at all, it's likely that the general topic you had in mind was poisons. Hydrogen cyanide is clearly a dangerous chemical. But despite its rather high toxicity, over 600,000 metric tons of the substance is made annually in the U.S. alone. It is an important component of a number of industrial reactions, so let's consider the ways in which HCN itself can be made.

There are several chemical processes that can be used to produce HCN for commercial purposes, and all involve the reaction of ammonia with some type of hydrocarbon. The primary method used in the United States is called the Andrussow process, and it involves the reaction of ammonia, methane and oxygen (from air) according to the following chemical equation.

$$2NH_3 + 2CH_4 + 3O_2 \rightarrow 2HCN + 6H_2O$$

In order to allow this reaction to be used on a commercial scale, it is carried out with either a platinum or a platinum-rhodium catalyst. A high temperature of 1,100°C is also needed to get this reaction to occur at a useful rate.

From what we have just learned, it is clear that this reaction is not a single step process, so the catalyst must lower the activation energy of the rate limiting step.

In addition to enhancing the reaction rate, though, engineers involved in the HCN production process also have to be aware of the importance of separating and purifying the desired product. Looking at the overall project provides the perspective shown in Figure 12.1. In this block diagram, we look at more than just the reaction to consider the isolation of HCN and other possible chemicals – for example sodium cyanide that is often produced by reacting the HCN with NaOH,

$$HCN + NaOH \rightarrow NaCN + H_2O$$

Doing so allows us to consider the concept of Green Chemistry, which suggests that chemical production should consider the entire process so that waste and pollutants can be minimized. Estimates vary about how many tons of waste and pollutants are produced in the production of HCN, but a number in the region of 25,000 tons per year is likely a reasonable guess. What considerations of the chemical reactions and separations used must be addressed to try to minimize these wastes? Because reaction kinetics and equilibrium are always involved in such questions, we will need to address both of these topics in relation to HCN production to be able to provide an answer.

A second process that can be used to produce HCN is called the Blausaure Methane Anlage, or BMA process. Once again ammonia is reacted with methane, but in this case no oxygen is added, so the second product is hydrogen rather than water.

$$NH_3 + CH_4 + \rightarrow HCN + 3H_2$$

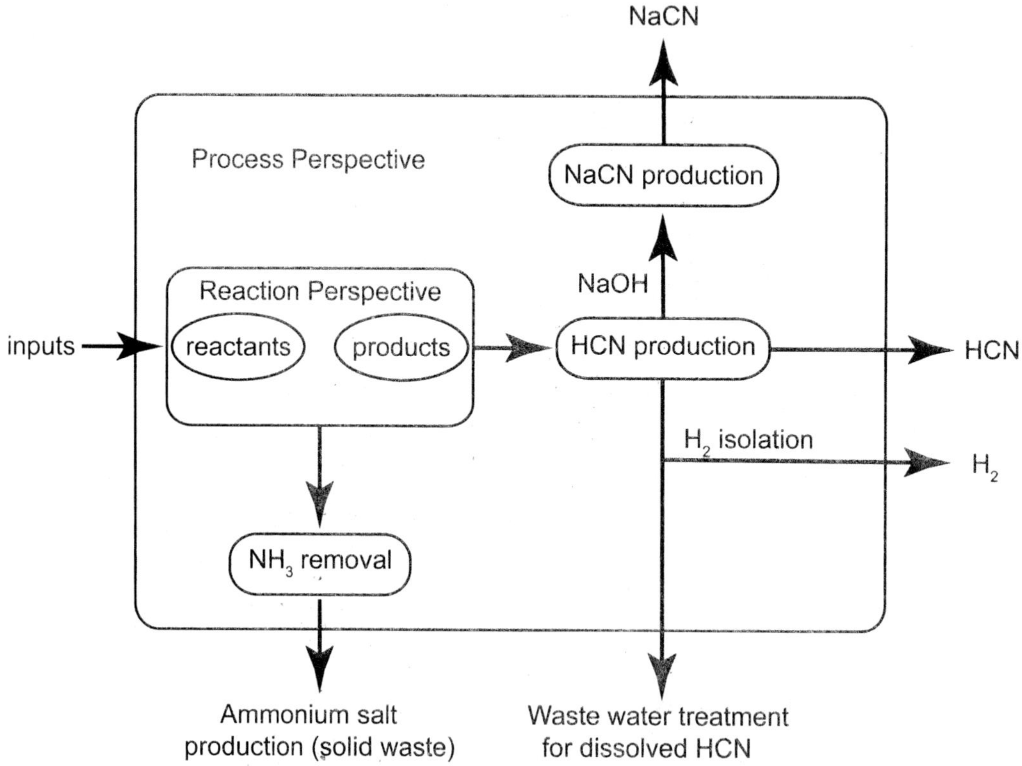

Figure 12.1 This flow diagram shows some of the considerations that enter into the thinking of a process engineer involved with the industrial production of HCN. Issues associated with product formation and waste disposal need to be examined in terms of both chemistry and economics.

One feature of this scheme is that the hydrogen can also be sold; so having a commercially viable byproduct plays a significant role in the decision to utilize the process. What types of chemical questions arise when making a choice between two viable processes?

In the case of HCN production, this question is even more complex, because a third process called the Fluohmic process can be used to produce HCN via another reaction.

$$3NH_3 + C_3H_8 \rightarrow 3HCN + 7H_2$$

This process requires more electricity to run the reactor, and we could certainly ask the question "why?" We should also note one additional method by which HCN is produced commercially. Much like the case of hydrogen production in the BMA or Fluohmic process, HCN is a commercially recoverable byproduct in the process used to make acrylonitrile. In that case, the side-reaction (a reaction that produces something other than the desired product) that takes place is

By now you may have spotted a pattern among these reactions. They differ only in the hydrocarbon and whether or not oxygen is used.

$$3NH_3 + C_3H_6 + 3O_2 \rightarrow 3HCN + 6H_2O.$$

With so many ways to produce HCN, we can obviously ask which one works best and why? We will need to look at this question from two perspectives – that of the chemist and that of the engineer.

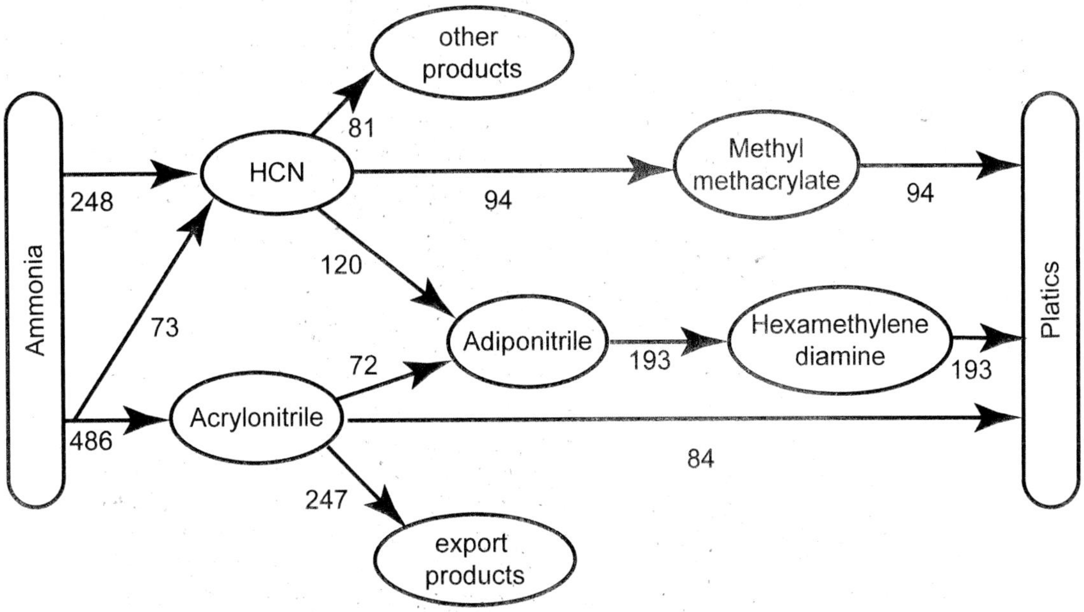

Figure 12.2 This diagram summarizes the production and consumption of HCN in the U.S. chemical industry. The majority of HCN ultimately finds its way into polymers of one type or another. Numbers on the various arrows represent typical annual production in thousands of metric tons (*i.e.*, millions of kilograms).

We also need to look at the reasons why HCN is produced at all. What are the products and materials that can be derived from all of this HCN? Figure 12.2 and Table 12.1 provide some insight. Nearly 75% of the HCN generated is subsequently used in the production of polymers. Sodium cyanide still finds use in the processing of gold at mining sites. What are the reactions that give rise to this usage, and can they be replaced by alternatives based on more environmentally benign materials?

Adiponitrile	41%
Acetone cyanohdrin(a)	32%
Sodium cyanide	14%
Methionine	4%
All other uses	9%

Table 12.1 Hydrogen cyanide is used in the production of several important industrial chemicals. The table shows the portion of total HCN production that is ultimately directed towards each of the most important product streams. (Acetone cyanohydrin is used in the production of methyl methacrylate.)

Chemical production facilities have to include a tremendous number of variables in their designs. From the perspective of chemistry, the role of kinetics and equilibrium provide some of the most important insight. We looked at kinetic factors in the previous chapter, so now let's turn our attention to the concept of equilibrium.

12.2 Chemical Equilibrium

In complex industrial processes like the manufacture of hydrogen cyanide, there are many variables that must be considered. The nature of the reactions used – including the equilibrium that ultimately dictates the efficiency of the reaction – represents the first issue that must be considered.

Figure 12.3 This photo sequence shows the water level in two glasses over the course of time. The glass on the left is covered, while that on the right is open to the air. In the left hand photo, the glasses are filled to the same level. By the time of the second photo, the water level in the right hand glass is visibly lower, while that in the left hand glass remains unchanged. The trend continues in the final photo. (The blocks are a calendar, and indicate the number of days that have passed between photos, but may not be legible here. They read "March 1," "March 11," and "March 17.")

Forward and Reverse Reactions

If you place a glass partially filled with water on a table and leave it overnight, in the morning you may find an empty glass, or at least find that the water level has decreased (Figure 12.3). The "missing" water has evaporated. Try the experiment again, this time covering the glass and you'll find water stays in the glass. The reason for the difference in observations is that in the closed system a dynamic equilibrium is established between liquid water and water vapor (Figure 12.4). At equilibrium, the rate at which water molecules leave the surface of the liquid equals the rate at which they return to the surface. The rate of evaporation will be determined by the distribution of molecular speeds in the liquid; only those molecules with sufficient kinetic energy will be able to overcome the intermolecular forces in the liquid and escape into the vapor phase. Assuming the water is at some fixed temperature, this means the rate of evaporation will be constant. What about the rate of condensation? Each time a gas phase molecule strikes the surface of the liquid, it will have some probability of "sticking" to that surface and returning to the liquid phase. And the rate at which gas molecules strike the surface of the liquid will be proportional to the number of gas molecules present (or to the gas pressure). If only liquid is present initially, the rate of condensation will be zero. But since evaporation is occurring, after a short time there will be some vapor in the container. And that means condensation will begin to occur. As long as the (constant) rate of evaporation is greater than the rate of condensation, the net effect will be to increase the pressure of the vapor phase. That, in turn, will increase the rate of condensation. Eventually the two rates will become equal, and after that point no net change in the amounts of liquid or vapor will be seen.

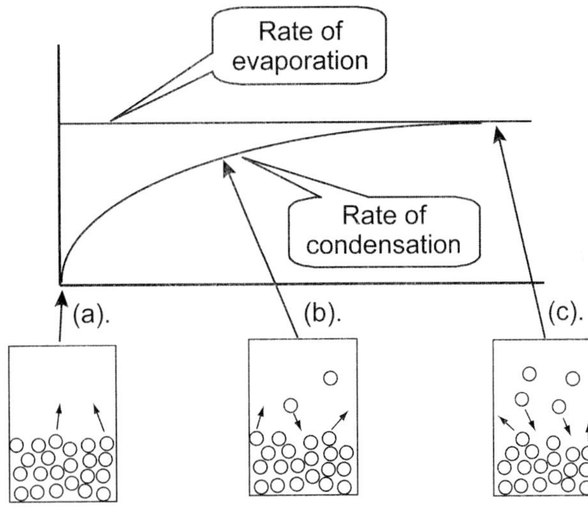

Figure 12.4 The equilibrium between liquid and vapor in a closed container is governed by the kinetics of evaporation and condensation. Here a liquid is placed into a closed container, with no vapor phase present. Energetic molecules at the surface of the liquid can move into the gas phase, as shown in (a). Once some vapor is present, gas molecules will occasionally strike and stick to the liquid surface, so condensation begins to compete with evaporation, as seen in (b). The rate of condensation rises over time, while that of evaporation stays constant. Eventually the two rates become equal. In this equilibrium state, both evaporation and condensation continue to occur, as shown in (c).

We call this situation a **dynamic equilibrium** because the evaporation and condensation do not stop at equilibrium. The rates of these competing processes become equal to one another, but they do *not* go to zero. So on the microscopic level, individual molecules continue to move from liquid to vapor phase and back again. But from a macroscopic perspective, the observable amounts of liquid and vapor no longer change.

Recall that in Chapter 8 we noted that this type of dynamic equilibrium is established when vapor pressure is measured.

Turning to chemical reactions, the manner in which we have written chemical equations thus far has been to consider only the *forward* direction of the reaction. In the *reverse* reaction, the substances we've been calling "products" react to form "reactants." In some cases, the reverse reaction may occur to a large degree and in other cases to an infinitesimal and immeasurable degree, but in principle it does occur in *every* chemical reaction. This means that in a closed system, any chemical reaction will eventually reach a dynamic equilibrium in a manner analogous to the one we described for evaporation.

At the moment we initiate a chemical reaction between two reactants there are no products present (Figure 12.5). We have already seen that the rate of a chemical reaction depends on the concentrations of reactants, so initially the rate of the reverse reaction is zero, because the concentration of its reactants (which are the products of the forward reaction) is zero. Thus, when we first mix the reactants the rate of the forward reaction is greater than that of the reverse reaction. Over time the concentration of the reactants decreases and that of the products increases. Changes in concentration are accompanied by changes in rate. The forward reaction slows down and the reverse reaction speeds up. Ultimately, the two rates become equal, chemical equilibrium is reached, and the observable concentrations of reactants and products both become constant. In any chemical equilibrium, *the rate of the forward reaction equals the rate of the reverse reaction.*

A graph such as this one inplies the extent of the reaction. In this case, one could argue the reaction does not proceed very far towards products. We will see how this type of observation can be made quantitative next.

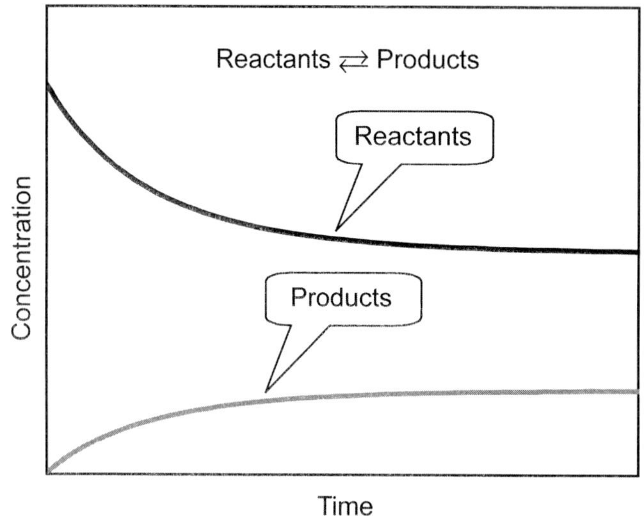

Figure 12.5 The graph depicts the approach of a chemical system to equilibrium. Initially only reactants are present; the forward reaction proceeds and forms products. As the product concentrations increase, the reverse reaction becomes increasingly important. Eventually the rates of the forward and reverse reactions become equal, and the measured concentrations of reactants and products all become constant at equilibrium.

Mathematical Relationships

The rate laws of the forward and reverse reactions suggest that a mathematical relationship can be written to describe equilibrium. For any reaction involving reactant R, and product P, we write the chemical equation at equilibrium with two arrows, one pointing in each direction, signifying the dynamic character of the process:

$$R \rightleftarrows P$$

For simplicity, we will assume that this reaction proceeds in each direction by a single elementary step. (The result of our derivation here does not actually depend on this assumption.) We can write the rate law for the forward direction of this reaction as

$$\text{rate}_{\text{for}} = k_{\text{for}}[\text{R}].$$

Likewise for the reverse reaction, we can write

$$\text{rate}_{\text{rev}} = k_{\text{rev}}[\text{P}].$$

At equilibrium we know these two rates must be equal, so

$$\text{rate}_{\text{for}} = \text{rate}_{\text{rev}}$$

and therefore

$$k_{\text{for}}[\text{R}]_{\text{eq}} = k_{\text{rev}}[\text{P}]_{\text{eq}}.$$

Since the rates are only equal if the system is at equilibrium, we explicitly designate the concentrations as *equilibrium* concentrations here. Both k_{for} and k_{rev} are rate constants, so we can manipulate this equation to obtain

$$\frac{k_{for}}{k_{rev}} = \frac{[P]_{eq}}{[R]_{eq}}$$

At a given temperature, therefore, $[P]_{eq}/[R]_{eq}$ is a constant. (Remember that rate constants vary with temperature). This expression will be central to our mathematical treatment of equilibrium in Section 12.3.

To place this idea in a more concrete setting, consider what happens when HCN is placed in water. As we will see in greater detail in Section 12.7, the HCN is an example of an acid that dissociates in water to form hydrogen ions and cyanide ions.

$$HCN(aq) \rightleftarrows H^+(aq) + CN^-(aq)$$

If we assume that the reactions in both directions are elementary processes (Section 11.6), the forward and reverse rate laws can be written.

$$\text{rate} = k_{for}[HCN]$$

$$\text{rate} = k_{rev}[H^+][CN^-]$$

and at equilibrium,

$$\frac{k_{for}}{k_{rev}} = \frac{[HCN]_{eq}}{[H^+]_{eq}[CN^-]_{eq}} = \text{a constant}$$

12.3 Equilibrium Constants

We mentioned earlier that cyanide is sometimes used in the extraction of gold from ores. The process is based on the fact that metallic gold will react with aqueous cyanide ions to form a soluble complex.

$$Au(s) + 2CN^-(aq) \rightarrow Au(CN)_2$$

The extent of this reaction, and even the availability of the cyanide itself, is typically dictated by a variety of chemical equilibria. If this reaction is being used as part of an industrial effort to recover gold, it is easy to imagine that it will be important to know just how much cyanide is available and how much gold should be obtained. The need for quantitative answers to questions such as this requires a mathematical model for equilibrium. We saw in the previous section that some sort of relationship exists between concentrations of reactants and products. This relationship, which was first proposed by Cato Maximilian Guldberg and Peter Waage in 1864, is often called the Law of Mass Action.

The Equilibrium (Mass Action) Expression

For a general chemical reaction,

$$aA + bB \rightleftarrows cC + dD$$

we can define the ratio

$$Q = \frac{[C]^c[D]^d}{[A]^a[B]^b} \quad (12.1)$$

for any situation, whether at equilibrium or not. We refer to this ratio as the **reaction quotient**, Q. Product concentrations appear in the numerator of the expression, and reactants appear in the denominator. Each concentration is raised to a power corresponding to the stoichiometric coefficient of that species in the balanced chemical equation for the equilibrium. At equilibrium, however, this ratio becomes the **equilibrium expression**, and the corresponding value of Q is called the **equilibrium constant**, K.

For the purpose of a definition such as this one, it is important to emphasize the fact that equilibrium is acheived by using the subscript "eq". This subscript is often left off and implied simply by the fact that "K" is being used rather than "Q".

$$K = \frac{[C]_{eq}^{c}[D]_{eq}^{d}}{[A]_{eq}^{a}[B]_{eq}^{b}} \tag{12.2}$$

To determine the value of K for a given reaction, we must measure the concentrations of the reactants and products at equilibrium. Note that, unlike rate expressions in chemical kinetics, *the equilibrium constant expression is always given directly by the reaction stoichiometry*. Let's write an equilibrium expression.

Example Problem 12.1

For the BMA process, the equilibrium below would be established if the products were not removed from the reaction vessel.

$$NH_3(g) + CH_4(g) \rightleftarrows HCN(g) + 3H_2(g)$$

Write the equilibrium expression for this reaction.

Strategy

We use the definition of the equilibrium expression. Write the products in the numerator and the reactants in the denominator. The exponent of the concentration is given by the stoichiometry of the chemical equation.

Solution

$$K = \frac{[HCN][H_2]^3}{[NH_3][CH_4]}$$

Check Your Understanding

Nitrogen oxides in the atmosphere may acidify rainwater. Write the equilibrium expression for the following reaction, in which gaseous nitrous acid forms in the atmosphere.

$$NO(g) + NO_2(g) + H_2O(g) \rightleftarrows 2HNO_2(g)$$

Homogeneous and Heterogeneous Equilibria

So far the equilibria we have examined are examples of **homogeneous equilibrium**, in which reactants and products are in the same phase, either gaseous or aqueous. If any of the reactants or products is in the solid or liquid phase, though, we will need to adjust our formulation of the equilibrium expression slightly. The concentration of a pure solid or liquid will not change as the reaction proceeds. Indeed, the phrase "concentration of a pure solid or liquid" should strike you as a bit odd; unlike gases or solutions, solids and liquids do not have variable concentrations. For example, solid $CaCO_3$ can be heated in a closed system to produce solid CaO and gaseous CO_2 (Figure 12.6).

$$CaCO_3(s) \rightleftarrows CaO(s) + CO_2(g)$$

Once equilibrium has been established at a particular temperature, the concentration of gaseous CO_2 can be expressed in terms of mol L^{-1}. If the system is heated, more CaO and CO_2 form, and, if the volume of the system is held constant, the concentration of CO_2 in mol L^{-1} increases as equilibrium is established at a higher temperature. But the *concentrations* of the solids do not change, because, as the numbers of moles of each changes, so do their volumes. The densities of solids change very little with a change in temperature, and the ratio of the number of moles (and mass) of $CaCO_3$, for example, to the volume of $CaCO_3$ remains constant. A solid's concentration does not change, nor does that of a liquid. This fact has led to a different way to describe the equilibrium expression in the case of **heterogeneous equilibrium**—equilibrium in a multiphase system. Because their concentrations are constant, reactants and products that are either liquid or solid do not appear in the expression, so the equilibrium expression for the system above would involve only the carbon dioxide concentration.

$$K = [CO_2]$$

In the following example, you can see how this works for precipitation reactions.

$$CaCO_3(s) \rightleftarrows CaO(s) + CO_2(g)$$

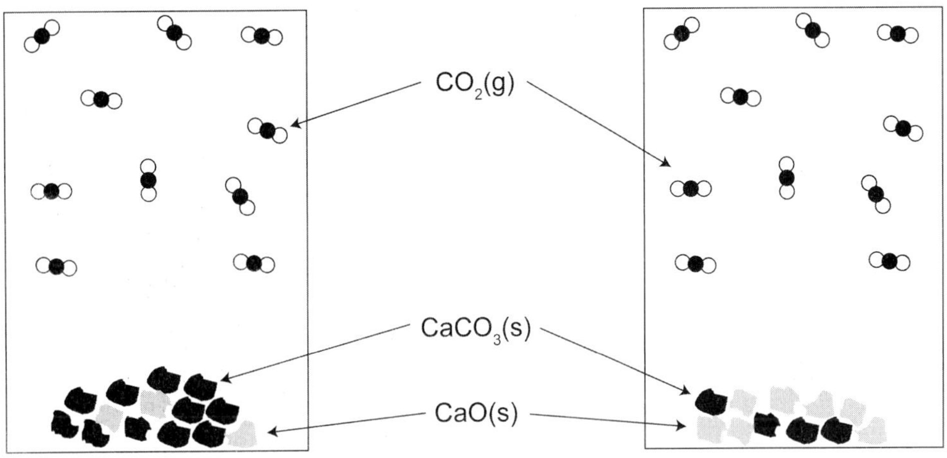

Note that to show enough CO_2 in this picture to make a comparison, the distance between particles is too small for any gas phase pressure that could be acheived.

Figure 12.6 Because solids do not have variable concentrations, the equilibrium constant for this reaction is just the concentration of CO_2. This means that the CO_2 pressure at a given temperature is independent of the amount s of $CaCO_3$ and CaO present. Here the CO_2 pressure is the same over two very different mixtures of the two solids.

Example Problem 12.2

Cyanide ions will precipitate out of solution if silver ions are also present as described by the equilibrium,

$$Ag^+(aq) + CN^-(aq) \rightleftarrows AgCN(s)$$

Write the equilibrium expressions for this reaction.

Strategy

Write the equilibrium expression as before, but do not include a term for the concentration of the solid product.

Solution

For the precipitation of silver cyanide,

$$K = \frac{1}{[Ag^+][CN^-]}$$

Check Your Understanding

Write the chemical equation for the precipitation of copper(II) phosphate from a solution containing copper(II) and phosphate ions, and then write the equilibrium expression.

Numerical Importance of the Equilibrium Expression

Being able to write an expression relating equilibrium concentrations in chemical reactions provides a powerful tool for understanding chemical equilibrium. While we will soon see how to evaluate these expressions for several classes of reactions, let's look first at numerical values of the equilibrium constant. For example, we could ask the question, "Which reaction is more likely to produce hydrogen gas at relatively low temperature – the BMA process or the Fluohmic process?" How can we use the equilibrium constant for each reaction to find the answer?

Some equilibrium problems are slightly more challenging if the exponent of the power of 10 exceeds 99 and causes errors on some calculators. Such problems can be handled using algebraic manipulations of the exponents.

Equilibrium constants range in value over many orders of magnitude. Values as small as 10^{-99} are known, as are values as high as 10^{99} or greater. What does the size of the equilibrium constant tell us about the reaction? If we look at the equilibrium expression for a reaction as an algebraic relationship, the answer is apparent.

$$K = \frac{[\text{products}]}{[\text{reactants}]}$$

Concentrations of products are in the numerator, and concentrations of reactants are in the denominator. For a very large value of K, the product concentrations must be large and the reactant concentrations must be small. By contrast, for the value of K to be small, concentrations of the reactants must be large and those of products must be small.

$$K = \frac{\Large[\text{products}]}{\scriptstyle[\text{reactants}]} = \text{large value} \qquad K = \frac{\scriptstyle[\text{products}]}{\Large[\text{reactants}]} = \text{small value}$$

Thus, the value of K can tell us in which direction a chemical reaction is likely to proceed. For large values of K, products are favored, while small values of K (meaning $K << 1$) are found in reactions that reach equilibrium with most of the reactants still intact. If K is on the order of one, the equilibrium has significant concentrations of both reactants and products. Example Problem 12.3 provides some practice interpreting the value of K.

Example Problem 12.3

We saw in Example Problem12.2 that cyanide ions will precipitate with silver. Bromide ions also precipitate with silver ions. The two pertinent equilibria are

$$Ag^+(aq) + CN^-(aq) \rightleftarrows AgCN(s) \qquad K = 1.68 \times 10^{18}$$

$$Ag^+(aq) + Br^-(aq) \rightleftarrows AgBr(s) \qquad K = 1.87 \times 10^{12}$$

Which ion is more likely to precipitate silver from a solution, assuming concentrations of cyanide and bromide ions are roughly equal?

Strategy

The likely extent of each reaction can be predicted by the size of the equilibrium constant. We must compare the two numbers to see which is more likely to precipitate the silver.

Solution

Both equilibrium constants are much greater than one, but the value for the reaction with cyanide ions is significantly larger than that for bromide. Thus silver cyanide has the greater tendency to form, and will be the more likely product.

Check Your Understanding

Silver sulfate precipitates according to the equation

$$2Ag^+(aq) + SO_4^{2-}(aq) \rightleftarrows Ag_2SO_4(s) \qquad K = 8.33 \times 10^4$$

Relative to cyanide and bromide ions, where would sulfate rate in terms of its ability to precipitate silver ions?

Mathematical Manipulation of Equilibrium Constants

When we work with precipitation reactions like those in the previous example, we need to look up values for the equilibrium constants. But the values we find in a reference book typically refer to the opposite process—solubility.

Reversing the Chemical Equation

When we write chemical equations for reactions that go to completion, the choice of "reactants" and "products" seems logically clear. But for an equilibrium reaction, the fact that both forward and backward reactions are potentially important makes the choice seem much more arbitrary. And indeed, we can write equations for equilibria in either direction. Neither choice is "wrong," provided we also realize that the equilibrium expression we use must match the equation we choose. Consider the precipitation of silver cyanide. When we write the equilibrium expression for both forms of the chemical equation,

$$Ag^+(aq) + CN^-(aq) \rightleftarrows AgCN(s) \qquad K = \frac{1}{[Ag^+][CN^-]}$$

$$AgCN(s) \rightleftarrows Ag^+(aq) + CN^-(aq) \qquad K = [Ag^+][CN^-]$$

we can see that the equilibrium expressions must be inverses of each other. Thus when we reverse the chemical equation, by switching reactants and products, we invert the equilibrium constant expression and its value.

$$K' = \frac{1}{K} \tag{12.3}$$

Adjusting the Stoichiometry of the Chemical Reaction

The stoichiometric coefficients play a direct role in the equilibrium expression, and if adjustments are made in the stoichiometry, they will affect the value of the equilibrium constant. For example, the reaction for the Andrussow process is sometimes written as either

$$2NH_3(g) + 2CH_4(g) + 3O_2(g) \rightleftarrows 2HCN(g) + 6H_2O(g)$$

or

$$NH_3(g) + CH_4(g) + 3/2\,O_2(g) \rightleftarrows HCN(g) + 3H_2O(g)$$

How is the equilibrium expression affected by this change in the stoichiometry? The two expressions are

$$K_1 = \frac{[HCN]^2[H_2O]^6}{[NH_3]^2[CH_4]^2[O_2]^3} \text{ and } K_2 = \frac{[HCN][H_2O]^3}{[NH_3][CH_4][O_2]^{3/2}}$$

As you can see, K_2 is the square root of K_1. Thus, dividing by two in the stoichiometry leads to a division by two in the *exponent* of the equilibrium expression, and we must look carefully at the way a chemical equation is written when we consider the value of the equilibrium constant.

Example Problem 12.4

Ammonia is the starting material for all industrial processes that produce HCN. Thus the production of ammonia from nitrogen and hydrogen is an important precursor to HCN production.

$$3\,N_2(g) + H_2(g) \rightleftarrows 2\,NH_3(g)$$

Write equilibrium expressions for

(a). the reaction as written.

(b). the reverse reaction.

(c). the reaction as written, but with the coefficients in the equation halved.

Strategy

Write the equilibrium expression for the reaction as written, and manipulate the expression as needed to match the other conditions. (If we are not confident with these manipulations, we can always write out the specified versions of the equilibrium equations and use them to obtain the needed equilibrium expressions.)

Solution

(a). $K = \dfrac{[NH_3]^2}{[N_2]^3[H_2]}$ (b). $K' = \dfrac{1}{K} = \dfrac{[N_2]^3[H_2]}{[NH_3]^2}$ (c). $K'' = K^{1/2} = \dfrac{[NH_3]}{[N_2]^{3/2}[H_2]}$

Check Your Understanding

Consider the following reaction in a closed system.

$$4\,NH_3(g) + 5\,O_2(g) \rightleftarrows 4\,NO(g) + 6\,H_2O(g)$$

Write the equilibrium expressions for

(a). the reaction as written.

(b). the reverse reaction.

(c). the reaction as written, when the coefficients in the equation are halved.

Equilibrium Constants for a Series of Reactions

In complicated chemical systems, a product of one chemical reaction can be a reactant in a subsequent reaction. Because of the nature of the equilibrium expression, if we know the equilibrium constant for the two separate reactions, we can determine an overall value for the combination of the two. Consider, for example, the stepwise reaction of phosphate ions with hydrogen ions.

1. $PO_4^{3-}(aq) + H^+(aq) \rightleftarrows HPO_4^{2-}(aq)$ $\qquad K_1 = \dfrac{[HPO_4^{2-}]}{[PO_4^{3-}][H^+]}$

2. $HPO_4^{2-}(aq) + H^+(aq) \rightleftarrows H_2PO_4^-(aq)$ $\qquad K_2 = \dfrac{[H_2PO_4^-]}{[HPO_4^{2-}][H^+]}$

3. $PO_4^{3-}(aq) + 2\,H^+(aq) \rightleftarrows H_2PO_4^-(aq)$ $\qquad K_3 = \dfrac{[H_2PO_4^-]}{[PO_4^{3-}][H^+]^2}$

The third equation is the sum of the first two equations. How are their equilibrium constants related? In order to obtain the chemical equation associated with K_3, we added the chemical equations for K_1 and K_2. To get the equilibrium constant for equation 3, however, we must multiply the expressions.

$$K_3 = K_1 \times K_2$$

$$K_3 = \frac{[HPO_4^{2-}]}{[PO_4^{3-}][H^+]} \times \frac{[H_2PO_4^-]}{[HPO_4^{2-}][H^+]} = \frac{[H_2PO_4^-]}{[PO_4^{3-}][H^+]^2}$$

The adjustment of K for the stoichiometry of the equation shown on page 471 is essentially a special case of this relationship.

This relationship holds true any time we add chemical equations to obtain a new equation: multiply the equilibrium constants to find the equilibrium constant for the reaction of interest.

Example Problem 12.5

Given the following equilibria.

$$CO_2(g) \rightleftarrows CO(g) + 1/2\ O_2(g)$$

$$H_2(g) + 1/2\ O_2(g) \rightleftarrows H_2O(g)$$

Determine the equilibrium expression for the sum of the two reactions.

Strategy

Write the equilibrium expressions for the two reactions. Multiply them to obtain the equilibrium expression for the sum of the two reactions. Check your answer by adding the two reactions and comparing the sum to the equilibrium expression you determined.

Solution

1. $CO_2(g) \rightleftarrows CO(g) + 1/2\ O_2(g)$ $\quad K_1 = \frac{[CO][O_2]^{1/2}}{[CO_2]}$

2. $H_2(g) + 1/2\ O_2(g) \rightleftarrows H_2O(g)$ $\quad K_2 = \frac{[H_2O]}{[H_2][O_2]^{1/2}}$

$$K_3 = \frac{[CO][O_2]^{1/2}}{[CO_2]} \times \frac{[H_2O]}{[H_2][O_2]^{1/2}} = \frac{[CO][H_2O]}{[CO_2][H_2]}$$

Discussion

The sum of the two reactions is $CO_2(g) + H_2(g) \rightleftarrows CO(g) + H_2O(g)$. So it is fairly easy to see that the equilibrium expression we obtained above is correct.

Check Your Understanding

Given the following equilibrium expressions for a pair of reactions in the gaseous phase, find the equilibrium expression for the sum of the two reactions, and write its chemical equation.

$$K_1 = \frac{[NO_2]^2}{[N_2][O_2]^2}$$

$$K_2 = \frac{[NO_2]^2}{[NO]^2[O_2]}$$

Units and the Equilibrium Constant

As you will see when we work with equilibrium constant expressions, the equilibrium constant can be used in various mathematical expressions, including as the argument in a natural log function. When a variable appears as an argument of a function, such as a natural log, it must be dimensionless. To produce equilibrium constants without units, we adjust the concentrations of all species by comparing them to a standard concentration of 1 *M*. Thus, the concentration of cyanide ion we use in the expressions we encounter will be $[CN^-] = [CN^-]_{eq} / 1\ M$. In this way the numbers we use in equilibrium constant calculations have no units attached, and the equilibrium constant will also be unitless. Because we choose 1 *M* as the standard concentration, the manipulation has no numerical consequences. It does, however, help avoid confusion later when we use the equilibrium constant in mathematical equations.

12.4 Equilibrium Concentrations

While equilibrium concentrations are measured to determine the value of the equilibrium constants in some lab experiments, other situations call for the prediction of equilibrium concentrations based on known equilibrium constants. We can look up the equilibrium constant for the dissociation of HCN in water, but how does this tell us the concentration of cyanide ion in a given solution? We can address this type of question by carrying out calculations using the equilibrium constant expression and the stoichiometry of the chemical reaction it describes.

It's always a good idea to obtain information experimentally, but being able to carry out calculations to make good predictions can often help dictate what methods of measurement should be used, for example.

Calculations based on chemical equilibria tend to be somewhat complicated, and often involve a large amount of data. So it is a good idea to approach these problems with a systematic method for handling the information. We'll illustrate how to approach some specific types of problems throughout this chapter. But first, it is worth noting three basic features of the strategy we'll employ in performing any equilibrium calculation.

- Write a balanced chemical equation for the relevant equilibrium or equilibria.
- Write the corresponding equilibrium expression(s).
- Create a table of concentrations for all of the reacting species. This will generally contain a column for each species involved in the equilibrium. The first row contains the initial concentrations, the second row contains the changes in these concentrations that occur as the system comes to equilibrium, and the third row contains the final equilibrium concentrations. Much of this information will not be known at the outset of the problem. But the table provides a convenient way of organizing all of the relevant information. As we work through a problem, more and more of the cells in this table will be filled in.

Equilibrium Concentrations From Initial Concentrations

If the BMA process was run in a batch so all the reactants and products remain present, what concentrations of each would be found at equilibrium? We can find answers to questions like this, and doing so represents an important starting point in the design of many industrial processes. To look at this type of problem, let's start with an algebraically simple example involving hydrogen and iodine. This will allow us to illustrate the use of the problem-solving framework that we will apply for equilibrium calculations. Once we've mastered that framework, we will return to questions related to the industrial production and uses of hydrogen cyanide.

Example Problem 12.6

When hydrogen gas reacts with iodine gas at elevated temperatures, an equilibrium is established.

$$H_2(g) + I_2(g) \rightleftarrows 2HI(g)$$

A student measured the equilibrium constant to be 59.3 at 400°C. If the student begins with a mixture that is 0.050 *M* in both hydrogen and iodine, what will be the equilibrium concentrations of reactants and products?

Strategy

It may help to begin by summarizing the information from the problem.

Given: $[H_2]_{initial} = [I_2]_{initial} = 0.050\ M;\ K = 59.3$ Find: $[H_2]_{eq}$, $[I_2]_{eq}$, and $[HI]_{eq}$

We'll use (and clarify) the three-point strategy outlined above. The chemical equation for the equilibrium is given in the problem. And from that equation we can easily write the equilibrium expression.

$$K = \frac{[HI]^2}{[H_2][I_2]} = 59.3$$

Next we construct the table of concentrations, with places to enter the initial concentrations, the changes in concentration, and the final (equilibrium) concentrations for the three species involved. At this point, only the given initial concentrations are known. (Since only H_2 and I_2 are present initially, the initial concentration of HI is zero.)

	H_2	I_2	HI
initial concentration	0.050 M	0.050 M	0 M
change in concentration			
final concentration			

The rest of the table will be filled in as we work our way through the solution.

At this point the problem seems daunting; we have three unknown equilibrium concentrations, and only one real equation to work with – the equilibrium expression. We will need to use the stoichiometry of the reaction to relate the changes in concentration to one another.

Solution

Since only H_2 and I_2 are present initially, we can easily see that their concentrations must decrease (and the HI concentration must increase) as the system goes to equilibrium. We also see from the balanced equation that each H_2 molecule that reacts must react with one I_2 molecule to form two HI molecules. This gives us the relationship we need to express the changes in concentration in terms of one variable. Let's call the change in $[H_2]$ '$-x$.' (The negative sign means the concentration decreases.) The 1:1 mole ratio between H_2 and I_2 means the change in $[I_2]$ must also be '$-x$.' And the 1:2 mole ratio between H_2 and HI means the change in [HI] must be '$+2x$.' This gives us the second row in the table.

	H_2	I_2	HI
initial concentration	0.050 M	0.050 M	0 M
change in concentration	$-x$	$-x$	$+2x$
final concentration			

Next we can fill in the third row. To do this, we use the simple relationship that the amount (concentration) of each species at equilibrium will be its initial concentration plus the change in concentration. So we just need to sum each column to get expressions for the third row.

	H_2	I_2	HI
initial concentration	0.050 M	0.050 M	0 M
change in concentration	$-x$	$-x$	$+2x$
final concentration	$0.050 - x$	$0.050 - x$	$2x$

Of course our aim is to find actual values for those three equilibrium concentrations, so we still need to find a way to solve for the variable we introduced (x). To do this, we return to the equilibrium expression. If we insert the expressions from the third row of our table into the equilibrium expression, we'll get the following equation.

$$K = \frac{[HI]^2}{[H_2][I_2]} = \frac{(2x)^2}{(0.050 - x)(0.050 - x)} = 59.3$$

Or

$$\frac{(2x)^2}{(0.050 - x)^2} = 59.3$$

This equation has only one unknown, so we should be able to solve it for x and then use that result to obtain values for the final concentrations. We could multiply the expression out and solve using the quadratic formula. Or we might save a little algebra by realizing that the left hand side is a perfect square. This lets us take the square root of both sides to get a linear equation, which we can easily solve.

$$\frac{2x}{0.050 - x} = 7.7$$

$$0.385 = 9.70x$$

$$x = 0.0397$$

Finally, we use this value of x to calculate all equilibrium concentrations.

$$[H_2] = [I_2] = 0.050 - 0.0397 = 0.010 \text{ M}$$

$$[HI] = 2x = 0.079 \text{ M}$$

Check Your Understanding

The same student measures the equilibrium constant for this reaction at a higher temperature and obtains a value of $K = 51.4$. An experiment is started with 0.010 M each of hydrogen and iodine. Predict the equilibrium concentrations of all species.

Two specific factors helped us to simplify the calculations in Example Problem 12.6. First, the stoichiometry yielded perfect squares in both the denominator and the numerator, *i.e.*, one mole of each of the two reactants yielded two moles of products. Second, we started with the same initial concentrations of each reactant. Had those initial amounts been different from each other, we would not have had a perfect square when we filled in the equilibrium expression. When we are not so fortunate in our reaction of interest we may have to use the quadratic equation to solve the resulting algebraic equation. While this may be a minor nuisance, it isn't really a major challenge. We consider this situation in the following example.

Example Problem 12.7

The reaction of chlorine gas with phosphorous trichloride to form phosphorous pentachloride has an equilibrium constant of 33 at 250°C. If an experiment is initiated with concentrations of 0.050 *M* PCl_3 and 0.015 *M* Cl_2 what are the equilibrium concentrations of all three gasses?

$$Cl_2(g) + PCl_3(g) \rightleftarrows PCl_5(g)$$

Strategy

The approach here is the same as in Example Problem 12.6. The only change occurs when we need to solve for x after we set up the equilibrium constant expression. The stoichiometry and the initial amounts in this case will force us to use the quadratic formula to solve for x.

Solution

Construct a table of concentrations, as we did in Example Problem 12.6. We'll insert the initial concentrations and the changes in concentration. Since the stoichiometry of this reaction is 1:1:1, the changes can be written as $-x$ for both Cl_2 and PCl_3 and $+x$ for PCl_5.

	Cl_2	PCl_3	PCl_5
initial concentration	0.015 M	0.050 M	0 M
change in concentration	$-x$	$-x$	$+x$
final concentration	$0.015 - x$	$0.050 - x$	x

Substitute into the equilibrium expression:

$$K = \frac{[PCl_5]}{[Cl_2][PCl_3]} = 33$$

$$\frac{x}{(0.015 - x)(0.050 - x)} = 33$$

To solve this we must use the quadratic formula, so we first multiply it out.

$$33x^2 - 3.145\ x + 0.025 = 0$$

Using the quadratic formula,

$$x = \frac{-b \pm \sqrt{b^2 - 4ac}}{2a}$$

$$= \frac{3.145 \pm \sqrt{9.89 - 3.3}}{66}$$

$$x = 0.087 \text{ or } x = 0.0088$$

It may seem disconcerting to get two numerical solutions to a problem like this. How do we decide which root is the correct one? The resulting equilibrium concentrations must be physically reasonable. Among other things, that means that none of them can be negative. But if we chose the first value for x (0.087), we would get negative concentrations for both Cl_2 and PCl_3. So the correct root must be the second one, which gives us the following results.

$$[PCl_5] = x = 0.0088\ M$$

$$[Cl_2] = 0.015 - x = 0.015 - 0.0088 = 0.006\ M$$

$$[PCl_3] = 0.050 - x = 0.050 - 0.0088 = 0.041\ M$$

Check Your Understanding

This reaction can also be considered as the decomposition of $PCl_5(g)$, for which, at the same temperature, the equilibrium constant is 0.30. If you begin an experiment with 0.040 M $PCl_5(g)$, what are the expected equilibrium concentrations of all species?

Mathematical Techniques for Equilibrium Calculations

The general approach we have put forward in these example problems will work for any equilibrium calculation. But we should point out that these examples were chosen at least in part because they can be solved with relative ease. Even our more complicated case required nothing more involved than using the quadratic formula. But not all equilibria lend themselves to such simple solutions. Consider the Andrussow process for the production of HCN. The overall equation is shown below. And since all species involved would be gases under the high temperature conditions needed for reaction, five concentration terms would appear in the equilibrium expression, along with exponents as high as six.

$$2NH_3 + 2CH_4 + 3O_2 \rightleftarrows 2HCN + 6H_2O$$

$$K = \frac{[HCN]^2[H_2O]^6}{[NH_3]^2[CH_4]^2[O_2]^3}$$

If we were to construct a problem similar to the ones we solved above but based on this equilibrium, we would be faced with solving an equation that included terms in x^8. Clearly this would present challenges well beyond solving a quadratic equation. Still, modern software does offer hope for solving such equations. Maple, Mathematica, and other packages provide tools for finding solutions to equations of this sort. We will not spend our time on such calculations here, though, because we would rather focus on the underlying chemical concepts. Being able to determine the equilibrium concentration of a desired reactant provides a useful starting point for exploring industrial processes like the production of hydrogen cyanide. But the equilibrium yield of a product may not be sufficient to make a process economically viable. In such cases, a process engineer needs to find ways to manipulate the equilibrium and allow higher levels of production or greater efficiency. How can equilibrium be manipulated? Our next section addresses this question.

12.5 LeChatelier's Principle

For the production of chemical commodities, chemical and industrial engineers choose reactions whose equilibrium constants favor products whenever possible. In some cases, though, this type of choice is simply not available. For example, the production of the ammonia feedstock for HCN synthesis does not favor the formation of ammonia. Despite this fact, the manufacture of ammonia is not only commercially viable, it is a major component of the chemical industry – chiefly because of the importance of ammonia to fertilizers and other agricultural chemicals. What are the variables that can be controlled in the plants and how do they fundamentally alter equilibrium?

Henry Louis LeChatelier came from a family of archeticts and engineers and carried out his work in chemistry from the 1880s to the early 1900s.

LeChatelier's Principle summarizes the ways that any equilibrium system responds to changes. *When a system at equilibrium is stressed, it responds by re-establishing equilibrium in a manner that partially offsets the applied stress.* There are three common means by which stresses are introduced in chemical equilibrium: changes in concentrations, changes in pressure, and changes in temperature. We will look at each of these in turn, and will think about them in the context of a single simple equilibrium system: the dimerization of NO_2 to form N_2O_4.

Effect on Equilibrium of a Change in Concentration of Reactant or Product

Chemical equilibrium results in a balance among the concentrations of the various species involved. So the most direct way to disturb a chemical system at equilibrium is to change the concentration of one or more reactants or products. Suppose we have a closed flask containing an equilibrium mixture of NO_2 and N_2O_4 gases. If we somehow inject additional NO_2 into the system, how will the equilibrium respond? The plot in Figure 12.7 shows the expected effect: additional N_2O_4 must be formed in order to reduce the applied stress to the equilibrium.

$$2NO_2 \rightleftarrows N_2O_4$$

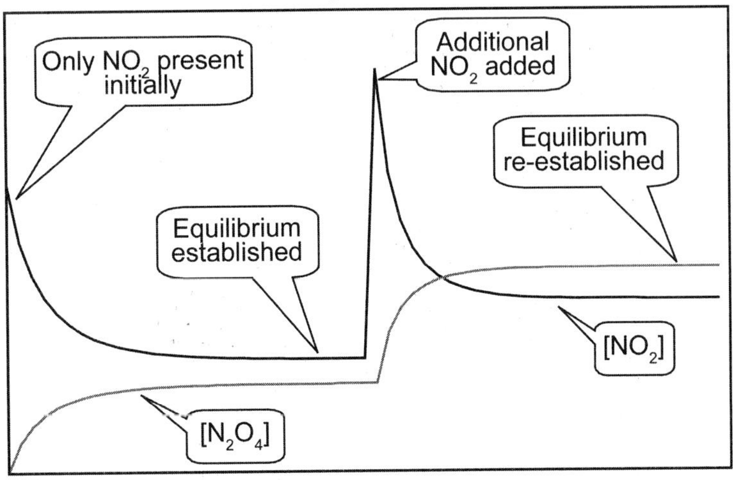

Figure 12.7 The graphs illustrate the response of an equilibrium to an increase in reactant concentration. Initially only NO_2 is present, and the formation of N_2O_4 establishes equilibrium. Midway through the time in the plot, a large amount of NO_2 is added to the system, perturbing the equilibrium. The system responds by forming additional N_2O_4 and establishing a new equilibrium.

But how does this response offset the applied stress? When we began our discussion of the equilibrium constant in Section 12.3, we first defined the reaction quotient, Q, and then considered the equilibrium constant as a special case of this ratio. Thus at equilibrium, we have the relationship, $Q = K$. But if we change the concentrations away from their equilibrium values, Q will no longer be equal to K. Immediately following an increase in the concentration of a reactant, as in Figure 12.7, Q must become smaller than K. So in order to establish a new equilibrium, the value of the ratio must increase. Thus the system must respond by increasing the concentrations in the numerator and decreasing the size of those in the denominator. Because products are in the numerator, this means that when $Q < K$ the reaction must shift toward products to reach equilibrium. Similar reasoning can establish that when $Q > K$ the reaction must shift towards reactants. Table 12.2 summarizes the way that an equilibrium system responds to the four possible types of concentration changes, and Example Problem 12.8 explores this further.

While LeChateliers principle is often used to gain qualitative insight into the effect of stresses on equilibrium, he himself was quite interested in the quantitative detail provided by mathematical treatments derived from looking at Q vs. K.

Type of concentration change	Resulting change in Q	Response of system
[products] increased	$Q > K$	more reactants formed
[products] decreased	$Q < K$	more products formed
[reactants] increased	$Q < K$	more products formed
[reactants] decreased	$Q > K$	more reactants formed

Table 12.2 The effects of concentration changes on equilibrium can all be rationalized by considering the reaction quotient Q and how it compares to the equilibrium constant K.

Example Problem 12.8

An experimental system in which hydrocyanic acid, HCN is in equilibrium with cyanide ion and hydrogen ion can be disturbed in several ways. Predict the change in the reaction quotient, Q, for each case and use that prediction to explain how the equilibrium is shifted by the stress.

(a). Sodium cyanide is added to the system.
(b). Additional hydrocyanic acid is added to the system.
(c). Sodium hydroxide is added to the system.

Strategy

In each case, determine what change the added chemical induces in the equilibrium concentrations. By noting whether the concentration change affects the numerator or denominator we can then judge whether the resulting reaction quotient is larger or smaller than K. Using the value of Q, we can predict the direction equilibrium will move to absorb the stress.

Solution

The equilibrium is

$$HCN(aq) \rightleftarrows CN^-(aq) + H^+(aq)$$

(a). Sodium cyanide is added.

The sodium ion is not involved in this equilibrium, but by adding CN^- ions, we increase the concentration of a product. This, in turn increases the value of *Q*, so $Q > K$. Equilibrium must shift towards reactants, so additional HCN will be formed.

(b). Additional hydrocyanic acid is added.
Because HCN is a reactant, the increase in concentration of the acid decreases the value of *Q*. *Q* is thus less than *K* so equilibrium shifts towards products. Some of the added HCN will dissociate, and the concentrations of CN^- and H^+ will both increase.

(c). Sodium hydroxide is added.
Again the sodium ion does not play a significant role in this perturbation, and at first glance it might seem like the hydroxide ion doesn't either. But if we use a little bit of the chemical intuition you've developed by this point in the course, we should realize that OH^- and H^+ ions will react with one another to form water. This will decrease the concentration of H^+. Because the concentration of a product is lowered, the size of the numerator decreases, and $Q < K$. Thus the equilibrium shifts to form more products.

Check Your Understanding

Describe two ways you could make this equilibrium shift towards reactants that haven't already been discussed in this problem. Explain your reasoning.

Effect on Equilibrium of a Change in Pressure, when Gases are Present

Several of the industrial processes that we have been focusing on involve gasses. We will look at them in the example problems for this part of the chapter – but let's consider a familiar example of the effect of pressure first. Think about carbon dioxide dissolving in water. A chemical reaction occurs, and equilibrium is established.

$$CO_2(g) + H_2O(l) \rightleftarrows H_2CO_3(aq)$$

An immediate example might be on your desk as you read this text—your can of soda. The carbonation of soft drinks results from dissolving gaseous CO_2 into water at high pressures. What happens when you open a soft drink can? The volume available to the carbon dioxide gas becomes large, so the pressure decreases. As pressure decreases, the soft drink fizzes, showing that carbon dioxide is leaving the solution. How does this observation conform to LeChatelier's Principle?

LeChatelier's Principle states that the equilibrium will shift in a way that tends to offset the applied stress. In this case, we have lowered the pressure of CO_2 above the soda. How could this pressure be increased? By having more dissolved gas leave solution (the soda) and enter the gas phase, thereby contributing to the pressure. When the soft drink fizzes, that is precisely what the system is doing—shifting in the direction that produces more gas. Since the container is open to the atmosphere, the CO_2 pressure never really rises, and gas continues to be released until the soda "goes flat."

This effect can be generalized for any chemical reaction that includes gases. If the number of moles of gases differs from reactants to products, a shift in pressure (due to a volume change) will result in a change in the equilibrium position. Carbon dioxide dissolving in water is an example of this, because, as the stoichiometry shows, there are more moles of gas phase reactants (one mole of CO_2) than there are moles of gas phase products (no moles of gas). If the total number of moles of gas reactants equals the total number of moles of gas products, inducing a pressure change produces no change in the position of equilibrium.

Note that changing the pressure by adding a nonreactive gas would not affect the equilibrium. Such changes do not vary the partial pressures of the gases involved in the equilibrium

Now apply the same reasoning to the equilibrium between NO_2 and N_2O_4. Two moles of gaseous NO_2 combine to form one mole of gaseous N_2O_4, so if pressure is applied, the system responds by producing more N_2O_4. This is illustrated in Figure 12.8.

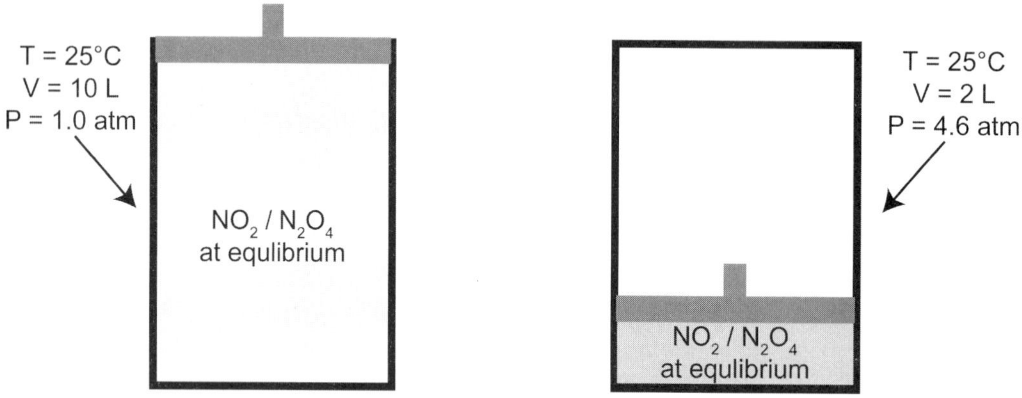

Figure 12.8 The effect of a pressure increase on the NO_2/N_2O_4 equilibrium is illustrated. A system is initially equilibrated in a 10-L volume, with a total pressure of 1.0 atm. Decreasing the volume to 2 L would cause an immediate pressure increase to 5.0 atm. The equilibrium responds by producing more N_2O_4, which reduces the total number of moles of gas present and reduces the extent of the pressure increase.

Example Problem 12.9

The reactions listed below are used in various HCN production processes. Assuming each system is initially at equilibrium, predict the direction the reaction will go to respond to the indicated stress.

(a). $NH_3(g) + CH_4(g) \rightleftarrows HCN(g) + 3H_2(g)$; pressure is increased.

(b). $2NH_3(g) + 2CH_4\ (g) + 3O_2(g) \rightleftarrows 2HCN(g) + 6H_2O(g)$; pressure is decreased.

Strategy

Count the total number of moles of gas on each side of the equation. If pressure is increased, the reaction that produces fewer moles of gas is favored; if pressure is decreased, the reaction that produces more moles of gas is favored.

Solution

(a). There are two moles of gas on the left, and four moles of gas on the right. If pressure is increased, the equilibrium will move to the left.

(b). There are seven moles of gas on the left, and eight moles of gas on the right. Decreasing pressure favors the reaction that produces more moles of gas, so the reaction to the right is favored.

Check Your Understanding

In each of the following, determine the direction the equilibrium will shift to absorb the stress.

(a). $3\ N_2(g)+ H_2(g) \rightleftarrows 2\ NH_3(g)$; pressure is decreased.

(b). $CO_2(g) + H_2(g) \rightleftarrows CO(g) + H_2O(g)$; pressure is increased.

Effect of Temperature Changes on Chemical Equilibrium

When we write chemical equations, we typically don't include any information about temperature. So it might seem hard to predict how temperature might impact an equilibrium. To get a sense of how temperature will affect a system, the key idea is to think of a temperature change as a flow of heat into or out of the reacting system. When put in those terms, you might recognize that the link between chemical reactions and heat comes from thermodynamics. As long as we know whether a reaction is exothermic or endothermic, we can predict how it will respond to temperature changes.

For an exothermic reaction, heat is given off by the system. We might say that this heat is a product of the reaction. We can note this explicitly in a form of the chemical equation:

Adding the word heat to an equation helps us figure out what direction the reaction will shift, but we must always remember that heat is not a thing that the products (in this case) have.

$$\text{Exothermic:} \qquad \text{reactants} \rightleftarrows \text{products} + \text{heat} \qquad (12.4)$$

Once we do this, the situation begins to resemble our discussion of the addition or removal of a product from the reaction mixture. When the temperature is increased, we stress the equilibrium by adding heat. By LeChatelier's principle, the system will respond to offset the stress and absorb the excess heat, moving the equilibrium towards reactants. On the other hand, when the temperature is lowered, heat flows out from the system, and the system will respond by generating additional heat—moving towards products.

The equilibrium between NO_2 and N_2O_4 that we have been considering must be an exothermic reaction, because a bond is formed between the two nitrogen atoms and no bonds are broken. Thus we can predict that formation of N_2O_4 is favored at lower temperatures, and Figure 12.9 confirms this.

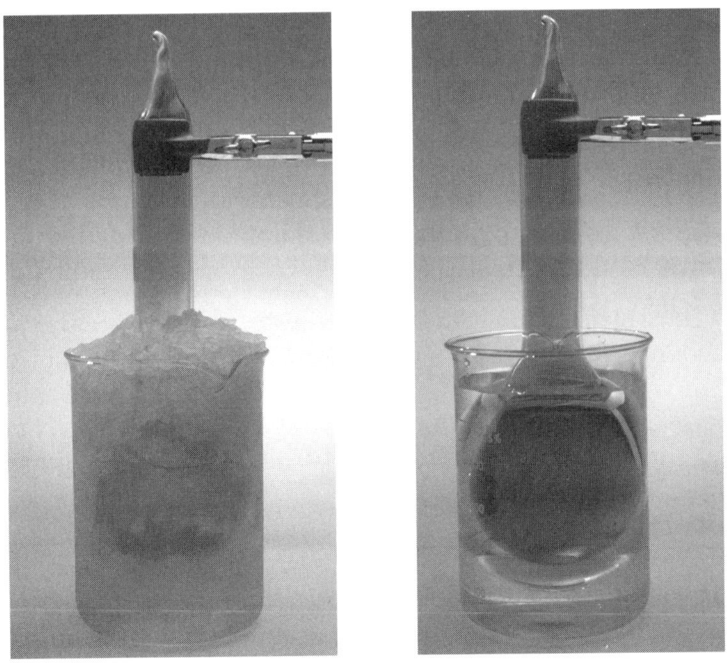

Figure 12.9 The flask in these photos contains a mixture of NO_2, which has a brownish color) and N_2O_4, which is colorless. In the left hand photo, the flask is held in an ice bath and the contents have a pale color. In the right hand photo, the same flask is held in a water bath at 50°C, and the color is markedly darker. This deepening of the color indicates that more NO_2 is present at the higher temperature, as predicted by LeChatelier's principle.

For an endothermic reaction, we can write:

$$\text{Endothermic:} \quad \text{reactants} + \text{heat} \rightleftarrows \text{products} \qquad (12.5)$$

Using similar reasoning, if the temperature is increased, the endothermic system will move in the direction that absorbs heat—towards the products. And if the temperature is decreased, a shift towards reactants liberates heat to offset this stress. We summarize the possible responses of chemical reactions to temperature changes in Table 12.3.

Type of temperature change	Response of system
T increased - exothermic reaction	more reactants formed
T decreased - exothermic reaction	more products formed
T increased - endothermic reaction	more products formed
T decreased - endothermic reaction	more reactants formed

Table 12.3 The effect of a temperature changes on a chemical system at equilibrium depends on whether the reaction is exothermic or endothermic. Unlike concentration or pressure changes, temperature changes also alter the value of the equilibrium constant.

Unlike changes in concentrations of reactants or products, in which the equilibrium constant is unchanged, a temperature shift also changes the *value* of the equilibrium constant. (As for rate constants, we are again confronted with a "constant" that varies as a function of some parameter!) The fact that equilibrium constants vary with temperature, however, plays an important role in the way chemical reactions occur in industrial settings. In many cases, high temperatures are required for kinetic reasons, but what if the equilibrium constant is then made less favorable? In such cases, chemists and engineers work together to find combinations of factors to make a reaction commercially viable. Nearly every adjustment made affects the reaction via LeChatelier's principle.

Effect of a Catalyst on Equilibrium

Equilibrium is established when the rate of the forward reaction equals the rate of the reverse reaction, although the position of equilibrium tells us nothing, in general, about the rate at which equilibrium was attained. What happens when a catalyst is added to a system at equilibrium? A catalyst increases the rates of both the forward and reverse reactions equally, and thereby does not affect the equilibrium position or concentrations.

12.6 Solubility Equilibria

Earlier we noted the fact that when cyanide ions encounter silver ions a precipitate forms. There are relatively few cations that precipitate cyanide, but other anions like phosphate or carbonate precipitate much more readily with a wide range of cations. For engineers and chemists that devise industrial processes, solubility often plays a vital role in both the isolation of desired products and the handling of waste streams. Because industrial processes are often carried out at high temperatures, the characteristics of solubility must be placed within the context of the equilibrium (and the equilibrium constant) of the reactions at those temperatures.

Solubility Product Constant

In Section 3.3, we introduced the term "solubility" as a measure of how much of a particular solvent can be dissolved. This idea is frequently translated into the solubility rules for ionic substances, which we introduced in Chapter 3. Those rules would classify both silver bromide and silver cyanide as insoluble. But in Example 12.3 we implied that there was a measurable difference in solubility between those two compounds. So we need to refine our notion of what it means for a compound to be "insoluble." We would be more accurate by referring to salts like silver bromide and cyanide as **sparingly soluble**. Given enough time and a constantly refreshing solvent (in other words, *not* an equilibrium condition), even the most insoluble salt will dissolve—after all, the dissolving of mountains by rainfall represents such a process, albeit one that requires hundreds of thousands of years. So, just how soluble is an "insoluble" compound?

Defining the Solubility Product Constant

Let's consider silver bromide and cyanide as examples. If either of these substances is placed in water, and the system is allowed to sit undisturbed, a dynamic equilibrium will be established between the salt and its constituent ions. The equilibria that describe these reactions, and those of any solubility process, are heterogeneous. So the equilibrium constant expression does not contain the concentration of the solid salt; only the product of the concentrations of the dissolved ions will appear. Because this form is universally employed for solubility equilibrium constant expressions, the constant has been named the **solubility product constant**, K_{sp}. Thus for the two salts we are considering, we can write the following expressions.

$$AgCN(s) \rightleftarrows Ag^{+}(aq) + CN^{-}(aq) \qquad K_{sp} = [Ag^{+}][CN^{-}] = 5.97\text{x}10^{-17}$$

$$AgBr(s) \rightleftarrows Ag^{+}(aq) + Br^{-}(aq) \qquad K_{sp} = [Ag^{+}][Br^{-}] = 5.35\text{x}10^{-13}$$

As we just learned, the small value for these equilibrium constants implies reactants are favored - so the salt does not dissolve very much.

Solubility product constants are important for a wide variety of applications, and they have been determined for many sparingly soluble salts. Table 12.4 lists some common salts and their K_{sp} values, and a more extensive table is in Appendix H.

Salt	K_{sp}
$CaCO_3$	4.96×10^{-9}
$FeCO_3$	3.07×10^{-11}
$PbCO_3$	1.46×10^{-13}
Ag_2CO_3	8.45×10^{-12}
$Ca_3(PO_4)_2$	2.07×10^{-33}
$Mg_3(PO_4)_2$	9.86×10^{-25}
Ag_3PO_4	8.88×10^{-17}
AgCl	1.77×10^{-10}
AgCN	5.97×10^{-17}
AgBr	5.35×10^{-13}
ZnS	2.93×10^{-25}
PtS	9.91×10^{-74}

Table 12.4 Values for a few compounds illustrate the wide range over which solubility product constants can vary.

Example Problem 12.10

Write the solubility product constant expression for the dissolution of calcium fluoride.

Strategy

Identify the formula for the salt, and write the chemical equation that represents the solubility equilibrium. Write the solubility product constant expression based on the chemical equation.

Solution

Because fluorine forms a -1 ion and calcium forms a +2 ion, calcium fluoride is CaF_2.

$$CaF_2(s) \rightleftarrows Ca^{2+}(aq) + 2\ F^-(aq)$$

so,

$$K_{sp} = [Ca^{2+}][F^-]^2$$

Check Your Understanding

Write the solubility product constant expressions for the following sparingly soluble salts: (a). manganese(II) hydroxide; (b). copper(II) sulfide; (c). copper(I) iodide; (d). aluminum sulfate.

The Relationship Between K_{sp} and Solubility

We have referred to the solubility of silver salts regularly in this chapter. Back in Chapter 3, we defined solubility in terms of the mass of solute that dissolves in 100 g of solute, but this was just one possible choice of concentration units. **Molar solubility** is the concentration of a solid present in a saturated solution, expressed in molarity. Based on the units for concentration presented in Chapter 4, we can convert between these two different expressions for solubility. Molar solubility is easily determined from K_{sp}, as you can see in Example Problem12.11.

Example Problem 12.11

What is the molar solubility of calcium phosphate, given that $K_{sp} = 2.07 \times 10^{-33}$? If the density of the saturated solution is 1.00 g/cm^3, what is the solubility in grams of $Ca_3(PO_4)_2$ per 100 grams of solvent?

Strategy

This problem asks us to find the equilibrium concentration of ions in a saturated solution. So it is actually very similar to ones we solved earlier in this chapter. Hence we will use the same general approach. To find molar solubility, begin by creating an equilibrium table for a solution with unknown equilibrium concentrations, and solve the resulting equilibrium expression for the unknown. Convert the concentration units to grams of solute per 100 grams of solvent, using the molar mass of the solute and the density of the water. (Since the solution will be very dilute, we will assume that its density is the same as that of water.)

Solution

The relevant equilibrium is the dissolution of calcium phosphate.

$$Ca_3(PO_4)_2(s) \rightleftarrows 3Ca^{2+}(aq) + 2PO_4^{3-}(aq)$$

We will need to work with the solubility product expression for this reaction, and the value of K_{sp} is given.

$$K_{sp} = [Ca^{2+}]^3[PO_4^{3-}]^2 = 2.07 \times 10^{-33}$$

We envision the process as starting with only solid present. Dissolution of some of the solid leads to the presence of ions in the saturated solution at equilibrium. In setting up our table, we must be sure to properly account for the 3:2 stoichiometry.

	$Ca_3(PO_4)_2(s)$	$Ca^{2+}(aq)$	$PO_4^{3-}(aq)$
initial concentration	solid	0 M	0 M
change in concentration	solid	$+3x$	$+2x$
final concentration	solid	$3x$	$2x$

Substitute into the equilibrium expression:

$$K_{sp} = 2.07 \times 10^{-33} = [Ca^{2+}]^3[PO_4^{3-}]^2 = (3x)^3(2x)^2$$

We can solve this for x.

$$2.07 \times 10^{-33} = 108\,x^5$$

$$x = 1.14 \times 10^{-7}$$

This means that the molar solubility of calcium phosphate is 1.14×10^{-7} M, because for every mole of $Ca_3(PO_4)_2(s)$ that dissolves, three moles of calcium ion and two moles of phosphate ion are released into solution.

Now convert units to find the solubility in terms of mass.

$$\text{solubility} = \frac{1.14 \times 10^{-7}\ \text{mol}\ Ca_3(PO_4)_2}{\text{L}} \times \frac{1\ \text{L}}{10^3\ \text{cm}^3} \times \frac{1\ \text{cm}^3}{1.0\ \text{g}} \times \frac{310\ \text{g}\ Ca_3(PO_4)_2}{1\ \text{mol}\ Ca_3(PO_4)_2}$$

$$= 3.54 \times 10^{-8}\ \text{g solute/ g}\ H_2O$$

Multiplying this by 100 gives us a solubility of 3.54×10^{-6} g $Ca_3(PO_4)_2$ per 100 g H_2O.

Check Your Understanding
Calculate the molar solubility of silver cyanide. What does this correspond to in grams of solute per 100 grams of solvent?

Common Ion Effect

When we look at the solubility product of silver bromide or silver cyanide we presume that an equilibrium is established from the salt in pure water. What if additional cyanide ion were introduced into the system by adding NaCN? How does the presence of an ion that is part of an equilibrium, but introduced into a system from another source, influence the solubility of a salt? This question is an important one in industrial settings because most commodities require several steps in their manufacture. The answer can be deduced from LeChatelier's Principle. Ions that are part of an equilibrium, but are introduced from an 'outside' source, are called **common ions**, because they are common to the solution and the solid. If we were to add a common ion (CN^-) to a solution of AgCN already at equilibrium, the resulting shift would be in a direction that consumes the added ion, forming additional solid AgCN. This is the **common ion effect** (Figure 12.10). If instead we add a sparingly soluble solid to a solution in which common ions are already present, their presence will suppress solubility of the solid (because they tend to stress the system towards the production of reactants). We explore this idea in Example Problem 12.12.

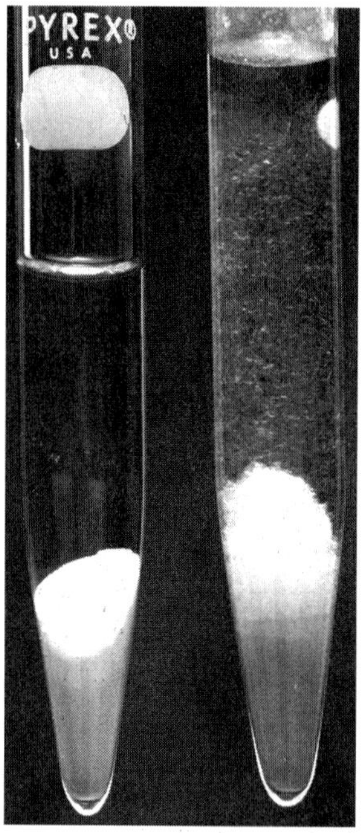

Figure 12.10 The test tube on the left contains a saturated solution of AgCN. Adding aqueous NaCN causes additional solid AgCN to precipitate due to the common ion effect, as seen in the tube on the right.

Example Problem 12.12
The K_{sp} of $Ca_3(PO_4)_2$ is 2.07×10^{-33}. Determine its molar solubility in 0.10 *M* $(NH_4)_3PO_4$. Compare your answer to the molar solubility of $Ca_3(PO_4)_2$ in water, which we calculated in Example Problem 12.11.

Strategy

$(NH_4)_3PO_4$ is very soluble in water, so $[PO_4^{3+}]$ in 0.10 *M* $(NH_4)_3PO_4$ is 0.10 *M*. When solid $Ca_3(PO_4)_2$ is added and equilibrium established, molar solubility will be found from the amount of $[Ca^{2+}]$ in solution. Write the reaction of interest and set up an equilibrium table as usual. Because the value of K_{sp} is very small, we will be able to use a simple approximation to solve for the molar solubility.

Solution

Let x = moles of $Ca_3(PO_4)_2$ that dissolve per liter (i.e., the molar solubility that we want to calculate). The initial concentration of Ca^{2+} is zero, but that of PO_4^{3-} is now 0.10 M. This gives us the following equilibrium table.

$$Ca_3(PO_4)_2(aq) \rightleftarrows 3\ Ca^{2+}(aq) + 2\ PO_4^{3+}(aq)$$

	$Ca_3(PO_4)_2(s)$	$Ca^{2+}(aq)$	$PO_4^{3-}(aq)$
initial concentration	solid	0 M	0.10 M
change in concentration	solid	$+3x$	$+2x$
final concentration	solid	$3x$	$0.10 + 2x$

Substitute into the solubility product expression.

$$K_{sp} = 2.07\ \text{x}10^{-33} = [Ca^{2+}]^3[PO_4^{3-}]^2 = (3x)^3(0.10 + 2x)^2$$

This looks daunting: multiplying out would give us a term involving x^5. But we can make use of a clever approximation. We know from the last example problem that the solubility in pure water was on the order of 10^{-7} M. And in this case the presence of the common ion should decrease the solubility considerably from that level. So we know that x will be less than 10^{-7}. This means that we can assume the "+2x" piece of the last term can be neglected in comparison to the 0.10.

$$0.10 + 2x \approx 0.10$$

This simplifies the problem considerably.

$$2.07\ \text{x}10^{-33} = (3x)^3(0.10)^2$$

$$x = 1.97\ \text{x}\ 10^{-11}\ M$$

In the previous example problem, we calculated the molar solubility of $Ca_3(PO_4)_2$ in water as 1.14 x 10^{-7} M. The presence of a high concentration of phosphate ion in solution has depressed the solubility to 1.97 x 10^{-11} M.

Discussion

Whenever we make a simplifying approximation of the sort used here, it is important to verify that the eventual solution is consistent with the underlying assumption. In this case we assumed that $2x$ was negligible compared to 0.10. And since we found x to be on the order of 10^{-11}, that assumption was certainly justifiable. We could have used appropriate tools to find a numerical solution to the actual equation, but the added work would not have improved our result by any discernible amount. Note that the underlying reason why this assumption could be made was the fact that the K_{sp} value was a very small number.

> **Check Your Understanding**
> CaF_2 has a solubility product of 1.7 x 10^{-10}. Calculate its solubility in water and in 0.15 *M* NaF.

Reliability of Using Molar Concentrations

We need to be aware that the calculations we've been carrying out have been simplified in various ways to make them compatible with the level of this course. These simplifications are reasonable under many typical circumstances, but can introduce significant errors under other conditions. One of the fundamental simplifications we have made is that we have been using the molarity of the ionic species in all our calculations of solubility equilibrium problems. In some cases, however, the utility of molarity dwindles. When ion concentrations are high, the interactions between oppositely charged particles, even other ions that are not part of a particular equilibrium, alters ion behavior. To take these alterations into account, we use the **activity** of an ion, the *effective* concentration of the ion. Activity is not a particularly intuitive modification of the molarity of a species, and we will not undertake any calculations using it here. We must note, however, that the types of calculations we have presented here do have limitations. In industrial chemistry, the models used to understand the synthetic process often require higher accuracy than is afforded by using molarity, so computer codes frequently offer the option of using activity instead.

Activities are often obtained from molarities by mulitplying by an activity coefficient whose value is usually obtained experimentally.

12.7 Acids and Bases

As we have seen in several examples in this chapter, HCN acts as an acid in water solutions. We introduced the essential concepts of acids and bases in Chapter 3, but the context of equilibrium allows us to consider an important distinction among them. Specifically, the strength of acids and bases is defined by a chemical equilibrium and is commonly measured utilizing pH, a concept we will define in this section.

The Brønsted-Lowry Theory of Acids and Bases

Our initial definition of acids and bases in Chapter 3 involved the formation of hydronium ions (for acids) or hydroxide ions (for bases). This definition is attributed to Svante Arrhenius, but it can be expanded to include non-aqueous solutions, among other things. Formulated independently in 1923 by two chemists, Johannes Brønsted in Denmark and Thomas Lowry in England, the Brønsted-Lowry definition does just this. By the Brønsted-Lowry definition, an acid is a proton (H^+) donor, and a base is a proton acceptor.

Brønsted received his undergraduate degree in Chemical Engineering in 1899.

This more general definition of an acid must include Arrhenius acids and bases, as well. That is to say that HCl, for example, must fit both the Arrhenius and Brønsted-Lowry theories. To see whether it does, let's look at what happens when HCl reacts with water. As hydrogen chloride gas dissolves in water, dipole-dipole forces come into play, and the partially positive hydrogen of the HCl molecule is attracted to the negative oxygen end of a water molecule. The attraction is strong enough to result in transfer of the hydrogen ion from HCl to the water molecule. The result is a **hydrated** proton, or hydronium ion, H_3O^+.

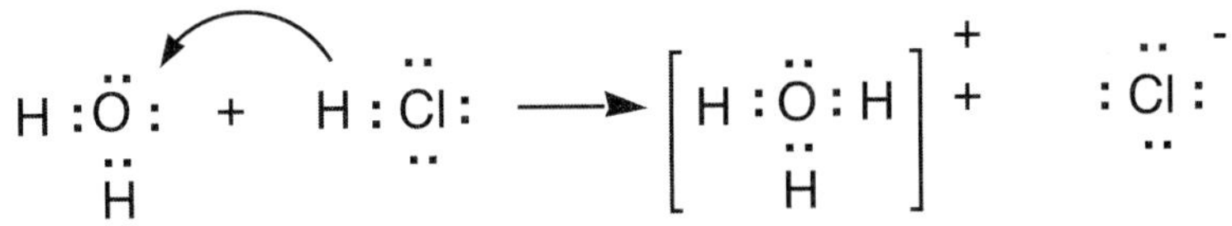

HCl has *donated* a proton to the water molecule, which accepted the proton, and we can write an equation to reflect this.

$$HCl(g) + H_2O \rightarrow H_3O^+(aq) + Cl^-(aq)$$

In the Brønsted-Lowry theory, a base is a proton acceptor. The $OH^-(aq)$ ion produced by the ionization of NaOH(s) is a base, because it can accept a proton from any proton donor.

$$OH^-(aq) + H^+(aq) \rightarrow H_2O$$

The Role of Water in the Brønsted-Lowry Theory

Next let's consider the ionization of HCN in water.

$$HCN(aq) + H_2O \rightleftarrows H_3O^+(aq) + CN^-(aq)$$

The HCN molecule donates a proton, so it is clearly an acid. But the water molecule accepts that proton. Does this make water a base? By the Brønsted-Lowry theory, it does. But by the same theory, water can also react as an acid, as you can see in the following reaction between ammonia and water.

$$NH_3(g) + H_2O \rightarrow NH_4^+(aq) + OH^-(aq)$$

The water molecule (an acid) donates a proton to the ammonia molecule (a base), which accepts the proton. So water can act as both acid and base, depending on the identity of the other reactant. We use the term *amphoteric* to refer to a substance that can be either an acid or a base.

The term amphiprotic is finding increasing use in chemistry and is slowly replacing the term amphoteric.

We also must be careful to note that the reaction of ammonia with water does not go to completion. The reaction above is actually in dynamic equilibrium.

$$NH_3(g) + H_2O \rightleftarrows NH_4^+(aq) + OH^-(aq)$$

Because the reverse reaction also involves a donated proton (this time from the ammonium ion), both the forward and reverse reactions are acid-base reactions.

$$\underset{base_1}{NH_3(g)} + \underset{acid_2}{H_2O} \rightleftarrows \underset{acid_1}{NH_4^+(aq)} + \underset{base_2}{OH^-(aq)}$$

The equation contains two pairs of acids and bases: NH_3 — NH_4^+ and H_2O — OH^-. These are called **conjugate acid-base pairs.** A **conjugate acid** of a base is the acid formed when the base accepts a proton; similarly, a **conjugate base** of an acid is the base formed when the acid donates a proton. NH_3 is the conjugate base of NH_4^+, and NH_4^+ is the conjugate acid of NH_3; H_2O is the conjugate acid of OH^-, and OH^- is the conjugate base of H_2O. We can link the pairs using brackets under the equation to highlight these relationships.

$$\underset{base_1}{NH_3(g)} + \underset{acid_2}{H_2O} \rightleftarrows \underset{acid_1}{NH_4^+(aq)} + \underset{base_2}{OH^-(aq)}$$

Example Problem 12.13

We have noted that HCN is an acid, called hydrocyanic acid. Write the equilibrium for its reaction in water and identify the conjugate acid-base pairs.

Strategy

Because hydrocyanic acid is behaving as a Brønsted-Lowry acid, it will require some species to accept a proton. Here the proton acceptor is water, as is often the case. This allows us to write the equilibrium for the HCN solution. Once it is written we identify the conjugate pairs by noting which species differ by only the addition or subtraction of a proton.

Solution

$$\underset{\text{acid}_1}{HCN} + \underset{\text{base}_2}{H_2O} \rightleftarrows \underset{\text{base}_1}{CN^-(aq)} + \underset{\text{acid}_2}{H_3O^+(aq)}$$

Check Your Understanding

Identify the conjugate acid-base pairs in the reaction between carbonate ion and water to form hydrogen carbonate and hydroxide ions.

One of the advantages of the Brønsted-Lowry theory is that it can describe acid-base reactions in a nonaqueous medium like liquid NH_3. In liquid NH_3, HCl(g) donates a proton to form NH_4^+ and Cl^-. No water is involved, so there are no H_3O^+ or OH^- ions formed. Ammonium and chloride ions are solvated by ammonia molecules. (Instead of "aq", we might use "am" to show solvation by ammonia molecules.)

$$HCl(g) + NH_3 \rightarrow NH_4^+(am) + Cl^-(am)$$

Weak Acids

There are many acids that, like HCN, do not react completely in water. Collectively these compounds are referred to as weak acids. These compounds lead to a conjugate acid-base system in which the acid and base on the right are stronger than their conjugate acid and base on the left, and only a small percentage of the acid molecules (here denoted as HA) ionize.

$$\underset{\text{weaker acid}}{HA(aq)} + \underset{\text{weaker base}}{H_2O} \rightleftarrows \underset{\text{stronger acid}}{H_3O^+(aq)} + \underset{\text{stronger base}}{A^-(aq)}$$

The equilibrium expression for this reaction is

$$K = \frac{[H_3O^+][A^-]}{[HA][H_2O]}$$

Much like we noted with solubility equilibrium in the last section, when a weak acid ionizes, the concentration of H_2O is essentially constant, so $[H_2O]$ can be combined with K to give a new constant, the **acid ionization constant, K_a**.

$$K[H_2O] = K_a = \frac{[H_3O^+][A^-]}{[HA]}$$

Table 12.6 lists acid ionization constants at 25°C for some common monoprotic acids.

Name of acid	Formula	K_a
Organic acids		
Formic	$HCOOH$	1.8×10^{-4}
Acetic	CH_3COOH	1.8×10^{-5}
Propanic	CH_3CH_2COOH	1.3×10^{-5}
Butanoic	$CH_3CH_2CH_2COOH$	1.5×10^{-5}
Chloroacetic	$ClCH_2COOH$	1.4×10^{-5}
Inorganic acids		
Hydrofluoric	HF	6.3×10^{-4}
Chlorous	$HClO_2$	1.1×10^{-2}
Carbonic	H_2CO_3	4.4×10^{-7}
Hypochlorous	$HClO$	2.9×10^{-8}
Hydrocyanic	HCN	6.2×10^{-10}

Table 12.6 Acid ionization constants are shown for some common acids at 25°C. Larger values of K_a indicate stronger acids.

It's fairly clear that these reactions are just a special case of the sort of equilibrium that we have noted throughout this chapter. When any acid dissolves in water, hydronium ions are always formed as one of the resulting products. This species is so common that an additional way to describe its concentration has been devised. In order to avoid dealing with small numbers in scientific notation, we often use the pH scale, in which pH is defined as the negative logarithm of the hydronium ion concentration.

$$pH = -\log[H_3O^+] \qquad (12.6)$$

Consider a solution with a hydronium ion concentration of 0.1*M* (or 10^{-1} *M*). This solution would have a pH = $-\log(10^{-1})$. The logarithm of 10^{-1} is –1. So the pH is 1. Applying the same logic to other hydronium ion concentrations in aqueous solution, the result is the pH scale (Table 12.7). From the table you can see that if the concentration of the hydronium ion is 1×10^{-x}, the pH = x. Special instruments, called pH meters, often allow us to determine the pH directly in the laboratory.

pH is a topic that is taught widely in elementary school, where you may have picked up some inaccurate "facts". For example, pH can be less than 0 and more than 14 under certain conditions.

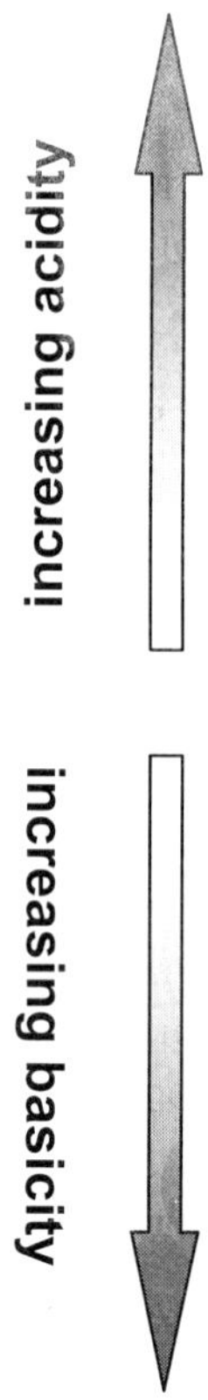

pH	$[H_3O^+]$, mol / L	pH of some common substances
0	1.0	1.0 *M* HCl
1	10^{-1}	HCl in the human stomach
2	10^{-2}	
3	10^{-3}	vinegar - $CH_3COOH(aq)$
4	10^{-4}	soft drinks
5	10^{-5}	
6	10^{-6}	milk
7	10^{-7}	pure water, blood
8	10^{-8}	sea water
9	10^{-9}	
10	10^{-10}	milk of magnesia - $Mg(OH)_2(aq)$
11	10^{-11}	household ammonia
12	10^{-12}	
13	10^{-13}	
14	10^{-14}	1.0 M NaOH

Table 12.7 The pH scale provides an easy way to measure the relative acidity or basicity of aqueous solutions, including many common substances from our daily lives. Pure water has a pH of 7.0. So solutions with pH < 7 are acidic and those with pH > 7 are basic. Because pH is measured on a logarithmic scale, it can describe H_3O^+ concentrations that vary over many orders of magnitude without needing to use scientific notation. Note that pH values less than zero or greater than fourteen are also possible.

If we know the K_a of an acid and its initial concentration, we can calculate the hydronium ion concentration and therefore the pH of the solution, as shown in Example Problem 12.14.

Example Problem 12.14

The K_a of acetic acid, CH_3COOH, is 1.8×10^{-5}. Calculate the pH of a 0.10 M solution of acetic acid.

Strategy

We begin as usual, by setting up an equilibrium table and determine the concentrations of all species. Once we know the concentration of H_3O^+, we can find the pH.

Solution

The initial concentration of acetic acid is 0.10 M and that of the ions is zero, leading to the following equilibrium table.

$$CH_3COOH(aq) + H_2O(l) \rightleftarrows H_3O^+(aq) + CH_3COO^-(aq)$$

	$CH_3COOH(aq)$	H_3O^+	CH_3COO^-
initial concentration	0.10	0	0
change in concentration	$-x$	$+x$	$+x$
final concentration	$0.10 - x$	x	x

Write the equilibrium constant expression, and substitute values from the table.

$$K_a = \frac{[H_3O^+][CH_3COO^-]}{[CH_3COOH]}$$

$$K_a = \frac{(x)(x)}{(0.10\text{-}x)} = \frac{x^2}{(0.10\text{-}x)} = 1.8 \times 10^{-5}$$

We should expect that the extent of dissociation should be small for a weak acid. So we can simplify the calculation if we assume that x is small enough that $(0.10 - x) \approx 0.10$. With that assumption, we can rewrite the equation above in a way that will not require us to use the quadratic formula.

$$K_a \cong \frac{x^2}{0.10} = 1.8 \times 10^{-5}$$

Solve this for x.

$$x^2 = 1.8 \times 10^{-5}(0.10) = 1.8 \times 10^{-6}$$

$$x = 1.3 \times 10^{-3}\ M$$

Therefore the $[H_3O^+] = [CH_3COO^-] = 1.3 \times 10^{-3} M$. We should confirm that the assumption we made was reasonable. We assumed that $(0.10 - x) \approx 0.10$. So we can test this by subtracting the value obtained for x from the initial concentration of CH_3COOH.

$$0.10\ M - 1.3 \times 10^{-3}\ M = 0.0987 \approx 0.10\ M$$

Our initial concentration was only given to two significant figures, so our simplifying approximation is reasonably valid. Since we now know the $[H_3O^+]$, the pH = -log $(1.3 \times 10^{-3}) = 2.9$.

Check Your Understanding

Using the approximation method shown above, determine the percentage ionization of a 0.10-M solution of butanoic acid. Is the approximation valid in this case?

12.8 Free Energy and Chemical Equilibrium

We've put equilibrium into perspective with the kinetics of the forward and reverse reactions. What is the relationship between free energy and equilibrium? *Equilibrium is a state of minimum free energy*. All systems move towards equilibrium because doing so will lower the free energy, and when they reach equilibrium the free energy no longer changes, so $\Delta G = 0$ at equilbrium.

Graphical Perspective

The graphs in Figure 12.11 show free energy as a function of progress along the reaction pathway. (This is similar in some ways to the potential energy diagrams we saw in Chapter 11.) While the vertical axis is the Gibbs free energy, keep in mind that we cannot determine an absolute value for the free energy of a system at any point. Instead, we can only measure *changes* in free energy. The purpose of the graph, therefore, is to help you understand that free energy decreases as you move toward equilibrium from either direction (reactants or products). *A chemical system tends to move spontaneously towards equilibrium.* When equilibrium is reached, the change in free energy will be zero.

Figure 12.11 also reminds us that $\Delta G°$ is the difference between the free energy content of the reactants and the products. For the reaction depicted in the lower panel of the figure, $\Delta G°$ would be positive because $G°_{products} > G°_{reactants}$. In Chapter 10 we asserted that a reaction with $\Delta G° > 0$ will not proceed spontaneously. Now we see that this was a bit of a simplification. All chemical reactions will proceed to the point of minimum free energy. While there is variability from one reaction to the next, the free energy minimum along the "progress of reaction" tends to lie closer to the state (reactants or products) that is lower in free energy. Thus, for a reaction where $\Delta G° < 0$, the equilibrium position lies closer to the products, and we tend to view such a reaction as proceeding towards products. For a reaction where $\Delta G° > 0$, on the other hand, the equilibrium position is usually achieved when a majority of reactants are still present. We tend to view these reactions as not proceeding spontaneously towards products, because they do not move in that direction very far before equilibrium is reached.

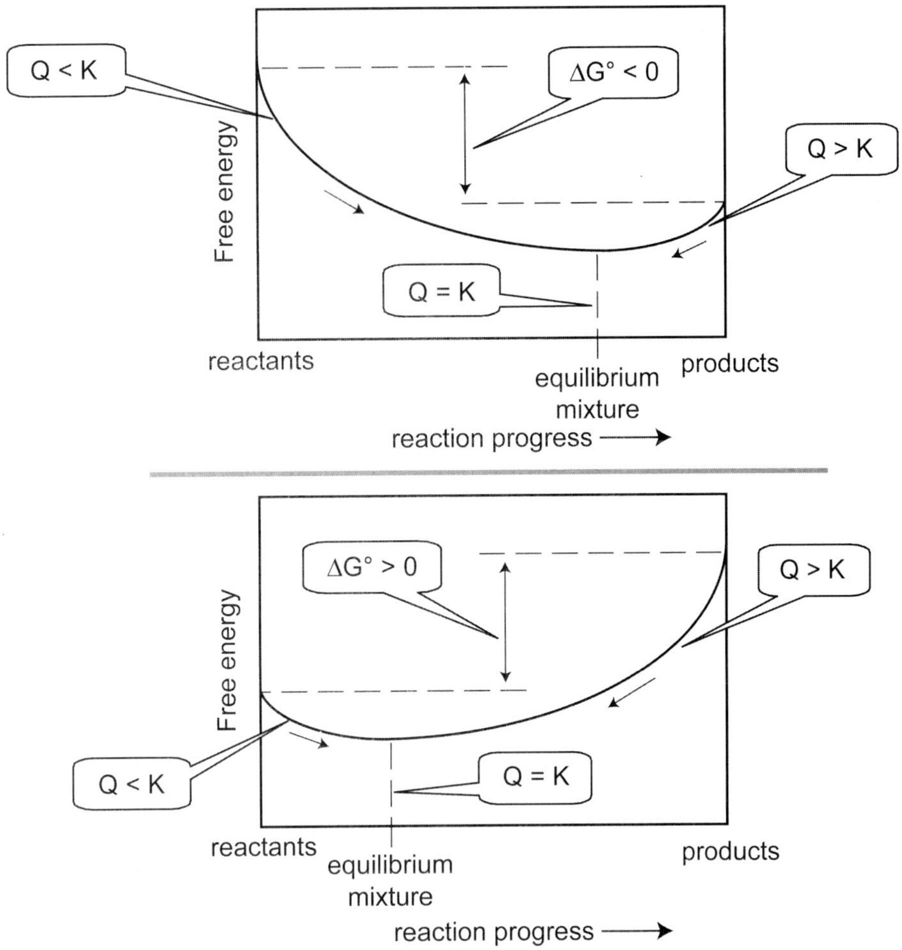

Finding a minimum on a curve is readily accomplished using derivatives, and derivations involving Free Energy commonly invoke calculus.

Figure 12.11 Chemical reactions always proceed towards a minimum in free energy. The upper plot illustrates what this means for a reaction with $\Delta G° < 0$. In this case the equilibrium constant will be greater than one, and equilibrium will lie towards products. A mixture of reactants will react to form products, while a mixture heavy in products will react "backwards" to form reactants. The lower figure shows the case for a reaction with $\Delta G° > 0$. Here the equilibrium constant will be less than one, and the situation will be reversed. In either case, equilibrium will be the point at which free energy is minimized.

Free Energy and Non-standard Conditions

To obtain a mathematical relationship between $\Delta G°$ and the equilibrium constant, K, we must first realize that the equilibrium state does not arise from standard conditions. That is, we cannot expect all reactants and products to be present at one atmosphere pressure and one molar concentration for every equilibrium reaction! The derivation of the mathematical relationship that adjusts for non-standard conditions is one you will learn if you continue to study chemistry. For now we simply need the expression itself.

$$\Delta G = \Delta G° + RT \ln Q \qquad (12.7)$$

Here R is the universal gas constant, T is the temperature, and Q is the reaction quotient. Recall from Section 12.3, however, that Q becomes K as the system reaches equilibrium. Combining this with the fact that $\Delta G=0$ at equilibrium means Equation 12.7 becomes:

$$0 = \Delta G° + RT \ln K.$$

Rearranging, we obtain

$$\Delta G° = -RT \ln K. \qquad (12.8)$$

With this equation we can calculate an equilibrium constant if we know the standard free energy change for the desired reaction. (And we can often obtain that by using tabulated data). Example Problem 12.15 applies Equation 12.8 to the conversion of methane to methanol.

Example Problem 12.15

Currently, much research is being directed towards finding ways to convert methane gas (CH_4) into methanol (CH_3OH), which is a liquid at room temperature. So far this process is not yet economically viable. Using tabulated thermodynamic data, calculate the equilibrium constant for the following reaction at 25°C.

$$CH_4\,(g) + 1/2\,O_2\,(g) \rightleftarrows CH_3OH(l)$$

Comment on whether or not the equilibrium position of this reaction is the likely source of the problem for the commercialization of the process.

Strategy

To solve this problem we must use the relationship between ΔG° and the equilibrium constant given in Equation 12.8. We can use tabulated values of the free energy of formation for the reactants and products to calculate ΔG° and then obtain K.

Solution

$$\Delta G° = \Delta G°_f\,[CH_3OH(l)] - \{\Delta G°_f[CH_4(g)] + 1/2\Delta G°_f[O_2(g)]\}$$

$$= -\,166.4 \text{ kJ mol}^{-1} - (-50.8 \text{ kJ mol}^{-1}) - (0 \text{ kJ mol}^{-1})$$

$$= -115.6 \text{ kJ mol}^{-1}$$

When using Equation 12.8, we must take care to make sure the energy units of ΔG° and R are consistent.

$$-115.6 \times 10^3 \text{ J mol}^{-1} = -(8.314 \text{ J mol}^{-1} \text{ K}^{-1})(298 \text{ K}) \ln K$$

so $\ln K = 46.66$ and $K = e^{46.66} = 1.83 \times 10^{20}$

Discussion

This reaction can be thought of as the partial oxidation of methane. The problem that scientists and engineers face with this process is that methane also has a strong tendency to completely oxidize to carbon dioxide and water. This issue is explored further in the "Check Your Understanding" problem.

Check Your Understanding

Calculate the equilibrium constant for the complete combustion of methane

$$CH_4(g) + 2O_2(g) \rightarrow CO_2\,(g) + 2H_2O(l)$$

Compare the equilibrium constant for the complete combustion with that for the partial oxidation calculated previously. Does this result help explain why it is so difficult to carry out the partial oxidation process?

Example Problem 12.17 shows that the equilibrium constant can be enormous if the free energy change is large and negative. Conversely, a reaction with a small equilibrium constant will have a standard free energy change that is large and positive. Only if the magnitude of ΔG° is small do we find equilibrium constants near one.

12.9 Insight into Borates and Boric Acid

Hydrogen cyanide is not the only small molecule that finds utility in the manufacture of the molecular giants, polymers. One of the most popular chemistry experiments for polymer chemistry – the synthesis of "slime" - involves the use of borax as a cross-linker, a molecule that forms bonds between two polymer chains. Borax and boric acid have many uses, both industrially and in the household. And like HCN, boric acid is an example of a weak acid.

Borax has the chemical formula $Na_2B_4O_7 \cdot 10H_2O$ and is found and mined in places such as Death Valley, California. As a chemical compound, it gained fame for its use as a laundry soap with the brand name "20-Mule Team Borax." But from an industrial perspective, the conversion of borax to boric acid through the following reaction is particularly important.

Death Valley is below sea-level so the hard work of bringing the borax out of the valley and over the mountains required large mule teams. There's even a 20 mule team museum in Boron, CA.

$$Na_2B_4O_7 \cdot 10H_2O + H_2SO_4 \rightarrow 4B(OH)_3 + Na_2SO_4 + 5\ H_2O$$

Once produced, boric acid finds a number of industrial applications. The largest area for growth in the use of borates in general – including boric acid – is in the manufacture of fiberglass. Fiberglass is used in both insulation and textiles (for items like surfboards) and borates are needed to control the temperature at which the glass melts during production. They also help manage variables like viscosity and surface tension (see Chapter 8), which allow the melted glass to be pulled into fibers.

Borates have long been used with glass – including the glassware you probably use in the chemistry laboratory. Borosilicate glasses are important in the lab and in the kitchen because of their resistance to heat-induced deformation. And because of that same property they are now used in other heat-producing items including halogen headlights (See Chapter 6 for a discussion of the characteristics of halogen lamps) and cathode ray tubes (used in traditional television sets or computer monitors).

As additives in the polymer industry, borates are primarily valued as fire retardants. In this case, the relevant chemical compounds are zinc borates, which have the important property that they tend to retain their waters of hydration at relatively high temperatures. This fact likely plays a role in the mechanism of retarding fires. Fires involving plastics containing zinc borates are not only slower to spread, but they also produce less smoke.

This section provides only a few examples of the uses of borates and boric acid. It serves as an additional example of how a small chemical plays a role in a number of ways, much like cyanides do. We can now continue to another topic where small molecules play an important industrial role – in the control of corrosion, as we'll see in Chapter 13.

Focus on Problem Solving

Question

You are designing an instant heat pack for a food application. Your design includes a non-toxic ionic salt that dissolves exothermically in water. You need to control the amount of heat released – and initial prototypes get too warm, even with a fairly small amount of salt used. What else could you do to reduce the heat released in your product? Identify any key limitations that would have to be met.

This is a conceptual question that requires that we identify both the chemical properties that are important and the practical ones. For a food application, for example, issues like toxicity become important. Any adjustment that can be made must still be safe for the consumer.

Answer

The first chemical principle to recognize is that dissolving, like any chemical process, continues until equilibrium is reached. For an ionic salt, we expect that the solid will dissociate into ions in solution. The resulting equilibrium can be adjusted by adding a common ion instead of using pure water for the solvent. The key constraints will include making sure that the solvent with the common ion could be produced economically. The common ion must not increase the toxicity of the product noticeably, and it has to be successful in reducing the amount of solid that dissolves. It might seem more practical to simply reduce the amount of solid used, but if the pack is to stay warm for some length of time there may be a need for a minimum amount of salt initially.

Chapter 12
Problems & Exercises

Insight into Hydrogen Cyanide

12.1 How many metric tons of HCN are produced in the US annually? How many moles of HCN is this.

12.2 If all of the HCN in the US was produced by the Andrussow, what mass of ammonia would be consumed?

12.3 How does the BMA process differ from the Andrussow process for HCN production? Speculate on added dangers of the BMA versus the Andrussow process.

12.4 What are the inputs and outputs that a process engineer has to consider in the production of HCN? How do they differ from the inputs and outputs that a chemical engineer must consider?

12.5 The production of HCN is accomplished partially as a side product for another industrial product, acrylonitrile. Why is this chemical important industrially?

12.6 Use the web to find where HCN is produced in the US. Is the plant you locate near another chemical production plant?

12.7 Use the web to find the chemical reactions associated with the processing of gold with sodium cyanide. Is gold processed in any other way so that NaCN might be replaced someday?

Chemical Equilibrium

12.8 On your desk is a glass half-filled with water and a square glass plate to use as a cover for the glass.

(a). How would you set up a dynamic equilibrium using the glass, water, and glass plate?

(b). Once equilibrium is established in (a), describe the processes that are occurring at the molecular level.

12.9 In the figure, white fish are placed in one aquarium and gray fish in an adjoining aquarium. The two tanks are separated by a removable partition that is initially closed.

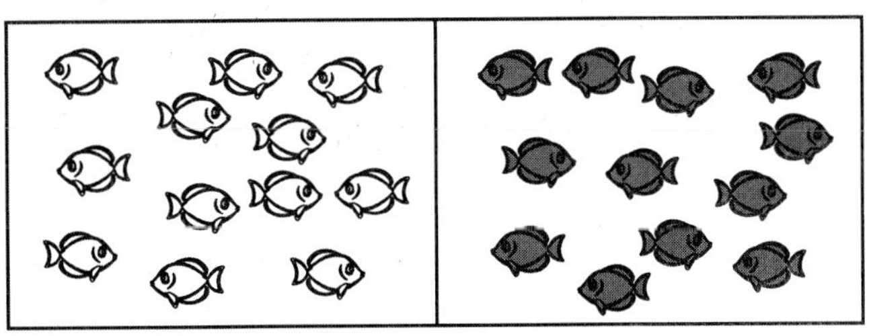

(a). Describe what happens in the first few minutes after the partition is opened.
(b). What would you expect to see several hours later?
(c). How is this system an analogy for dynamic chemical equilibrium?

12.10 At a particular temperature, iodine vapor, I_2, is added to a bulb containing $H_2(g)$. Describe what is happening to the concentrations of reactants on the molecular level

(a). at the moment I_2 enters the reaction bulb;
(b). as reaction proceeds;
(c). at equilibrium.

12.11 For the system in the preceding problem, show the equilibrium condition in terms of the rates of the forward and reverse reactions.

12.12 An equilibrium involving the carbonate and bicarbonate ions exists in natural waters:

$$HCO_3^-(aq) \rightleftarrows H^+(aq) + CO_3^{2-}(aq)$$

Assuming reactions in both directions are elementary processes,
(a). write rate expressions for the forward and reverse reactions;
(b). write a constant expression for the equilibrium, based on the rates of the forward and reverse reactions.

12.13 The graph represents the progress of the reaction, $A(g) \rightleftarrows B(g)$, which starts with only the reactant A present. Label the two lines to show which represents A and which represents B. Discuss what is occurring at each of the times indicated ((a), (b), and (c)).

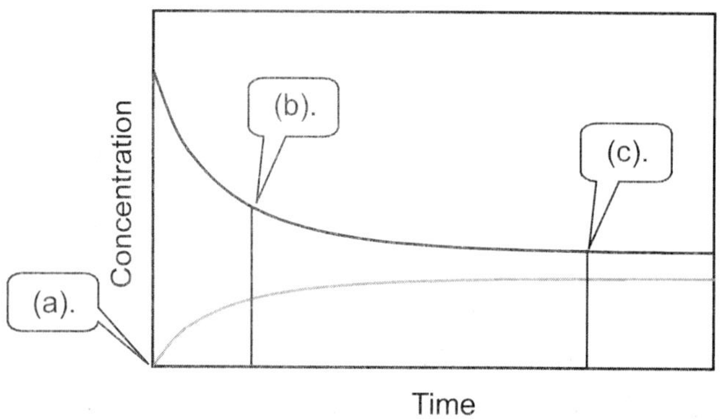

12.14 A small quantity of a soluble salt is placed in water. An equilibrium between dissolved and undissolved salt may or may not be attained. Explain.

12.15 In the following equilibrium in lake water, which of the following inferences can be drawn from the equation alone?

$$HCO_3^-(aq) \rightleftarrows H^+(aq) + CO_3^{2-}(aq)$$

(a). The rate of the forward reaction equals the rate of the reverse reaction.
(b). The equilibrium concentrations of all species are equal.
(c). Equilibrium was attained by starting with only $HCO_3^-(aq)$ in solution.

Equilibrium Constants

12.16 Write equilibrium (mass action) expressions for each of the following reactions.

(a). $H_2(g) + I_2(g) \rightleftarrows 2HI$

(b). $2\ NO(g) + O_2(g) \rightleftarrows 2\ NO_2(g)$

(c). $N_2(g) + 3H_2(g) \rightleftarrows 2NH_3(g)$

(d). $CO(g) + NO_2(g) \rightleftarrows CO_2(g) + NO(g)$

(e). $2CO(g) + O_2(g) \rightleftarrows 2CO_2(g)$

12.17 Write equilibrium (mass action) expressions for each of the following reactions.

(a). $2\ NOBr(g) \rightleftarrows 2\ NO(g) + Br_2(g)$

(b). $4\ HCl(g) + O_2(g) \rightleftarrows 2\ H_2O(g) + Cl_2(g)$

(c). $SO_2(g) + 1/2\ O_2(g) \rightleftarrows SO_3(g)$

(d). $CH_4(g) + 2\ O_2(g) \rightleftarrows CO_2(g) + 2\ H_2O(g)$

(e). $C_2H_5OH(g)\ + 3\ O_2(g) \rightleftarrows 2\ CO_2(g)\ + 3\ H_2O(g)$

12.18 What is the difference between *homogeneous* equilibrium and *heterogeneous* equilibrium?

12.19 Write equilibrium expressions for each of the following heterogeneous equilibria.

(a). $CaCO_3(s) \rightleftarrows Ca^{2+}(aq) + CO_3^{2-}(aq)$

(b). $AgCl(s) \rightleftarrows Ag^+(aq)\ + Cl^-(aq)$

(c). $Mg_3(PO_4)_2(s) \rightleftarrows 3Mg^{2+}(aq) + 2PO_4^{2-}(aq)$

(d). $Zn(s) + Cu^{2+}(aq) \rightleftarrows Cu(s) + Zn^{2+}(aq)$

12.20 Write equilibrium expressions for each of the following equilibria.

(a). $2\ C(s) + O_2(g) \rightleftarrows 2\ CO(g)$

(b). $Zn^{2+}(aq) + H_2S(g) \rightleftarrows ZnS(s) + 2H^+(aq)$

(c). $HCl(g) + H_2O(l) \rightleftarrows H_3O^+(aq) + Cl^-(aq)$

(d). $H_2(g) + 1/2\ O_2(g) \rightleftarrows H_2O(g)$

12.21 For each of the following, are products or reactants favored?

(a). $AgCl(s) \rightleftarrows Ag^+(aq)\ + Cl^-(aq)$ $K = 1.7 \times 10^{-10}$

(b). $Ca^{2+}(aq) + CO_3^{2-}(aq) \rightleftarrows CaCO_3(s)$ $K = 2.2 \times 10^{8}$

(c). $N_2O_4 \rightleftarrows 2\ NO_2$ $K = 1.10$

12.22 Which of the following is more likely to precipitate the hydroxide ion?

(a). $Cu(OH)_2(s) \rightleftarrows Cu^{2+}(aq) + 2OH^-(aq)$ $K = 2.2 \times 10^{-20}$

(b). $Ca(OH)_2(s) \rightleftarrows Ca^{2+}(aq) + 2OH^-(aq)$ $K = 5.5 \times 10^{-6}$

12.23 Which of the following is more likely to precipitate the sulfate ion?

(a). $PbSO_4(s) \rightleftarrows Pb^{2+}(aq) + SO_4^{2-}(aq)$ $K = 1.6 \times 10^{-8}$

(b). $CaSO_4(s) \rightleftarrows Ca^{2+}(aq) + SO_4^{2-}(aq)$ $K = 9.1 \times 10^{-6}$

12.24 The reaction, $3H_2(g)\ +\ N_2(g) \rightleftarrows 2\ NH_3(g)$, has the following equilibrium constants at the temperatures given.

at $T = 25°C$, $K = 2.8 \times 10^4$

at $T = 500°C$, $K = 2.4 \times 10^{-7}$

(a). At which temperature are reactants favored?

(b). At which temperature are products favored?

(c). What can you say about the reaction if the equilibrium constant is 1.2 at 127°C?

12.25 For each of the following equations, write the equilibrium expression for the reverse reaction.

(a). $2C(s) + O_2(g) \rightleftarrows 2CO(g)$

(b). $AgCl(s) \rightleftarrows Ag^+(aq) + Cl^-(aq)$

12.26 Write the equilibrium expressions for the following pairs of reactions.

(a). $2\ SO_2(g) + O_2(g) \rightleftarrows 2\ SO_3(g)$; $SO_2(g) + 1/2\ O_2(g) \rightleftarrows SO_3(g)$

(b). $3\ H_2(g) + N_2(g) \rightleftarrows 2\ NH_3(g)$; $3/2H_2(g) + 1/2N_2(g) \rightleftarrows NH_3(g)$

12.27 Using the following equations, determine

(a). equilibrium expressions, K_1 and K_2

(b). the equation for the reaction that is the sum of the two equations

(c). the equilibrium expression, K_3, for the sum of the two equations

1. $CO_3^{2-}(aq) + H^+(aq) \rightleftarrows HCO_3^-(aq)$ $\quad K_1 = ?$
2. $HCO_3^-(aq) + H^+(aq) \rightleftarrows H_2CO_3(aq)$ $\quad K_2 = ?$

12.28 (a). In Exercise 12.27, if $K_1 = 2.3 \times 10^6$ and $K_2 = 1.8 \times 10^{10}$, what is the value of K_3?
(b). What is the equilibrium constant, K_3' for the reverse reaction?

12.29 In Exercise 12.28(a), which reaction has the greater tendency to go toward completion?

12.30 Using the following equations,

$$HAsO_4^{2-}(aq) \rightleftarrows AsO_4^{3-}(aq) + H^+(aq) \qquad K_1 = 4.0 \times 10^{-12}$$

$$HAsO_4^{2-}(aq) + H^+(aq) \rightleftarrows H_2AsO_4^-(aq) \qquad K_2 = 5.9 \times 10^6$$

determine the equilibrium constant for the following reaction.

$$H_2AsO_4^-(aq) \rightleftarrows AsO_4^{3-}(aq) + 2\ H^+(aq)$$

12.31 In Exercise 12.30, which reaction has the greater tendency to go to completion as written, reaction 1 or reaction 2?

12.32 The following data were collected for the reaction, $H_2(g) + I_2(g) \rightleftarrows 2\ HI(g)$, at equilibrium at 25°C.

$[H_2] = 0.10$ mol L^{-1}, $[I_2] = 0.20$ mol L^{-1}, $[HI] = 4.0$ mol L^{-1}

Calculate the equilibrium constant for this reaction.

12.33 The following data were collected for a system at equilibrium at 140°C. Calculate the equilibrium constant for the reaction, $3\ H_2(g) + N_2(g) \rightleftarrows 2\ NH_3(g)$ at this temperature.

$[H_2] = 0.10$ mol L^{-1}, $[N_2] = 1.1$ mol L^{-1}, $[NH_3] = 3.6 \times 10^{-2}$ mol L^{-1}

12.34 The equilibrium constant for the reaction, $3\ H_2(g) + N_2(g) \rightleftarrows 2\ NH_3(g)$, at a given temperature is 1.4×10^{-7}. Calculate the equilibrium concentration of ammonia, if the $[H_2] = 1.2 \times 10^{-2}$ mol L^{-1} and the $[N_2] = 3.2 \times 10^{-3}$ mol L^{-1}.

Equilibrium Concentrations

12.35 0.100 mol each of compounds A and B were placed in a 1-L vessel and allowed to react, producing compound C, by the following reaction.

$$A(g) + B(g) \rightleftarrows 2\,C(g)$$

At equilibrium, 0.050 mol of C was present. What is the equilibrium constant for this reaction?

12.36 Hydrogen iodide is placed in a closed system, and equilibrium is established at 500 K. If the initial concentration of HI was 0.500 mol L^{-1}, what are the equilibrium concentrations of all species?

$$H_2(g) + I_2(g) \rightleftarrows 2HI(g) \qquad K = 160$$

12.37 0.100 mole of oxygen gas, O_2, is placed in a closed container, and the system is brought to equilibrium at 600 K.

$$O_2(g) \rightleftarrows 2\,O(g) \qquad K = 2.8 \times 10^{-39}$$

What are the equilibrium concentrations of O and O_2?

12.38 The reaction of $Cl_2(g)$ with $PCl_3(g)$ to produce $PCl_5(g)$ has an equilibrium constant of 50 at a particular temperature. If the initial concentration of PCl_3 is 0.012 mol L^{-1} and the initial concentration of Cl_2 is 0.125 mol L^{-1}, what are the equilibrium concentrations of all species?

12.39 The following reaction establishes equilibrium at 2000 K.

$$N_2(g) + O_2(g) \rightleftarrows 2\,NO \qquad K = 4.1 \times 10^{-4}$$

If the reaction began with 0.10 mol L^{-1} of N_2 and 0.10 mol L^{-1} of O_2, what were the equilibrium concentrations of all species?

12.40 The reaction in Exercise 12.39 was repeated. This time the reaction began when only NO was injected into the reaction container. If 0.20 mol L^{-1} NO was injected, what were the equilibrium concentrations of all species?

12.41 In the reaction in Exercise 12.39, another trial was carried out. The reaction was begun with initial concentration of N_2 equal to the initial concentration of NO. Each had a concentration of 0.10 mol L^{-1}. What were the equilibrium concentrations of all species?

12.42 The experiment in Exercise 12.39 was redesigned. 0.15 mole each of N_2 and O_2 were injected into a 1.0-L container at 2500 K. The equilibrium constant at 2500 K is 3.6×10^{-3}. What was the composition of the reaction mixture after equilibrium was attained?

12.43 Again the experiment in Exercise 12.39 was redesigned. This time 0.15 mol each of N_2 and O_2 were injected into a 5.0 L container at 2500 K, at which the equilibrium constant is 3.6×10^{-3}. What was the composition of the reaction mixture at equilibrium?

12.44 At a particular temperature, the equilibrium constant, *K*, for the dissociation of N_2O_4 into NO_2 is 133. If the initial concentration of N_2O_4 is 0.100 mol L^{-1}, what are the concentrations of all species at equilibrium?

$$N_2O_4 \rightleftarrows 2NO_2$$

12.45 A student is simulating the carbonic acid-hydrogen carbonate equilibrium in a lake.

$$H_2CO_3(aq) \rightleftarrows H^+(aq) + HCO_3^-(aq). \qquad K = 4.3 \times 10^{-7}$$

She starts with 0.10 *M* carbonic acid. What are the concentrations of all species at equilibrium?

12.46 Since carbonic acid undergoes a second ionization, the student in Exercise 12.38 is concerned that the hydrogen ion concentration she calculated in that Exercise is not correct. She looks up the equilibrium constant for the reaction

$$HCO_3^-(aq) \rightleftarrows H^+(aq) + CO_3^{2-}(aq).$$

Upon finding that the equilibrium constant for this reaction is 5.6×10^{-11}, she decides that her answer in 12.38 is correct. Explain her reasoning.

12.47 Since calcium carbonate is a sink for CO_3^{2+} in a lake, the student in Exercise 12.38 decides to go a step further and examine the equilibrium between carbonate ion and $CaCO_3$. The reaction is

$$Ca^{2+}(aq) + CO_3^{2-}(aq) \rightleftarrows CaCO_3(s)$$

The equilibrium constant for this reaction is 3.6×10^8. If the initial calcium ion concentration is 0.02 *M* and the carbonate concentration is 0.03 *M*, what are the equilibrium concentrations of the ions?

LeChatelier's Principle

12.48 The following reaction is in equilibrium in lake water.

$$HCO_3^-(aq) + H^+(aq) \rightleftarrows H_2CO_3(aq)$$

Predict the change in the reaction quotient, *Q*, for each disturbance below, and use that prediction to explain how the equilibrium is shifted by the stress.

(a). $NaHCO_3$ is added to the lake.
(b). H_2CO_3 is added.
(c). NaOH is added.

12.49 In the following equilibrium in a closed system, indicate how the equilibrium is shifted by the indicated stress.

$$HF(g) + H_2O(l) \rightleftarrows H_3O^+(aq) + F^-(aq)$$

(a). Additional HF(g) is added to the system.
(b). Water is added.
(c). $Ca(NO_3)_2$ solution is added, and CaF_2 precipitates.
(d). The pressure is increased.
(e). KOH is added.
(f). A catalyst is added.

12.50 In each of the reactions below, how does the equilibrium respond to an increase in pressure?

(a). $2\ SO_2(g) + O_2(g) \rightleftarrows 2\ SO_3(g)$

(b). $H_2(g) + I_2(g) \rightleftarrows 2\ HI(g)$

(c). $3\ H_2(g) + N_2(g) \rightleftarrows 2\ NH_3(g)$

12.51 In each of the reactions below, tell how the equilibrium responds to an increase in pressure.

(a). $CaCO_3(s) \rightleftarrows CaO(s) + CO_2(g)$

(b). $N_2O_4(g) \rightleftarrows 2NO_2(g)$

(c). $HCO_3^-(aq) + H^+(aq) \rightleftarrows H_2CO_3(aq)$

12.52 How would a temperature increase affect each of the equilibria below?

(a). $AgNO_3^-(s) \rightleftarrows Ag^+(aq) + NO_3^-(aq)$, $\Delta H = 22.6$ kJ mol^{-1}

(b). $C(s) + O_2 \rightleftarrows 2\,CO(g)$, $\Delta H = -209$ kJ mol^{-1}

12.53 The forward reaction in the equilibrium below is very slow at room temperature.

$$H_2(g) \rightleftarrows 2\,H(g)$$

A catalyst is added, and equilibrium is rapidly attained. How does the catalyst affect the equilibrium concentrations of reactant and product at room temperature?

12.54 The following equilibrium is established in a closed container.

$$NO(g) + 1/2\,O_2(g) \rightleftarrows NO_2(g) + \text{heat}$$

How is the concentration of NO affected by each of the following changes?
(a). NO_2 is removed from the system.
(b). The temperature is raised.
(c). The concentration of O_2 is increased.
(d). The volume of the container is decreased.

12.55 The following equilibrium is established in a closed container.

$$C(s) + O_2(g) \rightleftarrows CO_2(g) \qquad \Delta H° = -393 \text{ kJ mol}^-$$

How does the equilibrium shift to absorb each of the following stresses?
(a). The quantity of solid carbon is increased.
(b). A small quantity of water is added, and CO_2 dissolves in it.
(c). The system is cooled.
(d). The volume of the container is increased.

Solubility Equilibria

12.56 Write the K_{sp} expression for each of the following equilibria for the dissolution of the salt in water.

(a). $AgI(s) \rightleftarrows Ag^+(aq) + I^-(aq)$

(b). $PbI_2(s) \rightleftarrows Pb^{2+}(aq) + 2\,I^-(aq)$

(c). $Hg_2I_2 \rightleftarrows Hg_2^{++}(aq) + 2\,I^-(aq)$

(d). $Cu(IO_3)_2 \rightleftarrows Cu^{2+}(aq) + 2\,IO_3^-(aq)$

12.57 Write an equilibrium equation for the dissolution of each of the following in water and its K_{sp} expression.

(a). AgBr (b). $Cu(OH)_2$ (c). $BaSO_4$ (d). $PbCrO_4$ (e). $Mg_3(PO_4)_2$

12.58 What is the molar solubility of AgCl, if $K_{sp} = 1.7 \times 10^{-10}$?

12.59 The Safe Drinking Water Act of 1974 placed the maximum permitted concentration of silver ion at 0.05 ppm. What is the concentration of Ag^+ in parts per million in a saturated solution of AgCl? (1 ppm = 1 mg of solute / L of solution.)

12.60 In Exercise 12.59, what is the allowed concentration of AgCl in g per 100 g of water?

12.61 Lead chromate has been used as a pigment, although both the lead ion and the chromate ion are toxic. What is the molar solubility of $PbCrO_4$ in water?

12.62 Because barium sulfate is opaque to X-rays, it is suspended in water and taken internally to make the gastrointestinal tract visible on an X-ray photograph. Although barium ion is quite toxic, barium sulfate's K_{sp} of 1.1×10^{-10} gives it such low solubility that it can be safely consumed.

(a). What is the molar solubility of $BaSO_4$.

(b). What is its solubility in grams per 100 g of water?

12.63 An important source of mercury is the ore, cinnabar, HgS. A red solid, its solubility in water is 6.2×10^{-27} mol L^{-1}.

(a). Calculate its K_{sp}.

(b). What is its solubility in grams per 100 g of water?

12.64 Lead(II) phosphate has a solubility of 1.4×10^{-5} g per 100 g of water. What is its K_{sp}?

12.65 Cadmium sulfide, CdS, is a bright yellow pigment. Its solubility in cold water is 1.3×10^{-4} g in 100 g of cold water. Calculate its K_{sp}.

12.66 The K_{sp} of copper(II) hydroxide is 2.2×10^{-20}. Calculate its solubility in (a). pure water, and (b). 0.01 *M* NaOH.

12.67 The K_{sp} of copper(II) phosphate is 1.3×10^{-37}. Calculate its molar solubility in (a). pure water, and (b). 0.10 *M* Na_3PO_4.

12.68 In each of the following, use the concentrations of each ion to compute the reaction quotient, Q. If Q > K, a precipitate will form. If Q < K, no precipitate forms, because the concentrations of reacting species are too low. In which of the following will a precipitate form? The K_{sp} of AgCl is 1.7×10^{-10} (Remember that dilution changes the concentrations of ions!)

(a). 10 mL of 1.0×10^{-5} *M* Ag^+ are added to 10 mL of 1.0×10^{-5} *M* Cl^-.

(b). 50 mL of 1.0×10^{-2} *M* Ag^+ are added to 50 mL of 1.0×10^{-2} *M* Cl^-.

12.69 Copper(II) ion forms a complex ion with ammonia. If one copper(II) ion reacts with 4 ammonia molecules, write the formula for the complex ion that results.

12.70 A student is studying a solution of the sparingly soluble salt Ag_2SO_4, for which the K_{sp} is 1.4×10^{-5}. The solubility of Ag_2SO_4 is high enough that silver ions often encounter sulfate ions in solution. What would be the effect on the measurement of the solubility of Ag_2SO_4 if a large number of ion pairs exist in the solution? Defend your answer.

Acids and Bases

12.71 Write the formula of the conjugate base of each of the following acids.

(a). HNO_3 (b). H_2O (c). HSO_4^- (d). H_2CO_3 (e). H_3O^+

12.72 Write the formula of the conjugate acid of each of the following bases.

(a). OH^- (b). NH_3 (c). CH_3NH_2 (d). HPO_4^{2-} (e). CO_3^{2-}

12.73 In each of the following reactions, identify the conjugate acid-base pairs, and draw brackets linking them under the equation (see Example 12.15).

(a). $HOCl + H_2O \rightleftarrows H_3O^+ + OCl^-$

(b). $OH^- + H_3O^+ \rightarrow H_2O + H_2O$

(c). $H_3O^+ + NH_3 \rightarrow NH_4^+ + H_2O$

(d). $CH_3COOH + CH_3NH_2 \rightarrow CH_3COO^- + CH_3NH_3^+$

12.74 What makes a weak acid or base different from a strong acid or base?

12.75 Write chemical equations and equilibrium expressions for the reactions of each of the following weak acids with water.

(a). acetic acid, CH_3COOH

(b). propanoic acid, C_2H_5COOH

(c). hydrofluoric acid, HF

(d). chlorous acid, HClO

(e). carbonic acid, H_2CO_3

12.76 Write chemical equations and equilibrium expressions for the reactions of each of the following weak bases with water.

(a). ammonia, NH_3

(b). methylamine, CH_3NH_2

(c). acetate ion, CH_3COO^-

(d). hydrogen carbonate ion, HCO_3^-

(e). aniline, $C_6H_5NH_2$

12.77 Hydrofluoric acid is a weak acid used in the building industry to etch patterns into glass for elegant windows. Because it dissolves glass, it is the only inorganic acid that must be stored in plastic containers. A 0.1 M solution of HF has a pH of 2.1. Calculate the $[H_3O^+]$ in this solution.

12.78 The pH of a 0.10 *M* solution of hypobromous acid, HBrO, is 6.4 at 25°C. What is the value of K_a for hypobromous acid?

12.79 Calculate the pH of a 0.10 *M* solution of propanoic acid and determine its percent ionization.

12.80 Find the pH of 0.115 *M* solution of $NH_3(aq)$.

12.81 Acrylic acid is used in the polymer industry in the production of acrylates. Its K_a is 5.6×10^{-5}. What is the pH of a 0.11 *M* solution of acrylic acid, CH_2: CHCOOH?

12.82 Calculate the pH of a 0.00100 *M* solution of acetic acid.

12.83 Morphine, an opiate derived from the opium poppy (genus *Papaver*), has the molecular formula $C_7H_{19}NO_3$. It is a weakly basic amine, with a K_b of 1.6×10^{-6}. What is the pH of a 0.0045 *M* solution of morphine?

12.84 In a particular experiment, the equilibrium constant for the reaction, $Cl_2(g) + NO_2(g) \rightleftarrows Cl_2NO_2(g)$ is measured to be 2.8.

(a). Based on this measurement, calculate ΔG° for this reaction.

(b). Calculate ΔG° using data from Appendix E at the back of the book, and discuss the agreement between your two calculations.

12.85 For the reaction, $PCl_5(g)$ Ä $PCl_3(g) + Cl_2(g)$ the equilibrium constant is measured to be 7.7×10^{-3} at some temperature. (a) Based on this information predict if $\Delta G°$ is positive or negative. Explain your answer. (b) Calculate $\Delta G°$ from this information.

12.86 Use the values from Appendix E at the back of the book to calculate the equilibrium constant, K_p, for the following gaseous reactions,

(a). $H_2(g) + Cl_2(g) \rightleftarrows 2HCl(g)$

(b). $CH_4(g) + H_2O(g) \rightleftarrows CO(g) + 3H_2(g)$

(c). $SO_2(g) + Cl_2(g) \rightleftarrows SO_2Cl_2(g)$

(d). $2HCl(g) + F_2(g) \rightleftarrows 2HF(g) + Cl_2(g)$

12.87 A group of students working on chemistry homework calculates the equilibrium constants for three reactions. Reaction 1 has $\Delta G° = -50$ kJ, reaction 2 has $\Delta G° = -30$ kJ and reaction 3 has $\Delta G° = -10$ kJ. The students are surprised when their answers for each reaction are very close to K=1. Suggest what error the students must have made and calculate the actual values of K for these reactions.

Additional Problems

12.88 The following are injected into a 1-L flask: 0.10 mol H_2, 0.10 mol I_2, and 0.10 mol of HI. Calculate the equilibrium concentrations of all species for the reaction at 227°C, where $K = 160$.

$$H_2(g) + I_2(g) \rightleftarrows 2HI$$

12.89 Solid $CaCO_3$ is placed in a closed container and heated to 800°C. At equilibrium, $K = 2.5 \times 10^{-3}$. What is the equilibrium concentration of CO_2 due to the following equilibrium?

$$CaCO_3(s) \rightleftarrows CaO(s) + CO_2(g)$$

12.90 Stalactites form in caves as mineral-rich water containing calcium and carbonate ions drips from the ceiling of the cave. What must be the concentration of calcium ion for $CaCO_3$ to form at 25°C?

12.91 Your laboratory instructor gives you 100 mL of a saturated solution containing 50 g of NaCl in water. Look up the solubility of NaCl in the CRC *Handbook of Chemistry and Physics*. If you dilute the solution by a factor of 10, will all the NaCl dissolve?

12.92 For an experiment, you need the K_{sp} for copper(II) oxalate, and it is not in the K_{sp} list in your textbook. Look up the solubility of copper(II) oxalate in the CRC *Handbook of Chemistry and Physics*, and calculate its K_{sp}. (Check its formula, as well, in the *Handbook*.)

12.93 Methanol, CH_3OH, can be produced by the reaction of CO with H_2, with the liberation of heat. All species in the reaction are gaseous. What effect will each of the following have on the equilibrium concentration of CO?

(a). pressure is increased
(b). volume of the reaction container is decreased
(c). heat is added
(d). the concentration of CO is increased
(e). some methanol is removed from the container
(f). H_2 is added

12.94 The following equilibrium is established in a closed container.

$$C(s) + O_2(g) \rightleftarrows CO_2(g) \qquad \Delta H° = -393 \text{ kJ mol}^{-1}$$

Set up a table like the one below, and indicate whether equilibrium concentrations increase, decrease, or remain the same.

Change applied to system	C(s)	O_2(g)	CO_2(g)
pressure is decreased			
temperature is increased			
carbon is added			
oxygen is added			
carbon dioxide is added			
volume of the system is increased			
a catalyst is added			
heat source is removed			

12.95 Using the kinetic molecular theory, explain why an increase in pressure produces more N_2O_4 in the following system.

$$N_2O_4(g) \rightleftarrows 2NO_2(g)$$

12.96 The solubility of KCl is 34.7 g in 100 g of water at 20°C. Calculate its molar solubility in (a). pure water, and (b). 0.10 M NaCl.

12.97 To prepare community wastewater to be released to a lake, phosphate ions (among others) must be removed. $Al_2(SO_4)_3$ can be used for this purpose, precipitating $AlPO_4$. The K_{sp} of $AlPO_4$ at 25°C is 6.3 x 10^{-19}. If $AlPO_4$ is precipitated, and the precipitate removed, what concentration of PO_4^{3-} remains in solution in the water released to the lake?

12.98 A commercially bottled drinking water contains 10 ppm of NaCl. Look up the solubility of NaCl in the CRC *Handbook of Chemistry and Physics*, and calculate its solubility in ppm. By what factor is the concentration of NaCl in the bottled drinking water related to its solubility?

12.99 Copper(II) iodate has a solubility of 0.136 g per 100 g of water. Calculate its molar solubility in water and its K_{sp}.

12.100 In Exercise 12.99, what do you predict would happen to an aqueous copper(II) iodate equilibrium if sulfide ion, S^{2-}, were added? The K_{sp} of CuS is 6.3 x 10^{-36}.

Focus on Problem Solving

12.101 You have three white solids. What experiment could you carry out to be able rank them in order of increasing solubility.

12.102 You find a bottle in the laboratory. You know it is an acid and the label says it is a 1.05 M solution but doesn't say what acid it is. Suppose the label is correct for the concentration, how could you determine what acid it is? What would have to be measured and what would you have to look up?

12.103 You find a bottle of acid in the laboratory and it is labeled $H_2C_2O_4$. There is no concentration given and you don't have the reagents to carry out a titration. What other measurement could be made and what information would you need to look up to determine the concentration of the acid?

12.104 You need to find the equilibrium constant for a reaction that may exist in an industrial process. Describe how you can estimate this value using only thermodynamic data.

Chapter 13: Electrochemistry

Electrochemistry deals directly with electrons and their movement. Whether the electrons move under the influence of the driving forces of natural chemical processes or by the application of an external electric potential, their movement brings about a whole new set of opportunities and challenges. We introduce the chapter with the topic of corrosion, one result of electrochemistry that costs the U.S. economy an estimated $300 billion per year.

13.1 Insight into Corrosion

Over the history of the U.S. space program, NASA has used the Kennedy Space Center (KSC) on the coast of Florida as its primary launch site. In stark contrast, the site of the space center used by the cosmonauts of Russia is constructed in the middle of the Moyunqum Desert. While there are many factors that are considered when choosing a site for such facilities, the contrast between the environments in these two places is striking. The placement of the KSC so close to the ocean means rust has long been an important issue for the American space program. Rust is such a significant concern to NASA that they founded the Beach Corrosion Test Site in the 1960's to study the effect of the atmosphere on structures used in launching rockets. When rust forms on steel, it weakens the metal, and weaker materials are more prone to failure. NASA, therefore, studies rust and **corrosion**, the degradation of metals by chemical reactions with the environment. Corrosion generally involves the slow combination of oxygen with metals to form oxides.

For engineering applications it is important to remember that many metals corrode. Rust, the formation of iron oxide is only one, relatively familiar, example.

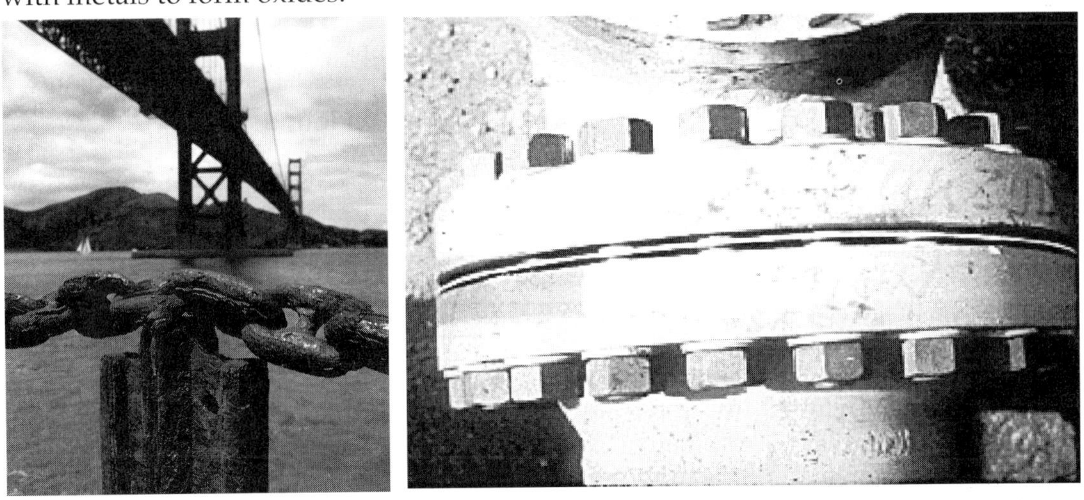

Figure 13.1 Corrosion occurs in a variety of forms. The chain on the left shows uniform corrosion, while the flanges on the right show corrosion only near the joint. This is called crevice corrosion, and can occur whenever small gaps exist between metal parts.

Research into corrosion has identified many different forms of corrosion according to the specifics of the physical environment in which it occurs. The rusting of automobile bodies is an example of **uniform corrosion**, and is one of the most visible forms of corrosion. What conditions are necessary for this process to occur? Another important form of corrosion is **galvanic corrosion**, which occurs only when two different metals are in contact with each other in the presence of an appropriate electrolyte. What is so special about the contact of two different metals? Other forms of corrosion tend to require specific conditions, yet many of these situations are common in designs for machinery. **Crevice corrosion** is a major problem in many large machines. When two pieces of metal touch each other, they tend to leave a small opening (unless the joint is covered with a coating, such as paint). At that crack, the metals are more likely to experience corrosion. How can we understand the tendency to corrode using basic principles of electrochemistry?

A puzzling feature about corrosion is the difference between iron (or steel) and other metals in their behavior towards corrosion. Aluminum, for example, has a greater tendency to corrode than does iron, but the corrosion of aluminum is not problematic. Why not? As we will see, the answer lies in the nature of the products of the corrosion reaction. What differences are there between the corrosive formation of aluminum oxide and iron(III) oxide? How can we picture corrosion at an atomic level to account for these differences?

While corrosion is an important example of electrochemical processes, we need to be careful not to imply that it alone defines electrochemistry. There are many important applications of electrochemistry that provide useful products for society rather than degrade them. The refining of some metal ores into useable materials and the production of batteries are two obvious examples of the practical utility of electrochemistry. But even in the production of batteries, we need to consider additional manufacturing steps to account for the need to reduce the impact of corrosion on the useful lifetime of the battery. Where does corrosion occur in a battery, and what can be done to combat it? The study of electrochemical principles will help us answer these questions.

13.2 Oxidation-Reduction Reactions and Galvanic Cells

What happens to a piece of steel that sits outside, unprotected? In most locations, it rusts. Would you expect to observe the same thing if that piece of steel were inside a house or in a desert? Perhaps not. There must be some special conditions that promote the reaction of iron with oxygen to form iron(III) oxide. We could design a set of experiments to study the formation of rust, but from a laboratory perspective rust formation is rather slow. To find out more about the basics of electrochemistry, let's begin with more easily observed reactions and apply what we learn to examples of corrosion.

Oxidation-Reduction and Half-reactions

In discussing batteries in the closing "Insight" section of Chapter 9, we mentioned that reactions involving the transfer of electrons are known as **oxidation-reduction reactions**. (The term is often inverted and shortened to **redox reactions**.) In order to delve more deeply into electrochemistry, we will first need to extend our understanding of this class of chemical reactions and define a few relevant terms. **Oxidation** is the loss of electrons from some chemical species, while **reduction** is the gain of electrons. By now you will recognize that electrons can't simply be "lost;" they have to go somewhere, in keeping with the notions of conservation of charge and matter. This leads directly to one of the most important principles of redox chemistry: the electrons "lost" in oxidation must always be "gained" in the simultaneous reduction of some other species. In other words, we can't have oxidation unless we also have reduction. To illustrate this idea, let's consider the reaction that takes place when we place a copper wire into a solution of silver nitrate—a popular classroom demonstration (Figure 13.2).

(a). (b). (c).

Figure 13.2 When a clean copper wire is placed into a colorless solution of silver nitrate, it is quickly apparent that a chemical reaction takes place. Crystals of silver metal form on the wire, and the solution takes on a blue color. In this reaction copper metal is oxidized to Cu^{2+}, while Ag^+ ions are reduced to silver metal.

Close inspection of the last photograph shows the formation of dendrites of silver. They are called dendrites because they branch, similar to tree limbs.

As you can see in the photographs, the original solution of silver nitrate is clear and colorless, and the copper wire is essentially smooth (Figure 13.2a). A short time after the wire is placed in the solution, though, the copper wire looks fuzzy, and the solution has become light blue (Figure 13.2b). These observations become more apparent over time, and in the final photo there is an obvious build up of slivers of silver metal while the solution is distinctly blue (Figure 13.2c). What has happened chemically? The blue color of the solution is indicative of the presence of aqueous Cu^{2+} ions. These ions must originate from the copper wire, since no other source of copper is available. So in forming the cations, copper atoms must have lost electrons. We say that copper has been oxidized, and we could write an equation to describe this change.

$$Cu(s) \rightarrow Cu^{2+}(aq) + 2e^-$$

But if cations are produced at the wire, where did the electrons go? The accumulation of silver on the copper wire provides us with the answer. The only apparent way to form metallic silver in this system is from the silver ions in the original solution. So we must conclude that silver cations in solution have accepted the electrons lost by the copper. We say that silver has been reduced, and again we can write an equation to show the change.

$$Ag^+(aq) + e^- \rightarrow Ag(s)$$

The two equations we have written describe what are called **half-reactions** for the oxidation of copper and the reduction of silver. Neither one can occur on its own, because oxidation and reduction must take place in concert with one another.

As you examine the half-reactions with this in mind, you may notice a small discrepancy. Copper loses two electrons in the oxidation half-reaction, while silver gains only one electron in the reduction half-reaction. This tells us that in order to conserve electrons, two silver ions will need to be reduced for every copper atom that is oxidized. We can multiply the reduction by two to make this explicit, giving us the following pair of half-reactions.

$$Cu(s) \rightarrow Cu^{2+}(aq) + 2e^-$$

$$2Ag^+(aq) + 2e^- \rightarrow 2\ Ag(s)$$

Here we can more readily see that the silver gains the two electrons that the copper loses. If we add the two half-reactions together, the electrons will cancel and we'll be left with the net ionic equation for the overall redox reaction.

$$2Ag^{+}(aq) + Cu(s) \rightarrow 2\ Ag(s) + Cu^{2+}\ (aq)$$

As we discussed back in Chapter 3, we could also write this as a molecular equation by including the spectator ions (NO_3^- in this case).

$$2AgNO_3(aq) + Cu(s) \rightarrow 2Ag(s) + Cu(NO_3)_2\ (aq)$$

Note that an agent in this sense is something that allows the process to occur. In everyday language, for example, an agent of change is something that helps change happen.

Looking at our example above, we could describe the process by saying that copper metal is oxidized by silver ions. Or we could say that silver ions are reduced by copper metal. Neither of these statements is any more or less correct than the other, since both processes must occur together. And the same idea can be generalized to give the following bit of terminology.

The species undergoing oxidation is referred to as a reducing agent.

The species undergoing reduction is referred to as an oxidizing agent.

Many students find this confusing, but it should make sense if you consider it carefully.

Building a Galvanic Cell

The half-reactions we introduced above provide us with a way of seeing the reaction as oxidation and reduction, but seem like a rather artificial idea. Both half-reactions took place in the same beaker, after all. But what would happen if we could set up these half-reactions in two separate containers? Consider the following experiment.

Nothing moves through the salt bridge - it dispenses ions on each end.

Place a piece of copper metal in a solution of copper(II) ions. (Figure 13.3(a); the spectator ion is not important—sulfate might be used, because $CuSO_4 \cdot 5H_2O$ is a common laboratory reagent.) In a separate beaker, place a piece of silver metal or a silver wire into a solution of silver ions. (Again, the spectator ion is not critical. We might use nitrate ions, from silver nitrate.) We observe no reactions in the individual beakers. Now connect the solutions by a conducting wire and connect a voltmeter between the copper and silver, as shown. If you have a good measuring device, you might notice a brief electrical current, but there is otherwise no observable change in the system. Why? The wire cannot allow the continuous flow of charge between the solutions, since the charged particles in the solutions are ions, not free electrons. Any initial flow of electrons from the copper-containing solution towards the silver-containing one is quickly stopped by the build-up of negative charge in the silver beaker. Connecting the two beakers in a second way, with a salt bridge, can alleviate the charge buildup (Figure 13.3(b)). A **salt bridge** contains a strong electrolyte that allows either cations or anions to migrate into the solution where they are needed to maintain charge neutrality. In this case, suppose the salt bridge is filled with NH_4Cl. Ammonium ions will flow into the silver beaker to offset the *removal* of $Ag^{+}(aq)$, and chloride ions will flow into the copper beaker to offset the *production* of $Cu^{2+}(aq)$ in the copper beaker. In Figure 13.3(b) we can see the various moving charges in this arrangement. The apparatus we have constructed in this way is an example of a *galvanic cell.*

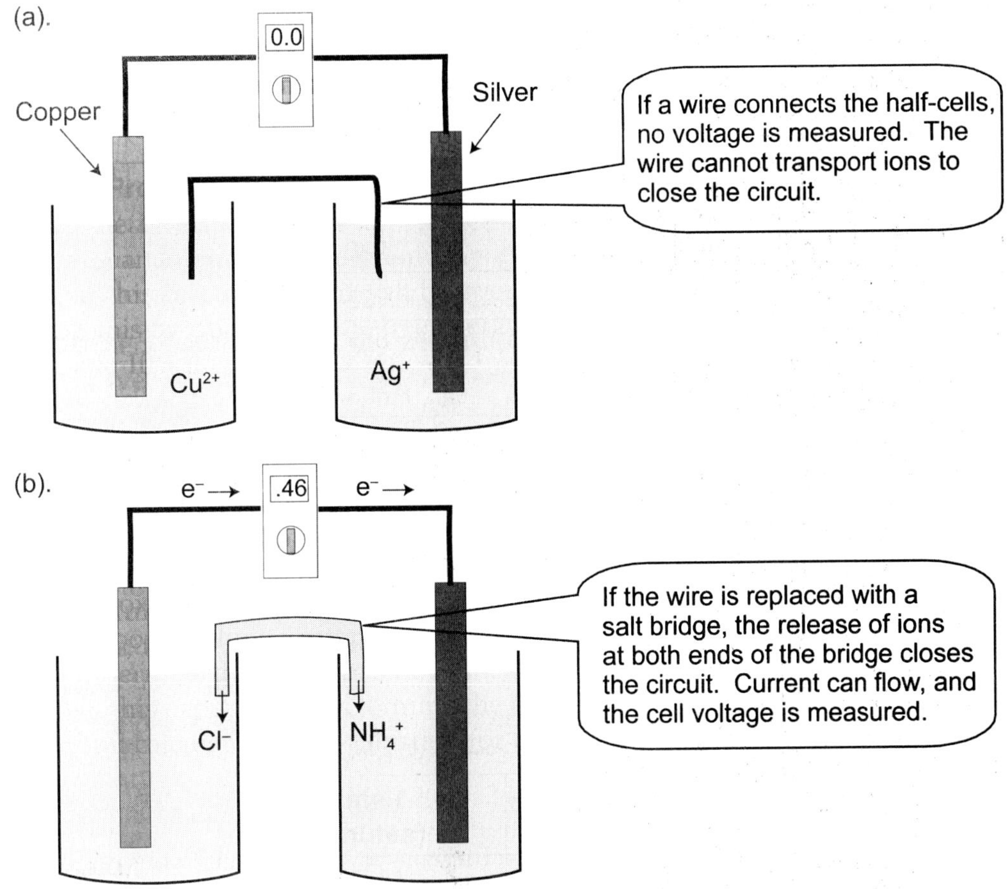

Figure 13.3 The salt bridge plays a crucial role in a galvanic cell. By allowing ions to flow into each half-cell, the bridge closes the circuit and allows current to flow. A wire can carry a current of electrons, but cannot transport the ions needed to complete the circuit.

Galvanic derives its name from the Italian physicist Luigi Galvani who conducted early experiments in the role of electricity in living systems (using frogs.)

To formally define the galvanic cell we must first note its most important aspect. By separating the two half-reactions, inserting a salt bridge between the solutions, and allowing electrons to flow along a wire connecting the metals, we have generated a sustainable electrical current. A **galvanic cell** is any electrochemical cell in which a spontaneous chemical reaction may be used to generate an electrical current. The observation of an electric current in such a cell led to the name *electrochemistry*. This connection of electricity to chemistry was vitally important in the development of modern thought about chemical principles.

Terminology for Galvanic Cells

There are an almost unlimited number of possible galvanic cells that could be built. All of them have common features, and terminology has been devised to describe these features. The electrically conducting sites at which either oxidation or reduction take place are called **electrodes**. Oxidation occurs at the **anode** and reduction occurs at the **cathode**. These terms are also used for other types of electrochemical cells, not just galvanic cells.

Because there are so many possible electrochemical cells, a shorthand notation for representing the essential chemistry has been devised. This **cell notation** lists the metals and ions involved in the reaction, and uses a double line, ||, to denote the salt bridge. The anode is always written on the left, and the cathode on the right.

Anode | electrolyte of anode || electrolyte of cathode | cathode

The previous example of copper and silver would be written as follows.

$Cu(s) \mid Cu^{2+}(aq)\ (1\ M) \parallel Ag^{+}(aq)\ (1\ M) \mid Ag(s)$

Once again we see that the spectator ions are not identified in this notation. Generally, however, the concentration of the electrolyte is included. In the example we include concentrations of 1 M. These values have special meaning because they are the assigned concentrations for the **standard state** of an electrochemical cell. If the electrochemical half-reaction includes the production or consumption of a gas, the standard state is a pressure of 1 atm. A standard state also implies that the electrode material is in its *thermodynamic* standard state, which is commonly the case, since most electrode materials are solids at room temperature.

Atomic Perspective of Galvanic Cells

To see the pervasive nature of corrosion, or to understand how electrochemistry can be harnessed as a battery, we need to understand the fundamental laws that are in play. We have noted that electrons flow in a properly constructed cell, but what causes this flow to occur? To understand this question we will "build" a galvanic cell again, but this time we will focus on what happens at the atomic level. If we construct the anode and cathode separately, what does the resulting system look like at each of the electrodes *before* we connect them with a salt bridge?

While it may seem a violation of your intuition, at the interface between the solid electrode and the solution surrounding it there is a build up of charge. We are accustomed to thinking that chemical systems are electrically neutral, but this constraint is true only within a single phase, not necessarily between phases. So, a system that does not yet have a complete circuit undergoes partial reaction, and at the anode some oxidation occurs and cations dissolve into solution, leaving behind a negative charge on the anode. At the cathode, reduction brings positively charged ions to the electrode, and the result is a build up of positive charge on that electrode. Thus before the circuit is complete, a form of equilibrium is quickly achieved where charge builds up on the individual electrodes, and it is locally offset by the ions in the solution as shown in Figure 13.4. We could consider this equilibrium state at each electrode *the half-reaction equilibrium*—but it is *not* an oxidation-reduction chemical equilibrium. Still, the buildup of charge on the electrodes is important, because it means there is the potential to do electrical work. This potential is called the **cell potential**, or the **electromotive force** (EMF). The EMF is important, because it is easily related to the maximum electrical work that can be obtained from an electrochemical cell,

$$w_{max} = qE \tag{13.1}$$

where q is the charge that is moved (electrons) and E is the EMF. This equation fits the scientific concept of work, because something is moved (charge) by the action of some force (the EMF).

How does the existence of EMF explain the tendency of metals to corrode? To answer this question, let's examine two types of corrosion.

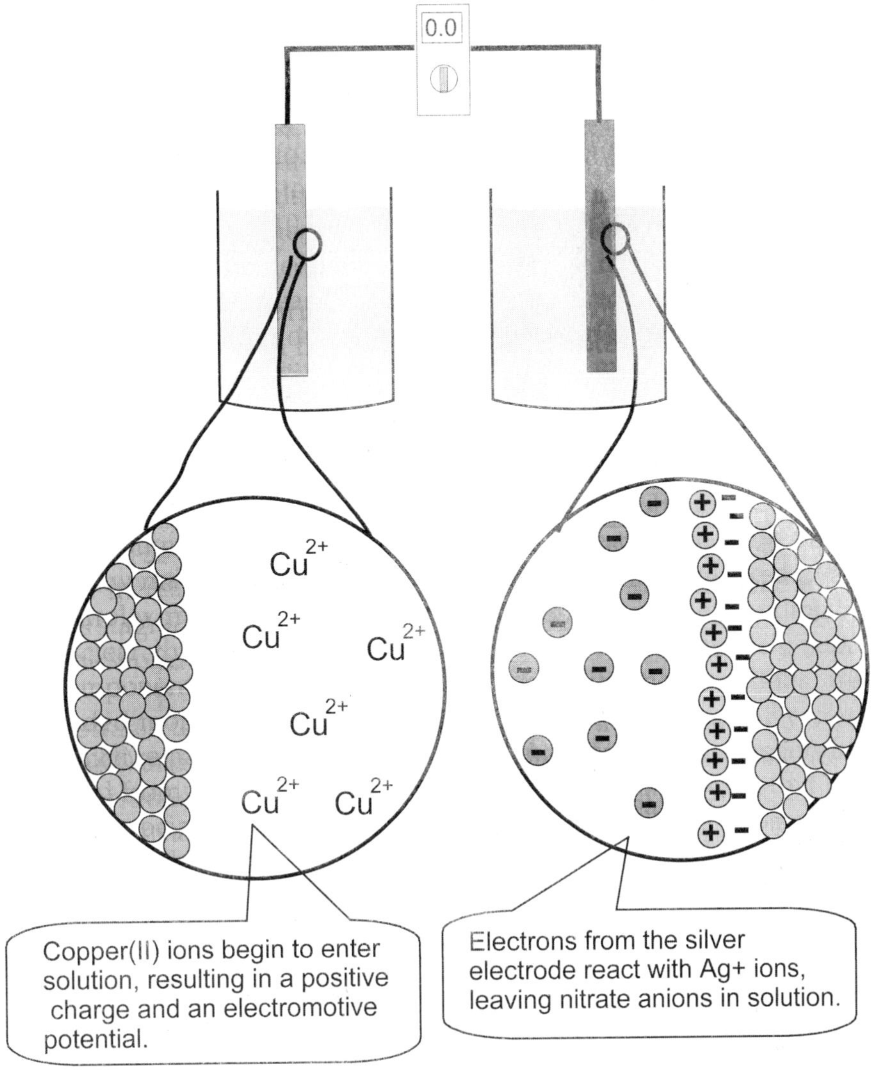

Figure 13.4 Without a salt bridge to close the circuit, local charges will build up around both electrodes. Neither electrode reaction can proceed to any significant extent, and no cell voltage can be measured.

Galvanic Corrosion and Uniform Corrosion

While common instances of corrosion are often subtler than laboratory-constructed cells, metals in contact with a solution are capable of establishing the sort of oxidation "half-reaction equilibrium" we've just described. If the solution contains a substance capable of undergoing reduction, a redox reaction may result. Under some circumstances, when two metals are in contact there is the possibility of *galvanic corrosion*. This form of corrosion resembles the galvanic cell in that oxidation and reduction take place in (at least slightly) different locations, but there is a connection of the metals that "completes the circuit". The separation of oxidation and reduction may be quite subtle, however as noted in Figure 13.5. The everyday "tin" can is mostly steel, though in some cases the steel is plated with tin. If there is a scratch in the tin and the underlying steel (mostly iron) is exposed to air and water, corrosion will be rapid. Why? Because the half-reaction equilibrium of the tin assists the process by which iron is oxidized, an example of galvanic corrosion. Pure steel will corrode more slowly, and zinc plated steel will corrode even more slowly.

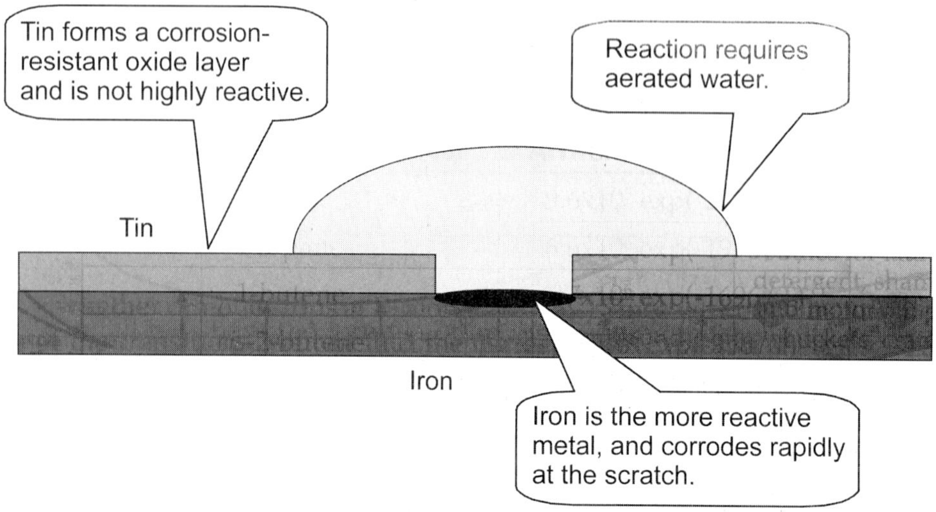

Figure 13.5 A "tin can" is usually tin-plated steel. If the tin coating is scratched to expose the underlying steel, iron in the steel will corrode rapidly.

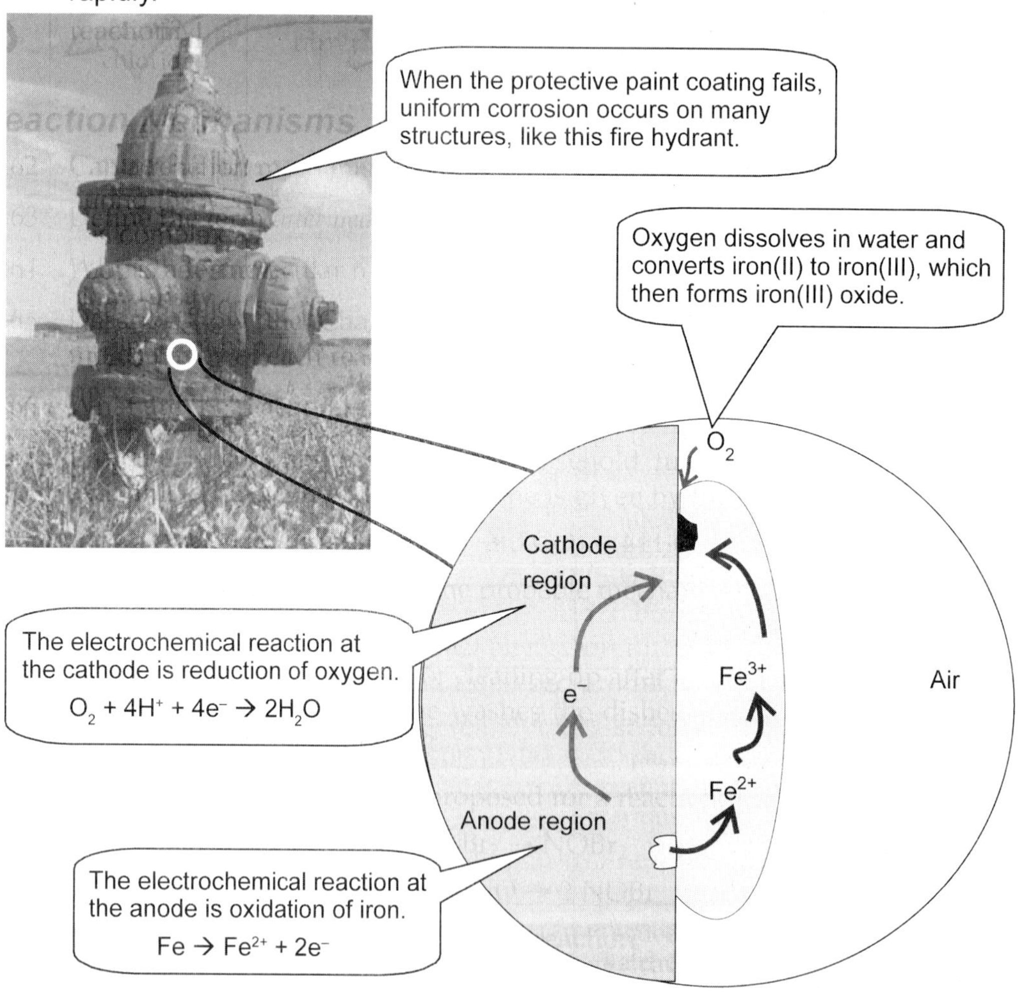

Figure 13.6 The rusting of iron exposed to the weather is an example of uniform corrosion. Iron is oxidized while oxygen from the air is reduced. Water is needed to allow ion mobility between the anodic and cathodic regions, and the presence of ionic salts speeds the reaction considerably.

A pile of nails left out in the yard rusts without contact with another metal (Figure 13.6). This implies that the second half-cell in this backyard redox reaction must involve a nonmetal. In *uniform corrosion*, rust can cover the surface of iron or steel. The second electrode is a second region of the iron itself, located some distance away from the first spot. The presence of conducting ions helps facilitate this process– so when the chloride ions of salt are present the rate of rusting is enhanced.

Both galvanic and uniform corrosion involve the concept of electromotive force or cell potential, but to understand them more thoroughly, we need to quantify the cell potential.

13.3 Cell Potentials

Our observation about the corrosive inclinations of various plated steels suggests we need to be able to put numbers on the cell potential. Just how strongly driven towards corrosion is steel that is connected to tin? Can the tendency of a metal such as zinc to protect steel be of practical use? For answers to these questions, let's turn to the concept of cell potential.

Measuring Cell Potential

If we take charged electrodes of a galvanic cell and connect them to a high impedance voltmeter, what will happen? Let's first think about what happens if we connect a voltmeter to an ordinary battery (which we will see soon is really just a well-engineered galvanic cell itself). As we see in Figure 13.7, the result depends on which terminal on the voltmeter is attached to which electrode. This observation reminds us that the voltmeter measures not only the size of the electrical potential, but also its polarity— the location of the negative charge (negative pole) and the positive charge (positive pole).

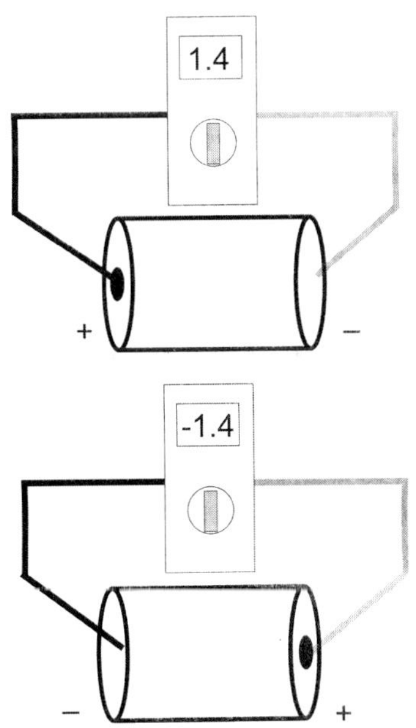

The positive terminal of the meter is connected with a black wire and the negative with a gray wire in these illustrations.

Figure 13.7 An electric potential has a fixed polarity as well as a voltage. Reversing the poles of a battery with respect to a voltmeter changes the sign on the measured voltage, but does not actually influence the electrochemical reaction taking place in the battery.

We can always change the terminal of the voltmeter to which we connect an electrode, so for the time being let's simply adjust the measurement to show a positive voltage every time. If we apply this technique to the example of copper and silver what do we observe? The cell Cu(s) | Cu^{2+} (1 M) || Ag^{+}(1 M) | Ag(s) has a potential of 0.463 V. If we take the same copper half-cell and connect it instead to a half cell that reduces iron(III) to iron(II), we find that the measured cell potential is 0.434 M. If this iron electrode is connected to the silver one the result is 0.029 V. These numbers are clearly related: 0.463 = 0.434 + 0.029. (Figure 13.8) This observation is important for two reasons. First, it shows a behavior of cell potentials that is akin to that of a state function. Second, it suggests that if we choose a specific standard electrode to which all other electrodes are compared, we can devise a practical system for the determination of cell potential.

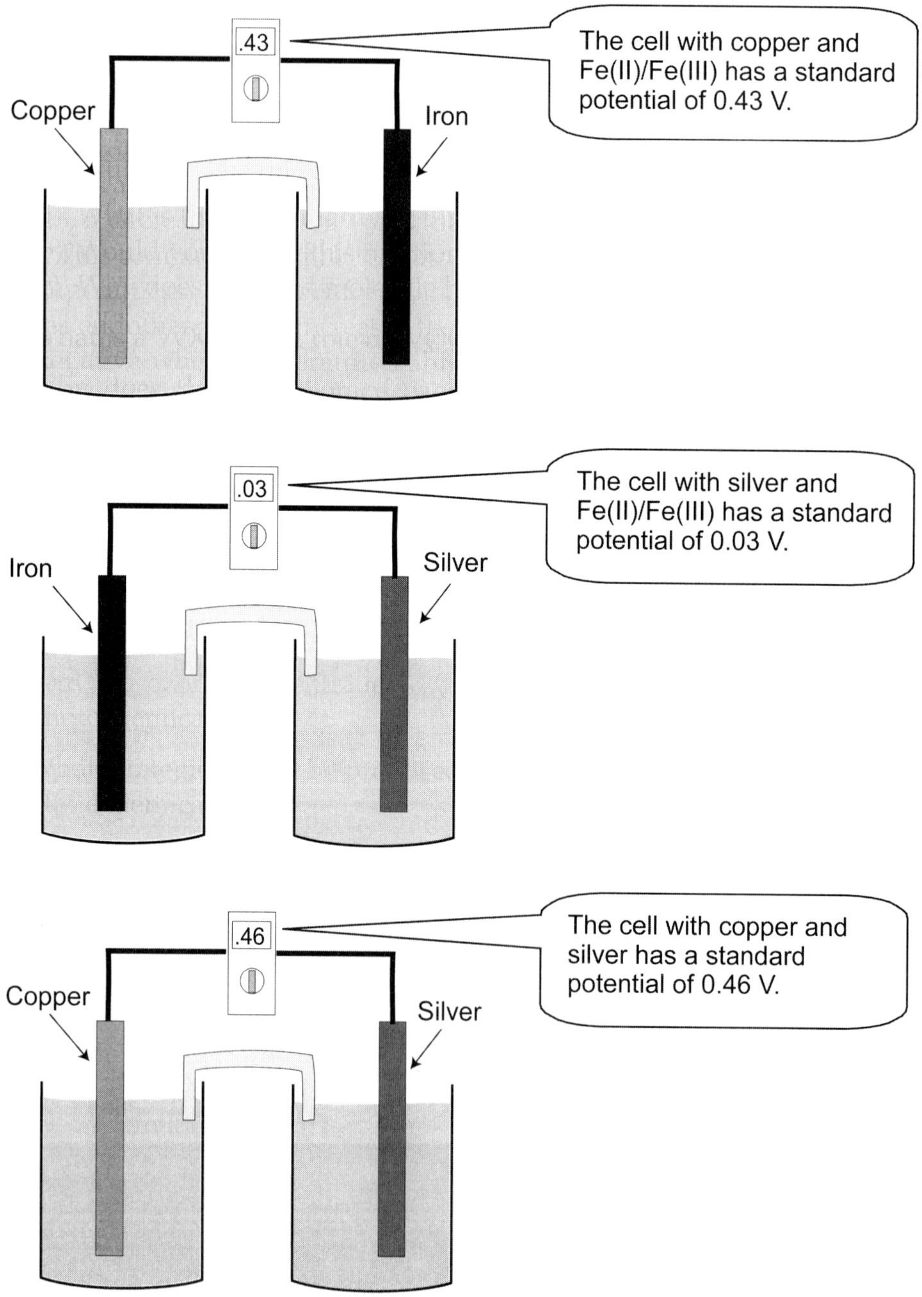

Figure 13.8 Measurement of standard cell voltages for various combinations of the same half-reactions suggests that a characteristic potential can be associated with a particular half-reaction.

The choice for the standard component in cell potential measurements is the standard hydrogen electrode (SHE), illustrated in Figure 13.9. A platinum wire or foil is the conducting source of electrons. Hydrogen gas is bubbled over the electrode at a pressure of 1 atm, and the electrolyte solution is 1 M HCl(aq). The resulting half-reaction is

$$2H^+(aq) + 2e^- \rightarrow H_2(g),$$

and the half-cell notation is

$$Pt(s) \mid H_2\ (g, 1\ atm) \mid H+\ (1\ M).$$

While the SHE is used as the standard for establishing cell potentials, it is a cumbersome device and other standards are often used in practical applications of electrochemistry.

By convention, this half-reaction is assigned a potential of 0.00 V. To determine the half-cell potential of any other electrode/electrolyte system, it can be connected to the SHE and the cell potential measured. Because the SHE is assigned a value of 0.00V, the observed cell potential is attributed to the other half-reaction.

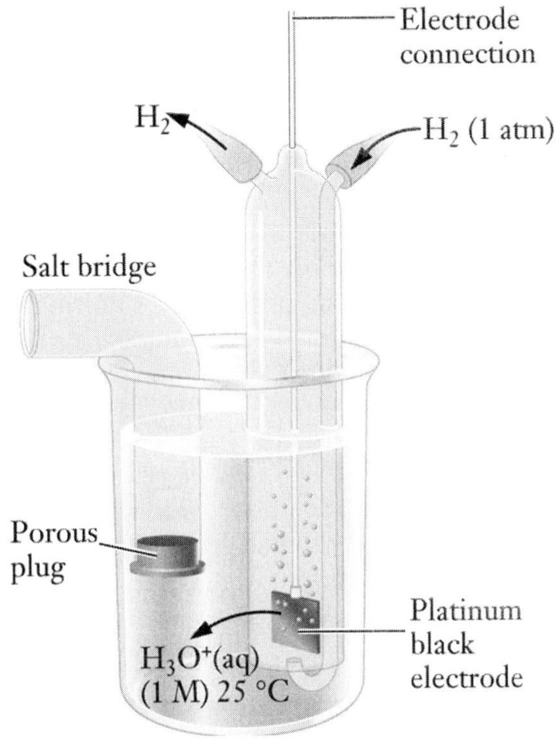

Figure 13.9 The standard hydrogen electrode has been chosen as the reference point for the scale of standard reduction potentials, and assigned a potential of zero. All other electrode potentials are measured relative to this standard.

We can now return to the issue of the direction of the electric potential. We have already set the SHE at a potential of 0.00 V, and we can further specify our observations by always connecting the SHE to the positive terminal of the voltmeter. When a variety of other electrodes are attached to the negative terminal of the voltmeter, both positive and negative values are observed. What is the connection of this observation to the chemistry that is taking place? The anode is the site of oxidation, so electrons are produced there. The anode has a negative charge in a galvanic cell. In an electric circuit containing several components, the negative pole of one component is connected to the positive pole of the next component. So, if the anode is connected to the positive terminal of the voltmeter, the voltmeter shows a positive reading. By contrast, at the cathode of a galvanic cell electrons are consumed in the reduction reaction. The cathode carries a positive charge. If a cathode is connected to the positive terminal we have violated the wiring convention of an electric circuit, and the voltmeter displays a negative voltage (Figure 13.10).

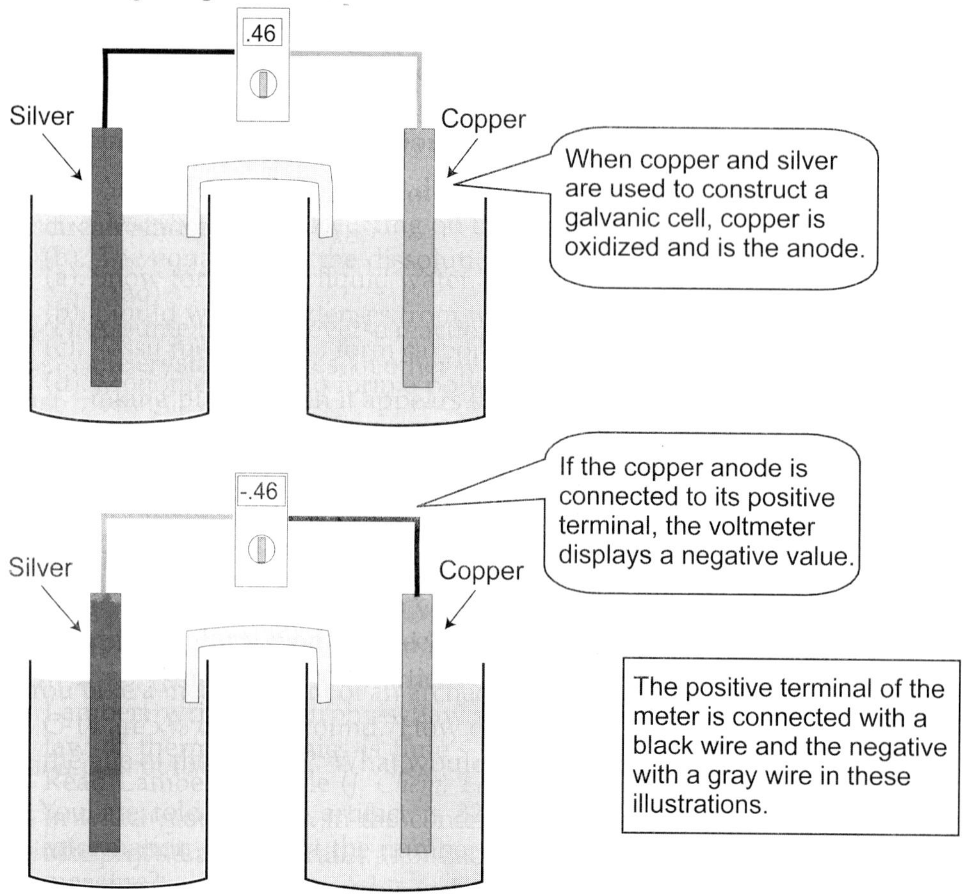

Figure 13.10 Just like a commercial battery, a galvanic cell has a fixed polarity. Electrons flow through the external circuit from the anode to the cathode. Reversing the leads of the voltmeter changes the sign of the reading, but does not influence the flow of current.

If we apply this reasoning to the measurement of standard potentials, where the standard hydrogen electrode is always connected to the positive terminal, the sign of the measured potential tells us the direction of the redox reaction. When the measured cell potential is positive, the SHE is the anode, and we know that H_2 is oxidized to H^+ (aq).

$$H_2(g) \rightarrow 2H^+(aq) + 2e$$

When the cell potential is negative, the SHE is the cathode, and H^+ is reduced to H_2. This reasoning allows us to use electrochemical measurements to assess the tendency of materials to undergo oxidation or reduction.

Standard Reduction Potentials

Just as some acids are stronger than others, the oxidizing or reducing strength of various chemicals has a range of values, as well. To help organize the oxidation-reduction trends of chemicals used in electrochemistry, all half-cell potentials are tabulated as reductions. Thus a table of **standard reduction potentials** shows the potential of any half-reaction, when connected to a SHE. Note that all electrodes must be constructed to meet the conditions of standard states, *i.e.*, (1) the materials are in thermodynamic standard states, (2) solutions have a concentration of 1 M and (3) gases have a pressure of 1 atm. A partial table of standard reduction potentials is given in Table 13.1, and a more extensive one appears in Appendix I. Several features of this table are important. First, although the half-reactions are all listed as reductions, in any electrochemical cell one half-reaction must be an oxidation. Second, some half-reactions have positive potentials while others have negative potentials. Recall that these measurements are taken with the SHE connected to the positive terminal. If the voltage is positive, the SHE is the anode—the oxidation site. So, a positive voltage in the standard reduction potential means the half-reaction proceeds as written (reduction occurs). If the value is negative, the half-reaction proceeds as an oxidation when connected to the SHE, because the SHE is serving as the cathode. (See Figure 13.10.)

Half-Reaction	Standard Reduction Potential (V)
$Zn^{2+} + 2e^- \rightarrow Zn$	-0.763
$Fe^{2+} + 2e^- \rightarrow Fe$	-0.44
$2H^+ + 2e^- \rightarrow H_2$	0.000
$Cu^{2+} + 2e^- \rightarrow Cu$	+0.337
$Fe^{3+} + e^- \rightarrow Fe^{2+}$	+0.771
$Ag^+ + e^- \rightarrow Ag$	+0.779

Table 13.1 Standard reduction potentials are shown for several of the half-reactions involved in the cells discussed in the text. A more extensive table of potentials appears in Appendix I.

These observations have implications about the tendency for oxidation and reduction of the chemicals involved in the half-reaction. A large, positive value for the standard reduction potential implies that the substance is readily reduced and is thereby a good **oxidizing agent**. Conversely, if a reduction potential has a relatively large (in magnitude) negative value, it indicates the reaction proceeds readily in reverse—the substance is oxidized. Thus, negative standard reduction potentials identify good **reducing agents**. The product of the half-reaction as written in the table is the reducing agent, because the reverse reaction is the oxidation. In either case the magnitude of the value is important in deciding what is oxidized and what is reduced. *The more positive value will be the reduction half-reaction in any galvanic cell.* The relationship between standard reduction potential and the direction of the redox reaction can be pictured by thinking of the reduction potentials along an axis as shown in Figure 13.11.

Ordering reactive tendencies based on standard reduction potentials is sometimes referred to as the electrochemical series.

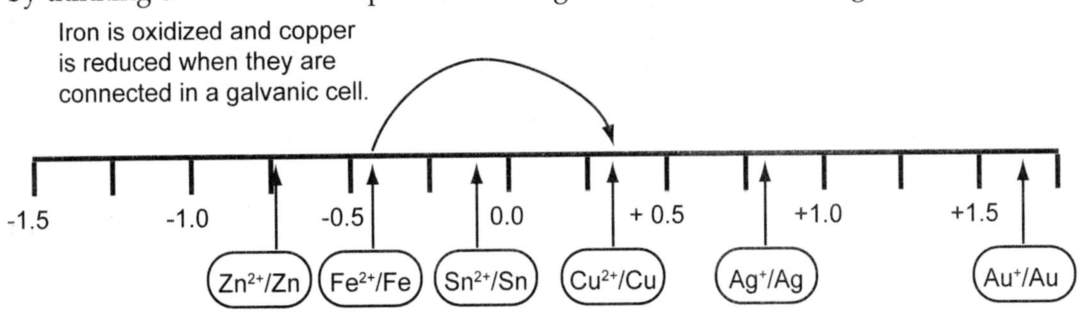

Figure 13.11 If we think of standard reduction potentials as being arranged horizontally, as shown here, we can easily identify the anode and cathode in a galvanic cell. For any pair of electrodes, the one appearing farther to the left on such a scale will be the anode and the one farther to the right will be the cathode in a galvanic cell.

With this information we can determine the standard cell potential for any pair of half-reactions by using the equation,

$$E^\circ_{cell} = E^\circ_{red} - E^\circ_{ox} \qquad (13.2)$$

The E°_{red} is the standard reduction potential of the cathode, and E°_{ox} is the standard reduction potential of the anode. The minus sign is needed because the oxidation is the opposite of the tabulated reduction potential. Example Problem 13.1 examines this concept in the context of materials that are important in galvanic corrosion.

Example Problem 13.1

Copper and iron (generally in the form of steel) are two of the many metals used in designing machines. (a). Using standard reduction potentials, identify the anode and the cathode, and determine the cell potential for a galvanic cell composed of copper and iron. Assume standard conditions. (b) Confirm that the potential of the galvanic cell, Cu(s) | Cu^{2+} (1 M) || Ag^{+}(1 M) | Ag(s), is 0.46 V.

Strategy

We must interpret the nature of an electrochemical system based on the information available in a table of standard reduction potentials. With two half-reactions there are only two possible outcomes—and one outcome yields a negative value for the cell potential. Since we know that a galvanic cell cannot have a negative value, we must determine the combination of half-reactions that provides a positive value for E°.

Solution

(a). Using Table 13.1 we find the two half-reactions,

$$Fe^{2+}(aq) + 2e^- \rightarrow Fe(s) \qquad E^\circ = -0.44\ V$$

$$Cu^{2+}(aq) + 2e^- \rightarrow Cu(s) \qquad E^\circ = 0.34\ V$$

In order for a combination of these two half-reactions to yield a positive potential, iron must be oxidized.

$$Fe(s) + Cu^{2+}(aq) \rightarrow Fe^{2+}(aq) + Cu(s) \qquad E^\circ_{cell} = ?$$

Using Equation 13.2,

$$E^\circ_{cell} = 0.34\ V - (-0.44\ V) = 0.77\ V$$

Copper is the cathode in this cell (because copper is reduced) and iron is the anode (because it is oxidized.)

(b). Find the two half-reactions in Table 13.1:

$$Ag^{+}(aq) + e^- \rightarrow Ag(s) \qquad E^\circ = 0.80\ V$$

$$Cu^{2+}(aq) + 2e^- \rightarrow Cu(s) \qquad E^\circ = 0.34\ V$$

From the values of the standard reduction potentials, we can see that copper is oxidized, and silver is reduced. We then use equation 13.2 to find the cell potential.

$$Cu(s) + 2Ag^{+}(aq)\ Cu^{2+}(aq) + 2Ag(s) \qquad E^\circ_{cell} = 0.46\ V$$

Discussion

The calculation of cell potential might seem more complicated if both reduction potentials are negative. If we keep in mind that the overall cell potential must be positive for a galvanic cell, then only one of the two choices will be correct, no matter what values are given.

Check Your Understanding

Suppose you wanted to make a 1.00 V cell using copper as one of the materials. Of the materials listed in Table 13.1, which would provide a cell potential closest to 1.00 volt when coupled with copper? What is the anode and what is the cathode in your cell?

Example Problem13.1, in which copper and iron are used to construct an electrochemical cell, points to the origin of galvanic corrosion. Thus far we have implied that the separation of the electrodes is something that is carefully constructed and uses a salt bridge. Deciding which electrode is the anode and which is the cathode, and identifying the cell reactions are relatively simple tasks. But in real world applications, the separation may be much more subtle.

As a starting point, let's consider the galvanic corrosion that can occur at a joint between copper and iron materials. The resulting degradation of the iron makes it obvious that iron acts as the anode. This means that the copper must be the cathode. But what is being reduced? It turns out that two reactions are involved at the cathode, and both involve the reduction of water. Which one will occur depends on the amount of oxygen present in the water and the cell potentials. If copper is not reacting, then is it even important that the material is copper? If it is water that is actually reacting, what would happen if some other metal such as zinc were present? Here is where the value of the standard reduction potentials becomes apparent.

Let's add the reduction of zinc to the list of half-reactions from Example Problem 13.1.

$$Fe^{2+}\,(aq) + 2e^- \rightarrow Fe(s) \qquad E° = -0.44\ V$$

$$Cu^{2+}\,(aq) + 2e^- \rightarrow Cu(s) \qquad E° = 0.34V$$

$$Zn^{2+}\,(aq) + 2e^- \rightarrow Zn(s) \qquad E° = -0.763\ V$$

If iron is paired with zinc in an electrochemical cell, which electrode will be the anode? The one that is the more negative—in this case zinc. So if zinc and iron are both present, iron does not corrode, because zinc serves as the anode. Copper by contrast is a better oxidizing agent than iron, and the iron will corrode via galvanic corrosion. That a material like zinc will prevent corrosion provides a common method of protecting iron. **Galvanized steel** includes a thin layer of zinc as a coating, making it notably less likely to rust.

An entire industry, the electroplating industry, has built up from the fact that coatings of metals on objects can provide useful properties, including corrosion resistance.

Non-standard Conditions

One thought that you may have about corrosion is that in real world applications we cannot expect standard conditions. What must we do in order to account for the differences that arise when standard conditions are not present? This is a very important question that has ties to thermodynamics. The equation that governs nonstandard conditions is called the **Nernst equation.**

$$E = E° - \frac{RT}{nF}\ln Q \qquad (13.3)$$

Here Q is the reaction quotient we worked with in Chapter 12: a ratio of concentrations of reactants and products, each raised to the power of its stoichiometric coefficients.

$$Q = \frac{[C]^c[D]^d}{[A]^a[B]^b} \qquad (13.4)$$

Among his many accomplishments, Faraday is responsible for many names used in electrochemistry. With help from a linguist friend William Whetwell he abandonned ideas such as eastode and westode and settled on cathode and anode.

F is Faraday's constant, and n is the number of electrons transferred in the redox reaction. Named after Michael Faraday, F has a value of 96,485 J $V^{-1}mol^{-1}$ or 96,485 C mol^{-1}. (Remember that 1J = 1 C V.) Equation 13.3 bears a resemblance to Equation 12.7, which we used to derive the relationship between ΔG° and the equilibrium constant. We can apply the Nernst equation to estimate the potential of the electrochemical system present in the corrosion of steel at more realistic concentrations. Example Problem 13.2 explores the use of the Nernst equation in an electrochemical cell modeling corrosion.

Example Problem 13.2

Suppose you worked for a company that designed the drive mechanism for large ships. The materials will obviously come into contact with environments that enhance the prospect of corrosion. To estimate the difficulties that corrosion might cause, you decide to build a model electrochemical cell utilizing electrolyte concentrations that might be present in your system when it is in service. Assume you have a cell that has an iron(II) concentration of 0.015 M, and the H^+ concentration is $1.0\text{x}10^{-3}$ M. The cell temperature is 38°C, and the pressure of hydrogen gas is maintained at 0.04 atm. What would the cell potential be under these conditions?

Strategy

This problem defines non-standard conditions that must be addressed using the Nernst equation. Virtually anytime you are given concentrations of electrolytes present in a cell (other than 1 M) you can expect to need this equation. This problem also presents the challenge of identifying the reactions used. Iron will be the anode, but we will need to scan the table of standard reduction potentials to identify the cathode reaction. The most likely suspect will be the reduction of H^+ to H_2. Once we know both half-reactions we can calculate the standard cell potential to fill in the appropriate values in the Nernst equation.

Solution

Anode reaction: $Fe^{2+}(aq) + 2e^- \rightarrow Fe(s)$ E° = - 0.44 V

Cathode reaction: $2H^+(aq) + 2e^- \rightarrow H_2(g)$ E° = 0.00 V

These reactions tell us two things: first, the standard cell potential will be

$$E° = 0.00\ V - (-0.44\ V) = 0.44\ V$$

and second, there are two electrons transferred in the overall redox reaction,

$$Fe(s) + 2H^+(aq) \rightarrow H_2(g) + Fe^{2+}(aq).$$

These facts, plus the values given in the problem and those of the constants allow us to use the Nernst equation to find the cell potential.

$$E = 0.44V - \frac{(8.314Jmol^{-1}K^{-1} \times 311K)}{(2 \times 9.6485 \times 10^4 JV^{-1}mol^{-1})} \ln\left[\frac{(0.015)^1(0.04)^1}{(0.0010)^2}\right] = 0.35V$$

Discussion

The correction made in the cell potential is rather small, even though we are far away from standard conditions. But what happens as the reaction proceeds? The concentration of H^+ gets smaller and Fe^{2+} concentration increases. So the potential is not a constant—the lnQ factor becomes larger (smaller number in the denominator, larger in the numerator) until the potential falls to zero (at least in principle). This rather complicated behavior is possible in a model that simplifies the real world application of corrosion, so we can see that for applications outside of the laboratory it can be vital to account for non-standard conditions.

Check Your Understanding

Suppose you want to construct a galvanic cell out of nickel and cadmium for a chemistry project. You take all the precautions necessary for dealing with a toxic heavy metal like cadmium, and you decide to use a 0.01 M solution of cadmium nitrate as one of the electrolytes. If you need a cell potential of 0.25V for this system, what concentration of $NiCl_2$ should you use? (For a follow-up web research project, compare this nickel-cadmium system with the one that is used in a commercial Ni-Cd battery.)

13.4 Cell Potentials and Equilibrium

Many metals are found in nature as oxide ores. Two common iron ores are hematite, Fe_2O_3, and magnetite, Fe_3O_4. Winning iron metal from its oxides in the refining process requires a great deal of energy. In the process of converting an oxide to the pure element, the entropy of the system decreases. The result is that refining is *not* a spontaneous process; the free energy change in the refining process is positive.

Corrosion can be thought of as the reverse of the process of refining a metal. When a metal corrodes, it is returning to its natural state, and it does so spontaneously. If we place an iron nail in a solution of aerated salt water, we know we will observe the formation of rust. Based on our experience from Chapter 10, the change in free energy of this process must be negative. Can we determine the free energy change associated with corrosion or with the operation of a galvanic cell?

Cell Potentials and Free Energy

Recall that free energy is related to the maximum possible amount of work that can be done by the system. In the case of a galvanic cell, the work done is electrical work, so we can state the relationship between free energy and the cell potential as

$$\Delta G° = -nFE°. \qquad (13.5)$$

Here we have replaced the charge, q, from Equation 13.1 with nF, because we often can determine moles more easily than charge in a chemistry-related problem. The minus sign is required, because a galvanic cell has a positive cell potential and spontaneously generates electrical work, and free energy must be negative (as required for a spontaneous process). This equation allows us to calculate the standard free energy change for an electrochemical reaction, provided we can look up standard reduction potentials. Let's see how this works.

Example Problem 13.3

Suppose we wish to study the possible galvanic corrosion between zinc and chromium, so we set up a cell, $Cr(s) \mid Cr^{2+}(aq) \parallel Zn^{2+}(aq) \mid Zn(s)$. What is the chemical reaction that takes place, and what is the standard free energy change of that reaction?

Strategy

To calculate the free energy change we must know two things: 1) the cell potential and 2) the number of electrons transferred in the reaction. Then we can simply use these values in Equation 13.5 to obtain the free energy change.

Solution

First we need the balanced chemical equation, which in this case can be written immediately as since two electrons are transferred (n =2):

$$Zn^{2+}(aq) + Cr(s) \rightarrow Cr^{2+}(aq) + Zn(s)$$

Now if we look up the standard reduction potentials, we find

$$Zn^{2+}(aq) + 2e^- \rightarrow Zn(s) \qquad E° = -0.763\ V$$

$$Cr^{2+}(aq) + 2e^- \rightarrow Cr(s) \qquad E° = -0.910\ V$$

According to Equation 13.2, the cell potential is

$$E°_{cell} = -0.763\ V - (-0.910\ V) = 0.147\ V$$

Inserting this into Equation 13.5,

$$\Delta G° = -nFE° = -2\ mol \times 9.6485 \times 10^4\ J\ V^{-1}mol^{-1} \times 0.147\ V = -2.84 \times 10^4\ J = -28.4\ kJ$$

Discussion

The most common error students make in this type of problem is forgetting to include n, the number of electrons transferred. Remember that even if you must multiply a chemical equation that represents the reaction given in the standard reduction potential by some value to match the number of electrons between oxidation and reduction, you do *not* multiply the standard reduction potential. Reduction potentials are not like thermochemical quantities in this regard.

Check Your Understanding

If this reaction could be reversed, what would be the value of the free energy change? What does this value tell you about the spontaneity of the reverse process?

The process of rusting is more complicated than this, so we must consider this number to be a rough approximation to the Free Energy change for the oxidation of iron to form rust.

How can we understand the spontaneity of corrosion using this type of calculation? What we must be able to do is identify the half-reactions involved and imagine them as a galvanic cell. Then the potential of this imaginary cell tells us the standard free energy change of the corrosion reaction. Let's return to iron rusting in aerated water, which is surely the most common example of corrosion. Iron is initially oxidized to its +2 state, and the oxygen in the water accomplishes the needed reduction according to the equation following half-reaction.

$$2H_2O + O_2 + 4e^- \rightarrow 4OH^-$$

Looking up values for these two reactions, we find that the cell potential of a galvanic cell comprised of them would be 0.81, so the free energy change is $\Delta G° = -313$ kJ. This large negative value tells us that equilibrium strongly favors the oxidation of iron metal.

Equilibrium Constants

We also learned in Chapter 12 that $\Delta G°$ is related to the equilibrium constant, K, by $\Delta G° = -RT \ln K$. By substituting the right hand side of this equation for $\Delta G°$ as given in Equation 13.5, we can obtain a relationship between cell potential and the equilibrium constant.

$$E° = \frac{RT}{nF} \ln K \qquad (13.6)$$

Here we have divided both sides of the equation by nF, to isolate the cell potential in the equation. This equation also looks much like the Nernst equation, Equation 13.4, and we can readily see how it arises. At equilibrium the free energy change is zero and the reaction quotient, Q, is the equilibrium constant, K.

We can gain some important insight into electrochemical reactions in general, and corrosion in particular, by replacing the natural logarithm with the common (base 10) log to obtain a slightly different form of the equation.

$$E° = \frac{2.303RT}{nF} \log K \qquad (13.7)$$

The factor of 2.303 is the result of changing to the common log function: ln(10) = 2.303. This change to the base 10 log function allows us think in terms of orders of magnitude. Both R and F are constants, so we can simplify Equation 13.7 by absorbing their values into the numerical term. And since we are most often interested in experiments carried out at the standard temperature of 25°C (298 K), we can insert that value as well. (Of course the resulting version of the equation will only be correct at this temperature.)

$$E° = \frac{0.0592V}{n} \log K$$

If we choose an electrochemical reaction that transfers only one electron at a potential of 1.0 V, and we determine the value of K, we obtain

$$K = 10^{\frac{(1)(1.0V)}{0.0592V}} = 7.8 \times 10^{16}$$

Thus, a modest cell potential of only one volt and only one electron transferred results in a very large value for K. Recall that large values of K indicate that equilibrium strongly favors products. In corrosion it is quite common to have reactions transferring two, three, or four electrons, which will tend to make the value of K even larger. This is illustrated in Figure 13.12.

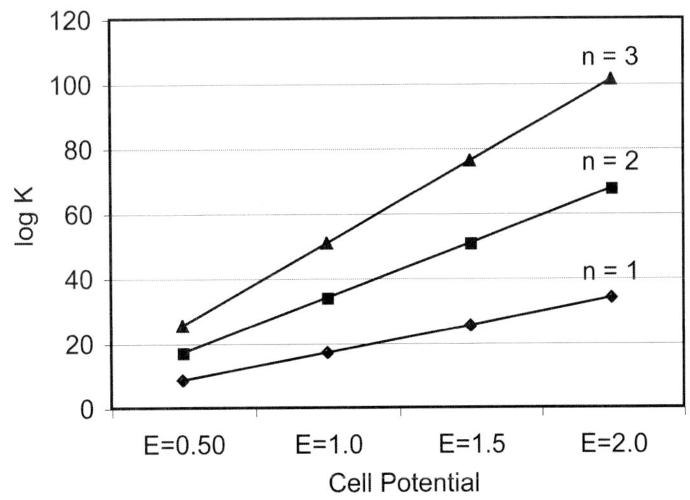

Figure 13.12 The variation of equilibrium constant with cell potential is shown. The different lines correspond to reactions involving the transfer of one, two, or three electrons, as indicated in the legend.

The vertical axis on this graph is the log of the equilibrium constant, K, so, for example, an increase of 20 is actually an increase in K of 10^{20}. Notice how either a greater cell potential or a larger number of electrons transferred quickly pushes the equilibrium constant to large values, indicating reactions that tend to proceed strongly towards products. From an equilibrium perspective, corrosion is likely to strongly favor the formation of the metal oxide products.

13.5 Batteries

Alessandro Volta, after whom the unit volt is named, was the first person to realize that the differences in chemical reactivity of metals could be used to generate electricity - his work was the precursor to the development of the battery..

The tendency of many electrochemical reactions to move towards complete formation of products presents a significant challenge in the case of corrosion, but it also provides opportunities for making use of these reactions in a positive way. Probably the most familiar example of this utility is the **battery**, a cell or series of cells that generates an electric current. Batteries are composed of many different materials and find many uses, but they share one common property—they provide a means by which we harness the electrical work of a galvanic cell and utilize it in a productive manner. We will see, however, that batteries have something else in common—they are susceptible to corrosion. So even when we intend to put electrochemistry to work for us, we need to think about corrosion. But first, let's look at batteries in a general sense, and then we will consider some specific examples.

Primary Cells

In our everyday experience, we know that some batteries are rechargeable while others are used once and thrown away. Single-use batteries that cannot be recharged are called **primary cells** or primary batteries. The most prevalent type of primary battery in use today is the *alkaline dry cell* battery. Flashlights, MP3 players, or hand-held video games all make widespread use of alkaline batteries. The manufacture of these batteries leads to stiff competition, as evidenced by the number of television commercials appearing for specific brands. Still, aside from some fairly minor engineering details, these alkaline batteries all rely on the same basic chemistry.

In an alkaline dry cell battery the anode is a zinc electrode and the oxidation half-reaction can be written

$$Zn(s) + 2\,OH^-(aq) \rightarrow Zn(OH)_2(s) + 2\,e^-.$$

The cathode is derived from manganese(IV) oxide and has the half-reaction

$$2MnO_2\,(s) + H_2O(l) + 2\,e^- \rightarrow Mn_2O_3(s) + 2OH^-\,(aq).$$

We can combine these two half-reactions to yield a net equation that represents the chemistry of an alkaline dry cell battery.

$$Zn(s) + 2MnO_2\,(s) + H_2O(l) \rightarrow Zn(OH)_2(s) + Mn_2O_3(s)$$

The essential design features of an alkaline battery are shown in Figure 13.13. The electrolyte used is KOH, but rather than having the electrolyte dissolved in liquid water it is present in the form of a paste or a gel—thus the term *dry cell*. The MnO_2 for the cathode is mixed with graphite to increase conductivity. The anode is a paste containing powdered zinc. (Powdering the zinc increases the surface area and improves performance.) The battery case also plays an important role in the design. Electrons generated by oxidation are "collected" by a piece of tin-coated brass connected to the bottom of the battery case. The remainder of the battery case is in contact with the cathode, but a protrusion on top makes it easier for a consumer to identify the positive terminal. All of these design elements are used to facilitate the basic chemistry that powers the battery.

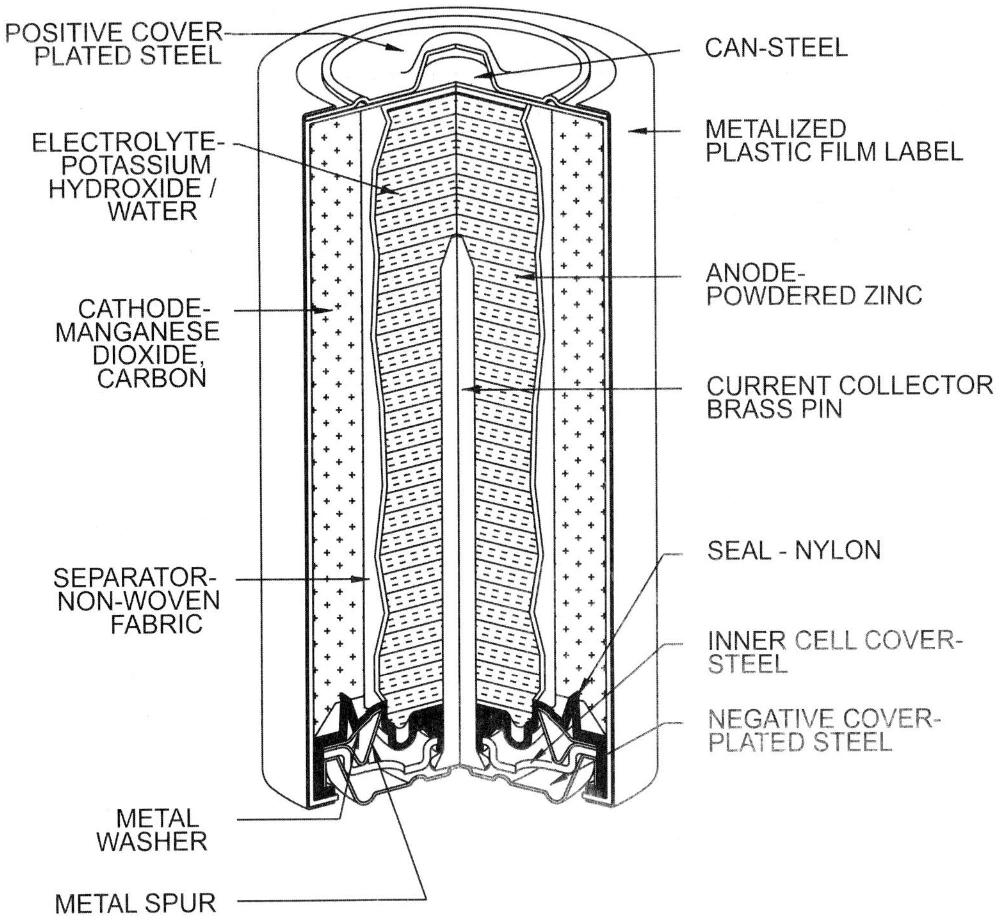

Figure 13.13 Construction of a typical alkaline battery is illustrated. (Figure courtesy of Eveready Battery Company.)

Battery research has long been important in US manufacturing. Thomas Edison held 147 patents for battery technology.

Other primary cells are also used as batteries for certain applications. In some cases the battery must be quite small. For a medical device, such as a heart pacemaker, such a battery should not only be small, but long lasting. Mercury batteries have come to fill this role. In the mercury battery, shown in Figure 13.14, zinc is the anode as in the alkaline dry cell.

$$Zn(s) + 2\ OH^-(aq) \rightarrow Zn(OH)_2(s) + 2\ e^-$$

The cathode, however, utilizes mercury(II) oxide.

$$HgO(s) + H_2O(l) + 2\ e^- \rightarrow Hg(l) + 2\ OH^-(aq)$$

A steel connection is used as the terminal for this reaction and is thus identified as the cathode of the battery. The important feature of this battery is that it provides a stable current and electrical potential for long periods, despite its small size.

Another intriguing form of alkaline battery with a zinc anode is the zinc/air battery, depicted in Figure 13.15. These batteries are sold as one-use, long lasting products for cellular phones. The cathode reaction in this battery is

$$1/2\ O_2\ (g) + H_2O(l) + 4\ e^- \rightarrow 2OH^-,$$

and the oxygen is obtained from air. With a significant portion of the working material obtained from the surrounding air, the prospect for lightweight batteries is intriguing for the zinc/air alkaline battery. There are, however, obstacles to the widespread use of this battery technology, because environmental factors, such as humidity, affect its long-term performance.

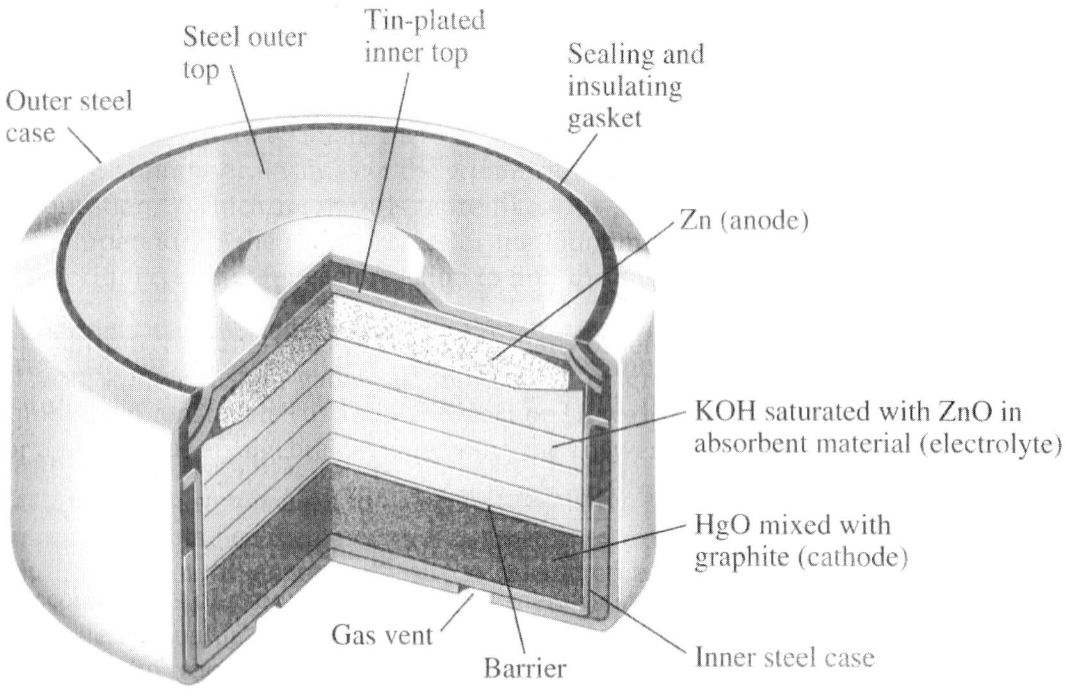

Figure 13.14 The mercury battery (also called a zinc-mercuric oxide cell) shown here has a voltage output that is extremely stable over time. These batteries are commonly used in devices where frequent battery changes would be a nuisance or hazard.

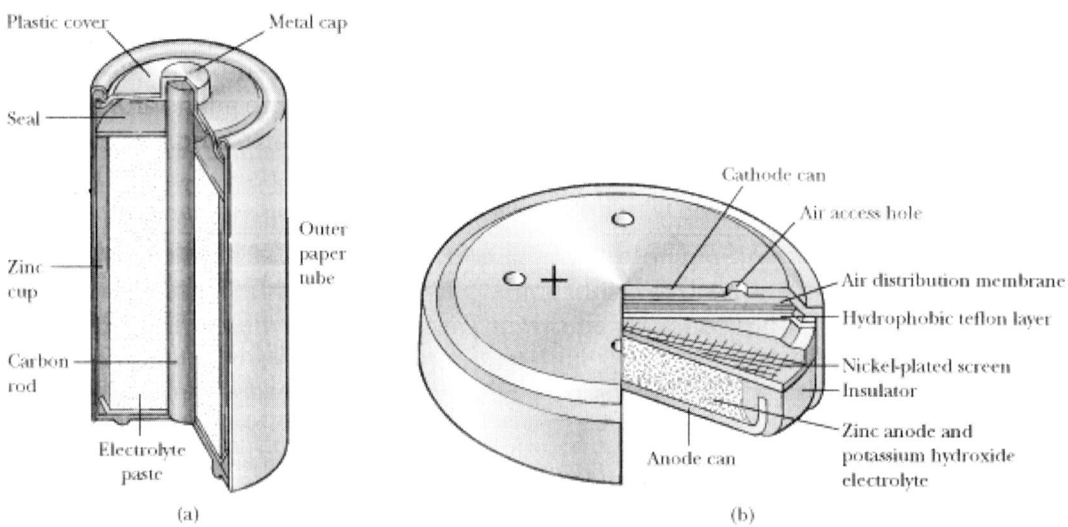

Figure 13.15 In a zinc-air battery, one of the reactants is oxygen from the surrounding air. As a result these batteries can offer a very attractive energy density. They can also be kept on hand in sealed bags and "charged" when needed by opening the bag to expose the battery to air.

Secondary Cells

Rechargeable batteries are increasingly common in consumer products. They represent examples of secondary cells or secondary batteries. With the increasing popularity of cellular phones and miniaturized computer devices such as palmtop and laptop computers, *nickel-cadmium*, or "ni-cad" batteries, shown in Figure 13.16, have become familiar to many consumers. The anode in the ni-cad battery is cadmium, reacting according to the following equation.

$$Cd(s) + 2\ OH^-(aq) \rightarrow Cd(OH)_2(s) + 2\ e^-$$

And the cathode reaction is best represented by the equation,

$$NiO(OH)(s) + H_2O(l) + e^- \rightarrow Ni(OH)_2(s) + OH^-\ (aq).$$

Ni-cad batteries can be expended and recharged many times, but they are sometimes susceptible to a performance-decreasing effect called "memory." The memory effect involves an inability to utilize all of the possible chemical energy of the battery unless it is completely discharged. So successive recharging ultimately results in shorter times before the battery appears to be "dead." The chemistry that gives rise to this effect involves the formation of a thin layer of material on the electrodes inside the battery that limit the oxidation-reduction reaction needed to generate electrical work.

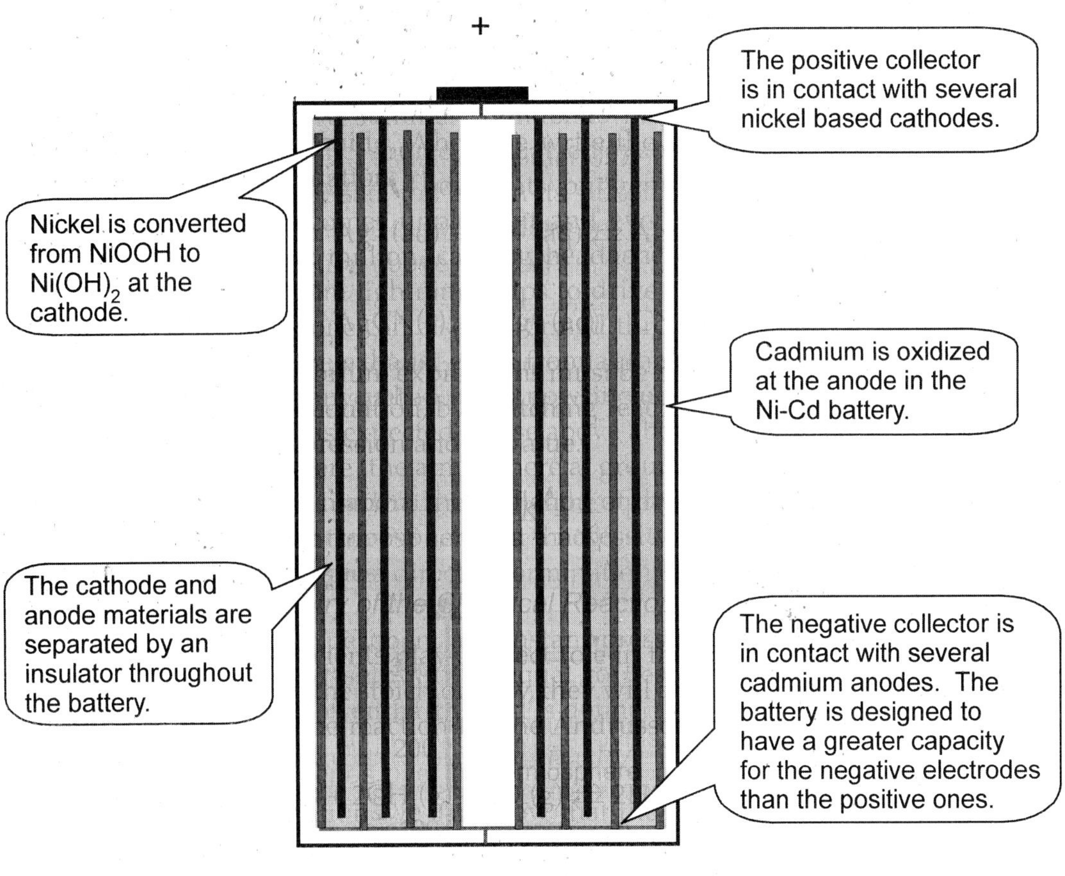

Figure 13.16 Important design features of a nickel-cadmium battery are shown. Transforming an electrochemical reaction into a useful consumer battery offers many engineering challenges.

Nickel-metal hydride batteries (Figure 13.17) find usage in many of the same devices as ni-cad cells. In this battery, the cathode reaction is the same, namely,

$$NiO(OH)(s) + H_2O(l) + e^- \rightarrow Ni(OH)_2(s) + OH^-\ (aq).$$

But the anode reaction is,

$$MH(s) + OH^-(aq) \rightarrow M + H_2O(l) + e^-,$$

where M stands for some metal or metal alloy. Some of the alloys used in commercially available nickel-metal hydride batteries contain as many as seven different metals, and the reason for the strong performance of these alloys is not yet well understood.

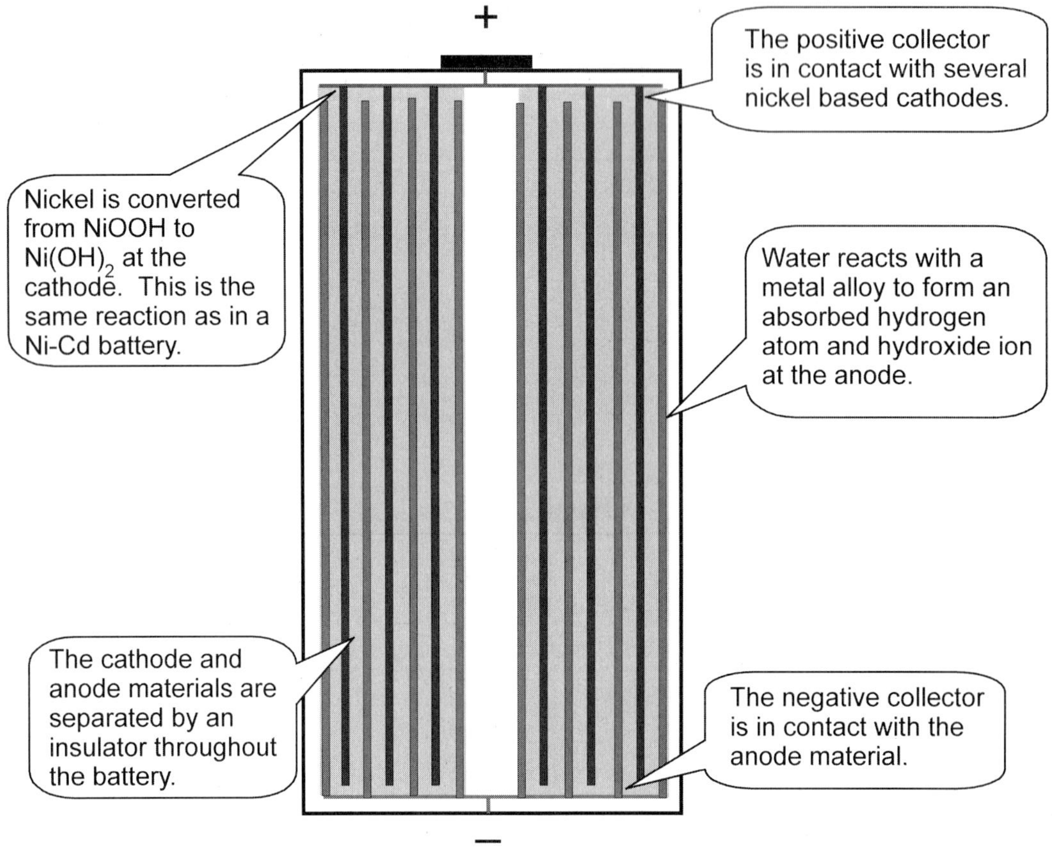

Figure 13.17 Nickel metal hydride batteries have become a popular option for rechargeable cells. The design is quite similar to the NiCd cell, but nickel metal hydride cells are less prone to memory effects over the life of the battery. Such advances in battery technology are driven by rising demand for portable electronic devices like digital cameras.

While these batteries are important in terms of their utility in personal electronics, they are not the most widely selling rechargeable batteries. That distinction still belongs to the *lead storage battery* in automobiles, which have been using the same technology for roughly 80 years. The anode of this battery is

$$Pb(s) + HSO_4^-(aq) \rightarrow PbSO_4(s) + H^+(aq) + 2\,e^-,$$

and the cathode is

$$PbO_2(s) + 3H^+(aq) + HSO_4^-(aq) + 2\,e^- \rightarrow PbSO_4(s) + 2H_2O(l).$$

While these equations do not show it explicitly, the electrolyte for this battery is sulfuric acid. We can infer this to be the case because the sulfate ion that might be present in the equations appears under the highly acidic conditions of the battery as hydrogen sulfate ions (HSO_4^-). The importance of the sulfuric acid in these batteries results in their sometimes being referred to as *lead-acid batteries*.

The cell designated by the reactions given above generates a potential of nearly 2.0 V. Thus to obtain a standard 12 V automobile battery there are six cells wired in series, so that their voltages are additive. Figure 13.18 shows a diagram of a lead storage battery. These batteries are rechargeable, because the lead sulfate product tends to adhere to the electrode surfaces. Passing electric current from the alternator of the automobile to drive the redox reactions in the non-spontaneous direction recharges the battery. Driving the reactions in a battery in the reverse direction to recharge it sometimes requires a larger voltage than the discharge voltage to assure that all chemical species are returned to their original state. Eventually, as the battery experiences mechanical shocks from bumping and jarring during normal use of a car, the lead sulfate falls away from the electrodes, and the battery cannot be recharged and must be replaced.

Lead-acid storage batteries account for 88% of the lead consumed in the US annually.

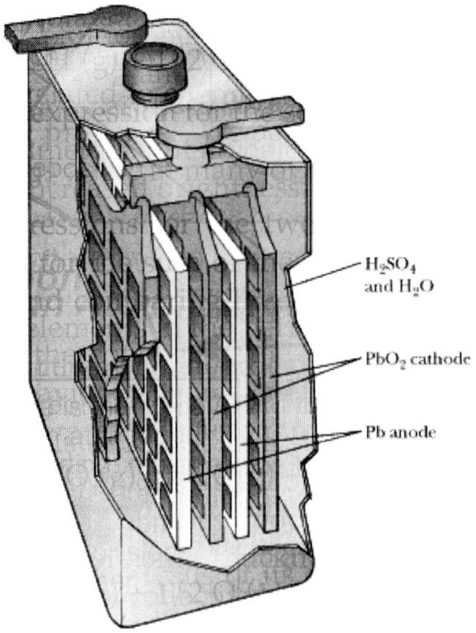

Figure 13.18 The lead-acid storage battery consists of Pb anodes alternating with PbO_2 cathodes, all immersed in sulfuric acid. While the underlying chemistry has remained the same over many years, engineering advances have drastically improved the lifetime and reliability of the modern lead-acid battery.

Limitations of Batteries

Battery manufacturing represents a major market, and the companies that produce products for this market carry out considerable research into the technology of batteries. Longer lasting, better performing, or lighter batteries are at a premium. Aside from the expenditure of the materials that compose the cathodes or anodes, the next most common chemical reason for the loss of performance of a battery is corrosion. Perhaps you forgot to remove the batteries from a radio or flashlight that you did not use for many months. When you looked into the battery compartment, you likely found that the batteries had corroded extensively, perhaps allowing fluids to leak out of the cells. Optimum conditions for corrosion are readily observed in batteries. In fact, older dry cell batteries were even more susceptible to corrosion effects than the common alkaline battery of today. How do manufacturers attempt to limit the performance diminishing effects of corrosion on batteries? One method is the protective plating of materials used in the battery by another electrochemical process called **electrolysis.**

13.6 Electrolysis

From our consideration of equilibrium, we know that chemical reactions may proceed in the direction of either reactants or products. In our previous discussions, however, we found ourselves merely observing the direction that nature would proceed to establish equilibrium. Perhaps, if the mathematical signs on ΔH and ΔS were the same, we could manipulate the direction of equilibrium by raising or lowering the temperature. In acid-base chemistry or in the formation of precipitates, however, it was not apparent that we could overcome free energy and make the reaction go the way we would like. In contrast, redox reactions *can be* manipulated in a fairly simple way. While the spontaneous direction of a redox reaction may be turned into a galvanic cell to provide an electrical current, we can use an external electric current to force a redox reaction in the non-spontaneous direction. **Electrolysis** is the process of passing an electric current through an ionic solution or molten salt to produce a chemical reaction.

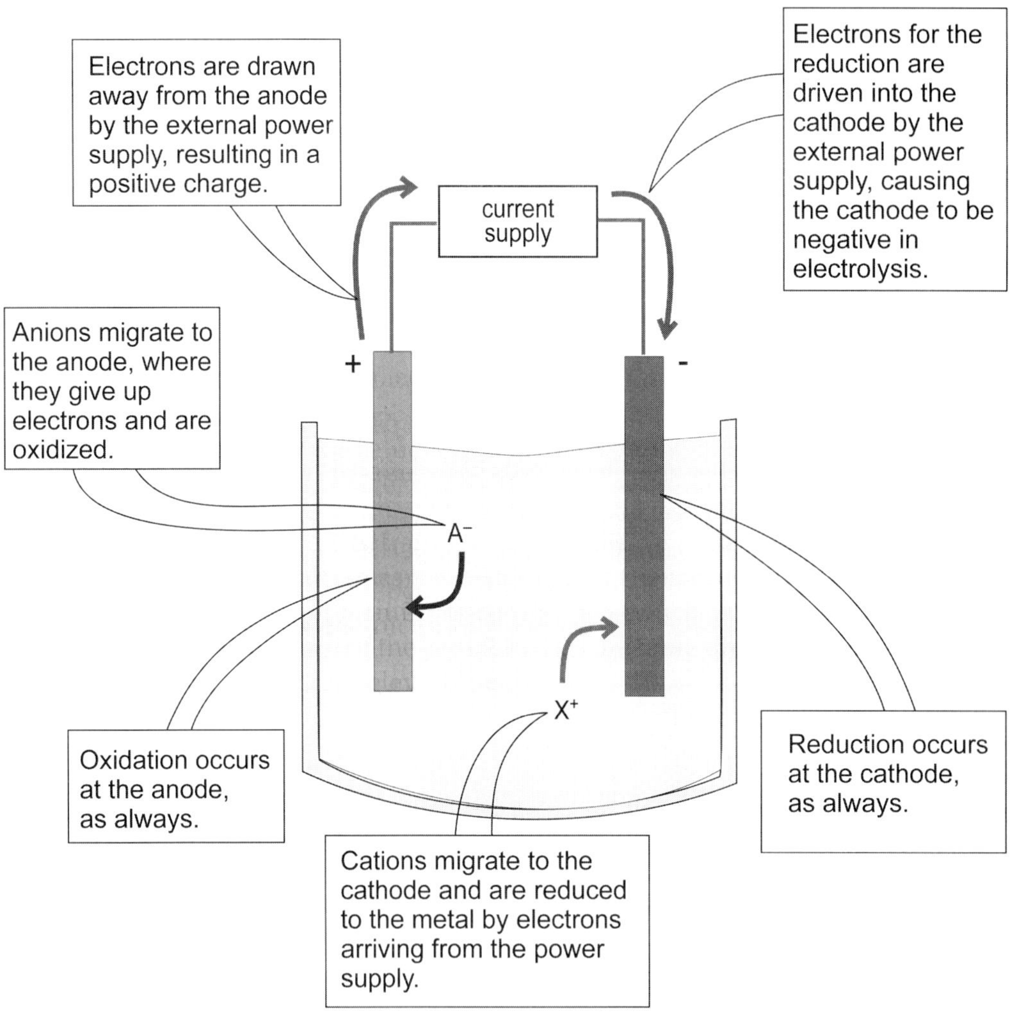

Figure 13.19 In electrolysis, and external source of current drives a redox reaction that would otherwise not be spontaneous. Electrolysis is useful for electroplating as well as for the purification of several metals.

Electrolytic cells can be divided into two categories based on the nature of the electrodes used. If the electrodes are chemically inert materials that simply provide a path for electrons, the process is called **passive electrolysis**. When the electrodes are part of the electrolysis reaction it is called **active electrolysis**. Passive electrolysis is used in industry to produce metals that corrode easily. Active electrolysis is used to plate materials to provide resistance to corrosion. We will consider one example of each form of electrolysis, but first we need to address the issue of electrical polarity in electrolysis cells.

Electrolysis and Polarity

Electrolysis changes the polarity of the electrodes in a system. Instead of the reactions that would occur spontaneously at the electrodes, the external power supply causes the reverse reactions to occur. Thus, for reduction to occur electrons are forced to the cathode in an electrolysis reaction. The cathode is always the site of reduction, but in electrolysis it is negatively charged, because the external power supply forces electrons (and their negative charges) to the location of the reaction. The anode is still the site of oxidation, but in electrolysis it is positively charged. In forcing the oxidation of the material, the external power supply pulls the electrons away from the region of the oxidation reaction. Figure 13.19 illustrates electrode charges in electrolysis.

Passive Electrolysis in the Refining of Aluminum

Aluminum metal finds widespread use as a structural material because it is resistant to catastrophic failure due to corrosion. This utility does not imply, however, that aluminum does not corrode. Pure aluminum metal reacts with oxygen very quickly, as we pointed out back in Chapter 1. However, when the corrosion occurs, a thin layer of Al_2O_3 is formed at the surface of the metal. The oxide surface is strongly bonded to the pure aluminum underneath, protecting it and allowing it to be useful in a variety of applications. The ready formation of Al_2O_3 can be predicted from the standard reduction potential of the Al(s) | Al^{3+} half-reaction. At –1.66 V, aluminum is a considerably stronger reducing agent than iron, whose Fe(s) | Fe^{2+} half-reaction has a potential of –0.44 V. If the reaction to form aluminum ions is thermodynamically favorable, the reverse reaction will not be spontaneous. (Recall from Chapter 1 that the refining of aluminum from its ore is so difficult that until the mid-19th century aluminum was considered a precious metal, even though large sources of Al_2O_3-containing ore were known.)

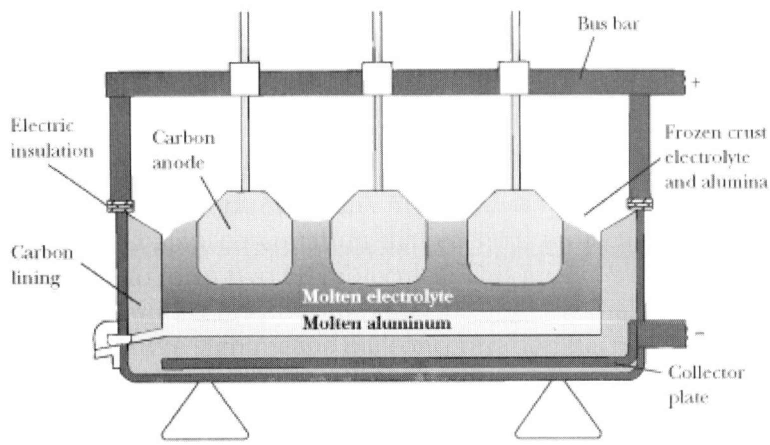

Figure 13.20 The Hall-Heroult process involves the electrolytic refining of aluminum from Al_2O_3. Carrying out the electrolysis is a bath of cryolite (Na_3AlF_6) lowers the required temperature significantly, making the process much more economical.

Electrolysis provides the means to overcome the non-spontaneous nature of the reaction to separate aluminum from oxygen. The industrial process used to accomplish this task was discovered simultaneously in the United States by Charles Martin Hall and in France by Paul Heroult in 1886. The Hall-Heroult process uses carbon electrodes as inert sites for passive electrolysis as illustrated in Figure 13.20. There are three components of this process that are important to its usefulness. The first is the inclusion of Na_3AlF_6, cryolite, in the reactor, which lowers the melting temperature from 2045°C for Al_2O_3 to a temperature closer to 1000°C for the mixture. Lower temperatures can be achieved more readily and therefore allow the refining of aluminum to be economically viable. Second, the process requires a very high current in order to produce substantial amounts of aluminum. Such vast amounts of electricity became available in the large hydroelectric projects in North America early in the 20th century. Finally, although the precise reactions are not known, the electrolysis reactions ultimately produce oxygen gas. This adds additional cost to the production of aluminum, because the oxygen slowly reacts with the carbon in the graphite rods to form carbon dioxide. The destruction of the graphite electrodes over time means that they must be replaced on a regular basis. Still, aluminum is a familiar material in everyday life because the electrolysis of aluminum-containing ore is both possible and economically viable. Because of the enormous electricity requirement for this process, recycling of aluminum is significantly less expensive than the recovery of aluminum from its ore. Recycling one can of aluminum saves enough electricity to light a 100-watt bulb by which we can study chemistry for four hours!

Cryolite is referred to as a flux - a substance that lowers the melting point of a material.

Active Electrolysis and Electroplating

In many devices, a thin coating of one metal on top of another is needed. In some cases this coating is cosmetic, while in others it is vital to some functionality of the coated piece. The process of depositing a thin coat of a metal using electricity is called **electroplating**. But where do we find electroplated products? One place is in the batteries we discussed in the last section. The electron collector that provides the external connection for the anode is electroplated. The metal coating gives this piece of a battery both corrosion protection and desirable conducting properties. Small parts such as this are plated in bulk by placing them in a barrel and then placing the barrel in the electrolyte solution. How does this process work? Let's consider the plating of silver to make parts for electrical devices.

You can see electroplating equipment at several manufacturers' web sites, including www.proplate.com

The solution from which silver is plated normally contains $CN^-(aq)$ ions, which are present to form complexes with the silver. The need for uniform coatings makes this step important. The anode and cathode reactions are as follows.

Anode: $Ag(s) + 2CN^-(aq) \rightarrow Ag(CN)_2^-(aq) + e^-$

Cathode: $Ag(CN)_2^-(aq) + e^- \rightarrow Ag(s) + 2CN^-(aq)$

That the reactions are the reverse of each other is common in electroplating operations. While a galvanic cell cannot be constructed out of such a combination—it would have a cell potential of zero—the fact that electrolysis is driven by an external current means the cell potential is not crucial. Most electroplating processes are designed to run at low voltages, however, so a zero cell potential is advantageous. The basic design of a barrel system for silver plating is illustrated in Figure 13.21. The barrel is rotated during the plating process so that all the small parts it contains obtain electrical contact with the cathode, thereby becoming part of the cathode themselves during the time of contact. The overall process normally involves several preliminary steps, not shown here, to prepare the parts for the silver plating. For many applications, an undercoating of copper is plated first, and then silver is plated over the copper. Silver plating is important industrially, because silver is a good conductor, is resistant to corrosion, and is generally considered attractive.

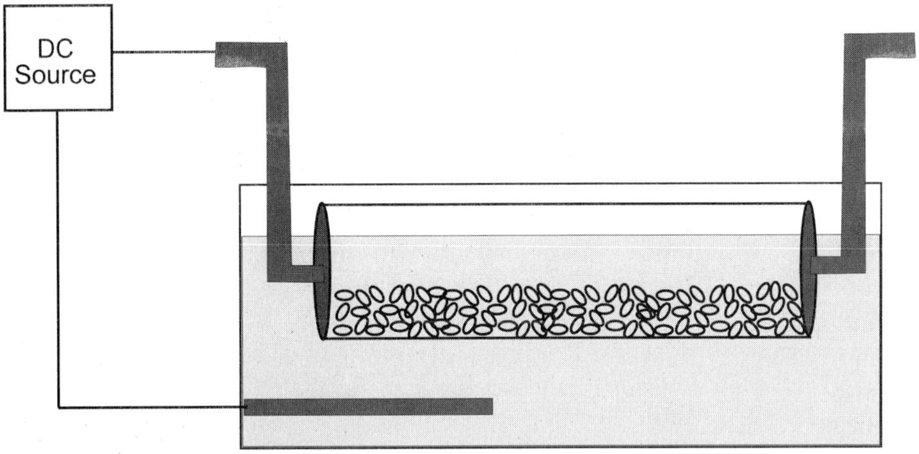

Figure 13.21 Barrel plating is often used to apply coatings to small parts. The barrel cathode rotates during electrolysis to assure that all of the parts come into contact with the electrode.

The use of plating to prevent corrosion in electrical apparatus shows how important an understanding of corrosion is. Because devices are built from a multitude of parts, there are inevitably places where different metals will come into contact. As we have already seen, this situation can lead to galvanic corrosion. To prevent galvanic corrosion, but not hinder the electrical performance of the device, parts are often plated. In some cases a limited area of a part is plated: nuts might be plated with silver on the inside of the threads only, for example. Doing so hinders galvanic corrosion between the nut and the bolt, because the metals that are in contact are corrosion resistant. Since silver coatings are relatively expensive, plating only the inside reduces the cost. The external area of the nut does not make contact with other metals, and therefore does not need plating.

13.7 Electrolysis and Stoichiometry

Silver plating the inside of a nut suggests the importance of knowing just how much material is being deposited on the surface. If the silver coating were to become too thick, the nut might not "fit" the bolt—the threads might become too narrow. So in electroplating, it can be vitally important to know precise amounts of materials. Fortunately, the flow of electrons in a current provides a ready source of information. Let's review some facts about electricity, and then apply those facts to electrolytic cells.

Electrolysis has important historical implications for chemistry as well. It was through electrolysis that the role of electrons in chemical bonding began to be elucidated.

Current and Charge

When the current is measured in an electric circuit, the observation is the flow of charge over a period of time. The base unit of current, the ampere (A), is a combined unit defined as a Coulomb per second: $1\ A = 1\ C\ s^{-1}$. Devices called amp-meters (or ammeters) measure current. If a known current passes through a circuit for a known time, the charge can be easily calculated.

$$\text{Charge} = \text{Current x Time}$$

$$Q = I \times t \tag{13.8}$$

Typically Q is in coulombs, I in coulombs/second, and t in seconds.

We have already introduced the Faraday constant which is the charge of one mole of electrons, F = 96,485 C mol^{-1}. So if we know the charge that passes through an electrolytic cell, we know the number of moles of electrons that pass. And assuming we know how many electrons were required to reduce each metal cation, it is simple to calculate the number of moles of material plated.

Example Problem 13.4

In a process called "flash" electroplating, a current of 2500 A is passed through an electrolytic cell for 5.0 minutes. How many moles of electrons are driven through the cell?

Strategy

Since we know both the current and the time that it was applied, we can use the relationship between charge and current in Equation 13.8 to obtain the charge. Then Faraday's constant will let us convert that charge into moles of electrons. We must be careful with units, however, because the ampere is charge per second and time is given here in minutes.

Solution

$$Q = 2500 \text{ A x } 300 \text{ s} = 7.50\text{x}10^5 \text{ C}$$

Now use Faraday's constant:

$$7.50\text{x}10^5\text{C} \times \left(\frac{1\text{mol e}^-}{96485 \text{ C}}\right) = 7.77\text{mol}$$

Discussion

This two-step manipulation is really a stoichiometry problem, as it allows us to find the number of moles involved in a chemical reaction. The central importance of moles in stoichiometry problems is well established at this point in our study of chemistry. Just as for other types of stoichiometry problems that we've considered, we sometimes need to use these relationships "in reverse." If, for example, you need to know how long to run a given current to obtain a desired amount of plated material, these relationships would still be used, only the known quantity would be moles of electrons.

Check Your Understanding

You need 0.56 mol of electrons to deposit a thin covering of silver on a part you are using for a prototype machine. How long would you need to run a current at 5.0 A to obtain this number of moles?

While current can be measured readily, the usage of electricity is often considered in terms of power. Electrical power is a rate of electricity consumption, and utilities charge based on consumption. The SI unit for power is the watt.

$$1 \text{ watt} = 1 \text{ J s}^{-1}$$

Thus wattage is the rate of energy usage in J s^{-1}. To determine energy used we multiply this rate by the time over which it occurs. To obtain numbers of a convenient magnitude, electrical utilities normally determine energy consumption in terms of kilowatt-hours, kwh. One kilowatt-hour is equal to 3.60×10^6 J. Back in Chapter 2 we noted the relationship between electrical potential and energy.

$$1\,\text{J} = 1\,\text{C V}$$

So a second perspective on electrolysis, one that is very important for an electroplating company because of its relationship to cost of production, is energy expenditure. From this relationship we can see why electroplating companies prefer low voltage applications. Example Problem 13.5 investigates the energy cost of electrolysis.

Treatment of wastes is another source of costs that we will not address with stoichiometry at this point - though the removal of wastes clearly involves stoichiometry.

Example Problem 13.5

Suppose a batch of parts is plated with copper in an electrolysis bath running at 0.15 V and 15.0 A for 2 hours. What is the energy cost of this process if the electric utility charges the company $0.05 per kwh?

Strategy

We can determine the energy expended because we know the current, time, and voltage. The current multiplied by the time gives us the charge, which when multiplied by the voltage yields the energy. Once we know the energy expenditure we can convert the value we calculate (in J) to kwh to obtain the cost of the electricity.

Solution

$$Q = I \times t = 15.0\text{ C s}^{-1} \times 7200\text{ s} = 1.08\times10^5\text{ C}$$

$$\text{Energy} = \text{charge} \times \text{voltage} = 1.08\times10^5\text{ C} \times 0.15\text{ V} = 1.62\times10^4\text{ J}$$

Now convert to kwh and determine the cost.

$$1.62\text{x}10^4\ \text{J}\times\left(\frac{1\text{ kwh}}{3.60\times10^6 J}\right)\times\left(\frac{\$0.05}{1kwh}\right)=\$0.00025$$

Discussion

This value seems small, and it points out the importance of running electroplating at low voltages. This example is not representative of costs in the electroplating industry, but it does help us to get a handle on some of the variables that industry must consider when implementing electrochemical techniques.

Check Your Understanding

If you need 25,000 A of current to carry out an electrolysis for 15 minutes, what voltage would you need if you use 1.7 kwh for the process?

Calculations Using Masses of Substances Involved in Electrolysis

Now that we have reviewed the important electrical concepts involved in quantitative applications of electrolysis, we can put these factors together with our customary approach to stoichiometry problems. In problems such as these, we generally encounter one of two questions. First, how much material is plated, given a specified current or electrical energy expenditure? And, second, how long must a given current be passed through the cell to yield a desired mass of plated material? Example Problems 13.6 and 13.7 illustrate these two types of problems.

Example Problem 13.6

An electrolysis cell that deposits gold (from $Au^+(aq)$) operates for 15 minutes at a current of 2.3 A. What mass of gold is deposited?

Strategy

As in any stoichiometry problem we need a balanced chemical equation, so we will start by writing the half-reaction for gold reduction. To determine the mass of gold deposited, we must calculate the number of moles of electrons used from the current and the time. We can use the half-reaction to obtain a mole ratio and convert moles of electrons into moles of gold. Once we have moles of gold, we convert to mass using the molar mass, as we have done many times in stoichiometry problems.

Solution

First write the balanced half-reaction.

$$Au^+(aq) + e^- \rightarrow Au(s)$$

Now calculate moles of electrons based on current and time.

$$Q = I \times t = (2.3\ C\ s^{-1})(900\ s) = 2.07 \times 10^3\ C$$

$$2.07 \times 10^3\ C \times \left(\frac{1\ mol\ e^-}{96485\ C}\right) = 2.15 \times 10^{-2}\ mol\ e^-$$

Now we note that the mole ratio of electrons to gold is 1:1, so we also have 2.15×10^{-2} mol of Au.

$$2.15 \times 10^{-2}\ mol\ Au \times 197\ g\ mol^{-1} = 4.23\ g\ Au$$

Discussion

The latter part of this problem looks familiar, as we have been carrying out stoichiometry problems similar to this since Chapter 3. The only changes from our previous experience were the use of a half-reaction (so we can see the electrons explicitly) and the use of some physical laws that govern electricity to obtain moles of electrons in the first place.

Check Your Understanding

Suppose a machine part requires 400 mg of copper to provide a coating that is thick enough for a certain electrical function. If you run an electrolysis using copper(II) ions for 10 minutes at a current of 0.5 A, will you meet the requirement?

Now we can consider an example where we approach the problem from the other direction.

Example Problem 13.7

Suppose you have a part that requires a tin coating. To achieve an adequate coat you know that you need to deposit 3.6 g of tin. If your electrolysis cell (using Sn^{2+}) runs at 2.0 A, how long must you operate the cell to obtain the desired coating?

Strategy

This problem is different from the last one, since it requires that we use the balanced half-reaction to relate moles of electrons and moles of a desired substance, in this case tin. Because we are given a mass of tin required, we can obtain the moles of tin, and that will tell us the moles of electrons required. We then determine the charge of those electrons with the Faraday constant, and we can use our understanding of electricity to get the time needed at the given current.

Solution

$$Sn^{2+}(aq) + 2\,e^- \rightarrow Sn(s)$$

$$3.6 \text{ g Sn} \times \left(\frac{1 \text{ mol Sn}}{118.7 \text{ g Sn}}\right) \times \left(\frac{2 \text{ mol e}^-}{1 \text{ mol Sn}}\right) \times \left(\frac{96485\text{C}}{1 \text{ mol e}^-}\right) = 5.85 \times 10^3\text{C}$$

Now recall that Q = I x t, so t = Q/I.

$$t = \frac{5.85 \times 10^3\text{C}}{2.0\text{Cs}^{-1}} = 2925\text{s} = 48.8\text{min}$$

Discussion

Perhaps this problem seems more complicated than some other stoichiometry problems, because we divided it into two steps. Still, the fundamental components of the problem are the same as always. We used a balanced chemical equation to relate moles of a substance (tin) to moles of electrons. Then we used the relationship, Q = I x t, to find the time.

Check Your Understanding

Suppose you need a coating of rhodium (a precious metal) on a part that is to be a mirror in a laser. Because rhodium is so expensive you wish to deposit only 4.5 mg on the surface of the part. If you run an electrolysis apparatus for 5.0 min, what current do you need to deposit the desired amount from a solution containing Rh^{3+} ions?

An additional type of calculation that engineers and chemists sometimes need to make in electroplating companies is to determine the amount of material based on a coating thickness. In such a case, the surface area of the part must also be known. Multiplying the coating thickness by the surface area provides a volume that, when multiplied by the density of the material, provides a mass. When considering the plating of parts of a battery to protect the battery from corrosion, these calculations are made more complicated by the fact that in barrel plating hundreds or even thousands of parts are plated at the same time. Still, many electroplating companies establish a significant portion of their business from the need to protect parts from the effects of corrosion, so this type of quantitative information is quite important.

Additionally, factors such as the rate of coating affect the quality of the product - so plating does not always use the most simple reactions to provide the metals. For example cyanide solutions are used when silver is plated to control the rate of deposition of silver metal.

13.8 Insight into Corrosion Prevention

The emphasis on corrosion throughout this chapter has revealed several important factors. First, the thermodynamics of corrosion, with its large, negative free energy for the reaction, suggests it will be a pervasive reaction. Second, we have seen that the nature of galvanic corrosion can influence what materials corrode, and in principle we should be able to use that information to protect a material like iron. Finally, we have noted that some materials, like aluminum, corrode readily, but the corrosion product (in the case of aluminum, Al_2O_3) ultimately forms a protective layer that eliminates further corrosion. These fundamental observations have been applied to reducing corrosion in a number of ways, some of which we will survey here.

Coatings

The most common means of preventing corrosion is to coat the material that might corrode. We have already noted the use of electroplating in the discussion of electrolysis. Another widely used coating practice is painting. How does painting affect corrosion? It protects the underlying material, such as iron, from exposure to water and oxygen. Paint coatings, however, are capable of being scratched, which can enhance rusting, as shown in Figure 13.22. Where the scratch exposes iron, the system is in contact with aerated iron, and the reduction of oxygen occurs at the iron surface, which acts as the cathode. Water that moves up under the paint is not aerated, so it only provides an electrolyte for the oxidation of iron, but the net result is still more corrosion.

Figure 13.22 The photo shows pronounced corrosion at an 'X' scratched into a paint layer on a metal plate. This was part of a NASA study assessing the usefulness of various anti-corrosion coatings. (http://corrosion.ksc.nasa.gov/)

Because paint is such a common means of combating corrosion, many paint companies add chemicals whose role in the mixture is to inhibit corrosion. Twenty rust inhibitors are listed by a professional society of industrial painters as environmentally approved additives for paints. While the specific chemical content may vary, most of these corrosion inhibitors contain one of the following ions: phosphate, borosilicate, chromate or phosphosilicate. The precise mode of corrosion inhibition depends on several factors, but all of these ions, when part of a paint coating, form compounds with oxidized iron that inhibit further formation of rust. Much like an oxide coating on the surface of aluminum prevents ongoing corrosion, these coatings stop the corrosion of the iron after only a small amount of material has reacted. This process is called **passivation**.

The business of paints and coatings is very competitive, with many chemical manufacturers specializing in compounds for this industry. The addition of a corrosion inhibitor is often only a minor factor in the overall success of the paint in protecting structural steel or iron. The ability of a paint to adhere to the metal is also quite important, as are factors that influence its ease of application and durability. Formulators, chemists whose job it is to devise improved mixtures of substances in products, spend millions of person-hours annually trying to devise improved paints that will better combat corrosion.

For automobiles, the process of applying paint has several intensely competitive factors. If you tour an automobile manufacturing plant, it's quite possible that the only part of the construction of the car that you won't see will be the painting! Ways to obtain paint coatings that protect against scratches and rust, but also appeal to consumers, are closely guarded industrial secrets. One common method of painting used in the automobile industry is electrospray painting, in which adhesion and coating of the paint is enhanced by holding the metal part at an electrical bias. Powdered pigments are sprayed on the part, and they adhere because of electrostatic attraction. Then the part is "baked," meaning it is heated past the melting point of the powdered pigment, and the resulting layer is both strongly adhered to the part and very durable. Formulating powdered pigments that include corrosion inhibiting chemicals remains an area of active research in the chemical coatings industry.

Cathodic Protection

The fact that some materials are more easily oxidized than iron provides a way to intentionally construct galvanic corrosion conditions that serve to protect the iron. If we choose a metal, such as magnesium, whose reduction potential is more negative than that of iron, the magnesium is oxidized, and iron is reduced.

$$Fe^{2+} + 2e^- \rightarrow Fe \qquad E° = -0.41\ V$$

$$Mg^{2+} + 2e^- \rightarrow Mg \qquad E° = -2.39\ V$$

If we wish to prevent iron from corroding, we can use a piece of magnesium as a **sacrificial anode**. Connecting the piece of magnesium to the iron forces the iron to be the cathode, thereby preventing the oxidation of iron. This process is called **cathodic protection**. By making the iron a cathode, we assure that it cannot be an anode, and so it will not corrode. To be effective, the sacrificial anode must be replaced periodically, but this method finds many uses, including the protection of iron or steel pipelines (Figure 13.23).

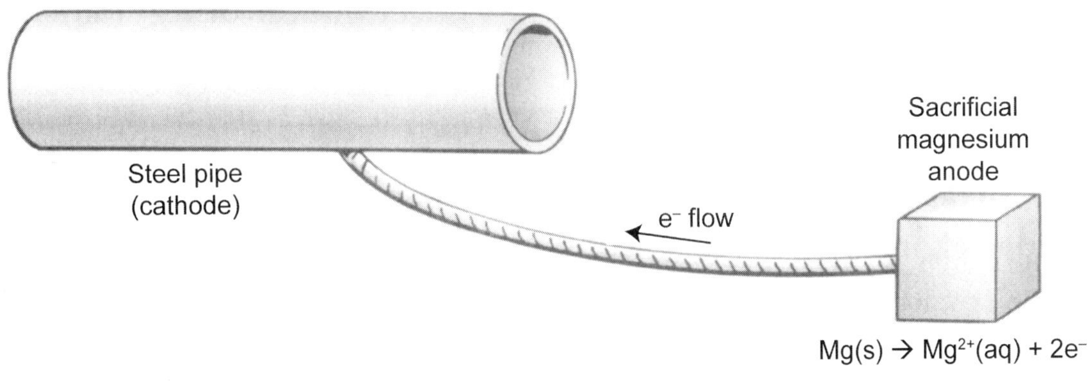

Figure 13.23 Sacrificial anodes are one effective method of corrosion prevention. An unprotected iron or steel pipe buried in the ground would be at high risk for corrosion. By connecting the buried pipe to a metal like magnesium, which is more easily oxidized, a galvanic cell is created with the pipe as the cathode. In the case of a buried pipe, the soil itself serves as the electrolyte. The anode is called "sacrificial" because it will be eaten away over time by oxidation. But replacing the anode – which may be nothing more than a metal block or stake – is much easier than replacing the buried pipe would be.

The fundamental concepts of cell potential are used in other forms of rust protection as well. One product to protect automobiles from rust uses wires that run from the battery to the steel body of the car, thus increasing the electric potential of the steel. The "cell voltage" of the corrosion cell on the body of the car is not sufficient to overcome the externally applied voltage, and the car does not rust.

Preventing Corrosion in Space

One issue that must be controlled if a manned mission to Mars is ever accomplished is corrosion control, so crews can stay in space with safe equipment for a trip of that duration.

Most of NASA's anti-corrosion efforts are concentrated on the launch site, where the high temperatures of the launch and the humid coastal atmosphere encourage corrosion. However, corrosion is a concern in other aspects of the space program, as well. Batteries used in the International Space Station (ISS) must be prevented from corroding in the ISS's artificial, earth-like atmosphere; NASA's mission of planetary exploration requires that scientists know the corrosivity of a planet's atmosphere before a craft can safely land. Venus, for example, has a highly corrosive atmosphere that makes lander design very difficult.

Since the large negative free energy change of corrosion is favored by thermodynamics, it will undoubtedly remain a problem in several areas of modern society. Many scientists are employed in research to find ways to hinder corrosion, either by slowing its kinetics, or by coupling additional chemical reactions to the corrosion reactions to make them less thermodynamically favorable. The details of such schemes can become quite complex, but are based on the general concepts of electrochemistry that we have explored in this chapter.

Focus on Problem Solving

Question

You are working on the design of a battery and need to have a small button plated with silver (because it has useful conductive properties and is corrosion resistant). You determine for the design that a coating of 1.5 mm will suffice. What would you have to look up to devise an initial cost estimate for this process?

This is an electrolysis question from a practical perspective. Cost versus coating thickness can be determined by finding out what materials are used to plate silver, what amount of silver is needed on the part and what is the approximate cost of the plating process itself. Several variables would have to be looked up.

Answer

The first, and probably most important, estimate would be to determine the amount of silver that would need to be deposited. The surface area of the unplated button would have to be measured and the amount of silver needed to cover it with a thickness of 1.5 mm. Once the amount of material is known, we can look at the chemistry of silver plating. There are some options – but the oxidation state from which the silver is plated is Ag^+ (in many cases complexes that contain Ag^+ are used to help make sure the coating is even.) From this fact, we could calculate the number of electrons needed to plate the desired amount of silver (one per atom of silver) and use that value to set the current parameters needed for the electroplating process. Thus, the things that would need to be looked up or measured include (1) the surface area of the button; (2) the chemical reactions used in the silver plating and the cost of the materials it uses; (3) the efficiency of the electroplating operation; (4) the cost of electricity where the plating will be carried out; (5) the cost of treatment and disposal of any wastes generated.

Chapter 13
Problems & Exercises

Corrosion

13.1 When you look at several older cars that are showing initial signs of rust formation, where do you expect to find the most rust? What does this observation imply about conditions that lead to corrosion?

13.2 What type of corrosion is associated with each of the following circumstances?
(a). two metals are touching each other
(b). a small crack separates the two metals
(c). a metal automobile body rusts in the atmosphere.

13.3 Using the internet, find three cases where corrosion was the cause of some sort of failure or malfunction of a device or a structure.

13.4 Using the internet, find three cases where concern about corrosion is used to help advertise a product or to convince people to buy a product.

Oxidation-Reduction Reactions and Galvanic Cells

13.5 For the following oxidation-reduction reactions, identify the half reactions and label them as oxidation or reduction.

(a). $Cu(s) + Ni^{2+}(aq) \rightarrow Ni(s) + Cu^{2+}(aq)$

(b). $2Fe^{3+}(aq) + 3Ba(s) \rightarrow 3Ba^{2+}(aq) + 2Fe(s)$

13.6 Explain why something that is oxidized is called a reducing agent.

13.7 What is the role of a salt bridge in the construction of a galvanic cell?

13.8 What half reaction takes place at the anode of an electrochemical cell? What half reaction takes place at the cathode?

13.9 If a salt bridge contains KNO_3 as its electrolyte, what ions diffuse into solution in the compartment of the anode of the galvanic cell? Explain your answer.

13.10 If a salt bridge contains KNO_3 as its electrolyte, what ions diffuse into solution in the compartment of the cathode of the galvanic cell? Explain your answer.

13.11 The following oxidation-reduction reactions are used in electrochemical cells. Write them using cell notation.

(a). $2Ag^{+}(aq)$ (0.50 M) + $Ni(s) \rightarrow 2Ag(s) + Ni^{2+}(aq)$ (0.20 M)

(b). $2Cu(s) + PtCl_6^{2-}$ (0.10 M) $\rightarrow 2Cu^{2+}$ (0.20 M) + $PtCl_4^{2-}$ (0.10 M) + $2Cl^{-}$ (0.40 M)

(c). $Pb(s) + SO_4^{2-}$ (0.30 M) + $2\,AgCl(s) \rightarrow PbSO_4(s) + 2\,Ag(s) + 2\,Cl^{-}$ (0.20 M)

(d). In a galvanic cell, one half cell contains 0.010 M HCl and a platinum electrode, over which H_2 is bubbled at a pressure of 1.0 atm. The other half cell is composed of a zinc electrode in a 0.125 M solution of $Zn(NO_3)_2$.

13.12 Given the following electrochemical reactions written in cell notation, write the balanced chemical equation, and identify the number of electrons transferred.

(a). $Ga(s) \mid Ga^{3+}(aq) \parallel Ag^{+}(aq) \mid Ag(s)$

(b). $Zn(s) \mid Zn^{2+}(aq) \parallel Cr^{3+}(aq) \mid Cr(s)$

(c). Fe(s), FeS(s) | S^{2-} (aq) || Sn^{2+} (aq) | Sn(s)

13.13 For the reactions in (a). and (b). in the preceding problem, there are no anions shown in the cell notation at all. Explain why this is not a concern.

13.14 Explain why the terms cell potential and electromotive force mean the same thing in electrochemical cells.

13.15 How does galvanic corrosion differ from uniform corrosion of iron?

13.16 How can each of the following serve as both the cathode and the anode of an electrochemical cell? (a). a single piece of metal or (b). two connected pieces of metal?

13.17 A student who has mercury amalgam fillings in some of her teeth is eating a piece of candy. She accidentally bites down on a piece of the aluminum foil wrapper and experiences a sharp sensation in her mouth. Explain what has happened in terms of electrochemistry.

Cell Potentials

13.18 Based on the cell potential measured for the cells,

Co(s) | Co^{2+} (aq) || Cu^{2+} (aq) | Cu(s) $E° = 0.614$ V

Fe(s) | Fe^{2+} (aq) || Cu^{2+} (aq) | Cu(s) $E° = 0.777$ V

what potential should you expect to find for the following cell?

Fe(s) | Fe^{2+} (aq) || Co^{2+} (aq) | Co(s)

13.19 If the reaction at a standard hydrogen electrode is $2H^+ + 2e^- \rightarrow H_2$, why is there a platinum foil present in the system?

13.20 Suppose we hook up two half cells, A and B, to a standard hydrogen electrode (SHE). For case A the voltage reads +0.51 V (using the convention that the SHE is connected to the positive terminal). In case B the voltage reads +0.10 V.
(a). In these two cases, is the SHE the cathode or the anode of the electrochemical cell?
(b). Does oxidation or reduction take place at the SHE?
(c). If half-cell A is connected to half-cell B and the positive terminal of the voltmeter is connected to A, will the voltmeter read a positive or negative value?
(d). What will be the numerical value of the potential for this cell?

13.21 If the SHE was assigned a value of 3.00 V rather than 0.00 V, what would happen to all of the values listed in the table of Standard Reduction Potentials?

13.22 Explain why a large negative value for the standard reduction potential indicates a half-cell that is likely to undergo oxidation.

13.23 For most tables of Standard Reduction Potentials that start from large positive values at the top and proceed through 0.0 V to negative values at the bottom, the alkali metals are normally at that bottom of the table. Use your chemical understanding of alkali metals and how they behave in bonding to explain why this is so.

13.24 Based on the table of standard reduction potentials, identify the metal in each pair that is more easily oxidized.

(a). Cu or Zn (b). Fe or Cd (c). Sn or Ni (d). K or Na

13.25 Using values from the tables of standard reduction potentials, calculate the cell potentials of the following cells.

(a). Ga(s) | Ga^{3+} (aq) || Ag^{+}(aq) | Ag(s)

(b). Zn(s) | Zn^{2+} (aq) || Cr^{3+} (aq) | Cr(s)

(c). Fe(s), FeS(s) | S^{2-} (aq) || Sn^{2+} (aq) | Sn(s)

13.26 Using values from the tables of standard reduction potentials, calculate the cell potentials of the following cells.

(a). Fe(s) | Fe^{2+} (aq) || Hg^{2+} (aq) | Hg(l)

(b). Pt(s) | Fe^{2+} (aq), Fe^{3+} (aq) || MnO_4^-(aq), H+ (aq), Mn^{2+} (aq) | Pt(s)

(c). Pt(s) | Cl_2(g) | Cl^-(aq) || Au^+(aq) | Au(s)

13.27 Using the standard reduction potentials given in Appendix I, devise galvanic cells with voltages of 3.0 V, 2.5 V, 2.0 V and 1.5 V.

13.28 Suppose you wish to make a galvanic cell with a voltage of nearly 2.0 V using a Zn(s) | Zn^{2+} (aq) half cell. What could be used as the other half cell? Which electrode would be the anode and which the cathode?

13.29 Use the Nernst equation to calculate the cell potentials of the following cells.

(a). $2Ag^+$(aq) (0.50 M) + Ni(s) → 2Ag(s) + Ni^{2+} (aq) (0.20 M)

(b). 2Cu(s) + $PtCl_6^{2-}$ (0.10 M) → $2Cu^{2+}$ (0.20 M) + $PtCl_4^{2-}$(0.10 M)+ $2Cl^-$ (0.40 M)

(c). Pb(s) + SO_4^{2-} (0.30 M) + 2AgCl(s) → $PbSO_4$(s) + 2Ag(s) + $2Cl^-$ (0.20 M)

13.30 Suppose you created a galvanic cell using the copper metal and silver metal example that was presented in the chapter. The silver nitrate solution you use is 0.45 M and the copper sulfate solution is 0.60 M. What is the cell potential when you first connect the two half reactions?

13.31 We noted that a tin-plated steel can will corrode more quickly than an unplated one. In cases of galvanic corrosion, one cannot expect standard conditions. Suppose you want to study the galvanic corrosion of tin-plated steel by constructing a cell with low concentrations of the ions. You have pieces of tin and iron. You also have a solution of tin(II)chloride that is 0.05 M and one of iron(II)nitrate that is 0.01 M.

(a). Describe the half reactions you construct for this experiment.

(b). Which half reaction will be the anode and which the cathode?

(c). Based on the solutions you have, calculate the cell potential for your experiment.

13.32 A student who attends a school located in the mountains is measuring cell potentials versus a SHE electrode and forgets to adjust the pressure so it is 1.0 atm. Instead, the pressure is 0.87 atm. If all solutions are 1.0 M, what percentage error would result in the measurements for the standard reduction potential for each of the following electrodes?

(a). Sn(s) | Sn^{2+} (aq)

(b). Zn(s) | Zn^{2+} (aq)

(c). Pt(s) | MnO_4^-(aq), H+ (aq), Mn^{2+} (aq)

13.33 You are a chemical engineer in a firm that manufactures computer chips. A new accountant, who understands budgets but not chemistry, sends out the following memo.

> "Because of the high price of gold, the company should change from gold-plated pins on our computer chips to pins made of copper, which is much cheaper."

Write a memo to your boss explaining why the accountant's choice of copper is a poor one and why gold should continue to be used in pin manufacturing.

13.34 Immediately after a person has his or her ears (or other body parts) pierced, a solid 14K gold post should be used to keep the hole open as it heals.

(a). Explain why solid gold is preferred over other metals, such as steel.
(b). Why do you think that a gold-plated steel post might not offer the same protection as the solid gold post?
(c). Surgical steel is often used as the metal post in a newly pierced ear, because it is less expensive than gold. Using the internet and other resources, find out the composition of surgical steel.

13.35 In May, 2000, a concrete pedestrian walkway collapsed in North Carolina, injuring over 100 people. Investigation revealed that $CaCl_2$ had been mixed into the grout that filled the holes around the steel reinforcing cables inside the concrete, resulting in corrosion of the cables, thus weakening the structure. Based on your understanding of corrosion, explain why the use of chloride compounds in steel-reinforced concrete is discouraged by the concrete industry.

13.36 A chemical engineering student is studying the effect of pH on the corrosion of iron. The following data are collected.

Corrosion rate (mm/year)	pH
0.03	13.00
0.06	12.00
0.16	11.00
0.22	10.00
0.25	9.00
0.25	8.00
0.25	7.00
0.25	6.00
0.25	5.00
0.27	4.00
0.37	3.00
0.38	2.80
0.63	2.80
1.00	2.80

Plot corrosion rate vs. pH. (Hint: label the x-axis with *decreasing* pH.)
(a). How would you describe the dependence of corrosion on pH?

(b). What change takes place in the reduction half reaction at a pH of about 3?
(c). How do you explain the shape of the curve at the lowest pH?
(Based on a graph in the *Kirk-Othmer Encyclopedia of Chemical Technology*, 4th ed, vol 7. New York: John Wiley & Sons, p. 559.)

Cell Potentials and Equilibrium

13.37 How is the relationship between maximum electrical and cell potential used to determine the free energy change for electrochemical cells?

13.38 Calculate the free energy change for the following cells.

(a). Ga(s) | Ga^{3+} (aq) || Ag^{+}(aq) | Ag(s)

(b). Zn(s) | Zn^{2+} (aq) || Cr^{3+} (aq) | Cr(s)

(c). Fe(s), FeS(s) | S^{2-} (aq) || Sn^{2+} (aq) | Sn(s)

13.39 Calculate the free energy change for the following reactions using the standard cell potentials for the half reactions that are involved.

(a). Fe(s) + Hg_2^{2+} (aq) → Fe^{2+} (aq) + 2 Hg(l)

(b). Fe^{3+} (aq) + Ag(s) + Cl-(aq) → Fe^{2+} (aq) + AgCl(s0

(c). 2 MnO_4^-(aq) + 5 Zn(s) + 16 H_3O^+(aq) → 2 Mn^{2+} (aq) + 5 Zn^{2+} (aq) + 24 H_2O

13.40 Suppose you cannot find a table of standard reduction potentials. You remember that the standard reduction potential of Cu^{2+} + $2e^-$ → Cu(s) is 0.34 0 V. Given that $\Delta G_f°(Cu^{2+}) = 65.49$ kJ mol^{-1} and that $\Delta G_f°(Ni^{2+}) = -45.6$ kJ mol^{-1}, determine the standard reduction potential of Ni | Ni^{2+} from this data.

13.41 Use the potential of the galvanic cell, Co(s) | Co^{2+} || Pb^{2+} | Pb(s), to determine $\Delta G_f°(Pb^{2+})$, given that $\Delta G_f°(Co^{2+}) = -54.4$ kJ mol^{-1}.

13.42 Determine whether or not the following reactions will proceed spontaneously to products as written. Explain your reasoning.

(a). N_2O(aq) + $2H^+$ + Zn(s) → N_2(g) + H_2O(l) + Zn^{2+} (aq)

(b). I_2(s) + H_2O_2(aq) + $2H_2O$(l) → 2 I^-(aq) + O_2(g) + $2H_3O^+$(aq)

(c). 3 PbO(s) + BrO_3^-(aq) → $3PbO_2$(s) + Br^-(aq)

13.43 Does the reaction, Cr(s) + 2 Cr^{3+} (aq) → 3 Cr^{2+} (aq) proceed spontaneously? What does this reaction imply about finding chromium(III) in solution if solid chromium is also present?

13.44 The equilibrium constant for a reaction is 3×10^{-15}. (a) Without carrying out any calculation, discuss whether ΔG° for the reaction is positive or negative. (b) Calculate ΔG° for this reaction.

13.45 Some calculators cannot display results of an antilog calculation if the power of ten is greater than 99. This shortcoming can come into play for determining equilibrium constants of redox reactions, which are sometimes quite large. Solve the following expressions for K.

(a). log K = 45.63 (b). log K = 25.00 (c). log K = 20.63

What is the relationship among the three expressions and the three answers? How can you use this relationship to solve problems that exceed 10^{99}, even if your calculator will not carry out the calculation directly?

13.46 Calculate the equilibrium constant for the following reactions using data from the standard reduction potential tables given in Appendix I.

(a). $Cl_2(g) + 2Br^-(aq) \rightarrow Br_2(g) + 2Cl^-(aq)$

(b). $Ni(s) + 2Ag^+(aq) \rightarrow 2Ag(s) + Ni^{2+}(aq)$

(c). $2Co(s) + 2O_3^{2-}(aq) + 3H_2O(l) \rightarrow 2Co(OH)_2(s) + S_2O_3^{2-}(aq) + 2OH^-(aq)$

13.47 Use the standard reduction potentials for the reactions $AgCl(s) + e^- \rightarrow Ag(s) + Cl^-(aq)$ and $Ag^+(aq) + e^- \rightarrow Ag(s)$ to calculate the value for K_{sp} for silver chloride. How does your answer compare with the tabulated value from Chapter 12?

13.48 Hydrogen peroxide is often stored in the refrigerator to help keep it from decomposing according to the reaction,

$$2H_2O_2(l) \rightarrow 2H_2O(l) + O_2(g).$$

Use information from the standard reduction potentials to determine the equilibrium constant of this reaction.

13.49 Calculate the equilibrium constant for the redox reactions that could occur for the following situations and use that value to explain whether or not any reaction will be observed.

(a). A piece of iron is placed in a 1.0 M solution of $NiCl_2(aq)$

(b). A copper wire is placed in a 1.0 M solution of $Pb(NO_3)_2(aq)$

(c). Chlorine gas is bubbled through a 1.0 M solution of NaOH

13.50 If a logarithmic scale had not been used for the graph of Figure 13.12, what would the plots look like?

13.51 Suppose you need a cell that will deliver a potential of exactly 1.52 V. Design and sketch a cell to provide this voltage, detailing the solutions, their concentrations, and the electrodes you will need. Write equations for all the possible reactions.

Batteries

13.52 What is the principal difference between a primary and a secondary cell?

13.53 Based on the chemistry that takes place, explain why an alkaline battery is called "alkaline."

13.54 If you put a 9-volt battery in a smoke detector in your home or apartment, you are not installing a single galvanic cell. Explain how and why this is so.

13.55 If alkaline batteries were not alkaline but rather acidic (as in the older dry cell batteries) what extra difficulties could you envision with corrosion, based on reactions that are part of the table of standard reduction potentials?

13.56 What would happen to the voltage of an alkaline battery if the zinc were replaced by steel? Assume the zinc reaction is simply $Zn \mid Zn^{2+}$ and that steel is iron.

13.57 Battery manufacturers often assess batteries in terms of their specific energy (or energy capacity). The weight capacity of a battery is defined as q x V / mass. Why would a battery manufacturer be interested in this quantity?

13.58 Suppose you had a fresh battery and wanted to measure its weight capacity, as described in the previous problem. Devise and describe an experiment that you could carry out to make this measurement. Do you think performing your experiment on a single battery would yield reliable results?

13.59 What product forms from the lead components of a lead storage battery? Why do mechanical shock (bumps) ultimately degrade the performance of the lead storage battery?

13.60 The electrolyte solution in a zinc/air battery (Figure 13.15) is aqueous KOH. In order for a zinc-air battery to obtain O_2 from the air there are tiny openings in the battery. However, these openings also allow water vapor to escape and enter the battery as the humidity changes. Optimum relative humidity for efficient zinc-air battery function, at which the KOH electrolyte is in equilibrium with water in the surrounding air, is 60% at 25°C. On the molecular level, describe what will happen to a zinc-air battery if used consistently in

(a). an arid environment.

(b). an extremely humid environment.

13.61 The graph below shows discharge curves for 4 button batteries under the same conditions and discharge load.

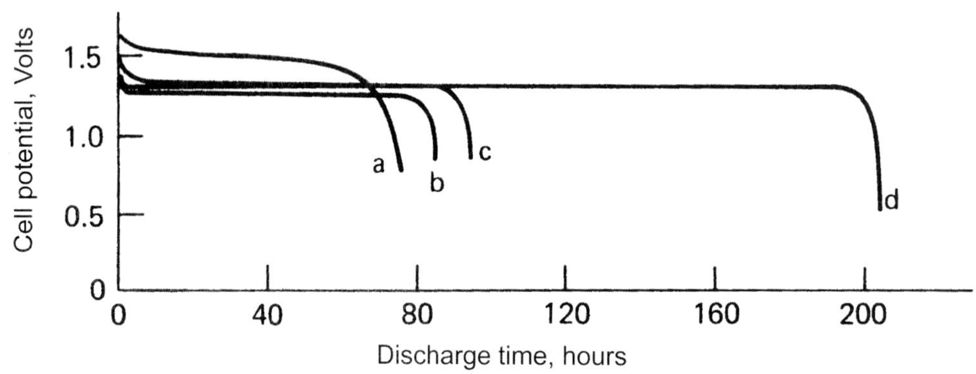

(a). Write a paragraph in which you interpret curve (a) and curve (d) to a customer seeking a battery for an electronic device.

(b). Both (b) and (d) represent zinc/air cells. Explain why there is a difference in these two curves. (Hints: Which electrode in a zinc/air cell has an unlimited life? Compare the approximate area of the other electrode for the two cells.)

13.62 Some batteries contain toxic materials that should not be disposed of in a sanitary waste dump. List the materials that have been mentioned in the text that are used in common, commercially available batteries. Using a resource such as Material Safety Data Sheets (MSDS), determine which battery discussed is the most environmentally friendly. (MSDS resources can be accessed on the web or in your chemistry department stockroom.)

13.63 On the internet, access the website of a major battery company, such as Rayovac, Duracell, or Eveready (Energizer). Search the site for a battery type not covered in this chapter. Find out all you can about the battery from engineering data or specification (spec) sheets. Print spec sheets that include graphs of discharge characteristics, operating temperature data, etc. Summarize and interpret your findings in a one-page report to which you attach the spec sheets.

Electrolysis

13.64 What is the difference between active and passive electrolysis? Based on the common meanings of the words active and passive, what part of electrolysis is the focus of the name?

13.65 Why is it easier to force an oxidation-reduction reaction to proceed in the non-spontaneous direction than it is to force an acid-base reaction in the non-spontaneous direction?

13.66 In any electrochemical cell, reduction occurs at the cathode. Why does the cathode carry a negative charge when electrolysis occurs rather than the positive charge it carries in a galvanic cell?

13.67 When aluminum is refined by electrolysis from its oxide ores is the process used active or passive electrolysis? Explain your answer.

13.68 What complication arises in the electrolytic refining of aluminum because of the production of oxygen?

13.69 In an electroplating operation the cell potential is sometimes 0 V. Why is a zero potential possible in electrolysis but not in a galvanic cell?

13.70 When copper wire is placed in a solution of silver nitrate we say the silver plates out onto the copper, but we don't call the process electroplating. Explain the difference.

13.71 In barrel plating, why is the barrel containing the small parts rotated?

13.72 When a coating is plated onto a metal, two different metals are in contact with each other. Why doesn't galvanic corrosion occur at this interface?

13.73 Use the internet to find electroplating companies that carry out silver plating. Popular impressions are that silver plating is mostly done for cosmetic reasons, making objects (like silverware) more attractive. Based on your internet research, do you believe this popular impression accurately reflects the plating industry?

13.74 Based on the reaction used for silver plating, suggest why there are environmental challenges associated with industrial plating companies. Use the internet to find out how some electroplating companies address their responsibilities to protect the environment.

Electrolysis and Stoichiometry

13.75 If a current of 15 A is run through an electrolysis cell for 2 hours, how many moles of electrons have moved?

13.76 Suppose somebody in a laboratory doesn't quite turn off the current in an electrolysis cell so that 4.2 mA of current continue to flow. If the cell is then left untouched for exactly 3 weeks, how many electrons (in moles) have flowed through the cell in that time?

13.77 If a barrel plating run consumes 200 A for 6 hours to carry out an electroplating application at 0.30 V, how many kilowatt-hours have been used in the run? If the voltage is 0.90V what is the power usage (in kwh)?

13.78 Suppose you are analyzing an electroplating run and want to calculate the charge that has been used. You know that the cell voltage is 0.25 V and that

10.3 kwh were expended. What was the charge? If the current of the apparatus is 3.0 A, how long was it running?

13.79 In a copper plating experiment in which copper metal is deposited from copper(II) ion solution, the system is run for 2.6 hours at a current of 12.0 A. What mass of copper is deposited?

13.80 Palladium is very expensive, and palladium coatings are normally quite thin. Suppose you have a part that requires 3.2 mg of palladium to have an adequate coating. How long would you need to run an electrolysis cell at 1.5 A to obtain this much palladium from a $PdCl_2$ solution?

13.81 Tin plated steel is used for "tin cans". Suppose in the production of sheets of tin plated steel, a line at a factory operates at a current of 100 A for 8 hours on a continuously fed sheet of unplated steel. What is the total mass of tin that has plated out in this operation?

13.82 Chrome is plated from solutions of chromium (III) ions. If a particular application requires 13.7 g of Cr in the coating, and the run takes 90 minutes, what current was used?

13.83 If a plating line that deposits nickel ($NiCl_2$ solutions) operates at a voltage of 0.40 V, with a current of 400 A, and a total mass of 49.0 kg of nickel is deposited. How many kwh were consumed in this process?

13.84 Suppose a student mistakenly thinks that copper is plated from solutions that contain copper (I) ions rather than copper (II) ions. How would this mistake affect the answer that student would get for the mass of copper plated out for a given time and current? (Is it too high, too low, and by what factor?)

13.85 When a lead storage battery is recharged by a current of 12.0 A for 15 minutes, what mass of $PbSO_4$ is consumed?

13.86 A small part with a surface area of 2.62 cm^2 is plated with a gold coating that is .0005 mm thick. The density of gold is 19.32 g cm^{-3}. What mass of gold is in this coating? If a barrel plating run has 5000 of these parts in a single barrel, how long would the run take if the current was 15.0 A? (Gold is effectively plated from $Au^+(aq)$)

13.87 A piece of metal that measures 1.3 cm by 0.83 cm will have a coating of rhodium plated on its surface to serve as a mirror. The rhodium thickness will be 0.0003 mm. If the operating current of the electrolysis is 1.3 A, how long will it need to be operated to obtain the desired coating? The density of rhodium is 12.4 g cm3. What mass of rhodium is deposited?

Corrosion Prevention

13.88 In corrosion prevention, what is meant by the term passivation?

13.89 Any coating that keeps a part free from environments that can cause rust is capable of helping reduce corrosion. What must paint manufactures do to produce a paint that is specifically designed to help prevent corrosion?

13.90 Spend 10 minutes searching on the internet for companies that sell corrosion inhibition additives for the paint industry. How many companies do you find in this short time? What does the result of your search suggest about the importance of corrosion in the modern economy?

13.91 Suppose you discovered a new, highly effective, corrosion-inhibiting molecule, but when your discovery is added to paint it tends to make the coating brittle.

What kinds of considerations would manufacturers of paint have to make about whether or not to purchase your product to add to their paints?

13.92 From where does the electro- prefix of the term electrospray technology for paint coatings arise?

13.93 Would nickel make an acceptable sacrificial anode to protect steel? Explain your answer.

13.94 Zinc is used as a coating for galvanized steel, where it helps prevent corrosion. Explain why zinc would also make an acceptable sacrificial anode for a steel pipeline. Are these two uses related in terms of the science that gives rise to their utility?

13.95 Sacrificial anodes are sometimes connected to steel using a copper wire. If the anode is completely corroded away and not replaced so that only the copper wire remains, what could happen at the location where the copper and iron meet? Explain your answer.

13.96 As a design engineer for NASA, you are responsible for the design of a Venus lander that will study the surface of Venus in great detail. Using the Internet and your background in chemical kinetics (Chapter 11), thermodynamics (Chapter 10), and electrochemistry, write a 2 –3 page paper in which you

(a) Examine the environmental conditions the lander will encounter on the surface of Venus;

(b) Detail the chemical effects that the atmospheric conditions on Venus could have on the lander;

(c) Suggest some specific methods of protecting the lander from Venus's harsh environment.

Focus on Problem Solving

13.97 For a voltage sensitive application, you are working on a battery that must have a working voltage of 0.85 V. The materials to be used have a standard cell potential of 0.97 V. What must be done to achieve the correct voltage? What information would need to be looked up?

13.98 For a battery application you need to get the largest possible voltage per cell that you can. If voltage were the only concern, what chemical reactions would you choose? Use the internet or your own background to speculate on the likely practical consequences of the choice driven by voltage alone.

13.99 An oxidation reduction reaction using Sn(s) to remove $N_2O(g)$ from a reaction vessel has been proposed. You cannot find any thermodynamic information on one product of the reaction, $NH_3OH^+(aq)$. How could you estimate the equilibrium constant of the reaction?

13.100 You are designing a system for a rather corrosive environment and are using an expensive alloy composed of primarily titanium. How would you choose a sacrificial anode to help protect this expensive component in your design? What information would have to be looked up?

13.101 You need to use a gold plated connector in a design to take advantage of the conductivity and corrosion resistance of gold. In order to justify this choice to the project director you must devise a cost projection for several levels of plating. What variables could be varied? What information would have to be looked up to provide the cost information requested?

International Table of Atomic Weights

Atomic Number	Symbol	Name	Atomic Weight	Atomic Number	Symbol	Name	Atomic Weight
89	Ac	Actinium	[227]	42	Mo	Molybdenum	95.94
13	Al	Aluminium	26.981538	60	Nd	Neodymium	144.24
95	Am	Americium	[243]	10	Ne	Neon	20.1797
51	Sb	Antimony	121.760	93	Np	Neptunium	[237]
18	Ar	Argon	39.948	28	Ni	Nickel	58.6934
33	As	Arsenic	74.92160	41	Nb	Niobium	92.90638
85	At	Astatine	[210]	7	N	Nitrogen	14.0067
56	Ba	Barium	137.327	102	No	Nobelium	[259]
97	Bk	Berkelium	[247]	76	Os	Osmium	190.23
4	Be	Beryllium	9.012182	8	O	Oxygen	15.9994
83	Bi	Bismuth	208.98038	46	Pd	Palladium	106.42
107	Bh	Bohrium	[264]	15	P	Phosphorus	30.973761
5	B	Boron	10.811	78	Pt	Platinum	195.078
35	Br	Bromine	79.904	94	Pu	Plutonium	[244]
48	Cd	Cadmium	112.411	84	Po	Polonium	[209]
55	Cs	Caesium	132.90545	19	K	Potassium	39.0983
20	Ca	Calcium	40.078	59	Pr	Praseodymium	140.90765
98	Cf	Californium	[251]	61	Pm	Promethium	[145]
6	C	Carbon	12.0107	91	Pa	Protactinium	231.03588
58	Ce	Cerium	140.116	88	Ra	Radium	[226]
17	Cl	Chlorine	35.453	86	Rn	Radon	[222]
24	Cr	Chromium	51.9961	75	Re	Rhenium	186.207
27	Co	Cobalt	58.933200	45	Rh	Rhodium	102.90550
29	Cu	Copper	63.546	37	Rb	Rubidium	85.4678
96	Cm	Curium	[247]	44	Ru	Ruthenium	101.07
110	Ds	Darmstadtium	[281]	104	Rf	Rutherfordium	[261]
105	Db	Dubnium	[262]	62	Sm	Samarium	150.36
66	Dy	Dysprosium	162.500	21	Sc	Scandium	44.955910
99	Es	Einsteinium	[252]	106	Sg	Seaborgium	[266]
68	Er	Erbium	167.259	34	Se	Selenium	78.96
63	Eu	Europium	151.964	14	Si	Silicon	28.0855
100	Fm	Fermium	[257]	47	Ag	Silver	107.8682
9	F	Fluorine	18.9984032	11	Na	Sodium	22.989770
87	Fr	Francium	[223]	38	Sr	Strontium	87.62
64	Gd	Gadolinium	157.25	16	S	Sulfur	32.065
31	Ga	Gallium	69.723	73	Ta	Tantalum	180.9479
32	Ge	Germanium	72.64	43	Tc	Technetium	[98]
79	Au	Gold	196.96655	52	Te	Tellurium	127.60
72	Hf	Hafnium	178.49	65	Tb	Terbium	158.92534
108	Hs	Hassium	[277]	81	Tl	Thallium	204.3833
2	He	Helium	4.002602	90	Th	Thorium	232.0381
67	Ho	Holmium	164.93032	69	Tm	Thulium	168.93421
1	H	Hydrogen	1.00794	50	Sn	Tin	118.710
49	In	Indium	114.818	22	Ti	Titanium	47.867
53	I	Iodine	126.90447	74	W	Tungsten	183.84
77	Ir	Iridium	192.217	112	Uub	Ununbium	[285]
26	Fe	Iron	55.845	116	Uuh	Ununhexium	
36	Kr	Krypton	83.798	118	Uuo	Ununoctium	
57	La	Lanthanum	138.9055	114	Uuq	Ununquadium	[289]
103	Lr	Lawrencium	[262]	111	Uuu	Unununium	[272]
82	Pb	Lead	207.2	92	U	Uranium	238.02891
3	Li	Lithium	6.941	23	V	Vanadium	50.9415
71	Lu	Lutetium	174.967	54	Xe	Xenon	131.293
12	Mg	Magnesium	24.3050	70	Yb	Ytterbium	173.04
25	Mn	Manganese	54.938049	39	Y	Yttrium	88.90585
109	Mt	Meitnerium	[268]	30	Zn	Zinc	65.409
101	Md	Mendelevium	[258]	40	Zr	Zirconium	91.224
80	Hg	Mercury	200.59				

Physical constants

Quantity	Symbol	Traditional Units	SI Units
Acceleration of gravity	g	980.6 cm/s	9.806 m/s
Atomic mass unit ($\frac{1}{12}$ the mass of ^{12}C atom)	amu or u	1.6606×10^{-24} g	1.6606×10^{-27} kg
Avogadro's number	N	6.0221367×10^{23} particles/mol	6.0221367×10^{23} particles/mol
Bohr radius	a_0	0.52918 Å 5.2918×10^{-9} cm	5.2918×10^{-11} m
Boltzmann constant	k	1.3807×10^{-16} erg/K	1.3807×10^{-23} J/K
Charge-to-mass ratio of electron	e/m	1.75882×10^{8} coulomb/g	1.75882×10^{11} C/kg
Electronic charge	e	1.60218×10^{-19} coulomb 4.8033×10^{-10} esu	1.60218×10^{-19} C
Electron rest mass	m_e	9.10940×10^{-28} g 0.00054858 amu	9.10940×10^{-31} kg
Faraday constant	F	96,485 coulombs/eq 23.06 kcal/volt · eq	96,485 C/mol e^- 96,485 J/V · mol e^-
Gas constant	R	$0.08206 \frac{L \cdot atm}{mol \cdot K}$ $1.987 \frac{cal}{mol \cdot K}$	$8.3145 \frac{kPa \cdot dm^3}{mol \cdot K}$ 8.3145 J/mol · K
Molar volume (STP)	V_m	22.414 L/mol	22.414×10^{-3} m^3/mol 22.414 dm^3/mol
Neutron rest mass	m_n	1.67495×10^{-24} g 1.008665 amu	1.67495×10^{-27} kg
Planck constant	h	6.6262×10^{-27} erg · s	6.6262×10^{-34} J · s
Proton rest mass	m_p	1.6726×10^{-24} g 1.007277 amu	1.6726×10^{-27} kg
Rydberg constant	R_∞	3.289×10^{15} cycles/s 2.1799×10^{-11} erg	1.0974×10^{7} m^{-1} 2.1799×10^{-18} J
Speed of light (in a vacuum)	c	$2.99792458 \times 10^{10}$ cm/s (186,281 miles/second)	2.99792458×10^{8} m/s

Electron configurations of atoms in their ground states

Z	Element	Configuration
1	H	$1s^1$
2	He	$1s^2$
3	Li	$[He]2s^1$
4	Be	$[He]2s^2$
5	B	$[He]2s^2\ 2p^1$
6	C	$[He]2s^2\ 2p^2$
7	N	$[He]2S^2\ 2p^3$
8	0	$[He]2S^2\ 2p^4$
9	F	$[He]2S^2\ 2p^5$
10	Ne	$[He]2s^22p^6$
11	Na	$[Ne]3s^1$
12	M g	$[Ne]3s^2$
13	AL	$[Ne]3s^23p^1$
14	Si	$[Ne]3s^23p^2$
15	P	$[Ne]3s^23p^3$
16	S	$[Ne]3s^23p^4$
17	CL	$[Ne]3s^23p^5$
18	Ar	$[Ne]3s^23p^6$
19	K	$[Ar]4s^1$
20	Ca	$[Ar]4s^2$
21	Sc	$[Ar]3d^14s^2$
22	Ti	$[Ar]3d^24s^2$
23	V	$[Ar]3d^34s^2$
24	Cr	$[Ar]3d^54s^1$
25	Mn	$[Ar]3d^54s^2$
26	Fe	$[Ar]3d^64s^2$
27	Co	$[Ar]3d^74s^2$
28	Ni	$[Ar]3d^84s^2$
29	Cu	$[Ar]3d^{10}4s^1$
30	Zn	$[Ar]3d^{10}4s^2$
31	Ga	$[Ar]3d^{10}4s^24p^1$
32	Ge	$[Ar]3d^{10}4s^24p^2$
33	As	$[Ar]3d^{10}4s^24p^3$
34	Se	$[Ar]3d^{10}4s^24p^4$
35	Br	$[Ar]3d^{10}4s^24p^5$
36	Kr	$[Ar]3d^{10}4s^24p^6$
37	Rb	$[Kr]\ 5s^1$
38	Sr	$[Kr]\ 5s^2$
39	Y	$[Kr]4d^15s^2$
40	Zr	$[Kr]4d^25s^2$
41	Nb	$[Kr]4d^45s^1$
42	Mo	$[Kr]4d^55s^1$
43	Tc	$[Kr]4d^55s^2$
44	Ru	$[Kr]4d^75s^1$
45	Rh	$[Kr]4d^85s^1$
46	Pd	$[Kr]4d^{10}$
47	Ag	$[Kr]4d^{10}5s^1$
48	Cd	$[Kr]4d^{10}5s^2$
49	In	$[Kr]4d^{10}5s^25p^1$
50	Sn	$[Kr]4d^{10}5s^25p^2$
51	Sb	$[Kr]4d^{10}5s^25p^3$
52	Te	$[Kr]4d^{10}5s^25p^4$
53	I	$[Kr]4d^{10}5s^25p^5$
54	Xe	$[Kr]4d^{10}5s^25p^6$
55	Cs	$[Xe]6s^1$
56	Ba	$[Xe]6s^2$
57	La	$[Xe]5d^16s^2$
58	Ce	$[Xe]4f^15d^16s^2$
59	Pr	$[Xe]4f^36s^2$
60	Nd	$[Xe]4f^46s^2$
61	Pm	$[Xe]4f^56s^2$
62	Sm	$[Xe]4f^66s^2$
63	Eu	$[Xe]4f^76s^2$
64	Gd	$[Xe]4f^75d^16s^2$
65	Tb	$[Xe]4f^96s^2$
66	Dy	$[Xe]4f^{10}6s^2$
67	Ho	$[Xe]4f^{11}6s^2$
68	Er	$[Xe]4f^{12}6s^2$
69	Tm	$[Xe]4f^{13}6s^2$
70	Yb	$[Xe]4f^{14}6s^2$
71	Lu	$[Xe]4^{14}5d^16s^2$
72	Hf	$[Xe]4f^{14}5d^26s^2$
73	Ta	$[Xe]4f^{14}5d^36s^2$
74	W	$[Xe]4f^{14}5d^46s^2$
75	Re	$[Xe]4f^{14}5d^56s^2$
76	Os	$[Xe]4f^{14}5d^66s^2$
77	Ir	$[Xe]4f^{14}5d^76s^2$
78	Pt	$[Xe]4f^{14}5d^96s^1$
79	Au	$[Xe]4f^{14}5d^{10}6s^1$
80	Hg	$[Xe]4f^{14}5d^{10}6s^2$
81	TI	$[Xe]4f^{14}5d^{10}6s^26p^1$
82	Pb	$[Xe]4f^{14}5d^{10}6s^26p^2$
83	Bi	$[Xe]4f^{14}5d^{10}6s^26p^3$
84	Po	$[Xe]4f^{14}5d^{10}6s^26p^4$
85	At	$[Xe]4f^{14}5d^{10}6s^26p^5$
86	Rn	$[Xe]4f^{14}5d^{10}6s^26p^6$
87	Fr	$[Rn]7s^1$
88	Ra	$[Rn]7s^2$
89	Ac	$[Rn]6d^17s^2$
90	Th	$[Rn]6d^27s^2$
91	Pa	$[Rn]5f^26d^17s^2$
92	U	$[Rn]5f^36d^17s^2$
93	Np	$[Rn]5f^46d^17s^2$
94	Pu	$[Rn]5f^67s^2$
95	Am	$[Rn]5f^77s^2$
96	C m	$[Rn]5f^76d^17s^2$
97	Bk	$[Rn]5f^97s^2$
98	Cf	$[Rn]5f^{10}7s^2$
99	Es	$[Rn]5f^{11}7s^2$
100	Fm	$[Rn]5f^{12}7s^2$
101	Md	$[Rn]5f^{13}7s^2$
102	No	$[Rn]5f^{14}7s^2$
103	Lr	$[Rn]5f^{14}6d^17s^2$
104	Rf	$[Rn]5f^{14}6d^27s^2$
105	Db	$[Rn]5f^{14}6d^37s^2$
106	Sg	$[Rn]5f^{14}6d^47s^2$
107	Bh	$[Rn]5f^{14}6d^57s^2$
108	Hs	$[Rn]5f^{14}6d^67s^2$
109	Mt	$[Rn]5f^{14}6d^77s^2$

Specific heats and heat capacities of some common substances

Substance	Specific Heat (J/g · °C)	Molar Heat Capacity (J/mol · °C)
Al(s)	0.900	24.3
Ca(s)	0.653	26.2
Cu(s)	0.385	24.5
Fe(s)	0.444	24.8
Hg(ℓ)	0.138	27.7
H_2O(s), ice	2.09	37.7
H_2O(ℓ), water	4.18	75.3
H_2O(g), steam	2.03	36.4
C_6H_6(ℓ), benzene	1.74	136
C_6H_6(g), benzene	1.04	81.6
C_2H_5OH(ℓ), ethanol	2.46	113
C_2H_5OH(g), ethanol	0.954	420
$(C_2H_5)_2O$(ℓ), diethyl ether	3.74	172
$(C_2H_5)_2O$(g), diethyl ether	2.35	108

Heats and temperatures of phase transitions for some common substances

Substance	mp (°C)	Heat of Fusion (J/g)	ΔH_{fus} (kJ/mol)	bp (°C)	Heat of Vaporization (J/g)	ΔH_{vap} (kJ/mol)
Al	658	395	10.6	2467	10520	284
Ca	851	233	9.33	1487	4030	162
Cu	1083	205	13.0	2595	4790	305
H_2O	0.0	334	6.02	100	2260	40.7
Fe	1530	267	14.9	2735	6340	354
Hg	−39	11	23.3	357	292	58.6
CH_4	−182	58.6	0.92	−164	—	—
C_2H_5OH	−117	109	5.02	78.3	855	39.3
C_6H_6	5.48	127	9.92	80.1	395	30.8
$(C_2H_5)_2O$	−116	97.9	7.66	35	351	26.0

Vapor pressure of water as selected temperatures

Temperature (°C)	Vapor Pressure (torr)	Temperature (°C)	Vapor Pressure (torr)	Temperature (°C)	Vapor Pressure (torr)	Temperature (°C)	Vapor Pressure (torr)
−10	2.1	21	18.7	51	97.2	81	369.7
−9	2.3	22	19.8	52	102.1	82	384.9
−8	2.5	23	21.1	53	107.2	83	400.6
−7	2.7	24	22.4	54	112.5	84	416.8
−6	2.9	25	23.8	55	118.0	85	433.6
−5	3.2	26	25.2	56	123.8	86	450.9
−4	3.4	27	26.7	57	129.8	87	468.7
−3	3.7	28	28.3	58	136.1	88	487.1
−2	4.0	29	30.0	59	142.6	89	506.1
−1	4.3	30	31.8	60	149.4	90	525.8
0	4.6	31	33.7	61	156.4	91	546.1
1	4.9	32	35.7	62	163.8	92	567.0
2	5.3	33	37.7	63	171.4	93	588.6
3	5.7	34	39.9	64	179.3	94	610.9
4	6.1	35	42.2	65	187.5	95	633.9
5	6.5	36	44.6	66	196.1	96	657.6
6	7.0	37	47.1	67	205.0	97	682.1
7	7.5	38	49.7	68	214.2	98	707.3
8	8.0	39	52.4	69	223.7	99	733.2
9	8.6	40	55.3	70	233.7	100	760.0
10	9.2	41	58.3	71	243.9	101	787.6
11	9.8	42	61.5	72	254.6	102	815.9
12	10.5	43	64.8	73	265.7	103	845.1
13	11.2	44	68.3	74	277.2	104	875.1
14	12.0	45	71.9	75	289.1	105	906.1
15	12.8	46	75.7	76	301.4	106	937.9
16	13.6	47	79.6	77	314.1	107	970.6
17	14.5	48	83.7	78	327.3	108	1004.4
18	15.5	49	88.0	79	341.0	109	1038.9
19	16.5	50	92.5	80	355.1	110	1074.6
20	17.5						

Selected thermodynamic data at 298.15 K

Species	ΔH_f° (kJ mol^{-1})	S° (J mol^{-1} K^{-1})	ΔG_f° (kJ mol^{-1})
Aluminium			
Al(s)	0	28.3	0
$AlCl_3$(s)	-704.2	110.7	-628.9
Al_2O_3(s)	-1676	50.92	-1582
Barium			
$BaCl_2$(s)	-860.1	126	-810.9
$BaSO_4$(s)	-1465	132	-1353
Beryllium			
Be(s)	0	9.54	0
$Be(OH)_2$(s)	-907.1	-	-
Bromine			
Br(g)	111.8	174.9	82.4
Br_2(l)	0	152.23	0
Br_2(g)	30.91	245.4	3.14
BrF_3(g)	-255.6	292.4	-229.5
HBr(g)	-36.4	198.59	-53.43
Calcium			
Ca(s)	0	41.6	0
Ca(g)	192.6	154.8	158.9
Ca^{2+}(g)	1920	-	-
CaC_2(s)	-62.8	70.3	-67.8
$CaCO_3$(s)	-1207	92.9	-1129
$CaCl_2$(s)	-795.0	114	-750.2
CaF_2(s)	-1215	68.87	-1162
CaH_2(s)	-189	42	-150
CaO(s)	-635.5	40	-604.2
CaS(s)	-482.4	56.5	-477.4
$Ca(OH)_2$(s)	-986.6	76.1	-896.8
$Ca(OH)_2$(aq)	-1002.8	76.15	-867.6
$CaSO_4$(s)	-1433	107	-1320
Carbon			
C (s,graphite)	0	5.740	0
C (s,diamond)	1.897	2.38	2.900
C(g)	716.7	158.0	671.3
CCl_4(l)	-135.4	216.4	-65.27
CCl_4(g)	-103	309.7	-60.63
$CHCl_3$(l)	-134.5	202	-73.72
$CHCl_3$(g)	-103.1	295.6	-70.37
CH_4(g)	-74.81	186.2	-50.75
C_2H_2(g)	226.7	200.8	209.2
C_2H_4(g)	52.26	219.5	68.12
C_2H_6(g)	-84.86	229.5	-32.9
C_3H_6(g)	20.41	266.9	62.75
C_3H_8(g)	-103.8	269.9	-23.49
C_6H_6(l)	49.03	172.8	124.5
C_8H_{18}(l)	-268.8	-	-
C_2H_5OH(l)	-277.7	161	-174.9
HCOOH(s)	-409.64	163.2	-355.718
$H_2C_2O_4$(s)	-825.968	45.61	-697.224
C_2H_5OH(g)	-235.1	282.6	-168.6
CO(g)	-110.5	197.6	-137.2
CO_2(g)	-393.5	213.6	-394.4
CS_2(g)	117.4	237.7	67.15
$COCl_2$(g)	-223.0	289.2	-210.5
Cesium			
Cs^+(aq)	-248	133	-282
CsF(aq)	-568.6	123	-558.5
Chlorine			
Cl_2(g)	0	223.0	0
Cl(g)	121.7	165.1	105.7
Cl^-(g)	-226	-	-
Cl^-(aq)	-167.29	56.48	-131.26
ClO(g)	101.22		
ClO_2(g)	104.60	249.4	123.4
HCl(g)	-92.31	186.8	-95.30
HCl(aq)	-167.4	55.10	-131.2
Chromium			
Cr(s)	0	23.8	0
$(NH_4)_2Cr_2O_7$(s)	-1807	-	-
Copper			
Cu(s)	0	33.15	0
CuO(s)	-157	42.63	-130
Flourine			
F^-(g)	-322	-	-
F^-(aq)	-332.6	-	-278.8
F(g)	78.99	158.6	61.92
F_2(g)	0	202.7	0
HF(g)	-271	173.7	-273
HF(aq)	-320.8	-	-296.8
Hydrogen			
H(g)	218.0	114.6	203.3
H_2(g)	0	130.6	0
H_2O(l)	-285.8	69.91	-237.2
H_2O(g)	-241.8	188.7	-228.6
H_2O_2(l)	-187.8	109.6	-120.4
Iodine			
I(g)	106.6	180.66	70.16
I_2(s)	0	116.1	0
I_2(g)	62.44	260.6	19.36
ICl(g)	17.78	247.4	-5.52
HI (g)	26.5	206.5	1.72
Iron			
Fe (s)	0	27.3	0
$FeCl_2$(s)	-340.67	117.9	-302.3
FeO(s)	-272	-	-
Fe_2O_3 (s,hematite)	-824.2	87.40	-742.2
Fe_3O_4 (s,magnetite)	-1118	146	-1015
FeS_2(s)	-177.5	122.2	-166.7
$Fe(CO)_5$(l)	-774.0	338	-705.4
$Fe(CO)_5$(g)	-733.8	445.2	-697.3
Lead			
Pb(s)	0	64.81	0
$PbCl_2$(s)	-359.4	136	-314.1
PbO(s,yellow)	-217.3	68.70	-187.9
$Pb(OH)_2$(s)	-515.9	88	-420.9
PbS(s)	-100.4	91.2	-98.7

Selected thermodynamic data at 298.15 K (continued)

Species	ΔH_f° (kJ mol^{-1})	S° (J mol^{-1} K^{-1})	ΔG_f° (kJ mol^{-1})
Lithium			
Li(s)	0	28.0	0
LiOH(s)	-487.23	50	-443.9
LiOH(aq)	-508.4	4	-451.1
Magnesium			
Mg(s)	0	32.5	0
Mg^{2+}(aq)	-454.668	-138.1	-455.57
$MgCl_2$(s)	-641.8	89.5	-592.3
MgO(s)	-601.8	27	-569.6
$Mg(OH)_2$(s)	-924.7	63.14	-833.7
MgS(s)	-347	-	-
Mercury			
Hg(l)	0	76.02	0
Hg_2Cl_2(s)	-224	146	-179
HgO(s, red)	-90.83	70.29	-58.56
HgS(s, red)	-8.2	82.4	-50.6
Nickel			
Ni(s)	0	30.1	0
$Ni(CO)_4$(g)	-602.9	410.4	-587.3
NiO(s)	-244	38.6	-216
Nitrogen			
N_2(g)	0	191.5	0
N(g)	472.704	153.19	455.579
NH_3 (g)	-46.11	192.3	-16.5
N_2H_4(l)	50.63	121.2	149.2
$(NH_4)_3AsO_4$(aq)	-1268	-	-
NH_4Cl(s)	-314.4	94.6	-201.5
NH_4Cl(aq)	-300.2	-	-
NH_4I(s)	-201.4	117	-113
NH_4NO_3(s)	-365.6	151.1	-184.0
NO(g)	90.25	210.7	86.57
NO_2(g)	33.2	240.0	51.30
N_2O(g)	82.05	219.7	104.2
N_2O_4(g)	9.16	304.2	97.82
N_2O_5(g)	11	356	115
N_2O_5(s)	-43.1	178	114
NOCl(g)	52.59	264	66.36
HNO_3(l)	-174.1	155.6	-80.79
HNO_3(g)	-135.1	266.2	-74.77
HNO_3 (aq)	-206.6	146	-110.5
Oxygen			
O_2(g)	0	205.0	0
O(g)	249.2	161.0	231.8
O_3(g)	143	238.8	163
OF_2(g)	23	246.6	41
Phosphorous			
P(g)	314.6	163.1	278.3
P_4(s, white)	0	177	0
P_4(s, red)	-73.6	91.2	-48.5
PCl_3(g)	-306.4	311.7	-286.3
PCl_5(g)	-398.9	353	-324.6
PH_3(g)	5.4	210.1	13
P_4O_{10}(s)	-2984	228.9	-2698
H_3PO_4(s)	-1281	110.5	-1119
Potassium			
K(s)	0	63.6	0
KCl(s)	-436.5	82.6	-408.8
$KClO_3$(s)	-391.2	143.1	-289.9
KI(s)	-327.9	106.4	-323.0
KOH(s)	-424.7	78.91	-378.9
KOH(aq)	-481.2	92.0	-439.6
Rubidium			
Rb(s)	0	76.78	0
RbOH(aq)	-481.16	110.75	-441.24
Silicon			
Si(s)	0	18.8	0
$SiBr_4$(l)	-457.3	277.8	-443.9
SiC(s)	-65.3	16.6	-62.8
$SiCl_4$(g)	-657.0	330.6	-617.0
SiH_4(g)	34.3	204.5	56.9
SiF_4(g)	-1615	282.4	-1573
SiI_4(g)	- 132	-	-
SiO_2(s)	-910.9	41.84	-856.7
H_2SiO_3(s)	-1189	134	-1092
Na_2SiO_3(s)	-1079	-	-
H_2SiF_6(aq)	-2331	-	-
Silver			
Ag(s)	0	42.55	0
Ag^+(aq)	105.79	73.86	77.12
AgCl(s)	-126.904	96.085	-109.8
Sodium			
Na(s)	0	51.0	0
Na(g)	108.7	153.6	78.11
Na^+(g)	601	-	-
NaBr(s)	-359.9	-	-
NaCl(s)	-411.0	72.38	-384
NaCl(aq)	-407.1	115.5	-393.0
Na_2CO_3(s)	-1131	136	-1048
NaOH(s)	-426.7	-	-
NaOH(aq)	-469.6	49.8	-419.2
Sulfur			
S(s, rhombic)	0	31.8	0
S(g)	278.8	167.8	238.3
S_2C1_2(g)	-18	331	-31.8
SF_6(g)	-1209	291.7	-1105
H_2S(g)	-20.6	205.7	-33.6
SO_2(g)	-296.8	248.1	-300.2
SO_3(g)	-395.6	256.6	-371.1
$SOCl_2$(l)	-206	-	-
SO_2Cl_2(l)	-389	-	-
H_2SO_4(l)	-814.0	156.9	-690.1
H_2SO_4(aq)	-907.5	17	-742.0
Tin			
Sn(s, white)	0	51.55	0
Sn(s, grey)	-2.09	44.1	0.13
$SnCl_2$(s)	-350	-	-
$SnC1_4$(l)	-511.3	258.6	-440.2
$SnC1_4$(g)	-471.5	366	-432.2
SnO_2(s)	-580.7	52.3	-519.7
Titanium			
$TiCl_4$(l)	-804.2	252.3	-737.2
$TiCl_4$(g)	-763.2	354.8	-726.8
Tungsten			
W(s)	0	32.6	0
WO_3(s)	-842.9	75.90	-764.1
Zinc			
ZnO(s)	-348.3	43.64	-318.3
ZnS(s)	-205.6	57.7	-201.3

Ionization constants for weak acids at 25°C

Acid	Formula and Ionization Equation	K_a
Acetic	$CH_3COOH \rightleftharpoons H^+ + CH_3COO^-$	1.8×10^{-5}
Arsenic	$H_3AsO_4 \rightleftharpoons H^+ + H_2AsO_4^-$	$2.5 \times 10^{-4} = K_{a1}$
	$H_2AsO_4^- \rightleftharpoons H^+ + HAsO_4^{2-}$	$5.6 \times 10^{-8} = K_{a2}$
	$HAsO_4^{2-} \rightleftharpoons H^+ + AsO_4^{3-}$	$3.0 \times 10^{-13} = K_{a3}$
Arsenous	$H_3AsO_3 \rightleftharpoons H^+ + H_2AsO_3^-$	$6.0 \times 10^{-10} = K_{a1}$
	$H_2AsO_3^- \rightleftharpoons H^+ + HAsO_3^{2-}$	$3.0 \times 10^{-14} = K_{a2}$
Benzoic	$C_6H_5COOH \rightleftharpoons H^+ + C_6H_5COO^-$	6.3×10^{-5}
Boric*	$B(OH)_3 \rightleftharpoons H^+ + BO(OH)_2^-$	$7.3 \times 10^{-10} = K_{a1}$
	$BO(OH)_2^- \rightleftharpoons H^+ + BO_2(OH)^{2-}$	$1.8 \times 10^{-13} = K_{a2}$
	$BO_2(OH)^{2-} \rightleftharpoons H^+ + BO_3^{3-}$	$1.6 \times 10^{-14} = K_{a3}$
Carbonic	$H_2CO_3 \rightleftharpoons H^+ + HCO_3^-$	$4.2 \times 10^{-7} = K_{a1}$
	$HCO_3^- \rightleftharpoons H^+ + CO_3^{2-}$	$4.8 \times 10^{-11} = K_{a2}$
Citric	$C_3H_5O(COOH)_3 \rightleftharpoons H^+ + C_4H_5O_3(COOH)_2^-$	$7.4 \times 10^{-3} = K_{a1}$
	$C_4H_5O_3(COOH)_2^- \rightleftharpoons H^+ + C_5H_5O_5COOH^{2-}$	$1.7 \times 10^{-5} = K_{a2}$
	$C_5H_5O_5COOH^{2-} \rightleftharpoons H^+ + C_6H_5O_7^{3-}$	$7.4 \times 10^{-7} = K_{a3}$
Cyanic	$HOCN \rightleftharpoons H^+ + OCN^-$	3.5×10^{-4}
Formic	$HCOOH \rightleftharpoons H^+ + HCOO^-$	1.8×10^{-4}
Hydrazoic	$HN_3 \rightleftharpoons H^+ + N_3^-$	1.9×10^{-5}
Hydrocyanic	$HCN \rightleftharpoons H^+ + CN^-$	4.0×10^{-10}
Hydrofluoric	$HF \rightleftharpoons H^+ + F^-$	7.2×10^{-4}
Hydrogen peroxide	$H_2O_2 \rightleftharpoons H^+ + HO_2^-$	2.4×10^{-12}
Hydrosulfuric	$H_2S \rightleftharpoons H^+ + HS^-$	$1.0 \times 10^{-7} = K_{a1}$
	$HS^- \rightleftharpoons H^+ + S^{2-}$	$1.0 \times 10^{-19} = K_{a2}$
Hypobromous	$HOBr \rightleftharpoons H^+ + OBr^-$	2.5×10^{-9}
Hypochlorous	$HOCl \rightleftharpoons H^+ + OCl^-$	3.5×10^{-8}
Hypoiodous	$HOI \rightleftharpoons H^+ + OI^-$	2.3×10^{-11}
Nitrous	$HNO_2 \rightleftharpoons H^+ + NO_2^-$	4.5×10^{-4}
Oxalic	$(COOH)_2 \rightleftharpoons H^+ + COOCOOH^-$	$5.9 \times 10^{-2} = K_{a1}$
	$COOCOOH^- \rightleftharpoons H^+ + (COO)_2^{2-}$	$6.4 \times 10^{-5} = K_{a2}$
Phenol	$HC_6H_5O \rightleftharpoons H^+ + C_6H_5O^-$	1.3×10^{-10}
Phosphoric	$H_3PO_4 \rightleftharpoons H^+ + H_2PO_4^-$	$7.5 \times 10^{-3} = K_{a1}$
	$H_2PO_4^- \rightleftharpoons H^+ + HPO_4^{2-}$	$6.2 \times 10^{-8} = K_{a2}$
	$HPO_4^{2-} \rightleftharpoons H^+ + PO_4^{3-}$	$3.6 \times 10^{-13} = K_{a3}$
Phosphorous	$H_3PO_3 \rightleftharpoons H^+ + H_2PO_3^-$	$1.6 \times 10^{-2} = K_{a1}$
	$H_2PO_3^- \rightleftharpoons H^+ + HPO_3^{2-}$	$7.0 \times 10^{-7} = K_{a2}$
Selenic	$H_2SeO_4 \rightleftharpoons H^+ + HSeO_4^-$	Very large $= K_{a1}$
	$HSeO_4^- \rightleftharpoons H^+ + SeO_4^{2-}$	$1.2 \times 10^{-2} = K_{a2}$
Selenous	$H_2SeO_3 \rightleftharpoons H^+ + HSeO_3^-$	$2.7 \times 10^{-3} = K_{a1}$
	$HSeO_3^- \rightleftharpoons H^+ + SeO_3^{2-}$	$2.5 \times 10^{-7} = K_{a2}$
Sulfuric	$H_2SO_4 \rightleftharpoons H^+ + HSO_4^-$	Very large $= K_{a1}$
	$HSO_4^- \rightleftharpoons H^+ + SO_4^{2-}$	$1.2 \times 10^{-2} = K_{a2}$
Sulfurous	$H_2SO_3 \rightleftharpoons H^+ + HSO_3^-$	$1.2 \times 10^{-2} = K_{a1}$
	$HSO_3^- \rightleftharpoons H^+ + SO_3^{2-}$	$6.2 \times 10^{-8} = K_{a2}$
Tellurous	$H_2TeO_3 \rightleftharpoons H^+ + HTeO_3^-$	$2 \times 10^{-3} = K_{a1}$
	$HTeO_3^- \rightleftharpoons H^+ + TeO_3^{2-}$	$1 \times 10^{-8} = K_{a2}$

**Boric acid acts as a Lewis acid in aqueous solution.*

Ionization constants for weak bases at 25°C

Base	Formula and Ionization Equation				K_b
Ammonia	NH_3	$+ H_2O$	$\rightleftharpoons NH_4^+$	$+ OH^-$	1.8×10^{-5}
Aniline	$C_6H_5NH_2$	$+ H_2O$	$\rightleftharpoons C_6H_5NH_3^+$	$+ OH^-$	4.2×10^{-10}
Dimethylamine	$(CH_3)_2NH$	$+ H_2O$	$\rightleftharpoons (CH_3)_2NH_2^+$	$+ OH^-$	7.4×10^{-4}
Ethylenediamine	$(CH_2)_2(NH_2)_2$	$+ H_2O$	$\rightleftharpoons (CH_2)_2(NH_2)_2H^+$	$+ OH^-$	$8.5 \times 10^{-5} = K_{b1}$
	$(CH_2)_2(NH_2)_2H^+$	$+ H_2O$	$\rightleftharpoons (CH_2)_2(NH_2)_2H_2^{2+}$	$+ OH^-$	$2.7 \times 10^{-8} = K_{b2}$
Hydrazine	N_2H_4	$+ H_2O$	$\rightleftharpoons N_2H_5^+$	$+ OH^-$	$8.5 \times 10^{-7} = K_{b1}$
	$N_2H_5^+$	$+ H_2O$	$\rightleftharpoons N_2H_6^{2+}$	$+ OH^-$	$8.9 \times 10^{-16} = K_{b2}$
Hydroxylamine	NH_2OH	$+ H_2O$	$\rightleftharpoons NH_3OH^+$	$+ OH^-$	6.6×10^{-9}
Methylamine	CH_3NH_2	$+ H_2O$	$\rightleftharpoons CH_3NH_3^+$	$+ OH^-$	5.0×10^{-4}
Pyridine	C_5H_5N	$+ H_2O$	$\rightleftharpoons C_5H_5NH^+$	$+ OH^-$	1.5×10^{-9}
Trimethylamine	$(CH_3)_3N$	$+ H_2O$	$\rightleftharpoons (CH_3)_3NH^+$	$+ OH^-$	7.4×10^{-5}

Solubility product constants for some inorganic compounds at 25°C

Substance	K_{sp}
Aluminum compounds	
$AlAsO_4$	1.6×10^{-16}
$Al(OH)_3$	1.9×10^{-33}
$AlPO_4$	1.3×10^{-20}
Antimony compounds	
Sb_2S_3	1.6×10^{-93}
Barium compounds	
$Ba_3(AsO_4)_2$	1.1×10^{-13}
$BaCO_3$	8.1×10^{-9}
$BaC_2O_4 \cdot 2H_2O$*	1.1×10^{-7}
$BaCrO_4$	2.0×10^{-10}
BaF_2	1.7×10^{-6}
$Ba(OH)_2 \cdot 8H_2O$*	5.0×10^{-3}
$Ba_3(PO_4)_2$	1.3×10^{-29}
$BaSeO_4$	2.8×10^{-11}
$BaSO_3$	8.0×10^{-7}
$BaSO_4$	1.1×10^{-10}
Bismuth compounds	
$BiOCl$	7.0×10^{-9}
$BiO(OH)$	1.0×10^{-12}
$Bi(OH)_3$	3.2×10^{-40}
BiI_3	8.1×10^{-19}
$BiPO_4$	1.3×10^{-23}
Bi_2S_3	1.6×10^{-72}
Cadmium compounds	
$Cd_3(AsO_4)_2$	2.2×10^{-32}
$CdCO_3$	2.5×10^{-14}
$Cd(CN)_2$	1.0×10^{-8}
$Cd_2[Fe(CN)_6]$	3.2×10^{-17}
$Cd(OH)_2$	1.2×10^{-14}
CdS	3.6×10^{-29}
Calcium compounds	
$Ca_3(AsO_4)_2$	6.8×10^{-19}
$CaCO_3$	4.8×10^{-9}
$CaCrO_4$	7.1×10^{-4}
$CaC_2O_4 \cdot H_2O$*	2.3×10^{-9}
CaF_2	3.9×10^{-11}
$Ca(OH)_2$	7.9×10^{-6}
$CaHPO_4$	2.7×10^{-7}
$Ca(H_2PO_4)_2$	1.0×10^{-3}
$Ca_3(PO_4)_2$	1.0×10^{-25}
$CaSO_3 \cdot 2H_2O$*	1.3×10^{-8}
$CaSO_4 \cdot 2H_2O$*	2.4×10^{-5}
Chromium compounds	
$CrAsO_4$	7.8×10^{-21}
$Cr(OH)_3$	6.7×10^{-31}
$CrPO_4$	2.4×10^{-23}
Cobalt compounds	
$Co_3(AsO_4)_2$	7.6×10^{-29}
$CoCO_3$	8.0×10^{-13}
$Co(OH)_2$	2.5×10^{-16}
CoS (α)	5.9×10^{-21}
CoS (β)	8.7×10^{-23}
$Co(OH)_3$	4.0×10^{-45}
Co_2S_3	2.6×10^{-124}
Copper compounds	
$CuBr$	5.3×10^{-9}
$CuCl$	1.9×10^{-7}
$CuCN$	3.2×10^{-20}
Cu_2O ($Cu^+ + OH^-$)†	1.0×10^{-14}
CuI	5.1×10^{-12}
Cu_2S	1.6×10^{-48}
$CuSCN$	1.6×10^{-11}
$Cu_3(AsO_4)_2$	7.6×10^{-36}
$CuCO_3$	2.5×10^{-10}
$Cu_2[Fe(CN)_6]$	1.3×10^{-16}
$Cu(OH)_2$	1.6×10^{-19}
CuS	8.7×10^{-36}
Gold compounds	
$AuBr$	5.0×10^{-17}
$AuCl$	2.0×10^{-13}
AuI	1.6×10^{-23}
$AuBr_3$	4.0×10^{-36}
$AuCl_3$	3.2×10^{-25}
$Au(OH)_3$	1.0×10^{-53}
AuI_3	1.0×10^{-46}
Iron compounds	
$FeCO_3$	3.5×10^{-11}
$Fe(OH)_2$	7.9×10^{-15}
FeS	4.9×10^{-18}
$Fe_4[Fe(CN)_6]_3$	3.0×10^{-41}
$Fe(OH)_3$	6.3×10^{-38}
Fe_2S_3	1.4×10^{-88}
Lead compounds	
$Pb_3(AsO_4)_2$	4.1×10^{-36}
$PbBr_2$	6.3×10^{-6}

Solubility product constants for some inorganic compounds at 25°C (continued)

Substance	K_{sp}	Substance	K_{sp}
Lead compounds (cont.)		***Nickel compounds (cont.)***	
$PbCO_3$	1.5×10^{-13}	NiS (α)	3.0×10^{-21}
$PbCl_2$	1.7×10^{-5}	NiS (β)	1.0×10^{-26}
$PbCrO_4$	1.8×10^{-14}	NiS (γ)	2.0×10^{-28}
PbF_2	3.7×10^{-8}	***Silver compounds***	
$Pb(OH)_2$	2.8×10^{-16}	Ag_3AsO_4	1.1×10^{-20}
PbI_2	8.7×10^{-9}	AgBr	3.3×10^{-13}
$Pb_3(PO_4)_2$	3.0×10^{-44}	Ag_2CO_3	8.1×10^{-12}
$PbSeO_4$	1.5×10^{-7}	AgCl	1.8×10^{-10}
$PbSO_4$	1.8×10^{-8}	Ag_2CrO_4	9.0×10^{-12}
PbS	8.4×10^{-28}	AgCN	1.2×10^{-16}
Magnesium compounds		$Ag_4[Fe(CN)_6]$	1.6×10^{-41}
$Mg_3(AsO_4)_2$	2.1×10^{-20}	Ag_2O ($Ag^+ + OH^-$)†	2.0×10^{-8}
$MgCO_3 \cdot 3H_2O$*	4.0×10^{-5}	AgI	1.5×10^{-16}
MgC_2O_4	8.6×10^{-5}	Ag_3PO_4	1.3×10^{-20}
MgF_2	6.4×10^{-9}	Ag_2SO_3	1.5×10^{-14}
$Mg(OH)_2$	1.5×10^{-11}	Ag_2SO_4	1.7×10^{-5}
$MgNH_4PO_4$	2.5×10^{-12}	Ag_2S	1.0×10^{-49}
Manganese compounds		AgSCN	1.0×10^{-12}
$Mn_3(AsO_4)_2$	1.9×10^{-11}	***Strontium compounds***	
$MnCO_3$	1.8×10^{-11}	$Sr_3(AsO_4)_2$	1.3×10^{-18}
$Mn(OH)_2$	4.6×10^{-14}	$SrCO_3$	9.4×10^{-10}
MnS	5.1×10^{-15}	$SrC_2O_4 \cdot 2H_2O$*	5.6×10^{-8}
$Mn(OH)_3$	$\approx 1.0 \times 10^{-36}$	$SrCrO_4$	3.6×10^{-5}
Mercury compounds		$Sr(OH)_2 \cdot 8H_2O$*	3.2×10^{-4}
Hg_2Br_2	1.3×10^{-22}	$Sr_3(PO_4)_2$	1.0×10^{-31}
Hg_2CO_3	8.9×10^{-17}	$SrSO_3$	4.0×10^{-8}
Hg_2Cl_2	1.1×10^{-18}	$SrSO_4$	2.8×10^{-7}
Hg_2CrO_4	5.0×10^{-9}	***Tin compounds***	
Hg_2I_2	4.5×10^{-29}	$Sn(OH)_2$	2.0×10^{-26}
$Hg_2O \cdot H_2O$*		SnI_2	1.0×10^{-4}
($Hg_2^{2+} + 2OH^-$)†	1.6×10^{-23}	SnS	1.0×10^{-28}
Hg_2SO_4	6.8×10^{-7}	$Sn(OH)_4$	1.0×10^{-57}
Hg_2S	5.8×10^{-44}	SnS_2	1.0×10^{-70}
$Hg(CN)_2$	3.0×10^{-23}	***Zinc compounds***	
$Hg(OH)_2$	2.5×10^{-26}	$Zn_3(AsO_4)_2$	1.1×10^{-27}
HgI_2	4.0×10^{-29}	$ZnCO_3$	1.5×10^{-11}
HgS	3.0×10^{-53}	$Zn(CN)_2$	8.0×10^{-12}
Nickel compounds		$Zn_2[Fe(CN)_6]$	4.1×10^{-16}
$Ni_3(AsO_4)_2$	1.9×10^{-26}	$Zn(OH)_2$	4.5×10^{-17}
$NiCO_3$	6.6×10^{-9}	$Zn_3(PO_4)_2$	9.1×10^{-33}
$Ni(CN)_2$	3.0×10^{-23}	ZnS	1.1×10^{-21}
$Ni(OH)_2$	2.8×10^{-16}		

$[H_2O]$ does not appear in equilibrium constants for equilibria in aqueous solution in general, so it does not *appear in the K_{sp} expressions for hydrated solids.*

†*Very small amounts of oxides dissolve in water to give the ions indicated in parentheses. These solid hydroxides are unstable and decompose to oxides as rapidly as they are formed.*

Standard reduction potentials in aqueous solution at 25°C

Acidic Solution		Standard Reduction Potential, E^0 (volts)
$Li^+(aq) + e^- \longrightarrow Li(s)$		−3.045
$K^+(aq) + e^- \longrightarrow K(s)$		−2.925
$Rb^+(aq) + e^- \longrightarrow Rb(s)$		−2.925
$Ba^{2+}(aq) + 2e^- \longrightarrow Ba(s)$		−2.90
$Sr^{2+}(aq) + 2e^- \longrightarrow Sr(s)$		−2.89
$Ca^{2+}(aq) + 2e^- \longrightarrow Ca(s)$		−2.87
$Na^+(aq) + e^- \longrightarrow Na(s)$		−2.714
$Mg^{2+}(aq) + 2e^- \longrightarrow Mg(s)$		−2.37
$H_2(g) + 2e^- \longrightarrow 2H^-(aq)$		−2.25
$Al^{3+}(aq) + 3e^- \longrightarrow Al(s)$		−1.66
$Zr^{4+}(aq) + 4e^- \longrightarrow Zr(s)$		−1.53
$ZnS(s) + 2e^- \longrightarrow Zn(s) + S^{2-}(aq)$		−1.44
$CdS(s) + 2e^- \longrightarrow Cd(s) + S^{2-}(aq)$		−1.21
$V^{2+}(aq) + 2e^- \longrightarrow V(s)$		−1.18
$Mn^{2+}(aq) + 2e^- \longrightarrow Mn(s)$		−1.18
$FeS(s) + 2e^- \longrightarrow Fe(s) + S^{2-}(aq)$		−1.01
$Cr^{2+}(aq) + 2e^- \longrightarrow Cr(s)$		−0.91
$Zn^{2+}(aq) + 2e^- \longrightarrow Zn(s)$		−0.763
$Cr^{3+}(aq) + 3e^- \longrightarrow Cr(s)$		−0.74
$HgS(s) + 2H^+(aq) + 2e^- \longrightarrow Hg(\ell) + H_2S(g)$		−0.72
$Ga^{3+}(aq) + 3e^- \longrightarrow Ga(s)$		−0.53
$2CO_2(g) + 2H^+(aq) + 2e^- \longrightarrow (COOH)_2(aq)$		−0.49
$Fe^{2+}(aq) + 2e^- \longrightarrow Fe(s)$		−0.44
$Cr^{3+}(aq) + e^- \longrightarrow Cr^{2+}(aq)$		−0.41
$Cd^{2+}(aq) + 2e^- \longrightarrow Cd(s)$		−0.403
$Se(s) + 2H^+(aq) + 2e^- \longrightarrow H_2Se(aq)$		−0.40
$PbSO_4(s) + 2e^- \longrightarrow Pb(s) + SO_4^{2-}(aq)$		−0.356
$Tl^+(aq) + e^- \longrightarrow Tl(s)$		−0.34
$Co^{2+}(aq) + 2e^- \longrightarrow Co(s)$		−0.28
$Ni^{2+}(aq) + 2e^- \longrightarrow Ni(s)$		−0.25
$[SnF_6]^{2-}(aq) + 4e^- \longrightarrow Sn(s) + 6F^-(aq)$		−0.25
$AgI(s) + e^- \longrightarrow Ag(s) + I^-(aq)$		−0.15
$Sn^{2+}(aq) + 2e^- \longrightarrow Sn(s)$		−0.14
$Pb^{2+}(aq) + 2e^- \longrightarrow Pb(s)$		−0.126
$N_2O(g) + 6H^+(aq) + H_2O + 4e^- \longrightarrow 2NH_3OH^+(aq)$		−0.05
$2H^+(aq) + 2e^- \longrightarrow H_2(g)$	(reference electrode)	0.000
$AgBr(s) + e^- \longrightarrow Ag(s) + Br^-(aq)$		0.10
$S(s) + 2H^+(aq) + 2e^- \longrightarrow H_2S(aq)$		0.14
$Sn^{4+}(aq) + 2e^- \longrightarrow Sn^{2+}(aq)$		0.15
$Cu^{2+}(aq) + e^- \longrightarrow Cu^+(aq)$		0.153
$SO_4^{2-}(aq) + 4H^+(aq) + 2e^- \longrightarrow H_2SO_3(aq) + H_2O$		0.17

Standard reduction potentials in aqueous solution at 25°C (continued)

Acidic Solution	Standard Reduction Potential, E^0 (volts)
$SO_4^{2-}(aq) + 4H^+(aq) + 2e^- \longrightarrow SO_2(g) + 2H_2O$	0.20
$AgCl(s) + e^- \longrightarrow Ag(s) + Cl^-(aq)$	0.222
$Hg_2Cl_2(s) + 2e^- \longrightarrow 2Hg(\ell) + 2Cl^-(aq)$	0.27
$Cu^{2+}(aq) + 2e^- \longrightarrow Cu(s)$	0.337
$[RhCl_6]^{3-}(aq) + 3e^- \longrightarrow Rh(s) + 6Cl^-(aq)$	0.44
$Cu^+(aq) + e^- \longrightarrow Cu(s)$	0.521
$TeO_2(s) + 4H^+(aq) + 4e^- \longrightarrow Te(s) + 2H_2O$	0.529
$I_2(s) + 2e^- \longrightarrow 2I^-(aq)$	0.535
$H_3AsO_4(aq) + 2H^+(aq) + 2e^- \longrightarrow H_3AsO_3(aq) + H_2O$	0.58
$[PtCl_6]^{2-}(aq) + 2e^- \longrightarrow [PtCl_4]^{2-}(aq) + 2Cl^-(aq)$	0.68
$O_2(g) + 2H^+(aq) + 2e^- \longrightarrow H_2O_2(aq)$	0.682
$[PtCl_4]^{2-}(aq) + 2e^- \longrightarrow Pt(s) + 4Cl^-(aq)$	0.73
$SbCl_6^-(aq) + 2e^- \longrightarrow SbCl_4^-(aq) + 2Cl^-(aq)$	0.75
$Fe^{3+}(aq) + e^- \longrightarrow Fe^{2+}(aq)$	0.771
$Hg_2^{2+}(aq) + 2e^- \longrightarrow 2Hg(\ell)$	0.789
$Ag^+(aq) + e^- \longrightarrow Ag(s)$	0.7994
$Hg^{2+}(aq) + 2e^- \longrightarrow Hg(\ell)$	0.855
$2Hg^{2+}(aq) + 2e^- \longrightarrow Hg_2^{2+}(aq)$	0.920
$NO_3^-(aq) + 3H^+(aq) + 2e^- \longrightarrow HNO_2(aq) + H_2O$	0.94
$NO_3^-(aq) + 4H^+(aq) + 3e^- \longrightarrow NO(g) + 2H_2O$	0.96
$Pd^{2+}(aq) + 2e^- \longrightarrow Pd(s)$	0.987
$AuCl_4^-(aq) + 3e^- \longrightarrow Au(s) + 4Cl^-(aq)$	1.00
$Br_2(\ell) + 2e^- \longrightarrow 2Br^-(aq)$	1.08
$ClO_4^-(aq) + 2H^+(aq) + 2e^- \longrightarrow ClO_3^-(aq) + H_2O$	1.19
$IO_3^-(aq) + 6H^+(aq) + 5e^- \longrightarrow \frac{1}{2}I_2(aq) + 3H_2O$	1.195
$Pt^{2+}(aq) + 2e^- \longrightarrow Pt(s)$	1.2
$O_2(g) + 4H^+(aq) + 4e^- \longrightarrow 2H_2O$	1.229
$MnO_2(s) + 4H^+(aq) + 2e^- \longrightarrow Mn^{2+}(aq) + 2H_2O$	1.23
$N_2H_5^+(aq) + 3H^+(aq) + 2e^- \longrightarrow 2NH_4^+(aq)$	1.24
$Cr_2O_7^{2-}(aq) + 14H^+(aq) + 6e^- \longrightarrow 2Cr^{3+}(aq) + 7H_2O$	1.33
$Cl_2(g) + 2e^- \longrightarrow 2Cl^-(aq)$	1.360
$BrO_3^-(aq) + 6H^+(aq) + 6e^- \longrightarrow Br^-(aq) + 3H_2O$	1.44
$ClO_3^-(aq) + 6H^+(aq) + 5e^- \longrightarrow \frac{1}{2}Cl_2(g) + 3H_2O$	1.47
$Au^{3+}(aq) + 3e^- \longrightarrow Au(s)$	1.50
$MnO_4^-(aq) + 8H^+(aq) + 5e^- \longrightarrow Mn^{2+}(aq) + 4H_2O$	1.507
$NaBiO_3(s) + 6H^+(aq) + 2e^- \longrightarrow Bi^{3+}(aq) + Na^+(aq) + 3H_2O$	1.6
$Ce^{4+}(aq) + e^- \longrightarrow Ce^{3+}(aq)$	1.61
$2HOCl(aq) + 2H^+(aq) + 2e^- \longrightarrow Cl_2(g) + 2H_2O$	1.63
$Au^+(aq) + e^- \longrightarrow Au(s)$	1.68
$PbO_2(s) + SO_4^{2-}(aq) + 4H^+(aq) + 2e^- \longrightarrow PbSO_4(s) + 2H_2O$	1.685
$NiO_2(s) + 4H^+(aq) + 2e^- \longrightarrow Ni^{2+}(aq) + 2H_2O$	1.7
$H_2O_2(aq) + 2H^+(aq) + 2e^- \longrightarrow 2H_2O$	1.77
$Pb^{4+}(aq) + 2e^- \longrightarrow Pb^{2+}(aq)$	1.8
$Co^{3+}(aq) + e^- \longrightarrow Co^{2+}(aq)$	1.82
$F_2(g) + 2e^- \longrightarrow 2F^-(aq)$	2.87

Standard reduction potentials in aqueous solution at 25°C (continued)

Basic Solution	**Standard Reduction Potential, E^0 (volts)**
$SiO_3^{2-}(aq) + 3H_2O + 4e^- \longrightarrow Si(s) + 6OH^-(aq)$	−1.70
$Cr(OH)_3(s) + 3e^- \longrightarrow Cr(s) + 3OH^-(aq)$	−1.30
$[Zn(CN)_4]^{2-}(aq) + 2e^- \longrightarrow Zn(s) + 4CN^-(aq)$	−1.26
$Zn(OH)_2(s) + 2e^- \longrightarrow Zn(s) + 2OH^-(aq)$	−1.245
$[Zn(OH)_4]^{2-}(aq) + 2e^- \longrightarrow Zn(s) + 4OH^-(aq)$	−1.22
$N_2(g) + 4H_2O + 4e^- \longrightarrow N_2H_4(aq) + 4OH^-(aq)$	−1.15
$SO_4^{2-}(aq) + H_2O + 2e^- \longrightarrow SO_3^{2-}(aq) + 2OH^-(aq)$	−0.93
$Fe(OH)_2(s) + 2e^- \longrightarrow Fe(s) + 2OH^-(aq)$	−0.877
$2NO_3^-(aq) + 2H_2O + 2e^- \longrightarrow N_2O_4(g) + 4OH^-(aq)$	−0.85
$2H_2O + 2e^- \longrightarrow H_2(g) + 2OH^-(aq)$	−0.828
$Fe(OH)_3(s) + e^- \longrightarrow Fe(OH)_2(s) + OH^-(aq)$	−0.56
$S(s) + 2e^- \longrightarrow S^{2-}(aq)$	−0.48
$Cu(OH)_2(s) + 2e^- \longrightarrow Cu(s) + 2OH^-(aq)$	−0.36
$CrO_4^{2-}(aq) + 4H_2O + 3e^- \longrightarrow Cr(OH)_3(s) + 5OH^-(aq)$	−0.12
$MnO_2(s) + 2H_2O + 2e^- \longrightarrow Mn(OH)_2(s) + 2OH^-(aq)$	−0.05
$NO_3^-(aq) + H_2O + 2e^- \longrightarrow NO_2^-(aq) + 2OH^-(aq)$	0.01
$O_2(g) + H_2O + 2e^- \longrightarrow OOH^-(aq) + OH^-(aq)$	0.076
$HgO(s) + H_2O + 2e^- \longrightarrow Hg(\ell) + 2OH^-(aq)$	0.0984
$[Co(NH_3)_6]^{3+}(aq) + e^- \longrightarrow [Co(NH_3)_6]^{2+}(aq)$	0.10
$N_2H_4(aq) + 2H_2O + 2e^- \longrightarrow 2NH_3(aq) + 2OH^-(aq)$	0.10
$2NO_2^-(aq) + 3H_2O + 4e^- \longrightarrow N_2O(g) + 6OH^-(aq)$	0.15
$Ag_2O(s) + H_2O + 2e^- \longrightarrow 2Ag(s) + 2OH^-(aq)$	0.34
$ClO_4^-(aq) + H_2O + 2e^- \longrightarrow ClO_3^-(aq) + 2OH^-(aq)$	0.36
$O_2(g) + 2H_2O + 4e^- \longrightarrow 4OH^-(aq)$	0.40
$Ag_2CrO_4(s) + 2e^- \longrightarrow 2Ag(s) + CrO_4^{2-}(aq)$	0.446
$NiO_2(s) + 2H_2O + 2e^- \longrightarrow Ni(OH)_2(s) + 2OH^-(aq)$	0.49
$MnO_4^-(aq) + e^- \longrightarrow MnO_4^{2-}(aq)$	0.564
$MnO_4^-(aq) + 2H_2O + 3e^- \longrightarrow MnO_2(s) + 4OH^-(aq)$	0.588
$ClO_3^-(aq) + 3H_2O + 6e^- \longrightarrow Cl^-(aq) + 6OH^-(aq)$	0.62
$2NH_2OH(aq) + 2e^- \longrightarrow N_2H_4(aq) + 2OH^-(aq)$	0.74
$OOH^-(aq) + H_2O + 2e^- \longrightarrow 3OH^-(aq)$	0.88
$ClO^-(aq) + H_2O + 2e^- \longrightarrow Cl^-(aq) + 2OH^-(aq)$	0.89

Answers to Check Your Understanding Exercises

Chapter 1

1.1 See Figure 1.5.

1.2 (a). 3 (b). 1

1.3 (a). 16.10 m (b). 0.005 g

1.4 \$299; The amount purchased might still need to be a multiple of five gallons, so it might need to exceed the actual amount needed.

1.5 4.3×10^6 W

1.6 3.34×10^4 g, or 33.4 kg

1.7

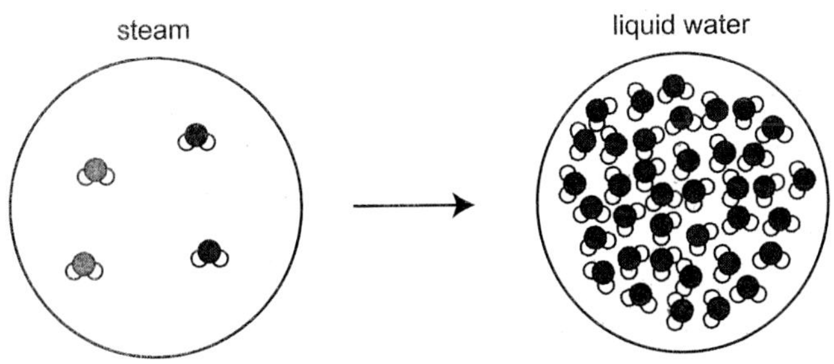

Chapter 2

2.1 19.90% ^{10}B, 80.10% ^{11}B

2.2 8 carbon atoms, 12 hydrogen atoms, 4 nitrogen atoms; C_2H_3N

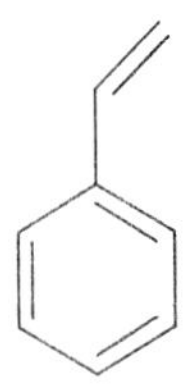

2.3

2.4 C_6H_9NO

2.5 (a). carbon disulfide (b). sulfur hexafluoride (c). dichlorine heptoxide

2.6 (a). copper(II) sulfate (b). silver phosphate (c). vanadium(V) oxide

Chapter 3

3.1 $C_3H_8 + O_2 \rightarrow 3CH_2O + H_2O$

3.2 (a), (b), and (c) are all soluble; (d). is insoluble

3.3 Molecular: $2HCl(aq) + Ca(OH)_2(aq) \rightarrow CaCl_2(aq) + 2H_2O(l)$
Total ionic:
$2H^+(aq) + 2Cl^-(aq) + Ca^{2+}(aq) + 2OH^-(aq) \rightarrow Ca^{2+}(aq) + 2Cl^-(aq) + 2H_2O(l)$
Net ionic: $H^+(aq) + OH^-(aq) \rightarrow H_2O(l)$
(These could also be written with H_3O^+ instead of H^+.

3.4 Molecular: $Na_2SO_4(aq) + Pb(NO_3)_2(aq) \rightarrow PbSO_4(s) + 2NaNO_3$
Total ionic:
$2Na^+(aq) + SO_4(aq) + Pb^{2+}(aq) + 2NO_3^-(aq) \rightarrow PbSO_4(s) + 2Na^+(aq) + 2NO_3^-(aq)$
Net ionic: $SO_4(aq) + Pb^{2+}(aq) \rightarrow PbSO_4(s)$

3.5 (a). 98.079 g mol^{-1} (b). 63.0128 g mol^{-1} (c). 95.0578 g mol^{-1}

3.6 70.5 mL

3.7 7×10^6 tons

3.8 $C_3H_5N_3O_9$

3.9 $C_4H_8N_8O_8$

3.10 59 moles

3.11 0.216 M

Chapter 4

4.1 13.5 mol H_2O
4.2 3.24 g H_2
4.3 96 g SO_2
4.4 > $4.5x10^3$ moles H_2
4.5 Same result as in example: S_8 is the limiting reactant
4.6 20.8 g B_2H_6
4.7 The remaining fuel is 262.2 g of aluminum.
4.8 80.9% yield
4.9 63.3 mL
4.10 0.828 M NaOH

Chapter 5

5.1 82.1 torr
5.2 330 K
5.3 0.77 ft^3
5.4 V = 29.6 L, P_{O2} = 0.348 atm, P_{N2} = 1.78 atm
5.5 6.4 g SO_2, 3.4 g SO_3; P_{SO2} = 0.98 atm, P_{SO3} = 0.42 atm
5.6 0.96 g $NaHCO_3$
5.7 9.67 L SO_2
5.8 P_{ideal} = 0.817 atm, P_{VdW} = 0.817 atm

Chapter 6

6.1 $6.2x10^{14}\ s^{-1}$
6.2 $1.7x10^{-19}$ J
6.3 $2.6x10^{-20}$ J
6.4 97.3 nm
6.5 *$1s^22s^22p^63s^23p^2$*
6.6 *[Xe] $6s^24f^{14}5d^{10}$*

Chapter 7

7.1 The 1st IE will be small, the 2nd IE wil be slightly larger, and the 3rd IE will be extremely high.
7.2 KCl; singly charged ions and smaller ions than in RbBr.
7.3 S-Br

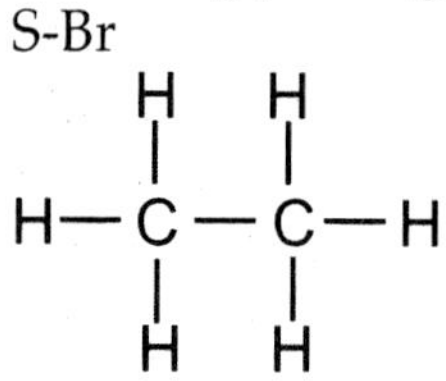

7.4

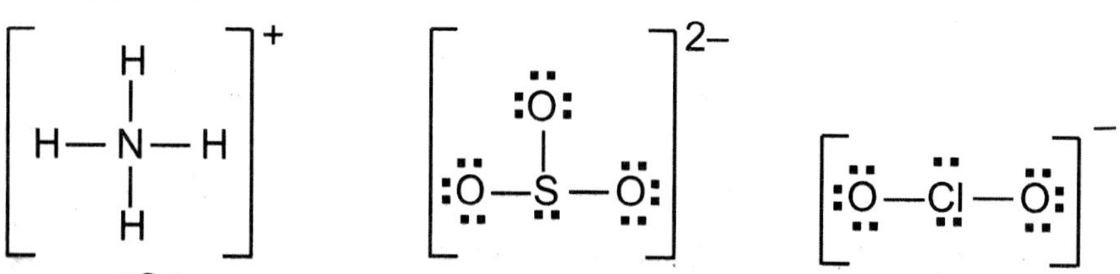

7.5

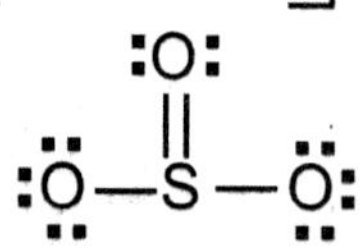

7.6
7.7 octahedral
7.8 (a). T-shape (b). linear
7.9 (a). trigonal planar (b). linear

Chapter 8

8.1 Values are given in text: 68% for bcc, 52% for sc

8.2 The dopant should be an element with more than four valence electrons. Phosphorous is the most likely choice. Band diagram will look like the right hand side of Figure 8.12.

8.3 Boiling point sequence should be: $Ne < CH_4 < CO < NH_3$

8.4 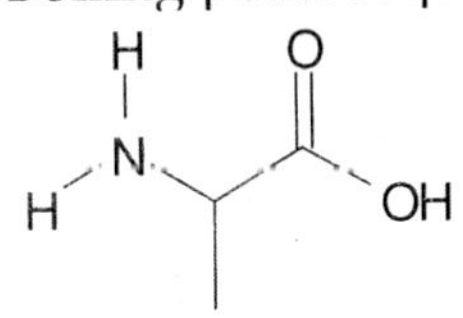

Chapter 9

9.1 +223 J

9.2 16.3°C

9.3 24.4 J mol^{-1} $°C^{-1}$

9.4 64°C

9.5 29.3 kJ/g

9.6 $-3.28x10^4$ J

9.7 1.9 kJ mol^{-1}

9.8 –252.58 kJ

9.9 $-1.215x10^3$ kJ

9.10 41.2 g NO, 19.2 g N_2

Chapter 10

10.1 266.8 J mol^{-1} K^{-1}

10.2 Low temperatures. The process is enthalpy driven, since it is exothermic but decreases the entropy of the system. Enthalpy dominates at low temperatures.

10.3 8.8 J mol^{-1} K^{-1}

10.4 1278 kJ mol^{-1}

Chapter 11

11.1 NO_2: $8x10^{-6}$ mol L^{-1} s^{-1}, O_2: $2x10^{-6}$ mol L^{-1} s^{-1}

11.2 (a). First order in H_2, first order in Br_2, second order overall
(b). First order in N_2O_5, first order overall

11.3 No, as long as we do the calculation correctly the results will agree within the experimental error.

11.4 Units on rate constants depend on the overall reaction order.

11.5 870 seconds

11.6 First order, $k = 0.0023$

11.7 $1.6x10^{-5}$ s^{-1}

11.8 41.6 kJ mol^{-1}

11.9 (a). Yes, it does match the actually stoichiometry, and does not involve any steps that involve collisions of three or more molecules. (b). N_2O_3 and NO are both intermediates in this mechanism. (c). The first two steps are unimolecular, and the third step is bimolecular.

Chapter 12

12.1 $$K = \frac{[HNO_2]^2}{[NO][NO_2][H_2O]}$$

12.2 $3Cu^{2+} + 2PO_4^{3-} \rightarrow Cu_3(PO_4)_2$; $K = \dfrac{1}{[Cu^{2+}]^3[PO_4^{3-}]^2}$

12.3 The equilibrium constant for sulfate ion is much smaller than the others, so sulfate will not precipitate silver as well as wither cyanide or bromide.

12.4 (a). $K = \dfrac{[NO]^4[H_2O]^6}{[NH_3]^4[O_2]^5}$ (b). $K = \dfrac{[NH_3]^4[O_2]^5}{[NO]^4[H_2O]^6}$ (c). $K = \dfrac{[NH_3]^2[O_2]^{5/2}}{[NO]^2[H_2O]^3}$

12.5 $K = \dfrac{[NO_2]^4}{[N_2][O_2]^3[NO]^2}$; $N_2 + 3O_2 + 2NO \rightarrow 4NO_2$

12.6 $[H_2] = [I_2]$=0.0022 M, [HI] = .0156 M

12.7 $[Cl_2] = [PCl_3]$ = 0.036 M, $[PCl_5]$ = 0.0043 M

12.8 Possibilities include removing HCN from the system somehow or adding additional acid, which would increase $[H^+]$.

12.9 (a). shift towards reactants (b). no change

12.10 (a). $K_{sp} = [Mn^{2+}][OH^-]^2$ (b). $K_{sp} = [Cu^{2+}][S^{2-}]$ (c). $K_{sp} = [Cu^+][I^-]$ (d). $K_{sp} = [Al^{3+}]^2[SO_4^{2-}]^3$

12.11 $1.1\text{x}10^{-8}$ M; $1.5\text{x}10^{-7}$ g/100 g H_2O

12.12 $3.5\text{x}10^{-4}$ M in H_2O; $7.6\text{x}10^{-9}$ M in 0.15 M NaF

12.13 Hydrogen carbonate ion is the conjugate acid of carbonate, and hydroxide ion is the conjugate base of water.

12.14 1.2% ionized; yes, the approximation is reasonably valid here

12.15 ln K = 330, which means that K for this reaction would be on the order of 10^{143}. This extremely large number indicates that the complete oxidation is the very strongly preferred pathway.

Chapter 13

13.1 Zinc; copper is the cathode and zinc is the anode.

13.2 The data give $[Ni^{2+}]$ = 19 M, which is too high to be practical.

13.3 Reversing the reaction would change the sign of ΔG, giving +28.4 kJ. That large positive value indicates that the reverse process is not spontaneous.

13.4 $1.08\text{x}10^4$ s, which is about 3 hours

13.5 0.272 V

13.6 No. The stated current and time give only 0.003 mol e^-, and we need 0.126 mol, so the coating will be far too thin.

13.7 0.021 A

Index

A

B

H

I

Z